An Introduction to

Physical Geography and the Environment

Second edition

Edited by

Joseph Holden

School of Geography, University of Leeds

PEARSON
Prentice
Hall

Harlow, England • London • New York • Boston • San Francisco • Toronto • Sydney • Singapore • Hong Kong
Tokyo • Seoul • Taipei • New Delhi • Cape Town • Madrid • Mexico City • Amsterdam • Munich • Paris • Milan

Pearson Education Limited
Edinburgh Gate
Harlow
Essex CM20 2JE
England

and Associated Companies throughout the world

Visit us on the World Wide Web at:
www.pearsoned.co.uk

First published 2005
Second edition published 2008

ISBN: 978-0-13-175304-4

British Library Cataloguing-in-Publication Data
A catalogue record for this book is available from the British Library

Library of Congress Cataloging-in-Publication Data
An introduction to physical geography and the environment / edited by Joseph Holden.
–– 2nd ed.
      p. cm.
    ISBN 978-0-13-175304-4 (pbk.)
1. Physical geography.   I. Holden, Joseph, 1975-   II. Title: Physical geography and the environment.
    GB54. 5. 158 2008   2008013319
    910' .2––dc22

10 9 8 7 6 5 4 3 2
12 11 10 09

Typeset in 9.95/13 Minion by 73
Printed and bound by Rotolito Lombarda, Italy

BRIEF CONTENTS

CONTENTS

Contents

Contents

Supporting resources

Visit www.pearsoned.co.uk/holden to find valuable online resources

Companion Website for students

- Video-clips allow you to see landscapes and processes in action
- Multiple choice questions to test your learning
- Models for practical learning
- Annotated weblinks for every chapter
- Comprehensive bibliography
- An online glossary to explain key terms
- Flashcards to test your understanding of key terms

For instructors

- PowerPoint slides that can be downloaded and used for presentations
- Gallery of selected diagrams and photographs from the book
- Fieldwork exercises

Also: The Companion Website provides the following features:

- Search tool to help locate specific items of content
- E-mail results and profile tools to send results of quizzes to instructors
- Online help and support to assist with website usage and troubleshooting

For more information please contact your local Pearson Education sales representative or visit www.pearsoned.co.uk/holden

GUIDED TOUR

NAVIGATION AND SETTING THE SCENE

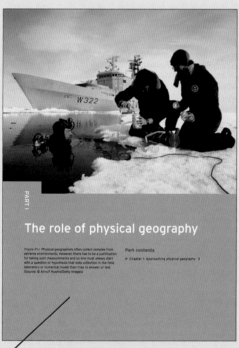

PART 1

The role of physical geography

Figure P1.1 Physical geographers often collect samples from extreme environments. However, there has to be a justification for taking such measurements and so one must always start with a question or hypothesis that data collection in the field, laboratory or numerical model then tries to answer or test. (Source: © Arnulf Husmo/Getty Images)

Part contents
➤ Chapter 1: Approaching physical geography 3

The book is divided into six **parts**, each with a part opener, describing the main themes and links between chapters within that part.

CHAPTER 21

Climate change: an unprecedented environmental challenge

John Grace
School of GeoSciences, University of Edinburgh

Learning objectives

After reading this chapter you should be able to:

➤ discuss the evidence for anthropogenic global warming
➤ understand how the use of fossil fuels has impacted upon the climate
➤ describe how the carbon cycle has been perturbed
➤ appreciate how humankind has created environmental problems and perceive how they may be solved

21.1 Introduction

Environmental change on a global scale first became a matter of public concern in the 1960s. Before then, the perceived environmental problem was urban pollution, which affected human health and the quality of life of many people. Although urban pollution became acute during the Industrial Revolution, it was not new. The smelting of toxic metals such as copper and lead was a health hazard in ancient Rome, as revealed by analysis of hair samples from the preserved corpses of Roman soldiers found in bogs, and

from traces of metal in Greenland ice cores. Coal was used in London in the thirteenth century. Coal contains not only carbon but also 1–4% sulphur and traces of heavy metals, and therefore its combustion releases a multitude of pollutants as well as carbon dioxide (CO_2). With the onset of the Industrial Revolution in western Europe, around 1780, the use of coal increased dramatically and cities such as London became heavily polluted with smog, a mixture of fog and smoke. Domestic coal burning was a major contributor to smog, and the industrial regions around Birmingham in England became known as the Black Country; even non-industrial Edinburgh was known as Auld Reekie, referring to the smell of coal burning.

Diseases such as bronchitis and tuberculosis were widespread following the Industrial Revolution, and nearly a quarter of deaths in Victorian Britain (1837–1901) were from lung diseases. In one week of December in 1952, 4000 Londoners were killed by a particularly severe episode of smog. The ensuing public outcry resulted in the Clean Air Act of 1957, which restricted coal burning and resulted in the use of cleaner energy sources such as oil, gas and electricity. Other coal-burning cities of the world such as Pittsburgh in the United States have a similar history. Problems were greatly exacerbated by the growth in use of the

Learning Objectives introduce topics covered and help you to focus on what you should have learned by the end of the chapter.

AIDING YOUR UNDERSTANDING

Figures, diagrams and photos feature throughout the text to illustrate key points and clarify topics discussed.

17.8 Fluvial-dominated coastal environments

Figure 17.37 Classification of fine-grained deltas proposed by Galloway (1975). The end-member types are fluvial-, tide- and wave-dominated. (Source: after Galloway, 1975)

distribution through the delta wedge is then a general fining in the vertical direction (Reading and Collinson, 1996). Based on the relative magnitude of river, wave and tide power, Galloway (1975) proposed a classification scheme of deltas with three end-members (Figure 17.37).

Fluvial-dominated deltas are associated with large catchments, river discharge into protected seas with minimal nearshore wave energy, and a small tidal prism. The freshwater river effluent is generally less dense than the saltwater in the receiving basin, and the river water flows out on top of the receiving water. If the outflowing river water is denser than the water in the receiving basin, for example due to extremely high suspended sediment concentrations, the river water will flow out along the seabed. Buoyancy limits the mixing between river and basin waters, allowing the river sediment to be transported further into the receiving basin before settling to the bed. As a result, the morphology of river-dominated deltas is characterized by pronounced seaward protrusions, a classic example of which is the so-called 'bird foot' of the Mississippi Delta (Figure 17.38).

Wave-dominated deltas are typically found in open-coast settings with a steep shoreface gradient, causing the deltaic coastline to be exposed to energetic waves. The action of

Figure 17.38 The Mississippi Delta is shaped like a 'bird's foot' and is the classic example of a river-dominated delta. The scene shown in this false-colour image covers an area of 54 km by 57 km and represents the currently active delta front of the Mississippi known as the Balize.

495

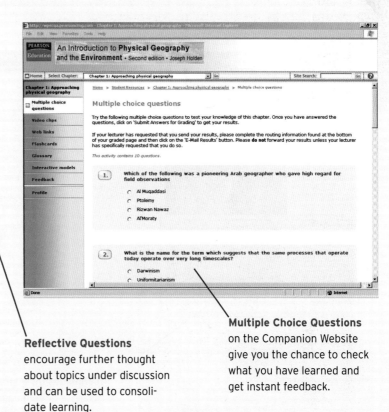

Chapter summaries recap and reinforce the key points to take away from the chapter. They also provide a useful revision tool.

Reflective Questions encourage further thought about topics under discussion and can be used to consolidate learning.

Multiple Choice Questions on the Companion Website give you the chance to check what you have learned and get instant feedback.

PHYSICAL GEOGRAPHY AND ENVIRONMENTAL ISSUES IN ACTION

Six categories of **boxed features** explore and illustrate topics and concepts through real world examples.

Fundamental principles offer further explanation of core concepts.

Techniques give detailed coverage of techniques and tools used within physical geography.

Case studies include a range of international examples and illustrations to add a real world relevance to topics discussed.

Environmental change explores environmental issues that shape the world around us.

New directions introduce the latest thinking and discoveries.

Video-clips allow you to see landscapes and processes in action.

Hazards apply the topics discussed to contemporary environmental problems.

GOING FURTHER

Further reading and **web resources** offer sources of additional information for those who wish to explore a topic further.

Annotated weblinks provide guidance and ideas about where to find useful further online resources and information.

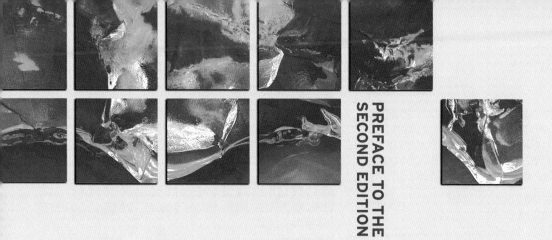

It has been four years since I wrote the preface to the first edition. Research in physical geography continues to advance and make a difference to the way humans interact with and manage the environment. It is therefore important to update textbooks to reflect the latest advances. This second edition has been updated with new discoveries, examples and techniques that were not known at the time of writing the first edition. There are new boxed features, new figures and three new chapters.

This edition contains a new chapter on coastal processes written by Gerhard Masselink. While the first edition contained three chapters on climate and weather and a further chapter on Quaternary climate change (last 2.4 million years), this new edition contains an additional two chapters on contemporary climate change written by John Grace. Climate change is such an important topic in today's changing world that it is now high on the political and public agenda and deserves considerable attention within a textbook on physical geography. The first of the new climate change chapters deals with recent and potential future climate change processes and impacts. The second new chapter on climate change deals with the role of vegetation in the climate system. Vegetation has acted as a key control to the Earth's climate system by absorbing carbon dioxide and producing oxygen. Humans have significantly modified the Earth's climate system through burning of fossil fuels, but also through the modification of the Earth's vegetation system. There are many ways in which the biosphere and atmosphere interact and this new chapter deals with these interactions.

Each of the contributors has worked hard to bring their material up to date. Unlike most other geography textbooks which are often written by just one or two people, this book is different in that each chapter is written by a specialist in their given field. Therefore it has been possible to produce a book that contains an extremely rich source of information and an impressively wide set of examples from around the world. The book draws on the experience and expertise of 16 contributors. They are internationally recognized environmental scientists who, together, have researched Earth system processes in every environment on the planet. With such a large number of contributors we can be sure that this book provides an extremely valuable resource for the budding physical geographer.

The contributors to this text will each have their own individual life stories about how they came to do the subject they study and these experiences will influence the way they think about the subject and the way they do research. For me there are many people whom I met during my schooldays and times at Cambridge, Durham and Leeds University who made an impact. They should all be assured that the way they did their research, the way they talked about and taught their subject and the way they interacted with others all made an impact upon me. Of those who have influenced my thinking are many students with whom I have worked in the lecture hall, the computer room, my office or the field. Teaching and learning are two-way processes. A scientist or academic who is not willing to learn from students will lack something from their repertoire of skills and understanding. Indeed, true academics remain students all their lives; they learn from their research and their interactions with other students, be they undergraduates, postgraduates, the general public, policy-makers or other academics and researchers.

When I was a small child my father, Henry Holden, would show me all the wonders of nature and explain what he knew about them with great excitement. He was not an academic geographer or scientist but he was, in a way, a truly modern geographer. He had a keenness for looking at changing landscapes such as the beaches of Northumberland, the shape of hillslopes and mountains, the marvels of the seasons and the atmosphere, the movements of water, the distribution of plants and animals, and the phenomenon of ice and snow. He wanted to understand and explain how all these things came to be and why and how they change.

During past centuries geographers have often been restricted to looking at pattern and distribution within individual places and regions rather than explaining things by examination of processes or looking at the inter-linkages between different places and different environments. However, it is now the question of *'why?'* which geographers pursue more than ever before and it is the study of environmental processes that forms the core of modern physical geography. I hope this book serves to help answer some of the *'why?'* questions that you may have, but I also hope that it helps to make you ask questions that you had never previously considered.

Scope of the book

This text provides an introduction to the major subjects of physical geography. It is aimed at those embarking on a university course and will help bridge the gap between studies at high school and higher education. The book should be of value throughout a university degree as it provides a baseline of understanding and additionally it directs the reader to resources that encourage them to develop their studies further. It focuses on understanding processes and is comprehensively illustrated in order to aid such process understanding. Physical geography is of wide interest and immense importance. It deals with the processes associated with climate, landforms, oceans and ecosystems of the world. The Earth has always been subject to changes in these systems and studying physical geography allows us to understand how Earth systems have come to operate as they do today. It also provides us with insights into how they may operate in the future. The impacts that humans have made on the Earth's environments are ever increasing as the world's population approaches seven billion. Thus the Earth's systems will change in the future both naturally and in a forced way through human action. However, it will be crucial to understand, manage and cope with such change and this can be achieved only by understanding the processes of physical geography.

Tools used in the book

In addition to providing a rich source of information the book uses a number of educational tools to aid understanding:

- The book is split into six parts, each with a **part opener** that describes the main themes of that part of the book and the links between the chapters within that part.
- **Learning objectives** clearly outline the purpose and aims of a particular chapter to help locate the reader within the book.
- **Boxed features** explore and illustrate topics and concepts through real-world examples. Scattered throughout every chapter, these insightful applications are differentiated into the following types:
  - fundamental principles,
  - techniques,
  - case studies,
  - environmental change,
  - new directions,
  - hazards.

- **Reflective questions** invite the reader to think about, and further explore, what they have just read. Useful for consolidating learning, these are found at the end of each major section of every chapter.
- The **comprehensive glossary** serves as an additional resource to help clarify concepts discussed within the book. Key words defined in the glossary are highlighted in the text.
- A **summary** draws together the key ideas of the chapter.
- An **annotated list** of further reading aims to inspire and enable deeper exploration into a topic.
- Useful **web resources** are listed that can be accessed in order to take the subject further.

Companion website

The book also has a dedicated **website** at www.pearsoned.co.uk/holden on which there is a suite of other educational resources for both students and lecturers alike.

Lecturer resources

- **PowerPoint slides:** a set of slides for every chapter comprising bulleted outlines of core topics and the key figures and images from the main text.
- **Field exercise ideas:** suggestions for a variety of field trips and associated activities.

Student resources

- **Digital video-clips:** a unique set of short films allowing students to see the landscapes and processes in action.
- **Multiple-choice questions:** a set of interactive questions for every chapter that allow students to test and consolidate their understanding.
- **Comprehensive glossary:** unique to this text, a handy resource to assist learning.
- **Interactive models for practical learning:** these models give students the opportunity to explore and understand environmental processes and the principles of modelling.
- **Bibliography:** an annotated list of further reading material.
- **Annotated weblinks:** several hundred annotated additional websites for students to further explore a topic.
- **Flashcards:** designed for students to test their knowledge of key terms for each chapter.

I hope that you are able to use the rich resources that this book provides and that the chapters engage you with their subjects and help inspire you to contribute to the subjects that come under the umbrella of physical geography.

Joseph Holden
January 2008

CONTRIBUTORS

Dr Pippa J. Chapman, University of Leeds

Dr David J. Gilvear, University of Stirling

Professor John Grace, University of Edinburgh

Dr Kate V. Heal, University of Edinburgh

Professor Joseph Holden, University of Leeds

Dr Timothy D. James, Swansea University

Professor Mike Kirkby, University of Leeds

Professor Michael D. Krom, University of Leeds

Dr John G. Lockwood, formerly of University of Leeds

Dr Gerhard Masselink, University of Plymouth

Professor John McClatchey, North Highland College

Professor Adrian T. McDonald, University of Leeds

Professor Tavi Murray, Swansea University

Professor Kevin G. Taylor, Manchester Metropolitan University

Professor David S.G. Thomas, University of Oxford

Dr Hilary S.C. Thomas, University of Wales, Newport

EDITOR'S ACKNOWLEDGEMENTS

I have been inspired and guided by many important figures throughout my career. There are many people at the places where I have studied and the places I have worked to whom I owe thanks. Many of these figures probably do not realize it but they influenced me a great deal. I would particularly like to thank the following: Alan Angus, Adrian Bailey, Lee Brown, Tim Burt, John Charlton, Nick Cox, Robin Glasscock, Frank Hodges, Christopher Keylock, Mike Kirkby, Stuart Lane, Adrian McDonald, Vincent McNeany, Sue Owens, Keith Richards, Ian Simmons, Ian Snowdon, Tom Spencer, Derek Talbot, Steve Trudgill and Jeff Warburton.

I am grateful to the contributors for their efforts and outstanding expertise in producing this second edition. They have worked hard to update the chapters and provide a comprehensive and up-to-date introduction to physical geography and the environment. I am indebted to many colleagues, friends, family members and students for help of various sorts. These include Eve Holden, Patricia Holden, Henry Holden, Vincent Holden, Clare Holden and a number of people who provided photographs and are acknowledged in the figure captions. Thanks also to Suada Khan, Sarah Smalley, David Appleyard and Alison Manson for help with work on some of the figures and supplementary materials.

I would also like to thank all the reviewers who read and commented in detail upon the chapters and suggested revisions that could be incorporated within this second edition. My kind appreciation also goes to the whole editorial and production team at Pearson Education including Anita Atkinson, Maggie Wells, Caterina Pellegrino and Sarah Busby, with particular thanks to Stephanie Poulter and Andrew Taylor who provided significant experience, expertise and support in helping to produce this second edition.

Finally I thank Eve for putting up with my long office hours, her continuous encouragement and for taking such wonderful care of our beautiful daughter Justina.

PUBLISHER'S ACKNOWLEDGEMENTS

The publishers would again like to thank Joe Holden for his fabulous and skilful steering of this complex project to its high-quality conclusion.

We would also like to thank the reviewers who have been involved in the development of this edition of the book. We are very grateful for their insightful comments and recommendations. Our thanks go to:

Dr. Ivar Berthling, Norwegian University of Science and Technology, Norway
Dion J. Wiseman, Brandon University, Canada
Randy W. Dirszowsky, Laurentian University, Canada
Dr. Robin Edwards, Trinity College, University of Dublin, Ireland
Dr. Jon French, University College London, UK
Professor Jack Hardisty, University of Hull, UK
Dr. Diane Horn, Birkbeck College, University of London, UK
Dr. Suzana Ilic, Lancaster University, UK
Dr. Jörg Kaduk, University of Leicester, UK
Professor Martin Kent, University of Plymouth, UK
Dr. Peter G. Knight, Keele University, UK
Professor Dr. Anders Lundberg, University of Bergen, Norway
Dr. Richard Pope, University of Derby, UK

We are grateful to the following for permission to reproduce copyright material:

Figure PI.1 © Arnulf Husmo/Getty Images; Figure 1.10 reprinted from *Sedimentary Geology*, 85, Nicholas J. Clifford, Formation of riffle-pool sequences: field evidence for an autogenetic process, p. 48, Copyright 1993, with permission from Elsevier; Figures 1.13, 1.14 and 1.15 after Gildor, H. and Tziperman, E., A sea-ice climate switch mechanism for the 100-kyr glacial cycles, *Journal of Geophysical Research*, 106, pp. 9117-33, 2001. Copyright 2001 American Geophysical Union. Modified by permission of American Geophysical Union; Figure 1.16 after Magnitude and frequency of forces in geomorphic processes in *Journal of Geology*, 68, The University of Chicago Press (Wolman, M.G. and Miller, W.P. 1960); Figure 1.19 courtesy of Natalie Suckall; Figure PII.1 Rob Reichenfeid © Dorling Kindersley, Figure 2.20 Max Alexander/Dorling Kindersley © Archaelogical Receipts Fund, Figure 8.9 Lloyd Park © Dorling Kindersley, Figure 8.15 Joe Cornish/DK Images, Figure 9.12 DK Images and Figure 20.16 photos courtesy of Dorling Kindersley; Figure 2.4 John Mead/Science Photo Library Ltd; Figure 2.5 reprinted by permission of Martin Jakobbson; Figure 2.6 from Magnetic survey off the west coast of North America, 30° N. latitude to 42° N. latitude, *Geological Society of America Bulletin*, 72, reprinted by permission of the Geological Society of America (Mason, R.G. and Raff, A.D. 1961); Figure 2.9 Krafft/Explorer/Science Photo Library Ltd; Figure 2.10 Landform Slides/GeoScience Features Picture Library; Figure 2.11 from Schematic cross section of ocean crust in Ophiolites, *Scientific American*, August, reprinted by permission of Andrew M. Tomko III (Bass, G.I. 1982); Figure 2.16 courtesy of Dr. W. Murphy; Figure 2.18 Getty Images/Stone; Figure 2.21 courtesy of Photolibrary Group; Figure 2.23 Peter Menzell/Science Photo Library Ltd; Figure 3.1 courtesy of Planetary Visions; Figure 3.3 adapted from *The Oceans, 1st Edition*, reprinted by permission of Pearson Education, Inc. (Sverdrup, H.U. *et al*. 1942); Figure 3.4 after K.A. Sverdrup *et al., Introduction to the World's Oceans,* copyright 2003, McGraw Hill Education, reproduced with permission of The McGraw-Hill Companies; Figure 3.5 from *Invitation to Oceanography, 1st Edition,* reprinted by permission of Jones and Bartlett Publishers (Pinet, P.R. 1996); Figure 3.12 Landsat-7 image courtesy of the Remote Sensing Group, Plymouth Marine Laboratory; Figure 3.15a Dr Ken MacDonald/Science Photo Library Ltd; Figure 3.15b Al Giddings/AL Giddings Images Inc; Figure 3.17 Steve Terrill/Corbis; Figures 3.23a,b &c from Future coral reef habitat marginality: climate change in the Pacific basin in *Coral Reefs,* 22, part of figure 2 on p. 555, copyright 2003, Springer-Verlag, with kind permission from Springer Science and Business Media (Guinotte, J.M., Buddemeier, R.W. and Kleypas, J.A. 2003); Figure 4.1 redrawn from *Climate Change 2001: The Scientific Basis. Contribution of Working Group 1 to the Third Assessment Report of the Intergovernmental Panel on Climate Change,* edited by Houghton *et al.,* pub Cambridge University Press, reprinted by permission of the Intergovernmental Panel on Climate Change (IPCC 2001); Figure 4.3 from *The Economics of Climate Change,*

Cambridge University Press (Stern, N. 2007); Figure 4.6 from *Man's Energy and World Climate*, Institute for Energy Analysis Report, Permission granted by Oak Ridge Associated Universities (ORAU), (Rotty, R.M. and Mitchell, J.M., Jr. 1974); Figure 4.8 from *Causes of Climate*, Edward Arnold, reprinted by permission of the author (Lockwood, J.G. 1979) and *Climate and Life*, Academic Press (Budyko, M.I. 1974); Figure 4.9 from The annual radiation balance of the earth-atmosphere system during 1969–70 from Nimbus 3 measurements in *Journal of Atmospheric Sciences*, 30, reprinted by permission of the American Meteorological Society (Raschke, K. *et al.* 1973); Figure 4.10 from *Causes of Climate*, pub Edward Arnold (Lockwood, J.G. 1979); Figure 4.11a after H. Riehl, *Introduction to the Atmosphere*, copyright 1965, McGraw Hill Education, Reproduced with permission of The McGraw-Hill Companies; Figure 4.11b from *Evaporation into the Atmosphere*, Figure 4.2, copyright © 1982 by D. Reidal Publishing Company, Dordrecht, Holland, Kluwer Academic Publishers BV, with kind permission of Springer Science and Business Media (Brutsaert, W. 1982); Figure 4.12 Sipa Press/Rex Features; Figures 4.13, 5.13 and 6.15 from *Contemporary Climatology, 2nd Edition*, pub Longman reprinted by permission of Pearson Education Limited (Henderson-Sellers, A. and Robinson, P.J. 1999); Figures 4.16 and 4.17 from Northern hemisphere monthly mean 500 millibar and 1000–500 millibar thickness charts and some derived statistics (1951–66), *Geophysical Memoirs*, 117, © Crown Copyright 2004, published by the Met Office, reprinted by permission of the Met Office (Moffit, B.J. and Ratcliffe, R.A.S. 1972); Figure 4.18 from The Indian Monsoon – a review, *Weather*, 20, reprinted by permission of Royal Meteorological Society (Lockwood, J.G. 1965); Figure 4.19 from Learning from El Niño, *Weather*, 53, reprinted by permission of Royal Meteorological Society (Philander, G. 1998); Figure 4.20 from The 1997/98 El Niño, *Weather*, 53, reprinted by permission of Royal Meteorological Society (Slingo, J. 1998); Figure 4.21 material taken from *Global Warming: The Complete Briefing, 1st Edition*, by J. Houghton, published by Lion Hudson plc 1994. Used by permission; Figure 4.22 from Impact of global warming on evapotranspiration, *Weather*, 48, reprinted by permission of Royal Meteorological Society (Lockwood, J.G. 1993); Figures 4.23, 21.2, 21.3, 21.4, 21.5, 21.6 and 22.16b from IPCC, 2007: Summary for Policymakers in: *Climate Change 2007: The Physical Science Basis. Contribution of Working Group 1 to the Fourth Assessment Report of the Intergovernmental Panel on Climate Change* [S. Solomon *et al.* (eds.)]. Cambridge University Press, Cambridge, United Kingdom and New York, NY, USA, figures SPM.2, SPM.3, SPM.1, SPM.4, SPM.5, SPM.7 and SPM.6 respectively; Figure 5.3 from: White *et al.*, *Environmental Systems: An Introductory Text, 2nd Edition*, D.N. Mottershead, S.J. Harrison and I. White, Copyright © 1992, Taylor and Francis Books Ltd. (Nelson Thornes), Reproduced by permission of Taylor & Francis Books UK; Figures 5.4 and 5.8 from *Meteorology Today, Introduction to Weather, Climate and the Environment 6th edition* by AHRENS. 2000. Reprinted with permission of Brookes/Cole, a division of Thomson Learning: www.thomsonrights.com. Fax 800 730-2215; Figures 5.5 and 5.7 adapted from *General Climatology, 3rd Edition*, reprinted by permission of Pearson Education, Inc. (Critchfield, H.J. 1974); Figure 5.17 © Warren Faidley/Weatherstock®.com; Figure 5.11 AP/PA Photos; Figure 5.18 Ron Kuntz/Corbis; Figure 5.22 from *Geography: An Integrated Approach, 2nd Edition*, pub Thomas Nelson and Sons Ltd., reprinted by permission of Nelson Thornes Ltd. (Waugh, D. 1995); Figures 6.4 and 24.2 courtesy of Alona Armstrong; Figures 6.5 and 6.23 from *Boundary Layer Climates*, Routledge (Oke, T.R. 1987); Figure 6.8 Barnabas Kindersley © Dorling Kindersley; Figure 6.10 Kent Wood/Science Photo Library Ltd; Figure 6.14 from Satellite images of the distribution of extremely low temperatures in the Scottish Highlands, *Meteorological Magazine*, 116, © Crown Copyright Met Office, reprinted by permission of the Met Office (McClatchey, J. *et al.* 1987); Figure 6.19 National Geographic/Getty Images; Figure 6.20 from The distinction between canopy and boundary layer urban heat islands, *Atmosphere*, 14, No. 4, Figure 5, reprinted by permission of the Canadian Meteorological and Oceanographic Society (Oke, T.R. 1976); Figure 6.22 reprinted with permission from Planning Advisory Service Report No. 376, *Improving Street Climate Through Urban Design* by Charles Thurow (Chicago: APA, 1983), p. 23; Figure 6.24 Reprinted from *Energy and Buildings*, 15, E. Jauregui, Influence of a large urban park on temperature and convective precipitation in a tropical city, p. 7, Copyright 1990–1991, with permission from Elsevier; Figure 6.25 courtesy of Photolibrary Group; Figure 6.26 courtesy of National Environment Agency of Singapore; Figures 7.6a-d and Figure 7.13 reproduced from National Soil Resources Institute, *Soils and their use in Northern England, Soil Survey of England and Wales, Rothamsted Experimental Stations, Harpenden, Herts*, © Cranfield University 1984, reprinted by permission of Cranfield University, National Soil Resources Institute (Jarvis *et al.* 1984); Figure 7.7 courtesy of Dr. J.S. Conway; Table 7.8 adapted from *The Nature and Properties of Soils, 12th Edition*, reprinted by permission of Pearson Education, Inc. (Brady, N.C. and Weil, R.R. 1999); Table 7.10 from *World Map of the Status of Human Induced Soil Degradation – An Explanatory Note. Global Assessment of Oil Degradation (GLASOD), 2nd Revised Edition*, ISRIC, Wageningen and UNEP, Nairobi, reprinted by permission of ISRIC (Oldeman, L.R. *et al.* 1991); Figures 7.9 and 7.10 courtesy of E.A. Fitzpatrick; Figure 7.11 from *Soils and Geomorphology, 3rd Edition*, figure 10.5, p. 274, reprinted with permission from Oxford University Press (Birkland, P.W. 1999); Figure 7.14 reproduced from National Soil Resources Institute, National Soil Map, © Cranfield University and for the Controller of HMSO, reprinted by permission of Cranfield University, National Soil Resources Institute; Figure 7.20 from: *Fundamentals of the Physical Environment, 3rd Edition*, P. Smithson *et al.*, Copyright © 2002, Routledge, Reproduced by permission of Taylor & Francis Books UK; Figure 7.22 from *The Nature and Properties of Soils, 13th Edition*, reprinted by permission of Pearson Education, Inc. (Brady, N.C. and Weil, R.R. 2002); Figure 7.24 Ron Giling/STILL Pictures The Whole Earth Photo Library; Figure 7.25 from *Status of UK Critical Loads - Critical Loads Methods Data and Maps,* February, reprinted by permission of Centre for Ecology and Hydrology (CEH, 2003); Figure 7.28 adapted from Revised estimates of the carbon mitigation potential of UK agricultural land, *Soil Use and Management,* 16, Blackwell Publishing (Smith, P. *et al.* 2000); Figure 8.3 from *UMD Global Land Cover Classification, 1 Kilometer, 1.0,* 1981–1994, Department of Geography, University of Maryland, College Park, Maryland, reprinted with permission from Dr. Matthew C. Hansen (Hansen, M. *et al.* 1998); Table 8.1 from *Vegetation of the Earth and Ecological Systems of the Geobiosphere, 3rd Edition,* Springer-Verlag, New York, with kind permission of Springer Science and Business Media (Walter, H. 1985); Figure 8.4 from *Die Ökozonen der Erde. 2. Aufl. UTB1514,* reprinted by permission of Verlag Eugen Ulmer (Schultz, J. 1995); Figures 8.6 and 8.8 courtesy of Nicholas Berry; Figure 8.7 Staffan Widstrand/Nature Picture Library; Figure 8.12 Stephen J. Krasemann/Photo Researchers Inc.; Figure 8.13 Carsten Peter/National Geographic; Figure 8.19 from *Vegetationszonen und Klima, 6 Aufl.,* p. 382, reprinted by permission of Verlag Eugen Ulmer (Walter, H. 1990); Figure 8.23 courtesy of Ed Reschke/Peter Arnold Inc/Still Pictures; Figure 8.24 Alan & Sandy Carey/Photodisc; Figure 8.25 Simon Fraser/Science Photo Library; Figure 8.26 from *Walter's Vegetation of the Earth: The Ecological Systems of the Geobiosphere, 4th Completely Revised and Enlarged Edition,* pub Springer-Verlag, reprinted by permission of Verlag Eugen Ulmer (Breckle, S.W. 2002); Figure 8.24 Alan & Sandy Carey/Photodisc; Figure 8.27 courtesy of Wouter Buytaert; Table 9.2 from *An Introduction to Applied Biogeography,* Cambridge University Press (Spellerberg, I.F. and Sawyer, J.W.D. 1999); Table 9.3 adapted from *Biodiversity and Conservation,* p. 111, Routledge (Jeffries, M.J. 1997); Table 9.4 after Man's attitude towards vegetation in *Man's Impact on Vegetation* edited by W. Holzner *et al.,* with kind permission of Springer Science and Business Media (Westhoff, V. 1983); Figure 9.10 Klein/Hubert/Bois/STILL Pictures; Figure 9.11 was published in *The History of British Mammals* by D. Yalden (ed), Copyright Elsevier, 1999; Figure 9.14 redrawn from Climatic Modelling of the Queensland fruit fly, Copyright CSIRO Australia; Figure 9.17 from *Urban Habitats,* Figure 3.2, Routledge, (Wheater, C.P. 1999); Figure 10.10 from A biodiversity concept map, Michael G. Robinson, from the web site of the 1999 Summer Biology Institute on Diversity, produced by the Leadership Program for Teachers at the Woodrow Wilson National Fellowship Foundation, http://woodrow.org/teachers/bi/1999/projects/group9/Robinson/conceptmap.html, used by permission of the Woodrow Wilson National Fellowship Foundation, 2004 (Woodrow Wilson National Fellowship Foundation 1999); Box 10.1 and Box 10.3 courtesy of Anne Goodenough, University of Gloucestershire; Figure 10.11 from Mechanisms of succession in natural communities and their role in community stability and organization, *The American Naturalist,* 111, University of Chicago Press (Connell, Joseph H. and Slayter, Ralph O. 1977); Figure 10.15 from Complexity and fragility in ecological networks, *Proceedings of the Royal Society of London B,* 268, pp. 1–7, Figure 1, reprinted by permission of The Royal Society (Solé, R.V. and Montoya, J.M. 2001); Figures 10.16 and 10.17 Copyright Anne Goodenough; Figure 10.18 courtesy of Kenfig National Nature Reserve; Figure 11.6 Robert Harding World Imagery/Tony Waltham; Figure 11.7 G.R. Dick Roberts/Natural Science Photos; Figure 12.2 from *Sand and Sandstone,* © 1972, 1987 by Springer-Verlag New York Inc., reprinted by permission of Springer-Verlag GmbH & Co. KG (Pettijohn, F.J., Potter, P.E. and Siever, R. 1987); Figure 12.5 from *Earth, 2nd Edition,* W.H. Freeman and Company (Press, F. and Siever, R. 1986); Figure 12.6 adapted from *Sedimentology and Sedimentary Basins,* Blackwell Publishing (Leeder, M.R. 1999); Figure 12.7 courtesy of Andrew Thomas; Figures 12.8 and 12.17 adapted from *Sedimentary Petrology: An Introduction,* Blackwell Publishing (Tucker, M.E. 1981); Figure 12.9 modified from diagrams in *Sedimentology and Stratigraphy,* Blackwell Publishing (Nichols, G. 1999). Reproduced with permission from Nichols (1999) and *The Depositional Environments, Sedimentology, Book 2,* reprinted by permission of Pearson Education Limited (King, C. 1991); Figures 12.10, 12.11, 12.13, 12.16 and 12.22 modified from diagrams in *Sedimentology and Stratigraphy,* Blackwell Publishing (Nichols, G. 1999). Reproduced with permission from Nichols (1999) and *The Depositional Environments, Sedimentology, Book 2,* reprinted by permission of Pearson Education Limited (King, C. 1991); Figure 12.14 courtesy of Chris Perry; Figure 12.18 from The role of alluvial fans in arid zone fluvial systems in *Arid Zone Geomorphology: Processes, Forms and Change in Drylands, 2nd Edition* edited by D.S.G. Thomas. Copyright 1997. © John Wiley & Sons Ltd. Reproduced with permission (Harvey, A.M. 1997); Figure 12.19 courtesy of Marli Bryant Miller; Figure 12.23 was published *in Journal of Environmental Radioactivity,* 32, P.N. Owens, D.E. Walling, and Q. He, The behaviour of bomb deprived Caesium-147 fallout in catchment soils, 169–191, Copyright Elsevier 1996; Figure 12.24 courtesy of Cathy Delaney; Figure 12.29 courtesy of Philip N. Owens; Figure 13.4 courtesy of the Hong Kong Observatory of Hong Kong Special Administrative Region; Figure 13.19 courtesy of Katy Gell; Figures 13.20 a & b courtesy of Martin Dawson; Figure 13.23 from: *Introducing Groundwater,* Michael Price, Copyright © 2002 Taylor and Francis Books Ltd. (Nelson Thornes). Reproduced by permission of Taylor and Francis Books UK; Figure 13.26 reprinted by permission of the National River Flow Archive, CEH Wallingford, and the Environment Agency, South Wessex area; Figure 13.27 from *Principles of Hydrology,* McGraw-Hill Publishing Company. Reproduced with the kind permission of the McGraw-Hill Publishing Company (Ward, R.C. and Robinson, M. 2000); Figure 13.29 and text in Box 13.4 courtesy of Sara Alexander, University of Leeds, river data for Figure 13.29 kindly supplied by the GRDC (Global Runoff Data Centre); Figure 13.30 Mark Edwards/STILL

Publisher's acknowledgements

Pictures The Whole Earth Photo Library; Figure 14.4 ©
Dominic Habron; Figure 14.5 Bernard Edmaier/Science Photo
Library Ltd; Figure 14.11 courtesy of Despina Psarra; Fig-
ure 14.26 from River channelisation, traditional engineering
methods, physical consequences and alternative practices,
Progress in Physical Geography, 9, reprinted by permission of
Arnold Publishers (Brookes, A. 1985); Figure 14.28b courtesy
of Charles Perfect; Figure 15.3 from Solutes in river systems
in *Solute Processes* edited by S.T. Trudgill. Copyright 1986.
© John Wiley & Sons Ltd. Reproduced with permission (Walling,
D.E. and Webb, B.W. 1986); Table 15.1 after *Global Environment:
Water, Air and Geochemical Cycles*, adapted by permission of
Pearson Education, Inc. (Berner, E.K. and Berner, R.A. 1996);
Figure 15.7 and Table 15.2 adapted from Hydrochemical
processes in *Contemporary Hydrology* edited by R.L. Wilby.
Copyright 1997. © John Wiley & Sons Ltd. Reproduced with
permission (Soulsby, C. 1997); Figure 15.8 was published in
Journal of Hydrology, 151, P.J. Chapman *et al.*, Hydrochemical
changes along stormflow pathways in a small moorland head-
water catchment in Mid-Wales, UK, 241-65, Copyright Elsevier
1993; Figure 15.10 after Runoff processes and solutional
denudation rates on humid temperate hill slopes in *Solute
Processes* edited by S.T. Trudgill. Copyright 1986. © John Wiley
& Sons Ltd. Reproduced with permission (Burt, T.P. 1986);
Figure 15.13 from *Biogeochemistry*, 68, 1-19, The influence of
catchment characteristics on the seasonality or carbon and
nitrogen species concentrations in upland rivers of Northern
Scotland, M.J. Clark, M.S. Cresser, R. Smart, P.J. Chapman and
A.C. Edwards, 2004, Kluwer Academic Publishers. With kind
permission of Springer Science and Business Media; Figure
15.14 was published in *Journal of Hydrology*, 178, D. Caissie, T.L.
Pollock, and R.A. Cunjak, Variation in stream water chemistry
and hydrograph separation in a small drainage basin, 137-57,
Copyright Elsevier 1996; Figure 15.19 was published in
Environmental Pollution, 88, D.A. Turnbull and J.R. Bevan, The
impact of airport de-icing on a river: the case of the Ouseburn,
Newcastle upon Tyne, 321-32, Copyright Elsevier 1995; Figure
15.20 adapted from *Biogeochemistry*, 70, 369-402, Trends in
dissolved organic carbon in UK rivers and lakes, F. Worrall
et al. 2004, Kluwer Academic Publishers. With kind permission
of Springer Science and Business Media; Figure 15.21 adapted
from Lake acidification in Galloway: a palaeoecological test of
competing hypotheses, *Nature*, 314, reprinted by permission of
Macmillan Magazines Ltd. (Battarbee, R.W. *et al.* 1985); Figure
15.22a and Figure 15.22b adapted from Water quality implications of an observed trend of rainfall and runoff, *Journal of the
Institution of Water and Environmental Management*, 5,
reprinted by permission of Chartered Institute of Water and
Environmental Management (Curran, J.C. and Robertson, M.
1991); Figure 15.23 courtesy of A.M. Hardie; Figure 15.24
adapted from Pathways of major elements from land to ocean
through rivers in *River Inputs to Ocean Systems* edited by J.M.
Martin, reprinted by permission of United Nations Environment Programme (Meybeck, M. 1980); Figure 15.29 adapted

with permission from Pharmaceuticals, hormones, and other
organic wastewater contaminants in US streams, 1999-2000:
A national reconnaissance, *Environmental Science and
Technology*, 36, Copyright 2002 American Chemical Society
(Kolpin, D.W. *et al.* 2002); Figure 15.30 after Hornberger, G.M.,
Cosby, B.J. and Wright, R.F., Historical reconstructions and
future forecasts of regional surface water acidification in
southernmost Norway, *Water Resources Research*, 25,
pp. 2009-18, 1989. Copyright 1989 American Geophysical
Union. Modified by permission of the American Geophysical
Union; Figure 15.32 after Maier, H.R. and Dandy, G.C., The use
of artificial neural networks for the prediction of water quality
parameters, *Water Resources Research*, 32, pp. 1013-22, 1996.
Copyright 1996 American Geophysical Union. Modified by
permission of the American Geophysical Union; Figure 16.5
after *World Atlas of Desertification*, *1st Edition*, pub Edward
Arnold, reprinted by permission of United Nations Environ-
ment Program (UNEP 1992); Table 16.4 adapted from Arid
environments: their nature and context in *Arid Zone Geomorphology: Process, Form and Change in Drylands* edited by
D.S.G. Thomas, Copyright 1997. © John Wiley & Sons Ltd.
Reproduced with permission (Thomas, D.S.G. 1997);
Figure 16.12 was published in *Ecosystems of the World*, Vol. 12A,
Hot Deserts and Arid Shrublands edited by M. Evenari *et al.*, A.
Shmida, M. Evenari, and I. Noy-Meir, Hot desert ecosystems: an
integrated view, 379-88, Copyright Elsevier 1986; Figure 16.14
after Langbein, J.B., and Schumm, S.A., Yield of sediment in
relation to mean annual precipitation, *Transactions of the
American Geophysical Union*, 39, pp. 1076-84, 1958. Copyright
1958 American Geophysical Union. Modified by permission of
the American Geophysical Union; Figure 16.17 courtesy of
Sebastian Engelstaedter; Figure 16.19 after *Atlas of World
Physical Features*, Copyright © 1972, Reprinted with permission of John Wiley & Sons, Inc. (Snead, R.E. 1972) and after
Sand seas and aeolian bedforms in *Arid Zone Geomorphology:
Process, Form and Change in Drylands* edited by D.S.G.
Thomas, Copyright 1997. © John Wiley & Sons Ltd. Reproduced with permission (Thomas, D.S.G. 1997); Figure 16.20
after Dune and dunefield development on Padre Island, Texas,
with implications for interdune deposition and water table
controlled accumulation, *Journal of Sedimentary Petrology*, 62,
reprinted by permission of Society for Sedimentary Geology
(Kocurek, G. *et al.* 1990); Figure 16.28 after Sand deserts
during the last glacial maximum and climatic optimum,
Nature, 272, reprinted by permission of Macmillan Magazines
Ltd. (Sarnthein, M. 1978); Figure 16.29 after *World Atlas of
Desertification*, *2nd Edition*, pub Edward Arnold, reprinted
by permission of United Nations Environment Program
(UNEP 1997); Figure 17.2 from Church, J.A. and White, N.J., A
20th century acceleration in global sea-level rise, *Geophysical
Research Letters*, 33, L01602, doi:10.1029/2005GL024826,
Copyright 2006 American Geophysical Union. Reproduced by
permission of American Geophysical Union; Table 17.3 and
Figure 17.36 from: *Coastal Systems*, Simon Haslett, Copyright

© 2000 Routledge, Reproduced by permission of Taylor & Francis Books UK; Figure 17.6 was published in *Atlas of Oceans, Wind and Wave Climate,* I.R. Young and G.J. Holland, Copyright Elsevier 1996; Figures 17.12a, c, e and 17.37 from: *Fundamentals of the Physical Environment, 2nd Edition,* D. Briggs *et al.,* Copyright © 1997, Routledge, Reproduced by permission of Taylor & Francis Books UK; Figure 17.21 reprinted from Open University, *Waves, Tides and Shallow-water Processes,* 1989, Butterworth-Heineman, with permission from Elsevier and Open University; Figure 17.22 was published in *Sedimentary Geology,* 80, R. Boyd, R. Dalrymple, and B.A. Zaitlin, Classification of clastic coastal depositional environments, 139–50, Copyright Elsevier 1992; Figure 17.23 from Danish-German-Dutch Wadden Environments in *The Physical Geography of Western Europe* edited by E.A. Koster, Oxford University Press, reprinted with permission from Oxford University Press (Hofstede, J., 2005); Figure 17.24 modified from *Geographical Variation in Coastal Development,* 2nd Edition © Longman, Reprinted by permission of Pearson Education Limited (Davies, J.L. 1980); Figures 17.25a and 17.25b from Wave dominated coasts in *Coastal Evolution: Late Quaternary Shoreline Morphodynamics* edited by R.W.G. Carter and C.D. Woodroffe, Cambridge University Press (Roy, P.S., Cowell, P.J., Ferland, M.A. and Thorn, B.G. 1994); Figure 17.29 from Aagard, T. and Masselink, G. (1999) The surf-zone. In: Short, A.D. (ed) *Handbook of Beach and Shore Morphodynamics,* pp. 72–118, © John Wiley & Sons Limited. Reproduced with permission; Figure 17.30 and Figure 17.42c courtesy of Aart Kroon; Figures 17.32, 17.33 and 17.34 from Estuarine facies models: conceptual basis and stratigraphic implications, *Journal of Sedimentary Petrology,* 62, reprinted by permission of Society of Sedimentary Geology (Dalrymple, R.W., Zaitilin, B.A. and Boyd, R. 1992); Figure 17.39 adapted from *A Celebration of the World's Barrier Islands,* Columbia University Press, NY (Pilkey, O.H. 2003); Figure 17.41 after *Geomorphology of Rocky Coasts,* Copyright 1992, © John Wiley & Sons Ltd. Reproduced with permission (Sunamura, T. 1992); Figure 18.1 after West Antarctic collapse – How likely?, *Episodes,* 13, No. 3, reprinted by permission of Episodes (Alley, A.B. 1991); Figure 18.2 courtesy of Jon Bamber and Adrian Luckman; Figure 18.3 modified from *Glacier Ice,* reprinted by permission of University of Washington Press (Post, A. and LaChappelle, E.R. 2000); Table 18.1 after *Glaciers and the Landscape,* Edward Arnold, © David E. Sudgen and Brian S. John 1976, reproduced by permission of Hodder Arnold (Sudgen, D.E. and John, B.S. 1976); Table 18.2 was published in *The Physics of Glaciers,* 3rd Edition, edited by W.S.B. Paterson, Copyright Pergamon 1994; Figure 18.4 courtesy of Lonnie G. Thompson, The Ohio State University; Figure 18.5b after Acceleration of Greenland ice mass loss in spring 2004, *Nature,* 443, Reprinted by permission of MacMillan Publishers Ltd (Velicogna, I. and Wahr, J. 2006); Figure 18.6 after The thermal regime of sub-polar glaciers mapped by multi-frequency radio-echo sounding, *Journal of Glaciology,* 42, reprinted from the *Journal of Glaciology* with permission of the International Glaciological Society and the author (Björnsson, H. *et al.* 1996); Figure 18.11a with permission from U.S. Geological Survey; Figure 18.11b courtesy Mats Wibe Lund/epa/Corbis; Figure 18.17b, Figure 18.19a and Figure 18.19b reprinted by permission of Norwegian Polar Institute; Figure 18.18a courtesy of NASA; Figure 18.18b was published in *Icarus,* 144, S.M. Clifford *et al.,* The state and future of Mars polar science and exploration, 210–42, Copyright Elsevier 2000; Figure 18.19c and Figure 18.19d reprinted by permission of A.K. Pedersen, Geological Museum of Copenhagen and Danish Lithospheric Centre; Figure 18.20, figure on left, from Snowfall-driven growth in East Antarctic ice sheet mitigates recent sea-level rise, *Science,* 308 (5730), reprinted by permission of AAAS (Davis, C.H., Li, Y. H. and McConnell, J.R. 2005); Figure 18.20, figure on right, from Recent ice sheet growth in the interior of Greenland, *Science,* 310 (5750), reprinted by permission of AAAS (Johannessen, O.M., Khvorostovsky, K. and Miles, M.W. *et al.* 2005); Figure 18.22 from Boulder shapes and grain-size distributions of debris as indicators of transport paths through a glacier and till genesis, *Sedimentology,* 25, Blackwell Publishing (Boulton, G.S. 1978); Figure 18.24 from Bimodal distribution of rock and mineral fragments in basal tills in *Till, A Symposium* edited by R.P. Goldthwait, Ohio State University Press, reprinted by permission of the authors (Dreimanis, A. and Vagners, U.J. 1971); Figure 18.25 was published in *Quaternary Science Reviews,* 10, J.K. Hart and G.S. Boulton, The interrelation of glaciotectonic and glaciodepositional processes within the glacial environment, 335–50, Copyright Elsevier 1991; Figure 18.25, photo (a) courtesy of Sarah J. Fuller; Figure 18.25, photo (c) courtesy of Andy J. Evans; Figure 18.30 Mike Potts/Nature Picture Library; Figure 18.31b from *Miscellaneous Report 28, Canada's Heritage of Glacial Features,* reproduced with the permission of the Minister of Public Works and Government Services Canada, 2007 and Courtesy of Natural Resources Canada, Geological Survey of Canada (Prest, V.K. 1983); Figure 18.34 Reprinted from *Sedimentary Geology,* 62, J. Menzies, Subglacial hydraulic conditions and their possible impact on subglacial bed formation, 125–50, Copyright 1989, with permission from Elsevier; Figure 18.35a reprinted from Drumlin formation by subglacial meltwater erosion, *Canadian Journal of Earth Science,* 24, reprinted by permission of NRC Research Press (Shaw, J. and Sharpe, D.R. 1987); Figure 18.35b reproduced with permission from John Shaw, A qualitative view of sub-ice-sheet landscape evolution, *Progress in Physical Geography,* 18, 159–184, Copyright © Sage Publications Ltd., 1994, by permission of Sage Publications Ltd.; Figure 19.2 courtesy of Larry D. Hinzman/University of Alaska Fairbanks; Figure 19.3 after Permafrost in *The Heritage of Engineering Geology: the first hundred years, Geological Society of America Centennial Special Volume 3* edited by G.A. Kiersch, reprinted by permission of the Geological Society of America (Péwé, T.L. 1991); Figures 19.4, 19.5 and Table 19.1 from *The Frozen Earth, Fundamentals of Geocryology,* Cambridge University Press (Williams,

P.J. and Smith, M.W. 1989); Figure 19.6b courtesy of Science Photo Library; Figure 19.7 from Ground surface temperature reconstruction based on data from a deep borehole in permafrost at Janssonhaugen, Svalbard, *Annals of Glaciology*, 31, reprinted from the *Annals of Glaciology* with permission of the International Glaciological Society and the author (Isaksen, K. *et al.* 2000); Figure 19.8 from *The Periglacial Environment, Permafrost and Man, Commission on College Geography Resource Paper No. 14*, Figure 5, reprinted by permission of the Association of American Geographers (Price, L.W. 1972); Figure 19.9 from The effect of thermal abrasion on the temperature of the permafrost in the coastal zone of the Laptev Sea, *Proceedings of the Second International Conference on Permafrost, Yakutsk, USSR*, reprinted by permission of National Academy of Sciences, courtesy of the National Academies Press, Washington, D.C. (Molochuskin, E.N. 1973); Figure 19.10 after Permafrost hydrology in North America, *Atmosphere-Ocean*, 24, Figure 10c, reprinted by permission of the Canadian Meteorological and Oceanographic Society (Woo, M.K. 1986); Figure 19.12 from: *The Ice Age in Britain*, B.W. Sparkes and R.G. West, Copyright © 1972 Methuen, Reproduced with permission from Taylor & Francis Books UK; Figure 19.13 courtesy of Julian Murton; Figures 19.14, 19.17 and 19.21 a,b and c from *The Periglacial Environment, 2nd Edition*, Longman, reproduced by permission of Hugh French (French, H.M. 1996); Figure 19.15b Peter Dunwiddle/Visuals Unlimited; Figure 19.18a Paolo Koch/Science Photo Library Ltd; Figure 19.19 from Pingo growth and subpingo water lenses, western Arctic coast, Canada, *Proceedings of the Fourth International CoFigureerence on Permafrost*, reprinted by permission of the National Academies Press (Mackay, J.R. 1983); Figure 19.20 from Pingos, springs and permafrost in Spitzbergen, originally published in *Norsk Polarinstitutt Arbok*, reprinted by permission of the Norwegian Polar Institute (Liestøl, O. 1977); Figure 19.22 from Surface soil displacements in sorted circles, Western Spitzbergen, *Proceedings of the 5th International Conference on Permafrost, Vol. 1*, reprinted by permission of Tapir Academic Press (Hallet, B. *et al.* 1988); Table 19.2 modified from Recent developments of mountain slopes in Karkevagge and surroundings, northern Sweden in *Geografiska Annaler*, 42, Blackwell Publishing (Rapp, A. 1960); Figure PVI.1 Seth Resnick/Getty Images; Figure 20.3 courtesy of Alice Milner; Figure 20.6e from *Quaternary of the Thames, 1st Edition*, Fig. 1.2, © 1994 Joint Nature Conservation Committee, Chapman and Hall, with kind permission of Springer Science and Business Media (Bridgland, D.R. 1994); Figure 20.8 courtesy of P. Williamson/University of Bern; Figure 20.9 modified from The Greenland Ice-Core Project (GRIP): reducing uncertainty in climate change?, *NERC News*, April 26–30, reprinted by permission of the author (Peel, D.A. 1994); Figure 20.10 after http://www.cru.uea.ac.uk/cru/info/thc/, Crown Copyright, Met Office, Hadley Centre, reprinted by permission of Met Office; Figures 20.13, 20.21 and Table 20.2 from *Reconstructing Quaternary Environments, 2nd Edition*, © Longman Group Ltd.

1984. This edition © Addison Wesley Longman Limited 1997, reprinted by permission of Pearson Education Ltd. (Lowe, J.J. and Walker, M.J.C. 1997); Figure 20.14 courtesy of Ken Hamblin; Figure 20.15 after Alluvial architecture of the Quaternary Rhin-Meuse river, *Geologie en Mjinbouw*, 72, Figure 2.42, © 1993 Kluwer Academic Publishers, With kind permission of Springer Science and Business Media (Ruegg, G.H.J. 1994); Figure 20.17 from Large-scale patterns of dune type, spacing and orientation in the Australian continental dunefield, *Australian Geographer*, 19, Taylor & Francis Ltd., http://www.informaworld.com (Wasson, R.J. *et al.* 1988); Figure 20.18 courtesy of John Corr; Figure 20.19 from Holocene vegetation and climates of Europe in *Global Climates Since the Last Glacial Maximum* edited by H.E. Wright, Jr. *et al.*, University of Minnesota Press, Copyright 1993 by the regents of the University of Minnesota (Huntley, B. and Prentice, K. 1993); Figure 20.20 from illustration by Katharine Sutliff in Quaternary refugia and persistence of biodiversity, *Science,* 297, 20 September, reprinted by permission of AAAS (Taberlet, P. and Cheddadi, R. 2002); Figure 20.24 after New long continental pollen record from a tropical intermontane basin: Late Pliocene and Pleistocene history from a 540 m core, *Episodes*, 14, reprinted by permission of Episodes (Hooghiemstra, H. and Sarmiento, G. 1991); Figure 20.25 from *A Climate Modeling Primer*, Copyright 1997, © John Wiley & Sons Ltd. Reproduced with permission (McGuffie, K. and Henderson-Sellers, A. 1997); Figure 21.7 from G.A. Meehl *et al.*, 2007: Global Climate Projections. In: *Climate Change 2007: The Physical Science Basis. Contribution of Working Group I to the Fourth Assessment Report of the Intergovernmental Panel on Climate Change* [S. Solomon *et al.* (eds.)]. Cambridge University Press, Cambridge, United Kingdom and New York, NY, USA, Figure 10.19, with permission from IPCC; Figure 21.10 was published in *Atmospheric Environment*, 40, Korbetis Malamo, Reay David S. and Grace John, New directions: rich in CO_2, 3219–220, Copyright Elsevier 2006; Figure 21.11 from *The Ozone Hole Tour: Part II: Recent Ozone Loss Over Antarctica*, reprinted by permission of Centre for Atmospheric Science, Department of Chemistry, University of Cambridge; Table 21.1, Figure 22.1 and Table 22.1 from Global climate and the distribution of plant biomes, *Philosophical transactions of the Royal Society B*, 359, pp. 1465–1476, reprinted by permission of The Royal Society (Woodward, F.I., Lomas, M.R. and Kelly, C.K. 2004); Figure 22.2 from *A History of Atmospheric CO_2 and its Effects on Plants, Animals and Ecosystems,* copyright 2005, Springer-Verlag, New York, with kind permission from Springer Science and Business Media (Ehleringer, J. *et al.* 2005); Figure 22.4 from Millennial-scale dynamics of Southern Amazonian rain forests, *Science,* 290, figure 2, reprinted by permission of AAAS (Mayle, F.E., Burbridge, R. and Killeen, T.J. 2000); Figure 22.5 from Height growth response of tree line spruce to recent climate warming across the forest-tundra of eastern Canada in *Journal of Ecology*, 92, Blackwell Publishing (Gamache, I. and Payette, S. 2004); Figure 22.6 from A global change-induced biome shift in the Montseny mountains (NE Spain), *Global Change Biology*, 9, Blackwell Publishing (Penuelas, J. and Boada, M. 2003); Figure 22.7 from European phonological response to climate change matches the warming pattern, *Global Change Biology*, 12, Blackwell Publishing (Menzel, A., Sparks, T.H., Estrella, N. *et al.* 2006); Figures 22.9 and 22.10 from Global response of terrestrial ecosystem structure and function to CO_2 and climate exchange: results from six dynamic global vegetation models, *Global Change Biology*, 7, Blackwell Publishing (Cramer, W., Bondeau, A., Woodward, F.I. *et al.* 2001); Figure 22.11 from Climate-carbon cycle feedback analysis: results from the C4MIP model intercomparison, *Journal of Climate*, 19, American Meteorological Society (Friedlingston, P., Cox, P., Betts, R. *et al.* 2006); Figure 22.12 The human footprint in the carbon cycle of temperate and boreal forests, *Nature*, 447, Reprinted by permission of MacMillan Publishers Ltd (Magnani, F. *et al.* 2007); Figure 22.14 after Galloway, J.N. *et al.*, Nitrogen fixation: anthropogenic enhancement and environmental response, *Global Biogeochemical Cycles*, 9, pp. 235-52, 1995. Copyright 1995 American Geophysical Union. Modified by permission of the American Geophysical Union; Figure 22.15 from Positive feedbacks amongst forest fragmentation, drought and climate change in the Amazon, *Conservation Biology*, 15, Blackwell Publishing (Laurence, W.F. and Williamson, B. 2001); Figures 23.2a, 23.2b and 23.9 reprinted by permission of Environment Agency; Figures 23.5, 23.10 a & b, 23.11 a & b, 23.13, 23.14, Table 23.2 and Table 23.3 from *Remote Sensing and Image Interpretation, 4th Edition*, Copyright © 2000 John Wiley & Sons, Inc. This material is used by permission of John Wiley & Sons, Inc. (Lillesand, T.M. and Kiefer, R.W. 2000); Figure 23.7 redrawn from *An Introduction to Remote Sensing for Environmental Monitoring, Report No. 1*, pub Institute of Environmental Studies, University of Wisconsin, reprinted by permission of the authors (Scherz, J.P. and Stevens, A.R. 1970); Figure 23.12 after P.R. Wolf and B.A. Dewitt, *Elements of Photogrammetry with Applications in GIS, 3rd Edition,* copyright 2000, McGraw-Hill Education, Reproduced with permission of The McGraw-Hill Companies; Figure 23.18 courtesy of Tavi Murray; Figure 23.31a, Figure 23.31b and Figure 23.32, images obtained from MIMAS Landmap Service. http://landmap.ac.uk. Copyright University of Manchester/University College London 2001. Original Landsat 7 distributed by Infoterra International; Figure 24.1 courtesy of Issac Agyemang; Figure 24.2 courtesy of Alona Armstrong; Figure 24.5 adapted and Reproduced by permission of SAGE Publications, London, Los Angeles, New Delhi and Singapore from MacDonald (2000) in Reflective Practice in Geography Teaching; Figure 24.6 Mickey Gibson/Animals Animals/Photolibrary.com; Figure 24.7 courtesy of B.C. Ministry of Environment; Figure 24.8 Christopher Boisvieux/Corbis; Table 24.4 adapted from *A Flood Risk Management Strategy for the River Severn at Gloucester*, BBR No: BWA 000 6359, © Environment Agency. All rights reserved. Figure 24.14 Fritz Hoffmann/The Image Works/TopFoto.

In some instances we have been unable to trace the owners of copyright material, and we would appreciate any information that would enable us to do so.

The role of physical geography

Figure PI.1 Physical geographers often collect samples from extreme environments. However, there has to be a justification for taking such measurements and so one must always start with a question or hypothesis that data collection in the field, laboratory or numerical model then tries to answer or test. (Source: © Arnulf Husmo/Getty Images)

Part contents

Scope

Science, and with it the subject of physical geography, has evolved over time. Part I deals with how physical geography has developed in order to help explain why we go about studying it today in many different ways. This part contains one chapter which deals with the ways in which we approach physical geography. It describes the basic frameworks for studying science and explains the roles of fieldwork, laboratory work and modelling. The chapter describes the advantages and disadvantages of a range of approaches that we should be aware of when studying physical geography. It therefore sets the scene for the rest of the book by providing the reader with an appropriate grounding in the nature of the subject.

What do we mean by physical geography?

Physical geography involves the understanding of interactions of processes involving the Earth's climate system, oceans, landforms, animals, plants and people. This understanding requires the synthesis of environmental dynamics (linking the physical systems together) and the synthesis of societal and environmental dynamics (relating human actions to the physical environment). Of interest to physical geographers are the mechanisms that maintain flows of energy and matter across the Earth. There are components of study which include processes associated with plate tectonics, geomorphology, climatology, glaciology and hydrology that shape the surface of the Earth; the collection of climatic and atmospheric processes acting as one of the ultimate controls on the landscape and biosphere; and the ecological and biogeographical patterns that characterize the living portion of the Earth. Physical geography involves the application of technology to study these components and changes within them. For example, remote sensing provides an aid to monitoring the world's constantly changing natural and human landscapes, the oceans, atmosphere and biosphere.

Geographers often say that they study the 'why of where'. By this they mean that their geographical curiosity is grounded by an interest in explaining the underlying processes that result in the patterns of natural phenomena and the ways in which humans interact with, and alter, these processes and patterns. In addition to a spatial context, change over time is also a central theme to physical geography.

It is important to be aware of the ways in which physical geographers study physical geography. Some kind of philosophical basis of enquiry is essential in order to allow fair comparisons of results and interpretation of conclusions between different research areas. The scientific methods discussed in Chapter 1 help to form this philosophical foundation. The underlying method does not necessarily mean that all research is done using the same techniques; indeed physical geography utilizes a variety of tools to help understand, measure, observe and predict environmental processes. However, by maintaining a philosophical basis, it reminds us to question the approach we take. In recent years, emphasis has shifted from a position where science represents the ultimate authority informing society, to a realization that science is itself influenced by society, and that many other sources of knowledge must be equally considered. Consideration of the advantages and limitations of a given approach is therefore vital so that we can assess the reliability and usefulness of the conclusions attained.

Why do we study physical geography?

Physical geography affects most aspects of our daily lives and is fundamental to our human existence. Physical geography allows us to be conscious of the nature of natural and human-induced environmental change and to understand the implications of such change for human life, landscapes and the biosphere. Indeed the variety of scales at which geographers approach their subject permits them to adapt readily to the relevant scale of enquiry required by policy-makers, helping to bridge the gap between science and society. We therefore study physical geography in order to make a difference to people's lives.

CHAPTER 1

Approaching physical geography

Joseph Holden

School of Geography, University of Leeds

Learning objectives

After reading this chapter you should be able to:

➤ describe the historical basis of physical geography

➤ understand the basic scientific method

➤ evaluate scientific methods for different types of research in physical geography

➤ appreciate the advantages and limitations of different approaches to physical geography

1.1 Introduction

Physical geography affects most aspects of our daily lives. It is fundamental to human existence. For example, it determines water availability and water quality, weather and climate, soil systems, potential for agriculture, the risk of landslides or other hazards, and if and how we can travel from one place to the next. Figure 1.1 illustrates the major components of physical geography. It deals with the Earth's climate system, which results from a combination of atmospheric, oceanic, land, ice and ecological processes. It also deals with a wide range of processes that affect the landscape of the Earth. For example, plate tectonic processes are responsible for mountain building, the movement of the continents, ocean floor spreading, ecological isolation and changing climate. In addition the landscape is worn down by weathering and erosion processes, many of which are driven by gravity and water (in solid, liquid and gas form). Water also transports nutrients from soils to plants and from rocks and soils to rivers and into the oceans. It transports nutrients and energy around the globe through the oceans and the atmosphere. It moves sediments across hillslopes, catchments and seas. Understanding the variety of processes that link the components shown in Figure 1.1 (atmosphere, oceans, landforms and biosphere) at global and small scales enables improved prediction of future change of the Earth's environmental systems.

A range of tools are available to physical geographers in order to help us understand, measure, observe and predict environmental processes. These include tried and tested methods along with new technologies such as advanced probes and laboratory methods or geophysical and remote sensing tools that allow us to measure the Earth's features and processes remotely. For example, there are now acoustic Doppler velocity probes that allow us to collect data on the speed of water in three directions at the same time (vertically up and down, laterally upstream and downstream and laterally across a river or water body), taking

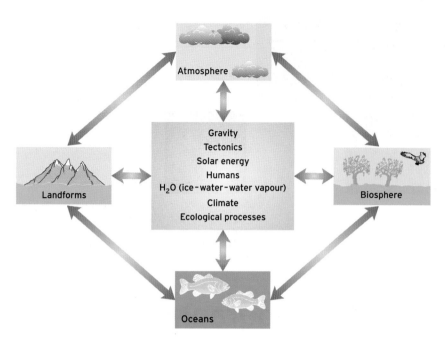

Figure 1.1 Components of the changing environment. The atmosphere, oceans, biosphere and landforms all interact with each other. Important links within the system include H_2O in its different forms, tectonics, ecological processes and humans.

measurements thousands of times per second. This allows us to understand how water actually flows around obstacles and thereby to develop better engineering structures such as bridges or flood protection that are less likely to be eroded or damaged. Another example of a recently developed tool now available to physical geographers is Light Detection And Ranging (LiDAR). This tool uses instruments on board aircraft to measure very accurately the topography of the Earth's surface and also the structure of the vegetation or built environment on top of the Earth's surface. This can be done with a resolution of just a few centimetres and therefore allows lots of important data to be collected over very large areas without the need for a time-consuming field survey, which is particularly advantageous in areas that are difficult to access on foot.

Fieldwork, laboratory work and numerical modelling are all important components of the method of physical geography today. However, each particular approach and method has its limitations. No matter what type of measurement device or approach is used, it is often how it is used and why it is being used in that way that is important. In other words, scientific approaches have a philosophical underpinning which can be evaluated. There are a range of approaches to science and physical geography and each approach has advantages and disadvantages. It is therefore necessary to understand these methods and their limitations so that we can: (i) evaluate which are the most appropriate methods to use for a given environmental investigation and (ii) fully evaluate the implications of any given research finding in physical geography.

The approaches that physical geographers have used have varied through time as the subject has developed. In order to understand contemporary practice in physical geography it is therefore necessary to know something about the history and development of the subject. This chapter will briefly describe the way physical geography has developed. It will then move on to discuss how the scientific method has been applied by physical geographers to studies of the environment. The remaining parts of the chapter will look at the principles of and approaches to (i) fieldwork, (ii) laboratory research and (iii) numerical modelling, all of which are important methods of physical geography. Prominent themes will emerge from these areas of discussion including the need to couple processes operating on different time and space scales and ideas of equilibrium, magnitude–frequency and feedback.

1.2 Historical development of physical geography

1.2.1 Physical geography before 1800

The ancient Greek and Roman geographers concerned themselves with topographic descriptions of places and their history, calculations about the circumference of the Earth and the production of maps, and a philosophical interest in the relations between humans and the environment. Between the time of the Roman Empire and the sixteenth century, European science progressed very slowly. Often scholars rejected anything that seemed to go against the teachings of the Christian Church. In the Middle East, however, Arab geographers such as Al Muqaddasi (who lived between AD 945 and 988) were pioneering fieldwork whereby observations were given precedence. Indeed Al Muqaddasi stated that he would not present anything unless he had seen it with his own eyes. Such Arab geographers maintained the Greek and Roman techniques and developed new ones. Arab traders travelled throughout Asia, Africa and the Indian Ocean and added a great deal of geographical knowledge to update the classical sources. Any European geographical work was trivial in comparison with the huge amount published by Islamic writers of the Middle Ages. Exploration and learning also flourished in China with advanced triangulation techniques allowing exceptionally good quality maps of the region to be produced from the first century AD onwards. Indeed official Chinese historical texts contained a geographical section from this period onwards, which was often an enormous compilation of changes in place-names and local administrative divisions controlled by the ruling dynasty, descriptions of mountain ranges, river systems, taxable products and so on.

While science was slow to progress in Europe before the sixteenth century, with the **Renaissance** came a renewed interest in the geographical knowledge of the ancients (which the Arab and Chinese scientists had already advanced significantly) and a willingness to test and refine their theories. The European explorations of the fifteenth and sixteenth centuries were part of a major period of invention and discovery. Table 1.1 lists some of the important discoveries made during this period. Improvements in measuring devices such as timekeepers and in mapping and printing techniques were combined with a new geographical knowledge about the world. Indeed many of these new technologies had roots in the pursuit of geographical knowledge. For example, methods for accurately keeping time were developed when stable navigation systems that could determine the longitude (east–west position) of a ship were invented. As the Earth is constantly rotating, knowing the time while making an altitude measurement to a known star or the Sun provided data to accurately calculate longitude. The experiences of the explorers had begun to overturn traditional views of those thought to be authority figures (such as leaders of the Christian Church and the theories of the ancient Greeks). For example, new continents were being discovered and the layout of landmasses across the Earth was being determined. A fundamental importance (as recognized much earlier by Al Muqaddasi) was beginning to be placed on the role of real-world experience. This meant that determining whether or

Table 1.1 Major discoveries of the fifteenth and sixteenth centuries

1410	A translation of Ptolemy's *Geography* was published in Europe (Ptolemy was an Egyptian astronomer and geographer who lived from AD 87 to 150)
1418	Prince Henry the Navigator established the Sagres Research Institute
1455	Gutenberg invented the printing press
1492	Columbus discovered the New World (although some suggest that there were earlier Norse settlements in North America and that there were original migrations from Asia and Europe around 14 000 years ago)
1498	Vasco da Gama sailed around the Cape of Good Hope to India
1500	Cabral discovered Brazil
1504	Columbus correctly predicted the total eclipse of the moon
1505	Portugal established trading posts in East Africa
1510	Henlein of Nuremberg invented spring-powered clocks permitting smaller (and portable) clocks and watches
1519	Magellan's ships began a circumnavigation of the Earth
1543	Copernicus suggested the Earth and other planets revolved around the Sun
1543	Vasalius produced a detailed description of human anatomy
1556	Tartaglia (Venetian mathematician) showed how to fix position and survey land by compass bearing and distance
1569	Mercator created his map of the world using a projection technique that bears his name
1581	Galileo concluded that the time for a lamp (pendulum) to swing does not depend on the angle through which it swings. This observation eventually led to the development of pendulum clocks
1590	Zacharias and Hans Janssen combined double convex lenses in a tube, producing the first telescope
1592	Galileo developed a type of thermometer based on air

not there was a Southern Ocean landmass could only be established through experience and not by just reading the works of Aristotle. This triumph of experience over authority was a central theme of the development of science during this period. However, it was because geography was inextricably linked to exploration, patriotism and colonization that it became a popular subject before 1800. It was therefore considered an important subject by the society of the time. Geographers were making the key advances in discovering new lands, mapping them, changing people's perception of the shape and size of features of the Earth and bringing potential 'wealth' to nations that conquered and colonized others.

1.2.2 Physical geography between 1800 and 1950

1.2.2.1 Uniformitarianism

Prior to the early nineteenth century the prevailing belief of the western world had been that the Earth was created in 4004 BC. The landscapes of the Earth were thought to be a result of catastrophic events. For example, it was thought that river valleys were scoured out during the biblical flood and that peatlands were remnants of the slime left behind after the flood receded. However, the increasing scientific knowledge acquired between the sixteenth and the end of the nineteenth centuries began to lead to different views developing. One new idea that emerged, for example, was that the Earth's landscapes gradually changed over time rather than being affected by sudden catastrophic events. Indeed one of the most persistent and influential themes to affect physical geography and especially geomorphology was the *Theory of the Earth* published by James Hutton in 1795 and clarified by Playfair (1802) in his *Illustrations of the Huttonian theory of the Earth*. Hutton and Playfair were scientists who examined the Earth's landscapes and tried to understand their formation. Hutton's theory rejected catastrophic forces as the explanation for environmental features and gave rise to a school of thought known as **uniformitarianism** (Gregory, 1985). The central tenet of this concept is that present-day processes that we can observe should be used to inform our understanding of past processes that we cannot observe. In other words, many of the processes we can see today are probably the same as those that occurred in the past and so we can infer what went on in the past from understanding contemporary environmental processes. Uniformitarianism propagated the idea that 'the present is the key to the past'. Although this idea was very satisfactory in terms of the processes for understanding the past, of course it cannot be assumed that the rates at which processes operate today (e.g. weathering of rock) are the same as those that occurred in the past.

Nevertheless it was still recognized that given enough time a stream could carve a valley, ice could erode rock, and sediment could accumulate and form new landforms. It would take tens of thousands of years to weather the rock to produce the features shown in Figure 1.2(a)–(c) and indeed Hutton speculated that millions of years would have been required to shape the Earth into its contemporary form. It was not until the early 1900s and the discovery of radioactivity that estimates of the age of the Earth became more reliable. Radioactive elements such as uranium and strontium are unstable and decay at a steady rate. Uranium^{-238}, for example, decays into lead^{-206}. Comparing the ratio of these two elements allows us to determine how much time has passed since the uranium sample was pure when the rock solidified. Radioactive decay also gives off heat and we can determine the rate of Earth cooling to determine a time when it formed. The Earth is in fact around 4.6 billion years old. The oldest rocks that have been found on the Earth date to about 3.9 billion years ago.

1.2.2.2 Darwin, Davis and Gilbert

Charles Darwin, who was a botanist who collected and organized specimens, read some of the writings on uniformitarianism and extended these ideas to biology. Darwin's *The origin of species* published in 1859 was hugely influential in the field of science and in society in general. Indeed it has often been referred to as the 'book that shook the world'. The book outlined how there could be a relatively gradual change in the characteristics of successive generations of a species and that higher plants and animals evolved slowly over time from lower beings. This evolution occurred as a result of competition within local interacting communities (see Chapters 9 and 10). Darwin's book helped throw the idea that there was a complete difference between humans and the animal world into turmoil as he reinforced the suggestion that humans evolved from lower beings. Prior to this it was believed in the western world, based on biblical works, that humans were created superior to other beings. With the idea that humans could have evolved from lower beings came the undermining of traditional religious opinions. However, although some religious leaders did embrace Darwinism at the time, the theories were very different from those that had come before. These ideas radically shook a society where, because of the increasing availability of printed books and papers, intellectual knowledge was being transferred in greater quantity than ever before.

Darwin's ideas therefore influenced most areas of science at the time. The idea of 'change through time' was reflected

(a)

(b)

(c)

Figure 1.2 Erosion features: (a) a weathered rock face; (b) rugged cliffs eroded by coastal processes; (c) an isolated rock weathered by water, wind, ice and chemical action over thousands of years.

in evolutionary attitudes to the study of landforms following Darwin's own 1842 study of the evolution of coral islands which was particularly influential in relation to the 'cycle of erosion' idea promoted by W.M. Davis (Gregory, 1985). The approach recommended by Davis, who was a very revered geomorphologist, dominated approaches to physical geography from the late nineteenth century

through until the 1950s. Davis suggested in 1889 that the normal cycle of erosion could be used to classify any landscape according to the stage that it had reached in the erosion cycle. He termed this the 'cycle of life', which was a rather biological metaphor for landform development. Figure 1.3 shows the **Davisian cycles of erosion**. A youthful uplifted landscape begins to be dissected by rivers. As the

(a) Young

Recently uplifted with new incision

(b) Mature

Deep and widespread valley incision

(c) Old

Almost flat, featureless peneplane, with the landscape eroded away

Figure 1.3 The Davisian cycles of erosion: (a) young uplifted stage with very limited incision; (b) a mature stage with deep valley incision and complex topography; (c) an old eroded landscape with few topographic features. (Source: after Davis, 1889)

landscape matures these valleys become wider and more gently sloping until eventually all that remains is a flat, old landscape (a **peneplain**). In addition there were random or chance elements to the normal cycle. These were volcanic activity, glacial activity and climatic change. The great success of the Davisian approach, dominating popular physical geography for 60 years, was due to the fact that it was simple and applicable by students to a wide range of landscapes. As a result of these ideas people then tried to determine the history of an area by establishing which stage of the Davisian cycle it was in. This approach was also known as **denudation chronology**. While Davis had based his ideas on the Appalachians in the United States, the Davisian ideas were applied by many to help interpret landscapes across the globe (e.g. Cotton, 1922, applied the ideas to parts of New Zealand and Wooldridge and Linton, 1939, produced a Davisian interpretation of south-east England). In plant geography and ecology a similar influence was being

expressed by Clements in his concept of succession (see Chapter 8). It is notable, however, that two themes of Darwin's work (struggle and selection/randomness and chance) did not have an immediate impact on physical geography (Stoddard, 1966). Indeed the unique contribution of Darwin's theory, which was 'random variation' whereby random change could occur to species from one generation to the next, did not really appear in work by physical geographers until the 1960s (Gregory, 1985). Nevertheless the theme of evolution provided an historical perspective to physical geography which still dominates geomorphology, studies of soils, biogeography and climatology.

An alternative approach that was advocated at the same time as the Davisian approach was that of G.K. Gilbert. Gilbert was an explorer of the American West. Gilbert wanted to understand *why* particular landforms developed rather than just classify them as being youthful or mature. In order to understand landform development he recognized the importance of describing physical processes and deriving systems of laws that determined how a landform could change. He attempted to apply quantitative methods to geological investigations. His ideas, however, were not taken on board during an era dominated by the descriptive techniques offered by Davis. It was not until the 1950s that physical geography came to revisit his approach and that Gilbert's ideas finally won favour. Until the 1950s, therefore, physical geography was largely descriptive and was concerned with regions. It was concerned with the evolution of environments and their classification. There were virtually no measurements of environmental processes involved and if you look at geography books from that period you will see that they are structured by regions and simply describe regional climates, landscapes, resources and trade (e.g. L. Dudley Stamp's 1949 book *The world: A general geography*).

1.2.3 Physical geography since 1950

1.2.3.1 The quantitative revolution

In the 1950s European and North American geography was forced to change. It was realized that describing places and putting boundaries around them, where in fact real boundaries did not actually exist, was no longer a useful approach. The 1950s were a time of increasing globalization when more people began to travel by air to far-flung destinations and when television began to show programmes made around the world, thereby opening up people's experience and views of the world. Global trade was increasing and mass-produced items such as refrigerators, cars and plastic became much-wanted goods. It became more common for people to own goods that were made in other countries (e.g. Europeans buying Ford cars made in the United States). It therefore became evident that there were increasing human and physical interlinkages between regions. It was also a period of modernity in which there was a societal commitment to order and rationality, and to science as the driving force for future developments and improvements in infrastructure and lifestyles. Physical geography needed to maintain its academic status and it could no longer do so within a society that now had a 'professional' science (see below). The Davisian cycles of erosion could not be verified from a scientific perspective and furthermore they did not *explain* observations. It was too difficult to measure such slow processes over such large spatial scales. Arthur Strahler, a geomorphologist particularly interested in rivers and landform change, proposed that a new dynamic basis for physical geography should be developed based on physical real-world measurements. It was also at this time that Hack (1960), a physical geographer, went into the Appalachians (coincidentally the very heart of the Davisian theory) and realized that landscapes were more delicately adjusted and that there was some form of equilibrium between rivers and landscapes. Box 1.1 describes these equilibrium approaches and their limitations.

It was also during this time that the work of G.K. Gilbert was revisited and his approach eventually embraced. This was largely due to the pioneering studies of the hydrologist Robert Horton and the development of his ideas by Strahler and his graduate students, who included Stanley Schumm, Marie Morisawa, Mark Melton and Richard Chorley. The 1950s are often referred to as the time of a quantitative revolution in geography due to the move away from description and towards measurement. Work began to concentrate on smaller spatial scales where processes could be measured during short-term studies.

1.2.3.2 Functional geomorphology

However, although quantitative techniques were being employed these were not necessarily those that Gilbert had proposed. The measurements that were being performed in the 1950s and 1960s often did not allow us to evaluate or understand physical processes properly. They tended to be quantitative descriptions rather than the measurement of processes. For example, in 1953 Leopold and Maddock, physical geographers who studied rivers, published results

of a survey of streams and rivers in the central and south-west United States. They found that stream width, depth and velocity increased in proportion to the discharge to the power of a given number (e.g. width is proportional to discharge to the power of 0.5; see Chapter 14). As the discharge increased downstream the equations suggested that channel width, mean depth and mean velocity should all increase. These equations could be used to make predictions about the discharge or **hydraulic geometry** of rivers across the world (see Chapter 14).

There are two problems with this approach. The first is that the relationships determined are purely statistical relationships (or functional relationships). In other words, they are just a result of the average value of the width, depth, velocity and discharge of all the rivers that were measured, but this does not explain *why* channel dimensions vary in such a way with discharge. These sorts of statistical relationship do not explain the physical processes. The second problem is that such functional geomorphological approaches are often not applicable to areas other than the area for which they were determined. This is because local factors can influence the development of a landform (such as geology or tree roots on a river bank holding the bank together and preventing it from eroding) so that it does not conform to the statistical average. Indeed sometimes it is the unusual cases that we are more interested in rather than the average. It was for these reasons that Yatsu (1992), a Japanese physical geographer with expertise in rock weathering, accused Strahler himself of 'crying wine and selling vinegar'. This means that he thought Strahler had advocated a new physical geography founded in physically based process measurement (the 'wine') but that Strahler was actually not measuring the physical processes. Hence Strahler was actually advocating a poorer type of physical geography (he was selling only 'vinegar'). Strahler and others around him made measurements but these were measurements of the wrong type. They were not physically based process measurements. It was Yatsu in 1966 who stated: '[Physical geographers] have often asked the what, where and when of things and they have seldom asked how . . . and they have never asked why. It is a great mystery why they have never asked why.' It was not until the mid-1970s that physical geography more fully adopted the idea of measuring processes in order to understand and explain the world. It was during this period that physical geography managed to embrace 'basic scientific methods' (see below). Nowadays the emphasis is very much on understanding processes (e.g. how vegetation interacts with soils and soil development – Chapters 7 and 8; how vegetation and climate interact – Chapter 22; or

FUNDAMENTAL PRINCIPLES

EQUILIBRIUM CONCEPTS IN PHYSICAL GEOGRAPHY

When Hack (1960) completed a field visit to the Appalachian Mountains he realized that rather than there being one long Davisian erosion process whereby rivers wore away mountains over time, there is in fact a more dynamic set of processes operating. He rejuvenated Gilbert's concept of 'dynamic equilibrium'. He suggested that every slope and every channel in an erosional system are adjusted to each other and that relief and form can be explained in spatial terms rather than historic ones. This work suggested that river profiles were never exactly concave. Instead, when sediment from a hillslope builds up in a river it has to steepen itself in order to move that sediment. Once removed, the river may become less steep in profile. In other words, the rivers and slopes would adjust to each other in an attempt to be in equilibrium.

Of course, the nature of equilibrium investigated depends on the timescale under investigation.

Figure 1.4 shows forms of equilibrium over three timescales (Schumm and Lichty, 1965). Note that over short timescales it may be possible to identify a static equilibrium (no change over time) or a steady-state equilibrium (short-term fluctuations about a longer-term mean value) while over longer time periods the equilibrium might be dynamic (shorter-term fluctuations with a longer-term mean value that is changing).

However, the concept of equilibrium has always been somewhat confusing because different people have chosen to identify different types of equilibrium and because the precise meaning is time dependent. Indeed, equilibrium may be just as generalized and untestable as the Davisian cycle of erosion it was meant to replace. Often it depends on where and when you measure something as to whether it will show equilibrium. Figure 1.5 illustrates this very simply for two systems that in the long term are behaving differently. Because the measurements were done at the times shown in Figure 1.5, it was not possible to identify the nature of the long-term

trend and in fact different trends have been determined from those that are actually occurring in the long term.

Figure 1.5 The timescale for human measurement makes it very difficult to identify long-term trends and the nature of equilibrium. Here the measurements are taken at two times (t_1 and t_2) for each case. However, because of the timing of the measurements we have incorrectly identified the nature of the long-term change in each case. In (a) we have established a downward trend where there is no long-term trend and in (b) we have identified no change while the long-term trend is upward.

Figure 1.4 Equilibrium over three timescales: (a) dynamic equilibrium; (b) steady-state equilibrium; (c) static equilibrium.

BOX 1.1

how glaciers erode their beds and move over the land surface – Chapter 18). However, as the following sections will demonstrate, although a basic scientific method may be adopted, there are actually many types of scientific method proposed and used, and each method can be criticized: this should be borne in mind when reading through the rest of this textbook and during your studies in general.

Reflective questions

➤ How did Darwin's *Origin of species* impact physical geography?

➤ What were the main differences between physical geography before 1950 and that since 1950?

➤ Why was there a change in approach to physical geography in the 1950s?

1.3 Scientific methods

Science uses a number of techniques and has adopted several philosophical approaches. It is important for physical geographers to be aware of such methods so that we can evaluate the advantages and limitations of the way in which we are approaching a research topic. The following section provides information on the key principles of scientific method.

1.3.1 The positivist method

Modern science is grounded in observation. It places special emphasis on empirical (derived from experiment and observation) measurements over theoretical statements (human ideas about how a system might work). Standard scientific method involves the formulation of hypotheses (e.g. 'the Earth's climate has warmed over the past 100 years') and the collection of information to test these hypotheses (e.g. comparing temperature measurements). This helps us explain why physical geographers (and scientists in general) are preoccupied with experimentation and measurement.

One of the key elements of scientific method is that of **causal inference**. This is the idea that every event has some sort of cause and so causal inference is the process by which we link observations under this assumption. However, it is

rare for causal inference not to be affected by what we know and think already. Therefore there must be some theoretical basis for research ideas. At the same time, however, the theories are tested so that observations still have greater status than theories.

One of the key approaches for linking theory and observation is known as **positivism**. Positivism is a traditional philosophy of science which has its origins with the philosopher Auguste Comte in the 1820s. Its idea was simply to stick to what we can observe and measure and ensure that science was separated from religious explanations for phenomena. It uses repeatable research methods so that the same tests can be performed again as a quality control. There are different forms of positivist approach. Two important ones are **logical positivism** (or logical empiricism) and **critical rationalism**.

In logical positivism scientists use **inductive** reasoning whereby unordered knowledge is defined, measured and classified into ordered knowledge. Once it is ordered, then regularities may be discovered and these might suggest a natural law. It uses experiments to acquire knowledge. For example, we might measure water table depth and find that it seems to control the rate of birch tree seedling growth. If we find that in areas with low water tables, such as sandy soils, seedling growth is slow compared with clay soils with higher water tables then we may develop a theory that water table depth controls the rate of tree growth.

This was criticized, however, by the philosopher Karl Popper, who suggested that you could not derive a law based on this technique. This is because the failure to do infinite experiments means that there could always be a case that does not fit the law. It may just be that you have not yet found the case that does not obey the law. Instead Popper argued that critical rationalist approaches should be adopted whereby you start with a theory that leads to the formation of a hypothesis. Then you test the hypothesis to the limit in an attempt to falsify the hypothesis. In other words, you try to disprove the hypothesis (rather than try to prove it). This is based on **deductive** reasoning. Figure 1.6 provides a conceptual diagram of this approach. If a theory survives falsification after testing it can still not be proved. You can only say that it has not been disproved. This is the basis for most scientific approaches today. For instance, one theory may be that all sheep have four legs. You test this by observing sheep. If you find every sheep has four legs then you can say that the evidence corroborates the theory and *probably* all sheep have four legs. You cannot be certain, however, that this is the case and it is impossible to verify the hypothesis (particularly as not all sheep have

Figure 1.6 Schematic diagram of the typical experimental approach used by critical rationalists.

been born yet). However, if you find a three-legged sheep then you can say that not all sheep have four legs. In this case you have falsified the hypothesis and now you must refine the hypothesis and test a new theory. The data collection, analysis and hypothesis testing strategy can involve field measurements, laboratory tests or even numerical modelling.

1.3.2 Critique of the positivist method

While physical geography has now had a 'scientific' basis since the 1950s there are a number of criticisms of the use of traditional science within environmental subjects. These include the following:

1. The idea that it is impossible to do completely objective science: this is because there is always a human who is doing the experiment and that person will always apply some sort of meaning to the research. In other words, there is an interaction between the enquirer and the research they are carrying out which is ignored by positivism. Indeed, sometimes a scientist might be motivated by a range of external forces when doing their experiments. For example, a scientist may be trying to obtain results in order to produce some data to show to those that

funded the research and also to help make sure that the scientist can be awarded further funding for more research in the future. This removes a great deal of independence and objectivity from research.

2. 'Context stripping': this means that although experiments might be well controlled and repeatable, because they involve simplification (e.g. manipulation of water table causing a change in seedling growth) the fact that there may be more complex interactions is often ignored. In this way positivism is **reductionist** because it assumes one can 'close' a system and look at the relationship between two variables in that system while holding all others constant. For example, if we think about the seedling growth example discussed above then water tables may themselves be controlled by soil texture and topographic setting. Seedling growth may also be more dominantly affected by soil nutrient availability rather than the water table. The water table position itself may control other factors, such as soil temperature. Thus, simple positivist experiments may lack the environmental context required for understanding the processes that are operating.

3. Positivist approaches tend to separate out grand theories from local contexts: positivist approaches often produce statistical generalizations that are statistically meaningful but that are inapplicable to individual places (e.g. a general equation based on data from 100 sites that links slope and the frequency of meanders on a river, but which does not seem to work for a particular site being studied).

4. By actually measuring something we may be changing it: for example, if we wanted to measure the erosion on a hillslope (see Chapter 11) we might dig a pit to collect the sediment. However, by digging that pit we may be changing the rate of erosion on the slope. Similarly if we wanted to measure the velocity of water in a river we might place a probe into the river to measure it. However, by placing that probe into the river we may be changing the velocity of the water around the probe compared with the probe not being there (Figure 1.7).

5. Not all things are necessarily measurable, yet they may still be important.

1.3.3 Realism as an alternative positivist approach

In the 1990s realism gained popularity in physical geography as an alternative positivist approach that tried to get around some of the critiques discussed above. This approach attempts to reintroduce philosophy into method.

Figure 1.7 Investigating flow in a sandy stream. The presence of the flow meter in the water causes changes to the water movement that the flow meter is trying to measure.

In one version of realism that is often cited in the physical geography literature (e.g. Richards *et al.*, 1997) there are three levels into which any phenomenon can be structured (Figure 1.8): (i) mechanisms (underlying processes that cause things to happen); (ii) events that they produce depending on circumstances; and (iii) the empirical

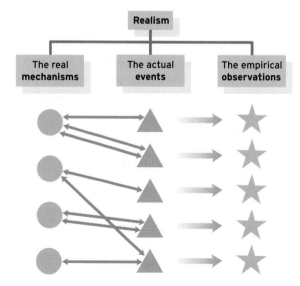

Figure 1.8 Facets of realism. Different combinations of mechanisms result in different events, some of which can then be observed or measured.

observations of those events made by humans. This approach accepts that in complex open environmental systems, interacting mechanisms may not always produce an event. Events occur only when the mechanisms interact at the right place at the right time.

Figure 1.8 shows how events are associated with the coming together of mechanisms, which are themselves influenced by other events. An event may be the shattering of a piece of rock. The underlying mechanisms to produce that event will include fluctuating temperatures above and below freezing, the presence of water in rock, and expansion and contraction of water within the rock upon freezing and thawing. However, the rock shattering will occur only if the rock contains water and if the rock has been weakened sufficiently by previous weathering. Therefore, while there may be fluctuating temperatures that cause freezing and thawing of water within a rock, it will only shatter (the event) when the right combination of processes (mechanisms) occurs at the right place and right time.

Different combinations of mechanisms result in different events. For example, a meander bend may form at a given point only if: (i) there is a river flowing past the point; (ii) the right turbulent flows, sediment erosion and transport mechanisms are operating; (iii) the soil material is readily erodible to allow a meander bend to form; and (iv) local vegetation material or geology does not restrict the meander bend formation. In this case the exact size, shape and location of the meander bend will be a result of general meander bend-producing processes and local factors. In addition this also suggests that the existing landform itself changes the way processes interact and thereby influence how the landform will develop in the future.

This approach brings together the idea of general laws and local events. It helps us understand how local factors influence the mechanisms to produce an individual example. In this way it is different from traditional positivist approaches which are just interested in generalizations and iron out any irregular or unusual cases. In realism the case study itself becomes of interest in helping us to understand the world around us. This is because what causes change is the interaction of mechanisms with particular places. This is particularly useful for geography, which has a tradition of examining case studies (see below). In addition, the realist approach allows us to evaluate different results at different scales. Often great scientific progress can be made when rules that emerge at one scale of investigation are used to question and critique those that emerge at other scales (Lane, 2001).

1.3.4 No such thing as a single scientific method in physical geography?

More recently, there have been moves by physical geographers to state that while it matters what approach they decide to take, it should not matter that their approach is different from another person's approach. In fact Lane (2001) argued that one of the ways science moves forward is by trying to solve disagreements between one set of findings and another set of findings that have been produced by a different method. It tries to make sense of disagreements. One of the important elements of physical geography is that it is examined by using a wide variety of techniques and approaches which makes for a rich subject and a diversity of findings. If everyone had the same method and approach to physical geography then this might actually hinder scientific progress, discovery and innovation.

A good example of this is provided by Chapter 20 which discusses the environmental changes during the past 2.4 million years. In order to investigate such changes it has been necessary to develop a wide range of techniques including climate, ocean circulation and ecosystem modelling, sediment dating techniques, analysis of different types of evidence (e.g. pollen records, fossils, gas bubbles trapped deep inside ice sheets and the contents of ocean floor sediments) and landform interpretation. Indeed much of physical geography is based on 'historical science'. Historical science in this sense means taking measurements that involve historical inference and developing an explanation of phenomena where it is not possible to measure directly the processes involved. An example would be determining the location of an ice sheet 23 000 years ago. To do this would require 'proxy measures' (physical measurements that are based on present evidence of past conditions). The approach is often to build up as much evidence as possible from many different types of sources (e.g. landform shape, radiocarbon dating, fossil plants) so that the hypothesis becomes more acceptable as the multiple lines of evidence are compiled.

It is often the mixture of backgrounds that physical geographers have, and the mixture of ideas about what they want to study, how much they want to achieve, and who they have been collaborating with, that defines how they will approach a new problem given to them. What is important is that you are aware of the advantages and limitations of the various approaches to physical geography so that you can be ready to incorporate them into the way you and others interpret the research findings of

investigations in physical geography. The following section describes some of the advantages and disadvantages of field, laboratory and modelling approaches.

Reflective questions

➤ What are the main advantages and limitations of a traditional positivist approach?

➤ What are the differences between logical positivism and critical rationalism and how would your method be different if you had a critical rationalist approach rather than a logical positivist approach?

➤ Why is realist theory so attractive to physical geographers?

➤ What are the benefits and disadvantages of having several types of scientific method that physical geographers use?

1.4 The field, the laboratory and the model

Fieldwork, laboratory investigation and numerical modelling are not necessarily independent of each other. Laboratory analysis often requires fieldwork, and numerical models typically require laboratory or field data for validation. Throughout this textbook the findings that are discussed and the processes that are explained will have been investigated using a mixture of field, laboratory and numerical modelling approaches. The following sections identify some of the main types of issues involved with choosing a particular method.

1.4.1 Approaching fieldwork

In physical geography, fieldwork is a fundamental component of research. It is a means of obtaining experience of the environment. However, there is a more specific set of reasons for doing fieldwork rather than other types of investigation. Harré (2002) suggested there were 12 uses of an experiment. These are listed in the first two columns of Table 1.2. We can see that there are a wide variety of reasons and purposes behind the experimental method. However,

Table 1.2 Harré's purposes for an experiment and the role of intensive and extensive fieldwork and of modelling. ✓ and ✗ indicate whether they fit this role

Harré's role of an experiment	Fieldwork		Numerical models
	Intensive/case study	Extensive case study	
As formal aspects of method			
1 To explore the characteristics of a naturally occurring process	✓ Detailed environmental processes examined	✓ General environmental relationships	✓ Traditional role of modelling to examine a range of scenarios
2 To decide between rival hypotheses	✓ Choose between potential processes	✓ Choose between potential influences	✓ Parts of model are tested to see which are the best hypotheses
3 To find the form of a law inductively	✗ Hard to find laws from case studies	✓ An important aspect	✗ Not possible because you need laws before you can get a model to work
4 As models to simulate an otherwise unresearchable process	✗	✗	✓ One of the most useful aspects of modelling
5 To exploit an accidental occurrence	✓ Good for investigating unusual cases	✗ Not included as by definition it is not accidental	✓ Used more to understand potential accidental occurrences such as a landslide
6 To provide negative or null results	✓ We can falsify from one case study and the advantage of this approach is understanding why it is false	✓	✗ Not possible as we do not know whether the negative result is just because we use models or whether it is a real falsification
In the development of the content of a theory			
7 Through finding the hidden mechanism of a known effect	✓ Main justification for case studies	✗ Hard to investigate mechanisms behind data	✗ We cannot find hidden mechanisms if they are not included in the original model

(continued)

Table 1.2 Harré's purposes for an experiment and the role of intensive and extensive fieldwork and of modelling. ✓ and ✗ indicate whether they fit this role (*continued*)

Harré's role of an experiment	Fieldwork		Numerical models
	Intensive/case study	Extensive case study	
8 By proving something exists	✓ Demonstrating something exists	✗ Demonstrates relationships but not phenomena	✓ They can corroborate findings of other methods
9 Through the decomposition of an apparently simple phenomenon	✓ Main justification for case studies	✗ Generally not enough detailed knowledge available	✗ Not possible
10 Through demonstration of underlying unity within apparent variety	✗ More likely to identify difference	✓ Identifies generalizations	✓ They can identify general patterns
In the development of technique			
11 By developing accuracy and care in manipulation	✓ Design for particular applications	✓ Statistical methods	✗ Not possible
12 Demonstrating the power and versatility of apparatus	✓ Develop/test new technique at case study site	✗	✗ Not possible

(Source: after Harré, 2002; Lane, 2003)

the type of experiment and purpose of the experimental method will vary depending on whether the research is extensive or intensive.

1.4.1.1 Extensive research

Extensive research consists of a large number of samples over many places with empirical observations as the basis for theoretical development. Extensive research methods involve measuring a large number of samples so that we can make statements about the entire population. For example, we might measure 1 000 000 of the sand grains on 1000 beaches. From this we then attempt to make a statement about all the sand grains on beaches across the world. In other words, we infer things from the sample for the entire population. It is just like asking 1000 people which party they would vote for in an election and then assuming that those 1000 people were representative of the entire country. However, we qualify our statements by saying what the probability is of a statement being correct. For example, we may state that we are 95% certain that the results have not occurred purely by chance. It is not possible to be 100% certain unless the entire population has been measured, which in physical geography is not easy! There is a great deal of literature on extensive sampling methods (e.g. Burt *et al.,* 2006).

1.4.1.2 Intensive research and case studies

Intensive research may be restricted to one or two places in which theoretical developments are used as the context for empirical observations and whereby process and mechanisms can be investigated. Case studies are intensive field research methods and are used widely in physical geography. They can involve sets of measurements taken of natural conditions, or indeed of conditions that humans have disturbed (e.g. an agricultural field compared with a natural hillside). It is very common for researchers to do field experiments at case study sites whereby you manipulate a factor (e.g. you change the vegetation cover) and measure how this affects other factors (e.g. water infiltration). In many ways the approach adopted in a field experiment could be the same as in a laboratory (see below) but the environmental context would be different as you have less control over the other factors that operate, but at the same time the situation is likely to be more representative of the environment you are studying.

There are a number of problems with using case studies. Firstly, if you spend a lot of time in just one place it must be asked whether the findings are relevant to other places. In most cases they will not be. Therefore, we have to ask what is actually the point of doing intensive research at just one place (e.g. on one glacier in New Zealand)? Case studies provide us with a better way of looking at how things are related. In extensive studies it is often hard to establish whether particular variables are really related to each other or whether other factors have affected them. In small-scale case studies, however, we may be able to see how one variable affects another. Furthermore, a case study fits into the realist perspective outlined above which states that we are interested in how general relationships occur in particular places. Therefore case studies are used to focus on individual events in order to identify how the processes are coming together to create that event (Richards, 1996). This helps us identify the causes of an event (e.g. a landslide). In other words, by using a case study in the field we can establish how processes operate and how they come together at a particular place and time to produce an event. Box 1.2 provides an example of a typical piece of realist field research which allows us to develop an alternative framework for thinking about the environment. This is one in which landform and process are inextricably linked

by a series of feedback mechanisms. It demonstrates how small events can have a major impact if they occur at the right time.

Table 1.2 shows how case studies and extensive field studies fit with Harré's roles for experiments. By using this table it is possible to identify the reasons why we might do fieldwork. However, it should be noted that Harré's table is heavily biased towards the experimental sciences, and of course much of physical geography is based on historical science.

1.4.2 Approaching laboratory work

Physical geographers use the laboratory (Figure 1.11) for two reasons: (i) analysis of something sampled in the field in order to derive its properties (e.g. concentration of carbon dioxide in an air sample); and (ii) experimentation in order to see how something behaves under controlled conditions. Of course with laboratory work care is needed in order to ensure that samples are not contaminated and that they are stored correctly so that they are not affected by laboratory conditions (e.g. storing ice cores in a freezer at the correct temperature). Laboratory work also requires very careful research design so that we can be sure that a 'fair test' has been carried out.

A good example comes from determining whether sulphate deposited in rainwater controls the release of dissolved organic carbon from soils. Sulphate has historically been released in many parts of the world where industrialization has occurred due to the burning of fossil fuels (Chapters 7 and 21). This led to rainwater with high concentrations of sulphate resulting in more acidic rain ('acid rain'). More recently, however, in western Europe and North America the concentrations of sulphate in rainwater have been falling. At the same time the concentration of dissolved organic carbon in river waters has been increasing (Worrall and Burt, 2004). Dissolved organic carbon is a hazard because it leads to the water becoming discoloured (brown) and also makes it difficult to treat to make it safe for drinking. In order to test whether the decrease in sulphate deposition is a causative factor in dissolved organic carbon release, a laboratory experiment can be set up. This can be done for three soil types as some soils might behave differently from others.

In the laboratory this requires a number of samples of each of the soil types in order that we have enough replicates to be sure that one result was not just an unusual result compared with others. It also requires that we establish whether sulphate in rainwater has any effect at all.

NEW DIRECTIONS

FORM-PROCESS FEEDBACK INTERACTIONS: THE EXAMPLE OF HOW ONE PEBBLE CAN CAUSE A WHOLE RIVER TO MEANDER

Clifford (1993) investigated riffle-pool sequences in the River Quarme, a tributary of the River Exe in south-west England. Riffles are stony, shallow parts of rivers with fast-moving water whereas pools are deeper slow-moving parts (see Figure 14.15 in Chapter 14). Often rivers have an alternating sequence of riffles and pools along their length. Clifford was interested in finding out why riffle-pool sequences actually formed. A 350 m reach was chosen for study. A detailed survey of the river bed topography was carried out and an array of velocity probes was used to provide detailed data on flow velocity. Data from the probes were then analysed in a numerical model which had been constructed based on existing theory about the conservation of mass and momentum.

The results suggested that an initial randomly located pebble causes the flow of water around the pebble to change (Figure 1.9a). This flow of water speeds up as it is forced to rise above the pebble, and this causes it to have increased ability to carry up more particles from the river bed. Therefore just downstream of the pebble the change in the flow of water causes the bed to erode and creates a scour hole. This hole itself now causes the flow structure of the water above the river bed to change. When the flow slows down (after its acceleration over the pebble) it can no longer hold all of the sediment it carries and thus deposition occurs. Thus there is an area of erosion and an area of deposition (Figure 1.9b). From a small pebble a much larger pool and

riffle sequence has been formed owing to the feedback between the landform (the pebble) and processes of water movement, erosion and deposition which have in turn affected the nature of the river bedform. Because there is now a pool and riffle sequence causing a much larger oscillation of water along the river bed, further pools and riffles

(a)

(b)

(c)

Figure 1.9 Flow around an obstacle on the river bed causes erosion and deposition around the obstacle (a). This in turn affects the flow at a larger scale which in turn affects the turbulent flow structure of the water, the erosion and the deposition processes, causing pool and riffle development and meander bend development (b) and (c). The whole process then becomes self-maintaining.

develop upstream and downstream because the flow is oscillating. Once riffles have formed on the river bed, water wants to move via an easier path around the obstacles and starts to create a new route for the river. This means water tries to take the easiest route past the riffle, which might mean it bends around it. This is the start of meander bend formation (Figure 1.9c). Thus, from one small landform there has been an interaction of processes which have changed the landform of the river at both a small and much larger scale. These landforms in turn alter the river flow, erosion and deposition processes. Thus rivers have a 'memory' of the forms and processes that have been operating in the recent past as these influence the forms and processes today, which in turn will influence the landforms and processes (at a variety of scales) in the future. This form-process feedback interaction provides an alternative way of approaching the study of the physical environment (Figure 1.10).

Figure 1.10 Flow-form feedback interaction. Bedforms interact with water flow processes which in turn interact with sediment erosion, deposition and transport processes. These processes then feed back to cause changes in bedforms. At the same time bedforms affect the sediment transport processes and there are thus complex interactions. (Source: after Clifford, 1993)

BOX 1.2

Figure 1.11 A physical geography laboratory with bottles ready for measurement.

Therefore we would need to put some of the soils into distilled water and some into rainwater with a sulphate solution. The distilled water tests act as a **control** with zero sulphate. We could then measure how much dissolved organic carbon is released from the soil samples. A weak sulphate solution and a strong sulphate solution could also be tested. In this case, there would be nine **treatments** (three sulphate solutions × three soil types). We would also need replicates of each of the treatments in order to establish how much variation there was. So if we said we wanted eight replicates per treatment this would result in 9 × 8 samples being required, a total of 72 (Figure 1.12). Thus, the number of samples can increase dramatically if you want to include more variables (e.g. another soil type, test the effect of water pH on dissolved organic carbon release and so on). Therefore in order to avoid a lot of wasted resources it is very important to make sure that the research design is carefully considered.

	Peat	Podzol	Brown earth
Distilled (control)			
Low-concentration sulphate solution			
High-concentration sulphate solution			

Figure 1.12 Treatments and replicates for a laboratory soil test. Note that there are eight replicates for each treatment. Treatments include distilled water, low-sulphate-concentration water and high-concentration water each on three soil types. There are therefore 72 separate tests for this simple experiment.

A range of critiques of laboratory work have been put forward, including the difficulty in scaling up results from the laboratory to the field and, like models (see below), the simplicity of the experiments ignores complex feedback processes and interlinkages that occur in the real world.

1.4.3 Approaching numerical modelling

Table 1.2 shows how numerical models fit into Harré's reasons for experimentation. Models can be used to help us understand whole systems, tell us which parts of the system are most important (by changing each part of a model separately you can see which changes result in the biggest differences overall), help us make predictions and simulate what might happen if we varied something in the landscape or climate (e.g. the effect of changing land management on catchment runoff and water quality). A good example of this type of modelling approach is given in Box 1.3. Models are ways of simplifying reality. For example, a map is a model of part of the Earth. There are distinctions between different types of model:

- conceptual;
- statistical;
- probabilistic;
- deterministic.

Conceptual models (e.g. a map, a flow chart) express the ideas about how processes in a system work. This is the starting point for any modelling, as without an idea about how things work you cannot do anything else. It is important to ensure that the concepts are correct before embarking on a numerical model.

Statistical models are used where data have been collected that allow relationships to be statistically established and then predictions to be made. For example, by measuring the diameter of 100 trees in each of 20 plots of a tropical rainforest it may be possible to predict how altitude or aspect influences tree growth. It may therefore be possible to make predictions about tree size in other areas of the rainforest just by determining the aspect or altitude. Note that this is a classic example of functional research.

Probabilistic models assume that there is some sort of random behaviour that is part of the system and assign a likelihood to events or data (e.g. 95% probability that this is the correct output). This is often expressed by a ranked numerical value or an estimate of best case, worst case or most likely.

Deterministic models assume a cause and effect link via a process mechanism. The models assume there is only one possible result (which is known) for each alternative course

CASE STUDIES

MODELLING PROCESSES WE CANNOT MEASURE

Gildor and Tziperman (2001) wanted to experiment with causes of past climatic change using a model. They developed a simple box model which is shown schematically in Figure 1.13 which represents the oceans, land and atmosphere. The land and oceans could be covered with variable amounts of ice. When there is more ice, the global climate (atmosphere) is cooler and also more of the Sun's energy is reflected back out to space by the ice. Rocks and water absorb more energy than ice thereby

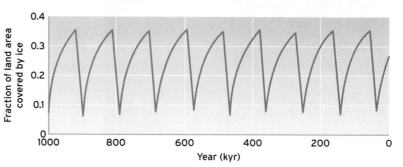

Figure 1.14 Land ice extent predicted by Gildor and Tziperman's (2001) model in the northern polar box as a proportion of total land area. (Source: after Gildor and Tziperman, 2001)

Figure 1.13 Box model of Gildor and Tziperman (2001): (a) side view of atmospheric boxes; (b) side view of ocean boxes with shaded regions representing sea ice cover; (c) top view of land and ocean with sea and land ice cover. (Source: after Gildor and Tziperman, 2001)

making the Earth's climate warmer (see Chapter 4).

The ocean model was made up of four surface and four deep water boxes as shown in Figure 1.13(b). The two polar boxes of the ocean may be covered by sea ice to variable extent. Each of the four atmosphere boxes had land, ocean, land ice or sea ice at its base, each with a specified reflection of the Sun's energy. The land part of the polar boxes may be covered by glaciers that can be of variable extent. Each model then used simple energy and mass balance equations.

When Gildor and Tziperman ran the model, which involved working through all of the equations at different time steps, they obtained an interesting variability on a timescale of 100 000 years. A result is shown in Figure 1.14 for 1 million years of model predictions. This result shows the land ice in the northern polar box as a fraction of land area. There is a sawtooth pattern whereby land ice cover gradually grows and

then quickly melts every 100 000 years. This is a similar pattern to the real ice cover over the past 2.4 million years as determined by records from deep ocean sediments (see Chapter 20).

Figure 1.15 shows the results of different parts of the model for a 200 000 year time series. Initially, the ocean has no ice cover (Figure 1.15b) and the atmosphere and ocean temperature in the northern box are mild (Figure 1.15c and g). However, glaciers start growing because there is enough moisture available to make snow accumulation greater than melting. The temperature of the northern atmosphere box is, while mild, still below freezing (Figure 1.15c). Thus, the reflection of the Sun's energy begins to increase and so the atmospheric and ocean temperatures begin to decline slowly (Figure 1.15c and g). Eventually, at 90 000 years, the sea surface temperature reaches the critical freezing temperature and sea ice begins to form very rapidly (Figure 1.15b). This further increases reflection of energy and decreases atmospheric temperatures. Sea ice can now grow more quickly. In less than

BOX 1.3 ➤

Figure 1.15 Model result for 200 000 years showing (a) northern land ice extent as a fraction of the polar box area (e.g. 0.1 = a tenth of the polar box is covered in land ice); (b) northern sea ice extent as a fraction of the polar ocean box area (e.g. 0.1 = a tenth of the polar box has sea ice cover); (c) northern atmospheric mean annual temperature, °C; (d) mean mid-latitude atmospheric temperature, °C; (e) deep-ocean mid-latitude temperature, °C; (f) source term and sink term (dashed) for northern land ice, 10^6 m^3 s^{-1}; (g) upper-ocean polar box temperature, °C; (h) thermohaline circulation discharge through the northern polar boxes, 10^6 m^3 s^{-1}. (Source: after Gildor and Tziperman, 2001)

20 years the sea ice cover in the polar box is almost total. However, the sea ice growth is self-limiting as it insulates the ocean below it from the cold atmosphere and thus there are no more net losses from the ocean and thus no more sea ice growth.

At this point air temperature is at a minimum, and land and sea ice cover at a maximum. However, because of the low air temperatures there is a significant reduction in moisture flux to the poles (i.e. evaporation of water from oceans and land is reduced, thereby reducing precipitation). Furthermore, the sea ice prevents

evaporation from the polar ocean and further reduces moisture sources for the land. This reduces precipitation for glaciers and now melting is greater than accumulation. The glaciers recede and reflection decreases. Hence atmospheric temperature starts to rise. The oceans begin to warm slowly, too. However, because there is still sea ice cover, not enough moisture is being delivered to land glaciers and they melt fairly rapidly. Once the glacier coverage decreases significantly, atmospheric temperatures rise rapidly and the sea ice begins to melt owing to heat from the mid-latitude oceans. The sea ice

melting is slow at first (Figure 1.15b; 90 000 to 70 000 years) until the deep ocean has warmed sufficiently (Figure 1.15e; 70 000 years). At this point sea ice melt is rapid and within 40 years the ocean is free of ice. The deep ocean warms during this phase because it is affected by vertical mixing from the upper ocean, which is affected by global warming due to the melting of land ice. Hence the deep ocean produces a delay in the response of sea ice. Now the lack of land and sea ice increases temperatures dramatically and with it the amount of precipitation available for land ice growth. Here the cycle reaches its starting point again and the land ice slowly starts to grow.

This model suggests that sea ice plays a critical role in climate switching and that ocean circulation changes and instabilities are not the trigger of rapid climate changes. Instead the ocean circulation responds to rapid changes in sea ice. Obviously the model suffers from low resolution because there are only eight ocean boxes and four atmospheric ones. Nevertheless, Gildor and Tziperman (2001) showed that the mechanism the model predicts is independent of other theories of global climate change such as changes in the Sun's energy reaching the Earth's surface (orbital forcing; see Chapter 20) and yet produces similar patterns of ice growth and retreat to that seen in the ocean sediments. Hence, this sort of simple modelling approach can point us in new directions for research. The next step after these sorts of model findings is to find evidence for the hypotheses they raise.

BOX 1.3

or action. The model contains no random components and hence each component and input are determined exactly. Both deterministic and probabilistic models are based on physical laws and obey rules of storage and energy transfer. There are a number of stages in developing these numerical models and these can be summarized as follows:

1. Start with a conceptual model based on theory and what is relevant.
2. Convert the theory into mathematical equations and rules.
3. Incorporate the equations into a simulation model (by connecting the equations together and giving them some sort of numbers as input data).
4. Apply the model to a real-world situation.
5. Compare the model predictions with real-world observations.
6. The model may now need some fine tuning by tweaking it so that it gives the best results for the majority of cases – note that this does not mean that it will work perfectly for all cases. It may even mean that while at first the model might work really well for one or two cases, the final model might not work as well for those cases. On average, however, it will work better overall when all cases are taken into account.
7. Now use the model to simulate and predict (as long as we know the limits of its predictive capability).

Of course numerical modelling is subject to a range of problems, including those associated with simplifying spatial and temporal scales in order to keep the model simple enough for a computer to run it. For example, if you wanted to run an atmospheric model that predicted how the air moved across the Earth's surface in response to pressure gradients, the surface topography and other forces (see Chapter 4), some topographic data would be needed. A decision would also have to be taken on how often predictions should be made. However, once topographic and air mass detail becomes very fine such as around $2\,\text{m} \times 2\,\text{m} \times 2\,\text{m}$ resolution then it becomes computationally difficult to do all of the calculations that move air from one $8\,\text{m}^3$ cell to the next and also to do it for every cell at the same time. Furthermore, repeating this every minute will lead to enormous amounts of computer power being required. Therefore we have to make decisions about which processes to include and what spatial and temporal scales to adopt. Therefore we must always remember when analysing numerical model predictions that the model itself is just a simplification of reality and that often many real-world processes have been removed.

It is also often difficult to apply models developed at one scale to situations at a very different scale (e.g. applying a valley glacier model to the whole of the Antarctic; see Chapter 18). There are often problems in dealing with the net effect of very small-scale processes (e.g. how does a catchment runoff model deal with rapid water movement through soil cracks? See Chapters 13 and 15). Other problems include the fact that models can become so tweaked that they no longer reflect the processes they were originally intended to represent (Beven, 1989). One of the most frequent problems with numerical modelling is in not having enough real-world data (or good enough quality data) to test the model and thus the need for field and laboratory work in order to help provide this. At the same time models can also help determine new areas for research. They often show which areas are the most important or which we know least about. This may be shown, for example, by changing the numbers in different parts of the model and investigating what effect that has on the overall model output. The part of the model that produces the biggest change in output overall can be called the most sensitive and therefore this part may be the most important to get right. Other parts of the model may have very little impact when they are changed and therefore are less important. Models may also be used to help us to predict or explain something that is actually impossible to measure (e.g. Box 1.3). In some cases it is not possible to compare the model predictions to real-world observations. Nevertheless, these sorts of models still provide us with useful information and understanding about the working of physical systems that may have not been possible to envisage without the numerical model.

Reflective questions

➤ In what ways have the various field, laboratory and modelling approaches discussed above been typical of logical positivist, critical rationalist or realist philosophies?

➤ Why is numerical modelling a useful tool to physical geographers?

1.5 Using physical geography for managing the environment

Many of the world's environmental problems lie in unsound human management of the environment (see Chapters 21, 22 and 24). Humans have the potential to recognize and respond to opportunities and to threats that are natural or caused by humans, and to perhaps avoid or mitigate them (see Chapters 21 and 24). Physical geography provides a long-term understanding of environmental change within which to place contemporary environmental change. Physical geographers study past and present climates (Chapters 4–6, 20–22), tectonics (Chapter 2), oceans (Chapter 3), glaciology (Chapters 18 and 19), landform development (Chapters 7, 11–20 and 23) and ecological and biogeographical processes (Chapters 8–10 and 22). They also study the interacting nature of these global and local processes. Physical geographers are involved with measuring environmental change. This can be done using a wide range of tools, many of which are discussed throughout this textbook (e.g. remote sensing – Chapter 23; water and sediment monitoring – Chapters 7, 11–21; vegetation survey – Chapters 8–10 and 22). In addition to monitoring environmental change we are able to test and refine theories about how processes actually operate. This allows us to understand more readily whether changes we see today are part of normal dynamic Earth processes or whether they are the result of human interaction with the environment.

Through fieldwork, laboratory investigation and numerical modelling geographers are able to understand a wide variety of environmental processes and how these processes interact. This improved understanding helps us to predict the likely effects of future environmental changes. This is because our predictive models rely on us first having a good conceptual model of how the system operates. Geographers are able to contribute to the debates surrounding global climate change and can help provide policy-makers with some answers as to how best to deal with certain problems. Rather than just investigating whether the increased flooding in a particular catchment is due to climatic change or land management, physical geographers are able to take a more holistic approach (looking at how the whole system responds rather than just one bit of it). A traditional approach would simply explore a problem (e.g. flooding) in relation to possible causes (e.g. land management, climatic change). This approach fails to address the linkages between problems that emerge within particular catchments and where any one solution to a problem (e.g. blocking land drains) may have positive (enhanced biodiversity) and negative (release of dissolved organic carbon into rivers) impacts upon other parts of the environment. In other words, rather than thinking about just one thing at a time in environmental management, we ought to be seeking to look at the full range of possible effects.

Furthermore, because of the history of physical geography, geographers have had experience of bringing together large-scale approaches with small-scale approaches, linking case studies with general context. In this way physical geographers can readily adapt to the relevant scale of enquiry required by policy-makers. They can also link together scales of approach through process understanding. Box 1.2 provides an example whereby small-scale processes are shown to impact on larger-scale processes in a way that suggests that landforms interact with the processes, causing feedbacks then to alter the landforms. Cause and effect are not clear and the alternative framework for thinking about environmental change is one in which form–process feedback interactions are considered. This is particularly important as we are finding that more and more environmental systems do not behave in a linear fashion. For example, Chapter 20 discusses how global warming can actually lead to global cooling because of a series of interacting feedback mechanisms. On a temporal scale, too, there have been approaches to physical geography that illustrated how the individual location, the history of that location, and therefore the exact timing of an event can be important in determining how a landscape will respond. Box 1.4 provides an example of this.

The fact that geographers have such a variety of approaches can be of enormous benefit to policy-makers. For example, while evaluating a river reach subject to erosion near houses, some geographers might examine the river sediment distribution to establish sediment transport processes and hence design some coarse stone block protection for the river banks. Other physical geographers might take a longer-term view of the river system and start to look for evidence of how the river has behaved over time. This might suggest a different management strategy. For example, if the geographers find evidence that the river has been subject to historically frequent avulsions (where the channel suddenly changes its position), then perhaps progressive abandonment of the area by those who live there would be a better strategy since no matter how much river bank engineering there is, that particular river may be prone to major channel change and homes would never be safe from the threat of flooding and erosion. Therefore using different skills and different scales of approach (long term and short term) geographers can aid environmental management decision-making and planning.

As a society we are just starting to learn about the problems with scientific research (as discussed above) and how it is approached. We are therefore moving away from a society that simply accepts science as a dominant source of knowledge to a position where society is just one of many different sources usefully contributing knowledge (Lane, 2001). Successful environmental management requires an enhanced engagement with

ENVIRONMENTAL CHANGE

MAGNITUDE-FREQUENCY CONCEPTS IN PHYSICAL GEOGRAPHY

In the environment there will be events that occur either very rarely or very frequently and those that occur somewhere in between. Typically small-magnitude and very large-magnitude events are very rare and so intermediate-sized events are most frequent. Wolman and Miller (1960) suggested that it is these intermediate-sized events that, in the long term, will do the most work (Figure 1.16). So for a river carrying sediment it might be the river discharge event that occurs twice a year that performs most of the sediment transport rather than the 1 in 100 year flood. At the time of the big 100 year flood it might seem as if there has been a great deal of sediment transport. However, in the long term it is the twice yearly events that have carried the most sediment downstream (see also Box 14.2 in Chapter 14).

Equally the response of a particular site to a flood event may vary depending on how long it has been since the last big flood event. This means that there may be a flood of a particular magnitude in a river but it may not carry as much sediment as an earlier flood of the same magnitude. This may be because the first flood flushed out all of the available sediment that had been building from the hillslopes and so the second flood did not have much sediment available to carry. In Figures 1.17 and 1.18 there is a plentiful supply of in-channel sediment so that the next flood in these cases will be able to transport large amounts of sediment down the valley. Therefore the timing of an event in relation to other events

Figure 1.18 A river has undercut the steep bank resulting in a small landslide and a plentiful supply of local sediment which can now be transported downstream.

matters. Furthermore different environments take different amounts of time to recover from a major event (such as a landslide or a flood). This is known as the **relaxation time**. In ecology, relaxation time is an important concept where disturbance can cause the population of species to decline. If there are a number of disturbances then the population will only recover to original levels if there is sufficient time between the disturbances for the recovery to take place (e.g. see Figure 10.2d in Chapter 10). Some environments are also more sensitive to change than others. Hence the environmental response to a particular process or event depends on where you are and at what time the event occurs in relation to other events.

Figure 1.16 The magnitude-frequency concept. In the long term the intermediate-size events that occur perhaps only once or twice a year will do the most work. (Source: after Wolman, M.G. and Miller, J.P., 1960, Magnitude and frequency of forces in geomorphic processes, *Journal of Geology*, **68**, pp. 54-74, Fig. 1. Reprinted by permission of the University of Chicago Press)

Figure 1.17 An upland river channel with lots of sediment available to be mobilized. How a landscape will respond to an event may depend on how long it has been since the last big event and how quickly a landscape recovers from events.

BOX 1.4

CASE STUDIES

STAKEHOLDERS AT THE CENTRE OF SCIENCE – DEVELOPING A STRATEGY FOR MOORLAND MANAGEMENT

Many upland environments such as the moorlands of the Paramo in the Andes or the hills of the United Kingdom are very sensitive to environmental disturbance but have been managed intensively over the past few hundred years (Holden *et al.*, 2007). Often there are conflicting interests between those who live and work in these moorland areas and those who have different views on how the landscape should be used and managed. There is often also a lack of scientific evidence about the impacts of some types of management and a lack of communication between scientists and the stakeholders involved. Furthermore some stakeholders such as water companies are worried that what other stakeholders (e.g. farmers, conservation bodies) might do to the land could impact their water quality.

In order to tackle moorland management in a way that fully engaged geographical scientists with stakeholders while at the same time enabling the different stakeholders to discuss the issues with each other a new project was initiated in UK moorlands (Dougill *et al.*, 2006). The project began by identifying the

Figure 1.19 Stakeholders and scientists discussing potential future management scenarios in the UK moorlands. (Source: photo courtesy of Natalie Suckall)

stakeholders for case study areas and then asking them (both independently and in groups) what they thought the major problems were and how they would like the moorlands to be managed in the future (Figure 1.19). This way the stakeholders, and not the scientists, decided the important topics to be examined. Then the geographers collected new data, used existing data and developed integrated models to understand how the management scenarios that stakeholders had suggested would alter the vegetation cover, the water quality, the flood risk, the carbon balance, biodiversity and economic livelihoods of the area. The results were then shown to the stakeholders who then reconsidered their management options in light of the results. The stakeholders revised management plans, the predictive models were run again and the new results shown to stakeholders. Each time results were shown, stakeholders were able to evaluate how their own planned management strategies would result in environmental change in the future. This meant the stakeholders were at the heart of the science being done and were at the heart of the dissemination of that science. To learn more about this project please see: http://www.env.leeds.ac.uk/sustainableuplands

BOX 1.5

the relationship between science and society and in human–environment interactions. This is where physical geographers should really be making headway in the world. Box 1.5 describes an example of where this has recently occurred. The last chapter of this book describes the considerations and processes involved in environmental management.

Reflective question

➤ For what reasons is it beneficial to have a range of approaches when considering environmental management?

1.6 Summary

This chapter has shown that some of the early roots of physical geography belonged to the time of exploration and colonization. Since then a number of key figures have influenced how physical geography was done, and these include Darwin and Davis. Until the 1950s geography was a subject of inventory and classification. It was essentially descriptive in nature. However, during the 1950s this position could no longer be upheld and physical geography began to embrace positivist science. Many positivist studies of the 1950s and 1960s lacked a focus on processes and instead were interested in statistical relationships and regularities. The aim was to generate generalizable equations and laws that could be used across the world. However, by the 1970s it was realized that this did not help us understand the world. For example, it was only by understanding the processes that we could explain the cases that did not fit the normal statistical relationships. Thus, experimental methods were adopted that tried to determine cause–effect relationships. Inductive methods were replaced by deductive methods following the Popperian ideal of falsification. In the 1990s realism was one of the philosophical approaches that physical geographers adopted in order to get around the problems with earlier forms of positivism. This allows geographers to recognize that all places are unique, yet there are underlying physical mechanisms that come together to produce an event. It is the place itself and the way the processes come together (and have come together at the place in the past) that matter. It is therefore valid to use a case study in fieldwork for physical geography as it is through case studies that we more adequately understand the operation of environmental processes. In a complex world where many processes are non-linear it is important to have the right framework for investigation. A framework that allows both intensive and extensive study and that appreciates that there are a range of feedbacks between a landform or an event and processes seems to be appropriate.

Fieldwork, laboratory work and numerical modelling are all important components of physical geography and many studies will adopt all three approaches. Each approach needs to be carefully designed and there are clear reasons for requiring each type of study. At the same time each approach has a number of problems and it is these that make science fallible. Behind every result presented and every graph plotted there was a human idea and a human reason for obtaining that result. Thus physical geography (and all science) has an element of subjectivity about it. A numerical model may produce huge amounts of data but that model can only be as good as the original conceptual model that was developed and can only produce results that are of equivalent or worse standard than the real-world input data it was given. Of course models can be used to simulate events that would otherwise be very difficult to measure and observe and they can be used to make predictions and guide future research directions.

This chapter has provided a somewhat different introductory chapter from those you will see in other physical geography textbooks. Rather than saying this is how we do things in physical geography it has tried to explain why we do certain things in the way that we do them. The aim of this chapter was really to encourage readers to think critically about the subject of physical geography and the approaches to it. In some ways, therefore, I am encouraging you to read the remaining chapters in this textbook with a critical eye. While the contributors have put together a comprehensive set of material on physical geography to provide you with a baseline knowledge and understanding of the physical environment, it is also evident that there is still much we do not understand about the natural environment and human interactions with it. In particular the environment is always changing and is dynamic in nature. Sometimes we can predict fairly well how a system will respond to change, but in most cases we cannot – particularly those systems that are global in nature and consist of a complex set of process interactions all operating on different spatial and temporal scales. I hope that this book will be the first step towards inspiring you to be one of those that helps us to understand the way in which the world works. Because there are so many new questions, so many processes we do not understand and so many hypotheses that are constantly being revised, there is a huge opportunity out there for you to make an impact on our understanding of the world. Will you take this opportunity and engage with physical geography?

Further reading

Clifford, N.J. and Valentine, G. (eds) (2003) *Key methods in geography.* Sage Publications, London.
This book contains useful and clear chapters on a range of topics including data collection, data analysis and numerical modelling.

Gildor, H. and Tziperman, E. (2001) A sea-ice climate switch mechanism for the 100-kyr glacial cycles. *Journal of Geophysical Research,* 106, 9117–9133.
This is the full paper providing details of the model described in Box 1.3.

Gregory, K.J. (2000) *The changing nature of physical geography.* Arnold, London.
This is a very good short introduction to the history of physical geography over the past 150 years; particularly focus on chapters 2 and 3.

Haines-Young, R.H. and Petch, J.R. (1986) *Physical geography; its nature and methods.* Harper and Row, London.
This book deals with a critical rationalist approach to physical geography and has important material on the major changes in physical geography.

Inkpen, R. (2004) *Science, philosophy and physical geography.* Routledge, London.
This book describes science–philosophy interactions and how there is not one simple single scientific method. Some complex material is described in a very accessible way by the author.

Jones, A., Duck, R., Reed, R. and Weyers, J. (2000) *Practical skills in environmental science.* Prentice Hall, Harlow.
This book has lots of great information about field and laboratory techniques, approaches to sampling and designing projects.

Montello, D.R. and Sutton, P.C. (2006) *An introduction to scientific research methods in geography.* Sage Publications, London.
This book provides a basic outline of scientific approaches and data analysis techniques.

Rhoads, B.L. and Thorn, C.E. (eds) (1996) *The scientific nature of geomorphology,* Proceedings of the 27th Binghampton Symposium in Geomorphology, 27–29 September 1996. John Wiley & Sons, Chichester.
This is a really excellent edited volume which explores the main issues that surround approaches to physical geography. Chapters 1, 2, 5 and 7 are particularly useful. Chapter 7, for example, outlines the role of case studies in physical geography.

Rogerson, P.A. (2006) *Statistical methods for geography.* Sage Publications, London.
A systematic approach to statistical methods and techniques with a companion website.

Schumm, S.A. (1991) *To diagnose the Earth; ten ways to be wrong.* Cambridge University Press, Cambridge.
Schumm's essay is a thoughtful discussion of methods in geomorphological work.

Woolgar, S. (1988) *Science: The very idea.* Methuen, London.
This is an important book on the problems of scientific method.

Web resources

Discoverer's Web
http://www.win.tue.nl/cs/fm/engels/discovery/
A comprehensive website is produced here that provides details on the many voyages of discovery undertaken over the ages and the key historical figures involved; clear and well-organized site.

Fundamentals of Physical Geography Online Textbook: Introduction to Physical Geography
http://www.physicalgeography.net/fundamentals/chapter1.html
This site provides a brief outline of the various components of physical geography and the history of development of the subject. It provides a useful recap of concepts raised in the chapter.

Geographers in the Web
http://www.colorado.edu/geography/giw/davis-wm/1909_ge/1909_ge_ch01.html
W.M. Davis (1901) provides an essay on the relationship between different facets of geography, including their continuity of thought and cause–effect relationships. The text is quite complex in parts but provides an example of early approaches. The site provides further essays written by a variety of geographers on the factors influencing the geographic approach.

Scientific Method
http://www.ezresult.com/article/Scientific_method
A detailed description of the scientific method and its foundations is given in this website. It contains numerous very useful links that provide greater detail of the various concepts involved. The link to the 'philosophy of science' page is particularly useful for understanding the basic tenet behind studies of the scientific method.

 Visit our website at **www.pearsoned.co.uk/holden** for further questions, annotated weblinks, interactive models and video-clips showing physical processes in action.

Continents and oceans

Figure PII.1 Billowing steam plumes from a volcano which is producing lava in the middle of the Pacific Ocean. Here land is being created to form the island of Hawaii. (Source: Rob Reichenfeld © Dorling Kindersley)

Part contents

Scope

The previous part introduced many of the concepts that underlie approaches to the study of physical geography. A theme emerged which is that physical geography has shifted over the years to an emphasis on understanding the processes and mechanisms that drive the Earth's environmental systems. Part II offers a foundation for gaining such an insight by describing the processes involved in some of the most important underlying elements of physical geography: plate tectonics and the oceans.

The distribution of the land surface and the shape and nature of the oceans, which themselves cover 71% of the Earth's surface, are ultimately controlled by plate tectonics. The realization in the 1960s that large-scale crustal plate movements were actually occurring over the Earth revolutionized geological and biogeographical study; it became possible to understand why hazards such as volcanoes and major earthquakes are located only in confined areas

and to explain the spatial distribution of mountain belts and flat plains. It is now possible to explain the three main types of rock we see on the Earth's surface, the topography of the ocean floor and the reasons behind geographical patterns of evolution. The development of plate tectonic theory and the concepts involved are approached in Chapter 2.

Throughout history, the oceans have often been perceived as a powerful and sometimes spontaneous natural force; only experienced fishermen had a sense of the behaviour of the sea. Although many of the secrets of the oceans remain guarded, our inquisitiveness as scientists has led to some spectacular discoveries. These include the pinnacle role played by the oceans in controlling the Earth's global and local climate system and the discovery of teeming life around volcanic vents in the ocean floor where there is no sunlight. The oceans are the major driving mechanism by which energy is circulated around the planet and their nature and importance are described in Chapter 3.

CHAPTER 2

Earth geology and tectonics

Michael D. Krom

Earth and Biosphere Institute, School of Earth and Environment, University of Leeds

Learning objectives

After reading this chapter you should be able to:

➤ describe how the theory of plate tectonics developed from ocean basin research

➤ understand the structure of the Earth and how plates are able to move across the Earth

➤ describe the nature of the three main types of rock found on Earth

➤ explain what happens at the boundaries of plates when they move apart or together in different ways

➤ describe how the theory of plate tectonics has enabled earth scientists to understand the processes of continental drift and the many geological features preserved on continents and underneath the oceans

2.1 Introduction

Until the mid-1960s, although geologists had assembled much detailed information on geological processes both on continents and in the ocean, there was no comprehensive theory to explain the interrelationship between these processes. It is no exaggeration to say that with the development of plate tectonics, the subject of geology was revolutionized. It is now possible to understand and explain such apparently widely divergent subjects as the location of volcanoes, where and why earthquakes occur and why mountain belts are located only in certain confined areas. We now understand the structure of the ocean including why the deepest part of the ocean is near the continents and not, as was once thought, in the centre of the ocean and why the crust under the ocean is much younger than most of the rocks on the continents. Many of the patterns seen in the evolution of fossils can now be explained and it has even been possible to explain why the animals of Australasia are dominated by marsupials; the Americas have a few and Africa and Asia have none at all.

The purpose of this chapter is to explain our present knowledge of the structure of the Earth, including both its interior and its outer layers, which we now know are divided into plates. The history of the development of global plate tectonic theory will be explained and the chapter concludes with a section explaining the location and development of many of the major features of the continents as we see them today. An understanding of plate tectonics is fundamental to understanding a wide range of other subjects in physical geography including biogeography, oceans and global landform development.

2.2 The Earth's structure

The Earth is roughly spherical, as even some of the ancient Greeks realized. It is now known that it is an oblate sphere, somewhat flattened at the poles with a radius of 6357 km at the poles and 6378 km at the equator. The flattening of the Earth is very small but the shape of the Earth is very close to that expected of a fluid rotating with the same velocity that the Earth travels through space. This suggests that, over a long timespan, the Earth behaves in some ways like a fluid.

2.2.1 The interior of the Earth

If the Earth were cut in half and the interior structure exposed, the layers would appear as in Figure 2.1. At the centre of the planet is a core that has a radius of 3470 km. The inner core is solid and very dense ($\sim$13 g cm^{-3}) with a thickness of 1390 km. It is magnetized, and has a temperature probably in the region of 3000°C. The solid inner core is surrounded by a transition zone about 700 km thick, which in turn is surrounded by a 1380 km thick layer of liquid material that together form the outer core. It is somewhat less hot than the inner core and also less dense ($\sim$11.5 g cm^{-3}). The core is thought to be made up of iron and nickel with a small amount of lighter elements such as silicon. The pressure at the core of the Earth is up to 3 million times the Earth's atmospheric pressure. The next layer is called the **mantle**, and contains the largest mass of any of the layers of the Earth, approximately 70% of the total mass. The outer part (180 km) of the mantle along with the overlying crust is called the **lithosphere** and is rigid; this floats on the more mobile **asthenosphere**. Beneath the asthenosphere is a transition zone (350–700 km deep) to the lower mantle, which is located from 700 km to 2900 km below the surface. The mantle is composed principally of magnesium–iron silicates and has a density of 3.35 g cm^{-3}. When the rocks in the mantle are subjected to differential pressure and heat, they flow slowly, rather like ice flows in a glacier (see Chapter 18).

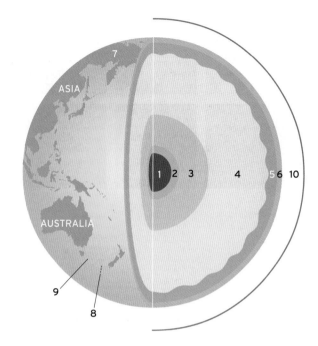

1 Inner core, solid, 1390 km radius
2 Transition zone, 700 km thick
3 Outer core, liquid, 1380 km thick
4 Lower mantle, semi-rigid, ~1200 km thick
5 Transition zone, 350-700 km deep
6 Asthenosphere, rigid, 180 km thick
7 Continental crust, rigid, 35-70 km thick
8 Oceanic crust, rigid, 6-10 km thick
9 Hydrosphere, ~4 km deep
10 Atmosphere, ~30 km+

Figure 2.1 The internal structure of the Earth showing the major layers and their known thicknesses. The core consists of an inner solid centre, a transition zone and an outer liquid zone. The mantle contains approximately 70% of the mass. The upper mantle consists of the asthenosphere and a transition zone. Above the asthenosphere lies the lithosphere, which is made up of the continental and ocean crust above which there is the hydrosphere and atmosphere.

2.2.2 The outer layers of the Earth

The Earth's outermost layer is the cold, rigid, yet thin layer that we are directly familiar with, known as the crust. The large continental land masses are formed primarily of granite-type rock, which has a high content of aluminium and silica. Quartz and sodium-rich feldspar are the two principal minerals present. The crust is comparatively thick ($\sim$35−70 km) and has a density of 2.8–2.9 g cm^{-3}. By contrast the rock underlying the oceans is mainly basalt, which is lower in silica and higher in iron and magnesium. The principal minerals in basalt are olivine, pyroxenes and feldspars rich in calcium. The oceanic crust is much thinner than the continental crust, being only 6–10 km thick, but has a higher density ($\sim$3 g cm^{-3}).

The boundary between the crust and the mantle is called the **Mohorovičić discontinuity**, named after its discoverer, and usually called simply the Moho. At one time it was thought that the Moho was the layer where the Earth's rigid crust moved relative to the mantle. However, it is now thought that the crust and the upper mantle together form the rigid upper layer known as the lithosphere (60–100 km thick) which floats on the asthenosphere and moves as the Earth's plates.

The difference between the average elevation of the continents and the oceans is determined principally by differences in the thickness and density of continental and oceanic crust (Figure 2.2). This is the principle of **isostasy**. A common analogy used to describe isostasy is that of an iceberg. The top of an iceberg is above sea level and is supported by the buoyancy of the displaced water below the surface. The deeper an iceberg extends below the surface, the higher the same iceberg reaches above the surface. Isostasy is, however, a dynamic situation. For example, when the major ice sheets, which until 10 000 years ago covered Europe and North America, melted (see Chapter 20) a significant weight was removed from the continents. These continents are still rebounding upwards as a result of slow flow processes in the upper mantle. An extreme example of this is in areas of west–central Sweden that are still moving upwards at a rate of 2 cm yr^{-1}. Areas can remain elevated where tectonic plates are colliding, as the crust thickens where two continental plates come together. This has

occurred in the Tibetan Plateau for example. It can also occur in areas of volcanic activity such as Yellowstone National Park, USA.

2.3 Rock type and formation

The rocks that are found at or near the Earth's surface can broadly be divided into three types: **igneous rocks**, **sedimentary rocks** and **metamorphic rocks**. There are many varieties of each of these three main types of rock and the exact type of rock that is formed can depend on factors such as temperature, pressure and the minerals present at the time of formation.

2.3.1 Igneous rock

Igneous rocks are formed from **molten rock** (rock that has melted) which cools and hardens. If the molten rock is erupted at a volcano, then the subsequent cooled and hardened rock is very fine grained with crystals that can only be seen under a microscope. A rock of this type is known as a basalt or a volcanic glass (e.g. see Figure 2.10). This type of rock generally cools quickly because it is exposed to the cooling air or water at the Earth's surface upon eruption. Basalt covers the ocean floor (which covers over two-thirds of the Earth's surface). If the rock cools more slowly then there is time for crystals to grow within the molten rock, and a coarse-grained rock is formed. Granites or gabbros are igneous rocks of this type containing large crystals (Figure 2.3a). The Sierra Nevada mountains, New Mexico, and Dartmoor, UK, are made up primarily of granite. In addition to grain

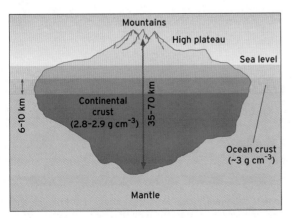

Figure 2.2 Diagrammatic cross-section of crust and upper mantle showing the principle of isostasy. This shows the less dense continental crust 'floating' on the upper mantle in a similar way to the manner in which an iceberg floats in seawater.

(a) Granite

(b) Sandstone

(c) Mica schist

Figure 2.3 Photographs and thin sections of three different rocks. (a) Granite is an igneous rock. Note the large crystals of pink and white feldspar (which can be seen in both the thin section and the hand specimen) which form as the molten rock cools slowly within the Earth's crust. (b) Sandstone is a sedimentary rock. The hand specimen shows layering formed when the sand was laid down under the sea. In the thin section it is possible to see the individual sand grains which show up as black or white grains under cross-polarized light. (c) Mica schist is a metamorphic rock. Note the fine layering seen in both the thin section and the hand specimen brought about by the recrystallization of the original minerals under conditions of high temperature and pressure. (Source: photos courtesy of Bob Finch, School of Earth and Environment, University of Leeds) size, igneous rocks are often divided into acid rocks (often light coloured) which are rocks formed from the melting of continental rocks and basic rocks (often dark coloured) formed by the melting of oceanic rocks.

2.3.2 Sedimentary rock

Sedimentary rocks are formed from the products of the chemical and/or physical weathering of rocks exposed at the Earth's surface. The sediment produced from such weathering can accumulate over time and eventually build up a deposit which, over time, can harden to form rock. Many sedimentary rocks are formed after the weathering products such as sand, silts or clays are transported by rivers and deposited downstream in coastal regions (see Chapter 12). Sediments are cemented together, and compacted and hardened over time by the weight and pressure of the sediments above them and by the precipitation of chemical cements such as calcium carbonate or silica. These processes result in rocks such as sandstones, siltstones or shales (Figure 2.3b). Sedimentary rocks often contain structures that represent a record of the physical conditions present when the rocks were deposited. In addition there are sedimentary rocks formed by the accumulation of the remains of either the skeletons or the organic remains of microscopic animals. These can be as fossils within other rocks or can represent most of the rock itself such as chalks, which are almost entirely the remains of coccoliths (dead microscopic plants). It is amazing to think that in some areas such as the south-east of England there are thick layers of the Earth's surface that are made up almost entirely from the skeletons (tests) of coccoliths (Figure 2.4). Other sedimentary rocks are formed when the concentration of a dissolved mineral in water is so great that mineral precipitates are formed. This can often happen when the water evaporates to leave behind the solid minerals. Halite or gypsum are formed, for example, when seawater is evaporated towards dryness.

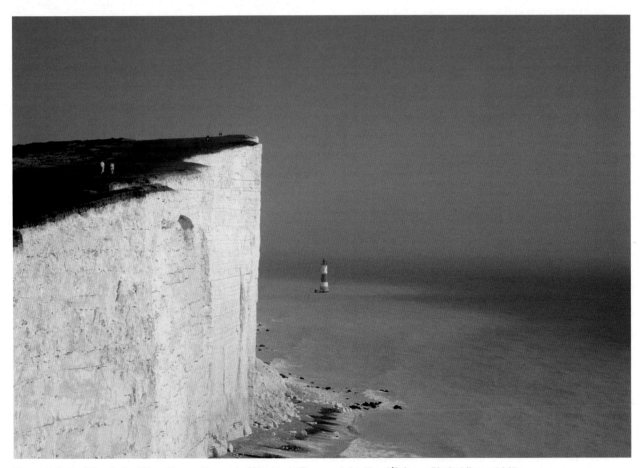

Figure 2.4 A white chalk cliff on the south coast of England. (Source: John Mead/Science Photo Library Ltd.)

2.3.3 Metamorphic rock

Metamorphic rocks form as a result of partial melting and recrystallization of existing sedimentary or igneous rocks. These changes usually take place where there is also high pressure such as under hundreds of metres of bedrock or where rock is crushed at the junction of tectonic plates (see below). As a result, many metamorphic rocks have a layered structure caused by this external pressure (Figure 2.3c). Metamorphic rocks tend to be harder than sedimentary rocks and are more resistant to weathering and erosion. For example, limestone and shale change to marble and slate when metamorphosed.

2.3.4 The rock cycle

Over time all rock types can convert into other forms and this has often been termed the rock cycle. Igneous and sedimentary rocks can become metamorphic rocks under pressure and heat. All rock types can erode to form the layers of sediment that can eventually become sedimentary rocks, and all rocks can be completely melted. When molten rock eventually cools and hardens at or near the Earth's surface it will form igneous rock.

Reflective question

➤ How are igneous, sedimentary and metamorphic rocks formed?

2.4 History of plate tectonics

2.4.1 Early ideas of global tectonics

As maps of the world became more complete a number of individuals began to examine in detail the shapes of the continents and their apparent relationships to one another. In a paper published in 1801, the German explorer Alexander von Humboldt noted that the bulge of South America seemed to fit into the bight of Africa (Figure 2.5). Francis Bacon, a British philosopher of the late sixteenth and early seventeenth centuries, is often erroneously credited with first noticing this fit. In fact he commented on the similar shape of the *west* coasts of Africa and South America in his *Novum Organum* written in 1620. Antonio Snider-Pelligrini in 1858 created a map that showed for the first time the positions of the American and

European–African continents before they broke up. It was not, however, until the early twentieth century that Alfred Wegener and Frank Taylor independently proposed that this fit was not coincidental and that the continents were in fact slowly drifting about the Earth's surface (Wegener, 1915, 1966). Taylor soon lost interest, but Wegener continued to develop and expand on his theory until his death in 1930. He suggested that starting in the Carboniferous period (250 million years ago), a large single continent, which he called **Pangaea**, slowly broke up and drifted apart creating the continents as we know them today (see Figure 2.24 below).

A wide range of evidence, apart from the fit of continents across the Atlantic, was advanced to support the theory of continental drift. Rare, identical fossils, such as the Cambrian coral-like organism *Archaeocyatha*, were found in rocks on different continents, now separated by thousands of kilometres of ocean. Glaciations that were known to have occurred in Carboniferous times appeared to have affected contiguous areas but only if the continents were fitted back together. Likewise mountain belts of similar age, rock types and tectonic history such as the mountains of Scandinavia, the Highlands of Scotland and the Appalachians of North America could all be fitted together within a reconstructed supercontinent.

Wegener's theory provoked considerable debate in the 1920s and for a time was quite widely accepted. However, most geologists simply could not believe it was possible to move the continents through the rigid crust of the ocean basins. Since there was no mechanism to drive this continental drift, the theory was not treated very seriously. It became a footnote in geology textbooks and was seldom taught seriously to students of the earth sciences.

2.4.2 Evidence that led directly to plate tectonic theory

The situation changed after the Second World War as a result of the major increase in our understanding of the **bathymetry** of the deep ocean and the rocks that underlie these basins (see Box 2.1 for the reason behind this increased knowledge of the deep ocean). For the first time scientists were able to construct detailed bathymetric maps of the sea floor beneath the oceans. It was found that there was a large and continuous mountain range running through the centre of many of the world's oceans. Detailed bathymetric surveys found that there was a valley in the centre of these mid-ocean mountain ranges that had the same shape as valleys caused by rifting on land. This was

Figure 2.5 Bathymetric map of the Atlantic Ocean showing how Africa and Europe have drifted apart from the Americas. Transform faults can be seen crossing the mid-ocean ridge. (Source: courtesy of Martin Jakobsen who created the image using data from GEBCO; http://www.ngdc.noaa.ngg/gebco/gebco.html)

interpreted as evidence for the pulling apart of the ocean basins at their centre and thus provided some of the first direct evidence in support of continental drift (Figure 2.5). It was also found that the deepest parts of the world's oceans were not, as had been expected, in the middle of the ocean but instead were located very close to the edge of the ocean, particularly within the Pacific. Prevailing theories had not predicted these major features and were unable to explain them.

Then in the 1960s, Harry Hess at Princeton University proposed that deep within the Earth's mantle, there are currents of low-density molten material that are heated by the Earth's natural radioactivity (Hess, 1962). These form convection cells within the mantle. When the upward-moving arms of these convection currents reach the rigid lithosphere, they move along it, cooling as they go until eventually they sink back into the Earth's interior. In the

regions where the upward-moving limb of mantle material breaks through the crust of the sea floor, underwater volcanoes would appear, developing a mountain range supported by the hotter mantle material below. Lava from these volcanoes erupted and then hardened to form new crust along the underwater mountain chain. As the limbs of the convection cell moved apart, they dragged the overlying ocean crust with them. Since the size of the Earth has remained constant, it is necessary to remove old crust at the same rate as it is being produced at the mid-ocean ridges. The great deep trenches around the Pacific were proposed as areas where the old crust dips down and disappears back into the Earth's mantle.

The second piece of evidence to explain these suboceanic features came from the study of **palaeomagnetism**. Most igneous rocks contain some particles of magnetite (an iron oxide, Fe_3O_4) which is strongly magnetic. Volcanic

CASE STUDIES

HOW THE EVIDENCE FOR THE THEORY OF PLATE TECTONICS DEVELOPED FROM THE COLD WAR BETWEEN THE SOVIET UNION AND THE UNITED STATES

It is a sad reflection on human civilization that many of the major advances in technology and knowledge have come about as a result of waging, or planning for, war. One dramatic example of this is evidence for the theory of plate tectonics. At the end of the Second World War, the wartime alliance between the (capitalist) United States and Britain and the (communist) Soviet Union (USSR) came to an abrupt end. The US Government became worried about the threat posed by the growing fleet of Soviet submarines, particularly because only a few years previously attacks on allied shipping by U-boats in the Atlantic had so nearly won the war for Germany. So the United States developed new sophisticated instruments and technology to study various aspects of the deep ocean basins. It also set up new oceanographic institutions and paid for the education and research expenses of a whole generation of oceanographers and marine scientists.

Detailed surveys were carried out of the bathymetry of the deep oceans to enable US submarines to navigate their way safely in the oceans. It was also necessary to know the most likely routes of passage for Soviet submarines and where they might hide. These surveys were carried out using the new, improved, echo-sounding equipment that was also

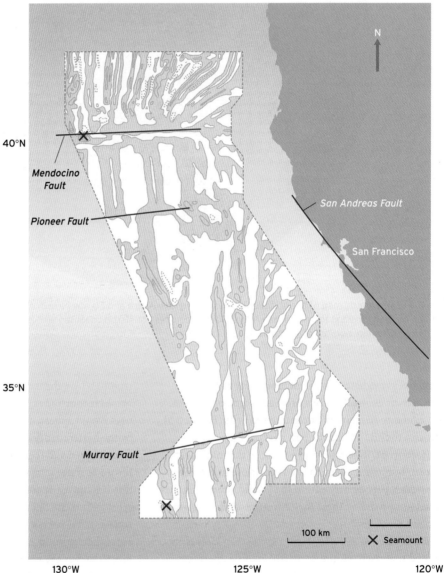

Figure 2.6 Magnetic anomalies of total magnetic field for the area off the western coast of California. Positive anomalies are shown in orange and negative anomalies in white. This was the first such map produced. (Source: after Mason and Raff, 1961)

BOX 2.1 ➤

used to find submarines underwater. The detailed maps of the oceans that show the presence of the mid-ocean ridges and deep trenches were made during this period (e.g. Figure 2.6).

During this explosion of oceanographic information, it was found that there was a depth zone in the ocean that had sound properties that allowed submarines to 'hide' from ships using echo sounders. Thus, at the same time, other technologies not based on sound waves were developed to search for submarines. One of these was underwater towed magnetometers. These very sensitive instruments were used to try to detect the distortion of the Earth's magnetic field caused by steel (magnetic) submarines. These magnetometers, however, were then used by marine geologists to examine the magnetic properties of the rocks

underlying the ocean. These surveys found magnetic stripes across the ocean floor that Vine and Matthews (1963) interpreted as crucial evidence for the formation of new oceanic crust at the mid-ocean ridges (Figure 2.6).

Furthermore, during the 1950s the United States and the Soviet Union tested their nuclear weapons in explosions at ground level, resulting in large amounts of radioactivity being released into the atmosphere. This radioactive fallout spread over the entire Earth and was found to be contaminating everything, including such sensitive substances as food. In particular it was found that the milk fed to children was contaminated with strontium-90. A campaign was launched by the Mothers' Union of North America and other groups to stop this atmospheric testing of

nuclear weapons. It is a tribute to the success of this campaign that a treaty banning the atmospheric testing of nuclear weapons was passed in 1963. From then on, during the remainder of the Cold War, all nuclear weapons testing was carried out underground. Naturally the US Government was keen to gain as much information as possible about the underground nuclear testing being carried out by the Soviet Union and later by China and other countries. So it set up a global network of seismic stations to monitor these tests. These seismic stations were able to locate and measure not only nuclear tests but also natural earthquakes. It is the data from these seismic stations that were crucial in defining the detailed geometry of the plate boundaries and the movement of the plates.

BOX 2.1

lavas such as the basalt which erupts at the mid-ocean ridges are high in magnetite and erupt at temperatures in excess of 1000°C. As the lava cools below 600°C, the **Curie point**, the particles of magnetite become oriented in the direction of the Earth's magnetic field, recording that field permanently relative to the rocks' location at the time they were erupted. If the Earth's magnetic field changes subsequent to the formation of the igneous rock, the alignment of these particles will not be affected. When geologists began to study this palaeomagnetism, they discovered to their surprise that the direction of the Earth's magnetic field reversed itself periodically. A compass needle that points towards the North Pole today would point south during a period of reversal. It is still not known why these reversals take place but they have occurred once or twice every million years for at least the past 75 million years. In fact, there is evidence of magnetic reversals into the Pre-Cambrian (i.e. more than 600 million years ago). A magnetic reversal is not sudden but is very quick in geological time, taking about 20 000 years for completion.

In 1960 **magnetometers** were towed over the sea floor off the western coast of North America, by scientists from Scripps Oceanographic Institute. A clear pattern of stripes appeared,

caused by changes in the polarity of the Earth's magnetic field locked into the crust of the sea floor (Figure 2.6). However, at that time, the bathymetric maps of that region of the Pacific Ocean were considered classified information by the US Government (see Box 2.1), and the scientists involved were not able to relate the magnetic patterns to the mid-ocean ridge, which was subsequently found to run through the area. It was not until Vine and Matthews (1963) proposed that these stripes represented a recording of polar magnetic reversals frozen into the sea floor that we had the evidence needed to demonstrate ocean floor spreading. As the molten basalt flowed out at the mid-ocean ridge, it solidified and retained the magnetic field existing at the time of formation. Sea floor spreading then moved this material away from the ridge on both sides to be replaced by more molten material (see Figure 2.8 below). Each time the Earth's magnetic field reversed it was preserved in the new basaltic lava crust. Vine and Matthews (1963) suggested that in such a case, there would be a symmetrical pattern of magnetic polarity stripes centred on the ridges and which would become progressively older as you travelled away from the ridge.

The third piece of evidence that appeared to create the new theory of plate tectonics came from the study of earthquakes. The global seismic network set up to monitor underground nuclear testing (see Box 2.1) gave geophysicists much detailed information on the epicentres of earthquakes and the direction of their movement. When maps were constructed of the worldwide distribution of earthquakes, it was found that there were narrow bands which were subject to frequent (and often destructive) earthquakes and relatively large areas of stable lithospheric crust in between. Isacks *et al.* (1968) looked not only at the precise epicentre of earthquakes (position and depth), but also at the direction of movement of the two layers of rock that caused these earthquakes. As a result of this work, it became obvious that there were a series of relatively rigid plates which were moving about on the Earth's surface colliding with one another. The continents, far from being crucial to this, appeared to be merely passengers being carried about on these moving plates. McKenzie and Parker (1967) put all this information together on a stringent geometrical basis and thus the theory of plate tectonics was born.

The final piece of evidence that confirmed plate tectonics came from the ocean drilling programme (ODP). Cores were drilled in a variety of locations in the ocean basins. It was found that the thickness of sediment increased from the mid-ocean ridges towards the edge of the basin. The thickest (and oldest) sedimentary cover was found nearest to the continents. However, at no place in the ocean were sediments found that were over 200 million years old. Thus the ocean floor is young compared with the continents.

Reflective question

➤ What are the key pieces of evidence that were used to develop the theory of plate tectonics?

2.5 The theory of plate tectonics

2.5.1 Lithospheric plates

The theory of plate tectonics states that the Earth's lithosphere is divided into a series of rigid plates which are outlined by the major earthquake belts of the world (Figure 2.7). The boundaries of the plates coincide with **mid-ocean ridges, transform faults,** trenches and actively growing mountain belts. At present, seven major lithospheric plates have been identified: the Pacific, Eurasian, African, Australian, North American, South American and Antarctic – as well as numerous smaller plates off South and Central America, in the Mediterranean area and along the north-west United States (Figure 2.7).

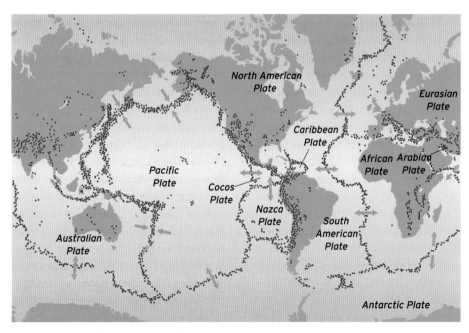

Figure 2.7 Major lithospheric plates of the world. The plates are delineated by the distribution of earthquakes over a period of nine years. The names of the plates and their relative motion are shown in the figure. (Source: based on data from NOAA)

The geological features found at a given plate boundary depend on the relative motion of the two plates that are in contact. Plate boundaries that are moving apart are called **divergent plate boundaries**. This is the place where new oceanic lithosphere is being created, the mid-ocean ridges. Plates slide past each other along major strike-slip faults called transform faults, such as the San Andreas Fault in California. When two plates move towards one another, at a **convergent plate boundary**, different features are formed. If one of the plates slides underneath the other, a **subduction** zone (the location where one plate is forced below another) is formed. Alternatively a mountain belt such as the Alps or Himalayas can be formed. Transform faults and convergent plate boundaries are the sites of major destructive earthquakes (e.g. Box 2.2 below).

It is now known that the driving forces that cause the motion of the plates are convection cells within the mantle. These convection cells transfer heat (and thus energy) from the very hot Earth's core towards the Earth's surface. The exact size and location of these cells are still matters of scientific debate. The maximum size of each cell should correspond to the size of the plates, although there may also be smaller cells. The upward-moving limbs of convection cells are called plumes or **hot spots** (see Section 2.6.4). Although these plumes ultimately cause the motion of the plates, the geometric relationship between the hot spots and the plates is not as simple as Hess first suggested. While there are examples where plumes are directly under spreading centres (e.g. Iceland), there are also hot spots that are far from present-day active divergent boundaries, such as Hawaii and Yellowstone. The motion of these convection cells moves the plates by viscous forces dragging the base of the lithosphere. In addition the upwelling magma at the mid-ocean ridges forces the plates apart as new crust is formed, while the plates are further dragged apart by the weight of the older, dense and cold part of plates which are furthest from the mid-ocean ridge and which also have a thick load of sediment as they sink back into the mantle.

2.5.2 Rates of plate movement

Vine and Matthews (1963) were the first to calculate both the rate and direction of plate motion over the past several million years. Knowing the age of the magnetic reversals and the position of the stripes relative to the mid-ocean ridges, it was possible to determine the rate of sea floor spreading. Using this method, it has been calculated that the sea floor moves away from the ridge system at a rate of $1-10$ cm yr^{-1}. More recently we have been able to locate positions on the Earth with extreme accuracy using laser beams fired from orbiting satellites (see Chapter 23). It has therefore been possible to confirm these estimated rates of plate motion by direct measurement of their present rates of movement determined by satellites. The direction of plate motion can be expressed only as a motion of one plate relative to the adjacent plate since all the plates on the Earth are moving and there is no fixed reference point that can be used.

While rates of $1-10$ cm yr^{-1} may appear to be slow, over geological timescales they are most definitely not. Consider, for example, the opening of the North Atlantic Ocean which, on the basis of geological evidence, started 200 million years ago. The present rate of movement of the North American Plate away from the Eurasian Plate is approximately 2.5 cm yr^{-1}. That corresponds to 1 km of movement in 40 000 years, resulting in a 5000 km separation over 200 million years. This distance corresponds quite closely to the actual present separation of these two plates. Plates, however, do not necessary move at a constant rate or indeed in a constant direction. If, for example, one plate with continental crust at its leading edge bumps up against a similar plate boundary on the adjacent plate, mountains are formed and the plates eventually stop moving relative to one another. This cessation of movement not only affects the two plates involved in the collision, but also has consequences on the motion of plates over a considerable fraction of the world.

Reflective questions

➤ What types of plate boundaries are there?

➤ Is the distance between London and New York increasing, decreasing or not changing?

2.6 Structural features related directly to motion of the plates

2.6.1 Divergent plate boundaries

2.6.1.1 Mid-ocean ridges

Mid-ocean ridges are the spreading centres where the lithosphere is created (Figure 2.8). The faster the spreading rate, the broader is the mountain range associated with that

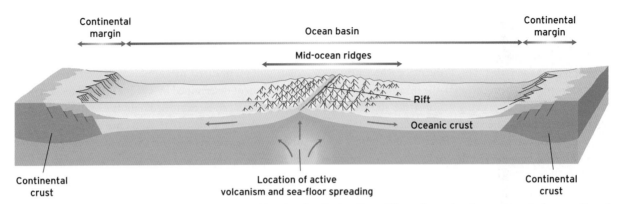

Figure 2.8 Diagrammatic cross-section of a mid-ocean ridge showing the location of the active volcanism and sea floor spreading at the ridge crest.

spreading. The best example of this is the East Pacific Rise, which is the mid-ocean ridge located between the Pacific Plate and the Nasca Plate where spreading rates are as high as 16.5 cm yr^{-1}. By contrast the Mid-Atlantic Ridge, with a spreading rate of 2–3 cm yr^{-1}, has steeper slopes characteristic of a slow spreading centre.

The axis of the mid-ocean ridge consists of a central valley or **graben**. Lava flows out onto the sea floor via long fissures that are orientated parallel to the ridge axis rather than from the more circular volcanic craters we see associated with other types of volcanism. As the oceanic crust is carried away from the active (and hot) central ridge, it gradually cools and sinks. Eventually, over the millions of years that the oceanic crust takes to be carried across the ocean basin, it becomes covered with deep ocean sediments and the original primary volcanic and tectonic features are blanketed over.

Mid-ocean ridges are where most of the Earth's volcanism occurs, producing about 20 km^3 of magma per year that solidifies to produce new ocean floor. The eruption of this magma occurs mostly underwater. The average depth of the mid-oceanic ridges is 2.5 km in regions. This makes it difficult to study such eruptions directly. As a result, much of the detailed knowledge we have of the processes that occur at the mid-ocean ridges has been obtained from studies of Iceland, the one location where there is a comparatively large island straddling the mid-ocean ridge (Figure 2.9). The rock that is erupted at these divergent plate boundaries is called mid-ocean ridge basalt. When mid-ocean ridge basalt erupts it is generally hot (1000°C) and very runny. Large volumes of basalt lava are formed at such volcanoes and the ease with which the lava flows when erupted on land leads to the formation of **shield volcanoes**. These are volcanoes with such gentle

Figure 2.9 Runny lava flowing from a volcanic eruption on Iceland. (Source: Krafft Explorer/Science Photo Library Ltd.)

slopes that sometimes they are barely discernible as mountains at all. Eruptions from these volcanoes are rarely explosive because any gas that is in the magma can easily escape through the runny liquid. Sometimes larger bubbles of gas do burst out of the magma and can produce spectacular fountains of fire. Eruptions can occur out of circular vents or quite frequently as walls of molten lava issuing from a linear crack in the Earth. When lava erupts underwater it cools rapidly, forming rounded structures called **pillow lavas**. Under such conditions the lava does not flow far from the vent where it was erupted (Figure 2.10).

The lava for these volcanoes comes from the partial melting of the upper mantle. When mantle rocks, which are themselves solid and move by **elastic creep**, move up towards the surface as the plates separate, pressure is released. This release of pressure causes the outer edges of the minerals to melt, forming a microscopic network of

Figure 2.10 Pillow lava, which has this characteristic shape, is formed when basaltic magma is erupted underwater. (Source: Landform Slides/GeoScience Features Picture Library)

Figure 2.11 The characteristic layers of ocean crust shown in this diagrammatic cross-section are found on land preserved within ophiolite suites such as in the Troodos Mountains in Cyprus. (Source: courtesy of Andrew M. Tomko III)

interconnected tubes and veins. The resulting fluid is so runny and the pressure so great that the lava is rapidly squeezed out and passes upwards into the lava chambers found below the mid-oceanic volcanoes and eventually out onto the surface in eruptions. The composition of this liquid is derived from, but not identical to, the mantle from which it was ultimately formed.

Our knowledge of these layers is derived to a large extent from study of those areas on land where oceanic crust has become exposed. These layers are called **ophiolites** and they are found in a number of locations. One of the best preserved sections of ophiolites is on the island of Cyprus. The Troodos Mountains contain a well-preserved geological section (Figure 2.11). On top are oceanic sediments (mainly chalk). Below this are pillow lavas and other volcanic rocks underlain by extensive areas of sheeted dykes which represent evidence for the cracks through which the lava flowed. The deepest rocks are **gabbros** and other rocks, which represent the lowest ocean crust and upper-mantle rocks.

Apart from this volcanic activity, the mid-ocean ridges are also the site of widespread hydrothermal activity. Seawater which seeps into the cracks and fissures in the basaltic lavas becomes superheated. As a result it rapidly rises and forms a series of hot vents along the rift valley of the ridge. During the passage of seawater through this subsurface plumbing system, the hot water interacts with the surrounding basaltic rock, causing chemical changes to occur. Sulphate ions, which are present in large quantities in seawater, are chemically reduced to sulphide, while at the same time trace metals including

manganese and iron are dissolved. The result of all these chemical changes is that the water re-emerges as a jet of hot water and carries with it a black precipitate of metal sulphides, which gives these features their popular name of **black smokers**. For a discussion of the unique biological **communities** associated with these hot vents see Chapter 3.

2.6.1.2 Rift zones on land

Rift zones are not limited to the ocean. They also occur on land and have in the past been responsible for the break-up of land masses. For such a feature to form on land, however, it is necessary for there to be considerable thinning of the continental crust first. The best example of this today is the Syrian–African Rift Valley (Figure 2.12). Initially localized faulting produced a sunken rift zone or graben. As the fault continued to pull apart there was a further thinning of the crust and deepening of the rift valley until eventually magma was erupted at the surface covering much of the Golan Heights. The magma in turn forced the crack to widen. As the rift continued to deepen, it eventually sank below sea level, as in the Dead Sea, which is at -399 m. In this example the rift to the south opened to the ocean, and seawater entered. This formed the present-day Red Sea. The floor of the Red Sea has both oceanic basalt and subsided blocks of continental-type crust on its floor. If this process continues, the crust will

Figure 2.12 The rift valley of the Jordan River can be seen in this image. The country of Israel is to the left of the river valley and the country of Jordan is to the right of the river valley. The Jordan River is 322 km long and originates in the Anti-Lebanon Mountains. The river flows southwards through the drained Hula Valley Basin into the freshwater Lake Tiberias (Sea of Galilee) which is the northernmost depression in this figure. The Dead Sea which is in the centre of this image is 70 km long and presently is 399 m below sea level. The Jordan River Rift Valley is a small part of the 6500 km long Syrian–African Rift. (Source: John Hall; Image Science and Analysis Laboratory, NASA-Johnson Space Center. 'The Gateway to Astronaut Photography of Earth.' <http://eol.jsc.nasa.gov/sseop/EFS/printinfo.pl?PHOTO=NM23-755-505>, accessed on March 14, 2008)

continue to thin and eventually magma will well up into the centre of the rift, forming a new mid-ocean ridge spreading centre.

The lava that is extruded from this type of volcano results in 'rivers and lakes' of lava which solidify into ropey lava also known by its Hawaiian name of pahoehoe (Figure 2.13) and a granular lumpy lava known technically as aa. The lava from such eruptions can flow for considerable distances and tends to infill existing landscapes. Once it begins to weather it can form very fertile soil.

2.6.2 Transform faults

Transform faults are found at the plate boundary where two plates slide along next to one another (Figure 2.14). Here the lithosphere is being neither created nor destroyed. The faults can exist as long boundaries between plates such as the Northern Anatolian fold in Turkey and the Altyn Tagh fault in China. These boundaries are characterized by frequent major and destructive earthquakes. The movement can be from a few centimetres in a comparatively small earthquake to 1–2 m in a large one.

Figure 2.13 Photograph of ropey lava also known as pahoehoe.

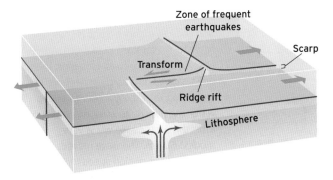

Figure 2.14 Diagrammatic cross-section of an oceanic transform fault showing the region of the fault with the most active seismicity. Where such transform faults occur on land they are the sites of the destructive earthquakes.

Transform faults can also occur in small sections in the ocean between spreading centres. The boundary of the divergent plate is not necessarily linear. For mechanical reasons the central graben and spreading centre are almost straight. Thus, in order to accommodate curves in the plate boundary, the central graben is displaced by transform faults. The opposite sides of a transform fault are two different plates moving in opposite directions, and the movement results in frequent shallow and often severe earthquakes. Where this fault zone extends beyond the active spreading centre, it is now within the boundary of a single plate and thus the two sides of the fault are moving in the same direction. The earthquakes that occur on this extended section of the fault are much smaller in magnitude and less destructive. The best known example of an active and dangerous transform fault is the San Andreas Fault. It extends from where the actively spreading East Pacific Rise intersects

the west coast of Mexico, through southern California and the San Francisco Bay region, to the Juan de Fuca Rift off the coast of Oregon and Washington states.

2.6.3 Convergent plate boundaries

2.6.3.1 Subduction and volcanism

When two plates consisting of oceanic lithosphere at their leading edge converge, one of the plates is driven under the other one. This subduction causes a deepening of the ocean at the boundary and a deep ocean trench generally forms. The only occasion when a trench is not formed at a subduction zone is when it is very close to the mid-ocean spreading centre. This is because the lithosphere will still be relatively warm and therefore less dense and more buoyant than the surrounding lithosphere. Hence there will be no propensity for a deep trench to form. At most subduction zones, however, the lithosphere is cool and dense and will readily subduct.

The **Wadati-Benioff** zone is a band of rock, 20 km thick, which dips from the trench region under the overlying plate. It contains the location of all the earthquake foci associated with the descending lithospheric plate. Initially the angle of dip is 45°, becoming steeper as the plate descends to greater depths. Deep-focus earthquakes associated with the descending plate have been observed to depths of 400 km in slowly moving plates and up to 700 km in faster plates. The major destructive earthquake and resulting **tsunami** that struck Indonesia and the eastern shore of the Indian Ocean occurred at a subduction zone (Box 2.2). Figure 2.16 (p. 48) shows the ruins of a building in El Salvador following an earthquake associated with subduction.

As the oceanic lithosphere is carried down into the mantle, it is heated. Water and other volatiles which were carried down with the plate are freed, producing a low-density mixture that rises to the surface on the overriding plate side of the trench. If the overriding plate is oceanic, then basaltic volcanoes are produced, which form into island arcs. The islands are generally situated about 100 km behind the oceanic trench.

When an oceanic plate collides with a plate that has continental crust at its leading edge, the oceanic plate is always the one that subducts beneath the continental one. The oceanic plate carries with it oceanic sediments, which are oldest and thickest at the plate boundary. Some of these sediments are scraped off the oceanic plate and are added directly to the continental plate (Figure 2.17, p. 48). The remaining sediment gets carried down with the descending

CASE STUDIES

EARTHQUAKE IN THE INDIAN OCEAN ON 26 DECEMBER 2004

On 26 December 2004 at 8 a.m. local time, the section of the plate boundary between the Indian and Burmese plates situated off the west coast of Sumatra, Indonesia (Aceh province), gave way catastrophically. This resulted in one of the largest earthquakes ever felt, with a magnitude of 9.3. It was the longest-duration earthquake ever recorded (500–600 seconds) and was large enough to cause the entire surface layers of the planet to vibrate by over a centimetre of vertical displacement. In total 1200 km of the fault moved. The slip did not happen instantaneously but took place over a period of several minutes. Seismograph and acoustic data (Figure 2.15) indicate that the first phase involved a rupture about 400 km long, 100 km wide at a depth of 30 km below the seabed. This is the longest rupture ever known to have been caused by a single earthquake. This first rupture, which was centred off Aceh, propagated at a speed of about 2.8 km s^{-1} or 10 000 km h^{-1} and lasted about 100 seconds. After a pause of about 100 seconds, a second major rupture occurred, continuing northwards towards the Andaman Islands. This rupture occurred more slowly than that in the south (2.1 km s^{-1}) and continued north for another 5 min to the point where the subduction boundary intersects with a transform fault. In addition there were several other sub-events, the largest of which would normally have been ranked as major earthquakes in their own right, with magnitudes of over 7.5. Subsequently when scientists examined the area of the fault line using high-

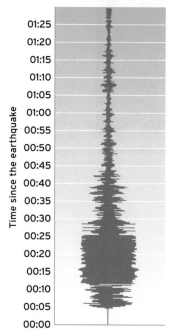

Figure 2.15 Seismograph of the magnitude 9.3 earthquake recorded on 26 December 2004. (Source: Modified from Rapid Earthquake Viewer, http://rev.seis.sc.edu/earthquakes/2004/12/26/00/58/50)

resolution side scan sonar, they found major disruption including several large landslides. In total there was 10 m of lateral movement and 4–5 m of vertical movement along the fault line. It has been estimated that the plate movements during the earthquake displaced ~30 km^3 of seawater radiating outwards along the entire 1200 km length of the rupture. The volume of sea floor upthrust raised global sea level by 0.1 mm. The total energy generated in this earthquake was 3.35×10^{18} joules (equivalent to 0.8 gigatonnes of TNT).

The largest earthquakes, such as the Sumatra–Andaman event, are almost always associated with major thrust events at subduction zones.

This earthquake was the third largest in the past 100 years but because of the location compared with other events, it had by far the largest loss of life. To put this into perspective, in Table 2.1 you can see the magnitude of several major famous and destructive earthquakes.

The reason for such a large loss of life in this earthquake was not the earthquake itself but the tsunami generated by the earthquake. Tsunami is a Japanese word for harbour wave. Japan is adjacent to a similar subduction zone as Indonesia and is thus liable to similar earthquake-generated phenomena. As with all tsunamis, this one behaved differently in deep water compared with shallow water. In deep water the maximum wave height 2 h after the earthquake was a very modest 60 cm. The wave moved, however, at very high speed (500–1000 km h^{-1}). In shallow water, near coastlines, the wave slows down but in doing so forms large destructive waves. In this case, a wave 24 m high struck the Indonesian coast, destroying towns and villages in its path. In total, 130 700 people were killed in Indonesia. The wave also propagated across the Indian Ocean and killed people in Sri Lanka (35 300), India (12 400), Thailand (5400) and many other countries. The tsunami even killed 8 people in Africa some 5000 miles to the west.

Could anything have been done to prevent such a large loss of life? Scientists at an international geophysics meeting a few months prior to the earthquake had forecast a major earthquake on this section of the fault but, like all such forecasts, it was impossible to say when it would occur or indeed how severe such an earthquake would be, although it was expected to be a 'big' one.

BOX 2.2 ➤

Table 2.1 A list of some of the largest earthquakes during the past 100 years, the estimated number of fatalities and whether a major tsunami was produced. Note that although the largest earthquakes are all associated with subduction zones and produce major tsunamis, the number of fatalities is most closely associated with the population density in the area affected

Earthquake location	Magnitude and date	Tsunami	Number of fatalities
Chile[+]	9.5, 1960	Yes	>2000
Sumatra-Andaman[+]	9.1–9.3, 2004	Yes	283 000
Prince William Sound, Alaska[+]	9.2, 1964	Yes	125
San Francisco[+]	7.8, 1906	No	~3000
Kashmir[+]	7.6, 2005	No	80 000
Tangshan, China	8.2, 1976	No	242 000[*]
Gansu, China	7.8, 1920	No	c.200 000[*]

[*]Government estimates. Aid agencies estimated the casualties as much higher.
[+]Data from **www.earthquake.usgs.gov**

Very little could have been done for the people of Aceh province. They had very little time between the earthquake being felt and the tsunami striking. In areas with a subduction fault situated off the coast, it is necessary for the local population simply to know that if there is a major earthquake felt, then they must make rapidly for high ground because the tsunami may be much more destructive than the earthquake that caused it. However, such major earthquakes happen in the same region very infrequently, possibly once every several hundred or even thousand years, so information about such events cannot generally be handed down by word of mouth. There is therefore a requirement for education of people by teachers and others in the area. In fact one of the few coastal areas to

evacuate before the tsunami arrived was on the Indonesian island of Simuelue. Here island folklore recounted an earthquake and tsunami in 1907. As a result of the 1907 story being well known by the islanders, they fled inland immediately after the initial shaking, knowing that a tsunami might be on its way.

Further away from the epicentre, it would have been possible to generate a tsunami warning. Such a system is in place around the Pacific Ocean (see Figure 2.7 for earthquake locations around the Pacific). If there is a major earthquake around the Pacific, the Tsunami Warning Center in Hawaii contacts regional authorities across the Pacific, who then have to get a warning to local residents. On 26 December 2004, the Tsunami Warning Center did realize very

quickly that a major tsunami was being generated in the Indian Ocean. However, the problem was that it simply did not have any mechanism to warn the people on the coasts of the Indian Ocean around Thailand, India and Sri Lanka of the danger that they faced.

So one is left with stories such as the one where Tilly Smith, a 10-year-old British girl, was on Maikhao beach in northern Phuket, Thailand. She was on the beach with her parents and saw the sea retreating dramatically from the beach. She had learned about tsunamis in geography lessons in school and recognized the warning signs of the retreating sea and frothing bubbles. As a result she and her parents warned others on the beach and they were able to evacuate to safety.

BOX 2.2

plate. The oceanic crust contains water that was trapped within it during its formation and from the pore waters of oceanic sediments. As the plate descends into the mantle, the water causes the basaltic rock to melt. The rising magma is also hot enough to start to melt the continental

crust of the overlying plate. The magma that results from this process is very sticky and would not flow at all were it not for the water it contains, which reduces its viscosity to the point where it continues to rise. As the ascending magma approaches the Earth's surface, the pressure is

Figure 2.16 Photograph of the destruction of a building in El Salvador caused by an earthquake. (Source: photo courtesy of Dr W. Murphy)

reduced. The water and other gases trapped in the magma begin to expand but cannot escape. As a result, a considerable overpressure builds up, which cannot escape because the volcanic vent is blocked with a plug formed from the cooling of the lava from a previous eruption. Eventually, the rising lava reaches the point where it breaks through this lava plug and erupts at the surface. The gases have expanded to such a degree that the lava is not so much a liquid, as you find in a mid-ocean ridge eruption, but in effect an incandescent mixture of foam and ash. The result can be a great and destructive explosion that can destroy large areas and kill many people.

Examples of these volcanoes that have resulted in great destructive eruptions include Santorini (Greece; see Box 2.3), Krakatoa (Indonesia), Vesuvius–Pompeii (Italy) and more recently Mount St Helens (USA). The lava can flow down the hillside as a boiling mud flow or as a stream of incandescent foam (called a **nuée ardente**). It can form pillars of gas and ash that can ascend as much as 40 km into the stratosphere. The dust and gases put into the atmosphere can have widespread effects on climate. After the Mount Tambora eruption in Indonesia in 1815, the world's temperature was lowered for over a year and the following year was known as the year without a summer because dust in the atmosphere reduced the amount of sunlight that reached the Earth. Once the initial destructive phase has passed, the sticky lava flows out of the mountain and builds its shape up into the characteristic steep-sided volcanoes such as Mount Vesuvius and Mount Fujiyama (Figure 2.18). Once the lava ceases to

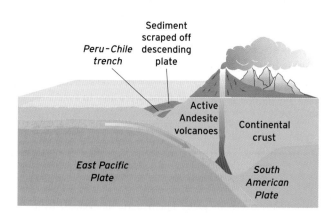

Figure 2.17 Diagrammatic cross-section of a convergent plate boundary showing the subduction of the oceanic plate at the continental margin producing a deep-sea trench and a volcanic belt close to the continental margin.

Figure 2.18 Mount Fujiyama, Japan, is an example of an andesitic volcano which is formed behind a subducting oceanic plate. Here the volcano is snow covered. In the foreground there is fog above a lake. These volcanoes are typically highly destructive when they erupt. (Source: Getty Images/Stone)

CASE STUDIES

THE DESTRUCTIVE POWER OF SANTORINI

Located 120 km north of Crete is the island of Thira, which is also known as Santorini (Figure 2.19). In 1628 BC it was the site of what was probably the largest volcanic eruption to take place during recorded history. The island is situated above the descending African Plate which sinks below the island of Crete. As such it is similar in type to the many volcanoes around the Pacific, such as Mount St Helens, Washington, USA, or Krakatoa, Indonesia. When the eruption of Santorini occurred it began as a typical but very large andesitic eruption. It has been calculated that during the eruption between 30 and 40 km^3 of molten rock and ash were ejected. The ash cloud was blown into the sky to a height of 35 km or more. Ash from the eruption of Santorini has been found throughout much of the eastern Mediterranean Basin including the islands of Kos and Rhodes, western Turkey and as far as the Nile Delta in Egypt, 800 km away (Stanley and Sheng, 1986).

What made this particular eruption so destructive was that, once the magma chamber was partially emptied, the crater (also known as a **caldera**) collapsed and the sea rushed in. The resulting interaction between seawater and the hot magma caused an explosion that was probably heard over much of the eastern Mediterranean.

By analogy with similar modern eruptions, particularly those in Indonesia, it is expected that the eruption would have resulted in a vast tsunami which has been calculated to have been possibly 200 m high. This tsunami would have devastated any coastal communities in the region. In particular, the explosion of Santorini may be largely to blame for the destruction of the Minoan civilization in Crete (Figure 2.20). All that remains of the original island after the cataclysmic explosion are the thin crescent-shaped islands around the original caldera (Figure 2.19). The small island in the middle of the caldera is the site of the volcanic vent itself, which is gradually growing out of the sea (Figure 2.21).

Such high-altitude ash clouds as produced by the Santorini eruption typically have severe effects on global climate. Locally this would have blocked out the sunlight almost completely and reduced global temperatures. Regionally it has been suggested that the ash cloud may have caused a failure in the monsoonal rains in East Africa, while globally it is likely to have caused widespread crop failure and hardship as global temperatures were lowered. The Minoan explosion has been dated by a prominent sulphuric acid peak in a south Greenland ice core, which was dated as 1645 BC (±20 years) (Hammmer et al., 1987), and high-altitude bristlecone pines in California and pines in Ireland suffered severe frost damage in the year 1672 BC (±2 years) (LaMarche and Hirschboeck, 1984; Baillie and Munroe, 1988). It has even been suggested that the explosion of Santorini may have been responsible for the biblical plagues in Egypt (Wilson, 1985).

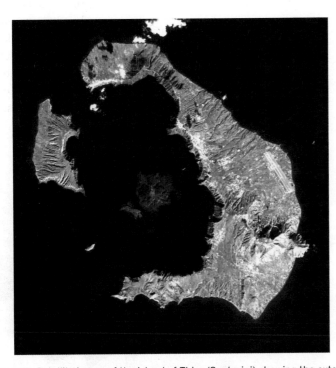

Figure 2.19 Satellite image of the island of Thira (Santorini) showing the extent of the caldera which was formed during the eruption of 1628 BC. The central cone is appearing again above the water in the centre of the caldera. (Source: NASA Jet Propulsion Laboratory (NASA JPL))

BOX 2.3 ➤

Figure 2.20 Knossos was the capital of the Minoan civilization on Crete which was essentially destroyed by the eruption of Santorini in 1628 BC. (Source: Max Alexander/Dorling Kindersley © Archaeological Receipts Fund)

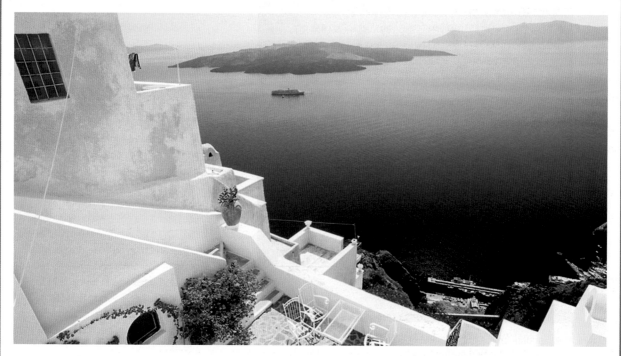

Figure 2.21 The volcanic cone appearing above the present sea level in the centre of the caldera of Thira (Santorini). (Source: Warwick Kent/Photolibrary.com)

BOX 2.3

flow, it cools in the volcanic vent. This volcanic plug allows a considerable overpressure to build up within the volcano, repeating the cycle until the next catastrophic eruption occurs.

The composition of the lava that erupts behind an oceanic trench depends on the precise nature of the continental rocks that are partially melted during their formation together with partially melted mantle material. One widespread form of this lava is called andesite, which is named after the Andes Mountains. It is intermediate in composition between basalt and granite. This process, over many cycles, is the ultimate origin of the rocks that make up the continents.

2.6.3.2 Mountain building

Although plate tectonics is very good at describing the principal features of the sea floor, and the changing distribution of the continents through geological time, it is not very good at describing the details of what happens when two continental plates collide (Figure 2.22). Indeed it might be argued that this is the reason why it took until the 1960s before plate tectonic theory was properly developed. There is no subduction when two plates carrying continental lithosphere at their leading edge collide. Examples of such boundaries today are in southern Europe where Italy has moved north to collide with Europe to form the Alps and in Asia where India has collided to form the Himalayas. In these regions there is no obvious plate boundary which can be defined either by earthquake epicentres or indeed geologically.

When two continents collide, both being buoyant, neither can sink under the other. The crust is thickened because it has been shortened and compressed as if squeezed in a vice. As a result the rocks are folded and deformed, crumpled and faulted and most dramatically uplifted to form great mountain belts. The thick crust that is a result of such plate collisions sticks up like an iceberg with high mountains and an even deeper root below floating on the dense mantle. As the high mountains are eroded, the root at the bottom moves up and exposes rocks and minerals that have been metamorphosed by the high temperatures and pressures which are found in the core of the mountain belts.

In addition to the high mountains that form at the boundary of the two plates, a number of other features are seen when two continents collide. These can include major horizontal faults. Faults of this type do not make mountains. They allow blocks to rotate or to move sideways out of the principal collision zone.

2.6.4 Hot spots

In addition to the comparatively simple pattern of volcanism associated with plate boundaries, there are areas where there are volcanoes which are not associated directly with plate boundaries. These features are called hot spots. They represent the surface expression of mantle plumes and are related to convective processes which originate at the core–mantle boundary deep in the Earth. At present 41 of these hot spots have been identified (Sverdrup *et al.*, 2004). Some are associated with ocean spreading centres and the large amount of lava that they provide has created islands such as Iceland. Hot spots can also form in mid-plate locations. In cases where there is a volcano situated immediately above the hot spot, then, as the plate moves over the hot spot, a series of volcanic islands or **seamounts** are formed. The best example of such a system is the Hawaiian island chain. The island of Hawaii is furthest to the east and is the site of the most active volcano on Earth at present (Figure 2.23). There are a series of islands, Maui, Molokai, Oahu and Kauai, in a line stretching more than 500 km to the west. Each island is made up of an extinct volcano, with the volcanoes getting progressively older as you travel west. Further west the islands are so old and eroded that they disappear below the ocean water surface to become seamounts. This also happens because, as the plate moves west, it cools and the ocean floor becomes deeper towards the ocean trench. During July and August 1996, there were a series of volcanic eruptions and earthquakes on the sea floor 30 km to the south-east of the big island of Hawaii, which seems to be associated with a new island forming.

Figure 2.22 Diagrammatic cross-section showing the type of plate boundary that is formed when two continental plates collide. Typically this results in multiple thrusting and folding of the rocks, double thickening of the continental crust, and high mountains and plateaux such as those found in the Himalayas and Tibet today.

(a)

(b)

Figure 2.23 (a) The islands of Hawaii are located above an oceanic hot spot. The island chain is caused by the interaction of that hot spot with the Pacific Plate, which is moving to the north-east. (b) A photograph showing red lava erupting into the blue sky from the gaping Kilauea caldera of Mauna Loa volcano in Hawaii. Cooling lava streams down the side of the black volcano (source: Peter Menzell/Science Photo Library Ltd).

Hot spots are also found in the middle of continental plates. One such hot spot is beneath Yellowstone National Park, and is responsible for the spectacular series of geysers, mud pots and other hydrothermal and volcanic features seen there.

Reflective questions

➤ What are the similarities and differences you would expect to see between the volcanic activity of a mid-ocean ridge when it is exposed on land (such as in Iceland) and the same feature at the bottom of the ocean?

➤ A majority of the most destructive earthquakes occur on transform faults. List the locations of all the earthquakes you have ever heard of and then look at Figure 2.7 to see what type of plate boundary they are associated with.

➤ What landforms would you associate with divergent and convergent plate boundaries and transform faults?

2.7 The history of the continents

Plate tectonics provides a theory to explain the nature and location of tectonic features on the present Earth. It also enables geologists to reconstruct how the present arrangement of continents came about. Using the magnetic record found within the oceanic crust and other direct evidence, it has been possible to reconstruct accurately the history of continental drift over the past 200 million years. When this reconstruction was carried out it was found that all the present continents were once bound together in a single land mass called Pangaea (Figure 2.24). This land mass was formed by the northern continent of Laurasia (which consisted of present-day North America and Eurasia) combined with the southern continents of South America, Africa, India, Australia and Antarctica (Gondwanaland) across the ancient Tethys Sea. This is the supercontinent proposed by Wegener (1966). The land masses appear to join most effectively if they are joined not at the present seashore but at the boundary of the continental slope at about 2000 m water depth.

When the continents moved together in this way, many hitherto inexplicable geological features become understandable. For example, it is possible to use the magnetism that is locked in rocks when they are formed to calculate how the position of the North Pole changed with time. These so-called polar wandering curves could be simply explained as the record of the tracks of the continents over the globe of the Earth. This movement of the continents resulted in them being found not only in

(a) Pangaea ~250 million years ago

(b) Gondwanaland (southern hemisphere) and Laurasia (northern hemisphere) ~180 million years ago

(c) Continents break up into recognizable form known today ~100 million years ago

Figure 2.24 The proposed break-up of a supercontinent called Pangaea into smaller continents which began around 200 million years ago.

different geometrical orientations but also at different latitudes. As a result, geological features such as **evaporites**, sandy desert features and coral reefs, which are common in the geological record of western Europe and North America in the Palaeozoic and Mesozoic (see Chapter 20), make sense since they were deposited when those continents were at lower latitudes than they are today.

One additional consequence of the break-up of Gondwana is that a series of passive continental margins were created. Passive margins are coastlines that are not directly related to present plate boundaries. Such coastlines, while not involved in earthquake or other tectonic activities, are important as being the base levels for continental denudation. They represent the locations where sediments eroded from the various continents that were part of Gondwana (e.g. Africa, India, South America, Australia and Antarctica) are deposited and build up.

The Earth is 4.6 billion years old (Eicher, 1976). It is thus a reasonable question to ask whether plate tectonics and continental drift were processes active earlier than 200 million years ago and, if so, how far back in time have these processes operated? There is no reason to believe that

these processes were not active well before 200 million years ago. However, the evidence we have available to us becomes rapidly more fragmentary the earlier we go back in time. There is no oceanic crust older than 200 million years and thus all evidence has to be obtained from continental rocks. There is no direct evidence from these rocks of ocean spreading or subduction. Transform faults have to be identified from geological and geophysical evidence. The most direct evidence is obtained by the study of ancient mountain belts, which are the result of only one type of particular plate boundary. Nevertheless geologists are confident that they can identify pre-Pangaea plate movements which began about 600 million years ago.

Reflective question

➤ Thinking about the themes of the other chapters of this book, what areas of physical geography have been changed in some way by the theory of plate tectonics?

2.8 Summary

The theory of plate tectonics totally revolutionized the way that environmental scientists looked at processes on the surface of the Earth and the geological record. The theory developed from data collected during the 1950s and early 1960s principally from oceanography and

marine geology. As a result of collecting data on the bathymetry of the ocean floor, the magnetism of oceanic rocks and the location of earthquakes and volcanoes, a pattern emerged that developed into the unifying theory of plate tectonics.

In this theory the Earth is made up of seven major lithospheric plates and a number of smaller ones which float on an asthenosphere. These plates move around the Earth. When these plates move apart, new ocean crust is formed. This occurs mainly at the mid-ocean ridges. When the plates collide with one another, one plate can descend beneath the other, in a subduction zone. These boundaries are characterized by a deep oceanic trench and by a series of highly explosive volcanoes. Alternatively, when two plates which have continental crust at their leading edge collide, then a mountain belt is formed such as the Alps or Himalayas. When two plates slide past each other, destructive earthquakes result. Because the only direct unequivocal imprint of plate tectonics is found in oceanic rocks and the oldest of those are only 200 million years old, it is not clear how long plate tectonics in its present form has operated on Earth. Geologists are confident, however, that they have found evidence of such movements back to 600 million years ago.

Further reading

Eicher, D.L. (1976) *Geologic time.* Prentice Hall, Englewood Cliffs, NJ.
This book gives a very readable account of how we understand geological time.

Hallam, A. (1973) *A revolution in earth sciences.* Clarendon Press, Oxford.
This is a relatively short book which summarizes the history of how plate tectonics developed.

Press, F., Siever, R., Grotzinger, J. and Jordan, T.H. (2004) *Understanding Earth.* W.H. Freeman, New York.
The authors present an introductory geology textbook which provides more detail on the general theory of plate tectonics.

Skinner, B.J., Porter, S.C. and Park, J. (2003) *The dynamic Earth: An introduction to physical geology.* John Wiley & Sons, New York.
This 5th edition provides much detailed information on physical geology and is a useful reference aid for those who want to learn about tectonics or other geological processes.

Wegener, A. (1966) *The origin of continents and oceans.* Methuen, London (translated from the 4th revised German edition of 1929 by J. Biram, with an introduction by B.C. King).
This book provides the original classic text describing continental drift before plate tectonics was first proposed.

Useful summary chapters of many aspects of plate tectonics are given in introductory oceanography textbooks that are also recommended as additional reading for Chapter 3. These include:

Pinet, P.R. (2000) *Invitation to oceanography.* Jones & Bartlett, Sudbury, MA (web-enhanced edition).

Segar, D.A. (1998) *Introduction to ocean sciences.* Wadsworth, Belmont, CA.

Sverdrup, K.A. and Armbrust, E.V. (2008) *An introduction to the world's oceans.* McGraw-Hill, London.

Thurman, H.V. (2004) *Introductory Oceanography.* Prentice Hall, Upper Saddle River, NJ.

Web resources

The ABCs of Plate Tectonics
http://webspinners.com/dlblanc/tectonic/ptABCs.php
A broad analysis of the basic principles of plate tectonics is presented here. However, in addition new more complex and controversial hypotheses concerning convection and landform formation are also discussed. (This is a personal website.)

This Dynamic Earth: The Story of Plate Tectonics (online edition)

http://pubs.usgs.gov/publications/text/dynamic.html
Here a US Geological Survey online textbook is presented, providing a clear but detailed discussion of the historical perspective, the development of the theory, the understanding of plate motions, 'hot spots', what drives plate movement and extraterrestrial tectonics. The site provides a very good description, with helpful illustrations throughout.

This Dynamic Planet: World Map of Volcanoes, Earthquakes, Impact Craters, and Plate Tectonics

http://pubs.usgs.gov/imap/2800/

A map showing the Earth's physiographic features, the current movements of the major plates and the location of volcanoes, earthquakes and impact craters. The map is fully explained and discussed in the adjoining article.

Nevada Seismological Laboratory Home Page

http://www.seismo.unr.edu/

This site provides information and images on tectonic activity.

NOAA Pacific Marine Environmental Laboratory (PMEL): Vents Program

http://www.pmel.noaa.gov/

Information about the work undertaken by the program in researching the effects of underwater hydrothermal venting systems.

Plate Tectonics

http://www.platetectonics.com/index.asp

This is a useful site containing a description of the major concepts in tectonics, a map of the world ocean floor and an article archive providing papers on plate tectonics and related geological topics.

San Andreas Fault

http://pubs.usgs.gov/gip/earthq3/

Information on where the fault is, what it is, the surface features that characterize it, the history of earthquakes and when the next one is due are presented here. The site is quite brief but provides some case study material.

USGS: Earthquakes Hazard Program

http://earthquake.usgs.gov/

An excellent set of resources is presented in this web resource with facts and information, animations, photos, and very useful links to other relevant sites related to earthquakes.

USGS Volcano Hazards Program

http://volcanoes.usgs.gov

This site contains an excellent set of resources with photos, diagrams, fact sheets and case studies of volcanoes and their hazards from around the world. There is also a section on Yellowstone National Park.

Visit our website at **www.pearsoned.co.uk/holden** for further questions, annotated weblinks, interactive models and video-clips showing physical processes in action.

Oceans

Michael D. Krom

Earth and Biosphere Institute, School of Earth and Environment, University of Leeds

Learning objectives

After reading this chapter you should be able to:

➤ appreciate the vast scale of the oceans and how the processes that control the shape of ocean basins are different from those that shape landscapes

➤ understand the processes that control and drive both the surface and deep water flows in the oceans

➤ understand the processes that control the nature and amount of sediment deposited within the oceans

➤ understand the factors that control primary productivity in the surface layers of the ocean and the distribution of animals in the sea including those that make up commercial fisheries

➤ describe the factors involved in the development of a commercial marine agriculture

3.1 Introduction

When astronauts first looked back on planet Earth from space (Figure 3.1) they were struck by its colour. The Earth looks blue from space because most of its surface is covered by seawater. The oceans not only cover most of the area of the planet but also contain almost all of the water on Earth. Furthermore, much of the oxygen we breathe is produced by phytoplankton that live in the surface layers of the ocean. These interactions between the ocean and the atmosphere play a large part in controlling both the climate and the weather at the Earth's surface (see Chapters 4, 5, 6, 20, 21 and 22). The oceans transfer energy across the planet in the form of ocean currents and waves. Much of the excess carbon dioxide resulting from human emissions is taken up by oceanic processes either temporarily or permanently. The study of oceanic processes is therefore crucial if we are to understand global climate change.

We also use the oceans as a source of food. However, present fishing methods have become so efficient that we are in danger of hugely depleting fish stocks over large parts of the ocean. We are only just beginning to develop the technology to grow marine organisms as an agriculture in the same way as we farm the land. Indeed this is probably the last major untapped food resource on the planet. At the same time we use the oceans as a repository for our waste, often assuming the oceans are so large that they have an infinite capacity to absorb our pollutants. Yet it is clear from the increased incidence of toxic plankton blooms and other undesirable effects that this is not true. For all these reasons it is important that we study the oceans and understand how they operate.

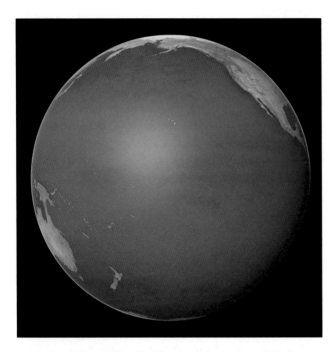

Figure 3.1 A view of the cloudless Earth from space. This view is centred over the Pacific and highlights how much of the Earth's surface is water covered. (Source: Planetary Visions)

3.2 The ocean basins

3.2.1 The scale of the oceans

There are a number of ways of expressing the total amount of water in the oceans. Seawater covers 361 million square kilometres (361×10^6 km^2) which represents 71% of the surface of the globe. The total volume of water is enormous: 1.37 thousand million cubic kilometres (1.37×10^9 km^3). Most of this water is contained in the three great oceans of the world: the Pacific, Atlantic and Indian Oceans (Table 3.1). Such numbers are so large that they are difficult to conceptualize. One way of looking at this

number is to consider a sphere the size of the Earth completely covered with water. The ocean water would then cover such a sphere to a uniform depth of almost 2700 m. That amount of water represents almost 93% of the total amount of water on the planet. Glacial and land ice make up a further 1.6% and groundwater 5% (see Chapter 4). Freshwater, in the form of rivers and lakes, the water that we as land animals require for our very existence, represents only 0.04% of the total amount of water on the planet. Even the largest lakes in the world are very small in area and particularly in volume compared with the world's oceans (Table 3.1). This freshwater is derived entirely from evaporation from the surface of the oceans.

3.2.2 Geological structure of the ocean basins

Geologists have suggested that oceans have been present on Earth for much of its 4.6 billion year history. Yet it came as a considerable surprise to discover in the 1960s that although there are rocks on land that were formed very early in the history of the Earth, the rocks that underlie the ocean are all comparatively recent. The oldest rocks in the ocean are only 200 million years old. Furthermore, the oldest rocks in the ocean were not found in the centre, as might be expected, but at the ocean margins. The explanation for this observation is plate tectonics (see Chapter 2). The ocean basins form when two plates spread apart and new crust is formed at the mid-ocean ridge. As the plates continue to diverge, the ocean basin gets wider (and older). As the oceanic crust cools it sinks, so the depth of water increases as one travels away from the mid-ocean ridges. The edge of the ocean basin, and generally its deepest part, is the bottom of the continental rise which is also where the oceanic lithospheric plate, composed mainly of basalt, is joined to the continental (granitic) plate. Eventually, as a result of the movement of plates over the Earth's surface, one or both of these passive margins

Table 3.1 The amount of water in each ocean and its depth with some large freshwater lakes

	Atlantic Ocean	Pacific Ocean	Indian Ocean	Lake Superior	Lake Baikal
Area (km^2)	106×10^6	180×10^6	75×10^6	82×10^3	32×10^3
Volume (km^3)	346×10^6	724×10^6	292×10^6	12×10^3	23×10^3
Mean depth (m)	3332	4028	3795	150	740

becomes a subduction zone and the oceanic crust plunges back into the Earth's mantle to be remelted and returned whence it came. Chapter 2 describes these tectonic features in more detail.

3.2.3 The depth and shape of the ocean basins

The deep ocean is nearly as rugged in its bathymetry as the terrain we see on land. There are high mountains, deep valleys and canyons as well as areas of flatter plains, with hills of varying heights rising from the plains below. Indeed the undersea mountains are in general longer, the valley floors wider and the canyons often deeper than the equivalent features on land. The principal difference is that while the dominant feature on land is erosion, with rain, wind and ice acting to erode the landscape and often to accentuate features (see for example Chapters 11 and 18), the net process underwater is sedimentation, which acts to fill in the valleys and to cover these features. However, the rate of sedimentation tends to be very slow so that topographic features formed near the mid-ocean ridge remain largely intact as the floor spreads away from the ridge.

Extending from the shoreline is the **continental shelf** (Figure 3.2). The width of the continental shelf can vary from a few tens of metres in areas such as parts of the western coast of North and South America to hundreds of kilometres in the North Sea and off the northern coast of Siberia. It is usually less than 150 m deep and is the area of the seabed where most of the coarse-grained sediment (sand and silt) derived from the physical erosion of the land is deposited. Because it is shallow and close to land, dissolved nutrients, which fertilize the sea and are needed by

phytoplankton to grow, are relatively abundant (see below). It is often an area of rich fisheries. Also, because it is underlain with continental rocks, it has been exploited for oil and other minerals.

The continental slopes beyond the shelf break are similar in relief to mountain ranges on continents. Their total height is generally 1–5 km above the **abyssal plain** below (Sverdrup *et al.*, 2004). The continental slope, and to a lesser extent the continental shelf, are often dissected by large submarine canyons. These canyons, which may originally have been formed as erosional features on land during the recent Ice Age when sea level was much lower (up to 130 m lower) than it is now, are often the locations for turbidity currents. These are flows of sediment-laden water, which periodically move down the slope at speeds in excess of 20 km h^{-1} and can cause considerable erosion on their way. Once these currents reach the continental rise, they form deep-sea fans, which have many of the characteristics of deltaic fans found in shallow water (see Chapter 17).

The continental slope and, beyond it, the abyssal plain, where depths often exceed 4000 m, are entirely underlain by basaltic oceanic crust. It is, however, untrue to think of these regions as flat and featureless. Seamounts (such as guyots or abyssal hills), all of which are volcanic in origin, dot the deep-sea floor. For example, in the Pacific Ocean alone, there are more than 20 000 seamounts, a few of which are tall enough to reach the surface as oceanic islands or **atolls**.

Mid-ocean ridges are a continuous set of mountain chains that extend for some 65 000 km. They are tectonically active with frequent volcanic eruptions and earthquakes. They are the origin of the basaltic rocks which underlie all the ocean basins (see Chapter 2). In many areas of the mid-ocean ridges there are also submarine hot springs (see Box 3.2 below).

The deepest valleys on Earth are the narrow, steep-sided oceanic trenches which characterize particularly the edge of the Pacific Ocean. These are the locations where oceanic plates plunge down into the mantle of the Earth. The Challenger Deep, which is the deepest part of the Marianas Trench in the western Pacific, is 11 020 m deep. By comparison Mount Everest is only 8848 m high.

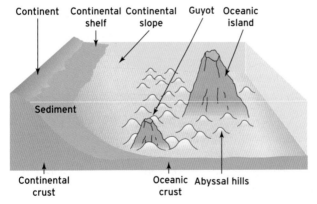

Figure 3.2 An idealized cross-section of the floor of an ocean basin from the coast of the continent. Abyssal hills and guyots are known to be volcanic in origin.

Reflective questions

➤ What are the main features of ocean basins?

➤ Why are the oldest rocks on the ocean floor not very old in geological terms?

3.3 Physical properties of the ocean

3.3.1 Salinity

In 1872, HMS *Challenger* sailed from Portsmouth on a voyage lasting three and a half years. This was the first systematic scientific investigation of the world's oceans and was as much of a voyage into the unknown as space probes to Mars and Venus are today. One of the most surprising things that the scientists on board discovered was that wherever you go in the world, the chemical composition of the ocean is *approximately* the same. The major salts that make up most of the salts in seawater are present in the same ratio to one another. Seawater is made up principally of sodium chloride and magnesium sulphate with significant amounts of potassium, calcium and bicarbonate (Table 3.2).

The chemicals that make up the salt in seawater were originally derived from the chemical weathering of rocks on land. They then flow as a dilute solution in river water into the sea. Once these chemicals reach the sea, they are involved in a variety of chemical, geological and biological reactions which ultimately remove them from seawater and deposit them in the sediments at the bottom of the ocean. So, for example, calcium can be removed by marine organisms to form their shells and skeletons. These are then deposited, sometimes in enormously thick and extensive deposits of calcium carbonate, such as the chalk deposits of south-east England. It is the balance between the processes of supply of salts by the rivers and their removal into the sediments of the oceans that maintains the particular chemical composition of the oceans that we find today.

Although the ratio of most chemical elements in seawater is constant over all the world's oceans, the actual concentration varies depending on where you are (Pinet, 2000). Over

Table 3.2 The concentration of the major elements present in seawater

Element	Concentration in seawater (mg l^{-1})
Chloride	19 500
Sodium	10 770
Magnesium	1 290
Sulphur (as sulphates)	905
Calcium	412
Potassium	380

large parts of the ocean, seawater salinity is 35.5 g of solid matter per kilogram of water. The concentration of salt in the ocean is generally expressed as parts per thousand (ppt) or in salinity units. However, surface water in the ocean can be diluted by the addition of freshwater from rain, river inflow or ice melting. Surface waters can also be concentrated by evaporation. The salinity of surface water in the climatic bands of the dry desert latitudes (20°N and S) is higher than in the wetter regions of the equator or the temperate latitudes (Figure 3.3) because there is the greatest net evaporation in those latitudes. It is perhaps not surprising that the salinity of enclosed basins can vary by more than that of the open ocean. The highest salinities in the ocean today are found in the Red Sea where the salinity reaches 41.5 ppt due to high evaporation. By contrast the waters of the upper Baltic Sea near Finland, which has large amounts of freshwater both raining upon it and flowing in from the adjacent land, can get as low as 5 ppt.

3.3.2 Temperature structure of the oceans

Oceans are very important in controlling the climate of the Earth. The surface of the ocean gains heat by radiation from the Sun, particularly in lower latitudes, and by conduction and convection from warm air flowing over the waves. Heat is lost by evaporation, by reradiation to space and by conduction to cold air above (Barry and Chorley, 2003). Measurements made over the Earth's surface show that more heat is gained than lost at the low latitudes, while more heat is lost than gained at the high latitudes. Because water has a very high specific heat (see Chapters 4 and 5), it acts as a major store of the energy of the Sun. The movement of ocean currents from the low latitudes to the high latitudes is very important in transferring energy to the colder regions and thus maintaining the Earth's temperatures in the present range, which is basically conducive to life as we know it. When ocean circulation changes occur, such as during El Niño events, then they can have a dramatic effect on the climate of the Earth. El Niño is the situation where unusually the entire central Pacific from Australia to Peru is covered with a relatively thin layer of warm water (see Chapter 4). As a result of this change, weather patterns alter not only adjacent to the current, such as rainfall and floods in the Atacama Desert and snowfall in Mexico City, but also over much of the world.

For most areas of the ocean, there is a warm surface layer. The thickness of this surface mixed layer varies with both season and location but is generally of the order of 100 m thick. This layer is underlain by waters of decreasing temperature (and often increasing salinity). The depth zone with

Figure 3.3 Surface salinity of the oceans across the Earth in parts per thousand. This is controlled by net precipitation (rainfall – evaporation) with a maximum at the same latitude as the major deserts and minima at the equator and towards the higher latitudes. (Source: Sverdrup, H.U., Fleming, Richard H., Johnson, Martin W. *Oceans*, 1st Edition, © 1942. Adapted by permission of Pearson Education, Inc., Upper Saddle River, NJ)

rapidly changing temperature is called the **thermocline** (Figure 3.4). Below this deep thermocline, the temperature is relatively uniform, often showing only a small further decrease to the ocean bottom. Figure 3.4 also shows a **halocline**, which is the layer where salinity concentration also rapidly changes.

Temperature and salinity together control the density of seawater. If the density of seawater increases with depth, then the water column is stable. If, however, there is more dense water on top of less dense water, then the situation is unstable. Vertical mixing will take place until the water column has a similar density over that depth interval. This is the process that drives the three-dimensional circulation of the oceans (see below).

Reflective questions

➤ Why would differences in water temperature or salinity encourage water movement?

➤ Why does ocean water contain salt?

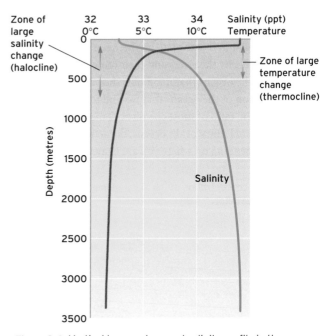

Figure 3.4 Vertical temperature and salinity profile in the ocean based on the north-east Pacific. There are large changes near the surface. (Source: after Sverdrup, K.A. *et al.*, *Introduction to the world's oceans* 7th edition, 2003, McGraw-Hill, Fig. 6.11. Reproduced with permission of the McGraw-Hill companies)

3.4 Ocean circulation

3.4.1 Surface currents

The first coherent body of knowledge that humans discovered about the oceans concerned the surface currents because such information was crucial for any voyages by ships driven by wind or oar power. The surface currents are driven by the prevailing winds. Except for the dramatic seasonal changes in those parts of the northern hemisphere affected directly by the monsoons, there is a nearly constant pattern for the winds blowing over the ocean that drives the large-scale surface currents. The trade winds that blow out of the south-east in the southern hemisphere and out of the north-east in the northern hemisphere (see Chapters 4 and 5) drive the Northern and Southern Equatorial Currents, which move in a westerly direction parallel to the equator (Figure 3.5). These currents are deflected by the continents when they reach the coastal areas.

In addition to the continents deflecting the water, the **Coriolis effect** (see Chapter 4) also causes water to deflect to the right in the northern hemisphere and to the left in the southern hemisphere. This forms the warm western boundary currents along the eastern coasts of North and South America, eastern Australia, eastern Asia and eastern Africa. In the northern hemisphere in the Atlantic, this warm western boundary current is better known as the Gulf Stream. The Gulf Stream forms the western and northern section of the North Atlantic subtropical **gyres**, which is one of the most dominant surface features of the world's oceans (Figure 3.5). These subtropical gyres are so called because their centres are located approximately 30°N or S of the equator, in the subtropics. In all of the world's oceans there is a westerly current in the temperate

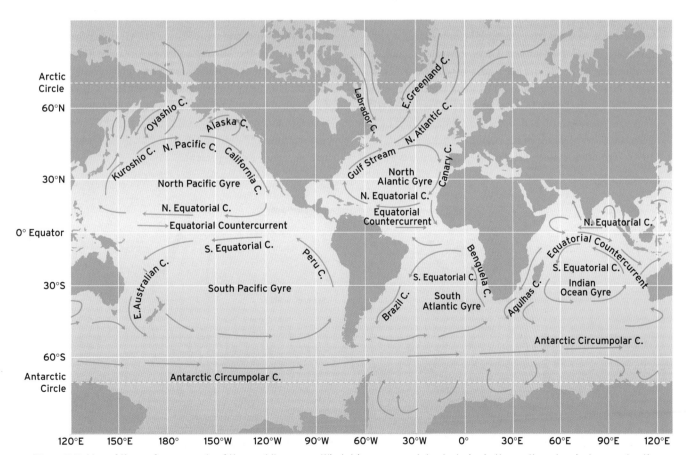

Figure 3.5 Map of the surface currents of the world's oceans. Wind-driven gyres rotate clockwise in the northern hemisphere and anti-clockwise in the southern hemisphere. (Source: after Pinet, P.R. *Invitation to oceanography*, 1st Edition, 1996: Jones and Bartlett Publishers, Sudbury, MA, www.jbpub.com. Reprinted with permission)

latitudes and the gyre is completed by a cold current flowing back towards the equator on the western side of the continents.

The other major feature of the Earth's surface circulation is the circumpolar currents which flow around Antarctica with a westerly current below 60°S and an easterly drift above 60°S. Because most of the land on Earth at present is in the northern hemisphere, there is no equivalent circumpolar current in the northern hemisphere.

3.4.2 The deep currents of the oceans

The large-scale deep circulation of the world's oceans is not driven by wind but by density variations. The density of seawater is controlled by its salinity and temperature. The resulting circulation is called the **thermohaline circulation.** When a body of water becomes denser than the water surrounding it, it sinks. In the world at present, the two most important regions where deep-water currents form are the North Atlantic–Arctic Ocean (see Box 3.1) and the Antarctic oceans (Figure 3.6). In the North Atlantic Ocean, relatively saline water from the Gulf Stream moves rapidly north into the Norwegian Sea. There, in winter, it cools to a temperature of 2–4°C and has a salinity of 34.9 ppt. This water, being denser than the surrounding water, then sinks away and forms the **North Atlantic Deep Water (NADW).** This water flows south and forms the major fraction of the deep water in the whole of the Atlantic (Dickson *et al.*, 1990). The other major source of deep cold water is formed along the edge of the Antarctic continent where a mass of even denser water with a temperature of −0.5°C and a salinity of 34.8 ppt is formed in winter. This **Antarctic Bottom Water (ABW),** which is the densest water formed in the oceans at present, then flows north under the NADW. Eventually the two water masses mix before flowing eastwards into the Indian and Pacific Oceans where the mixture forms a major fraction of their deep water. The major deep water masses in the oceans at present are formed in the Atlantic and/or the Southern Oceans. That is because there is no free access to the Arctic Ocean from either the Indian or Pacific Ocean.

Subsurface water can also be formed nearer to the equator. Near the equator, the upper boundary of the NADW is formed by water produced at about 40°S. This Antarctic Intermediate Water (AIW) is warmer (5°C) and less salty (34.4 ppt) than the NADW and hence remains above it. The other source of intermediate water (water that sinks to a depth of over 1000 m in the Atlantic) is Mediterranean Intermediate Water. This water is formed in the eastern basin of the Mediterranean. It has a much higher salinity (37.3 ppt) and temperature (13°C) than other water masses in the Atlantic and therefore once it flows out into the Atlantic at Gibraltar it remains recognizable as a water mass as far as Iceland and the West Indies. It is also of geological importance, since it is believed that water formed in this manner (by evaporation and then cooling) was present in the deep waters of the world's oceans when there were no ice caps present.

Figure 3.6 Vertical structure of water masses in the Atlantic Ocean. The Antarctic Bottom Water is densest and flows north from Antarctica. The North Atlantic Bottom Water flows south from Greenland above the Antarctic Bottom Water. Above these are intermediate water masses that are formed somewhat nearer to the equator. The surface layers extend to 300–500 m depth and are controlled by winds and other factors (see Figure 3.5).

FUNDAMENTAL PRINCIPLES

THE FORMATION OF NORTH ATLANTIC DEEP WATER AND THE OCEAN CONVEYOR BELT

Figure 3.7 shows the ocean conveyor belt. The Gulf Stream starts in the Caribbean and flows initially as a rather narrow, rapidly flowing stream of water off the coast of Florida and the southern states of the United States up to Cape Hatteras, North Carolina. From Cape Hatteras it leaves the coast and spreads out to form the North Atlantic Drift, which forms the northern part of the North Atlantic Subtropical Gyre (see also

(Figure 3.5). As it flows north, it gradually cools but the flow is still rapid enough to mean that the Gulf Stream and its equivalent in the western Pacific, the Kuroshio Current, are extremely important in the global transfer of heat from equatorial regions towards higher latitudes. The Gulf Stream flows around Iceland and into the Norwegian and Greenland Seas.

Here in winter, this relatively saline (34.9 ppt) and warm water cools to the point where it becomes denser than surface waters and falls away into the depths. Initial studies of the

rate of North Atlantic Deep Water (NADW) current formation were carried out in 1972–1973 during the Geochemical Ocean Section Study (GEOSECS) programme. Radioactive pollutants produced both in the atmosphere by nuclear bomb testing (^{3}H and ^{14}C) and by the discharge of nuclear waste at Windscale (Sellafield) in north-west England (^{137}Cs) were measured in the water column. Since it was known when these radioisotopes were first released into the environment, it was possible to determine from their distribution how far and fast they had moved, and thus

Figure 3.7 A schematic diagram of the main features of the global thermohaline circulation, also known as the oceanic conveyor belt.

BOX 3.1 ➤

the rate and amount of the NADW produced (Schlosser *et al.*, 1999). Scientists were surprised to discover that water downwelled as fast as an average of 0.1 to 1.5 m day^{-1}. More recently direct current measurements suggest that most of the NADW passes rather close to the Greenland coast as a western boundary current and then into the deep North Atlantic (Dickson *et al.*, 1990).

It has been suggested that global warming may cause sufficient fresh-water to be released from the melting of ice in the northern latitudes to slow

or even to stop the formation of the NADW by reducing salinity to the level where it is insufficiently dense to sink. If this occurred it would stop the Gulf Stream flowing north and thus cause the oceanic conveyor belt, which is transferring heat from the equator, to cease. This would have a dramatic cooling effect on the climate of north-west Europe, making it resemble that experienced at present in areas of north-east Canada such as Labrador. Such a scenario has been used to explain the re-cooling that occurred during the Younger Dryas (Broecker,

1998) at the end of the last Glacial Maximum (see Chapter 20) as well as the global climate anomaly 8200 years ago (Barber *et al.*, 1999). A series of current measurement and tracer studies are being carried out by a joint international programme of the US National Science Foundation and UK Natural Environment Research Council RAPID programme. The aim of this programme is to determine whether measurable and significant changes are taking place in the north-ward flow of water in the North Atlantic Drift.

BOX 3.1

Water that sinks away into the deep has to be balanced by water upwelling to the surface. This occurs many thousands of kilometres from the regions where deep water is formed. There are regions in the world, principally along the eastern margins of the major continental masses, where the surface currents are driven directly offshore. In order that volume is preserved, this water is replaced by upwelling water from below. These regions of upwelling are important because the upwelled water contains an abundance of plant nutrients, nitrates and phosphates, which, when they reach the surface layers where there is sufficient light, result in a vast bloom of phytoplankton which in turn sustains major fisheries. One example of such an upwelling region is off the coast of Peru. This is the area affected by El Niño, which results in the upwelling being interrupted by the covering of the warm surface waters from the west and the collapse of the fisheries.

3.4.3 The weather of the ocean

From 1971 to 1973 an international expedition was mounted to study intensively an area of ocean several hundred kilometres across in the western Atlantic. To the surprise of the scientists involved in this experiment, which was called the Mid-Ocean Dynamics Experiment (MODE), it was found that the pattern of currents was much more complex than the comparatively simple ocean currents described above. What we have described so far might be called the climate of the ocean. What the MODE scientists found was that the ocean also has weather. They found that there are a series of eddies and fronts in the ocean in much the same

way as there are cyclones, anticyclones and warm and cold fronts in the atmosphere (see Chapters 4 and 5). These eddies are characteristically of the order of 100 km in diameter. They can spin either clockwise or anticlockwise and can have a warm or a cold core. Some eddies are formed at the boundary between two major currents. Figure 3.8 shows eddies that form at the boundary between the warm Gulf Stream and the adjacent cold Labrador Current. These rings spin off and can survive for a few months to a couple of years before either dissipating or being incorporated back into the original current. Other eddies are caused by the interaction of a current with a promontory of land, such as at the entrance to the Mediterranean Sea (Figure 3.9), or by a current flowing over a seamount. Some eddies have been observed in the middle of the ocean. The cause of these is simply unknown at present. In practical terms these eddies can be important in transferring energy from one part of the ocean to another as well as being sites for enhanced or reduced biological productivity (Smith *et al.*, 1996).

Reflective questions

➤ How are the surface and deep ocean currents different? What drives these currents?

➤ If NADW formation was stopped as a result of global warming, what effects might it have on ocean currents in the rest of the world?

Figure 3.8 Colour satellite image of the western Atlantic showing eddy formation. The colours indicate mixing of waters of different temperatures with warmer waters in yellow-red and cooler waters in blue-green. The mixing of these waters occurs in 'swirls' or giant eddies. (Source: image courtesy of Liam Gurmley, MODIS Atmosphere Team, University of Wisconsin, Madison)

Figure 3.9 Colour satellite image of the straits of Gibraltar with Portugal and Spain to the north and Morocco to the south. The figure shows an eddy being formed as incoming water from the Atlantic 'relaxes' after passing through the narrow strait.

3.5 Sediments in the ocean

The continental shelves and other near-shore areas are usually underlain by sands and silts. Because we are living in a warm period with ice caps melting and sea level

rising (see Chapter 17), much of the new sediment eroded from the continents is being deposited in estuaries, while the shelf areas such as the North Sea are mainly underlain with relic sediments. However, there are regions, particularly opposite major rivers such as the Bramaputra–Ganges and the Amazon, where sediments are being deposited at present (Goodbred and Kuehl, 1999). These sediments can deposit to such depths that they become unstable and become sources for turbidity currents, which carry the sand and silt onto the continental rise and beyond.

There are four types of sediments that are found in the deep ocean. These are: **biogenous**, the remains of the skeletons of marine organisms; **lithogenous**, particles derived from the physical and chemical breakdown of rocks and minerals; **hydrogenous**, sediments derived from geochemical processes involving organic matter, and particularly iron and manganese oxides and seawater; and finally **cosmogenous**, particles derived from outer space.

The most abundant sediments on the ocean floor are sediments derived from biological remains (Figure 3.10). **Calcareous oozes** are found over large parts of the oceans and particularly the Atlantic Ocean. They are made up of the remains of minute organisms called plankton that live in the surface waters of the oceans. The most abundant organisms found in calcareous oozes are coccoliths, foraminifera (see Chapter 20) and pteropods. These oozes, which are made up principally of calcium carbonate, are found to a depth of 3500 m in the Pacific Ocean and to a somewhat greater depth in the Atlantic Ocean. Below that depth, the increasing concentration of carbon dioxide in the water and hydrostatic pressure cause the water to become undersaturated with respect to calcium carbonate, and the shells dissolve.

In some parts of the ocean, particularly around the Antarctic and central Pacific, the principal planktonic organisms growing in the surface waters do not have calcareous skeletons. They are diatoms in the Antarctic Ocean and radiolaria in the central Pacific. These organisms have skeletons made of silica. As a result, the sediments below these regions are called siliceous oozes.

Over the remaining areas of the deep ocean, the sediments are mainly lithogenous in origin. Red clays are derived mainly from dust transported by the prevailing winds and then deposited. As might be imagined, the rate of deposit of such sediments is very slow, ~1 mm per 1000 years. In some of these areas of slow deposition, manganese nodules form. We still know very little about exactly how these nodules form. We do know that they

Figure 3.10 (a) Map of sediment types on the ocean floor. Calcareous oozes are formed from plankton such as coccoliths, siliceous oozes are formed mainly from diatoms (Southern Ocean) and radiolaria (Tropical Ocean). Red clay is deposited mainly by dust storms shown occurring over the Red Sea in (b). Terrigeneous and continental shelf sediments are formed from land-derived material such as that seen entering the oceans from rivers such as the Yangtze in (c) which are then redistributed by oceanic currents; and also glacial marine sediments mainly derived from melting icebergs. (Source: (a) adapted from Davies and Garsline, 1976; (b) and (c) Jacques Desloitres, MODIS Land Rapid Response Team, NASA/GSFC)

grow exceedingly slowly, at a rate of between 1 and 200 mm per million years. It is probable that at least part of the manganese which forms the nodules is supplied by black smokers at the mid-ocean ridges (Box 3.2 below). The nodules are usually concentric and in addition to manganese contain significant amounts of iron, cobalt, copper and zinc. See also Chapter 12 for further information on ocean sediments.

3.6 Biological productivity

3.6.1 Photosynthesis in the ocean

The basic building blocks of all life in the sea, as on land, are the photosynthesizing plants. It may appear to the casual observer, whose main contact with the sea is a stroll along the beach, that the only plants in the sea are seaweeds, but that is far from true. By far the largest biomass of plants in the sea are phytoplankton. Phytoplankton, the grasses of the sea, are mainly single-celled plants known as algae. Although some plankton have the ability to move to some degree, they make no purposeful motion against the ocean currents and are carried from place to place suspended in the water. Phytoplankton, which are able to photosynthesize, vary in size from ultraplankton, which are less than 5 μm (0.005 mm) in diameter, to net plankton, which can be found in chains between 70 and 1000 μm (1 mm) in length. Groups of organisms belonging to the phytoplankton include diatoms, **dinoflagellates** and coccolithopores. Because they reproduce mainly by asexual division, which can occur every 12–24 h, under favourable conditions phytoplankton can rapidly bloom to form vast numbers of individuals. These can discolour the water and even, in the case of coccolith blooms, be seen from orbiting satellites.

The basic equation for photosynthesis is:

$$106CO_2 + 16NH_3 + H_3PO_4 + 106H_2O$$
$$\rightarrow (CH_2O)_{106}(NH_3)_{16}(H_3PO_4) + 106O_2 \qquad (3.1)$$

An additional requirement for this photosynthetic reaction to take place is light energy. It is the chlorophyll within the plant that is used to convert light energy into chemical energy, into the organic matter plants require for growth. The rather curious elemental ratio of $106C : 16N : 1P$ (carbon, nitrogen, phosphorus) is used because it has been found that almost all marine organic matter has this particular elemental ratio (Redfield *et al.*, 1963). The reverse reaction, respiration, involves the breakdown of organic matter to release energy. This is the energy all living organisms need for their life processes. All plants both photosynthesize and respire. This provides the food for animals. However, a healthy plant population will produce more organic matter by photosynthesis than it will respire. All animals only respire using the organic matter that they consume as fuel.

Plants are restricted to the surface layers of the ocean where there is sufficient light for them to photosynthesize. This layer is called the **photic zone**. As a rough guide, plankton can still grow in light that is 1% of the light intensity that reaches the sea surface. Some species of ultraplankton can successfully photosynthesize at a tenth of this light level. The penetration of sunlight in clear and turbid seawater is compared in Figure 3.11. In clear oceanic water the photic zone stretches down to approximately 100 m and in extreme cases such as the eastern Mediterranean to 200 m. As the turbidity of the water increases, owing to the presence of sediment or other particles, the depth of the photic zone decreases.

The amount of light available for photosynthesis also varies with time of day and with the weather. Clouds will reduce not only the amount of light but also the depth to which that light will penetrate. Light also varies with season and with latitude. These variations in light affect the amount and timing of productivity in the ocean. In the North Atlantic all the other requirements for primary productivity are present throughout the winter except for the fact that there is simply not enough light for the plankton to grow. In March, the light flux increases sufficiently for the

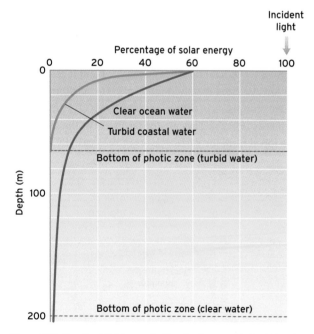

Figure 3.11 Depth of light penetration in clear and in turbid water. The photic zone represents that part of the water body into which solar radiation can penetrate allowing photosynthesis.

Figure 3.12 A near-true colour Landsat image of a coccolith bloom off the south-west coast of England. (Source: courtesy of the Remote Sensing group at the Plymouth Marine Laboratory)

plankton to start to grow. Plankton grow very fast, doubling their numbers over a 24 h period; this results in a plankton bloom (Figure 3.12).

3.6.2 Importance of nutrient supply to primary productivity

The presence of light does not guarantee plant life. Phytoplankton also need dissolved nutrients. The most important nutrients for plants in the sea are nitrates (NO_3^-) and phosphates (PO_4^{3-}). They are the fertilizers of the sea and are stripped from the surface layers of the ocean by the

plankton, which incorporate them into their tissues. When the plankton die, the organic matter is broken down and these nutrients are released back into the water. This can occur at depth or they can be returned to the water in the form of waste products of herbivores and carnivores. Figure 3.13 illustrates a typical depth profile for nutrients in ocean water. Vertical mixing brings these nutrients back to the surface.

The highest productivity in the oceans occurs in those areas where the supply of dissolved nutrients is greatest. The most productive areas of the ocean are the areas of coastal upwelling which have abundant life. Dissolved nutrients are supplied naturally from land by chemical weathering and by the breakdown of land plants and animals. Hence estuaries are areas of high productivity. This natural process is often amplified by humans, resulting in **eutrophication** (areas of unnaturally high primary productivity induced by high nutrient loading). The high productivity on the shallow coastal shelves is sustained not only by coastal runoff but also by mixing and recycling from the shallow sea floor. By contrast large areas of the world's oceans that do not have a direct supply of dissolved nutrients are, for all practical purposes, a biological desert.

Plant populations are limited by the lack of any essential nutrient: if the concentration of such a nutrient falls below the minimum required, the population growth stops until the nutrient is supplied. Most of the world's oceans have been found to be essentially limited equally by nitrates and by phosphates. Both need to be added in a ratio of 16 : 1 for natural growth to occur. There are some areas of the ocean, most notably the eastern Mediterranean, which are phosphorus limited. Apart from the major nutrients, N and P, plants also need micronutrients such as iron, copper, zinc and molybdenum. Recently it has been shown that large areas of the eastern central Pacific and in the Southern Ocean, which are so far away

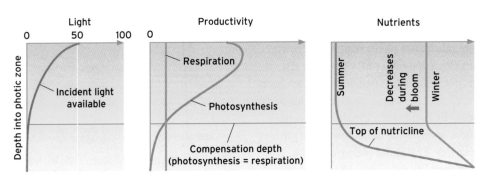

Figure 3.13 Typical nutrient and productivity profile for ocean water during spring plankton bloom.

from land that they are affected neither by runoff from land nor by dust input from the atmosphere, are iron limited (Boyd *et al.*, 2000).

Life in water requires carbon dioxide and oxygen, as does life on land. Carbon dioxide is required by plants for photosynthesis. It is contributed by respiration processes and also is absorbed by water from the atmosphere. Because seawater has the capacity to absorb large quantities of carbon dioxide, under natural conditions the amount of carbon dioxide in water is never limiting. Oxygen is also needed by all organisms that liberate energy for organic matter by normal aerobic respiration. There are groups of bacteria that are able to liberate energy from the breakdown of organic matter using other compounds such as nitrates or sulphates as the oxidizing agent. These bacteria respire **anaerobically**. Oxygen is supplied to surface water by exchange with the atmosphere and also as a by-product of photosynthesis. There are, however, no processes that supply oxygen to deep water, while processes of respiration consume oxygen. Thus in most areas of the ocean there are lower concentrations of dissolved oxygen than at the surface, as shown in Figure 3.14.

Plants and marine animals other than birds and mammals do not control their body temperatures. Their physiology is regulated by the temperature of the water and within limits metabolic processes occur more

rapidly in warm water than in cold water. It is therefore perhaps surprising that some of the most productive areas of the world's oceans are in the Antarctic Ocean and in the northern Atlantic Ocean in early spring. The reason for this is that the other factors discussed above, availability of light, nutrients and so on, are more important in controlling primary productivity than is temperature.

There is, however, one group of creatures that does not rely on light energy for survival of their food chain in the oceans. Instead they survive by chemical energy produced by inorganic reactions at the mid-ocean ridge. This is also the only ecosystem on the planet that remains unaffected by anthropogenic inputs and change. These systems are described in Box 3.2.

3.6.3 Animals of the sea

3.6.3.1 Food chains

In all the waters of every ocean, one organism preys on another. At the base of this food chain (see Chapter 10 for further explanation of food chains) are the phytoplankton, the primary producers, while at the top of the pyramid are the carnivores: sharks, tuna and of course also humans. On the first rung of this ladder are the herbivores which eat plant life directly (Figure 3.16). In the oceans by far the most numerous are the zooplankton. Within the zooplankton there are two types of organism:

1. Those animals that remain as floating animals throughout their life cycle. This includes animals such as salps and krill, the tiny shrimps that form the major food source for animals, including the blue whale.
2. The juvenile forms of many larger marine organisms spend part of their life cycle in a planktonic form. This enables **sessile** or quasi-sessile organisms to spread their offspring into new areas.

There may be many or few links in the food chain from the primary producers to the top carnivore (Figure 3.16). Food chains in the ocean are rarely simple. They are most likely to show complex branching and interactions. As a result, such interactions are better described as a food web.

In general as one moves upwards from the first **trophic level** (see Chapter 10 for information on trophic levels), the size of the organism increases and the number of organisms decreases. The larger numbers of small organisms at the

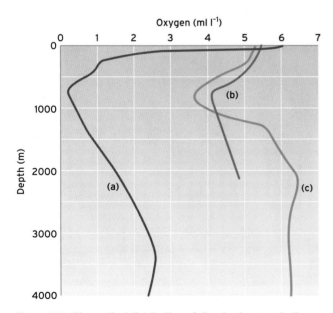

Figure 3.14 The vertical distribution of dissolved oxygen in three ocean waters: (a) the Pacific south of California; (b) the eastern Mediterranean; and (c) the Gulf Stream. (Source: adapted from Open University, 1989)

NEW
DIRECTIONS

MID-OCEAN RIDGE HOT SPRINGS

During the 1970s, geologists and geochemists who studied the processes occurring at the mid-ocean ridges hypothesized that apart from the volcanic eruptions and earthquakes, which they knew took place at these spreading centres, it was likely that there should be hot springs there as well. If such springs existed, they were expected to have unusual chemistries and possibly to have a significant effect on the chemical composition of the ocean as a whole. In March 1977, an expedition of scientists from Woods Hole Oceanographic Institute with the research submersible *Alvin* arrived over the Galapagos ridge in the eastern Pacific to locate and investigate the ridge area at a depth of 2000 m looking for vents. What they found when they finally located a hot spring, and dived down to investigate it, surprised and delighted them.

They did indeed find water issuing from the vents at temperatures of 17°C, which was significantly warmer that the 2°C found in the surrounding water. The hot water was seen to rise from the vent, producing shimmering upward-flowing streams rich in silica, barium, lithium, manganese, hydrogen sulphide, methane, hydrogen and sulphur. Since that time studies have been carried out of many other vents that have been found in the mid-ocean ridges in all of the world's oceans. Temperatures as high as 380°C have been recorded for the water issuing from these vents. This temperature was high enough to melt the plastic coating on the probes that were first used to measure temperature in these streams. It was also hot enough to soften the Perspex windows of the submersible *Alvin* itself! At many

of these areas, mounds and chimney-shaped vents were found which were tens of metres high and which ejected a continuous stream of black particles. These black smokers contain particles

(a)

of manganese, lead, cobalt, zinc, silver and other metals as well as sulphur and sulphide (Figure 3.15a). When such deposits are uplifted on to land, they result in commercially important

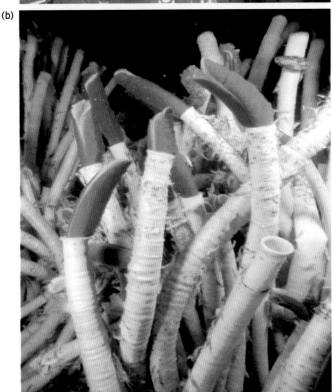

(b)

Figure 3.15 Mid-ocean ridge hot springs: (a) photograph of a black smoker vent from a deep-sea diving vehicle (source: Dr Ken MacDonald/Science Photo Library Ltd); (b) tube worms in a deep-sea hydrothermal vent community (source: Al Giddings).

BOX 3.2 ➤

massive sulphide deposits such as at Troodos, Cyprus.

However, probably most remarkable and surprising of all the discoveries at the vents was the presence of dense communities of large animals. Many of these animals looked very different from animals found elsewhere on Earth including giant tube worms, 3 m in length, with red tips (Figure 3.15b), clams with red blood and blind shrimps.

It was immediately clear that these vent communities were unlike any other biological community on Earth. They were not dependent on organic matter derived from the surface layers of the ocean and hence on food that was ultimately derived from the Sun's energy. The vent communities derive their metabolic energy from the chemical energy contained within the hot water in the springs. It has subsequently been shown that at the base of this remarkable food chain are groups of bacteria that derive their metabolic energy from the oxidation of the hydrogen sulphide, methane or hydrogen emitted with the vent water.

Detailed studies at the mid-ocean ridge in the centre of the equatorial Atlantic have found that the most abundant animal found at these vents is the blind vent shrimp, *Rimicaris*. These shrimps live very close to the hottest (and most toxic) vents. They live both by grazing chemosynthetic bacteria and by growing these bacteria within their shells (rather like humans growing potatoes and carrots on their backs). Because of this symbiotic relationship with these bacteria, the shrimps have to live right next to the outpouring waters where the highest concentrations of hydrogen sulphide are found. Although the shrimps are blind, they do have a heat-sensing organ on their backs which they use to avoid becoming boiled.

One intriguing question still to be answered is how the individuals in these communities manage to spread from one vent to the next. It is known that each individual vent has only a quite limited lifespan (~100 years) before the waters switch off and the vent ceases to flow. Since the next vent may be several kilometres away, it is necessary for the organisms to spread their offspring in such a way as to reach the new vent wherever it is. Scientists have now shown that for many vent organisms such as the shrimps and tube worms, the larvae are broadcast into the surrounding waters and live on phytoplankton and other organic debris until they find a new vent to colonize, or die in the attempt to find one.

Recently similar communities of animals which are dependent on chemical rather than solar energy for their energy source have also been found at vents of active volcanoes associated with subduction zones and with oceanic hot spots. These vent communities are often found in much shallower water. For example, hot vent fluids have been found with similar ecosystems in 240 m of water associated with volcanoes in the Mariana arc, western Pacific.

BOX 3.2

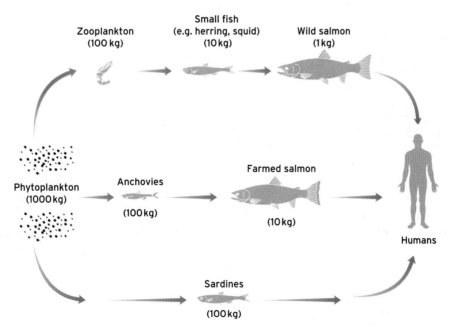

Figure 3.16 A simplified example of three oceanic food chains. Food chains are inefficient and lose around 90% of their energy between different food levels (see Chapter 10). Therefore 1000 kg of phytoplankton (photosynthetic cost) can produce 1 kg of wild salmon, 10 kg of farmed salmon and 100 kg of sardines.

lower trophic levels collectively have a much larger biomass than the smaller numbers of large organisms at the upper levels. It is estimated that the overall efficiency of energy transfer up each layer of an open-ocean trophic pyramid is about 10%. Thus in order to gain 1 kg of weight, a salmon must eat 10 kg of smaller fish, and the fish need to consume 100 kg of carnivorous zooplankton which in turn requires 1000 kg of phytoplankton. Thus it takes 1000 kg of primary producers to produce 1 kg of the top carnivore. The 90% energy loss at each level of the pyramid goes into the metabolic needs of the organisms at that level, such as energy required for moving, breathing, feeding and reproduction (see Chapter 10).

3.6.3.2 Fisheries

At present we harvest marine organisms principally by hunting them in boats using nets (Figure 3.17). The catch depends on custom, culture, economics and availability. Possibly more than with any other food, what one culture considers a great seafood delicacy, another will not eat either because of taste or in some cases because of religious taboos. Many commercial fisheries catch from both the higher trophic levels (e.g. cod, salmon, halibut and tuna) and the lower levels (e.g. herring, shellfish, anchovies and sardines). Harvesting fish high in the trophic pyramid is energy inefficient. Because of the complex interactions within the marine food web as well as our very poor knowledge of the marine biological system, depleting the stocks of fish at one trophic level has unpredictable effects on a number of organisms elsewhere in the food web. What is clear, however, is that with modern fishing methods, large boats and nets, sonar and other efficient locating equipment, we are succeeding in severely depleting many of the world's fish stocks (e.g. at Grand Banks, off Newfoundland, and the North Sea).

There is, however, hope. Most marine organisms have a very high **fecundity**. For example, each female gilthead seabream, a fish at the higher levels of the trophic pyramid in the Mediterranean, produces 1 million eggs per year. In nature only two of those offspring need to survive every several years to ensure the continuation of the species. In captivity, survival rates of 20% or more have been achieved from egg to adult fish, so that 200 000

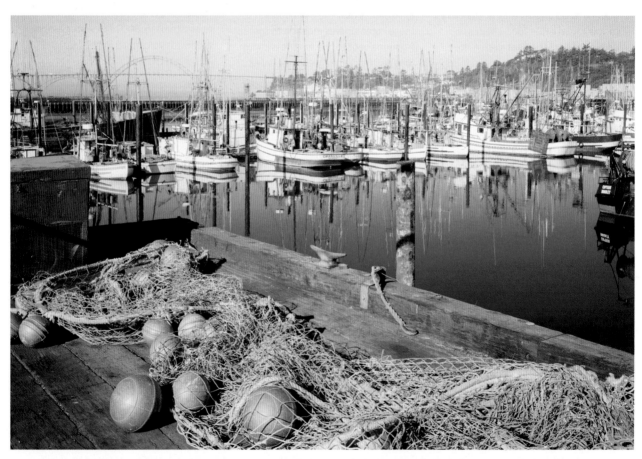

Figure 3.17 Fishery catch depends upon custom, culture, economics, technology and availability. (Source: Steve Terrill/Corbis)

offspring have survived. Thus it is possible, once over-fishing has stopped, for marine stocks to rebound to previous natural levels much faster than it is possible to restore natural systems on land.

3.6.3.3 Mariculture

Mariculture, the culturing of marine organisms as in agriculture, is probably the last major untapped food resource on Earth. Freshwater aquaculture began in China with the farming of carp some 4000 years ago. In medieval times in Europe almost every monastery had its own fish-ponds to provide the monks with fresh fish for Fridays. Despite this, the farming of marine fish is a comparatively recent technology. There are basically three types of mari-culture. Sea ranching is where the marine organism is restrained in cages or in the case of shellfish grown on ropes hung from rafts. No artificial food is provided in this type of culture and thus the overall yield of the system is constrained by the natural productivity of the area. An example of this type of mariculture is shown in Figure 3.18 which shows rafts in Ria Arosa, north-west Spain, with ropes seeded with mussels and oysters hung beneath them. The shellfish grow over a period of several months and are then harvested and sold.

By far the most widespread form of mariculture carried out at present is growth of fish in pens or nets. In such systems, shown in Figure 3.19, fish are put into a net slung beneath a set of floats when they are fingerlings. They are then fed an artificial diet and allowed to grow until they are large enough to be harvested. Sea-cages are now common in the sea lochs of Scotland, where the principal species being cultured is salmon. Salmon is also grown in Scandinavia, the United States and Canada. In the Mediterranean, gilthead seabream is the major species grown, while in Japan and South-East Asia, where most of the world's mariculture is carried out, a number of different species are cultured. In such systems the waste products of the culture, the 90% of the food that is not used for growth, is simply discharged into the environment and washed away. This can cause local problems of pollution such as the degradation of the coral reef in Eilat and possibly also encourage blooms of 'red tide' toxic dinoflagellates (red microscopic organisms rather like algae).

The third type of fish culture system is fishponds. These ponds, which are adjacent to the sea, are used to grow fish and shrimps, particularly in South-East Asia. Areas of natural lagoons have also been netted off and used to grow fish. The principal problem with such systems is that the accumulated metabolic wastes, particularly ammonia, are

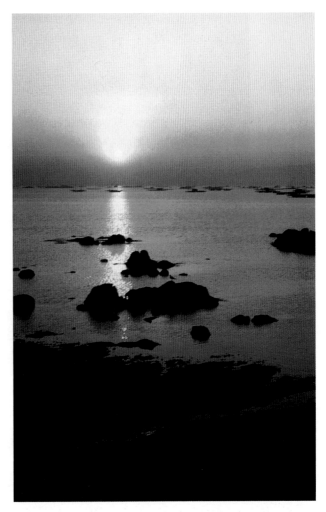

Figure 3.18 Sunset in Ria Arosa, Spain. On the horizon there are dozens of rafts which have ropes seeded with mussels and oysters hung beneath them.

toxic to the fish. The resulting water quality problems limit the number of fish that can be grown and hence the commercial success of the fish farms. An example of integrated mariculture systems that attempt to deal with the problems of pollution is provided in Box 3.3.

3.6.4 Pollution

Until recently it was considered relatively normal practice to discharge domestic and industrial wastes of various types into the ocean. This was done because the oceans are so vast that it was considered that the input would result in no significant change in the concentrations already there. This was the so-called 'dilution is the solution' method of waste disposal. However, we now realize that even the most remote parts of the world have been changed by our careless discharge of wastes. The snows in Greenland now contain

Figure 3.19 Fish-cages in a sea loch in Scotland. Fish are grown inside the cages.

CASE STUDIES

INTEGRATED MARICULTURE IN FISHPONDS

The farming of fish in fishponds, while being technically more difficult than culture in cages, principally because the fish swim about in their own metabolic waste products, has several advantages. It is possible to reduce the pollution into the environment by recycling or treating the waste. It is also possible to mimic the natural food chains by producing an integrated fish culture. Such systems are more energy efficient and can result in a greater profit since more than one species can be cultured.

The National Mariculture Centre, Eilat, Israel, has been in the forefront of developing such integrated fishpond culture systems. In these systems the seabream, a fish that lives naturally in the Mediterranean and is much sought after by the French and Italians, is grown. The first system developed was a semi-intensive system in which the fish were grown in earthen ponds. The nutrients excreted by the fish caused phytoplankton to grow. The phytoplankton removed most of the toxic ammonia from the fishpond water and added life-giving oxygen. Because of the particular conditions of the pond culture in Eilat, a large number of these phytoplankton were large benthic diatoms. These phytoplankton were then fed to oysters, which became a second successful cultured product. The commercial success of this system

was, however, limited because it was not possible to prevent the phytoplankton blooming and crashing, which caused large fluctuations in the water quality in the ponds and occasional mass mortalities of the fish.

In order to solve these problems a fishpond-seaweed system was developed. In this system fish are grown at high density in a concrete tank. The water from the fish tank is cleaned of the waste metabolic nutrients, most significantly ammonia and phosphates, by being passed through a tank in which seaweed is grown. It requires approximately three tanks of seaweed to clean the waste from one tank of fish. Seaweed, though edible by humans, does not have great commercial value. However, in this

BOX 3.3 ➤

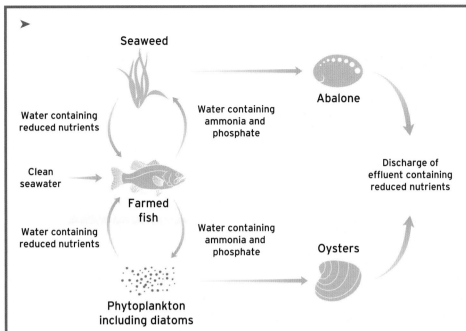

Seaweed

Water containing reduced nutrients

Water containing ammonia and phosphate

Abalone

Clean seawater

Farmed fish

Discharge of effluent containing reduced nutrients

Water containing reduced nutrients

Water containing ammonia and phosphate

Oysters

Phytoplankton including diatoms

Figure 3.20 Integrated mariculture systems. In the lower system waste nutrients (ammonia and phosphate) excreted by the fish are taken up by phytoplankton including diatoms which are then fed to oysters. In the upper system, the waste water is passed over seaweed (often ulva) which removes the nutrients. The seaweed is then fed to abalone. The treated waste water is then recycled to the fishponds or discharged. Such systems produce abalone, seaweed and oysters in addition to fish as commercial products while reducing the amount of nutrient waste being discharged into the environment.

system the seaweed is fed to abalone, which at present retail at $50 per kg. The waste faeces from the fish tank are put into a sedimentation tank and the water flowing over that is used to support a successful oyster culture. This system uses the natural sunlight that is abundant in Eilat to convert waste nutrients into seaweed, abalone and oysters, all of which represent additional commercial products from this aquaculture system (Figure 3.20).

An alternative system for an environmentally friendly intensive mariculture system that does not require abundant sunlight involves the use of bacterial biofilters to treat the wastes from the fishponds. In this system (Figure 3.21), the water from the fishpond containing ammonia and oxygen is passed over a nitrifying filter, which contains bacteria that convert ammonia to the nitrate in the presence of air. Although the nitrate is far less toxic than ammonia to fish, it is still undesirable to let it accumulate in the system. To remove the nitrate, some of the water is passed over a sedimentation pond containing anaerobic denitrifying bacteria that convert the nitrate to

nitrogen gas. These bacteria also remove phosphates from the reaction stream at the same time. Although this system produces only one commercial product, the fish, it does have several practical advantages. It is a completely closed system, which means that it can be operated far from the sea and potentially near the

market for the fish. Also, because the system is closed, it is commercially viable to heat the system. This means that fish can be grown in climates such as in Britain and northern Europe that are too cold to grow fish fast enough to be commercially viable under normal temperatures.

Figure 3.21 This pilot mariculture system uses bacterial biofilters to clean the waste nutrients from the water and enable the system to operate with no waste water discharge at all.

BOX 3.3

measurable lead from car exhausts while the penguins of Antarctica contain DDT residues in their body fats. Pollution of ocean waters can occur via direct discharge into the oceans of sewage and industrial effluent, from shipping (e.g. oil spills) and from changing rainfall chemistry. Pollution can also occur indirectly via runoff from rivers and land (e.g. sewage, fertilizers and pesticides, industrial effluent, sediment). This pollution can alter the nutrient balance in the water, add toxic chemicals to the system and alter the turbidity. A good example of the pollution problem is given in Box 3.4 for the Mediterranean Sea.

A further set of problems related to pollution results from the interactions of the oceans and atmosphere (see Chapter 4). Many atmospheric gases are taken up by the oceans via precipitation which contains the chemicals or via living matter near the surface of the oceans. Therefore the ocean chemistry may change because we have altered the composition of the

atmosphere. Box 3.5 describes the process of acidification of ocean waters through this process.

Reflective questions

➤ A turbid river containing abundant plant nutrients discharges into a coastal sea. The particles drop out of the water column in decreasing grain size until eventually only the finest particles remain in the water column. How might such a system control the primary productivity distribution on the coastal shelf?

➤ Under what conditions might each of the terms in the photosynthetic equation become limiting?

➤ What is mariculture and what types of mariculture are there?

HAZARDS

POLLUTION IN THE MEDITERRANEAN SEA

One of the most vulnerable regions of the world's oceans to anthropogenic pollution is the Mediterranean Sea. It is almost landlocked, with a narrow and comparatively shallow outlet to the Atlantic Ocean at Gibraltar. The population in the Mediterranean catchment area is approximately 120 million people living close to its shoreline and is subjected to an annual invasion of a further 120–200 million tourists who flock to the sandy beaches and clear blue waters. These tourists put a considerable strain on the local water resources and waste treatment facilities in many of the areas they visit. Many of the industries of southern Europe and elsewhere discharge their waste into the Mediterranean directly or via rivers such as the Po, the Rhone, the Nile and many smaller rivers. Some scientists allege that as a result

of the discharge of organic matter and nutrients down the Po, significant areas of the northern Adriatic have actually become **anoxic** at times in summer (Figure 3.22). Local sewage discharges into areas such as the Bay of Naples have long ago made consumption of local shellfish a hazardous experience, particularly for tourists who have lower immunities to the local bacteria. Especially vulnerable areas are the shallow lagoons adjacent to the Mediterranean such as the Venice Lagoon. These areas are particularly important because they provide the spawning ground for many of the commercial fisheries such as that for the gilthead seabream in the sea. Partially as result of pollution of these lagoons and partially as a result of overfishing, many of the commercial fisheries in the Mediterranean are under considerable stress.

Before 1967, the Mediterranean was the major shipping route for oil

tankers from the Middle East to the industrialized nations of Europe and North America. Even today there is still sufficient oil tanker traffic to pose a significant pollution hazard both from the illegal cleaning of tanks at sea and from occasional tanker accidents. Many beaches around the Mediterranean are still fouled by significant numbers of tar balls.

One major anthropogenic change, which affects the entire Mediterranean, and the effects of which are still comparatively poorly known, is the reduction in the natural flow of the rivers. All rivers in the area are utilized to some extent or other for irrigation, and for domestic and industrial use. In Israel, 98% of the water that falls is used, leaving almost no natural flow remaining in any of the local streams and rivers. The River Nile was dammed at Aswan in 1965. As a result the flow into the Mediterranean is less than 10% of what its natural flow used to be. It

BOX 3.4 ➤

Figure 3.22 Colour satellite image of the Mediterranean. Red is productive water with high chlorophyll and blue is unproductive water with low chlorophyll.

has recently been suggested that this may be altering the natural currents and flows in the eastern basin of the sea. To discuss and deal with these multinational, multidimensional environmental problems, the United Nations has set up the MEDPOL Programme. This programme is one of the few forums where all the nations

of the basin sit down together as equal participants. That includes the Israelis, Libyans and Syrians, as well as the Greeks, Turks and Cypriots. As a result of this programme, sewage and industrial discharges to the basin have been significantly reduced. Many countries have installed high-quality sewage treatment facilities, which

have not only reduced pollution but also enabled them to conserve and recycle valuable water resources. As a result of MEDPOL the amount of oil pollution in the basin has been significantly reduced. Although there are still many problems to be overcome, an encouraging start has been made.

BOX 3.4

ENVIRONMENTAL CHANGE

ACIDIFICATION OF THE SURFACE OCEAN

It is known that until approximately 2006, 50% of the carbon dioxide added to the atmosphere by

anthropogenic activities was removed from the atmosphere by natural buffer systems. The ocean system formed a very important part of this buffer system with excess carbon dioxide being

transferred into the deep ocean with deep water formation in places such as the North Atlantic and Southern Ocean. In addition some excess carbon dioxide was taken up by biological productivity

BOX 3.5 ➤

in surface waters and then transferred to the deep water via waste products from surface-grazing organisms.

There is now evidence that this natural buffer system has been used up and the carbon dioxide we emit into the atmosphere is largely accumulating in the atmosphere. One of the effects of this increased carbon dioxide in the atmosphere is that it will dissolve in the surface waters of the ocean and make them more acidic. Scientists are very concerned about this effect and are actively studying the possible consequences of acidification of the surface ocean.

One effect that has been predicted concerns the ability of marine organisms to build their shells from calcium carbonate. This is important not only for the seashells that are commonly found at beaches but also for coral reefs and for phytoplankton such as coccoliths which can form blooms in vast numbers under favourable conditions (see Figure 3.12). At present the surface waters of the ocean are close to the saturation point of calcium carbonate and natural processes such as photosynthesis can locally make the system supersaturated and facilitate the precipitation of calcium carbonate. However, if the surface water becomes more acid, then it will be more difficult for corals to grow. Indeed it has been predicted that all coral reefs will cease to grow and start to be dissolved within 30–50 years based on current estimates for increased carbon dioxide in the atmosphere (Figure 3.23).

(a)

(b)

(c)

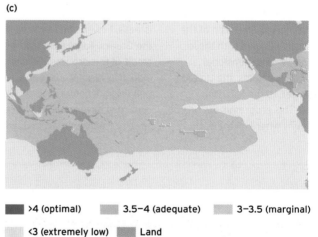

■ >4 (optimal)   ■ 3.5–4 (adequate)   ■ 3–3.5 (marginal)

■ <3 (extremely low)   ■ Land

Figure 3.23 Changes in growth potential of Pacific coral reefs as related to rising atmospheric CO_2 levels and declining pH and saturation state of surface ocean waters: (a) calculated pre-industrial values for 1870 with atmospheric CO_2 concentrations of 280 ppm; (b) projected values for 2020–2029 with CO_2 concentrations of 415 ppm; (c) projected values for 2060–2069 with CO_2 concentrations of 517 ppm. (a) was high pH conditions while (c) will be low pH conditions. (Source: after Guinotte *et al.*, 2003)

BOX 3.5

3.7 Summary

Oceans cover 71% of the surface of the Earth and a very considerable higher percentage of the habitable niches on planet Earth. The mean depth of the oceans is approximately 3500 m. The bottom of the ocean is far from flat. Most of the basic structure is controlled by plate tectonic processes. At the centre of the oceans are mid-ocean ridges. Away from the centre, the abyssal plains gradually deepen towards the edge of the ocean basin where in some locations they plunge into the ocean trenches. The continental slopes and shelves are of varying widths from hundreds of kilometres to less than a kilometre. In most areas of the ocean, the hills and valleys are covered by a drape of sediments derived from the continent or from biological of chemical processes in the ocean basins themselves.

While the surface currents of the world are driven by prevailing winds, the deeper circulation is controlled by the density of the water, which in turn is controlled by its temperature and salinity. The two most important deep water masses in the world are formed in the North Atlantic in winter (North Atlantic Deep Water) and at the edge of the Antarctic (Antarctic Bottom Water).

Most of the primary productivity in the world occurs in the surface waters of the oceans. The total amount is controlled by the depth to which usable light penetrates, the photic zone, and by the supply of dissolved nutrients (e.g. nitrogen and phosphorus). All organisms in the ocean, except those living at oceanic hot springs, depend on these phytoplankton via a food chain. We are just beginning to develop marine agricultures as a sustainable alternative to our current over-exploitation of fish stocks. In addition we are only beginning to realize and deal with the problems associated with pollution of the ocean waters.

Further reading

Pinet, P.R. (2000) *Invitation to oceanography*. Jones & Bartlett, Sudbury, MA (web-enhanced edition).
This book contains good sections on geology, chemistry and biology as well as current processes. There are also sections on tides and waves which are relevant to coastal processes discussed in Chapter 17.

Segar, D.A. (1998) *Introduction to ocean sciences*. Wadsworth, Belmont, CA.

Another well-written and colourful textbook with useful questions and good discussion of the history of oceanographic study.

Sverdrup, K.A. and Armbrust, E.V. (2008) *An introduction to the world's oceans*. McGraw-Hill, London.
This book develops the themes discussed in this chapter in more depth. It is a very well-illustrated text containing lots of weblinks and suggested further reading.

Thurman, H.V. (2004) *Introductory oceanography*. Prentice Hall, Upper Saddle River, NJ.
This is a clearly written oceans textbook containing some useful reflective questions and exercises.

Web resources

Fisheries Information Centre Home Page
http://www.fao.org/fi/default_all.asp
The website for the major programme on fisheries undertaken by the Food and Agricultural Organization (FAO) of the

United Nations. The site provides knowledge resources on the fishing industry, useful links and information on current fisheries issues/news. A very extensive report 'The State of World Fisheries and Aquaculture 2001' on the state of the world's fisheries and the major issues facing the industry is also provided at:

http://www.fao.org/docrep/005/y7300e/y7300e00.htm

Ocean Drilling Programme Australia: Ocean Sediments

http://www.odp.usyd.edu.au/odp_CD/oceplat/opindex.html

This is a brilliant site describing all the major characteristics of the different types of ocean sediments, including how they are created (their origin) and their global distribution, including diagrams, photos and ingenious student exercises.

Ocean Planet Home Page

http://seawifs.gsfc.nasa.gov/ocean_planet.html

An electronic online edition of the 'Ocean Planet' exhibition at the US National Museum of Natural History is provided here; including ocean science, an excellent section on threats to the ocean, human relationship with the oceans and an extensive list of related web resources.

Ocean World

http://oceanworld.tamu.edu/index.html

Supported by NASA through the TOPEX/Poseidon Education Project, this site provides educational information on oceans and climate, waves, currents and fisheries among other subjects. Everything is explained in a very simple manner but it is still very useful.

US National Weather Service: Ocean Prediction Center

http://www.opc.ncep.noaa.gov/

This is a site providing recent news on events and changes in ocean behaviour along with predictions on ocean conditions.

Visible Earth: NASA

http://visibleearth.nasa.gov/view_set.php?categoryID=817

NASA provides here a set of ocean images including ocean winds, circulation, temperatures and bathymetry.

 Visit our website at **www.pearsoned.co.uk/holden** for further questions, annotated weblinks, interactive models and video-clips showing physical processes in action.

Scope

Part III describes features of the Earth's climate and weather. The climate at any location is the most important environmental influence on the natural vegetation, the landscape and human activity at global and local scales. Chapter 4 offers an understanding into the science behind the processes occurring in the atmosphere, the most unstable and rapidly responding element of the climate system. Chapter 5 then outlines how these processes result in the nature of the climatic zones that cover the Earth. These broad global zones contain within them substantial variation at the local and regional scale and these smaller-scale processes form the focus of Chapter 6.

Climate change is an important topic and while some of the processes are introduced in Chapter 4, the topic of climate change is dealt with in more detail in three chapters (Chapters 20, 21 and 22) within Part VI of this book. Nevertheless Chapters 4, 5 and 6 still discuss the interactions of humans with climate and weather systems. Humans adapt to local and regional features of the Earth's climate system as well as modify them. For example, urban areas often have different local climates to the surrounding rural zones and humans have developed techniques for locally inducing rainfall by seeding clouds, and for reducing wind speed or snowfall by building shelter belts.

When we view the Earth's climate system as a whole its complexity seems unbounded: the physical mechanisms interact at a variety of scales within an open system, exchanging energy and matter in ways that often seem hidden from our grasp. However, it is possible to approach these mechanisms at a variety of levels. At the most simple level, we can observe a basic driving force behind these systems. This is the need to dissipate the energy that circulates through the system and that results in processes that transport energy and matter. It is possible to think of a system in terms of an energy or matter budget, with inputs, outputs, stores and transfers. As an example the global climate system principally reacts to the input of energy from the Sun. This input is concentrated at the equator and the large-scale atmospheric and oceanic circulation of the Earth is the mechanism by which this energy is transported away from the equator in an attempt to create a global balance. The Sun's energy is eventually released back out of the atmosphere into space so that the Earth's average temperature remains relatively stable. Although this is a very crude conception, it illustrates how we can simplify systems in order to help understand what drives the Earth system processes we pursue and how the same concepts can be used for a variety of subjects and over different scales of space and time.

CHAPTER 4

Atmospheric processes

John G. Lockwood

Formerly of School of Geography, University of Leeds

Learning objectives

After reading this chapter you should be able to:

➤ describe the major components of the climate system and their most significant interactions

➤ understand the basic properties of the atmosphere and the various flows of energy and matter that connect components of the climate system

➤ describe general wind circulation processes and understand concepts of Rossby waves, jet streams, El Niño and North Atlantic Oscillation

➤ understand the greenhouse effect and predict the likely impact of enhanced greenhouse effect on climate and be aware of some positive and negative feedback effects

4.1 Introduction

The term 'climate system' is often used to refer to an interactive system consisting of five major components: the atmosphere, the hydrosphere (e.g. oceans, lakes, rivers), the cryosphere (e.g. ice, snow, glaciers), the land surface and the biosphere (e.g. vegetation) (Figure 4.1). The climate system is driven by various external mechanisms, the most important of which is the Sun. In this chapter the main concern is with the atmosphere, which is the unstable and rapidly changing part of the climate system. In particular its composition, which is continually changing, together with the resulting climate changes have become of major international concern. The Stern Review (Stern, 2006) commented that an overwhelming body of scientific evidence indicates that the Earth's climate is rapidly changing, predominantly as a result of increases in so-called greenhouse gases caused by human activities. Greenhouse gases trap heat near the Earth's surface and cause the surface temperature to increase. Carbon dioxide (CO_2) and methane (CH_4) are among the most important anthropogenic greenhouse gases. Since pre-industrial times (around 1750), CO_2 concentrations have increased by just over one-third from 280 parts per million (ppm) to 379 ppm in 2005, predominantly as a result of burning fossil fuels, deforestation and other changes in land-use. This has been accompanied by rising concentrations of other greenhouse gases, particularly methane and nitrous oxide. Analysis of air bubbles trapped in Antarctic ice cores suggests that current levels of greenhouse gases are higher now than at any time in at least the past 650 000 years, and probably the past 800 000 years (IPCC, 2007a). Largely as a result of increasing atmospheric greenhouse gas concentrations, global mean temperatures have increased by 0.7°C

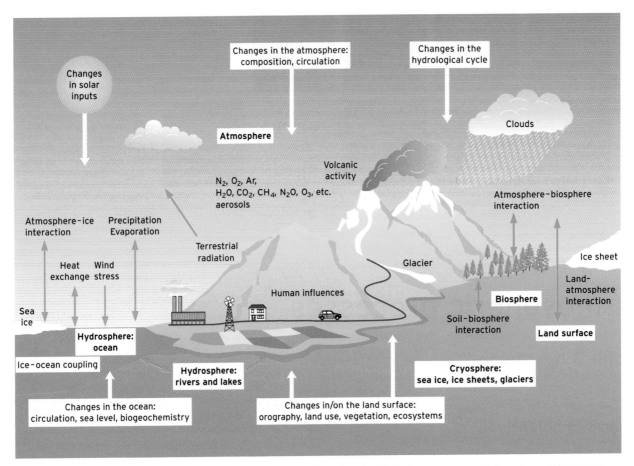

Figure 4.1 Schematic view of the components of the global climate system (bold), their processes and interactions (thin orange arrows) and some aspects that may change (thick white arrows). (Source: IPCC, 2001)

since around 1900. Over the past 30 years, global temperatures have risen rapidly and continuously at around 0.2°C per decade, bringing the global mean temperature to what is at or near the warmest level reached in the current inter-glacial period, which began around 12 000 years ago. IPCC (2007a) reported that a global assessment of data since 1970 has shown it is likely that anthropogenic warming has had a discernible influence on many physical and biological systems. Most climate model calculations show that a doubling of pre-industrial levels of greenhouse gases is very likely to commit the Earth to a rise of between 2 and 5°C in global mean temperatures (IPCC, 2007a). This level of greenhouse gases will probably be reached between 2030 and 2060.

Over the past 2.4 million years there have been numerous advances and retreats of major ice sheets across the northern parts of Europe and North America (see Chapter 20). It has often been assumed that since the last retreat of these ice sheets, around 10 000 years ago, global climate has undergone few variations. However, examination of both

proxy climate indicators (such as plant and animal remains) and direct climate observations indicates that the recent climate shows a surprisingly high degree of variability. It can therefore no longer be assumed that the global climate during the past 10 000 years has been either stable or benign. Furthermore, humans have interacted with atmospheric processes by burning fossil fuels and adding important trace gases to the atmosphere. These contribute to global and local climate changes. This chapter therefore seeks to explore the Earth's fundamental global atmospheric processes so that we may better understand, predict and mitigate the effects of such climate changes. It will start by introducing the basics of atmospheric processes before moving on to examine energy systems and moisture circulations. The chapter will then return to a more detailed treatment of atmospheric motion including processes linked with inter-annual variations in climates such as those associated with El Niño. Finally the important processes associated with the **greenhouse effect** will be discussed.

4.2 The basics of climate

Global atmospheric circulation consists of wind systems that can show marked annual and seasonal variations. They are one of the principal factors determining the distribution of climatic zones. The two major causes of the global wind circulation are inequalities in the distribution of solar **radiation** over the Earth's surface, particularly in a north–south direction (more received at the equator), and the Earth's rotation. Global wind and ocean current circulations redistribute heat from equatorial regions where it is in surplus to polar regions where it is in deficit. The solar radiation received at the Earth's surface and reradiated back out from the surface provides the energy to drive the global atmospheric circulation, while the Earth's rotation determines its shape.

Fundamental causes of seasonal differences in climate across the globe are the approximately spherical shape of the Earth and the way the Earth's axis of rotation is tilted at 23.5° in relation to a perpendicular to the plane of the Earth's orbit round the Sun. The spherical shape results in sharp north–south temperature differences. This is because, as Figure 4.2 shows, the same amount of solar radiation is spread over a larger area nearer the poles and also has a thicker layer of atmosphere to penetrate because of the angle at which the solar radiation is received. The Earth's tilt is responsible for month-by-month changes in the amount of solar radiation reaching each part of the planet, and hence the variations in the length of daylight throughout the year at different latitudes and the resulting seasonal

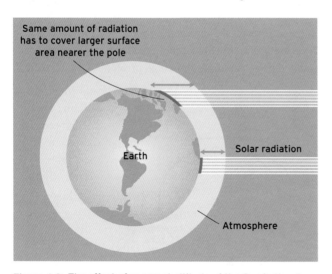

Figure 4.2 The effect of apparent altitude of the Sun in the sky upon solar radiation received at the Earth's surface. The Earth's 23.5° tilt means that the Sun appears overhead at midday at the Tropic of Cancer (23°27′N) on 21-22 June and the Tropic of Capricorn (23°27′S) on 22-23 December.

Same amount of radiation has to cover larger surface area nearer the pole

Earth

Solar radiation

Atmosphere

weather cycle. The 23.5° tilt accounts for the position of the tropics: the Tropic of Cancer at 23°27′N and the Tropic of Capricorn at 23°27′S. At the tropics the Sun is overhead at midday on the solstices, 21–22 June and 22–23 December respectively. The length of daylight throughout the year does not vary significantly from 12 h over the whole of the area between the tropics. At the spring (21 or 22 March) and autumnal (22 or 23 September) equinoxes, when the noon Sun is vertically overhead at the equator, day and night are of equal length everywhere across the Earth. This is not so at the winter and summer solstices, since each year the areas lying polewards of the Arctic (at 66°33′N) and Antarctic (at 66°33′S) Circles have at least one complete 24 h period of darkness at the winter solstice and one complete 24 h period of daylight at the summer solstice. At the poles themselves there is almost six months of darkness during winter followed by six months of daylight in summer.

The major controls of very long-term climatic change include palaeogeography, greenhouse gas concentrations, changing orbital parameters and varying ocean heat transport. On timescales of a few years, variations in ocean heat transport and atmospheric greenhouse gas concentrations are particularly effective in changing climate. The greenhouse effect is a natural process that keeps the Earth's surface around 30°C warmer than it would be otherwise. Without this effect, the Earth would be too cold to support life. Fourier realized in the 1820s that the atmosphere was more permeable to incoming solar radiation than outgoing infrared radiation and therefore trapped heat (Fourier, 1824). The link between greenhouse gases and climate change is illustrated in Figure 4.3, and is explored further later in the chapter.

Reflective question

➤ Why are there seasonal differences in climate across the globe?

4.3 The global atmospheric circulation

Modern observations of the atmosphere support a three-cell circulation model in each hemisphere as shown in Figure 4.4. Air rises near the equator where there is a heat surplus, flows poleward at high levels, sinks at about 30°N and S and returns at low levels to the equatorial regions. Climatologists know this simple circulation cell as the Hadley cell; it is

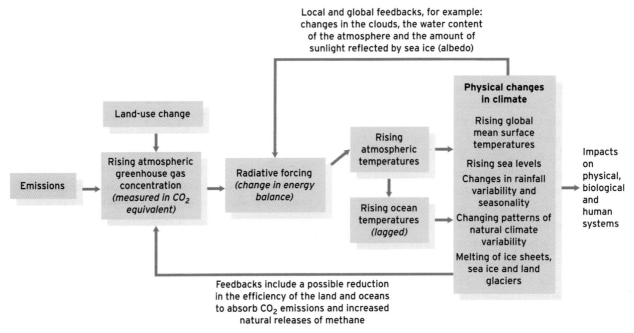

Figure 4.3 The link between greenhouse gases and climate change. (Source: after Stern, 2006)

restricted to tropical regions by the Earth's rotation (see Section 4.6.2). Over both poles where there is a heat deficit, a similar circulation with subsiding air and surface-level winds flowing back towards the equator exists, but they are diverted in an easterly direction by the Earth's rotation. Between the tropical Hadley cell and the polar cell there is, to complete the three-cell structure, a reverse cell with sinking air around 30°N and S, poleward-flowing westerly winds, and rising air along the boundary (the **polar front**) between the mid-latitude and polar cells. This middle-latitude cell is known as the Ferrel cell.

Between 20° and 30° latitude from the equator are areas of high pressure, such as the Azores anticyclone, where air

from the tropical Hadley cells descends (Figure 4.5). Since descending air warms (see Section 4.6.1) there is little chance of moisture condensation and generally these are regions of clear skies with high pressure and light winds. Almost all the major deserts of the world lie in these latitudes (see Chapters 8 and 16). The equatorward-flowing easterly winds on the equatorial sides of the subtropical anticyclones are known as **trade winds.** The trade winds meet in a trough of low pressure near the equator. This trough contains a convergence zone near the surface known as the intertropical convergence zone (ITCZ) where conditions are favourable for ascending motion, condensation of water vapour, cloudiness and high precipitation rates. It is

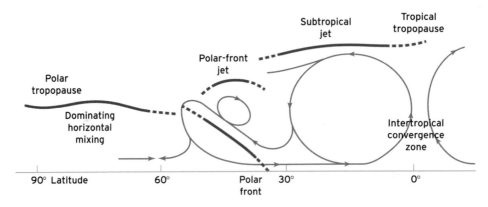

Figure 4.4 Schematic representation of atmospheric circulation cells and associated jet stream cores in winter. The tropical Hadley, the middle-latitude Ferrel and the polar cells are clearly visible. Surface winds are not directly north or south because of the Earth's rotation that forces the air to move in an easterly or westerly direction.

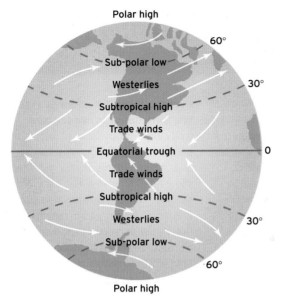

Polar high
60°
Sub-polar low
Westerlies
30°
Subtropical high
Trade winds
Equatorial trough
0
Trade winds
Subtropical high
Westerlies
30°
Sub-polar low
60°
Polar high

Figure 4.5 Idealized mean surface wind circulation.

only a few hundred kilometres in width and its position varies from day to day. Thus the ascending limb of the Hadley cell is limited to a few areas of cloud and rain in the equatorial trough, while the descending motion covers much of the subtropics giving widespread desert conditions. The Hadley cells are relatively stable; therefore Figure 4.4 is a reasonable representation. Compared with the Hadley cells, the middle-latitude atmosphere is highly disturbed and the Ferrel cell circulation shown in Figure 4.4 is highly schematic. At the surface the predominant features are a series of eastward-moving closed cyclonic and anticyclonic circulation systems. At high levels in the atmosphere these features are associated with large-scale open waves with **jet streams** imbedded in them. These waves and jet streams are discussed in more detail in Section 4.6.4.

Reflective question

➤ Which major circulation zone do you live in?

4.4 Radiative and energy systems

4.4.1 The nature of energy

Many meteorological processes involve flows of energy; it is therefore essential to have some understanding of the nature of energy. Energy may formally be defined as the capacity for doing work. It may exist in a variety of forms including heat, radiation, potential energy, kinetic energy, chemical

energy and electromagnetic energies. It is a property of matter capable of being transferred from one place to another, of changing the environment and is itself susceptible to change from one form to another. Except at the sub-atomic scales, energy is neither created nor destroyed and from this it follows that all forms of energy are exactly convertible to all other forms of energy, though not all transformations are equally likely. For example, the transformation of light into heat is common, and if sunlight falls on green plants it can be transformed into chemical energy via the formation of new compounds. The direct transformation of sunlight into kinetic energy of atmospheric motion is extremely uncommon. It is possible for any particular system to produce an exact energy budget, in which energy gained exactly equals the energy lost plus any change in storage of energy in the system.

Stated in simple climatological terms, heat is a form of energy and the indirect transformation of heat energy into various forms of mechanical energy is the process that drives the global circulation of the atmosphere. Thus the transformation of heat into mechanical energy is responsible for the formation of weather systems whose cumulative effects define the climate of a particular region. Friction continually degrades and destroys atmospheric motions by converting them back into heat, but this heat is in the form of low-grade energy, which cannot create fresh motion systems. Throughout Section 4.4 and Box 4.1 it will be demonstrated that the source of energy, which drives the atmospheric circulation, is sunlight, while the low-grade energy produced as a result of the motions is lost to space as long-wave infrared radiation.

4.4.2 Distinguishing between temperature and heat

It is important to distinguish between temperature and heat. Temperature is a measure of the mean kinetic energy (speed) per molecule of an object, while heat, or sensible heat as it is often called, is a measure of the total kinetic energy of all the molecules of that object. Thus the temperature of the air is simply a measure of the 'internal energy' of the air. This internal energy is associated with the random motion of the molecules. If two masses of gas are brought into intimate contact, this internal energy is rapidly shared between them and their temperatures become equal.

As the temperature decreases it is possible to imagine a state being reached when the molecules are at complete rest and there is no internal energy, a point on the temperature scale known as **absolute zero**. This has been determined as 273.15 degrees Celsius below the melting point of pure ice

ENERGY EXCHANGES AND ENTROPY

Energy transformations are best understood in terms of the first and second **laws of thermodynamics** and the concept of **entropy** (Lockwood, 1979). The first law of thermodynamics is a statement of the law of conservation of energy for a thermodynamic system (see Chapter 23). Entropy, however, is a measure of the unavailability of a system's heat energy for conversion into mechanical work. It is also a measure of the degradation or disorganization of the system: the higher the entropy, the more disorganized the system. Friction continually degrades and destroys atmospheric motions by converting them back into heat, but this heat is in the form of low-grade energy, which cannot create fresh motion systems.

The second law of thermodynamics asserts that the entropy of an isolated system increases with time so that it becomes more disordered. So if the system starts off in a state with some kind of organization, this organization will, in due course, become degraded, and the special features (e.g. wind motions) will become converted into 'useless' disorganized particle motions. Thus, over time, all organized atmospheric motions, unless continually renewed, will be destroyed by friction and vanish.

However, observation suggests that many systems do not increase their entropy through time. On medium and large scales, atmospheric motions show a high degree of stable organization. Since entropy is always tending to increase in these systems, they must be receiving new supplies of low entropy to maintain the stable, low-entropy conditions. It is therefore necessary to enquire about the source of this supply of low entropy. The answer to this may be illustrated by a consideration of energy flow in plants. Green plants take atmospheric CO_2, separate the oxygen from the carbon, and then use the carbon to build up their own substance and grow leaves, roots, fruits and so on. This procedure, photosynthesis, results in a large reduction in entropy. Green plants achieve this entropy reduction by using **short-wave radiation** from the Sun. The light from the Sun brings energy to the Earth in a comparatively low-entropy form, namely in the photons of visible light. The Earth does not retain this energy, but after some time reradiates it all back into space. However, the reradiated energy is in a high-entropy form that is relatively useless radiant heat or infrared photons. The sky is in a state of temperature imbalance: one region of it (occupied by the Sun) is at a very much higher temperature than the rest. The Earth receives energy from the hot Sun in a low-entropy form and reradiates it to the cold regions of the sky in a high-entropy form. The total amount of entropy exported by the Earth-atmosphere system to space is 22 times the amount of entropy imported by the incoming solar radiation at the top of the atmosphere. The gains in entropy represent the export, in the form of **long-wave radiation**, of the disorder continually being created in atmospheric flow patterns and surface ecological systems.

BOX 4.1

(0°C), and the Kelvin temperature scale is measured upwards from absolute zero in Celsius units, making 0°C equivalent to 273.15 K. The Kelvin temperature scale is often used in basic physical equations, particularly those concerned with radiation.

Heat is transferred from high- to low-temperature objects and this alters either the temperature or the state of the substance, or both. Thus, a heated body may acquire a higher temperature and this is known as **sensible heat** gain. Alternatively a heated body may change to a higher state (solid to liquid, or liquid to gas) and therefore acquire latent (or hidden) heat. For example, ice at 0°C can be heated and melt to form water. Once the melting process is complete the temperature of the water may still be 0°C. The extra heat was simply used to change the state of the solid to a liquid and this is known as **latent heat**.

One or more of the processes of **conduction, convection** or radiation affects the transfer of heat to or from a substance. Conduction is the process of heat transfer through matter by molecular impact from regions of high temperature to regions of low temperature without the transfer of matter itself. It is the process by which heat passes through solids (e.g. soils, rocks). The effects of conduction in fluids (liquids and gases) are usually negligible in comparison with those of convection. Convection is a mode of heat transfer in a fluid (e.g. atmosphere, oceans), involving the movement of substantial volumes of the substance itself. Examples are wind systems in the atmosphere and ocean currents.

Radiation is the final form of heat transfer and is transmitted by any object that is not at a temperature of absolute zero (0 K). This type of heat transfer is discussed in the following section.

4.4.3 Radiation

4.4.3.1 The nature of radiation

Radiation does not require any medium through which to travel and so it can pass through a vacuum with the speed of light. This is how radiation from the Sun reaches the Earth. Radiation can be regarded as having both a wave structure and a particle (**photon**) structure: see Box 23.1 in Chapter 23). Sometimes its properties are best described in terms of waves, while at other times in terms of a particle structure. Radiation is characterized by wavelength, of which there is a wide range or spectrum extending from very short X-rays through ultraviolet and visible light to infrared, microwaves and long radiowaves (see Figures 23.4 and 23.5 in Chapter 23).

The radiation laws contained in Box 4.2 describe the emission of radiation by an object. Further information about electromagnetic energy (radiation) is also provided in Chapter 23. These laws show that the amount of radiant energy emitted by any object is much greater as temperature increases. At the same time hotter objects also produce more radiation at shorter wavelengths than cooler objects

---

FUNDAMENTAL PRINCIPLES

RADIATION LAWS

A perfect **black body** is one that absorbs all the radiation falling on it and that emits, at any temperature, the maximum amount of radiant energy. This definition does not imply that the object must be black, because snow, which reflects most of the visible light falling on it, is an excellent black body in the infrared part of the spectrum. For a perfect all-wave black body the intensity of radiation emitted and the wavelength distribution depend only on the absolute temperature, which is the temperature measured in degrees Celsius from absolute zero, and in this case the Stefan–Boltzmann law (equation 4.1) applies. This law states that the flux of radiation from a black body is directly proportional to the fourth power of its absolute temperature, that is:

$$F = \sigma T^4 \qquad (4.1)$$

where F is the flux of radiation (watts per square metre), T is the absolute temperature (kelvin) and σ is a constant (10^{-8} W^{-2} K^{-4}).

A black body emits radiation with a range of frequencies, but with a maximum at frequency λ_{max}. It can be shown by the Wien displacement law that the wavelength of maximum energy λ_{max} is inversely proportional to the absolute temperature:

$$\lambda_{max} = \alpha / T \qquad (4.2)$$

where α is a constant (1.035×10^{11} K^{-1} s^{-1}).

Objects on the Earth's surface are commonly assumed to emit and absorb in the infrared region as a grey body: that is, as a body for which the Stefan–Boltzmann law takes the form:

$$F = \varepsilon \sigma T^4 \qquad (4.3)$$

where the constant of proportionality ε is defined as the infrared emissivity or, equivalently, the infrared absorptivity (one minus the infrared albedo). Typical infrared emissivities are in the range 0.90 to 0.98.

At normal temperatures many gases are not black bodies since they emit and absorb radiation only in selected wavelengths. Many molecules in the atmosphere possess spectra that allow them to emit and absorb thermal infrared radiation (4-100 μm); such gases include water vapour, CO_2 and ozone, but not the main constituents of the atmosphere, oxygen or nitrogen. These absorption properties are directly responsible for the natural greenhouse effect. Vigorous vertical mixing takes place in the **tropopause** as explained in Section 4.6.1. Therefore it is not the change in thermal infrared radiation at the Earth's surface that determines the strength of the greenhouse warming. Instead it is the change in the irradiance or radiative flux (F) at the tropopause that expresses the forcing of the climate system by radiation. Unless a gas molecule possesses strong absorption bands in the wavelength region of significant emission from the ground surface, it can have little effect on the radiative flux at the tropopause (Houghton, 1997). Thus global climate change is strongly affected by the atmospheric composition. The roles of the **troposphere** and tropopause are discussed in Section 4.6 (see Figure 4.13).

BOX 4.2

because the maximum wavelength emitted by an object is
inversely proportional to absolute temperature (K). The
high temperature of the radiating surface of the Sun (about
6000 K) results in over 99% of the solar energy being at
wavelengths of less than 4 μm (μm stands for micrometre
and is equivalent to one-millionth of a metre), whereas the
much lower temperature of the Earth's atmosphere and
surface (generally around 300 K or lower) yields most radiative energy in the 4–100 μm region.

It is therefore convenient to divide the entire atmospheric radiative regime into two parts: the solar (or
short-wave) regime and the terrestrial (or long-wave)
regime. A division of the spectrum at about 4 μm effectively
separates the two. The atmosphere is nearly transparent to
short-wave radiation from the hot Sun, of which large
amounts reach the Earth's surface. However, the atmosphere
readily absorbs longer-wave infrared radiation.

4.4.3.2 Global radiation

Global radiation is the sum of all short-wave radiation
received, both directly from the Sun (direct solar radiation)
and indirectly (diffuse radiation) from the sky and clouds,
on a horizontal surface (radiation pathways are shown in
Figure 4.6). Direct solar radiation casts shadows, and its
intensity on a horizontal surface varies with the angle of the
Sun above the horizon (solar angle). Diffuse radiation, in
contrast, does not cast shadows, is usually much less intense
than the direct component and only varies slightly
throughout daylight. Since diffuse radiation is the result
mainly of scattering and absorption in clouds, its intensity
depends on cloud structure and thickness, not just on solar
angle (see Figure 23.6 in Chapter 23).

Global radiation with a large direct component shows a
strong diurnal variation in intensity (e.g. Figure 4.7), with

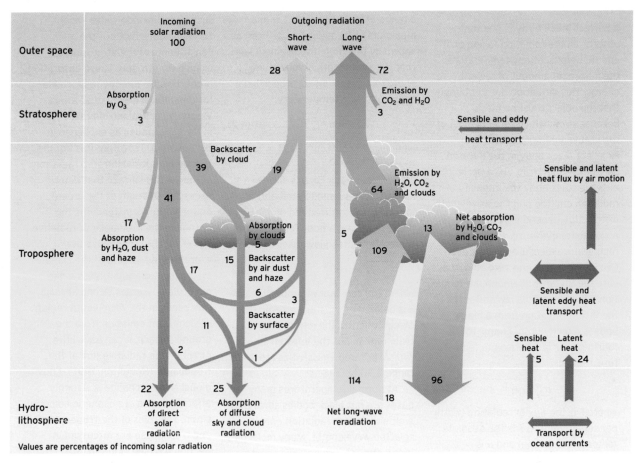

Figure 4.6 Schematic diagram showing the interactions that radiation undergoes in the atmosphere and at the land surface. Values
are percentages of incoming solar radiation. The stratosphere is the upper atmosphere and the troposphere the lower atmosphere. The
hydro-lithosphere refers to the land or oceans. (Source: after Rotty and Mitchell, 1974)

Figure 4.7 Solar irradiance (F) on a cloudless sky (16 July 1969) at Sutton Bonington (53°N, 1°W): S_t total flux; S_b direct flux on horizontal surface; S_d diffuse flux. Solar elevation is the angle of the Sun above the horizon. (Source: after Monteith and Unsworth, 1990)

maximum values when the Sun is at its greatest angle above the horizon around midday. This diurnal variation nearly vanishes on cloudy days when the diffuse component dominates the global radiation. The actual worldwide distribution of global radiation reflects astronomical factors and the distribution of cloud, which in turn is determined by the global wind circulation. Thus Figure 4.8 shows that the areas receiving most global radiation are found in the subtropics where there are clear skies because of the prevailing anticyclonic conditions associated with the descending air from the Hadley cell circulations illustrated in Figures 4.4 and 4.5.

4.4.3.3 Albedo

Radiation falling on a solid surface may be partly reflected, partly absorbed and partly transmitted. In contrast, water is translucent and light penetrates the surface layers of the oceans, while the atmosphere is nearly transparent to short-wave radiation. Radiation received from the surface of an object may result from either reflection by that surface, or radiation emitted (according to the radiation laws in Box 4.2) from the surface, or indeed both. **Albedo** is a measure of the reflecting power of a surface, being the fraction of the short-wave radiation received that is reflected by a surface. Albedo is sometimes called **reflectivity**, but this term properly refers to

Figure 4.8 Worldwide distribution of annually averaged global (direct plus diffuse) surface-received solar radiation (W m^{-2}). (Source: after Lockwood, 1979 and Budyko, 1974)

the reflected : received ratio for a specific wavelength. Reflectivity therefore varies with wavelength; for example, grass is green because it reflects much of the green light in solar radiation and absorbs most of the energy in other colours.

Albedo is important because it determines the fraction of the incoming solar radiation that is absorbed by the surface. This absorbed radiation is then available to heat the surface. Typical albedo values are contained in Table 4.1. The greater the albedo, the less solar radiation is absorbed. Albedo varies from 0.08 for dark, moist soils, to 0.86–0.95 for fresh snow. In the latter case nearly all the incident solar radiation is reflected and the solar warming of the surface is small. Clouds also have very high albedos, and therefore greatly restrict the amount of sunlight falling on the ground below them. The planetary albedo of the Earth as seen from space is made up of three main components. These are the light reflected from the actual land and sea surfaces of the Earth, the light reflected by clouds, and light scattered upwards by the atmosphere (Figure 4.6).

The global distribution of surface albedo is shown in Figure 4.9, where the extremely high albedo values of the ice fields of Antarctica should be noted and compared with the low albedo values over the equatorial rainforests.

Table 4.1 Values of albedo for various surfaces

Surface	Albedo (percentage of incoming short-wave radiation which is reflected)
Fresh, dry snow	80-95
Sea ice	30-40
Dry light sandy soils	35-45
Meadows	15-25
Dry steppe	20-30
Coniferous forest	10-15
Deciduous forest	15-20

Particularly high gradients of surface albedo are seen in both the Arctic and Antarctic where ice fields replace open water or ice-free land. High albedo values also occur over the dry sands of the subtropical deserts. Albedo over the oceans is, except for the high latitudes, nearly uniform at around 0.10.

Figure 4.9 Distribution of minimum albedo (%) from Nimbus 3 satellite measurements during 1969–1970. This is a good approximation to the albedo of the surface. The assumption is that the albedo of the Earth-atmosphere system is higher over each area in the presence of clouds than for a cloud-free atmosphere. Therefore estimates of the albedo of the surface and also locations of persistent cloud fields and of ice and snow can be made. The effects of changing cloud fields are removed by displaying only the lowest observed satellite albedo value in each area. (Source: after Raschke *et al.*, 1973)

4.4.3.4 Net radiation and energy balances

The full energy balance of, for example, a sample land surface (Figure 4.6) involves not only radiative fluxes, but also fluxes of energy in the form of sensible heat (heat that can be detected by a thermometer) and latent heat (e.g. energy contained in evaporated water vapour). The sensible heat fluxes may be both downwards into the soil and upwards into the atmosphere. The energy available for the sensible and latent heat fluxes depends on the net radiation. The **net radiation** at a surface is the difference between the total incoming and the total outgoing radiation. The net radiation indicates whether net heating (positive) or cooling (negative) is taking place; the net radiation will normally be negative at night, indicating cooling, but during the day it may be positive or negative depending on the balance of the incoming and the outgoing radiation. Low cloud blankets the Earth's surface and at night restricts the long-wave radiation loss, thereby causing only small net radiative losses and temperature decreases. If the sky is clear at night, the long-wave radiative loss is large, so is the net radiative loss, resulting in a large nocturnal temperature decrease. The term 'available energy' is often used to describe net radiation minus the sensible heat transfer into the soil.

Net radiation is defined as:

$$R_n = (1 - \alpha)Q - L_{out} + L_{in} \qquad (4.4)$$

where R_n is the net radiation, α is the albedo of the surface, Q is the incoming global radiation, L_{out} is the outgoing long-wave radiation from the surface and L_{in} is the incoming long-wave radiation from the atmosphere. For a normal surface (land or ocean) the sensible and latent heat fluxes are balanced by the net radiation. This gives an equation of the form:

$$R_n = \lambda E + H + S \qquad (4.5)$$

where λ is the latent heat of vaporization, E is the evaporation rate, H is the sensible heat flux into the atmosphere and S is the sensible heat flux into the soil or ocean.

4.4.3.5 The Bowen ratio – the partition of net radiation

The key issue over land is the partition of the net radiation into sensible and latent heat fluxes. One approach to this problem is to consider the Bowen ratio, which is defined as $H/\lambda E$ (sensible heat flux divided by latent heat flux). For normal moist surfaces, values of the ratio are in the range 0 to 1 (Figure 4.10). The value of the Bowen ratio varies widely over land. The Bowen ratio decreases with increasing

Figure 4.10 Values of the Bowen ratio at a freely evaporating surface for various surface temperatures. This diagram applies only to a freely evaporating surface. If the surface becomes dry, the value of the Bowen ratio will increase to above that appropriate for the surface temperature. (Source: after Lockwood, 1979)

temperature and the ratio reaches zero at about 32–34°C over moist surfaces. The ratio also depends on the availability of water for evaporation, which is controlled by soil moisture, vegetational controls, and input of dry air from above. Drier conditions result in an increase in the Bowen ratio.

If water is readily available for evaporation, because it is on the soil surface or the roots of vegetation have access to water in the soil, then as much as 75–80% of the total daytime transport of energy to the atmosphere occurs as latent heat. This high evaporation rate from a moist surface cools the surface and keeps daytime maximum temperatures below about 32–34°C. At this temperature all the net radiation is used for evaporation. Higher evaporation rates can occur only by extracting heat from the atmosphere and thereby cooling it. At the other extreme of the desert, where water for evaporation is very limited, most of the net radiation during daylight is transferred to the atmosphere as sensible heat, which warms the lower atmosphere. Under these circumstances the Bowen ratio has values above 1 and daylight surface temperatures can reach values considerably above 34°C since the surface is not cooled by evaporation. Because the desert surface is hot in daytime, the long-wave emission can be very large, significantly reducing the net radiation. The degree of wetness of a surface therefore exerts, through the Bowen ratio, a strong control over surface temperature and near-surface atmospheric temperature. For example, maximum temperatures observed over tropical oceans or moist tropical land surfaces rarely exceed 32–34°C, while over dry tropical deserts they may reach 50°C.

4.4.4 The atmospheric energy balance

The atmosphere continually loses more heat to space in the form of long-wave radiation than it gains from the absorption of short-wave radiation, and this is equivalent to a cooling of 1 or 2°C per day. The heat lost to space in the form of long-wave radiation is largely compensated for by the latent heat released by atmospheric water vapour that has evaporated from the surface, condensing to form precipitation. Typical annual global values for the energy balance of the atmosphere are:

- net radiative loss = -99.4 W m^{-2} (this must be balanced by the sensible and latent heat gained by the atmosphere);
- sensible heat gain = 20.4 W m^{-2};
- latent heat gain = 79.0 W m^{-2}.

Annual global precipitation is therefore linked closely to the atmospheric energy balance and can change only if there are associated changes in the global energy balance. The latent heat gain of 79 W m^{-2} is equivalent to an average annual global precipitation of about 1004 mm yr^{-1}, which has shown yearly variations of about ±50 mm (±5%) over the past 100 years.

On a global mean basis, to a good approximation, a balance between radiative cooling and condensational heating maintains the mean equilibrium atmospheric temperature. This has interesting and important consequences for the relationship between changes of global atmospheric temperature and associated changes in global precipitation. For example, a temperature increase initiated by a reduction in radiative cooling in the lower atmosphere, because of an increase in atmospheric CO_2, and therefore the greenhouse effect, can be achieved with a corresponding decrease in condensational heating and therefore a reduction in precipitation. In contrast, for a temperature increase initiated by an enhancement in condensational heating (increase in global precipitation), for instance, due to increased sea surface temperatures, and therefore evaporation, the atmosphere can adjust itself to a new steady state through an increase in radiative cooling. For both cases, although the new steady-state temperature is higher, the changes in global precipitation, and therefore in the intensity of the hydrological cycle, are in opposite directions. Thus an increase in temperature due to global warming may decrease or increase the rainfall. This is explored further in the next section and it is to the role of moisture in the atmosphere that we now turn our attention.

4.5 Moisture circulation systems

4.5.1 Moisture in the atmosphere and the hydrological cycle

The full cycle of events through which water passes in the surface–atmosphere system is best illustrated in terms of the hydrological cycle, which describes the circulation of water from the oceans, through the atmosphere back to the oceans, or to the land, and then to the oceans again by overland and subsurface routes. Water in the oceans evaporates under the influence of solar radiation and the resulting water vapour is transported by the atmospheric circulation (see Section 4.3) to the land areas, where precipitation may occur in a variety of forms including rain, hail and snow. Some of this precipitation will infiltrate into the soils and rocks below the surface from where it flows more slowly to river channels or sometimes directly to the sea (see Chapter 13). The water remaining on the surface and in the upper layers of the soil will partly evaporate back to the atmosphere. Since evaporation is linked to the surface energy balance, the hydrological cycle is directly linked to, and indeed is part of, global energy exchanges.

It is simplest to regard the hydrosphere as a global system consisting of four reservoirs linked by the hydrological cycle.

These are the world ocean, polar ice, terrestrial waters and atmospheric waters; estimates of the amount of water stored in these reservoirs are summarized in Table 4.2. The average residence times (Table 4.3) of water in the various reservoirs are of particular interest to climatologists. The water vapour in the air at any one time represents about one-fortieth part of the annual rainfall or the supply of water for about 10 days of rainfall. The Earth's atmosphere contains on average an amount of water vapour which, if it were all condensed and deposited on the surface of the Earth, would stand to a depth of about 25 mm. Therefore to maintain the precipitation amounts observed in many parts of the world, there must be a continual flow of very moist air into the area from the surroundings. The general wind circulation (Section 4.3) is therefore very important in explaining the observed global precipitation patterns. Particularly high precipitation regions are found where moist winds from oceanic areas either flow onto windward continental coasts, Europe for example, or converge into low-pressure troughs such as that found in the equatorial zone.

4.5.2 Global distribution of precipitation and evaporation

While on a global scale over an annual period, evaporation approximately equals precipitation, this is not so for either particular regions or seasonal timescales. Figure 4.11 shows estimated annual distributions of precipitation and evaporation. It is clear that the main areas of evaporation, mostly over the oceans, do not coincide with the major precipitation regions. This implies an atmospheric flux of water vapour from the oceans to the land masses. Indeed, the water forming precipitation over the continental land masses can be divided into two components. Firstly, there is the water component of the precipitation that originated from distant oceans and has been circulated over the land by low-level winds. Secondly, there is the component that is formed of locally evaporated water. In the middle-latitude continents, the first component dominates in winter and is associated with the middle-latitude westerlies and active frontal depressions, while the second is more important in summer, and tropical/equatorial areas, and is associated with convective storms. Evaporation rates (Figure 4.11b) over the land masses are relatively low, only about 50% of the global average (Brutsaert, 1982). There are a number of reasons for this, mostly linked to the moisture-holding capacity of the land surface soils, and the annual distribution of precipitation and net radiation. Many land surfaces are able to sustain typical evaporation rates of 2–3 mm day^{-1} for less than a month. In climates with marked wet and dry seasons, found particularly in the tropics, soil moisture soon becomes exhausted in the dry season and evaporation rates fall to very low values. Here vigorous evaporation is restricted to the wet season. In many areas in the subtropics, the wet season is short or non-existent, and annual evaporation rates are very low. Similar arguments apply in many middle-latitude regions, where precipitation is often plentiful in winter, but because of a Sun low in the sky associated with short daylight hours, net radiation is restricted and hence evaporation rates are low. Much of the winter precipitation either recharges soil

Table 4.2 Estimated volumes of principal components of the hydrosphere

	Surface area covered (10^6 km^2)	Present amount of water resources (10^3 km^3)	(%)
World ocean	361	1 370 000	93
Polar ice	16	24 000	2
Terrestrial waters	134	64 000*	5
Atmospheric waters	510	13	0.001

*Includes groundwater (almost 64 000), lakes (230), soil moisture (82).

Table 4.3 Average residence times of water

Water source	Residence time
Atmospheric waters	10 days
Terrestrial waters	
Rivers	2 weeks
Lakes	10 years
Soil moisture	2–50 weeks
Biological waters	A few weeks
Groundwaters	Up to 10 000 years
Polar ice	15 000 years
World oceans	3600 years

Figure 4.11 (a) Global distribution of annual precipitation (mm), simplified to show only major regimes; oceanic rainfall estimated. (Source: after Riehl, H., *Introduction to the atmosphere*, 1965, McGraw-Hill, Fig. 2.16. Reproduced with permission of the McGraw-Hill Companies) (b) Global distribution of annual evaporation (mm) based upon a number of recent estimates. Oceans are almost infinite moisture sources, while land areas frequently become dry, so there is a marked discontinuity between adjacent land and ocean regions. Oceanic evaporation shows a dependence upon ocean surface temperature. (Source: after Brutsaert, 1982)

moisture or forms river runoff. In contrast, in summer with long daylight hours and a Sun high in the sky, the net radiation values are high but soil moisture becomes seriously depleted, because of low precipitation rates, leading to restricted evaporation rates (Box 4.3).

In many summer and dry-season cases the actual land surface evaporation is considerably less than predicted by the energy balance equations with assumed moist land surfaces. The evaporation may be considered as being restricted by some form of resistance in the evaporative pathway from the soil to the atmosphere (Box 4.4). When the soil is moist the resistance is relatively low, but as the soil moisture decreases the resistance increases in response and limits the evaporative loss.

HAZARDS

DROUGHT

Droughts rank among the world's costliest natural disasters because they affect a very large number of people each year (Figure 4.12). Drought is a recurring phenomenon that has plagued civilization throughout history. A drought is considered to be a period of abnormally dry weather that causes serious hydrological imbalance in a specific region. However, the definitions of 'serious' and 'abnormally dry' depend on the nature of the local climate and the impact of the drought on the local society. A dry spell in a humid climate may be classified as a drought, while similar conditions in a semi-arid climate would be considered a wet period. This makes it difficult to produce a definition of drought that applies in a variety of climates.

The American Meteorological Society (1997) grouped drought definitions into four categories: meteorological or climatological, agricultural, hydrological and socio-economic. If atmospheric conditions result in the absence or reduction of precipitation over several months or years the result is a meteorological drought. A few weeks' dryness in the surface layers (vegetation root zone), which occurs at a critical time during the growing season, can result in an agricultural drought that severely reduces crop yields, even though deeper soil levels may be saturated. The onset of an agricultural drought may lag that of a meteorological drought, depending on the prior moisture status of the surface soil layers. **Precipitation deficits** over a prolonged period that affect surface or subsurface water supply, thus reducing stream flow, groundwater, reservoir and lake levels, will result in a hydrological drought, which will persist long after a meteorological drought has ended. Socio-economic drought associates the supply and demand of some economic good with elements of meteorological, agricultural and hydrological drought.

There are a number of quantitative definitions of drought based upon knowledge of precipitation, soil moisture, **potential evapotranspiration** or some combination of the above. A commonly used and widely accepted meteorological drought index is the Palmer Drought Severity Index (PDSI). It originated in the United States and is obtained from a simplistic model of the cumulative anomaly of moisture supply and demand at the land surface, which requires knowledge of both precipitation and potential evapotranspiration. Model results are adjusted to allow for the local climate before the final index is calculated. Regardless of the simple nature of PDSI, it was found by Dai *et al.* (2004) to be significantly correlated with the observed soil moisture content within the upper 1 m of soil during warm season months. They also found that PDSI was a good proxy for stream flow. Using a global data set of PDSI, Dai *et al.* (2004) found that very dry regions over global land areas have increased from ~12% to 30% since the 1970s, with a large jump in the early 1980s due to an El Niño-induced precipitation decrease and enhanced surface warming. IPCC (2007a) commented that it is likely that droughts have increased in many areas since the 1970s, and that it is more likely than not that there is a human contribution to this trend. During the past two to three decades, there has been a tendency for more extreme (either very dry or very wet) conditions over many regions, including the United States, Europe, east Asia, southern Africa and the Sahel.

Figure 4.12 Drought is a major hazard, sensitive to global warming, that disrupts many aspects of human existence ranging from agricultural production to water supply and river flow. The area covered by extreme drought is predicted to increase from 1% now to 30% by the end of the twenty-first century. (Source: Sipa Press/Rex Features)

BOX 4.3 ➤

> The UK Hadley Centre's global climate model (HadCM3) has been used, together with PDSI, to study drought on a global basis (Burke *et al.*, 2006). The model shows that between 1952 and 1998 on average 20% of the land surface was in drought at any one time. At decadal timescales, on a global basis, the model reproduced the observed drying since 1952 reported by Dai *et al.* (2004) and analysis showed that there was a significant influence of anthropogenic emissions of greenhouse gases and sulphate **aerosols** in the production of this drying trend. Future projections of drought in the twenty-first century using HadCM3 show regions of strong wetting and drying with a net overall global drying trend. The proportion of the land surface in extreme drought is predicted to increase from 1% for the present day to 30% by the end of the twenty-first century.
>
> **BOX 4.3**

4.5.3 The influence of vegetation on evaporation

Evaporative losses from land surfaces have three components, known as transpiration, **interception** loss and soil surface evaporation. While they are governed by the same physical laws, they do not necessarily behave in the same manner under similar conditions. Water is extracted from the soil by the roots of actively growing vegetation and transpired through small holes in the vegetation leaves called stomata. Transpiration is therefore strongly controlled by

NEW DIRECTIONS

SYSTEM RESISTANCES AND EVAPORATION

Evaporation from the ground surface depends on the net radiation, which supplies the energy for the latent heat of vaporization, and also the vertical gradient of water vapour pressure. Equations describing evaporation are normally expressed in terms of conceptual resistances, using Ohm's law for electricity as a direct analogue. Ohm's law gives the relationship between current (amps) in a circuit to the electrical potential (volts) and the resistance of the wire:

$$\text{Current} = \frac{\text{potential difference}}{\text{wire resistance}} \quad (4.6)$$

For entities transported from the surface through the atmosphere such as water vapour, heat and CO_2 this may be rewritten to read:

$$\text{Flux rate} = \frac{\text{concentration difference of property}}{\text{resistance to flow exerted by the system } (r)} \quad (4.7)$$

where r represents the appropriate system resistance with units of seconds per metre. In the case of evaporation the concentration difference may be expressed as the vertical gradient of vapour pressure.

In the case of **transpiration**, which is the removal of water from plants, two resistances are frequently used: one applies to the passage of water though the vegetation from the soil to the atmosphere, and the other applies to the passage of water vapour from the canopy surface to a standard level (often at 2 m) in the overlying atmosphere. The value of the first (the bulk canopy resistance) depends on such factors as the vegetation type, soil moisture, sunlight and atmospheric humidity. The value of the second (the aerodynamic resistance) depends on such factors as vegetation height, wind speed and so on.

Stomatal pores in leaves are necessary for the uptake of CO_2 from the ambient air, but at the same time water vapour escapes via transpiration since there is a common physical pathway of water vapour and CO_2 through the stomatal pores. The degree of opening of the pores can be considered as a compromise in the balance between limitation of water loss and admission of CO_2. The generally observed closure of stomata when CO_2 increases is an expression of this compromise. Thus for many vegetation types the bulk canopy resistance increases as atmospheric CO_2 increases, resulting in a fall in transpiration rates. As a result vegetation uses water more efficiently in an enhanced CO_2 atmosphere. Therefore some vegetation types could grow in drier climates than they do at present (Lockwood, 1993, 1999).

BOX 4.4

the nature and condition of the vegetation. During rainfall, water is intercepted in vegetation canopies and evaporated back to the atmosphere during and just after the rainfall event. This re-evaporated rainwater is termed interception loss. Provided there is rainwater on the plant leaves, the rate of interception loss is independent of the vegetation state but strongly dependent on atmospheric conditions. Lastly, water is evaporated directly from moisture held on or just below the soil surface. The sum of transpiration, interception loss and soil surface evaporation is termed evapotranspiration. Further discussion of evapotranspiration is provided in Chapter 13.

Reflective questions

➤ What are the fundamental differences in precipitation and evaporation likely to be during the year between northern and southern Europe?

➤ What reasons are there that actual evaporation from a desert surface might be lower than predicted by typical energy balance equations?

4.6 Motion in the atmosphere

4.6.1 Convective overturning

Convection is the dominant process for transferring heat, mostly in the form of latent heat, upwards from the surface. During daylight the surface of the Earth is warmed by short-wave radiation from the Sun. Air close to the surface is heated and rises, because on warming its density becomes lower than that of the surrounding cooler air. In this discussion it is useful to consider an isolated parcel of air which is warmed at the surface. Because pressure falls with height in the atmosphere, as the parcel rises it also expands so that the internal pressure of the parcel and the external pressure at the same level in the atmosphere are always equal. An expanding parcel of air chills because its molecules lose a little of their internal energy, which as explained earlier we sense as temperature, as they bounce off the retreating parcel walls.

Vertical motions in the atmosphere are usually rapid enough to make temperature changes occur much more quickly than any interchange of heat, by conductive or radiative processes, between the parcel and its surroundings. The temperature changes may then be said to be **adiabatic**: that is, there is no interchange of heat between the parcel and its surroundings. The rate of temperature fall with height (**dry adiabatic lapse rate**) in the rising air parcel

under these conditions is 9.8°C km^{-1}. A similar rate of warming is observed in a parcel of air sinking under adiabatic conditions. Cooling in a rising air parcel may cause the air to become saturated with water vapour, condensation to take place and clouds and precipitation to form. When water vapour condenses to form water droplets in the rising air parcel, it releases latent heat which warms the air slightly and therefore causes the rate of temperature decrease with height to fall below the dry adiabatic lapse rate. The new **lapse rate**, in the presence of condensation, is known as the **saturated adiabatic lapse rate**. Its exact value is variable since it depends on the rate of condensation. See Box 6.1 in Chapter 6 for further information on lapse rates.

Considerations of mass balance dictate that as some air masses rise, other air masses descend, so the lower atmosphere is continually turning over. This layer of the atmosphere where there is large-scale mixing, and which also contains most of the atmospheric water vapour as well as most clouds and weather phenomena, is known as the troposphere (Figure 4.13). The upper boundary of the troposphere is known as the tropopause, which has an altitude of about 5–6 km over polar regions and 15–16 km over equatorial regions. Temperature in the troposphere falls from around 15°C at the surface to around −57°C at the tropopause at a rate determined by these convective processes. Since condensation (to form clouds and rainfall) and the release of latent heat occur during the convective processes, the observed fall of temperature with height in the troposphere is less than the dry adiabatic lapse rate. Observations indicate that it is about 6°C km^{-1}, but it varies with location in the troposphere. The **stratosphere** extends from the top of the tropopause to about 50 km height (the stratopause). In the stratosphere the temperature either varies very little with height or increases with height with warmer air overlying cooler air. This is because the maximum temperatures are associated with the absorption of the Sun's short-wave radiation by ozone which is found at this altitude. In the thermosphere temperatures increase with altitude because of the absorption of extreme ultraviolet radiation (0.125–0.205 μm) by oxygen but really these 'temperatures' are theoretical because there is so little air (so few molecules – and remember that temperature is simply a measure of the mean kinetic energy of the molecules of an object).

4.6.2 The Earth's rotation and the winds

Over timescales of around a year or longer, the complete Earth–atmosphere system is almost in thermal equilibrium. Consequently the total global absorption of solar radiation by the Earth's surface and atmosphere must be balanced by

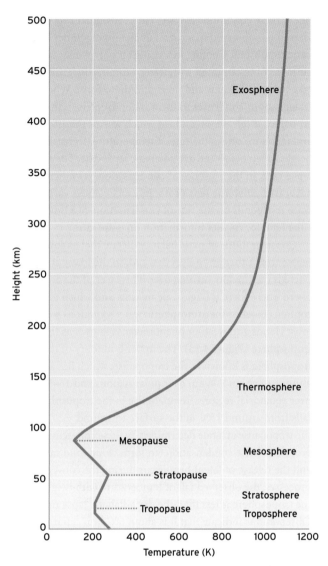

Figure 4.13 Temperature variation with height (lapse rate) in the atmosphere. Note the three regions of increased temperature where absorption occurs: (i) the surface (visible and near-infrared absorption); (ii) lower stratosphere (ultraviolet absorption by ozone); (iii) thermosphere (high-energy absorption). 0 K = −273.14°C, 273.14 K = 0°C. (Source: from Robinson and Henderson-Sellers, 1999)

latitudes. This very simple circulation both implies a rotating Earth and is further modified by the Earth's rotation.

The simplest atmospheric circulations would arise if the Earth always kept the same side towards the Sun. Under these circumstances the most likely atmospheric circulation would probably consist of rising air over an extremely hot, daylight face and sinking air over an extremely cold, night face. The diurnal cycle of heating and cooling obviously would not exist since it depends on the Earth's rotation. Surface winds everywhere would blow from the cold night face towards the hot daylight face, while upper flow patterns would be reversed. The climatic zonation would be totally different from anything observed today. Theoretical studies suggest that if this stationary Earth started to rotate, then as the rate of rotation increased, the atmospheric circulation patterns would be progressively modified until they approached those observed today (Lockwood, 1979). Even with the present rate of rotation two different atmospheric circulation patterns are possible. This is because if the Earth rotated towards the west instead of towards the east as at present, then the mid-latitude westerlies would become easterlies and depressions and anticyclones would rotate in the opposite sense. Under these circumstances, Europe would have a climate similar to that of present-day eastern Canada, while eastern Canada would enjoy the present-day, mild, western European climate.

Observations of air flow in the atmosphere show that, except very near to the surface and in equatorial regions, it is almost parallel to the **isobars** (contours of equal pressure). This seems odd because air should move at 90° to isobars from high to low pressure. Instead the Earth's rotation acts on the atmosphere through the so-called **Coriolis effect** (see Box 4.5). The force due to the horizontal pressure gradient balances the Coriolis effect and so the wind blows parallel to isobars (Figure 4.14). This wind, which exists

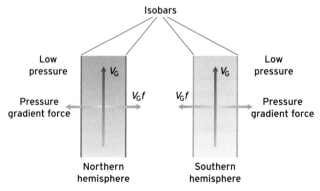

Figure 4.14 The geostrophic wind: V_G is the geostrophic wind; f is the horizontal component of the Coriolis force. The pressure gradient force tends to balance the Coriolis force. The pressure gradient force acts from high to low pressure.

the total global emission to space of infrared radiation. While most of the solar radiation is absorbed during daylight in the tropical regions of the world, infrared radiation is lost continually day and night to space from the whole of the Earth's surface and atmosphere, including the middle and polar latitudes. Heat energy is therefore transferred from the tropics where it is in surplus, to middle and high latitudes where it is in deficit, by the north–south circulations in the atmosphere and oceans. These circulation patterns were described earlier and take the form of a number of cells in the atmosphere, with rising air motion at low and middle latitudes, and sinking motions in subtropical and high

FUNDAMENTAL PRINCIPLES

THE CORIOLIS EFFECT

The concept of the Coriolis effect may be briefly explained as follows. Large-scale, horizontal winds flowing across the Earth's surface tend, if they are not deflected by horizontal pressure gradients, to move in a straight path in relation to a reference frame fixed in relation to the distant stars. The movement of winds over the Earth's surface is usually made with reference to the latitude and longitude grid which rotates with the Earth and thus

continually changes orientation in relation to the distant stars. Thus to an observer on the rotating Earth, who is unaware of the Earth's rotation, it appears that a force is acting on the winds that causes them to be deflected to the right (left) in the northern (southern) hemisphere (Figure 4.15). The observer will make the same 'mistake' whatever the initial wind direction. The observer may conclude that the deflection to the right has been brought about by a force acting

to the right of the wind. In the nineteenth century the French physicist Coriolis formalized this concept of an apparent force caused by the Earth's rotation, later called the Coriolis effect. Thus winds blowing towards the equator are deflected by the Coriolis effect towards the west (generating easterly winds), while poleward-flowing winds are deflected towards the east (generating westerlies). The magnitude of the Coriolis effect varies with latitude as shown in Figure 4.15(b).

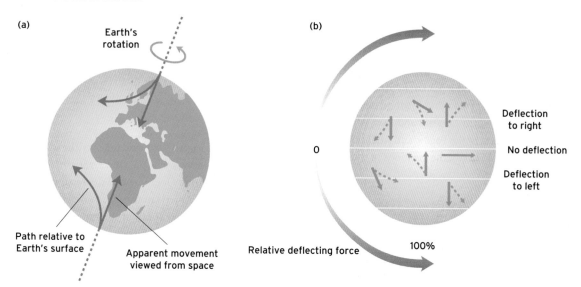

Figure 4.15 (a) The influence of the Coriolis effect on air movement. (b) The changing magnitude of the Coriolis effect with latitude.

BOX 4.5

above about 1 km from the surface, is known as the **geostrophic wind**. The speed of this wind is proportional to the pressure gradient (i.e. wind speeds are greater when the isobars are closer together) but is also determined by the strength of the Coriolis effect. As the Coriolis effect varies with latitude (Figure 4.15b) the geostrophic wind for the same pressure gradient will decrease towards the poles. The direction is such that the air flow is anticlockwise around low pressure in the northern hemisphere. The converse applies in the southern hemisphere where air flow is clock-

wise around low pressure. For high-pressure anticyclones the flow in both hemispheres is the opposite way round.

4.6.3 Rossby waves

Compared with the Hadley cells discussed in Section 4.3, the middle-latitude atmosphere is highly disturbed and the suggested circulation of the Ferrel cells in Figure 4.4 is largely schematic. At the surface, the predominant features are irregularly shaped, closed, eastward-drifting, cyclonic and anticyclonic systems, while higher up, smooth wave-shaped patterns are generally found. These waves are often termed **Rossby waves**, after C.G. Rossby who first investigated their principal properties in the late 1930s and early 1940s. Rossby waves are important because they strongly influence the formation and subsequent evolution of surface weather features. Middle-latitude frontal depressions tend to form and grow rapidly just downwind of upper troughs of Rossby waves, while surface anticyclones tend to develop just downwind of upper ridges. Surface depressions in particular tend to move in the direction of the upper flow with a speed which is directly proportional to the upper flow.

In a simple atmosphere, Rossby waves could arise anywhere in the middle-latitude atmosphere, as is observed in the predominantly ocean-covered southern temperate latitudes. In contrast, the northern temperate-latitude Rossby waves tend to be locked in certain preferred locations. These preferred locations arise because the atmospheric circulation is influenced not only by the differing thermal properties of land and sea, but also by high north–south-aligned mountain ranges such as the Rockies and Andes. In January, at about 5.5 km in the northern hemisphere, the dominant troughs are found near the eastern extremities of North America and Asia, while ridges lie over the eastern parts of the Pacific and Atlantic Oceans (Figure 4.16). Climatologically, the positions of the two

Figure 4.17 Monthly mean 500 millibar contours in the northern hemisphere for July, based on data from 1951 to 1966. (Source: after Moffit and Ratcliffe, 1972)

troughs are associated with cold air over the winter land masses and the position of the ridges with relatively warm sea surfaces. Atlantic Ocean frontal depressions form just downwind of the American trough, move eastwards with the upper winds and decay south of Iceland in the upper trough. In July, the mean 5.5 km flow pattern found in January over the Pacific has moved about 25° west and now lies over the warm North American continent, while there is a definite trough over the eastern Pacific (Figure 4.17).

4.6.4 Jet streams

Embedded in the Rossby waves are fast-moving bands of air known as jet streams. They are caused by sharp temperature gradients associated with frontal depressions. Separate powerful westerly jet streams, not associated with the middle-latitude frontal jet streams, are also found along the poleward borders of the tropical Hadley cells. Normally a jet stream is thousands of kilometres in length, hundreds of kilometres in width and several kilometres in depth. An arbitrary lower limit of 30 m s^{-1} (about 60 mph) is assigned to the speed of the wind along the axis of a jet stream. The global distribution of jet streams is briefly described below.

Figure 4.16 Monthly mean 500 millibar contours in the northern hemisphere for January, based on data from 1951 to 1966. (Source: after Moffit and Ratcliffe, 1972)

Figure 4.18 Schematic illustration of the major upper-troposphere features of the Asian monsoon in the northern winter (top panel) and summer (bottom panel). The 200 mb pressure level in the atmosphere is around 12 km altitude; the 150 mb level is at around 14 km. (Source: after Lockwood, 1965)

4.6.4.1 The subtropical westerly jet streams

The subtropical westerly jet streams are found near 30°N and S, corresponding to the poleward boundaries of the tropical Hadley cells, and often form continuous belts around the globe (Figures 4.4 and 4.18). The jet stream cores with the highest wind speeds are normally found near the 12 km level, just below the tropopause. The subtropical jet streams draw their high wind speeds partly from the circulations associated with the Hadley cells and are associated with the descending limbs of the cells. In winter the northern subtropical westerly jet stream divides north and south around the Tibetan Plateau. The two branches merge to the east of the plateau and form an immense upper convergence zone over China. In May and June, in association with the development of the south-west monsoon, the subtropical jet stream over northern India slowly weakens and disintegrates, causing the upper westerly flow to move northwards into central Asia. While this is occurring, an easterly jet stream, mainly at about 14 km altitude, builds up over the equatorial Indian Ocean and expands westward into Africa (Figure 4.18).

4.6.4.2 The equatorial easterly jet stream

During summer, the northern subtropical westerly jet stream is less well developed, especially over Asia, where it moves polewards and is replaced by an easterly jet stream at about 14 km altitude over the equatorial Indian Ocean. The formation of the equatorial easterly jet stream is connected with the formation of an upper-level high-pressure system over Tibet. In October the reverse process occurs: the equatorial easterly jet stream and the Tibetan high disintegrate, while the subtropical westerly jet stream reforms over northern India.

The equatorial easterly jet stream forms over Indonesia and disintegrates over northern Africa. The disintegration of the jet stream creates a region of subsidence around North Africa, which reinforces the subsiding limb of the Hadley cell and the associated Sahara Desert conditions. When it is well developed the equatorial easterly jet stream therefore has the effect of intensifying the summer aridity over the Sahara Desert and the Middle East.

4.6.4.3 Polar front jet streams

The polar front jet streams are associated with the warm and cold fronts of temperate-latitude depressions, and are found at approximately 10 km altitude. They show considerable day-to-day variation as surface-level fronts form and develop.

Reflective questions

➤ How does the Coriolis effect vary with latitude?

➤ To what extent does convection play a role in atmospheric processes?

4.7 The influence of oceans and ice on atmospheric processes

We have seen how the land surface can interact with atmospheric processes via albedo effects, the supply or retardation of moisture and topographic interactions. However, the

oceans also influence atmospheric processes, not only via albedo and moisture circulation but via other energy transfer processes too. Most substances become denser as they get colder. However, freshwater is unlike any other substance because it is densest at about 4°C. Therefore when the surface is cooled below that temperature, the coldest water stays on top as it is less dense (and hence more buoyant) than the water at 4°C. When water freezes, the ice floats on the top. This is because water expands when it freezes and becomes less dense. All other substances contract when they freeze.

By contrast, when saline water, which is normal for the world's oceans, is cooled it becomes denser, but does not reach its maximum density until near its freezing point, at about −2°C. Just above that temperature the ocean water will undergo marked convection, transferring cold water to the depths and warm water to the surface. Sea ice will form only when a layer of the ocean close to the surface has a relatively low salinity. The existence of this layer allows the temperature of the surface water to fall to freezing point, and ice to form, despite lower levels of the ocean having a higher temperature. For example, the perennial Arctic sea ice is formed in a relatively freshwater layer formed by runoff from rivers draining the surrounding continents. This is important because the river drainage of much of both Arctic Canada (Mackenzie River system) and Siberia (e.g. Lena River system) is into the Arctic Ocean, thus giving it a positive water balance and causing a net outflow of ocean water across 70°N. However, precipitation variations across either Arctic Canada or Siberia may cause variations in river runoff and thereby influence the amount of freshwater reaching the Arctic Ocean. Thus, through a range of feedback processes, if precipitation changed then the amount of sea ice formed might change and this may cause major northern hemisphere climatic changes linked to changes in albedo, ocean circulation and heat transfers between deep and shallow layers of the ocean (see Chapters 3 and 20).

Western Europe receives an almost continuous supply of heat and moisture from the North Atlantic. This is most noticeable in winter when north-western Europe is significantly milder than the temperature norm for its latitude. Indeed, the mildness of the winter climate of western Europe is one of the more spectacular latitudinal anomalies in the world climatic pattern; in January on the Norwegian coast the temperature anomaly may be as high as 22–26°C, gradually decreasing to 1°C southwards across the continent. This winter temperature anomaly arises not only from the prevailing westerly winds but also because at comparable latitudes the Atlantic Ocean is 4–5°C warmer than the

Pacific. The extension into very high latitudes and the northward narrowing of the northern North Atlantic have consequences on the Atlantic Ocean circulation which in turn has a series of unique effects on the climatic system. This is in complete contrast to the much more benign Pacific Ocean. Warm, saline surface water flows into the northern North Atlantic, after travelling from the Caribbean Sea, via the Gulf Stream and the North Atlantic Drift (see Chapter 3), warming western Europe. This is part of the so-called thermohaline circulation which is strongly developed in the Atlantic, but completely absent from the Pacific. Thus the atmosphere forms just one part of a complex interacting climate system that also incorporates the land, the oceans, ice cover and vegetation. It is because of such a system that there can also be inter-annual variations in atmospheric processes and it is to examples of these that we now turn.

Reflective questions

➤ Why is western Europe mild compared with eastern Canada?

➤ Why does ice form on the sea surface rather than on the sea floor?

4.8 Inter-annual climate variability

We are all very familiar with the marked diurnal and seasonal weather and climate cycles which are forced by daily and seasonal changes in solar radiation. However, the atmosphere–ocean system also oscillates with periods longer than annual. These oscillations exist, usually on a regional scale, with time periods from a few years to decades. These include the **El Niño Southern Oscillation (ENSO)** and the North Atlantic Oscillation (NAO). These are described in the following sections.

4.8.1 El Niño Southern Oscillation

Since the Sun is nearly overhead for the whole year near the equator, the seasonal cycle there is weak. This allows other cycles to dominate the equatorial ocean–atmosphere system. When atmospheric pressure is high in the Pacific Ocean, it tends to be low in the Indian Ocean from Africa to Australia.

The Southern Oscillation is dominated by an exchange of air between the south-east Pacific high and the Indonesian

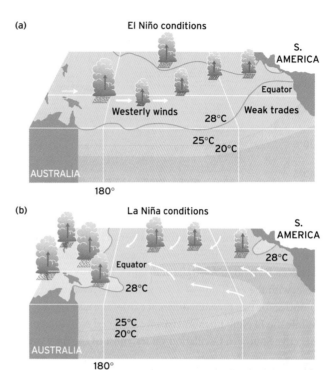

Figure 4.19 A schematic view of (a) El Niño and (b) La Niña. During La Niña, intense trade winds result in a cold equatorial Pacific in the east and warm waters in the west where moist air rises. The air subsides in the east, a region with little rainfall. During El Niño the trade winds along the equator relax, as does the temperature gradient of the ocean waters, and the warm surface waters of the west flow eastwards. (Source: after Philander, 1998)

equatorial low, with a period that varies between roughly 1 and 5 years (Philander, 1998). For several years during one phase of this oscillation, the trade winds are intense and converge into the warm-water regions of the western tropical Pacific, including the Indonesian islands, where rainfall is plentiful and sea-level pressure is low (Figure 4.19). At such times the ocean surface in the eastern tropical Pacific is cold and the associated atmosphere over the ocean and coastal equatorial South America is also cold and dry. More extreme forms of this 'normal' condition are known as La Niña. Every few years, for several months or longer, the trade winds relax, the zone of warmer surface waters and heavy precipitation shifts eastwards, and sea-level air pressure rises in the west while it falls in the east (Figure 4.19). This is El Niño. Under these conditions Indonesia experiences drought and associated forest fires, while coastal equatorial South America experiences heavy rainfall and floods. These ocean and atmospheric changes are part of an oscillating system known as the El Niño Southern Oscillation (ENSO), consisting of the warm ocean pool near the Indonesian Islands, the cold Pacific Ocean pool off

Figure 4.20 Schematic diagram showing the major impacts of El Niño during the period June to December 1997. (Source: after Slingo, 1998)

equatorial South America, and the atmospheric circulation over the equatorial Pacific. The Southern Oscillation may be defined in terms of the difference in sea-level pressure between Darwin in Australia and Tahiti. Records are available, with the exception of a few years and occasional months, for this pressure difference from the late 1800s. While the El Niño phenomena are associated with extreme negative Southern Oscillation values, most of the time there are continuous transitions from high to low values, with most values being positive. ENSO is important climatologically for two main reasons. Firstly, it is one of the most striking examples of inter-annual climate variability on a global scale. Secondly, in the Pacific it is associated with considerable fluctuations in rainfall and sea surface temperatures and also extreme weather events around the world (Figure 4.20 and Box 4.6).

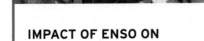

CASE STUDIES

IMPACT OF ENSO ON AUSTRALIAN VEGETATION

Australian rainfall is more variable than could be expected from similar climates elsewhere in the world, mainly due to the impact of ENSO. In Australia rainfall variability is typically one-third to one-half higher compared with stations with the same mean rainfall in areas not affected by ENSO. In particular, because of ENSO, much of Australia has extended periods of drought lasting 12 months or longer, followed by equally extended periods of intense rainfall. The dry periods are associated with ENSO events and the heavy rainfall with anti-ENSO events. Thus Australian vegetation should be suited to an environment of highly variable rainfall with frequent severe droughts or wet periods.

Among the characteristics of Australian vegetation that may be, at least in part, attributable to ENSO's influence on climate are the following:

- Absence of succulents (see Chapter 8): succulents are almost totally absent from Australian arid and semi-arid regions, because although adapted to arid climates and requiring little moisture, they need regular rainfall. Such plants are therefore unsuited to the high variability ENSO produces over much of Australia.

- Vegetation height: Australia has more trees at a given level of aridity than elsewhere. This is because intense rainfall events accompanying anti-ENSO periods produce deep water penetration into the soil. Larger trees, with vast root structures able to remove the stored rainwater from far below the surface, are favoured by the intense rainfall, relative to

an area where rainfall was lighter and more frequent.

- Drought tolerance/avoidance: longevity is a useful characteristic in a country of highly variable rainfall where the conditions for successful initial establishment will occur rarely.

- Fire resistance/dependence: much of the Australian flora is fire resistant or even dependent on fire for successful reproduction. Fires are common during drought periods, which are often associated with ENSO events.

- Fluctuating climax: the high inter-annual variability of annual rainfall in arid and semi-arid parts of Australia affected by ENSO means that vegetation appears adapted to the climate in such a way that demographic components are in a state of unstable equilibrium.

BOX 4.6

4.8.2 North Atlantic Oscillation

A major source of inter-annual variability in the atmospheric circulation over the North Atlantic and western Europe is the NAO, which is associated with changes in the strength of the oceanic surface westerlies. It is often measured by the difference of December to February atmospheric pressure between Ponta Delgado in the Azores (37.8°N, 25.5°W) and Stykkisholmur, Iceland (65.18°N, 22.7°W). Statistical analysis reveals that the NAO is the dominant mode of variability of the surface atmospheric circulation in the Atlantic and accounts for more than 36% of the variance of the mean December to March sea-level pressure field over the region from 20° to 80°N and 90°W to 40°E between 1899 and 1994. The oscillation is most marked during the winter. The oscillation is normally defined by an index which is the value of the sea-level air pressure difference from the Azores to Iceland. When the values are greater than the average the index is said to be positive. There can be great differences between winters with high and low values of the NAO index. Typically, when the index is high the Icelandic low pressure is strong, which increases the influence of cold Arctic air masses on the north-eastern seaboard of North America and enhances the westerlies carrying warmer, moister air masses into western Europe. Thus, NAO anomalies are related to downstream wintertime temperature and precipitation across Europe, Russia and Siberia. High-index winters are anomalously mild while low-index winters are anomalously cold.

Reflective question

➤ How can ENSO affect human activity across the world?

4.9 Interactions between radiation, atmospheric trace gases and clouds

4.9.1 The greenhouse effect

The gases nitrogen and oxygen which make up the bulk of the atmosphere neither absorb nor emit long-wave radiation. It is the water vapour, CO_2 and some other minor gases present in the atmosphere in much smaller quantities which absorb some of the long-wave radiation emitted by the Earth's surface and act as a partial blanket. To examine the effect of these absorbing trace gases on surface temperature it is useful to consider an atmosphere from which all cloud, water vapour, dust and other minor gases have been removed, leaving only nitrogen and oxygen. As the Sun's short-wave radiation passes through the atmosphere, about 6% is scattered back to space by atmospheric molecules and about 10% on average is reflected back to space from the land and ocean surfaces. The remaining 84% heats the surface. To balance this incoming radiant energy the Earth itself must radiate on average the same amount of energy back to space in the form of long-wave radiation. It can be calculated from radiation laws (Box 4.2) that to balance the absorbed solar energy by outgoing long-wave radiation, the average temperature of the Earth should be −6°C. This is much colder than is actually observed (15°C).

This calculation error occurs because the atmosphere readily absorbs infrared radiation emitted by the Earth's surface with the principal absorbers being water vapour (absorption wavelengths of 5.3–7.7 µm and beyond 20 µm), ozone (9.4–9.8 µm), CO_2 (13.1–16.9 µm) and all clouds (all wavelengths). Only about 9% of the infrared radiation from the ground surface escapes directly to space. The rest is absorbed by the atmosphere, which in turn re-radiates the absorbed infrared radiation, partly to space and partly back to the surface. This blanketing effect is known as the natural greenhouse effect and the gases are known as greenhouse gases. It is called 'natural' to distinguish it from the enhanced greenhouse effect due to gases added to the atmosphere by human activities such as the burning of fossil fuels and deforestation. It needs to be stressed that the natural greenhouse effect is a normal part of the climate of the Earth and that it has existed for nearly the whole of the atmosphere's history. Concern about the greenhouse effect arises over two issues: how the natural greenhouse effect may vary with time, and how human activities might modify and enhance the natural effect.

Human activities since the industrial revolution have increased atmospheric trace gases such as CO_2. Before the start of the industrial era, around 1750, atmospheric CO_2 concentration had been 280 ± 10 parts per million (ppm) for several thousand years. It has risen continuously since then, reaching 379 ppm in 2005, with a rate of increase over the past century that is unprecedented over at least the past 20 000 years (see also Chapter 21). The annual CO_2 growth rate was larger during the last 10 years (1995–2005 average 1.9 ppm per year) than it has been since the beginning of continuous direct atmospheric measurements (1960–2005 average 1.4 ppm per year). The present atmospheric CO_2 concentration has not been exceeded during the past 650 000 years, and possibly the past 20 million years. Several lines of evidence confirm that the recent and continuing increase of atmospheric CO_2 content is caused by human CO_2 emissions, and in particular fossil fuel burning. These human-induced CO_2 emissions enhance the already existing greenhouse effect, causing global warming and fundamental changes in climate (see Chapter 21). The 100-year linear global temperature trend is 0.74°C and the linear warming trend over the last 50 years of 0.13°C per decade is nearly twice that for the past 100 years (IPCC, 2007a).

4.9.2 A simple climate model of the enhanced greenhouse effect

It is possible to construct a simple climate model to illustrate the enhanced greenhouse effect due to the addition to the atmosphere of radiatively active trace gases such as CO_2. Figure 4.21 shows a mean temperature profile for the troposphere, calculated assuming convective equilibrium, which gives a lapse rate of 6°C km^{-1}. The upper atmospheric temperature is estimated by assuming that the stratosphere is in radiative equilibrium (its temperature is controlled by its radiation balance and not by vertical convection). The

height at which these two straight-line temperature profiles connect is the tropopause. Viewed from space, the Earth is observed to have a temperature of about −18°C. This is because most of the infrared radiation to space takes place from the middle atmosphere where the temperature is around −18°C; most of the greenhouse gases are below this level. As the amount of infrared-absorbing greenhouse gases mixed into the atmosphere is increased, it becomes more likely to absorb infrared radiation. Thus the effective level at which outgoing infrared radiation originates must rise to allow the radiation to escape to space at the same rate as before. However, because the tropospheric lapse rate does not change significantly from 6°C km^{-1} then the temperatures at a given height in the troposphere must be dragged up by the changes. Furthermore, since the height of the tropopause is increased the stratosphere must cool slightly. In particular, and of great importance, the surface temperature is also increased.

From this simple model, the doubling of the CO_2 content of the atmosphere in the absence of other effects would increase the tropospheric temperature by about 1.2°C. In reality various other effects could increase the predicted temperature changes.

4.9.3 Radiative interactions with clouds and sulphate aerosols

The terms 'forcing' and 'feedback' are frequently used in climatology. Forcings are processes that act as external agents to the climate system, such as changes in solar input to the Earth, the loading of the atmosphere with volcanic ash and aerosols, or rising levels of CO_2 gas. As a result of temperature changes caused by increasing concentrations of greenhouse gases, other changes may take place that could in turn influence temperature. For example, an increase in temperature could result in greater evaporation from the ocean, enhancing atmospheric humidity. Since water vapour is an active greenhouse gas, the increase in atmospheric humidity causes a further increase in temperature. This is an example of a positive feedback process. It also demonstrates that the intensity of the hydrological cycle is closely linked to the greenhouse effect.

4.9.3.1 Clouds

Conventionally, the radiative effect of clouds is discussed in terms of 'cloud radiative forcing', even though cloud effects are actually feedback processes. The radiative forcing of the Earth's climatic system is in part determined

Figure 4.21 The distribution of temperature in a convective atmosphere (solid line). The dashed line shows how the temperature increases when the amount of carbon dioxide present in the atmosphere is increased (for doubled carbon dioxide in the absence of other effects the increase in temperature is about 1.2°C). Also shown for the two cases are the average levels from which thermal infrared radiation leaving the atmosphere originates (about 6 km for the unperturbed atmosphere). (Source: after Houghton, 1994)

by the distribution of cloudiness, since clouds strongly influence the distribution of both short-wave and long-wave radiative fluxes within the atmosphere. At short wavelengths, clouds generally increase planetary albedo and so cool the planet by reflecting more solar radiation to space. However, for long-wave radiation, clouds generally add to the greenhouse effect as they absorb upward-moving infrared radiation and reradiate it back downwards. Thus cloudy days are normally cooler than clear-sky days, particularly in summer, because the short-wave radiative input to the surface is restricted. In contrast, cloudy nights are warmer than clear nights, particularly in winter, because the long-wave radiative loss from the surface to space is restricted (Figure 4.22). However,

observations from satellites show that the albedo cooling effect dominates, and the net effect of clouds at present is to cool the Earth.

4.9.3.2 Aerosols

Over the last few years, it has become evident that, when averaged over the global atmosphere, a substantial part of the enhanced greenhouse effect has been offset by a negative forcing due to the human-induced emission of aerosols (Figure 4.23). Aerosols have two effects, a direct one and also by modifying cloud properties. Aerosol particles reflect short-wave radiation and therefore increase atmospheric albedo. Aerosols also increase cloud albedo and influence cloud lifetime and precipitation. Sulphate aerosols, formed from sulphur dioxide emitted during the combustion of fossil fuels and partly responsible for producing acid rain, are regarded as particularly important. Sulphate aerosols located in the troposphere are rapidly washed out of the atmosphere by rainfall; hence they do not travel far from their industrial sources. The IPCC Report (2007a) suggested that anthropogenic contributions to aerosols (primarily sulphate, organic carbon, black carbon, nitrate and dust) together produce a cooling effect, with a total direct radiative forcing of -0.5 W m^{-2} and an indirect cloud albedo forcing of -0.7 W m^{-2}. The combined radiative forcing due to increases in CO_2, methane and nitrous oxide is estimated to be $+2.30$ W m^{-2}. Therefore IPCC (2007a) reports with a very high confidence that globally averaged net effect of humans activities since 1750 has been one of warming, with a radiative forcing of between $+0.6$ and $+2.4$ W m^{-2} (Figure 4.23).

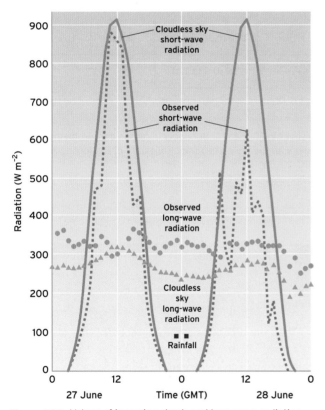

Figure 4.22 Values of incoming short- and long-wave radiation for actual and cloudless sky conditions for a site near Pateley Bridge, North Yorkshire, for two days in June, 1984. During the night of 27/28 June, 2 mm of rain fell in two events of 1 mm each. It is seen that, while cloud makes effectively no contribution to the downward-moving long-wave radiation across the midday period of 27 June (a relatively cloudless day with near-maximum values of incoming short-wave radiation), it contributes around 90 W m^{-2} throughout the rainfall of the night of 27/28 June and only slightly less during the relatively cloudy 28 June (with much reduced incoming short-wave radiation). (Source: after Lockwood, 1993)

Reflective question

➤ Sulphate aerosols originating from industrial areas cool the atmosphere but cause acid rain. Attempts to remove such aerosols from the atmosphere will accelerate global warming. What therefore should be the policy on industrial pollution?

RF terms		RF values (Wm⁻²)	Spatial scale	LOSU
Long-lived greenhouse gases	CO₂	1.66 (1.49 to 1.83)	Global	High
	N₂O CH₄ Halocarbons	0.48 (0.43 to 0.53) 0.16 (0.14 to 0.18) 0.34 (0.31 to 0.37)	Global	High
Ozone	Stratospheric Tropospheric	−0.05 (−0.15 to 0.05) 0.35 (0.25 to 0.65)	Continental to global	Medium
Stratospheric water vapour from methane (CH₄)		0.07 (0.02 to 0.12)	Global	Low
Surface albedo	Land-use Black carbon on snow	−0.2 (−0.4 to 0.0) 0.1 (0.0 to 0.2)	Local to continental	Medium to low
Total aerosol — Direct effect		−0.5 (−0.9 to −0.1)	Continental to global	Medium to low
Cloud albedo effect		−0.7 (−1.8 to −0.3)	Continental to global	Low
Linear contrails		0.01 (0.003 to 0.03)	Continental	Low
Solar irradiance		0.12 (0.06 to 0.30)	Global	Low
Total net anthropogenic		1.6 (0.6 to 2.4)		

Radiative forcing (Wm⁻²)

Figure 4.23 Global-average radiative forcing (RF) estimates and ranges in 2005 for anthropogenic carbon dioxide, methane, nitrous oxide and other important agents and mechanisms, together with the typical geographical extent of the forcing and the assessed level of scientific understanding (LOSU). (Source: IPCC, 2007a)

4.10 Summary

The climate system consists of a closely coupled atmosphere-ocean-ice-land-vegetation system. Short-wave radiation from the Sun is the main energy input that drives the climate system. The processes of absorption, reflection and reradiation of this energy, however, are of fundamental importance to atmospheric processes. There is a radiation surplus at the equator and a radiation deficit at the poles and so the net radiation leads to heat transfers. These transfers to rectify the equator/pole imbalance take place through atmospheric and oceanic motions.

The basic circulation of the atmosphere is dominated by a three-cell structure in each hemisphere which is controlled by a pressure gradient force. The Earth's rotation, however, has a profound influence on air movements across its surface, making circulation systems much more complex than they would otherwise be. At the surface in middle latitudes, the predominant features are closed, eastward-drifting, cyclonic and anticyclonic systems, while higher up in the troposphere smooth wave-shaped patterns are the general rule. There are normally five to eight upper (Rossby) waves circling the poles. They are important because frontal depressions tend to form and grow rapidly just downwind of upper troughs while surface anticyclones tend to develop just downwind of upper ridges. Embedded within the Rossby waves are strong, narrow currents of air known as jet streams.

Hydrological processes play an important part in atmospheric and climatic processes. For example, condensation is important in controlling the convective overturning of the atmosphere and clouds act as both cooling and warming agents across the planet. Atmospheric circulation patterns are naturally variable on an inter-annual scale. Among the more important oscillations are the El Niño Southern Oscillation and the North Atlantic

Oscillation. On longer timescales there is also natural climate variability.

A natural greenhouse effect operates because certain trace gases in the atmosphere such as water vapour and CO_2 act to absorb long-wave radiation emitted from the Earth's surface and reradiate it back downwards. This acts as a blanket. This effect is being enhanced by increased concentration of such greenhouse gases. However, there are also negative feedback effects from atmospheric pollution. Increased clouds and sulphate aerosols can increase atmospheric albedo and reflect more of the Sun's energy directly back into space and thus partly offset the greenhouse effect.

Further reading

Barry, R.G. and Chorley, R.J. (2003) *Atmosphere, weather and climate*. Routledge, London.
This book provides excellent further coverage of the processes discussed in this chapter and has been very popular with students for many years.

Bigg, G.R. (2001) Back to basics: the oceans and their interaction with the atmosphere. *Weather*, 56, 296–304.
This short paper provides a straightforward introduction to the principal ocean–atmosphere interactions.

Houghton, J.T. (2004) *Global warming: The complete briefing*, 3rd edition. Cambridge University Press, Cambridge.
This book describes the uncertainties surrounding predictions of the effects of global warming including positive and negative feedbacks.

IPCC (2001) *Climate change 2001: The scientific basis*. Contribution of Working Group 1 to the Third Assessment Report of the Intergovernmental Panel on Climate Change (ed. Houghton, J.T., Ding, Y., Griggs, D.J., Noguer, M., van der Linden, P.J., Dai, X., Maskell, K. and Johnson, C.A.). Cambridge University Press, Cambridge.
This is a comprehensive, internationally renowned, report on climate change.

IPCC (2007a) *Climate change 2007: The physical science basis*. Contribution of Working Group 1 to the Fourth Assessment Report of the Intergovernmental Panel on Climate Change.

Solomon, S., Qin, D., Manning, M. *et al.* (eds). Cambridge University Press, Cambridge.
The latest assessment of the Intergovernmental Panel on Climate Change demonstrating the lastest scientific understanding.

IPCC (2007b) *Climate change 2007: Climate change impacts, adaptation and vulnerability*. Contribution of Working Group 2 to the Fourth Assessment Report of the Intergovernmental Panel on Climate Change. Adger, N. *et al.* (eds). Cambridge University Press, Cambridge.
The latest assessment of the Intergovernmental Panel on Climate Change predicting the potential magnitude of change and impacts of climate change.

Lockwood, J.G. (1999) Is potential evapotranspiration and its relationship with actual evapotranspiration sensitive to elevated atmospheric CO_2 levels? *Climatic Change*, 41, 193–212.
A paper which provides further information on Box 4.4.

Nicholls, N. (1991) The El Niño-Southern Oscillation and Australian vegetation. *Vegetatio*, 91, 23–36.
Useful further information about the case study outlined in Box 4.6.

Robinson, P.J. and Henderson-Sellers, A. (1999) *Contemporary climatology*. Pearson Education, Harlow.
The authors provide a good overview of climate basics with some excellent diagrams.

Stern, N. (2006) *The economics of climate change: The Stern Review*. Cambridge University Press, Cambridge.
The Stern Review is an independent, rigorous and comprehensive analysis of the economic aspects of climate change.

Web resources

American Meteorological Society

http://www.ametsoc.org/

The official website of the American Meteorological Society; provides a useful search engine of various online journal articles on all aspects of atmospheric science.

Arctic Climatology and Meteorology

http://nsidc.org/arcticmet/glossary/

An extensive and useful glossary of terms involved in atmospheric processes.

Climatic Research Unit

http://www.cru.uea.ac.uk

Home page of the Climatic Research Unit, University of East Anglia. The site provides information on its current research projects, updates on important climate and meteorological data and fact sheets on a variety of climate-related subjects.

The Hadley Centre for Climate Prediction and Research

http://www.metoffice.gov.uk/research/hadleycentre/index.html

The site offers an insight into the work carried out by the Hadley Centre (a division of the UK Meteorological Office) on climate systems research. Information and brief discussion are provided on the nature of the climate system, climate monitoring, climate modelling, climate predictions and the carbon cycle with links to related sites.

Intergovernmental Panel on Climate Change

http://www.ipcc.ch/

These are the IPCC pages. The latest reports contain a massive amount of evidence on atmospheric processes and anthropogenic impacts. There are also less technical reports for policymakers.

NASA: Visible Earth

http://visibleearth.nasa.gov/view_set.php?categoryID=108

A search directory of images, visualizations and animations of the Earth from above, particularly the climate system.

National Oceanic and Atmospheric Administration: Climate

http://www.noaa.gov/climate.html

Archived and current climate data are provided including graphical representation of various time series of climate data. Links to reports and publications on climate issues can also be found.

NOVA Online: El Niño Tracking

http://www.pbs.org/wgbh/nova/elnino/

Contains basic useful information on El Niño and La Niña events: their relationship to global climate, their formation, and the related climatic effects in different parts of the world, including the current El Niño situation. The site also provides several links to other El Niño sites.

Online meteorology guide

http://ww2010.atmos.uiuc.edu/(Gh)/guides/mtr/fw/home.rxml

A comprehensive collection of web-based modules provided by the Department of Atmospheric Sciences (DAS) at the University of Illinois Urbana–Champaign (UIUC). The modules cover the forces influencing atmospheric circulation (pressure gradients and the Coriolis force), processes of precipitation formation, weather forecasting and extreme weather events.

UK Met Office

http://www.metoffice.gov.uk

UK Meteorological Office home page containing world weather news, UK weather forecasts, educational material on meteorology and climate change, global and local satellite imagery and Met Office research operations.

University Corporation for Atmospheric Research: Introduction to the Atmosphere

http://www.ucar.edu/learn/

This site provides a brief clear overview of the scientific properties associated with the atmosphere, the climate system, the greenhouse effect and the ozone layer with plenty of diagrams.

 Visit our website at **www.pearsoned.co.uk/holden** for further questions, annotated weblinks, interactive models and video-clips showing physical processes in action.

CHAPTER 5

Global climate and weather

John McClatchey

Environmental Research Institute, North Highland College (a UHI Millennium Institute Academic Partner), and Honorary Research Fellow, University of Nottingham

Learning objectives

After reading this chapter you should be able to:

➤ understand how major features of the general atmospheric circulation play a role in climate

➤ describe the types of climates that exist across the Earth

➤ identify the type of weather associated with different types of climate

➤ recognize the importance of seasonality in different climates

➤ understand why there is a distinct geographical distribution of climates

➤ explain why there are differences between the climates of the northern and southern hemispheres

5.1 Introduction

The climate at any place is the most important environmental influence on the natural vegetation, the landscape and human activity. For the natural world, three components of climate are crucial, these being the precipitation, the thermal conditions and the wind conditions. For example, in desert climates the landscape is usually dominated by **aeolian** (wind-related) features, while in much of the middle latitudes, fluvial (water-related) features dominate the landscape. The vegetation cover is also strongly controlled by the thermal conditions and by the availability of moisture.

The climate at any one location arises from the types of weather that are experienced there over a period of years. However, the climate is not just about average values of weather elements such as temperature or precipitation but includes the range and extremes of those elements and the frequencies of types of weather. For example, in desert regions the average monthly precipitation total may be almost meaningless, as precipitation might occur only on one or two days a month. Of much more importance is the number of days on which precipitation falls and the amounts that fall in each event. Of course in all climates there can be substantial local and regional variations from the average conditions (see Chapter 6) but despite such local variations, the types of weather and the weather systems in each major climate regime will be similar. Therefore in order to understand why a location has a particular climate, the starting point should always be to understand what types of weather are a feature of that climate. This chapter therefore starts by outlining the general controls of weather conditions and climate across the Earth, before discussing the features of the major global climate zones.

5.2 General controls of global climates

The Sun drives both weather systems and the global climate and, as a result, the latitude of any location plays an important role in determining the climate. The decline of the solar radiation input that takes place in moving from the equator to higher latitudes eventually leads to a negative net radiation balance at the top of the atmosphere (the balance between the incoming solar and outgoing terrestrial radiation). This occurs at close to 45°N and 45°S and would lead to a cooling at higher latitudes if energy were not transported there from lower latitudes by ocean currents and the general circulation of the atmosphere. Therefore, although important, solar radiation alone does not control the climate. The distribution of the oceans and continents, ocean currents and the general circulation of the atmosphere also play an important role (see Chapter 4). Hence despite having a similar solar radiation input, places at the same latitude may have mild winters in one case and very cold winters in another. For example, Scotland has much milder winters than Labrador at the same latitude in north-east Canada. Climate is often assessed by using a network of observation stations that record meteorological and climatological data. There are a number of types of such records that can be used to assess climate and these are discussed in Box 5.1.

TECHNIQUES

METEOROLOGICAL AND CLIMATOLOGICAL OBSERVING NETWORKS

It is important to be aware of the availability and type of climatological records that can be used to assess climate. There are a number of different types of observing station but the three principal types are synoptic stations, climatological stations and precipitation stations. They are not spread uniformly across the planet. There are very few in oceanic locations or in less developed countries, deserts, polar regions and other harsh or inaccessible places. Such stations tend to be concentrated in developed countries.

Synoptic stations

Synoptic observing stations provide surface weather element observa-tions as an input into numerical weather forecasting models. The principal elements recorded are temperature, humidity, wind speed, wind direction and atmospheric pres-sure (and change over past 3 hours). These elements are recorded at all synoptic stations, including automatic stations. For stations at which meteor-ological observers are based, obser-vations of cloud cover, cloud heights, visibility, precipitation, current weather and past (over past 3 hours) weather are also recorded but recent developments in equipment now allow visibility and some cloud infor-mation to be recorded at some auto-matic weather stations (Figure 5.1).

Synoptic stations record weather elements at least every 6 hours at the same time at every station in the world (at 00:00, 06:00, 12:00 and 18:00Z, the Z denoting Universal Time, equivalent to Greenwich Mean Time) and many stations make hourly observations. These synoptic stations are located to provide measurements representative of a wide area. Many are at airports or military aviation bases and therefore the local area surrounding the station is generally flat and open.

Solar radiation is recorded at only a few stations, mostly at what are termed 'agrometeorological stations' (used to provide information specifi-cally for agriculture). At some of those stations both the direct and diffuse solar radiation are recorded and at a few stations there will also be an observation of the net radiation which is the balance between the short- and long-wave radiation (see Chapter 4).

Climatological stations

Climatological observations are generally taken at least once per day,

BOX 5.1 ➤

Figure 5.1 An automatic weather station. The blue solar panel is used to charge the battery supply.

although at some sites it is twice per day. The time at which observations are taken at an 'ordinary' climatological station is decided by each country (e.g. 09:00 GMT in the United Kingdom). All climatological stations make an observation of the maximum and minimum temperatures and precipitation total over the previous 24 h. The 'average' temperature is just the sum of the maximum and minimum temperatures divided by 2. Note that for synoptic stations, however, a more representative mean can be obtained from the hourly values. Some stations record soil temperatures (at various depths)

and some record a night 'grass' (surface) minimum temperature. Some non-instrumental observations of weather phenomena are also recorded.

There are far more climatological stations than synoptic stations (which also record climatological observations), as they are used to provide an indication of the variation of climate within a region (Chapter 6). Across the United Kingdom there is about one climatological station per 500 km², although the density is much less in the Scottish Highlands and much greater in lowland England. This is

a high density of stations compared with most countries. The number of stations required to show the local variations of climate within a region depends on the topography and the proximity to the sea or very large lakes (such as the North American Great Lakes).

Precipitation stations

In addition to climatological stations there are also a large number of precipitation stations at which only precipitation is recorded. In areas of frequent winter snowfall, special snow gauges are used to obtain a rainfall equivalent measure. The United Kingdom has around 5000 precipitation stations (about one per 30 km²), a high density compared with most countries.

World Meteorological Organization standards

Both synoptic and climatological stations have to conform to certain standards set out by the World Meteorological Organization (WMO). These ensure that observations taken at any place in the world are directly comparable. For example, thermometers are sited in a screen (Figure 5.2) at between 1 and 2 m above a grass surface (this may be a snow surface during winter). This screen has to be above the surface as at night there can be differences of over 10°C between the (colder) grass (surface) minimum and screen minimum temperatures as a result of surface radiational cooling. The height at which wind is recorded is also important as the friction of the surface slows the wind and causes the wind direction to change. The standard height for an anemometer (wind speed recorder) and wind vane is 10 m (although a range between 8 and 12 m is acceptable).

BOX 5.1 ➤

Figure 5.2 A Stevenson screen used for housing thermometers.

A 'climate normal' (average) is established over a 30 year period for a standard station updated every 10 years (e.g. 1951-1980; 1961-1990, 1971-2000), which is the 30 year average of the individual monthly values. Often, however, meteorological and climatological observations are made as part of micrometeorological, ecological, hydrological and even geomorphological investigations. Such measurements may not conform to WMO standards but may be appropriate in terms of the particular investigation. However, non-standard observations cannot be directly compared with standard measurements and therefore they need to be used with care if used in any local climate assessment.

BOX 5.1

A key factor determining global climate is the presence of zones of ascent or descent of air. In areas of the world subject to high-pressure anticyclonic conditions, air is subsiding (descending) and is being warmed by compression. That warming leads to the formation of a **temperature inversion** at a few hundred to perhaps 2000 m above the surface whereby warmer air overlies colder air. This prevents any further rise of the lower colder air because it is more dense and less buoyant than the overlying warm air. This inversion therefore inhibits the growth of clouds, and as clouds need to be deep to give substantial falls of precipitation, areas in which anticyclones are located tend to be dry. Conversely, in areas of the world with low pressure, air is ascending (as a result of convergence in the lower atmosphere) and clouds can often grow to considerable depths. This convergence can enhance thermal convection or encourage **slantwise convection** (movement of air upwards and north or south) if there are noticeable thermal contrasts between regions of air (e.g. at fronts).

A map of average surface pressure (Figure 5.3) therefore provides an initial indication of likely areas of ascent (relatively wetter areas) and descent (relatively drier areas). In addition, because the middle and high latitudes have an overall energy deficit (more radiation is lost to space at the top of the atmosphere than is gained from solar radiation) the weather systems help transport energy from lower to higher latitudes. Therefore the average position of the weather systems can give an indication of regions for which air is generally being brought into from lower latitudes. These regions will be milder than might otherwise be expected and the circulation of air around pressure systems means that these regions will be on the southern and eastern sides of low-pressure systems in the northern hemisphere (northern and eastern sides in the southern hemisphere).

The type of weather experienced at a particular location depends on latitude, the position and movement of pressure systems and the presence or absence of areas of ascent or descent (subsidence) of air. The thermal climate will depend on latitude (height of the Sun in the sky), the principal prevailing wind directions (from warmer or colder areas) and the cloudiness (thick clouds reduce daytime maximum temperatures but keep night-time minimum temperatures higher). Cloud and precipitation

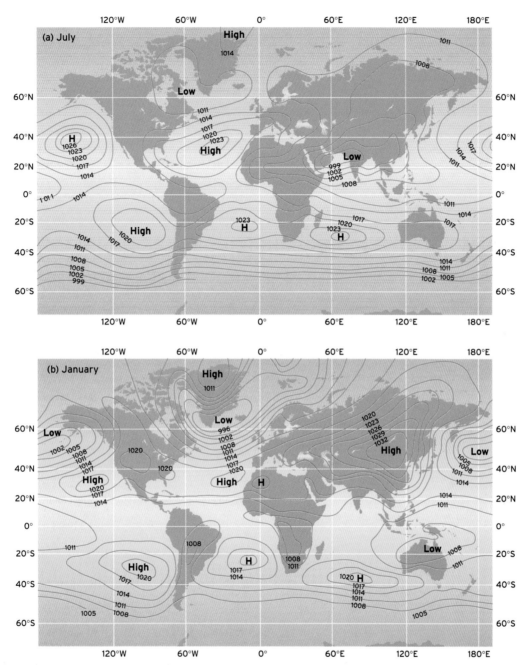

Figure 5.3 A map of average surface pressure over the Earth in (a) July and (b) January in hPa (or millibars). High-pressure areas represent zones of descending air (air at the surface is diverging away from that point) while low-pressure areas are zones of ascending air (air at the surface is converging towards that point). (Source: after White *et al.*, 1992, *Environmental Systems: An Introductory Text*, 2nd Edition, by D. N. Mottershead, S. J. Harrison & I. White, Fig 4.5, p. 84. Published by Taylor and Francis Books Ltd (Nelson Thornes) 1992. Reproduced by Permission of Taylor & Francis Books UK.)

depend on the presence or absence of ascending air (giving cloud) or subsidence (giving clear skies). Therefore, although the following discussion will be separated into latitudinal zones, there will be substantially different climates experienced within each latitudinal zone. A number of classifications of climate have been suggested,

with probably the most commonly used being Köppen's classification that was developed in the early twentieth century. The classification scheme was modified by Köppen's students (Geiger and Pohl, 1953) and is shown in Table 5.1. It is, however, important to realize that the boundaries between different types of climate are not

Table 5.1 Köppen's climate classification with additional modifications by Geiger and Pohl (1953)

1st	2nd	3rd	Basic description	Classification criteria[†]
	Letter code			
A			Humid tropical:	
	f		tropical wet (rainforest)	Wet all year
	w		tropical wet and dry (savanna)	Winter dry season
	m		tropical monsoon	Short dry season
B			Dry:	PET > ppn
				BS/BW boundary is one-half dry/humid boundary*
	S		semi-arid (steppe)	Mean annual $T \geq 18°C$
	W		arid (desert)	Mean annual $T < 18°C$
		h	hot and dry	Mean annual $T \geq 18°C$
		k	Cold and dry	Mean annual $T < 18°C$
C			Moist with mild winters:	Coolest month $T < 18°C$ and $> -3°C$
	w		dry winters	Wettest summer month 10 times rain of driest winter month
	s		dry summers	Driest month < 40 mm and wettest winter month $\geq$ 3 times driest month
	f		wet all year	Criteria for w or s not met
		a	summers long and hot	Warmest month > 22°C with more than 4 months > 10°C
		b	summers long and cool	All months below 22°C with more than 4 months > 10°C
		c	summers short and cool	All months < 22°C with 1 to 3 months > 10°C
D			Moist with cold winters:	Coldest month $\leq -3°C$ warmest month > 10°C
	w		dry winters	Same as Cw
	s		dry summers	Same as Cs
	f		wet all year	Same as Cf
		a	summers long and hot	Same as Cfa
		b	summers long and cool	Same as Cfb

Table 5.1 (*continued*)

1st	2nd	3rd	Basic description	Classification criteria[†]
		Letter code		
		c	summers short and cool	Same as Cfc
		d	as c with severe winters	Coldest month < 38°C
E			Polar climates:	Warmest month < 10°C
	T		tundra	Warmest month > 0°C but < 10°C
	F		ice cap	Warmest month ≤ 0°C

*ppn = annual precipitation; T = mean annual temperature; PET = potential evapotranspiration. Potential evapotranspiration greater than annual precipitation (PET > ppn); dry/humid boundary given by ppn = $2T + 28$ when 70% of rain in summer half-year; ppn = $2T$ when 70% of rain in winter half-year; ppn = $2T + 14$ when neither half-year has 70% of rain.

[†]Temperature limits given for warmest and coldest months in classification section are for average monthly temperatures.

(Source: Geiger and Pohl, 1953)

distinct and that climates merge into each other. Nevertheless a very general map of climatic zones is provided by Figure 5.4 based on Köppen's classification.

Reflective question

➤ What are the main factors controlling the location of climate zones?

5.3 The tropics and subtropics

5.3.1 Equatorial regions

Horizontal air movement arises from the pressure gradient force that is created if there is a pressure difference between different places. Over much of the Earth, once air moves as a result of a pressure gradient force, the Coriolis effect becomes apparent (see Chapter 4) and the winds deviate to the right in the northern hemisphere and to the left in the southern hemisphere. Close to the equator the Coriolis effect is negligible and therefore air follows the pressure gradient from high to low pressure. This is important as it means that near the equator, the weather is not dominated by the movement of large circulatory weather systems such as anticyclones and depressions. However, outflow from the large subtropical anticyclones (the north-east and south-east trade winds) does play an important role leading to

convergence of air into a region of somewhat lower pressure sometimes termed the equatorial trough.

This trough can be seen as a distinct intertropical convergence zone (ITCZ), particularly over the oceans, and is part of the Hadley cell circulation (see Chapter 4 for further details on the formation of the ITCZ). The ITCZ is not located along the equator but moves quite well north of the equator in some places in the northern hemisphere summer and a little south of the equator in the southern hemisphere summer, with an average position somewhat north of the equator (Figure 5.5). This movement is in response to seasonal differences in pressure gradients caused by changes in solar energy received during the year. The movement of the ITCZ north and south is greater over land because the seasonal cooling and warming are more pronounced over land (the oceans warm and cool at a much slower rate than land). The ITCZ is typically from 4° to 8°N (in March and September respectively) in the Pacific. There are, however, extensions of the ITCZ further south of the equator (Linacre and Geerts, 1997), particularly in the southern hemisphere summer. These are into southern Africa, east of the Andes in South America and a larger and more consistent extension called the South Pacific convergence zone, roughly south-south-west from the north-east of Australia (Figure 5.5).

Low-level convergence such as at the ITCZ leads to rising air and the formation of clouds. Therefore the weather close to the equator is dominated by frequent convectional clouds and rain. This convectional activity may be relatively random with many individual cumulus clouds being formed but there are also more organized cloud clusters. For example,

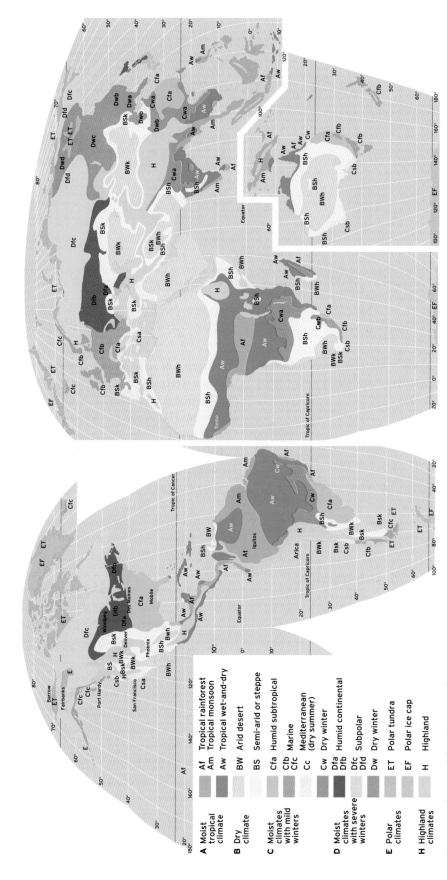

Figure 5.4 A map of the Köppen-Geiger-Pohl climate classification. (*Source: from Meteorology today: Introduction to weather, climate and the environment 6th edition by AHRENS, 2000. Reprinted with permission of Brooks/Cole, a division of Thomson Learning: www.thomsonrights.com, Fax 800-730-2215*)

Figure 5.5 Average position of the intertropical convergence zone in January and July. North-east trade winds and south-east trade winds converge at the ITCZ. Here solar heating and convergence of air results in ascension and instability. Thus precipitation rates are high. (Source: Critchfield, Howard J., *General Climatology*, 3rd edition, 3rd © 1974. Adapted (or electronically reproduced in case of e-use) by permission of Pearson Education, Inc., Upper Saddle River, NJ)

in some tropical coastal areas, more organized features sometimes occur diurnally as part of a land and sea breeze circulation (see Chapter 6). Equatorial climates are therefore dominated by the movement of the ITCZ with many places close to the equator having no distinct dry season. However, some equatorial areas show a clear peak in the monthly precipitation totals at each of the equinoxes with lower totals when the ITCZ is at its northern and southern limits (summer in the northern and southern hemisphere respectively). Total annual precipitation is high and in some parts of the Amazon and West Africa totals can be from 2500 to over 4000 mm.

As the Sun is always high in the sky, average temperatures in the equatorial regions are fairly constant throughout the year, although maximum temperatures may be a little lower during the wetter periods. Humid conditions keep night temperatures high, giving a relatively small diurnal range with average annual temperatures around 27°C. Typical monthly temperature and precipitation values for equatorial stations are given in Figure 5.6. Equatorial climates are not

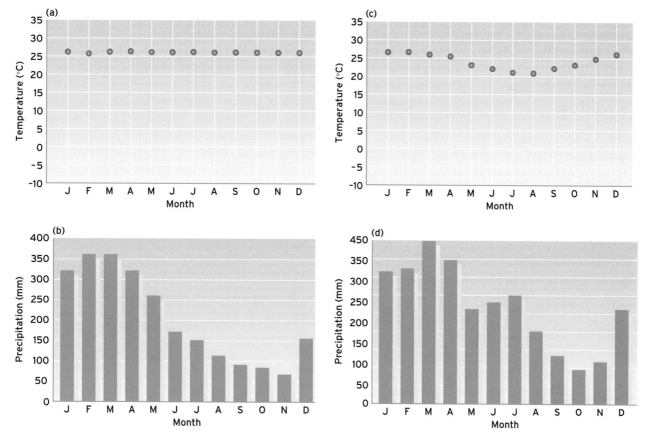

Figure 5.6 Example temperature and precipitation graphs for equatorial stations. Mean monthly temperatures (°C) and mean monthly rainfall (mm) for (a) and (b) Belem, Brazil (1.5°S), and (c) and (d) Tamavate, Madagascar (18°S).

exactly alike in every place, particularly over the continents of Africa and South America, as topography and proximity to the sea do play a role in altering thermal and precipitation regimes (see Chapter 6).

Moving north and south away from the equator into the trade wind belt, between about 5° and 20° latitude, the ITCZ still plays a key role in the climate but as the ITCZ is only close to these areas in the summer (in that hemisphere), there is now a distinct rainy (summer) and dry (winter) season. This greater seasonality gives a greater range of temperatures through the year, but still much less than in many middle and high latitudes, with the highest temperatures tending to be observed just before the rainy season. There can be quite large average diurnal ranges of temperature in places (15°C or more), particularly in the dry season. The trade winds also mean that there is a fairly steady but not particularly severe wind regime.

5.3.1.1 The easterly wave and tropical cyclones

One important type of more organized cloud is related to easterly waves in the trade winds. Such waves represent moving troughs of slightly lower pressure denoting zones of low-level convergence. This convergence is concentrated along the trough line resulting in a zone of heavier rain. These wave features (Figure 5.7) are important as they can develop into tropical depressions and tropical storms (Malkus, 1958) and some of those become tropical cyclones. These cyclones are called hurricanes in the Atlantic and typhoons in the west Pacific (Boxes 5.2 and 5.3). The WMO has produced a classification system that allows us to define when tropical depressions become severe enough to be called tropical storms or tropical cyclones (Table 5.2).

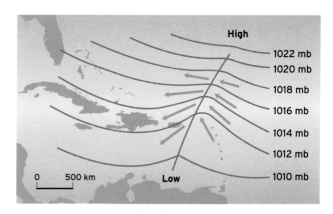

Figure 5.7 An easterly wave in the subtropical easterly flow. Easterly waves occur within the trade wind belt and represent waves of air (moving north and south) for air travelling in an easterly direction. (Source: after Critchfield, H.J. *General Climatology*, 3rd edition, 3rd © 1974. Adapted (or electronically reproduced in case of e-use) by permission of Pearson Education, Inc., Upper Saddle River, NJ)

Tropical cyclones require high sea surface temperatures (at least 27°C) and do not form over land as the energy of the system comes from the release of latent heat when the clouds are formed. The moisture evaporating from the sea is therefore a key factor in sustaining tropical cyclones and when that source of moisture is cut off (when hurricanes move inland) the intensity declines. Although tropical cyclones occasionally move inland, the majority are confined to the oceans. Even across the oceans the tracks of tropical cyclone systems show a distinct geographical distribution (Figure 5.8). Tropical cyclones do not occur close to the equator as there needs to be a sufficient Coriolis effect to help develop the circulation of the initial depression. As well as requiring high sea surface temperatures, the mid-troposphere has to be moist in order to allow tropical cyclone development. There must also be no strong temperature inversion inhibiting mixing of the

Table 5.2 World Meteorological Organization classification of tropical storms

	Wind speed		
	m s⁻¹	knots	Beaufort scale
Tropical depression	<17.2	<34	7
Moderate tropical storm	17.2-24.4	41-47	8-9
Severe tropical storm	24.5-32.6	48-63	10-11
Hurricane (tropical cyclone)	>32.6	>63	12

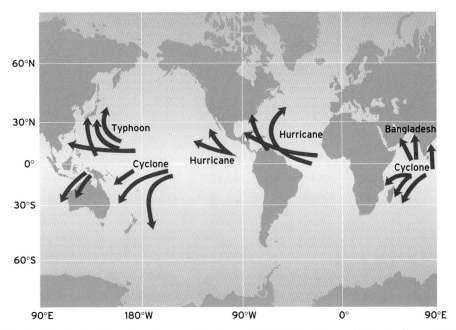

Figure 5.8 Typical tracks of tropical cyclones, hurricanes and typhoons. Note that hurricanes and typhoons are also tropical cyclones but that the names depend on their location (called hurricanes in the Atlantic, typhoons in the west Pacific). (Source: from *Meteorology today: Introduction to weather, climate and the environment* 6th edition by AHRENS, 2000. Reprinted with permission of Brooks/Cole, a division of Thomson Learning: www.thomsonrights.com, Fax 800-730-2215)

ENVIRONMENTAL CHANGE

ARE HURRICANES ON THE INCREASE?

The official hurricane season runs from 1 June to 30 November. In the US National Oceanic and Atmospheric Administration (NOAA) report on the 2005 Atlantic Hurricane season, a record 28 named storms were reported. Prior to this, the number of reported storms in any year since reliable records began in 1944 did not exceed 19 (although it is believed that there would have been 21 named storms in 1933 and there may have been other years before the 1940s when 20 or more storms might have occurred). Of the 28 named storms in 2005, 15 were hurricanes and 7 of

these were major hurricanes (categorized as 3 or more on the Saffir-Simpson scale - Table 5.3). In addition, 4 of those 7 major hurricanes reached category 5 status, the highest number of category 5 hurricanes in any year since records began. The 2005 season started early and as well as having the most costly and one of the most deadly hurricanes (Katrina) the season also had hurricane Wilma (another category 5 storm) in October which had the lowest central pressure ever recorded (882 hPa) for any Atlantic hurricane (the previous record was 888 hPa in hurricane Gilbert in 1988).

An alternative to counting the number of storms to determine how

stormy a year has been is to use the NOAA Accumulated Cyclone Energy Index. This is based on the cumulative strength and duration of each storm. Using that index 2005 was not the most active season, but the third most active season on record, behind 1950 and 1995.

Multi-decadal signal and the El Niño Southern Oscillation signal

Tropical cyclone activity in the Atlantic has been above normal since 1995. This has been largely in response to the active phase of the 'multi-decadal signal'. The multi-decadal signal relates to cycles of about 20-40 years in monsoon rains

BOX 5.2 ➤

Table 5.3 The Saffir–Simpson hurricane category scale

Hurricane category	Wind speed				Storm surge	
	mph	km h^{-1}	knots	m s^{-1}	ft	m
1	74-95	119-153	64-82	33-42	4-5	1.2-1.5
2	96-110	154-177	83-95	43-49	6-8	1.8-2.4
3	111-130	178-209	96-113	49.5-58	9-12	2.7-3.7
4	131-155	210-249	114-135	58.5-69	13-18	4.0-5.5
5	>155	>249	>135	>69	>18	>5.5

over West Africa and the Amazon Basin and in North Atlantic sea surface temperatures. The current phase of the cycle has involved lower **wind shear** (change of wind speed and/or direction with height) and warmer sea surface temperatures across the tropical Atlantic. In addition there have been weak low-level easterly (trade) winds and a westward expansion of upper-level easterly winds from Africa, the African Easterly Jet. These conditions are favourable to the development of hurricanes. This phase is expected to continue for the next decade or perhaps longer.

As well as the multi-decadal signal, El Niño Southern Oscillation (see Chapter 4) episodes occur roughly every three to five years, and generally last 9 to 15 months. El Niño events involve reduced upwelling and there-

fore lead to a warming of the surface ocean waters over the central equatorial Pacific. Strong upwelling gives La Niña events, cooling the surface waters. Changes in sea surface temperatures in this region alter the patterns of tropical convection across the central and east-central equatorial Pacific with warmer temperatures (El Niño) increasing convection and colder temperatures (La Niña) reducing convection (see Chapter 4). The La Niña episodes encourage upper atmosphere easterly winds and reduced wind shear at lower levels in the tropical Atlantic which as noted above are conditions that are favourable to Atlantic hurricanes. In contrast, El Niño events produce upper westerly winds and increased wind shear in the same region.

Although ENSO events can encourage (La Niña) or act against

the formation of Atlantic hurricanes (El Niño), the tropical multi-decadal signal is the dominant feature and can mask any influence of ENSO events. A multi-decadal signal favourable to hurricanes can be enhanced by a La Niña event and diminished by an El Niño. The influence of an El Niño was evident in 1997, 2002 and 2006, the only years with below-average hurricanes in the 11 years to 2006. However, the relationship is not simple, as the exceptional hurricane season of 2005 was not a La Niña year.

The average number of named storms per year since 1995 has been 13.0, compared with 8.6 during the preceding 25 years when the multi-decadal signal was in an inactive phase (Table 5.4). An average of 7.7 hurricanes and 3.6 major hurricanes per year since 1995 compares with 5

Table 5.4 Number of Atlantic hurricanes per year

	Average 1970-1994	Average 1995-2005	2005	2006
Storms named	8.6	13.0	28	9
Hurricanes	5.0	7.7	15	5
Major hurricanes	1.5	3.6	7	2

BOX 5.2 ➤

hurricanes and 1.5 major hurricanes per year between 1970 and 1994 (Table 5.4).

This box has indicated that there are many factors that combine to determine the numbers of hurricanes that form. However, the International Panel of Climate Change report has warned that global climate change will lead to more hurricanes over the next century (IPCC, 2007b). In fact they suggest that there is a greater than 66% chance that the intensity of hurricanes will also increase over the next few decades. However, as noted in Table 5.4, the number of hurricanes in 2006 was typical of the 1970-1994 average and was less than the average over the most recent period (1995-2005). Potential future increases in hurricanes and hurricane intensity and of typhoons in the Pacific must therefore be put into the context of multi-decadal and ENSO signals that can overlap and mask potential long-term change. There will be considerable variability in the frequency and intensity of tropical cyclones (hurricanes and typhoons) from year to year and it will therefore be difficult to discern a definite increase in either intensity or frequency of tropical cyclones for many years to come.

BOX 5.2

CASE STUDIES

HURRICANE KATRINA

In 2005, hurricane Katrina was the most costly in US history and also caused the highest number of deaths from a single hurricane since 1928. Tropical storm Katrina was designated on 24 August 2005 at which time it was located in the central Bahamas (Figure 5.9). Katrina began strengthening rapidly and became a Category 1 hurricane 24 km east-north-east of Fort Lauderdale at 17:00 EDT (21:00 UTC) on 25 August. At 18:30 EDT (22:30 UTC), the hurricane made landfall between Hallandale and North Miami Beaches with sustained winds estimated at 36 m s^{-1} gusting to over 40 m s^{-1}. Katrina moved south-west across the tip of the Florida Peninsula during the night but the landfall did little to reduce the intensity as the storm re-intensified as it moved back to sea over the warm waters of the Gulf. The sustained winds over Florida were never higher than 36 m s^{-1} but the heavy rain and gusty winds caused substantial damage and flooding, and 14 people lost their lives. By way of comparison, the 1987 October storm in southern England also had maximum sustained winds of around 36 m s^{-1}.

Katrina moved west after entering the Gulf of Mexico and then over the next few days gradually turned to the north-west and then north. The high sea surface temperatures and an upper-level anticyclone over the Gulf encouraged the rapid intensification, which led to Katrina attaining 'major hurricane' (Category 3) status on the afternoon of 26 August. Katrina continued to strengthen and by 07:00 CDT (12:00 UTC) on 28 August, hurricane Katrina reached Category 5 status with wind speeds of 72 m s^{-1} or more and a pressure of 908 hPa with the maximum sustained wind speeds of close to 78 m s^{-1} being reached at 10:00 CDT, remaining at that speed until the afternoon. At 16:00 CDT (21:00 UTC), Katrina's minimum central pressure dropped to 902 hPa, one of the lowest pressures ever recorded. At this time Katrina was at its peak strength with hurricane force winds extending outwards up to 168 km from its centre and tropical storm force winds (up to 33 m s^{-1}) extending outwards up to nearly 370 km. Sustained tropical storm force winds were already battering the south-east Louisiana coast and the 16:00 CDT (21:00 UTC) Bulletin from the National Hurricane Center warned of coastal storm surge flooding of 5.5 to 6.7 m above normal tide levels, locally as high as 8.5 m, and stated 'some levees in the Greater New Orleans area could be overtopped'.

At 04:00 CDT (09:00 UTC) on 29 August the hurricane's centre was 144 km south-south-east of New Orleans with winds of 67 m s^{-1} near the centre and gusts to hurricane force (33 m s^{-1}) along the coast. Just over 2 h later Katrina made landfall in Plaquemines Parish just south of Buras (between Grand Isle and the mouth of the Mississippi River) as a strong Category 3 hurricane (wind speeds about 57 m s^{-1} and a central pressure of 920 hPa). By 08:00 CDT (13:00 UTC), Katrina was

BOX 5.3 ➤

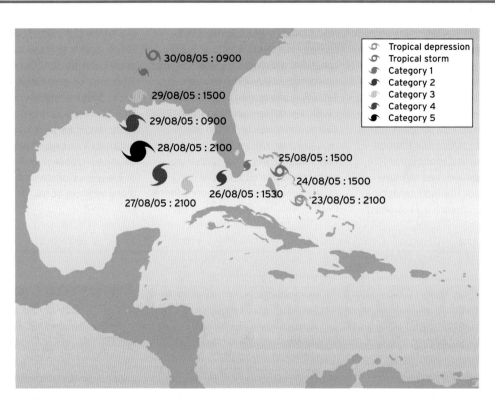

Figure 5.9 The storm track of hurricane Katrina.

only 64 km south-east of New Orleans with hurricane force winds extending up to 200 km from the centre of the storm. In the right front quadrant of the storm, which is where the strongest winds are generally found as there is an additive effect from the winds circulating round the hurricane with the direction of movement (Figure 5.10), Pascagoula Mississippi Civil Defense reported a wind gust to 53 m s^{-1} and Gulfport Emergency Operations Center reported sustained winds of 42 m s^{-1} with a gust to 45 m s^{-1}. By 10:00 CDT (15:00 UTC), the eye of Katrina was making its second northern Gulf coast landfall near the Louisiana-Mississippi border. The northern eyewall (Figure 5.10) was still reported to be very intense by WSR-88D radar data and the intensity was estimated to be near 54 m s^{-1}.

Katrina caused enormous damage to homes and businesses in both Louisiana and Mississippi estimated

Figure 5.10 A schematic diagram of a horizontal slice through the structure of a hurricane.

at around US$125 billion. The loss of human life was even more catastrophic with a death toll of 1833 with several hundred people still listed as missing (Graumann *et al.*, 2005). The majority of the deaths were in Louisiana (1577)

BOX 5.3 ➤

➤

and Mississippi (238) with 14 deaths in Florida and 2 each in Alabama and Georgia. This made Katrina the third deadliest hurricane since 1900, after the Galveston hurricane of 1900 (at least 8000 deaths) and the Lake Okeechobee Hurricane of 1928 (over 2500 deaths).

A detailed account of the hurricane with satellite images and a discussion of the historical perspective has been produced as a NOAA Technical Report (Graumann *et al.*, 2005) and much of the above summary of the progression of the hurricane is based upon that report.

While the wind damage caused by Katrina was significant, the bulk of the devastation was caused by flooding, largely due to the very substantial storm surge which peaked at 8.5 m at Pass Chritian, Mississippi. A surge of 7.3-8.5 m was estimated along the western Mississippi coast across a path of about 32 km. The surge was 5.2-6.7 m along the eastern Mississippi coast, 3.0-5.8 m along the Louisiana coast and 3.0-4.6 m along the Alabama coast.

A number of factors contributed to the extreme storm surge:

- the massive size of the storm;
- the strength of the system (Category 5) just prior to landfall;
- the 920 mb central pressure at landfall; and
- the shallow offshore waters.

In the delta country south-east of New Orleans, a number of towns were completely flooded with Plaquemines and St. Bernard Parishes particularly badly affected. The levee system protecting New Orleans was put under severe pressure due to the rise in the level of Lake Pontchartrain caused by the surge. As reported by (Graumann *et al.*, 2005) the damage and high-water marks indicate that the surge reached up to 19 km inland in some areas, especially along bays and rivers, and in New Orleans there were significant failures in the levee system on 30 August on the 17th Street Canal, Industrial Canal and London Avenue Canal levees. As much of New Orleans lies below sea level, the failure of the levees led to drainage of water into the city, leading to 80% of the city being underwater to depths of over 6 m (Figure 5.11). Graumann *et al.* (2005) also noted

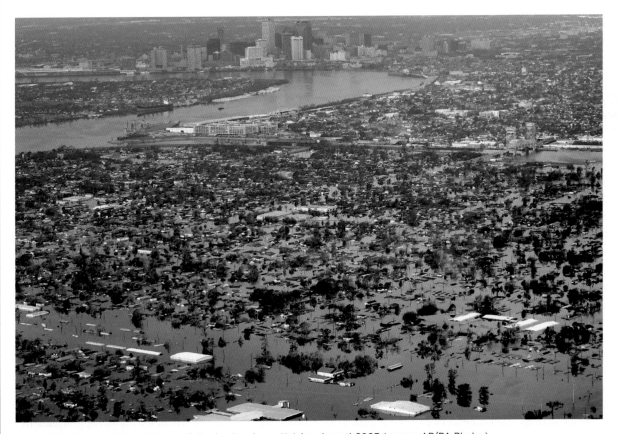

Figure 5.11 Flooding in New Orleans, following hurricane Katrina, August 2005 (source: AP/PA Photos).

BOX 5.3 ➤

that while much of the flood waters
had been cleared by 20 September,
the storm surge from hurricane Rita
on 23 September caused a new
breach in the repaired Industrial
Canal levee and many of the areas of
the city were flooded again.

Unfortunately, in some senses, this
was a disaster waiting to happen.
Much of New Orleans is below sea

level and the Mississippi Delta will
always be liable to flooding whether
due to river floods, extreme rainfall
events or storm surges. While levees
can be built up and strengthened,
deltas are notoriously unstable and
there is still the danger that levees
may be overtopped or undermined.
The disaster in New Orleans
suggests there is a need to adopt

the precautionary principle, as the
loss of life and the cost of damage
caused by Katrina provide a warning
about possible future events. If
global climate change increases the
intensity of hurricanes, the possibility of storm surges and extreme
rainfall events will increase, further
exacerbated by the projected rises in
sea level (IPCC, 2007b).

BOX 5.3

warm moist surface air with the mid-troposphere air. This
is why tropical cyclones are extremely rare in the South
Atlantic. In the South Atlantic there is a noticeable extension of the trade wind inversion from south-west Africa
towards Brazil and hence there is insufficient opportunity
for the build-up of a deeper moist layer. A strong temperature inversion also extends out into the Pacific from
northern Chile and Peru and the sea surface temperatures
are also relatively cool. This again inhibits the development
of tropical cyclones. However, the Pacific is much wider
than the Atlantic and therefore a deeper moist layer has
time to build up and as a result tropical cyclones are found
in the central and west Pacific south of the equator
(Brasher and Zheng, 1995). Figure 5.8 shows the tracks of
tropical cyclones in both the northern and southern hemispheres. It is also of note that there are fewer tropical
cyclones in the southern hemisphere as the summers are
somewhat cooler than those in the northern hemisphere
and therefore the area of the ocean with high sea surface
temperatures is smaller and confined to north of 20°S. In
the northern hemisphere, high sea surface temperatures of
27°C and above extend to 30°N.

Tropical cyclones give very high rainfall totals and
intensities. The rainfall falling over two to three days from
a single hurricane can even be close to the average annual
total in some places. The relatively low frequency of tropical storms, and the even lower frequency of tropical
cyclones at any individual location, mean that the standard
climate statistics do not provide any real indication of the
importance of these events in the climate of these areas.
They are high-magnitude, low-frequency events. In any
one year, two locations that have similar mean rainfall

totals may well have widely differing totals if one of the
locations had experienced a severe tropical storm or tropical cyclone.

5.3.1.2 Monsoons

One other important feature of the climate in the north-east
and south-east trade wind areas is that there are regions that
experience an exceptionally wet rainy season with a very
distinct dry season (Figure 5.12). These are the regions that
experience **monsoon** conditions (monsoon is from the
Arabic, meaning season). The largest and most intense
monsoon is in Asia but there are monsoon-type climates in
West and East Africa, and in Australia (Webster, 1981).
There is a much less well-defined monsoonal climate in
South America, partly as a result of the relatively cool sea
surface temperatures to the west of South America that
increase stability in the lower layers of the atmosphere.
The Andes also play a role as they obstruct the trade winds
and air descent in the lee of the mountains increases
atmospheric stability.

Monsoon regions are all subject to the switching of
the wind direction as the ITCZ moves north and south.
During the Asian winter, the winds are generally north-
easterly (although they may be north-westerly over
western India). At high levels in the troposphere (above
11 000 m) there is a distinct westerly jet stream (see
Chapter 4) located to the south of Tibet. This jet stream is
the southern and stronger branch of the subtropical westerly jet stream, the northern branch being located to the
north of Tibet. In spring, this southerly branch of the jet
stream weakens but remains south of Tibet while the

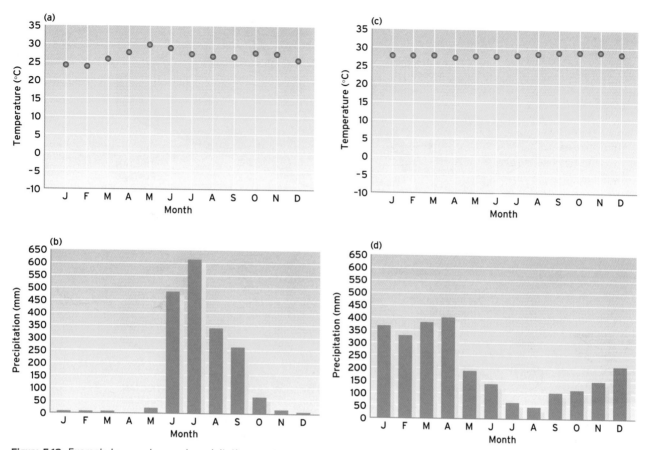

Figure 5.12 Example temperature and precipitation graphs for monsoon climate stations. Mean monthly temperatures (°C) and mean monthly rainfall (mm) for (a) and (b) Bombay, India (19°N), and (c) and (d) Manaus, Brazil (3°S).

northern branch strengthens and becomes extended. At the same time, the north of India is warming, with temperatures reaching a maximum in May. This warming creates a 'heat low' beginning a process that encourages the inflow of warm moist air from the south (Robinson and Henderson-Sellers, 1999).

In summer the upper-level westerly jet stream to the south of Tibet breaks down and then moves north across Tibet. As it moves across the mountains and the Tibetan Plateau, the high ground blocks the flow and the lower portion of the jet is deflected and re-established to the north. The strong convection over India (enhanced by the heat low) creates an outflow of air aloft and the southerly outflow develops into an easterly jet (under the influence of the Coriolis force). This upper air flow reversal from a westerly to an easterly jet is associated with the onset of the monsoon season in India and South-East Asia (Figure 5.13). As noted earlier, the south-east trades are located to the south of the ITCZ. In some parts of the world these winds

remain basically south-easterly as the ITCZ moves north during the northern hemisphere summer. Over India and Asia where the ITCZ moves much further from the equator than in most regions, the south-east trades move far enough north to become affected by the Coriolis force. As a result, they are deflected to the right and become south-westerly. The warm moist winds of the monsoon are therefore south-westerly. The return of the upper westerly jet takes place in October but the cessation of the monsoon rains is less distinct than their start. October and November also have the most frequent occurrences of tropical cyclones in the Bay of Bengal, and the rains from those cyclones give rise to a rainfall maximum at this time of year in south-east India.

Monsoon rains are not continuous throughout the monsoon summer period, as there are breaks between more active phases. In some places there is **orographic** enhancement of rainfall (forcing air to rise leading to further condensation, see Chapter 6) and the alignment of hills can also lead to increased low-level convergence

Figure 5.13 Upper air flow reversal and the onset of the Asian monsoon. (Source: after Robinson and Henderson-Sellers, 1999)

(creating zones of ascending air). In such places the rainfall can be exceptionally high, as for example in Assam (north-east India) where a number of places have exceptionally high rainfall totals (annual totals in excess of 10 000 mm). Cherrapunji, at 1340 m above sea level, is the most famous of the wet places. These exceptionally high totals compare with more typical values of between 1500 and 2000 mm close to the Bay of Bengal coast (~300 km to the south of Cherrapunji).

Despite such examples of orographic and convergence-induced enhancement of rainfall, the dominating influence on rainfall in monsoon regions is the large-scale circulation system creating the monsoon itself. The Asian monsoon therefore comes about as a complex interaction of the formation of the heat low, the changes in the upper air flow patterns, the movement of the ITCZ and the topographical barrier of the Himalayas and the Tibetan Plateau. It is also important to note that there is great variability from year to year. For example, El Niño years can be associated with the

failure of the monsoon rains, resulting in food shortages (Kumar *et al.*, 2006).

There is a less intense monsoon circulation over West Africa but the lack of a major mountain barrier allows a more steady movement of the ITCZ. However, as with Asia, the northward movement is sufficient to allow the Coriolis force to deflect the winds round to the south-west, bringing warm moist air in from the Atlantic Ocean. Northern Australia also has some monsoon rains but there is no topographical barrier to the south, and the land mass is not as large as in Asia and so the 'heat low' is not as intense.

5.3.2 The Sahel and desert margins

The influence of the poleward movement of the ITCZ declines further away from the equator. The effects of this are seen most clearly in the Sahel region of Africa that lies on the southern side of the Sahara Desert (Figure 5.14).

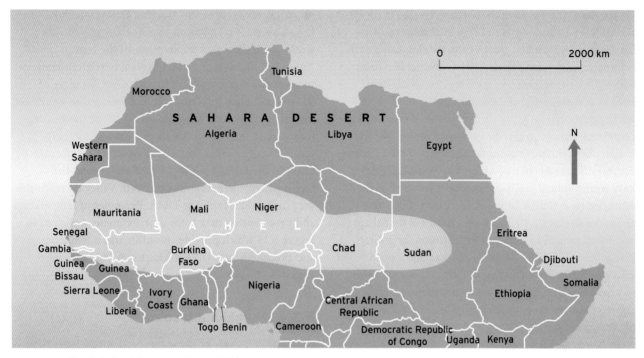

Figure 5.14 The Sahel and Sahara regions of Africa.

Most of the year the region is under the influence of the north-east trade winds blowing out of the subtropical anticyclone to the north. Temperatures are high during the day (possibly higher than 40°C in early summer) and there is a substantial diurnal range of temperature (10–15°C). The Sahel region still has a rainy season (when the ITCZ is at its furthest north), but the amounts are generally modest (300–600 mm) and in some years the rains fail to come (Nicholson *et al.*, 1996). During normal years, the rainfall is usually sufficient to sustain the vegetation and any grazing by animals. In dry years, however, there is an encroachment of the desert from the north (although that encroachment may be partly due to over-exploitation by people – see Chapter 16). There are several regions of the world with a similar climate to that of the Sahel, but they are not as extensive and the problems of drought are less severe. For example, in northern Argentina and south-east New South Wales in Australia the water supply is better because rivers flow into them from more humid areas.

5.3.3 Subtropical deserts

To understand the geographical pattern of these deserts, it is necessary to examine the positioning and intensity of the subtropical anticyclones that form the descending part of the Hadley cells north and south of the equator. These anticyclones are large features extending across the North and South Atlantic and Pacific Oceans centred at about 30°N and S. The intensity of the temperature inversions is greater on the eastern side of these anticyclones and the inversions are at lower altitudes: 300–500 m, as opposed to 1500–2000 m on the western extension of the anticyclones. In part this is due to stronger subsidence in the east but it is also due to the circulation of the ocean currents, with cool currents being present under the eastern end of these anticyclones. The lower sea surface temperatures in these currents help to increase atmospheric stability and reduce convection.

The driest hot deserts are therefore found in the western coastal regions of the continents where the subtropical anticyclones are most intense (see Figures 5.4 and 16.5). In the southern hemisphere they are found in Namibia in south-west Africa (Namib Desert), in western Australia (Great Sandy and Gibson Deserts) and in northern Chile (the Atacama Desert). In the northern hemisphere they are found in southern California (Sonoran and Mojave Deserts), in Africa (Sahara Desert), in Arabia (Arabian Desert) and in north-west India and southern Pakistan (Great Indian Desert). The extent of the hot desert region from the western Sahara, through to the Arabian Desert east of the Red Sea and then again in southern Iran, Pakistan and north-west India, is not mirrored elsewhere in the world, except in Australia.

This very large extent of the hot desert region in Africa, the Middle East and southern Pakistan arises as the large land mass allows a much greater eastward elongation of the subtropical anticyclone from the east Atlantic. This helps damp down convection over these regions. In addition, unlike the western side of the Pacific and Atlantic where warm ocean currents bring warm moist air to the east of the continents, warm currents do not move polewards in the Indian Ocean but move along the equator and then cross the equator off the east coast of Africa. The Australian deserts have greater annual precipitation totals than some of the other deserts at similar latitudes; the driest regions have annual totals close to 90 mm, while stations in the Sahara may have less than 15 mm. Part of the reason for greater precipitation is that the Australian anticyclone is not a constant feature but an average of individual anticyclones moving eastwards across the continent, allowing occasional inflow of moister air from the oceans to the north and south of the continent.

In North and South America there is only a relatively small area of desert due to major mountain barriers to the east: the southern Rockies and Mexican mountain ranges in the north and the Andes in the south. However, as the deserts in North and South America are situated to the lee of the mountains, the descent of the trade winds as they cross the mountains further dries the air, thereby intensifying the aridity. It is also important to note that variations in weather patterns over time can lead to extension or contraction of desert regions (Tucker et al., 1991).

The main features of the weather in most desert regions are the wind and the high daytime temperatures (typically over 45°C in the summer in Libya). The wind increases aridity and causes considerable aeolian erosion (by sand and other particles carried in the wind) in some places (see Chapters 12 and 16). The dry air and clear skies of the anticyclones give large diurnal ranges of temperature (as much as 20°C in some places). During winter, night temperatures can even drop below freezing in parts of these deserts.

In South America the northern parts of the desert are narrow and rainfall increases rapidly inland. This is due to the sea breeze circulation that can trigger thunderstorms if it reaches the edge of the Andes where forced ascent of the moist air can penetrate the temperature inversion and trigger the potential instability aloft. South of 10°S the desert widens and even the western Andes are dry. The aridity is most noticeable in the Atacama Desert in northern Chile. While it does rain on a few days per year in most deserts (even in the Sahara),

rainfall is very rare in the Atacama Desert which has the lowest annual precipitation totals of any place in the world (with the possible exception of the central Antarctic). Close to the coast, typical maximum temperatures in summer may be 25°C with diurnal ranges of 5–10°C and although temperatures do increase somewhat inland they are never extreme owing to the moderating influence of the ocean. The desert areas in southern California are similar, but inland from the Californian coast, in the lee of both the coastal ranges and the Sierra Nevada, temperatures in Death Valley do reach the very high values found in Libya. Examples of temperature and precipitation data from desert climates are given in Figure 5.15 which provides monthly mean temperature and precipitation totals for Khartoum (Sudan) and Baghdad (Iraq).

5.3.4 Humid subtropics

On the western side of the subtropical anticyclones there is a deep moist layer and convective activity is stronger than on the eastern side giving a higher likelihood of the development of rain clouds. A particular feature of this type of climate is the hot (often over 32°C), very humid summers associated with the tropical maritime air. These uncomfortable conditions can occasionally be interrupted if cooler air moves in from higher latitudes as part of the return flow of the Hadley or Ferrel circulation cells (see Chapter 4). However, if a ridge in the upper westerlies becomes established, very hot and humid conditions can last for weeks. The winters are generally mild, although there can be outbreaks of cold polar air into these regions. These unusually cold periods can cause major damage to sensitive crops such as citrus fruits and coffee beans (e.g. in Florida or south-east Brazil). Further into the continental interiors winters become more severe and at the same latitude winters in China are colder than winters in the United States. Annual precipitation totals are typically between 1100 and 1700 mm in this climate regime. Examples of areas with this type of climate include south-east Australia (Figure 5.16a and b), Taiwan and south-east Brazil, Paraguay, Uruguay and north-east Argentina in South America and the south-east states in the United States (e.g. Florida, Georgia, Alabama, Louisiana; Figure 5.16c and d), North and South Carolina. There is also a small zone of this type of climate in the east of southern Africa. The climate of eastern China is similar except that in central and southern China there is a winter minimum of precipitation (increasingly evident towards the north) rather than the more evenly spread rainfall in, for example,

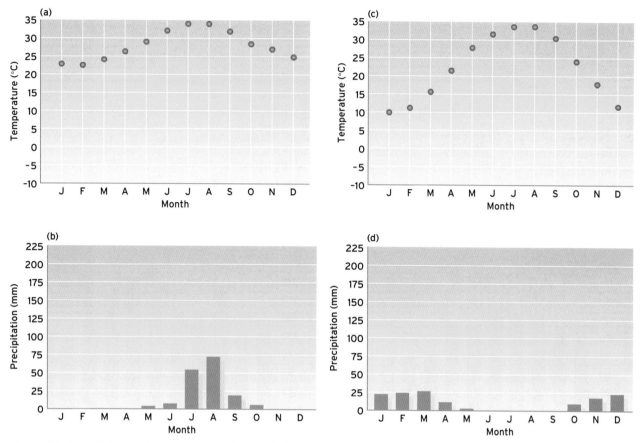

Figure 5.15 Example temperature and precipitation graphs for desert stations. Mean monthly temperatures (°C) and mean monthly rainfall (mm) for (a) and (b) Khartoum, Sudan (15.5°N), and (c) and (d) Baghdad, Iraq (33.5°N).

Georgia and Alabama. This winter minimum is a result of cool dry winds circulating around the winter Siberia high-pressure region. Areas with a winter precipitation minimum can also be found in Africa (e.g. east Zimbabwe) and South America (south-east Brazil and northern Paraguay).

Humid subtropical climates do not generally suffer from severe winds but the more coastal areas (in both the United States and China) can be hit by tropical cyclones (hurricanes) as they turn northwards and westwards. In Texas, Oklahoma and Kansas, tornadoes are another local-ized feature of the climate which is important. Here warm moist air from the Gulf of Mexico moves northwards inland and initially becomes trapped under a temperature inversion in the westerly winds aloft. If the temperature inversion is penetrated, substantial instability is released, leading to the growth of very substantial storm clouds, some of which will have associated tornadoes (Box 5.4). Although rare at any one place, these tornadoes are an important feature of the climate of this part of the United States.

Reflective questions

➤ What is the principal feature influencing the climate in equatorial regions and how does it affect temperature and rainfall?

➤ What are the conditions necessary for tropical cyclones to form and why are there not as many tropical cyclones in the southern hemisphere as the northern hemisphere?

➤ What factors make the Asian monsoon a much more marked feature than in other parts of the world at similar latitudes?

➤ Why are the deserts located where they are and what factors help increase aridity in many of these deserts?

➤ What is the dominant feature of summer weather in humid subtropical regions?

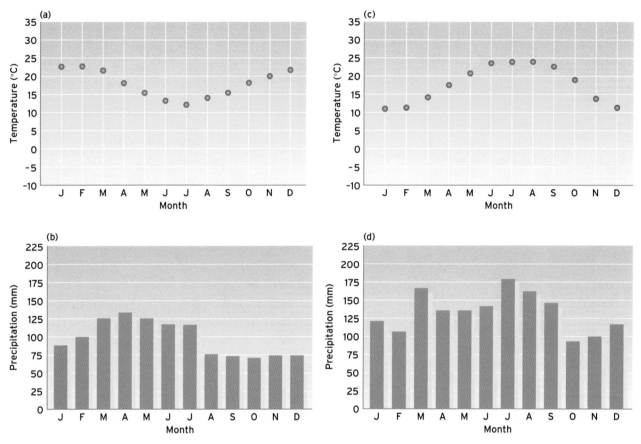

Figure 5.16 Example temperature and precipitation graphs for humid subtropical climate stations. Mean monthly temperatures (°C) and mean monthly rainfall (mm) for (a) and (b) Sydney, Australia (34°S), and (c) and (d) New Orleans, USA (30°N).

HAZARDS

TORNADOES

Tornadoes are violently rotating columns of air extending to the ground (Figure 5.17). They are capable of causing great damage with wind speeds of over 300 km h^{-1} (Figure 5.18). A tornado can be either very narrow and only a few metres across or very large with some over 500 m wide. They can often travel long distances, with some causing damage over 75 km. Tornadoes are often thought of as a phenomenon of the mid-west United States. However, they occur all over the world (e.g. Holden and Wright, 2004) and even in the United Kingdom the Tornado and Storm Research Organisation (TORRO) reports an average of 33 tornadoes per year (mainly in southern Britain, rarely in Northern Ireland or Scotland). Nevertheless, the largest ones tend to be concentrated in the Plains of the United States where atmospheric conditions often occur that suit their formation.

Tornadoes form when temperature and wind flow patterns in the atmosphere can cause enough moisture, instability, lift and wind shear for tornadoes to form in association

BOX 5.4 ➤

Figure 5.17 A tornado in Texas rampaging across fields while menacing a country road. Tornadoes form when two air masses of different temperatures and humidity meet. If the lower layers of the atmosphere are unstable, a strong upward movement of warmer air is formed. This starts to spiral as it rises, and intensifies. Only a small percentage of these systems develop into the narrow, violent funnels of tornadoes. Wind speeds can reach up to 400 km h^{-1} and they can damage an area 1 mile (1.6 km) wide and 50 miles (80 km) long. Tornadoes come in many shapes and sizes. (Source: © Warren Faidley/Weatherstock®.com)

Figure 5.18 Damage to homes on the end of a cul-de-sac demolished by a tornado in Grandview, Missouri in 2003. (Source: Ron Kuntz/Corbis)

with thunderstorms (Figure 5.19). The most destructive and deadly tornadoes occur from supercells which are rotating thunderstorms with a well-defined radar circulation called a mesocyclone. As well as tornadoes, supercells can also produce damaging hail and strong winds. All thunderstorms tend to produce lightning and heavy precipitation and in supercells the lightning is often more frequent and the precipitation can lead to flash floods. The rotating in the storms is due to wind shear which is when the wind direction changes and the wind speed increases with height. This kind of wind shear and instability usually exists only ahead of a cold front and depression system. The rotation of a tornado partly stems from updrafts and downdrafts caused by the unstable air interacting with the wind shear. Cyclonically flowing air which is already slowly spinning to the left (in the northern hemisphere) converges towards the centre of the thunderstorm, causing it to spin faster due to the conservation of angular momentum. This is a similar process to one you can try for yourself. If you sit on an office chair and spin round with your arms outstretched you spin slowly. If you then pull your arms towards your chest you will suddenly start spinning faster. It is this conservation of angular momentum that creates the very high wind speeds within tornadoes.

Most tornadoes rotate *cyclonically* (counterclockwise in the northern

BOX 5.4 ➤

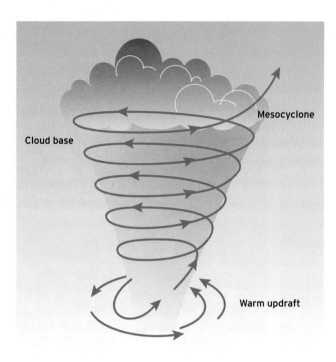

Figure 5.19 Formation of a tornado. Warm rising air meets cooler air aloft creating turbulence. These winds start to rotate because of wind shear which is when the wind direction changes and the wind speed increases with height. The larger mesocyclone which then develops aloft can generate sufficient strength to extend a funnel cloud down to the ground.

hemisphere and clockwise in the southern hemisphere) but the Coriolis force does not play any real part in the rotation as the size of tornadoes is too small for the Coriolis force to have any real effect. In fact there have been observations of anticyclonic rotating tornadoes, usually in the form of waterspouts (which are essentially tornadoes over water and also mainly have a cyclonic rotation), non-supercell land tornadoes, or anticyclonic whirls around the rim of a supercell's mesocyclone.

Tropical cyclones (hurricanes and typhoons) can also spawn tornadoes and while not all tropical cyclones that move across land will have tornadoes, some have major tornado outbreaks. The size of the tornado outbreak does not appear to be associated with the intensity of the tropical cyclone.

Tornado intensity is measured by the Fujita scale (F-scale named after Theodore Fujita), which was replaced in February 2007 by the enhanced Fujita scale (EF-scale) (Table 5.5).

Table 5.5 Enhanced F-scale for tornado damage (1 mph = 1.6 km h^{-1})

	Fujita scale		Derived EF-scale		Operational EF-scale[*]	
F number	Fastest 1/4-mile (mph)	3 second gust (mph)	EF number	3 second gust (mph)	EF number	3 second gust (mph)
0	40-72	45-78	0	65-85	0	65-85
1	73-112	79-117	1	86-109	1	86-110
2	113-157	118-161	2	110-137	2	111-135
3	158-207	162-209	3	138-167	3	136-165
4	208-260	210-261	4	168-199	4	166-200
5	261-318	262-317	5	200-234	5	Over 200

[*]Note that the EF-scale still is a set of wind estimates (not measurements) based on damage. It uses 3 s gusts estimated at the point of damage based on a judgement of eight levels of damage related to 28 indicators.

BOX 5.4

5.4 Mid- and high-latitude climates

In the tropics and subtropics the weather is often relatively predictable but the middle latitudes are dominated by weather systems that move across the planet. This makes the weather both much more variable and also much less predictable, particularly in areas close to the oceans. The middle latitudes of the southern hemisphere are dominated by oceans. The only land in the southern hemisphere middle latitudes is the southern tip of South Africa, the most southern parts of Australia and New Zealand, and central and southern Argentina and Chile. In the northern hemisphere, however, as well as the North Atlantic and Pacific Oceans, there are the major land masses of the North American, European and Central Asian continents. These span all of the middle latitudes and extend into the polar regions, and (in the case of North America and Asia) into the subtropics.

5.4.1 Depressions, fronts and anticyclones

The equatorial and subtropical climates are dominated by the weather systems associated with the thermally direct Hadley cell circulation (convection along the ITCZ and subsidence in the subtropical anticyclones) that transfers energy from lower latitudes. In the middle latitudes, however, there is also an energy transfer from lower latitudes to higher latitudes but this is not achieved by direct convection via heating at the surface. Instead the transfer is accomplished through the movements of large weather systems. However, just as in the tropics and subtropics, the key to understanding the development, movement and dissipation of mid-latitude weather systems is not what happens at the surface but what happens in the middle and upper troposphere.

The key factor is the positioning and movement of the westerly polar front jet stream. The jet stream is a 'thermal wind' related to sharp thermal gradients in the atmosphere. In the case of the polar front jet stream, the thermal gradient is created by the temperature difference between polar and tropical air where the two **air masses** meet (the polar front). A range of types of air masses exist and it is the interaction and modification of these air masses that may determine the weather conditions experienced in the middle and high latitudes. Further explanation and examples of air masses are discussed in Box 5.5. The polar front is just a steeper part of the normal low- to high-latitude temperature gradient (it is where there is a sharp transition between warm and cold air). The polar front jet stream is not fixed both in terms of its location as it moves north and south

and there are marked waves along its length which vary in amplitude with shallower waves giving what is called a zonal flow (west to east) and larger amplitude waves giving a meridional flow (a more marked north to south component as shown in Figure 5.20).

Air accelerates into anticyclonically curved waves creating a zone of upper-level divergence while the flow slows down as it enters a cyclonically curved wave creating upper-level convergence. If the upper-level divergence is greater than any low-level convergence this leads to a fall in surface pressure and creates a zone of ascending air. Such a situation is one of **cyclogenesis** (development of a depression). Conversely, if there is strong upper-level convergence and weaker lower-level divergence, there is a zone of

Zonal flow (generally west to east)

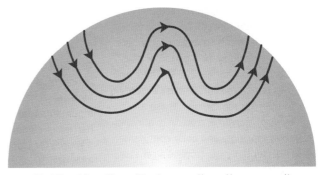

Meridional flow (flow with a large north-south component)

Cut-off high and cut-off lows (giving 'blocked' conditions)

Figure 5.20 Zonal and meridional flow.

subsiding air and this situation is one of **anticyclogenesis**. The thermal gradient in the upper westerlies is therefore of more significance in the development of weather systems than temperatures close to the surface.

Waves along the upper westerly polar front jet stream (Figure 5.21) can develop into frontal depressions. Air generally rises at fronts (by slantwise convection, as warm air is forced to rise above the cooler air it meets), leading to the formation of clouds (and precipitation) as the air cools on ascent. Fronts therefore mark areas of general precipitation (which may fall as snow in winter), although there are bands of more intense precipitation embedded within those areas. Some less active fronts may produce little or no precipitation.

L-Depression (low-pressure system with fronts)

Figure 5.21 Jet stream development. Waves along the jet stream can develop into frontal depressions.

Jet streams are not simple continuous features. They have marked entrances where the flow becomes more concentrated into a stronger jet. They also have exits

FUNDAMENTAL PRINCIPLES

AIR MASSES

The term 'air mass' is given to a body of air that has a very large horizontal extent and in which its potential temperature and moisture content are similar through most of the troposphere (close to the surface there may be differences). An air mass develops over a source region where it has remained for a period of days (Barry and Chorley, 2003). There are four basic types of air mass according to their source regions. These are tropical maritime, tropical continental, polar maritime and polar continental.

Essentially the tropical air masses are from low latitudes or the subtropics and polar air masses are from high latitudes. There are also extreme versions of polar air masses called Arctic maritime and Antarctic continental. Continental air masses are relatively dry and maritime air masses are relatively humid. The term 'relatively' has to be used as warm air holds much more moisture than cold air. A polar maritime air mass with 90% relative humidity holds 3.9 g of

water vapour per kg of air at 0°C while a tropical continental air mass with a 30% relative humidity has 8 g of water vapour per kg of air at 30°C.

Tropical maritime air is common in both hemispheres, but tropical continental air is less common as the only really large land mass in the subtropics is northern Africa and to a lesser extent in Australia. India is a source region for tropical continental air in winter but the intense winter high pressure over Siberia and the mountains to the north act as a barrier. In summer, central Asia can be a source for tropical continental air although strictly speaking it is in the middle latitudes. Polar maritime air is common in both hemispheres with source regions in the high-latitude oceans. As with all air masses, there is not a single set of characteristics defining this air mass as the values change according to the actual source region and the time of year. Polar continental air is found only in the northern hemisphere as there is no continent at higher latitudes in the southern hemisphere other than the

Antarctic where the air is classed as being Antarctic continental.

It is important to note that all air masses are modified by the underlying surface. If the surface is colder than the air mass then low-level stability will be increased. If it is warmer than the air mass, low-level stability will be decreased. The best example of this is Arctic maritime air moving south across the North Atlantic. The sea surface is much warmer than the air mass and this warms the lowest layers, decreasing stability and encouraging convection. This leads to the formation of frequent instability showers, a characteristic of this type of air mass. By studying source areas for air and tracking their movement over a particular region and predicting how the air masses might be modified by local conditions and how they might interact with other air masses, it is possible to provide short-range weather forecasts. Figure 5.22 provides an example of how air masses can affect the British weather.

BOX 5.5 ➤

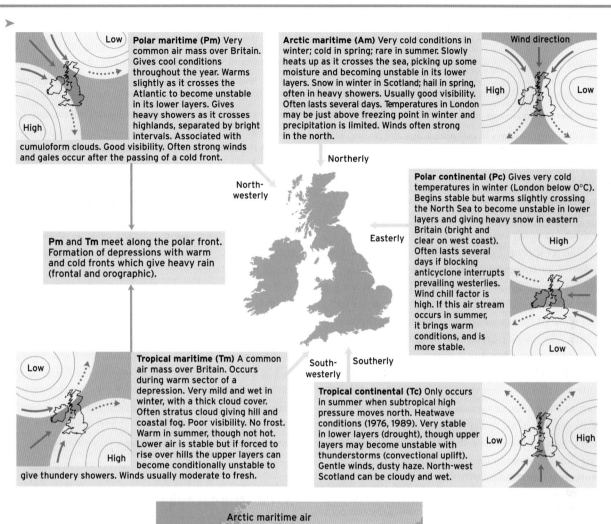

Polar maritime (Pm) Very common air mass over Britain. Gives cool conditions throughout the year. Warms slightly as it crosses the Atlantic to become unstable in its lower layers. Gives heavy showers as it crosses highlands, separated by bright intervals. Associated with cumuloform clouds. Good visibility. Often strong winds and gales occur after the passing of a cold front.

Arctic maritime (Am) Very cold conditions in winter; cold in spring; rare in summer. Slowly heats up as it crosses the sea, picking up some moisture and becoming unstable in its lower layers. Snow in winter in Scotland; hail in spring, often in heavy showers. Usually good visibility. Often lasts several days. Temperatures in London may be just above freezing point in winter and precipitation is limited. Winds often strong in the north.

Polar continental (Pc) Gives very cold temperatures in winter (London below 0°C). Begins stable but warms slightly crossing the North Sea to become unstable in lower layers and giving heavy snow in eastern Britain (bright and clear on west coast). Often lasts several days if blocking anticyclone interrupts prevailing westerlies. Wind chill factor is high. If this air stream occurs in summer, it brings warm conditions, and is more stable.

Pm and **Tm** meet along the polar front. Formation of depressions with warm and cold fronts which give heavy rain (frontal and orographic).

Tropical maritime (Tm) A common air mass over Britain. Occurs during warm sector of a depression. Very mild and wet in winter, with a thick cloud cover. Often stratus cloud giving hill and coastal fog. Poor visibility. No frost. Warm in summer, though not hot. Lower air is stable but if forced to rise over hills the upper layers can become conditionally unstable to give thundery showers. Winds usually moderate to fresh.

Tropical continental (Tc) Only occurs in summer when subtropical high pressure moves north. Heatwave conditions (1976, 1989). Very stable in lower layers (drought), though upper layers may become unstable with thunderstorms (convectional uplift). Gentle winds, dusty haze. North-west Scotland can be cloudy and wet.

Figure 5.22 Air masses affecting the British Isles. (Source: after Waugh, 1995)

BOX 5.5

139

CON - Convergence (leads to anticyclogenesis and descent of air)
DIV - Divergence (leads to cyclogenesis
    and ascent of air from below)

Figure 5.23 Jet stream entrance and exit, cyclogenesis and anticyclogenesis.

where the jet spreads out and the flow rate reduces. Associated with jet entrances and exits are marked zones where the formation of depressions and anticyclones (or anticyclonic ridges) is favoured through divergence or convergence of the upper winds (leading to cyclogenesis and anticyclogenesis respectively) (Figure 5.23). It is important to remember that if there is an upper-level outflow (divergence) the pressure will fall at the surface and air will therefore converge at lower levels and rise up to replace the ouflow. As a result there is a general (slow) upward motion of air in depressions in the conveyor belts moving through the system. In anticyclones or ridges the air flows inwards in the upper atmosphere, leading to an increase in pressure with descent and outflow at lower levels.

The descent of air in anticyclones increases the temperature (adiabatically) and dries out the air, which means anticyclones bring clear conditions. However, the descending air does not fall all the way to the land surface and due to the adiabatic warning a temperature inversion forms. While the air above may be clear, on some occasions the inversion can trap moister air below and a layer of cloud can form below the inversion, giving a condition termed 'anticyclonic gloom'. This is more common in winter than summer. This is because during summer solar radiation warms the air below the inversion sufficiently to cause the cloud to dissipate. On occasions the upper westerly flow can be interrupted by what is termed a blocking anticyclone generally formed by an intensification of a ridge in the upper westerlies into a closed circulation. This can lead to quite long periods of anticyclonic weather in areas where normally the weather would be characterized by the passage of mid-latitude depressions.

The ascent of air in depressions is concentrated to some extent along the warm and cold fronts. Once aloft, the air rotates to become more parallel with the upper-level flow (Figure 5.24). The ascent of the air is at perhaps 20 cm s^{-1}

compared with the 5–20 m s^{-1} that is typically found in large convective clouds. An ascent of air will eventually produce cloud as water vapour condenses out of the atmosphere in cooler conditions. The ascent of warm moist tropical air will quickly lead to the formation of layers of cloud (stratiform cloud) and continued ascent of air is likely to lead to precipitation. The ascent of air in depressions takes place over a much wider area and over a much greater time than is the case for a convective cloud, and therefore although precipitation rates (e.g. how heavy it is raining) may be less than those found with convective clouds, there can be substantial amounts of precipitation arising from a frontal depression. It is important to note that even within a frontal precipitation zone there will be areas of more intense precipitation so that fronts do not produce simple areas of steady precipitation.

Fronts tend to slope gently at a rate of 1 m vertical rise for every 80–150 m of lateral distance (slope of 1 : 80 to 1 : 150) with cold fronts being steeper than warm fronts. Over time the cold front tends to overtake the warm front, leading to what is termed an **occluded front** which is classified as warm or cold depending on whether the air ahead of the warm front is colder or warmer than the air following the cold front (Figure 5.25). The complex three-dimensional nature of depressions and fronts is important and is one reason why weather forecasting is a very complicated science.

5.4.2 Mid-latitude western continental margins

The Mediterranean-type climate is the first subtype of climate found polewards of the subtropical desert regions and is characterized by a mild wet winter half-year and a hot dry summer half-year. It is found in the far south-west of South Africa, in central Chile, on south-west-facing coastlines in the south of Australia, in California, as well as in the Mediterranean itself. In winter, mid-latitude frontal depressions bring rain to these areas, although in the Mediterranean itself, most of the depressions are not the frontal depressions of the Atlantic, the latter accounting for only 9% of Mediterranean depressions. A significant proportion of Mediterranean depressions develop as a result of dynamic effects on air flow over the Alps and Pyrenees that can lead to the formation of 'orographic' low-pressure areas (Barry and Chorley, 2003). These lows can develop frontal characteristics, particularly if the air flow across the mountains has a cold front embedded within it. This does not happen in the other Mediterranean climate zones of the world as they comprise only relatively narrow coastal areas. The

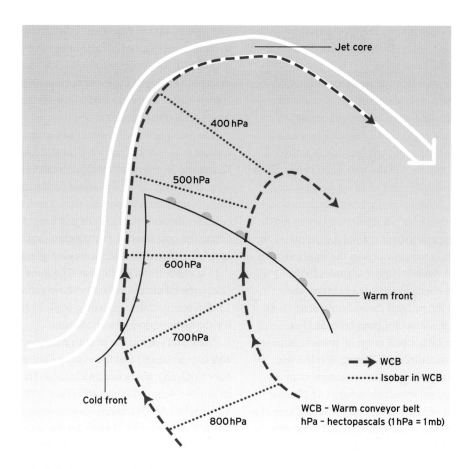

Figure 5.24 Conveyor belts in depressions. Slantwise convection (i.e. with a strong horizontal motion as well as conductive ascent) in the warm conveyor belt carries sensible and latent heat polewards.

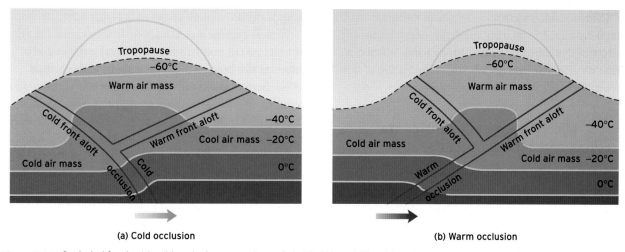

Figure 5.25 Occluded fronts: (a) cold occlusions are where air behind the cold front is colder than air ahead of the warm front; (b) warm occlusions are where air behind the cold front is less cold than air ahead of the warm front.

Mediterranean Sea provides the mechanism for extending the climate type much further into the continent.

Typically average winter temperatures in Mediterranean climates will be between 5 and 12°C with summer daytime maximum temperatures between 25 and 30°C. Rainfall totals will typically be between 400 and 750 mm with a distinct summer minimum. It is also of note that the summers in the Mediterranean climate zones of California and central Chile are drier than in the Mediterranean owing to the upwelling of cold water off those coasts. This stabilizes the air and inhibits convection.

Further polewards, mid-latitude western continental margins are most extensive in the northern hemisphere, although a similar climate regime is experienced by southern Chile, Tasmania and New Zealand. These climates have unusually mild winters for their latitude (e.g. the British Isles, western Europe and the west of Norway). These mild winters are particularly marked in the north-east Atlantic as the North Atlantic Drift pushes relatively warm water a long way north. There is a similar climate in much of New Zealand where again there is a warm current off the western coast. Generally temperatures in the southern hemisphere are lower than those at similar latitudes in the northern hemisphere owing to the large extent of the southern oceans and the paths of their ocean currents. As well as unusually mild winters for their latitude, these climates have a remarkably small range of annual temperature, have precipitation distributed throughout the year and there is considerable orographic enhancement of precipitation in the coastal mountain ranges (see Chapter 6). The mountain ranges of North and South America keep this climate confined to a relatively narrow coastal strip while in Europe the relatively low-lying ground from the Netherlands to Russia allows this climate type to extend to Poland in the east (Robinson and Henderson-Sellers, 1999). Average winter temperatures are typically between 2 and 8°C with average summer maximum temperatures between 15 and 25°C. Precipitation totals are generally in the range 500–1200 mm. The mid-latitude depressions that are a feature of this climate can bring strong winds which can cause considerable damage. Windiness is a feature of this type of climate, particularly in coastal areas.

5.4.3 Mid-latitude east continental margins and continental interiors

On the eastern side of North America and Asia, the eastern continental margin climates merge into the continental interior climates. Being within the mid-latitude westerly belt, the winds experienced on these continental margins have generally had a considerable passage across land. For that reason, the climate is more closely related to the continental interior than to the oceans, although some weather systems do come from the oceans to the east. Winters are much colder, with frequent snowfall, than those experienced in the subtropical humid climates to the south. These mid-latitude east continental margin climates do not exist in the southern

hemisphere as they require large land masses. Furthermore, in South America east of the Andes, where such a climate might exist, the Andes act as a block to the westerlies and descent in the lee of the mountains dries the air and creates a climate that is more like a mid-latitude continental interior (semi-arid) climate.

The main extent of the humid continental type of climate is in North America, China and eastern Russia. In China and North America this climate merges from the humid subtropical into a humid continental maritime margin, with increasingly severe winters, although at lower latitudes the summers are hot and long. As noted earlier, in China the cool winds circulating around the winter Siberian high pressure mean that there is less precipitation in winter than at other times of the year. The three winter months (December, January, February) have a total of close to 13 mm precipitation in Beijing while in New York the total for the same period is over 230 mm.

Summers are similar in both Asia (eastern China, Korea and central Japan) and the eastern United States (south of New England), being hot and humid. Average temperatures in July for Beijing, China (40°N) are 26°C and those in New York (41°N) 24.5°C. Mean winter temperatures are −4°C and 0°C respectively. In summer the June, July, August precipitation total is over 460 mm in Beijing while it is less than 320 mm in New York. Overall Beijing is drier (annual total close to 620 mm) than New York (over 1110 mm) owing to the aridity of Asia north and east of the Tibetan Plateau. Some examples of conditions experienced at mid-latitude climate stations are given in Figure 5.26.

Winters become increasingly colder as you move north or west into the mid-latitude continental interiors and summers also become milder and less humid (north of 45°N). Further north still (north of 50°N) the climate becomes subpolar with severe winters and relatively short summers (only three, or fewer, months with average temperatures above 10°C). For much of the mid-latitude continental interiors, precipitation is distributed throughout the year but generally with a distinct summer maximum. This summer precipitation is mainly in the form of convective showers although there are some weak frontal systems. Winter precipitation tends to fall as snow and as temperatures are low it often does not melt until the spring thaw. Total precipitation amounts are relatively low (below 500 mm), but the cold winter period and summer maximum of precipitation ensures that in most years there is sufficient moisture for plant growth. The main wheat-growing areas of North America have this type of climate. In Asia, however, the southern part of these mid-latitude continental interiors is

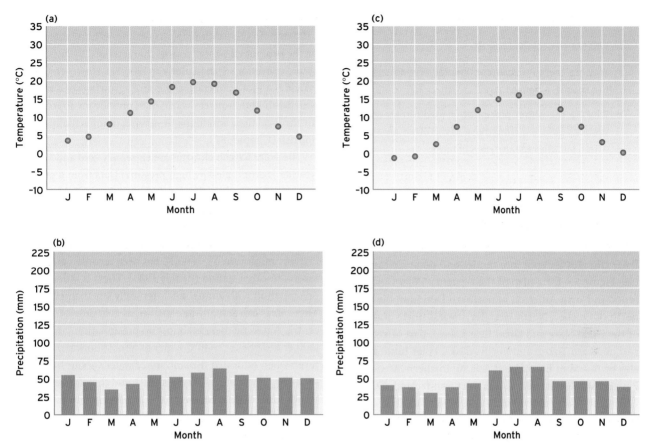

Figure 5.26 Example temperature and precipitation graphs for mid-latitude climate stations. Mean monthly temperatures (°C) and mean monthly rainfall (mm) for (a) and (b) Paris, France (49°N), and (c) and (d) Berlin, Germany (52.5°N).

semi-arid (as are those states in the United States just east of the Rockies). East of the Caspian Sea the climate becomes truly arid. These are cold desert regions in which winters are cold, although summers may still be warm. The Gobi Desert is an example of this type of desert.

The further north you go the shorter the summer and the growing season. Winters are cold with average temperatures below −12°C in the coldest month and below −25°C in northern regions. In the coldest regions of Siberia the average temperatures in the coldest month can even be as low as −50°C. There is a very large range in temperatures with the warmest months having average temperatures of over 21°C, and even in the coldest parts of Siberia, July average temperatures reach over 13°C (an annual range of over 60°C). In North America there can also be some extreme diurnal ranges in temperature, particularly in areas prone to Chinook winds (see Chapter 6) or if warm moist air from the south pushes much further north than usual. Diurnal changes in temperature have even exceeded 50°C. These regions are influenced by mid-latitude weather systems. However, in winter, high pressure dominates, especially in Siberia where pressures can reach over 1080 mb. Precipitation totals tend to fall as you move north within the mid- and high-latitude continents of the northern hemisphere, with totals below 400 mm in northern United States and southern Canada and falling below 300 mm further north. In eastern Siberia, annual totals can even be less than 150 mm.

Reflective questions

➤ Why are the waves in the upper westerlies important in encouraging the formation of depressions and anticyclones?

➤ Why are depressions associated with cloud formation and precipitation?

> ➤ What is an occluded front?
>
> ➤ Why are there no polar continental air masses in the southern hemisphere?
>
> ➤ Why are the winter precipitation Mediterranean systems different from those in the other areas with a Mediterranean-type climate?

5.5 Polar climates

Polar climates are split into tundra and polar ice cap types (see Figure 5.4 for a map of their extent). Polar tundra is found in North America, in northern Labrador, the far north of Quebec, North West Territories (north and east of the Great Slave and Great Bear Lakes), part of central and northern Yukon and north Alaska. Polar tundra can also be found in Europe and Asia in northern Sweden, Finland and Russia and also in northern Iceland. Coastal Greenland is also classed as having a polar tundra climate as is the northern part of Graham Land in the Antarctic. The polar ice caps are found in central Greenland and all of the Antarctic (except the northern part of Graham Land).

In the polar tundra climate the temperature of the warmest month will be above 0°C but below 10°C. Winter temperatures are generally extremely low (average temperature in January below −25°C), although in coastal Greenland winter temperatures are higher (−7°C in January). In North America and Asia the annual precipitation will typically be less than 300 mm and even below 120 mm in parts of north Alaska and northern Siberia, but in coastal Greenland annual totals are higher from 750 to over 1100 mm. In the tundra regions weather is dominated by the prolonged winter season characterized by anticyclonic conditions (particularly in Siberia). However, this provides a sharp contrast to the short growing season, which although not very warm does provide an opportunity for the local flora and fauna to survive, if not actually flourish (Chapter 8).

The polar ice cap climates are extremely cold and weather is dominated by high pressure. Data from a polar station at Ivigut, Greenland, are given in Figure 5.27. Summer temperatures are generally below 0°C and winter temperatures below −40°C. In parts of the Antarctic the mean annual temperature can be close to −50°C and an extreme minimum of −89.6°C has been recorded at the Vostok research station (21 July 1983). There is little precipitation with annual totals typically less than 100 mm. These areas can actually be classified as cold deserts. In the central Antarctic there is almost no precipitation. Air with a temperature below −40°C contains almost no water vapour (even when saturated the amount of water vapour held at such low temperatures is very small), and therefore even if clouds form there is unlikely to be any precipitation. Polar ice caps cool the air in contact with them and as a result there can be strong winds blowing off the centre of the ice caps towards the coasts. This climate type is therefore dominated by the cold and by the frequent strong winds that produce extreme wind chill.

Reflective question

> ➤ Why does Antarctica tend to be windy even though it is dominated by high-pressure conditions?

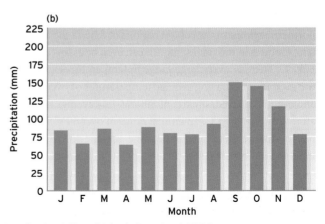

Figure 5.27 Example temperature and precipitation graphs for a polar climate station at Ivigut, Greenland (61°N).

5.6 Summary

This chapter has shown that the climate of any region is a result of the type and frequency of the weather systems found in that region. Zones of ascending and descending air such as the ITCZ (ascending), the subtropical anticyclones (descending) and the slantwise convection at the polar front (ascending) play an important role in climate. Equatorial climates are dominated by movements of the ITCZ whereas at higher latitudes in the tropics the easterly wave, monsoonal conditions and tropical cyclones may be more important, although it is still the movement of the ITCZ that partly controls these. Thus, some regions may have a fairly constant climate but are subject to occasional extreme events that have a substantial impact (e.g. regions prone to tropical cyclones). At still higher latitudes in the tropics, desert conditions are prevalent associated with the anticyclonic conditions related to the descending limb of the Hadley cell.

The distribution of land masses and oceans play an important role in global climates. Thus the southern hemisphere experiences different mid- and high-latitude climate conditions to the northern hemisphere owing to the lack of land masses within these regions. Air masses are important in the middle and high latitudes. They are distinguished by the source area from which they originate (continental or maritime, polar or tropical). The boundaries between air masses are known as fronts and represent sharp contrasts in temperature and moisture contents of the air. At fronts the warm air rises above the cooler air, often resulting in condensation and precipitation. Continental interiors tend to have different climates from those close to oceans even at the same latitude.

Individual climate elements and their changes through the year vary in their importance between different climates. For example, the distribution of precipitation during the year as well as annual precipitation totals is important because a region may not be arid if there is a winter rainfall maximum, even if it has a fairly low annual total and a hot summer. In some regions the heat and humidity of the summer are the dominating features of the climate while in others it may be precipitation totals or winter cold. There are also some climate types where there may be large differences between individual locations in one or more climate element (this is discussed further in Chapter 6). Finally it is important to recognize that climate types do not have distinct boundaries, as unless there is a major mountain barrier one type of climate usually merges gradually into another.

Further reading

Ahrens, C.D. (2003) *Meteorology today – An introduction to weather, climate, and the environment.* Brooks/Cole, Pacific Grove, CA.
This is an American textbook which provides a good clear overview and is very nicely illustrated with colourful figures. There are lots of reflective and essay-style questions and a useful interactive CD.

Barry, R.G. and Chorley, R.J. (2003) *Atmosphere, weather and climate.* Routledge, London.
This book contains useful chapters on air masses, fronts and depressions and on climates of temperate and tropical zones. It has been a very popular book over the years and is now in its eighth edition.

O'Hare, G., Sweeney, J. and Wilby, R. (2005) *Weather, Climate and Climate Change.* Prentice Hall, Harlow.
Excellent and accessible introduction to the area.

Linacre, E. and Geerts, B. (1997) *Climates and weather explained.* Routledge, London.
The presentation is good on the general principles and there is a large amount of material on winds at different scales.

McIlveen, J.F.R. (1998) *Fundamentals of weather and climate.* Stanley Thornes, Cheltenham.
This textbook goes into great depth and there is a lot of science (equations!). This should suit a range of interests, but those keen to get into real detail should look at this text.

Robinson, P.J. and Henderson-Sellers, A. (1999) *Contemporary climatology.* Pearson Education, Harlow.
This book contains good sections on tropical and mid-latitude climates.

Strangeways, I.C. (2006) *Precipitation: theory, measurement and distribution.* Cambridge University Press, Cambridge.
The book deals with precipitation formation as well as measurement methods.

Web resources

Earth Space Research Group: Indian Ocean Monsoon
http://www.crseo.ucsb.edu/esrg/IOM2/Start2_IOM.html
A good site describing what a monsoon is, the physical mechanisms of its formation and the uniqueness of the Indian monsoon (written by the Institute for Computational Earth Systems Science).

The Encyclopedia of the Atmospheric Environment: Climate
http://www.ace.mmu.ac.uk/eae/english.html
Very useful site that clearly and simply explains the nature of the global climate system (general circulation, ENSO events, wind belts, moons, etc.) and all the major global climate regions. It has a very clear and easy-to-use format.

Global Climate
http://apollo.lsc.vsc.edu/classes/met130/notes/chapter19/index.html
Lyndon State College Meteorology Department website provides an online textbook. This chapter on global climate offers classification and description of the different global climate types (including temperatures, precipitation rates, general characteristics and spatial distribution). It is a clear and accessible site, with appropriate diagrams.

Hurricane Intensity
http://www.magazine.noaa.gov/stories/mag184.htm
Article on the increase in hurricane intensity and natural variability.

Hurricane Katrina
http://cimss.ssec.wisc.edu/tropic/archive/2005/storms/katrina/katrina.html
This site contains movies, storm track details and other information about Katrina.

World Meteorological Organization Home Page
http://www.wmo.ch
Information on the many climate observation programmes undertaken by the World Meteorological Organization with links to other sites concerned with monitoring global climates.

Visit our website at **www.pearsoned.co.uk/holden** for further questions, annotated weblinks, interactive models and video-clips showing physical processes in action.

CHAPTER 6

Regional and local climates

John McClatchey

Environmental Research Institute, North Highland College (a UHI Millennium Institute Academic Partner) and Honorary Research Fellow, University of Nottingham

Learning objectives

After reading this chapter you should be able to:

➤ **understand how local factors can modify regional climate**

➤ **understand how altitude and topography control local and regional climates**

➤ **describe how large water bodies influence local and regional climates**

➤ **recognize how human activity can have a deliberate or inadvertent impact on local climate**

6.1 Introduction

The broad global climate zones discussed in Chapter 5 contain within them substantial variation at the local and regional scales. In some places the variation is part of a relatively gradual change from one climate type to another. For example, in the mid-west states of the United States, moving north from Tennessee through Kentucky to Illinois there is a steady change in climate (Figure 6.1). The whole of this large area north of the southern half of Missouri, Illinois, and Indiana and north of the whole of Ohio (i.e. the coverage of Figure 6.1) is within the humid continental

climate type. However, there are more gradual regional variations. Tennessee has a humid subtropical climate but from Illinois to the north into Canada the climate is described as humid continental (see Chapter 5). In moving northwards the winters become increasingly severe. Northern Iowa has summers that are noticeably cooler than the hot humid summers experienced in the south of Missouri. This cooling of the summers continues into Canada. North of Winnipeg the summers become much shorter with less than four months having temperatures above 10°C. Iowa, Wisconsin and Michigan are within the same climate subtype with at least four summer months with temperatures above 10°C. However, even within this area, there are marked regional variations in climate on top of the gradual change northwards. The most obvious of these differences are associated with the areas bordering the Great Lakes of Michigan and Superior.

In any climate region there can be very marked local variations in certain climate elements. For example, over a distance of little more than 100 km across Scotland there are places in the west where the total annual precipitation is nearly 10 times greater than that in parts of the east. However, on a global or even European scale the whole of Scotland falls well within the limits of a single climate type. The classification or description of a climate at a particular place therefore depends on the scale at which that climate

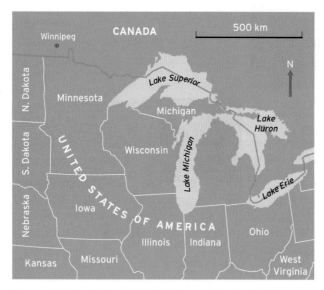

Figure 6.1 The mid-west United States and the Great Lakes.

is being considered. The smaller the geographical extent of the area of interest, the more important it becomes to have detailed climatological observations and statistics (see Box 5.1 in Chapter 5) in order to specify the climate of an individual location. It is also important to note that regional and local climates are a feature of the land and not the oceans because altitude, topography and proximity to the sea (or large bodies of water) are the key features that lead to the development of local and regional climates. These factors and associated processes will be discussed in this chapter as it is important to understand how both local and global processes interact at different scales. This interaction helps to explain the nature of local and regional climate across the Earth's surface.

6.2 Altitude and topography

The climate of mountains has always been of interest to scientists. Early studies helped establish that pressure and temperature fell with height and in the latter part of the nineteenth century a number of mountain observatories were established in Europe and North America to support astronomical studies and weather forecasting. Examples include Mount Washington (New Hampshire, established 1870), Sonnblick (Austria, 1886) and Ben Nevis (Scotland, 1883). Some of these mountain observatories closed after 10–20 years of observations (e.g. Ben Nevis, 1883–1904) but many are still in existence. The rate of fall of temperature with altitude (the **lapse rate**, see Box 6.1) varies in different parts of the world. The amount of solar radiation that can potentially be received by the ground surface actually increases with height. This is because less radiation has been absorbed or reflected by components of the atmosphere back into space. The lower the altitude, the thicker the layer of atmosphere that can reflect solar radiation back into space. However, whether solar radiation received at the ground surface actually increases with height in any particular mountain range depends upon local cloudiness. Harding (1979) reported that in the mountains of the United Kingdom, which tend to be cloudy, solar radiation decreased by 2.5 to 3 million J m^{-2} day^{-1} km^{-1} which was regarded as typical by Grace and Unsworth (1988). Nevertheless, even in those areas where received solar radiation increases with altitude, temperature is still likely to decline upwards because of adiabatic processes (see Table 6.1). In many places wind speed and precipitation increase with altitude but this is not true everywhere. Substantial mountain ranges can also act as a barrier to the movement of weather

Table 6.1 Predicted change of pressure and temperature with height based on a lapse rate of 6.5°C per m (pressure *P* in hPa or millibars and temperature *T* in °C)

Height (m) (above mean sea level)	Tropical		Mid-latitude		High-latitude	
	P	*T*	*P*	*T*	*P*	*T*
0	1013	25.0	1013	15.0	1013	5.0
1000	902	18.5	899	8.5	894	−1.5
2000	801	12.0	795	2.0	788	−8.0
3000	710	5.5	701	−4.5	691	−14.5
4000	627	−1.0	616	−11.0	605	−21.0
5000	552	−8.5	541	−17.5	527	−27.5

FUNDAMENTAL PRINCIPLES

LAPSE RATES

The rate at which temperature falls with increasing altitude is known as the **environmental lapse rate**. An air parcel will rise if it is warmer than the surrounding environment. Once the air parcel reaches the same temperature as the surrounding environment it will stop rising (Figure 6.2). When air rises (ascends) it expands. This is because the air pressure decreases. Conversely if air sinks (descends) it is compressed as the pressure increases. If no energy is added to or lost from that air as it rises (or falls), the changes in pressure and temperature are the result of what is termed an adiabatic process. When air rises and the pressure falls, the energy for the expansion of the air comes from the air itself. As temperature is a measure of the energy of air, if energy is removed by expansion of the air then the temperature of the air decreases. The reverse is true if air sinks and is compressed, in which case the temperature of the air increases.

If air expands adiabatically as it ascends the temperature of the air falls at a constant rate of 9.8°C km^{-1} (Figure 6.2). If air is compressed adiabatically as it sinks the temperature of the air increases at the same rate. This rate of temperature change is called the dry adiabatic lapse rate. The term 'dry' gives a clue as to why this is not the only rate of change of temperature with height, as it applies only if the atmosphere remains unsaturated. Air can hold only a certain amount of water vapour at any temperature and if the temperature falls or if more water is evaporated into the atmosphere then once saturation is reached water vapour will condense out of the atmosphere. When water condenses out of the air, energy is released. This energy is the latent heat of vaporization and is the energy released when gas water vapour condenses into liquid water. When water evaporates or ice melts, this uses up energy in order to change the state of the water but without changing the temperature of the water (see Chapter 4). When the reverse occurs energy is released. This energy release reduces the cooling rate of the air as it rises and expands. In the absence of any loss of total water content (gaseous, liquid or solid water) from the air, the rate of change of temperature with height is given by what is termed the saturated adiabatic lapse rate.

However, unlike the dry adiabatic lapse rate, the saturated adiabatic lapse rate depends on the amount of water vapour the air can hold at any temperature. As can be seen in Table 6.2 the amount of water vapour air can hold more than doubles for each 10°C increase in temperature. This means that the amount of latent heat released is much less at low temperatures than at high temperatures as less water vapour will condense out of the air at lower temperatures. The saturated adiabatic lapse rate therefore varies from around 0.3°C per 100 m close to the surface in the topics, where air temperatures are over 30°C, to close to the dry adiabatic rate at temperatures below −40°C. Temperatures of below −40°C are normally found at heights of between 5 and 10 km in mid-latitude regions.

When saturated air ascends, the temperature decreases at the

Figure 6.2 Air parcel buoyancy. A parcel of air warmed at ground level will rise if it is warmer than its surroundings and will cool at the environmental lapse rate (ELR). Note that top of the cloud occurs where the air parcel temperature is the same as its surroundings. DALR – dry adiabatic lapse rate (9.8°C/1000 m); SALR – saturated environmental lapse rate.

BOX 6.1 ➤

Table 6.2 Water vapour saturation vapour pressures

Temperature (°C)	−40	−30	−20	−10	0	10	20	30
Water vapour pressure (hPa)	0.19	0.51	1.25	2.86	6.11	12.27	23.37	42.43

saturated adiabatic lapse rate and the water that condenses out of the air forms a cloud. However, when saturated air descends, it will warm at the saturated adiabatic rate only if all of the liquid or solid water (cloud droplets or ice crystals) is re-evaporated back into the air. If any of the water has fallen out of the cloud as precipitation there will be less water to evaporate and the air will therefore warm more on descent than it cooled on ascent.

The difference between the dry and saturated adiabatic lapse rates is crucial in understanding why some atmospheric conditions are unstable (or **conditionally unstable**) (Figure 6.3). This instability occurs because if rising air becomes saturated, any further ascent will cause the air to cool at the saturated adiabatic rather than dry adiabatic lapse rate. Such saturated air will therefore be warmer than the surrounding air and as warmer air is less dense than colder air, the saturated

air will ascend further as a result of this density difference. This leads to strong upward convection and the growth of shower clouds. A situation is described as conditionally unstable if the environmental lapse rate is steeper than the saturated adiabatic lapse rate through the lower atmosphere (around 10 km). If clouds form and grow into this layer with a steep lapse rate they are then going to continue to grow to great depths. The instability is 'conditional' as clouds have to form and reach the height at which the cloud temperatures become warmer than the surrounding air. Although adiabatic processes are common in the atmosphere, particularly when there is widespread ascent or descent of air such as in frontal ascent or anticyclonic subsidence, energy can be gained or lost from air by a number of processes including the loss of energy associated with precipitation that falls out of the atmosphere.

Radiative exchanges can also be important in the atmosphere. While air is largely transparent to solar radiation, certain atmospheric gases absorb terrestrial (long-wave) infrared radiation (see Chapter 4). Liquid or solid water completely absorbs and reradiates long-wave radiation. This means that clouds play an important role in radiative exchanges. Radiational losses at the top of clouds can cause localized cooling. Radiational heating from the ground surface is also important. Strong heating can create steep lapse rates in the lowest layers of the atmosphere. These steep lapse rates can be greater than even the dry adiabatic lapse rate and are termed **super-adiabatic**. Such steep lapse rates cause rapid local convection which tends to mix the atmosphere. This effect is greatest in summer and at lower latitudes.

At night when long-wave infrared radiational emission from the surface

Figure 6.3 Atmospheric stability relationships between the ELR, DALR and the SALR: (a) stable; (b) conditional instability; (c) absolute instability.

BOX 6.1 ➤

cools the ground, the lowest few hundred metres of the atmosphere can be cooled as a result. This creates a temperature inversion perhaps a few hundred metres deep. This is where warmer air overlies cooler air so that it is no longer the case that temperature declines with altitude. Once warmer air overlies cooler air, the lower layer is trapped because it is denser. Thus, pollutants from fossil fuel combustion (e.g. fires, car engines) may not be able to escape from the lower air layer and thus a long-lasting 'smog' can develop. During these times public health can be at severe risk.

However, while adiabatic lapse rates are important in the atmosphere, other processes, such as mixing of air, affect temperatures. If rising air mixes with its surroundings its energy will be shared with the surroundings and the temperature changes will no longer be adiabatic.

BOX 6.1

systems and even smaller ranges of mountains and hills can give rise to noticeable differences in the weather (and hence in climate) on the lee side of those mountains. Hills and mountains are therefore important as they can create substantial regional and local modifications to the general climate of that part of the world. The following discussion will explain how individual climate elements are modified by hills and mountains, and how a specific regional climate may be developed on the leeward side of upland areas.

6.2.1 Pressure

The fall of air pressure with height is the most consistent feature of mountain climate. Up to around 3000 m the fall in pressure is close to 10 mb per 100 m. The rate of fall is more rapid in colder (denser) air and therefore pressures are higher at the same altitude in the tropics compared with the middle and high latitudes (Table 6.1).

6.2.2 Temperature

The change of temperature with altitude is known as the lapse rate. Box 6.1 describes the important characteristics of lapse rates and should be read in order to understand fully the following section. The values in Table 6.1 give an indication of the differences in the fall of pressure with altitude in tropical, mid-latitude and high-latitude regions, but they are based on an assumed lapse rate of 6.5°C per 1000 m. However, there can be substantial diurnal variations in lapse rates and these will influence altitudinal pressure gradients. During the day strong solar heating warms the air close to the ground, steepening the lapse rate in the first few hundred metres of the atmosphere. In the first few tens of metres the lapse rate can exceed the dry adiabatic lapse rate of 9.8°C per 1000 m (see Box 6.1), although the lapse rate will depend on how well mixed the atmosphere is. The stronger the wind, the greater the turbulent mixing and the closer the lapse rate will be to the dry adiabatic rate (as long as there is no condensation of water vapour). In cloudy conditions with strong winds, the atmosphere is so well mixed in the lowest few hundred metres that below the cloud base the lapse rate will approximate the dry adiabatic lapse rate. Within the cloud the lapse rate will be at the saturated adiabatic rate. At night, light winds and clear skies allow the ground to cool through long-wave radiation loss and a strong temperature inversion can form close to the ground (Figure 6.4). Katabatic drainage can further strengthen the night-time inversions (see Box 6.4 below). These diurnal variations in the fall of temperature with height are influenced by cloud cover and the presence or absence of vigorous mixing caused by strong winds. Together these play an important role in controlling temperatures of mountain regions.

As well as diurnal variations, the lapse rates also vary according to air mass. As air masses move they can be warmed or cooled by the underlying surface, leading to a steeper or a shallower lapse rate respectively. Hence Arctic or Antarctic maritime air masses, which are always warmed from below as they move into lower latitudes, have the steepest lapse rates, which are often close to the dry adiabatic lapse rate. Tropical air masses, however, are cooled from below as they move away from lower latitudes. This reduces the lapse rate. The affect of warming and cooling from below is illustrated in Figure 6.5 which shows an air mass being cooled as it moves across a cool lake and one being warmed as it moves from a cold lake to a warm land area. Temperature inversions are temporarily formed as the air is warmed or cooled from below. Polar maritime air may also have steep lapse rates if there is a large contrast between the air and sea surface temperatures. As the contrast between the air and sea surface temperatures varies with the seasons, there are some seasonal differences in lapse rates. The average lapse rates in mountains of the British Isles are steeper than in most mountainous regions as a result of the

Figure 6.4 Temperature inversion in the early morning. Cooler air lies below warmer air and a mist has formed close to the ground surface in the valley. (Source: photo courtesy of Alona Armstrong)

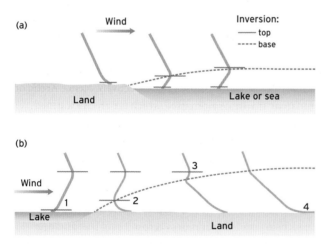

Figure 6.5 Temperature inversion: (a) the modification of an unstable temperature profile to give a surface-based inversion over a cold surface (e.g. lake/ice); (b) elevated inversion due to the advection of stable lake air across the shoreline to a warmer land area on a spring afternoon. (Source: after Oke, 1987)

frequency of polar and Arctic air masses and the relatively warm surrounding sea surface temperatures (Harding, 1978). Air with steep lapse rates is conditionally unstable (see Box 6.1) and instability showers are therefore common. Air with a shallow lapse rate is stable, damping down convective activity. However, such air can be subject to forced ascent over mountains or dynamically induced ascent such as in a mid-latitude depression. If the ascent is sufficient to cool the air to its dew point temperature, clouds will form and this can lead to enhancement of precipitation. Tropical continental air moving towards higher latitudes is cooled from below but in summer that cooling may only extend through a relatively shallow layer. Tropical continental air is dry and is formed over an area in which there was strong heating from below. Therefore above the layer of air that is cooled from below, the lapse rate will be much steeper. This means that there can be strong conditional instability. Convection that rises above the cooled layer can trigger the rapid formation of deep convectional clouds.

Anticyclonic conditions are also important in determining temperature variations with altitude. The large-scale descent of air in anticyclones creates a dry adiabatic lapse rate aloft, above a strong temperature inversion. For example, in the trade wind belt flowing out of the subtropical anticyclones, there is a marked temperature inversion,

Approximately 10 km

Figure 6.6 Lapse rate under anticyclonic conditions with a subsidence inversion.

Figure 6.7 Acceleration of winds over ridges and mountain tops where the air is compressed by a temperature inversion aloft. Because the space for the air to pass through is narrower then air is forced through at greater speeds.

generally at a few hundred metres above the surface on the eastern side of the anticyclone rising to 2000 m in the west. Mountain ranges such as the Atlas in Africa, the Andes in South America and the Sierra Nevada in California often penetrate well above these subtropical inversions. This leads to distinct changes in temperatures and therefore local climate on ascending such mountain ranges.

Subsidence inversions as shown in Figure 6.6 are a feature of all anticyclones. Air is forced to descend under the anticyclonic conditions but part of this descending air may be warmer than the air below it. Thus the temperature might vary with altitude as shown in Figure 6.6. Depending on the height of the inversion, mountain ranges in any part of the world may be above the height of the inversion base. This will lead to unusual conditions on those mountains with different lapse rates above and below the inversion. There may also be a very low humidity. The descending air in anticyclones leads to exceptionally low values of humidity close to the base of the inversion. However, as anticyclones found in the middle and high latitudes are much more transient than the subtropical anticyclones, on some days there will be a sharp change in the weather conditions experienced on ascending the mountains. Such changes will be incorporated into the averages of the climate elements.

6.2.3 Wind

As well as changes in temperature with height, the wind regime in mountains can be quite different from that at lower levels. It is not, however, altitude that is necessarily the key factor. It is the topography itself that is important. For example, the wind can be funnelled through valleys or even gaps between individual peaks, particularly if the orientation is in the direction of the prevailing wind. This will result in much greater local wind speeds. In addition,

individual peaks and exposed ridges will experience higher winds as there will be less surface friction acting to reduce wind speeds as the wind approaches. Friction reduces surface winds by about 30% compared with the 'free atmosphere'. In some circumstances, the compression of air between mountain summits and a temperature inversion aloft can lead to wind speeds above free atmosphere values (Figure 6.7). This frequently occurs around the Cairngorm Mountains in Scotland and at Mount Washington in New Hampshire where maximum gusts have been recorded at 76 and 103 m s^{-1} respectively. The wind regimes in mid-latitude mountains are also influenced by the westerly winds. The westerly winds are generally faster aloft. However, in the tropical and subtropical trade wind belts, the north-east and south-east trade winds generally weaken with height. Therefore mean wind speeds can be low on tropical and subtropical mountains. For example, typical wind speeds are 2 m s^{-1} during the period December to February at 4250 m in New Guinea and an annual mean of 5 m s^{-1} at 4760 m in Peru. In the Himalayas the monsoon circulation gives strong westerly winds through the winter half-year (October–May) with more moderate easterly winds in the summer (June–September). The westerly winds decrease from over 25 m s^{-1} at 9 km in the winter half-year to only 10 m s^{-1} by the end of May (being replaced by easterlies in the second half of June). These wind speed changes in the Himalayas emphasize the importance of the weather systems experienced in mountain regions in determining the wind regime.

One other local and regional climate feature of winds is a warm and dry wind that blows down lee slopes of hill and mountain ranges. These winds tend to warm and dry as a result of the compression and adiabatic warming of the air in the lee of the hills and mountains as it descends from higher levels. It is called a **Föhn wind**, although it has other names in different parts of the world such as in Canada where it is called the **Chinook**. The onset of the wind is typically accompanied by a sharp rise in temperature often with a substantial decrease in relative humidity. In Canada, in the lee of the Rockies, temperature rises of over 20°C have been recorded in just a few minutes. Evidence of smaller rapid rises in temperature has been found in Scotland and even with winds across the Pennines in England, where the hills are typically only 500–700 m high (Lockwood, 1962). Föhn winds in the Alps and other mountainous regions can cause rapid snow melt, greatly increasing avalanche risk and flooding (Barry, 1992).

It has already been noted that surface cooling can lead to katabatic drainage into valley bottoms. Over glaciers and particularly over the Antarctic ice sheets more substantial katabatic winds can form due to local cooling (e.g. Renfrew and Anderson, 2002). These winds, which can be extreme, will flow into hollows and valley bottoms and are a special case of mountain winds (see Box 6.4 below).

6.2.4 Precipitation

The amount of moisture the air can hold is strongly dependent on temperature and as temperatures fall with height, generally the moisture content of air does so too. It might therefore be expected that precipitation would also decrease with height as the moisture content declines. However, in the lowest 3000 m of the atmosphere this is certainly not the case. Air forced to rise over mountains cools at the dry adiabatic lapse rate until the dew point temperature is reached. At this point clouds form and temperature then decreases at the saturated adiabatic lapse rate if there is any further ascent of the air. Therefore, while temperature is reduced by adiabatic expansion as the pressure falls, the moisture content of the air does not change until saturation is reached. Hence, even in a very dry region, if air is forced to rise, it will eventually reach saturation. For example, if dry air (say 30% relative humidity) at 30°C is forced to rise, it will become saturated at about 2000 m and cloud will form. Unless there is no wind, the forced ascent of air over hills will therefore provide a supply of moisture from lower levels, even though there may be a decrease of vapour pressure with height in the free atmosphere. This forced ascent of air means that in moist airstreams, clouds will form over relatively low hills. Even in dry airstreams clouds will form if mountains are sufficiently high (Figure 6.8). The mere formation of clouds

Figure 6.8 Cloud formation over mountains at Applecross, in western Scotland. As the air is forced to rise over the mountain it expands adiabatically and saturation of the air occurs aloft. The water vapour can then condense to form clouds. (Source: Barnabas Kindersley © Dorling Kindersley)

will not, however, lead to precipitation. Precipitation has to be initiated. This occurs either through the formation of ice crystals in the upper parts of a cloud which then fall through the cloud leading to aggregation of crystals and production of supercooled water (**Bergeron process**), or by coalescence of smaller droplets onto larger droplets falling more quickly through the cloud. In a convective cloud, the maximum rate of precipitation will be close to the cloud base, as once rain falls out of the cloud the raindrops begin to evaporate. If there are strong updrafts even the raindrops may be transported upwards and, if that is happening, the zone of maximum precipitation may be above the cloud base (Figure 6.9).

Thunderstorms can often occur when the atmosphere is very unstable, producing very rapid falls of precipitation. Box 6.2 provides more details of these hazardous features of the atmosphere.

In the tropics and subtropics, precipitation is often as a result of convective activity and therefore the highest rainfall totals are found at typically between 1000 and 1500 m, at or just above the average cloud base. This is very common in the trade wind belts where the air above the trade wind inversion is very dry. For example, rainfall on Mauna Loa in Hawaii is over 5500 mm at 700 m but only 440 mm on the summit at 3298 m, well above the inversion (Barry, 1992). In the moister equatorial regions, rainfall generally tends to decrease with height.

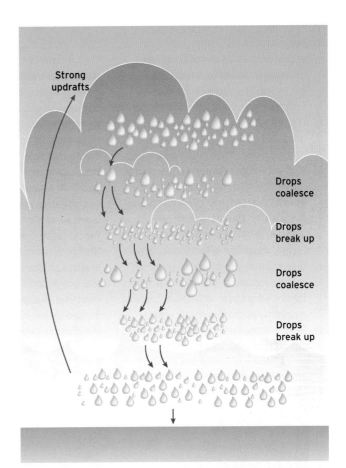

Figure 6.9 Raindrop formation where there are strong updrafts.

HAZARDS

THUNDERSTORMS

Thunderstorms form when significant condensation of water vapour occurs, resulting in the production of many water droplets and ice crystals. This happens when the atmosphere is in an unstable condition that supports fast upward motion. Although thunderstorms often happen during warm weather when heating of the ground surface causes sufficient moisture to accumulate in the lower atmosphere, and the warm surface causes there to be a steep adiabatic lapse rate, what is required for thunderstorm formation

is an unstable atmosphere (see Box 6.1) through a considerable depth of the atmosphere (possibly right up to the tropopause). The unstable atmosphere means that considerable energy is released as water condenses out of the atmosphere to form the deep cumulonimbus clouds that characterize thunderstorms. That energy can range from the equivalent of a small nuclear bomb (say 10 kt of TNT equivalent) in a small thunderstorm to over 100 times more in a severe thunderstorm. Thunderstorms occur anywhere in the world but are most

frequent in the tropics where they can be an almost daily occurrence. Thunderstorms are more common in summer in the mid-latitudes, although winter thunderstorms can occur, due to low-level convergence along a cold front, while at high latitudes thunderstorms are fairly rare, and form only in the summer. At high latitudes in winter, the air is so cold at the surface and throughout the atmosphere that there is insufficient moisture in the air to provide enough energy for a thunderstorm. Thunderstorms are usually accompanied by heavy rainfall, often

BOX 6.2 ➤

with strong winds and possibly hail. Thunder is caused by the explosive expansion of a narrow column of air which is heated by a lightning discharge (Figure 6.10). Therefore lightning precedes all thunder. Large cumulonimbus clouds form, often extending to great heights during thunderstorms, although the storm cloud is not normally larger than a few kilometres in diameter. Thunderstorms can be single or multicellular and a series may form a squall line with an associated gust front of strong winds. Supercell storms are severe storms characterized by wind shear with height creating a rotating updraft or mesocyclone. Severe tornadoes are associated with supercell storms (see Box 5.4).

Lightning occurs when a large charge is built up within a cloud and is then discharged. In tall cumulonimbus clouds electrical charge is built up as water droplets, hail and ice crystals collide with one another in the strong air movement (a bit like when you rub a balloon on your jumper you can create a charge that, when you put the balloon near your hair, makes your hair stand on end). The positive and negative electrical charges in the cloud separate from one another, the negative charges dropping to the lower part of the cloud and the positive charges staying in the middle and upper parts (Figure 6.11). Positive electrical charges also build upon the ground below. When the difference in the charges becomes large, a flow of electricity occurs in a discharge event. In-cloud lightning is most common but lightning does strike the ground or strikes from the ground to the cloud. A lightning strike occurs in less than a millionth of a second. The temperature of a lightning bolt can be hotter than the surface of the Sun. Although the lightning is extremely hot, the short duration means it is not necessarily fatal: hundreds of people are struck by lightning every year, but not all die. Lightning at the ground is the most common natural cause of forest fires, which can cause considerable damage and present a major hazard to both people and the environment, particularly after long periods of drought.

It is worth noting that large hail (which in supercell storms can be up to 10 cm in diameter, although large hail is more typically 2–4 cm across) can cause major damage to crops, property and vehicles. NOAA notes that hail causes damage to crops and property amounting to $1 billion per year in the United States.

Figure 6.10 Lightning. (Source: Kent Wood/Science Photo Library Ltd.)

Rain and hail

Figure 6.11 Electrical charge in a thunderstorm.

BOX 6.2

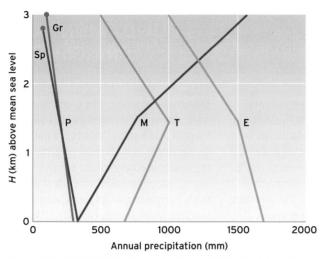

Figure 6.12 Precipitation changes with height in different parts of the world: E, equatorial; T, tropical; M, middle latitude; P, polar (Sp, Spitzbergen; Gr, Greenland). Only in the middle latitudes does precipitation increase with altitude over the 3000 m range shown. For equatorial and polar areas precipitation decreases with altitude and for tropical areas precipitation increases to about 1500 m and then declines above this height. (Source: after Lauscher, 1976)

For example, in equatorial Africa rainfall on mountains above 3000 m is only 10–30% of the highest totals which are observed lower down the mountains. In the middle latitudes, however, precipitation totals increase with altitude above 3000 m. Thus, there are distinct latitudinal differences in the change of precipitation with height in mountains (Figure 6.12).

The presence of mountains in the middle latitudes enhances precipitation in a number of ways. The most important effect is that low-level cloud is formed as air is forced to rise over the mountains (Box 6.3). Although convective precipitation can form a significant proportion of the rainfall totals in some mid-latitude locations, much of the precipitation arises from frontal activity associated with depressions. Orographic enhancement through the **feeder-seeder mechanism** (Box 6.3) can be substantial at warm fronts and in warm sectors, and to a lesser extent with cold fronts. The forced ascent of air over hills and mountains may also intensify vertical motions in depressions and troughs or even trigger conditional instability in polar or arctic airstreams. The general increase of wind speeds with height also ensures there is a supply of moist air brought in to replace any loss of water content through precipitation. As a result of orographic enhancement of precipitation, mid-latitude hills and mountains have unusually high annual precipitation totals (Box 6.3). In New Zealand, average precipitation

totals on the windward side of the Southern Alps can reach $10\,000$ mm yr^{-1}.

As well as enhancing precipitation on the windward side of mountains, there can be substantial reductions of precipitation in lower-lying areas to the lee of the mountains. These 'rain shadow' areas are found in many places such as northern Chile (south-east trade wind belt), Patagonia in Argentina and east of the Rocky Mountains in the United States (mid-latitude westerlies) and on a smaller scale in many hilly locations. This occurs because the air which can now descend down the lee side of the mountains warms adiabatically. This in combination with the earlier loss of moisture through precipitation on the windward and summit parts of the mountains makes the air less saturated and thus less likely to produce precipitation.

The other feature of mountains is that precipitation may fall as snow which can accumulate over time. Over the winter substantial amounts of snow can accumulate in mountains. Often snowmelt in the spring can produce large river flow peaks downstream even when there is no precipitation at the time. Precipitation in mountains is therefore not just enhanced but its hydrological impact may occur a number of months (or years) later. Given sufficient depth of snow (many metres) surviving over many summers it is possible for a glacier to form. At present nearly all mid-latitude mountain glaciers are retreating as annual melt is greater than snow accumulation. Despite that retreat, as the glacier surface ice is at 0°C even in mid-summer, glaciers have an impact on their local summer climate as they act as heat sinks. Glaciers cool air in contact with the surface and depending on the moisture content of the air they can act as either a local moisture source or sink. If air is warm then water vapour can **sublimate** (change directly from the solid to gaseous state) into the air above the glacier. If the air is cool and dry then water can condense out of the air onto the surface of the glacier which is then acting as a moisture sink.

Even in mountains where no glaciers exist, snow patches can survive over the summer. The Observatory Gully snow patch on Ben Nevis in the Scottish Highlands is an example of this, having only melted completely a few times over the past 120 years. Although snow may last for many years before melting, the Observatory Gully patch is too shallow to form a glacier. Like glaciers, such snow patches can also modify the microclimate close by. However, this is to a much smaller extent than glaciers. The snow can also have local ecological impacts creating a niche for certain alpine species. Again the local topography is important with snow

FUNDAMENTAL PRINCIPLES

OROGRAPHIC ENHANCEMENT OF PRECIPITATION

Precipitation totals increase with altitude in the tropics (up to around 1500 m) but the increase with altitude is greatest in the middle latitudes. This mid-latitude orographic enhancement of precipitation is related to the feeder–seeder mechanism. As air is forced to rise over the mountains the adiabatic cooling reduces air temperature to the point at which the air becomes saturated with water vapour. Further ascent leads to the formation of cloud, a common feature of mid-latitude mountains. This cloud acts to increase precipitation falling from higher 'seeder' clouds because the lower 'feeder' cloud droplets are swept into the precipitation falling through the cloud (Browning and Hill,

1981). This feeder–seeder enhancement of precipitation as illustrated in Figure 6.13 can be substantial at warm fronts and in warm sectors and to a lesser extent with cold fronts. This is most apparent in coastal mountain ranges such as the New Zealand Alps, the Coast Mountains of British Columbia and the western Highlands of Scotland.

Orographic enhancement of precipitation is difficult to measure as in the middle latitudes winter precipitation often falls as snow and the turbulent flow around mountains means that the deposition of snow can be into gullies and hollows. Standard measurements of precipitation using rain gauges are therefore impossible (see Chapter 13). In addition, orographic enhancement is not constant in all precipitation events.

Enhancement tends to be greatest at warm fronts and in warm sectors and therefore as weather events are not distributed evenly through time, there are seasonal differences in enhancement. In addition, even across a relatively narrow mountain range such as the Scottish Highlands, typically 80 km wide and only up to around 170 km at their widest (from the west coast to the east of the Cairngorms), there is a rapid decline in enhancement from west to east. While enhancement rates can be over 4.5 mm m^{-1} in the western Highlands, in the east of the Cairngorms the rate is only of the order of 1.33 mm m^{-1} (McClatchey, 1996). This enhancement gives annual totals of just over 2000 mm on the highest tops of the eastern Cairngorms, as compared with 6000 mm in parts of the western Highlands.

Figure 6.13 Schematic diagram of the feeder–seeder mechanism. The frontal cloud aloft produces precipitation which falls and hits the water droplets in the orographic cloud. These then combine to increase the overall precipitation totals on the mountain/hilltop.

BOX 6.3

patches forming and surviving in local shaded depressions or gullies in the mountains.

6.2.5 Frost hollows

Cold air tends to move downhill by katabatic drainage. This drainage of cold air into lower-lying areas can give unusually high occurrences of frost and low temperatures in certain locations. Such frost hollows are much more a feature of middle and high latitudes than other locations. This is because low temperatures occur when there is strong radiational cooling at the surface resulting in a

surface temperature inversion. As water vapour is a strong absorber of long-wave radiation, a significant amount of which it re-emits down to the surface, the relatively moist air in most subtropical and tropical regions (deserts are the exception) reduces the amount of this surface radiational cooling. Geiger (1965) described an extreme example of local surface cooling at the Gstettneralm sinkhole in Austria where temperature inversions from the bottom to the top (about 150 m) can be over 27°C and extreme minima of below −50°C have been recorded in the valley bottom. An example of an area prone to low temperatures in Europe is given in Box 6.4. Low-lying areas with well-drained soils

KATABATIC DRAINAGE AND EXTREME TEMPERATURE MINIMA

The density of air is inversely proportional to temperature ($\rho \propto 1/T$; density increases as temperature falls). In light winds, mechanical mixing of the air is very limited, and at night there is a lack of convective turbulence that is normally created by solar radiation warming the ground during the day. As a result the air close to the surface is cooled during the night as the surface temperature drops. This cooling is greatest when the sky is cloud free and the air is dry. As this cooled air is

now denser than the air aloft, if it is on a slope, the air can start to move downslope in what is called katabatic drainage. This flow of air is not fast and is not like that of water but more like the flow of something like porridge. Very rarely genuine katabatic winds can occur but this really only happens in the Antarctic when cold air flows off the main ice sheets (which can be at over 3000 m above sea level) down to the coastal ice shelves. In most parts of the world katabatic drainage is slow and the cold air can pond up behind restrictions in the flow such as walls (on a small scale) and on a larger scale where a wide valley becomes

constricted at a lower point in the valley.

Anticyclonic conditions are most likely to give rise to stronger katabatic drainage, as winds tend to be light and the sky cloud free (calm and clear conditions). Such nighttime conditions are sometimes called radiation nights as the surface has its maximum radiational loss under such conditions. Anticyclones have a marked temperature inversion aloft formed by the subsidence of air from higher levels and on such radiation nights the surface temperature inversion, formed by the cooling at the surface, can extend up to the anticyclonic inversion aloft.

An example of an exceptionally deep surface temperature inversion was observed in the Cairngorm Mountains, Scotland, in 1982. On 8 January 1982, a surface minimum temperature of −31.3°C was recorded at Grantown-on-Spey (the minimum air temperature was −26.8°C) while at the summit of Cairngorm (1247 m above Grantown-on-Spey) the minimum temperature was −12.6°C, a temperature inversion of over 18°C from the valley surface. Even lower temperatures were recorded in the Spey Valley on 10 January 1982 but no observation was available from Cairngorm summit on that date. A map of the temperatures on 10 January 1982 (Figure 6.14) using remote sensing (see Chapter 23) shows clearly how there are particular areas (frost hollows) that have the lowest minimum temperatures (McClatchey et al., 1987).

Figure 6.14 Satellite-derived temperatures in the Scottish Highlands for 10 January 1982. Severe low temperatures can be seen in the low-lying hollows whereas warmer temperatures are found on the mountain tops. This is an example of katabatic drainage. (Source: after McClatchey et al., 1987)

BOX 6.4

159

such as sands or gravels, or thin soils on chalk, are more likely to experience increased frequencies of frosts or unusually low temperatures. Outside the tropics and subtropics, the lowest surface air temperatures are, however, almost always recorded when the ground is snow covered as the snow insulates the air from the soil heat flux (Robinson and Henderson-Sellers, 1999). In addition the cold air above snow surfaces contains less water vapour and therefore the surface radiational cooling will be greater.

Reflective questions

➤ While an average temperature lapse rate for the troposphere may be around 6.5°C km⁻¹, under what circumstances would the lapse rate in mountains (i) show an increase in temperature with height from the valleys to well up the hillsides; (ii) be close to the dry adiabatic lapse rate; (iii) show a fall in temperature followed by a rise and then a further steep fall in temperature on ascent?

➤ Why do rainfall totals in the middle latitudes increase much more greatly with height than those in the tropics?

➤ Why is a general increase in wind speed with height important with regard to orographic enhancement of precipitation?

➤ Why is the change of wind speed with height in tropical and subtropical mountains different from that in mid-latitude mountains?

➤ Why is snow cover important in the occurrence of extreme temperature minima?

➤ What conditions are needed for thunderstorms to form and what hazards do they pose?

6.3 Influence of water bodies

Unlike land surfaces, water bodies have little diurnal change in surface temperature except in very shallow water close to the water's edge. Surface temperatures are fairly constant for a number of reasons. Solar radiation is transmitted through water to a considerable depth and is not absorbed at the surface as is the case for land surfaces. The high **specific heat** of water (the energy required to increase water temperature) means that it requires more energy to be absorbed for

any given temperature change than other substances. Furthermore the surface layers of water bodies tend to be well mixed, which helps spread any temperature change through a substantial depth of water. In addition, energy at the surface is used largely for the latent heat needed for evaporation rather than sensible heat that would cause a change in water temperature.

Over the land there are much more substantial diurnal changes in air temperature particularly in the summer half-year when solar radiation is stronger. In the middle latitudes, sea surface temperatures (and the air in the layers close to the surface) are therefore cooler than land surfaces during the day in the summer half-year. They are warmer than land surfaces at night. The same is true in the high latitudes but in the winter half-year the temperature of the snow-covered land may remain colder than sea surface temperatures during both day and night. Such differences in local temperature result in sea and land breezes as shown in Figure 6.15. Features like this can also develop where there are large inland bodies of water, such as the Great Lakes in North America, when they are called lake breezes.

Sea breezes form only when there are light wind conditions (typically anticyclonic conditions) as the stronger

Figure 6.15 Circulation of sea and land breezes for (a) and (b) day-time and (c) night-time. (Source: after Robinson and Henderson-Sellers, 1999)

winds of more active systems help reduce land–sea temperature differences through vigorous mixing of air. Summer sea breezes are a feature of many coastal areas and typically have speeds between 2 and 5 m s^{-1}. Sea breezes exist from the surface to 2 km above ground and may penetrate 30 km or more inland (occasionally as far as 100 km). The sea (or lake) breeze brings cooler (occasionally up to 10°C cooler) more humid air inland and a shallow sea breeze front may be evident. Uplift takes place along this front and can trigger the development of cumulus cloud (Figure 6.16). These clouds can be carried seawards by the counterflow present aloft. Occasionally sea breezes from different directions can converge enhancing uplift, which can lead to convective showers. The sea breeze dies off as night falls and is often replaced by a weak land breeze (Figure 6.15).

In the introduction, the regional change in climate across the mid-west United States was highlighted as an example of how, although within a single climate type, there would be differences in local climate. A further illustration of this can be seen in the temperatures and rainfall of Milwaukee and

(a)

(b)

Figure 6.17 Mean monthly temperature (a) and precipitation (b) at Milwaukee (43°N) and Madison (43°N). Milwaukee is close to Lake Michigan and so has milder winters owing to the influence of the warmer water. The lake water in winter is warmer than the surrounding land because it cools more slowly than land. Thus air over the lake becomes warmed by the lake water. The lake also acts as a moisture source and so winters in Milwaukee are wetter than those in Madison that is too far away from Lake Michigan to be affected.

Madison in Wisconsin, USA (Madison is about 120 km west of Milwaukee). Both are in the same climate type but Milwaukee comes under the influence of Lake Michigan, giving it slightly milder winters and slightly drier summers than Madison. This is illustrated by climate data shown in Figure 6.17.

Reflective question

➤ Why do sea and land breezes form?

6.4 Human influences

6.4.1 Shelter belts

It was noted earlier that coastal areas are subject to stronger winds than inland areas when winds are blowing off the sea. This is a result of reduced friction over the relatively smooth

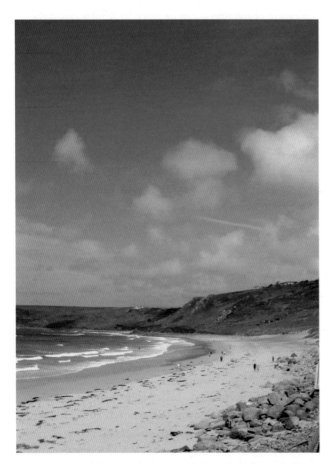

Figure 6.16 Cumulus clouds develop at the coast where air rises over the land and migrates seawards.

Figure 6.18 The role of shelter belts in reducing wind speed downwind. Low- and medium-density belts offer better shelter than high-density belts (e.g. walls) for a greater distance away from the belt. This is because the small amount of air flowing through the low-density belts cushions the air flowing over the top of the belt. Without this the air flowing over the top immediately subsides causing fast-moving eddies to form and thus rendering high-density belts less useful. (Source: after Nägeli, 1946)

sea surface. This suggests that any alterations in surface roughness have an impact on the local wind. However, unless there is a permanent change to a new surface (as from sea to land) individual roughness elements such as trees or buildings will have an effect for only a relatively short distance downwind. If a line of trees or hedges is planted upwind of a field sown with sensitive crops it is possible to reduce the local wind speeds in the field to provide some protection. The same thing can be done around a garden to provide shelter. The ideal shelter belt is slightly permeable as the lower layer of air acts somewhat like a cushion and extends the reduction in wind speed over a longer distance as shown in Figure 6.18. An impermeable shelter belt will produce a greater reduction in wind speed close to the belt. However, recovery of the wind speed to upwind values is more rapid for impermeable belts, taking place over a distance of about 10–15 times the height of the

barrier. Low-density shelter belts have an impact 15–20 times the height of the barrier downwind with medium-density shelter belts having the greatest impact of up to 20–25 times the height of the barrier.

The use of shelter belts is best when there is a particular wind direction from which damaging winds come. Examples can be found in southern France where shelter belts are planted to protect crops from winds that can come down the Rhône Valley from the north. Snow fences set back from roads and railway lines are also a type of shelter belt as they are used to reduce air flow to allow snow to fall to the ground before the air reaches the road (Figure 6.19). This keeps roads and rails more clear of snow than would otherwise be the case. In many ski resorts snow fences are also used to reduce local air flow and help allow snow to accumulate on the pistes.

6.4.2 Urban climates

The fabric of towns and cities substantially alters surface characteristics compared with surrounding rural areas. The urban surface is much rougher than most vegetation. An indication of roughness can be given by what is called the **roughness length**. This is of the order of 5–20 cm for agricultural crops but up to 10 m for tall buildings. There are also important changes to the radiation and energy fluxes in urban areas. The urban fabric (stonework, road materials, roofs and so on) strongly absorbs solar radiation. There is also a substantial release of energy into the urban atmosphere as a result of humans heating their environment (domestic and industrial). This is especially the case for mid- and high-latitude cities in winter. The urban atmosphere is also affected by air pollution with increased levels of carbon monoxide, oxides of nitrogen and various hydrocarbons. Although the local climatic impact of urban areas is always present, it is reduced in strong wind conditions as the vigorous mixing spreads any impact through a greater depth of the atmosphere and rapidly transports effects away from the urban area. The greatest impact of urban areas on the local climate is therefore found during light wind conditions. While the most important impact of urban areas on climate is linked to air pollution, there are local climate changes in both the temperature and wind regimes experienced by towns and cities.

The most commonly discussed climate modification in urban areas is the **urban heat island** effect. This is so called as the urban area is an 'island' of warmer air within the surrounding cooler rural air. The urban heat island occurs mainly at night when the urban atmosphere cools more slowly than that in the surrounding rural areas. It is more

Figure 6.19 A snow fence acting as a shelter belt near a railway line across a grassland in the Rocky Mountains. (Source: National Geographic/Getty Images)

strongly developed in generally light wind conditions with clear skies when long-wave radiational loss is greatest (at night). There appear to be critical wind speeds above which the urban heat island disappears. Examples are 12 m s^{-1} for London, 11 m s^{-1} for Montreal and 4–7 m s^{-1} for Reading (England) and these critical speeds are related to size of the urban area.

Oke (1976) identified two parts to the modification of the urban atmosphere: the **urban canopy layer** and the **urban boundary layer** (Figure 6.20). Oke (1987) also suggested how an urban heat island would develop these two layers. In the case of the urban canopy layer he suggested that the following would all play a role: (i) greater absorption or direct solar radiation due to 'canyon geometry' (width of the road plus the height of the buildings on either side); (ii) greater daytime heat storage due to properties of urban materials; (iii) anthropogenic heat release from buildings (largely due to heating losses); and (iv) decreased evaporation. In the urban boundary layer, however, he suggested entrainment of air from the canopy layer, anthropogenic heat from roofs and chimneys, and downward flux of sensible heat

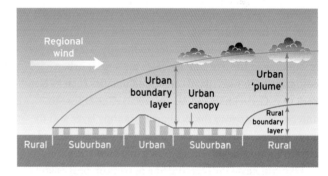

Figure 6.20 The urban canopy and boundary layer. The canopy layer consists of the spaces around the local buildings beneath the mean height of the buildings, whereas the boundary layer is the layer affected by the urban environment. (Source: after Oke, 1976)

(see Chapter 4) from the overlying stable layer would be the principal causes of the urban boundary layer heat island.

Even in relatively light wind conditions, the turbulence caused by air flow over buildings is enough to maintain a well-mixed atmosphere (up to an altitude of a few hundred metres in the largest cities). This mixing establishes an

High █████████ Low
Heat loss

Figure 6.21 Heat loss from an urban area detected using infrared sensors on an aircraft. It is possible to see which individual buildings are in need of increased insulation.

adiabatic lapse rate and as a result a stable layer is formed aloft. The heat island formed in this well-mixed urban boundary layer is strongly related to the city size and types of building in the city and therefore the largest urban boundary layer heat islands are found in cities such as New York (e.g. Gedzelman and Austin, 2003). The failure to differentiate between the local urban canopy layer and the more general urban boundary layer heat islands can lead to erroneous conclusions being drawn in observational studies.

Oke and East (1971) recorded a maximum heat island of up to 12°C in Montreal in winter which is unusual for mid-latitude cities where the maximum heat island is normally observed in summer. However, in high-latitude cities the anthropogenic heat losses in winter can be very large (owing to the extra internal heating switched on in offices and homes), leading heat island maxima to occur at that time of year. Even in London, a mid-latitude city with relatively mild winters, anthropogenic heat losses reach over 200 W m^{-2} in winter which is greater than typical solar radiation amounts at that time of year. Infrared technology is now being attached to aircraft to examine which buildings emit most

heat and to determine those areas most in need of additional insulation (Figure 6.21).

While urban heat islands are at their maximum under light wind conditions, in strong winds the buildings can create powerful gusts as the wind is forced to flow round tall buildings. In a zone close to tall buildings gusts may reach 2.5 or even 3 times the mean wind speed upwind of the building (Figure 6.22). The **Venturi effect** occurs when winds are forced to funnel between two buildings increasing localized wind speeds (Figure 6.22a). These winds can cause difficulty in walking and opening doors and may put severe stress on the buildings. Transverse currents can be generated when buildings are at right angles to the wind. Here pressure differences between the upwind and downwind sides of the buildings can lead to unexpected strong gusts (Figure 6.22b). However, overall large urban areas tend to reduce overall mean wind speeds as their greater roughness slows wind speeds to below their rural upwind values. This modification of wind flow around buildings also plays a role in the dispersion of air pollution and can lead to areas when the local air pollution is unusually high as a result of the trapping of

(a) Venturi effect

(b) Transverse currents

(−)

(−)

(+)

(+)

Zones of increased wind speed
(up to 2.5 to 3 times regional wind)

+ Increased pressure
− Decreased pressure

Figure 6.22 Wind regime around buildings: (a) flow can be funnelled into narrow passages between buildings resulting in the Venturi effect; (b) transverse currents can develop when buildings are at right angles to the wind. (Source: after Thurow, 1983)

pollution between buildings. Although there is some evidence that convectional rainfall can be enhanced by urban areas (but perhaps downwind of the city), there is no real indication that urban areas have any influence on precipitation events. Therefore, the climate modifications created by urban areas are largely the canopy and urban boundary layer heat islands and the modification of the wind regime. Box 6.5 describes techniques that you can use for measuring the urban heat island. In addition there is increased air pollution in cities but levels at any one place are related to emissions, dispersion and local geography and not just to any change in the local climate. Urban areas do also have some

TECHNIQUES

OBSERVING URBAN HEAT ISLANDS

Urban heat islands provide students with an opportunity to make their own observations of human influence on the local climate. The urban heat island (measured by the difference between temperatures in the surrounding rural areas and those within the town or city) is most clearly marked at night and as such is not a warming of the urban atmosphere but rather a reduced cooling rate in comparison to surrounding rural areas (Figure 6.23).

The best conditions under which to observe the urban heat island are on nights with light winds and clear skies. The light winds are important as stronger winds will mix the atmosphere and the surface cooling which takes place at night (through long-wave radiation loss) will be spread through a greater depth of the atmosphere. The lack of cloud cover is important as the water droplets in clouds mean that clouds act as black body radiators at terrestrial temperatures and so the long-wave radiation loss from the surface at night will be largely compensated for by the long-wave radiation emitted from the cold base. In fact, well-mixed air in the **surface boundary layer** of the atmosphere will tend to have an adiabatic lapse rate in the absence of strong surface heating (through solar radiation warming) or cooling (long-wave radiation loss to the atmosphere). It is

BOX 6.5 ➤

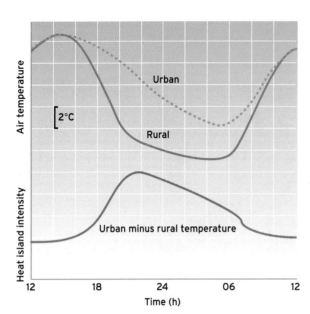

Figure 6.23 Typical temporal variation of urban and rural air temperatures under clear skies and weak air flow. (Source: after Oke, 1987)

therefore the case that on nights with strong winds and a complete low-level cloud cover, the lapse rate up to the cloud base will be close to the dry adiabatic lapse rate (this is also the case during the day if the low-level cloud cover is fairly dense limiting strong solar heating).

In making observations on what are termed 'radiation nights' (nights with light winds and clear skies), it is important to be aware that certain lower-lying spots are often prone to record somewhat lower temperatures than those in the surrounding areas (see Section 6.2.5). Thus in taking an observation at any site to assess the urban heat island effect it is important to note that the relative elevation of the site and the opportunity for colder air to drain away from, or towards, the site will have an impact on the observation.

Early research on the urban heat island made use of thermometers attached to vehicles. These vehicles were usually driven on a route into the town or city and then returned by the same route. Assuming the

weather conditions remained the same over the period of the two transects (inwards and outwards), the first set of observations on the inward transect would record higher temperatures at each location than the return transect as cooling would be taking place throughout the period. Observations of the urban heat island typically show a relatively sharp rise in the value of the urban heat island in moving into the built-up area of the town or city. There then tends to be a slower rise into the city centre.

It is important to note that in making observations of the urban heat island any individual, or a group of individuals, will generally be making their observations within the urban canopy layer (that layer of air within what are termed the 'street canyons') and not the urban boundary layer. Observation of the urban boundary layer requires the use of meteorological masts or remote sensing techniques. There are a number of suggested causes of the canopy layer heat island (Table 6.3) and the importance of these will vary depending on the particular building, the building structure, orientation with respect to the wind, traffic volume and the canyon geometry (topography of the buildings and urban structures).

Rather than studying a long transect from the rural surroundings to the centre of the town or city, an interesting study can be made by taking observations at locations which have clear differences in urban features. For example, quite rapid changes in temperatures can be observed in moving from urban parks into built-up areas (Figure 6.24). Urban heat island investigations can make interesting projects but the research design, detailed observation of the environment and proper and consistent exposure of the thermometer(s) are vital.

Table 6.3 Suggested causes of canopy layer urban heat island

Urban feature	Effect on canopy layer urban heat island
Air pollution	Increased absorption and re-emission
Canyon geometry	Reduced view of sky, reduced wind speed
Buildings	Direct heating of the local environment
Vehicles	Direct heating of the local environment
Construction materials	Increased thermal admittance and water proofing

BOX 6.5 ➤

Figure 6.24 Isotherms (heat contours) around an urban park, Mexico City, with clear and calm air. (Source: after Jauregi, 1990–91)

Examples of investigations could include:

- observing any temperature differences between relatively narrow and broad street canyons;
- looking at differences between deep and shallow street canyons of the same width;
- recording differences between temperatures in parkland (on lit paths) and nearby built-up areas;
- vehicle traverses from rural areas into and out of the town or city.

In undertaking any study of the urban heat island canopy layer, it is important to make detailed observations of the local environment (e.g. building type, materials, canyon geometry) as well as taking temperature observations. Any change in weather conditions will also influence observed temperatures and it is crucial that there is a proper and consistent exposure of the thermometer. Thermometer exposure refers to the way in which the thermometer is presented to the environment. At fixed sites ther-

mometers are in screens designed to ensure that the thermometer records the air temperature as it is shielded from solar and terrestrial radiation (see Figure 5.2 in Chapter 5). Ideally all thermometer exposures should provide radiation shielding and the observations can be improved if air is drawn past the thermometer (**aspiration**). Errors in exposure can invalidate observations. An interesting discussion on exposure is provided by Perry *et al.* (2007). The temperature observations must also be made at the same height above the surface, as in some places temperature lapse rates can be steep at night. The difference between the ground surface and the minimum temperature recorded by a thermometer inside a Stevenson screen at 1.2 m height can be over 10°C on a calm clear night with a snow-covered surface.

Before undertaking any urban heat island investigation there are certain safety issues that need to be addressed. Personal safety is important and as investigations will take place at night it is important that

such a study should not be undertaken by one person on their own. Observations made using thermometers mounted on a vehicle provide increased safety but it is important that the driver is not involved in making any observation of either temperatures or the nature of the environment through which the vehicle is being driven. The driver's attention must be fully given to the road, pedestrians and to all types of other vehicles. If a vehicle is not used, high-visibility clothing should be worn and observations should be made by two or more persons remaining together. Again for safety, when on foot the observers should keep to reasonably well-lit areas and should avoid areas which are known to be unsafe.

In assessing the results it will be important to examine whether the geographical location (how close to the centre of the town or city) and the nature of the local environment at each location (Table 6.3) help provide any explanation for observed differences. It is also important to be aware that the strength of the urban heat island varies over the course of a night and temperature differences between different locations may be the result of the different times at which the observations were made (Chow and Roth, 2006). Any differences in temperatures between different locations may also provide some indication of the rate at which any pollution is dispersed as poor mixing of air both vertically and horizontally will encourage greater pollution concentrations and a higher local air temperature can be related to areas of calm air where mixing is reduced. Air pollution in towns and cities is largely due to that from vehicle exhausts and high vehicle volumes combined with poor local mixing can lead to higher pollution concentrations in certain urban streets.

BOX 6.5

influence on local atmospheric moisture conditions related to lower evapotranspiration (Deosthali, 2000).

6.4.3 Atmospheric pollution and haze

Most air pollution events are localized in urban or industrial areas where traffic emissions or pollutants from factories occur. There can be periods when the atmospheric conditions of urban areas are a danger to human health (Figure 6.25). In regions with high solar radiation such as Los Angeles, Athens and Mexico City, the ultraviolet radiation reacts with the uncombusted hydrocarbons from vehicle emissions and produces a photochemical smog which irritates the eyes, nose and throat. However, often these conditions are localized and do not persist for long periods. Recently, however, in some places there have been haze pollution events that have lasted months and spread over many hundreds of kilometres.

Many of the large haze pollution events have resulted from forest fires. Forest fires can occur both through natural action and by human intervention. Forest fires in Indonesia are very largely the result of human activity as fire is used to clear land for agricultural purposes. Fire is cheap and, as well as reducing vegetation cover, enriches what are often very poor soils. Indigenous tribes, such as the Dayak people in Kalimantan, have traditionally used **shifting cultivation** (slash and burn) techniques and their use has been in tune with the natural environment with strict traditional rules of using fire. Unfortunately, the large number of settlers who came from other islands and new plantation companies do not follow rules that help the long-term maintenance of the environment. Plantation companies (or people hoping to profit from providing services to them) have a particular responsibility as they are largely aware of potential environmental damage and yet place a higher value on their own profits. The Indonesian Government has banned the use of fire for clearing land for a number of years but fires continue to be lit as Indonesia expands its wood pulp, palm oil and rubber industries. In addition to fires, the extensive logging of the rainforests, particularly selective logging, plus other land-use changes have played an important role in making the Indonesian rainforests more susceptible to fire. Rainforests are humid and fires do not naturally take hold. Selective logging and other agricultural land-uses open up the forest and allow it to dry out more easily. The forest fires have caused massive damage within Indonesia but due to the smoke from the fires, damage has also been caused to neighbouring countries, such as Malaysia and Singapore. The haze from the fires in 1986 covered the South-East Asian region for weeks, particularly affecting Malaysia and Singapore, causing health problems, disruption of shipping and aviation, and also caused the temporary closure of international airports. As a result there are significant economic losses as well as ecological damage. A useful history of Indonesian fires is presented by Gellert (1998) and the Indonesian fires of 2006 are covered in more detail in Box 6.6.

Figure 6.25 Smog over Mexico City. Smog blocks out the sunlight and promotes respiratory and other health problems. (Source: Photolibrary Group.)

CASE STUDIES

SOUTH-EAST ASIAN HAZE OF 2006

A severe haze occurred from September to late October 2006 across South-East Asia (Figure 6.26). This event was caused by uncontrolled burning associated with land clearance in Indonesia. The haze affected Mayalsia, Singapore and southern Thailand and may have reached South Korea. The haze added to any local sources of pollution and so was an even greater problem in industrial and high-density urban areas. As with many air pollution events, the local topography played a role, especially in the highly urbanized and industrialized Klang Valley of Malaysia. Air quality improved in late

Table 6.4 Pollution Standards Index

PSI value	PSI descriptor
0–50	Good
51–100	Moderate
101–200	Unhealthy
201–300	Very unhealthy
Greater than 300	Hazardous

October after heavy rainfall helped put out the forest fires.

Singapore has a good air quality monitoring network and there is a telemetric network of air monitoring stations strategically located in different parts of Singapore. These stations measure sulphur dioxide, nitrogen dioxide, ozone, carbon monoxide and particulate matter called PM10 (particulate matter of 10 microns or smaller in size). The

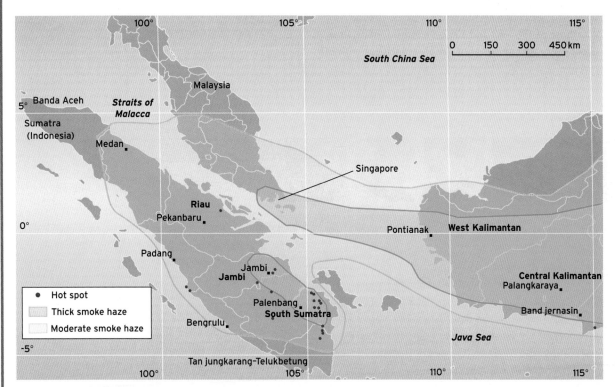

Figure 6.26 Haze pollution across South-East Asia on 19 October 2006. (Source: National Environment Agency of Singapore)

BOX 6.6 ➤

169

Singapore Government uses the 'Pollutant Standards Index' (PSI), an index developed by the US Environmental Protection Agency to provide accurate, timely and easily understandable information about daily levels of air pollution. The index is based on the pollutants listed above. The PSI value is calculated and graded according to Table 6.4. A PSI value is assigned to each pollutant through a linear function relating the pollutant level to its PSI sub-index. Once all sub-indices are calculated, the maximum sub-index is given as the overall PSI value. During the 2006 haze event the PSI value in Singapore reached 150 on 7 October (well into the unhealthy level) as a result of the PM10 from the Indonesian forest fires. Even higher values were recorded in Malaysia. The pollution led the Singapore Ministry of Education to recommend the suspension of outdoor activities.

BOX 6.6

As well as human activity, natural climate cycles can increase the susceptibility of forests to fires. For example, during El Niño events there tends to be reduced rainfall over Indonesia which leads to drier than usual vegetation. Severe fires in Indonesia took place during the two strongest El Niño events of recent years, 1982–1983 and 1997–1998, and hence these fires are commonly cited as impacts of El Niño events. However, while there is some evidence of a connection between El Niño and forest fires for earlier strong El Niño events, there were no fires as severe as those of 1982–83 and 1997–98. Additionally, fires now occur every year and are not simply coincident with El Niño events.

Reflective questions

➤ Why is a permeable barrier more effective as a shelter belt than an impermeable barrier?

➤ Why is it important to take care over what is being observed in any study of the urban heat island?

➤ If mean wind speeds are generally lower in cities, why can gusts be higher?

6.5 Summary

Local climate variations vary from extremely localized microclimates to more generalized regional climates within a general climate type. Regional and local geographies are the key factors in determining the magnitude and importance of these climates. A particular climate classification type can therefore be affected more locally by a number of different influences, some of which will depend on particular weather conditions or on the time of year. The most obvious and rapid climate gradients result from changes in altitude. With increasing altitude atmospheric pressure decreases, and associated with this are decreases in temperature. Precipitation totals often increase substantially with altitude and the wind regime in hills and mountains can be much more severe than that for nearby low-lying ground. Small topographic features can also be important. At night, radiational cooling takes place, reducing the temperature of the air close to the ground. This cooler air is slightly denser and tends to move down slopes to lower ground and into depressions (this is termed katabatic drainage). There are also microclimate variations introduced by vegetation and these include both forests and deliberately planted shelter belts.

The climate near sea coasts is also modified as a result of the influence of the sea, particularly in middle and high latitudes. Coastal areas are often subject to stronger winds than more inland areas. Winds over the sea are stronger as the frictional reduction of wind speed close to the surface is much less than over land. Therefore, if the wind direction is from the sea to land, coastal areas are subject to stronger winds. In addition, there can often be substantial differences between air temperatures over land and sea partly due to the high specific heat of water but also due to the ability of solar radiation to penetrate to considerable depths before being fully absorbed. This can lead to cooling sea breezes in coastal regions

particularly in summer (see below). Lakes can also have an influence on the climate, but except in the case of very large lakes (such as the North American Great Lakes that behave like inland seas) any effect is limited to a very narrow strip around the edge of the lake.

Human activity can also alter climate on a regional and even on a global scale. Examples include the haze pollution caused by forest fires, and emissions of CO_2 into the atmosphere through the burning of fossil fuels (see Chapters 21 and 22). However, there are more localized changes to climate that are a result of human activity. The most obvious example is the creation of a distinct urban climate in built-up areas. Urban areas affect the urban atmosphere and in lighter winds strong heat islands can develop. The urban atmosphere can be split into a canopy (between building) layer and a boundary (above roof) layer and each will have its own heat island. Buildings also modify wind flow. Towns and cities tend to have lower wind speeds but greater gustiness than surrounding rural areas.

Further reading

Barry, R.G. (1992) *Mountain weather and climate.* Routledge, London.
Excellent book on topographic controls of local, regional and global climate with a whole section dedicated to case studies.

Chandler, T.J. (1965) *The climate of London.* Hutchinson, London.
While this book is only about London, as a case study it is very useful because it illustrates the full range of possible modifications of climate that occur in urban environments. It is therefore applicable across the world.

Geiger, R., Aron, R.H. and Todhunter, P. (2003) *The climate near the ground.* Rowman and Littlefield, Oxford.
This is republication of a book from 1965 which is still of great relevance today. It details a whole range of ground–air

interactions ranging from soil and vegetation energy balances to the deposition of dew on ponds. It is a very detailed text.

Oke, T.R. (1987) *Boundary layer climates.* Routledge, London.
This is a fairly technical and physically based textbook with good detail and excellent explanations of differences between processes over vegetated and non-vegetated surfaces.

Robinson, P.J. and Henderson-Sellers, A. (1999) *Contemporary climatology.* Pearson Education, Harlow.
Chapters 10 and 11 are particularly relevant to the material discussed above.

Rosenberg, N.J., Blad, B.L. and Verma, S.B. (1983) *Microclimate: The biological environment.* John Wiley & Sons, New York.
Only chapter 9 covering the subject of shelter belts is relevant here.

Web resources

Heat Island Group

http://eetd.lbl.gov/HeatIsland/

The site describes the different causes of the urban heat island effect, the use of current climate data for urban heat islands research, and methods for reduction of the effect.

NASA: Urban Climatology and Air Quality

http://weather.msfc.nasa.gov/urban/

A NASA project, aiming to investigate and model the urban heat island effect, the effect of urban land cover change on air quality and the overall effects of urban development on surface energy budget characteristics across urban landscapes. The site provides a lot of basic information.

Urban Heat Islands

http://cimss.ssec.wisc.edu/wxwise/heatisl.html

A description of the urban heat island effect and the factors that cause it.

 Visit our website at **www.pearsoned.co.uk/holden** for further questions, annotated weblinks, interactive models and video-clips showing physical processes in action.

Soils, biogeography and ecology

Figure PIV.1 At different levels from the soil to the top of the tree canopy there are interactions of species, matter and energy. The concepts and processes associated with soils, biogeography and ecology are described in the four chapters of Part IV.

Part contents

Scope

In Part IV we turn primarily to the living components of our planet. However, it is first necessary to grasp the role played by soil as the interface between the atmosphere, lithosphere, hydrosphere and biosphere in which we exist. It represents a fundamental component of the terrestrial biosphere and performs a wide range of essential functions that sustain life. Soil acts as a medium for plant growth, providing support and essential nutrients (which in turn support animal life); provides a reservoir for water and water quality maintenance; recycles dead plants and animals; and provides a habitat for organisms and the raw materials that create the Earth's landscapes. The nature and properties of the soil therefore have an enormous impact on the wider environment. These properties are determined by the processes that maintain the transfers, exchanges, inputs and outputs of material through the soil system. Furthermore, soil is not an unlimited resource. It can be degraded, lost and improved by natural or human activities. An appreciation of these issues is imperative if we want to preserve our soil in a healthy state and improve our understanding of the relationship between soil and the biosphere. Chapter 7 attempts to provide such a basis.

The critical environmental factors that control soil formation, such as climate, topography and time, are also major influences on the biosphere. At the same time the biosphere exerts an influence on soil formation and soil processes. The various features of the biosphere can, in fact, be associated with all aspects of geography, from the climate system, oceanography, geology, hydrology, social issues and even global tectonics. These interactions result in distinctive regions of plants and animals (biomes) that differ depending upon the controlling variables, and often phase into one another along a gradient of change. Chapter 8 discusses the characteristics of these various biomes. It is not enough, however, simply to describe their features, but it is necessary to explain their nature and form. Chapter 9 is therefore concerned with studying the processes behind the spatial distribution of plants and animals and their change over time.

Any study in biogeography is inextricably linked to ecosystem processes which consider not only transfers of energy and matter but also individual species characteristics. Chapter 10 delves deeper into the characteristic features that underpin any understanding of modern ecology, such as the nature and role of ecosystems, habitats, communities, life strategies and the environmental niche at all scales down to an individual tree. By studying ecosystem processes, we become aware of the very dynamic nature of ecosystems and the checks and balances operating as drivers of ecological change. The planet's biomes are not static. Closer observation shows that important links between plants, animals and soils are related to processes involving the movement of energy and organic and inorganic materials through the system so that ecosystems are in a constant state of change.

The alteration and fragmentation of major biomes by non-sustainable practices of exploitation have been features of the human impact on the environment for thousands of years. However, the increasing density of human populations and improvements in technological capabilities put added pressure on the biosphere and the soil medium that supports it. The diversity and complexity of the systems discussed within Part IV can mask their fragility with respect to human disturbance. The requirement of environmental managers to balance the preservation of ecosystem stability with the human needs for the ecosystem therefore necessitates an understanding of the major ecological, biogeographical and environmental processes that characterize the living portion of our planet. Vegetation and soils also play an important role in the climate system and there are two-way interactions between climate and vegetation. However, discussion of these processes will be dealt with mainly in Part VI of this book which deals with environmental change. More specifically the whole of Chapter 22 is devoted to the topic of vegetation and environmental change.

CHAPTER 7

Soil and the environment

Pippa J. Chapman
School of Geography, University of Leeds

Learning objectives

After reading this chapter you should be able to:

➤ describe the components that make up soil and understand how they affect the physical and chemical properties of soil

➤ describe the processes of soil formation and the factors that control soil development, and understand how these lead to the features seen in soil profiles

➤ compare and contrast the different methods of soil classification

➤ define the important physical properties of soil and explain important cation-clay mineral interactions and the processes that control soil pH and relate these concepts to soil fertility

➤ appreciate the environmental importance of soil and understand the threats posed to soil and soil processes by human activities

7.1 Introduction

Soil is a major component of the Earth's **ecosystems** (see Chapters 8, 9, 10 and 22) and forms at the interface of the atmosphere (air), lithosphere (rocks), **biosphere** (plants and animals) and hydrosphere (water). Soil can be defined as a complex medium, consisting of inorganic materials (such as clay minerals), organic matter (living and dead), water and air, variously organized and subject to dynamic processes and interactions. As well as being a major component of the natural system, soil has a key role in the use and management of the environment by humans, where it performs a wide range of essential functions:

- It is a medium for plant growth, providing plants with support, essential nutrients, water and air. Plant life, in turn, supports animal life.
- It acts as a reservoir for water, influencing the quantity of water in our rivers, lakes and aquifers (see Chapter 13).
- It has a filtering and transforming role for materials added to the soil (see Chapter 15). Thus, soil is often able to protect the quality of our air and water.
- It recycles dead plants and animals into nutrients needed by all living things.
- It provides a habitat for organisms. A handful of soil may be home to billions of organisms, belonging to thousands of species.
- It provides raw materials such as clays, gravels, sands and minerals as well as fuels such as peat. It also provides a physical base for the foundations of buildings and roads.

Soil is, therefore, essential to the maintenance of the environment. Without soil, the biosphere in which we live could not function. As Doran and Parkin (1994) stated: 'The thin layer of soil covering the Earth's surface represents the difference between survival and extinction for most terrestrial life.'

Soil, however, is not an unlimited resource and can be lost, degraded or improved by natural processes and human activities. As most soils take thousands or even millions of years to form, they cannot be replaced if they are washed away or polluted. Understanding the nature and distribution of soils and the processes operating within soil is therefore essential if we want to preserve our soil in a healthy state and improve our understanding of ecosystem dynamics. This chapter begins by describing the components of soil before examining soil formation processes. These processes impact the physical and chemical properties of soil, which are also described. The chapter concludes by examining a range of impacts of human activities on soils and soil processes.

7.2 The components of soil

Soil comprises four major components: the inorganic or mineral fraction, organic matter, water and air. The relative proportion of these four components greatly influences the physical and chemical properties of a soil. In a soil, the four components are mixed in a complex way. Figure 7.1 shows that approximately half the soil volume of a typical **topsoil** consists of solid material (inorganic and organic); the other half consists of voids or **pore spaces** between the solid particles. Most of the solid material is inorganic mineral matter. However, the influence of the organic component on soil properties is often far greater than its

Figure 7.1 The components of soil. This diagram shows the composition by volume of a typical topsoil. The dashed line between water and air indicates that the proportion of these two components varies with soil moisture. Water and air make up the pore spaces whereas the solid material is made up from mineral and organic matter.

small proportion suggests. Most agricultural soils contain between 1 and 10% organic matter and are referred to as inorganic or mineral soils because of the low organic content.

Air and water fill the pore spaces between solid soil particles. The relative proportion of air and water fluctuates greatly and they are inversely related to each other. Following rainfall, water fills the pore spaces, expelling much of the air. As water gradually drains away or is used by plants, air refills the pores.

7.2.1 Mineral particles

In most soils the mineral fraction predominates. Mineral particles are derived from the **weathering** of **parent material**. Weathering (see Chapter 11) refers to the breakdown of rocks and minerals by the action of physical and chemical processes. The larger mineral particles, which include boulders, stones, gravel and coarse sands, are generally rock fragments, whereas smaller particles are usually made of a single mineral. There are two major types of mineral particles: **primary minerals**, which are minerals that have changed little since they were formed in magma, such as quartz, feldspars and micas; and **secondary minerals**, which are formed from the breakdown and chemical weathering of less resistant primary minerals such as clays and oxides of iron and aluminium. The composition and size of mineral particles have a great influence on the physical and chemical properties of a soil (Brady and Weil, 2002).

7.2.2 Soil organic matter

Soil organic matter can be divided into three categories: (i) decomposing residues of plant and animal debris referred to as **litter**; (ii) resistant organic matter known as **humus**; and (iii) living organisms and plant roots collectively referred to as the **biomass**. Fresh plant and animal litter is progressively decomposed by soil microorganisms to a more or less stable end product called humus which is resistant to further decomposition. During the breakdown of organic matter by microorganisms, plant nutrients are released, particularly nitrogen, phosphorus and sulphur. This process is called **mineralization**. The balance between inputs of plant and animal materials and losses by decomposition determines the amount of organic matter in the soil. Usually, organic matter represents between 2 and 6% by volume, but it has a large impact upon the soil. Where decomposition processes are drastically slowed, however, such as in waterlogged conditions, a surface accumulation of only partially decomposed material builds up to form depths of several metres. These organic soils are known as **peat** (see Chapter 12).

Soil organic matter is a very important component of soils because it: (i) is the main food for soil organisms; (ii) binds mineral particles together and therefore stabilizes the soil's structure and protects it from erosion; (iii) improves water holding capacity; (iv) improves **porosity** and aeration and therefore aids the growth of plants; and (v) is a major source of nutrients and therefore influences soil fertility.

7.2.3 Soil water

Water is essential to the ecological functioning of soils. Plant and soil organisms depend on water to survive. Water is also a major driving force in soil formation as it is required for parent material weathering. All chemical weathering processes depend on the presence of water. Water together with soil air fills up the pore spaces between the mineral and organic components of the soil. Soil water, however, is not pure; it contains dissolved organic and inorganic substances and is known as the **soil solution**. When some compounds dissolve in the soil solution, the atoms become separated as **ions**. An ion is an atom, or group of atoms, bearing an electrical charge. For example, when table salt (sodium chloride), which has the chemical formula NaCl, dissolves in water the atoms separate and form ions (see Figure 7.2). The sodium ions have a single positive charge and are indicated by the symbol Na^+, whereas chloride has a single negative charge as indicated by the symbol Cl^-. Positively charged ions are referred to as **cations** and negatively charged ions as **anions**. In soil solution important cations and anions include those in Table 7.1.

An important function of the soil solution is to ensure the continual supply of some of these cations and anions to the plant roots. Soil solution is also the main agent of **translocation**, carrying dissolved ions, including pollutant ions, and small particles through the soil to surface and groundwaters.

Water is held in soil by the attraction of water molecules to each other and to soil particles. Water exists as one of three states in the soil (Figure 7.3). The amount held in each state changes over time, which affects the amount of water available to plants and the potential movement of nutrients and pollutants within the soil. When all the soil pores are filled with water from rainfall, the soil is described as **saturated** (Figure 7.3a). The soil, however, does not stay in this state for very long as, under the action of gravity, water will start to drain out of the larger pores and is replaced by air (Ward and Robinson, 2000). This water is called **gravitational water** and when all of it has drained away the soil is said to be at **field capacity** (Figure 7.3b). The small pores retain water against the force of gravity: this water is

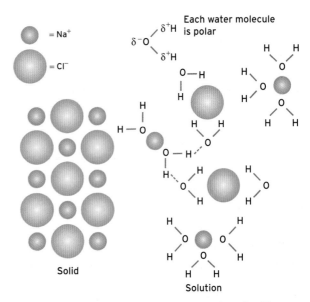

Figure 7.2 The dissolution of table salt (NaCl) in water. Since water (H_2O) is a polar molecule (one end has a different charge from the other end), the negative end is attracted to positive ions and the positive end is attracted to negative ions. Table salt completely dissolves (dissociates) in water as the water molecules keep the sodium (Na^+) and chloride (Cl^-) ions apart and stop them from reforming the solid salt. The figure shows how the water molecules are linked to the sodium and chloride ions.

known as **capillary water** and represents the majority of water that is *available* for plant uptake. This water remains in the soil because the combined attraction of (i) the water molecules to each other and (ii) water to the soil particles is greater than the gravitational force. Capillary water moves within the soil from zones of higher potential (wet areas) to lower potential (dry areas) (see Chapter 13). The most common movement is towards plant roots and the soil surface, where it is lost by evaporation and transpiration. The final type of water is **hygroscopic water**, which is held

Table 7.1 Some important cations and anions in soil solution

Cations		Anions	
H^+	Hydrogen	Cl^-	Chloride
Na^+	Sodium	NO_3^-	Nitrate
K^+	Potassium	SO_4^{2-}	Sulphate
Ca^{2+}	Calcium	HCO_3^-	Bicarbonate
Mg^{2+}	Magnesium	OH^-	Hydroxide
Al^{3+}	Aluminium		

Figure 7.3 Soil water states at: (a) saturation, when all pore spaces are filled with water; (b) field capacity, when the smaller pores are filled with water and the larger pores are filled with air; (c) permanent wilting point, when plants can no longer exert sufficient suction to withdraw the water that is tightly held around the soil particles.

Figure 7.4 The general relationship between soil texture and soil water availability.

as a tight film around individual soil particles (Figure 7.3c). This water is *unavailable* to plants as the attraction between the water and the soil particles is greater than the 'sucking power' of plant roots. A soil in which all the water is hygroscopic will appear dry although some water still remains. The drier the soil, the harder the plant has to work to obtain the remaining water held in progressively smaller pores. Eventually there comes a point when plants cannot withdraw the tightly held water from the soil and this is known as the permanent **wilting point** (Figure 7.3c). The water retained in the soil between the states of field capacity and the wilting point is known as the plant available water or **available water**.

The amount of water held in each state is related to a number of factors including **soil texture, soil structure** (see below) and organic matter content. Since soil water occurs as films around soil particles, if there are many small particles (i.e. clay size particles) in a soil it will hold more water due to the greater soil surface area per unit volume. However, much of the water held in soils with a high proportion of clay size particles is unavailable to plants (i.e. hygroscopic water) because it is held in very small pores. The general relationship between soil texture and soil water availability is shown in Figure 7.4. Soil structure influences the nature and abundance of soil pores and soil permeability and therefore the rate at which water drains through the soil. For example, if a soil has a lot of well-connected pores and is very permeable, water will percolate rapidly through it. Organic matter increases the soil's moisture holding capacity and indirectly affects water content through its influence on soil structure and total pore space (White, 1997).

7.2.4 Soil air

Soil air occupies pores that are not filled with water. Soil animals, plant roots and most microorganisms use oxygen and release carbon dioxide when they respire. In order to

maintain biological activity, oxygen needs to move into the soil and carbon dioxide must move out of the soil. This ventilation of the soil is known as **aeration**. Aeration is affected primarily by the pore size distribution, pore continuity, the soil water content and the rate of oxygen consumption by respiring organisms. As soil air is 'compartmentalized' by the presence of water and intervening soil particles, the composition of the soil air differs from that of atmospheric air. The composition of soil air varies considerably from place to place in the soil. Generally it also has higher moisture and carbon dioxide and lower oxygen concentrations than the atmosphere (see Table 7.2). Carbon dioxide concentrations are often several hundred times higher than in the atmosphere. However, the composition of soil air is constantly changing with marked diurnal and seasonal fluctuations. These changes are often associated with the differences in biological activity between night and day or between summer and winter. There will be less respiration on a cold winter's night in a

Table 7.2 The composition (% by volume) of soil air relative to the open atmosphere. There tends to be less oxygen and more carbon dioxide in soil air than in the atmosphere because of the respiration of microorganisms in the soil which uses oxygen and produces carbon dioxide

	Atmosphere	Soil air
Nitrogen (N_2)	79.01	79.0
Oxygen (O_2)	20.96	18.0-20.8
Carbon dioxide (CO_2)	0.035	0.15-1.0

temperate zone than on a warm summer's day and so carbon dioxide concentrations in soil air may be much lower in winter.

Reflective questions

➤ What is the difference between primary and secondary minerals?

➤ Why does soil organic matter have such an important impact on the properties of soil?

➤ What factors affect the soil's moisture holding capacity and why?

➤ Why is the concentration of carbon dioxide in soil air higher than that in the air above the soil?

7.3 Soil profile

Soils are described by the characteristics of their **soil profile**. This consists of a vertical section through the soil from the ground surface down to the parent material. It is made up of a series of distinctive horizontal layers known as **soil horizons**. This horizontal alignment is mainly due to the translocation of materials by the movement of water through the soil. The removal of solid or dissolved material from one horizon is called **eluviation**, while the deposition in another horizon is referred to as **illuviation**.

The soil horizons are given letters according to their genesis and their relative position in the profile. The major horizons are shown in Figure 7.5. Note that not all the horizons described here are present in every soil. The O horizon is a surface layer dominated by the accumulation of fresh or partially decomposed organic matter. The A horizon can occur at or near the surface (beneath the O horizon) and contains a mixture of mineral and organic (mainly humus) material and is therefore usually darker than the horizons below. Beneath this occurs the E horizon or elluvial horizon. As the E horizon is a zone of depletion (e.g. of clay, organic matter, iron) it is usually a pale, ashy colour. E horizons are common in high-rainfall areas, especially in soils developed under forests. The underlying B horizon is often a zone of accumulation (e.g. of clay, iron, organic matter, carbonates) often referred to as the illuvial horizon. In some soils, the accumulation of iron oxides in the B horizon gives it a reddish colour. The A, E and B horizons are sometimes referred to as the **solum** (from the Latin for soil or land). It is in the solum

Figure 7.5 A hypothetical mineral soil profile showing the relative position of the major horizons that may be present in a well-drained soil in the temperate humid region. Not all the horizons described here are present in every soil profile, and the relative depths vary.

that the soil-forming processes are active and that plant roots and animal life are largely confined. The B horizon usually grades into the C horizon, which largely comprises unconsolidated weathered parent material known as the regolith. Although the regolith is affected by physical and chemical processes it is little affected by biological activity and therefore not part of the soil solum. If unweathered rock exists below the C horizon it is called bedrock and is designated the R horizon.

In some soil profiles, the soil horizons are very distinct in colour, with sharp boundaries, whereas in other soils the colour change between horizons may be very gradual, and the boundaries difficult to locate. However, colour is just one of the many physical, chemical and biological characteristics by which one horizon may differ from the horizon above or below it.

The informal terms 'topsoil' and 'subsoil' are often used to describe soil. Topsoil refers to the upper portion of the soil (usually the A horizon or plough horizon) and is the part most important for plant growth. The subsoil refers to the part of the soil below the topsoil (plough depth) and usually relates to the B horizon.

Reflective questions

➤ Draw a diagram of the typical soil horizons and explain their importance in determining the properties of soil.

➤ What is the difference between eluviation and illuviation?

7.4 Soil formation processes

7.4.1 Pedogenesis

The process of soil formation, called **pedogenesis**, takes place over hundreds and thousands of years. The soil is an open system, which allows input of materials to the soil, the loss of materials from the soil and internal transfers and reorganization of these materials within the system. Soil horizons develop as a result of a number of processes occurring within the soil, which can be classified into the following categories: additions, removals, mixing, translocations and transformations.

The main input of soil material comes from the parent material of the soil. Mineral particles are released from the parent material by weathering at the base of the soil, and contribute to the lower layers of the soil. Significant inputs of material come from surface accumulation, particularly of organic matter. Inputs also include solutes and particles carried by precipitation and the wind, energy from the Sun and gases from the atmosphere.

The main losses from the soil occur through wind and water erosion and **leaching**. Leaching is the removal of soil material in solution and is most active under conditions of high rainfall and rapid drainage. The percolating water carries soluble substances downwards through the soil profile, depositing some in lower layers but removing the most soluble entirely (see Chapter 15). Removals also include the loss of gases and uptake of solutes by plants.

Mixing of organic and inorganic components is an important process that is carried out by soil animals, microbes and plant roots, freezing and thawing of water, and shrinking and swelling of the soil. Humans also cause physical mixing of the soil by ploughing. Chemical and biological processes can also transform soil components. Organic compounds decay and some minerals dissolve

while others precipitate. These transformations result in the development of soil structure and a change in colour from that of the parent material. Translocation of material within the soil profile often occurs in response to gradients of water potential (e.g. suction) and chemical concentrations within soil pores. Suspended and dissolved substances may move up or down through the soil profile.

The net result of these processes occurring over a long period of time is the formation of different soil horizons. However, the processes that dominate at a particular site are dependent on the environmental conditions at that site. In areas where rainfall exceeds evapotranspiration, net water movement is down through the soil (Figure 7.6a). The extent of leaching is often indicated by the acidity of the soil (Jarvis *et al.*, 1984). In many freely draining soils, clay is carried from the upper horizons by percolating water to lower horizons and this is known as clay eluviation, or lessi-vage (Figure 7.6b). The clay is redeposited as skins or coats on the surfaces of aggregates or in pores and around stones. Soil horizons characterized by clay accumulation are described as **argillic**. Clay eluviation tends to produce a group of soils known as acid brown earths or **luvisols** (Figure 7.7).

Podzolization may occur in soils where there is intense leaching and translocation of material (Figure 7.6c). Organic acids complex with iron and aluminium compounds that are transported downwards from the E horizon by percolating water and deposited in the B horizon. Podzolization occurs on freely drained sites under forests and heath plants and the end product of this process is a soil called a podzol (Figure 7.8), the characteristics of which are the presence of an organic layer, a leached E horizon and an accumulation of iron, aluminium and humic material in the B horizon. These soils are not very productive for agriculture because they are acidic and the free drainage results in leaching of fertilizers away from plant roots.

In many locations waterlogging leads to the reduction, mobilization and removal or redeposition of iron compounds in the soil (Figure 7.6d). The reduction of ferric (Fe^{3+}) to the more mobile, grey ferrous (Fe^{2+}) iron compound (see Chapter 15) by microorganisms is known as **gleying**. The soil loses the brown/red colour of ferric oxide and becomes grey or bluish. Alternate phases of reduction and oxidation due to fluctuations in the water content result in soil having a mottled appearance with brown/red iron oxide spots or streaks occurring along root channels and larger pores as shown in Figure 7.9.

Laterization (ferralitization) occurs in tropical and subtropical soils where high temperatures and heavy rain result in

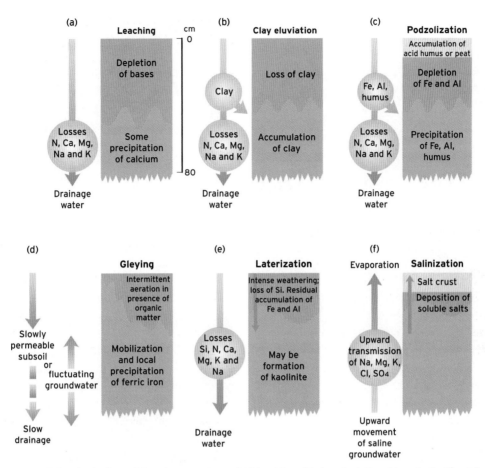

Figure 7.6 The movement of water in the soil-forming processes of (a) leaching, (b) clay eluviation, (c) podzolization, (d) gleying, (e) laterization and (f) salinization. (Source: (a)-(d) Reproduced from National Soil Resources Institute, Bulletin 10, Jarvis *et al.* (1984) 'Soils and their use in Northern England', Soil Survey of England and Wales, Rothamsted Experimental Stations, Harpenden, Herts, Fig. 14, p. 47. © Cranfield University 1984. No part of this publication may be reproduced without the express permission of Cranfield University; (e) and (f) adapted from McRae, 1988)

intense weathering and leaching (Figures 7.6e and 7.10). Almost all the by-products of weathering are leached out of the soil leading to the development of horizons depleted in base cations (e.g. calcium, magnesium, potassium and sodium) and enriched in silica and oxides of aluminium and iron (McRae, 1988). The red colour of these soils is due to the presence of haematite and goethite (Figure 7.10). Conversely in areas where evapotranspiration exceeds rainfall, such as arid and semi-arid areas, water is drawn to the soil surface and as water evaporates salts are precipitated at or near the surface (Figure 7.6f; Chapters 12, 13 (Box 13.5) and 16). This salinization process is almost the complete opposite of leaching.

7.4.2 Factors affecting soil formation

The major processes involved in soil formation described above are controlled by local and regional environmental factors. In the late 1800s, Dokuchaiev, a Russian scientist, was one of the first to recognize that soils do not occur by chance but usually form a pattern in the landscape and develop as a result of the interplay of climate, parent material, organisms and time. Building on this work in the 1930s and 1940s, Hans Jenny suggested that relief was an additional important factor (Jenny, 1941).

7.4.2.1 Climate

Climate is perhaps the most influential factor affecting soil-forming processes as it determines the moisture and temperature regimes under which a soil develops. In addition, climate is influential in determining vegetation distribution (see Chapters 8 and 22; Figure 22.1). Rainwater is involved in most of the physical, chemical and biological processes that occur within the soil, and particularly weathering and

Figure 7.7 An argillic brown earth from Kent, UK. (Source: photo courtesy of John Conway)

Figure 7.9 A stagnohumic gley. (Source: photo courtesy of E.A. Fitzpatrick)

Figure 7.8 A podzol from the Upper Wye catchment, Wales.

leaching. To be effective, however, water must pass downwards through the whole of the soil profile and into the regolith. The amount of precipitation that percolates downwards through the soil is mainly related to total annual precipitation and rate of evaporation (from vegetation and soil), although topography and permeability of the parent material are also important factors. Overall, percolating

water stimulates weathering processes, helps to differentiate the soil into horizons and influences soil depth.

The main effect of temperature on soils is to influence the rate of soil formation via mineral weathering and organic matter decomposition. For every 10°C rise in temperature, the speed of chemical reactions increases by a factor of two or three; biological activity doubles and evaporation of water increases. As rates of chemical weathering are greatest under conditions of high temperature and humidity, soils in tropical areas are often several metres deep while those in polar regions are shallow and poorly developed (Figure 7.11). In addition, soils are influenced by microclimates that are related to altitude and aspect.

7.4.2.2 Parent material

Soils may develop on the weathered surfaces of exposed, consolidated *in situ* rock surfaces, or unconsolidated superficial material that has been transported and deposited by gravity, water, ice or wind. Parent material influences soil formation through the process of weathering and then through the influence of the weathered material on soil processes. Rock types influence the rate of weathering through their mineralogical composition and the surface area of the rock exposed. The larger the exposed surface area, the faster the rate of weathering. Some minerals are more susceptible to weathering than others. Goldich (1938) proposed a 'stability series' for the silicate minerals as shown in Figure 7.12. This arrangement of minerals is identical to Bowen's reaction series, where the silicate minerals are placed in their order of crystallization. The minerals that crystallize first form under much higher temperatures than those that crystallize last. Consequently, the minerals that crystallize first, such as olivine and pyroxene, are not as stable at the Earth's surface, where the temperature and pressure are very different from the environment in which they form. In contrast, quartz, which crystallizes last, is the most resistant to weathering (Figure 7.12).

Knowledge of rock mineralogy allows rocks to be placed in their order of susceptibility to weathering. Hard igneous rocks and Carboniferous and Jurassic sandstones weather slowly to give shallow, stony, coarse-textured soils. In contrast, softer Permo-Jurassic sandstones weather more rapidly to give deeper, less stony, loamy or sandy soils. The soils that develop on all these parent rocks are generally acidic owing to the low base cation content of these rocks or the bases are leached from the soil faster than they are

Figure 7.10 A red laterite soil formed under a wet, humid and warm tropical climate. (Source: photo courtesy of E.A. Fitzpatrick)

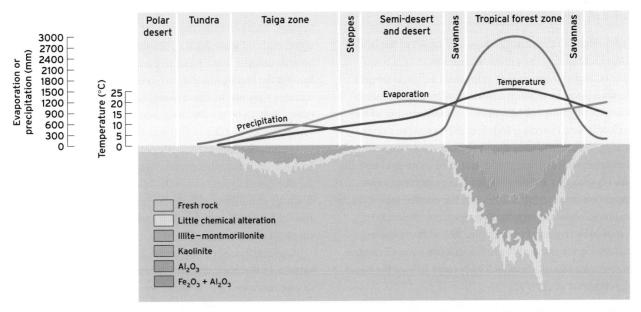

Figure 7.11 Schematic representation of the variation of soil depth with climate and biome from the equator to the north polar region. See Chapter 8 for information on each of the named biomes. Soils are deeper in the wet humid tropics and in the temperate zone and most shallow in dry or very cold locations. The weathering products of aluminium and iron oxides are also shown. (Source: after Strakhov, 1967, as adapted in Birkland, 1999, Fig. 10.5, p. 274)

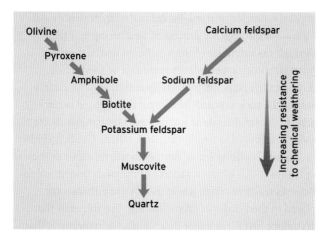

Figure 7.12 Weathering sequence for common rock-forming minerals. The sequence ranks the silicate minerals according to the general ease of weathering.

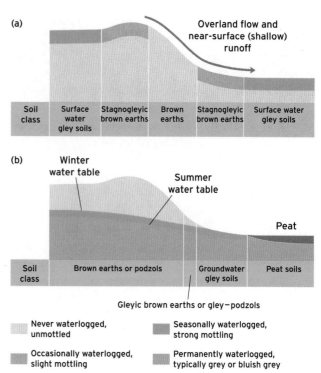

Figure 7.13 Relationship between slope, hydrology and soil formation on (a) slowly permeable parent material and (b) permeable parent material. (Source: Reproduced from National Soil Resources Institute, Bulletin 10, Jarvis *et al.* (1984) 'Soils and their use in Northern England', Soil Survey of England and Wales, Rothamsted Experimental Stations, Harpenden, Herts. Fig. 16, p. 52 © Cranfield University 1984. No part of this publication may be reproduced without the express written permission of Cranfield University)

replenished by weathering. Carboniferous, Permo-Triassic and Jurrassic clays, siltstones, mudstones and shales are all fine-grained rocks which weather to give silty or clayey soils which are generally slowly permeable. The weathering products of chalk and limestone are very soluble, and therefore soil depths are often shallow, particularly on steeper slopes. At the foot of the slope, where deeper soils form, they are well drained and base rich. Further information about parent material weathering can be found in Chapter 11 and the further reading at the end of this chapter.

7.4.2.3 Relief

Relief relates to the altitude, slope and aspect of the landscape and can hasten or delay the influences of climatic factors. Slope steepness is an important factor, as steeper slopes reduce the amount of water infiltrating and percolating through the soil and allow increased erosion of the surface layers. Therefore soils formed on steeper slopes tend to be thin, coarse textured and poorly developed compared with soils on gentler slopes or more level terrain (see Chapter 11). However, weathering rates tend to be greater on steeper slopes, although the weathering products do not accumulate very deeply as they are efficiently removed by erosion. For example, 90% of the dissolved material in the rivers of the Amazon Basin comes from the steep Andes Mountains which only cover 12% of the Basin (Gaillardet *et al.*, 1997).

On slopes with less permeable parent material, surface waterlogging causes gleying on flat ground, whereas the soils on steeper slopes are drier as most precipitation runs off the surface or through the upper horizons to lower

ground. This produces the pattern of soil distribution illustrated in Figure 7.13(a). On slopes with very permeable parent material water tends to penetrate to the subsoil, leaving the higher ground and steep slope well drained, whereas soils on the lower slopes and valley bottoms are more likely to be affected by groundwater as shown in Figure 7.13(b). Milne (1935) was the first to use the term soil **catena** for topographically determined soil profiles in East Africa. Where there is no change in the geology along the slope, soil differences in the catena are brought about by drainage conditions, differential transport, eroded material and the leaching, translocation and redeposition of mobile chemical constituents.

Aspect affects the solar energy received at the ground surface. In the northern hemisphere, south-facing slopes receive more and are therefore warmer and generally lower in moisture than north-facing slopes. Consequently, soils on the south slopes tend to be drier, less densely vegetated, and thus lower in organic matter. These differences are reversed in the southern hemisphere.

Altitude influences climate (see Chapter 6). Temperature declines with altitude and precipitation tends to increase with altitude in the middle latitudes. This leads to an excess of rainfall over evaporation and as a result leaching rates are high and waterlogging occurs where the drainage is poor. The lower temperatures also lead to a reduction in biological activity and therefore slower decomposition of organic matter. This in turn leads to the accumulation of thick organic horizons at the surface and ultimately to the formation of peat.

7.4.2.4 Organisms

Organisms include plants, animals, microorganisms and humans. Vegetation extracts water and nutrients from the soil and under natural conditions returns most of the nutrients it uses to the soil in litter. The type of vegetation influences the type and amount of litter that is returned to the soil. Different soil types support different vegetation communities. Vegetation also protects the soil from water and wind erosion by intercepting rainfall, decreasing the velocity of runoff, binding soil particles together, improving soil structure and porosity, and providing a litter cover which protects the soil surface against raindrop splash.

Earthworms and other small animals such as moles mix and aerate the soil as they burrow through the soil. Earthworms have been found to increase the infiltration rate of fine-textured soils and contribute towards increasing the stability of the soil structure by intermixing organic matter with mineral particles (Curtis *et al.*, 1976). Soil organisms, including fungi, bacteria and single-celled protozoa, play a major role in the decomposition of organic matter. The end product is humus. Humans influence soil formation through manipulation of vegetation, agricultural practices such as drainage and irrigation, the additions of fertilizers, lime and pesticides, and urban and industrial development.

7.4.2.5 Time

Over time soil is continually forming from the parent material, under the influence of the climate, topography, vegetation and soil organisms. Soil genesis is a long process; the formation of a layer 30 cm thick takes from 1000 to 10 000 years. During this time, the properties of the soil continually change. This is manifest by changes in the soil profile including the number of horizons, their depth and their degree of differentiation. When the rate of change of a soil property with time is negligible, the soil is said to be in steady state. However, in reality soil rarely reaches this state

because of changes in one of the environmental factors. For example, changes in the world's climate over geological time accompanied by changes in sea level, erosion and deposition have produced large changes in the distribution of vegetation and parent material. Therefore most soils have not developed under a single set of environmental factors but have undergone successive waves of pedogenesis. The most recent large change in climate resulted in alternating glacial and interglacial periods of the Quaternary (see Chapter 20). In high and middle latitudes, glaciation removed the majority of soils and covered large areas with drift material. Therefore, soil development in these areas began again on new surfaces after the final retreat of the ice about 10 000 years ago.

7.4.2.6 Combined influences

It can be seen that the five factors influencing soil formation do not operate as single independent factors. Climate influences vegetation and human activities and is itself affected by topography. Vegetation is influenced by climate and parent material. The combined influence of the five factors produces a set of soil-forming processes, which results in the world's distinctive soil profiles. Not all soils develop the same amount or combination of horizons and therefore specific combinations of horizons are used to classify soils. Box 7.1 provides details on soil classification schemes.

Reflective questions

➤ What are the similarities and differences between the soil-forming processes of leaching and clay eluviation?

➤ Can you explain why soils are considered as open systems and how this influences soil formation? (Drawing a diagram may help with your answer.)

➤ Can you describe the process of podzolization? What materials are removed? Why does this happen? Where are they being redeposited?

➤ Why are soils in tropical areas often several metres deep while those in polar regions are shallow and poorly developed?

➤ How does soil development vary down a slope of relatively uniform parent material?

TECHNIQUES

SOIL CLASSIFICATION

Not all soils develop the same amount or combination of horizons and therefore specific combinations of horizons are used to classify soils. There are a number of different soil classification systems used throughout the world, many of which are summarized by FitzPatrick (1983) and Gerrard (2000). The two most commonly used are the soil taxonomy (classification) of the United States Department of Agriculture (USDA) and the system used by the Food and Agriculture Organization – United Nations Educational, Scientific, and Cultural Organization (FAO-UNESCO).

The USDA soil taxonomy scheme is a hierarchical classification with soils divided into: (1) order, (2) suborder, (3) great group, (4) subgroup, (5) family and (6) series. There are 11 orders that are differentiated by the presence or absence of diagnostic horizons, features that show the dominant set of soil-forming processes that have taken place or chemical properties. This is essentially a subjective process as there are no fixed principles involved. A brief summary of the characteristics of the soil orders is presented in Table 7.3. Suborders are differentiated using criteria that vary from order to order. The number of subgroups ranges from two to seven

per order. In the differentiation of the great groups, the whole assemblage of horizons is considered, together with a number of diagnostic features. Great groups are divided into subgroups by the addition of adjectives to the great groups' names. Further subdivision into families occurs on the basis of physical and chemical properties. The final level of subdivision, the series, is achieved on the basis of the locality in which that type of soil was first recognized. It has no real value in terms of soil classification but is used in soil mapping at more detailed scales.

The FAO-UNESCO (1974) scheme was designed for the production of the Soil Map of the World. It is now a very

Table 7.3 Characteristics of the soil orders of the soil taxonomy scheme of soil classification

Order	Characteristics
Alfisols	Generally possess an argillic horizon and a moderate to high base saturation. They are common in humid temperate environments
Andisols	Form on parent material of volcanic origin, especially ash, and are often very fertile
Aridisols	Occur in both cold and hot dry areas where effective rainfall is low and often have a high salt content
Entisols	Recently formed soils with limited development of horizons
Histosols	Organic-rich soils
Inceptisols	Soils of humid regions with altered horizons that have lost material by leaching but still contain some weather-able minerals
Mollisols	Dark-coloured, base-rich soil of temperate grasslands
Oxisols	Red, yellow or grey soils of tropical and subtropical regions. They have strongly weathered horizons enriched in silica, clay and oxides of aluminium and iron and are acidic with a low nutrient status
Spodosols	Podzolic soils that have an organic surface horizon, a bleached grey to white eluvial horizon and a B horizon enriched in organic matter and oxides of iron and aluminium. These soils are coarse textured, highly leached and acidic, with a low nutrient status
Ultisols	Highly weathered, leached, acidic and have a low nutrient status. They usually have an argillic horizon and the B horizons of well-drained soils are red/yellow due to the accumulation of iron oxides. They have developed over a long period of time in humid, warm, temperate and tropical regions
Vertisols	Swelling clay soils with deep, wide cracks

BOX 7.1 ➤

➤

widely used scheme. It has 28 major soil groups that are subdivided into 153 units. The major group names come from a number of linguistic routes and many have been used before in other classification schemes, while others have been newly devised. Approximate equivalents with the soil taxonomy orders are shown in Table 7.4. Many of the soil characteristics used to define the major soil groups are morphological, such as texture, structure and colour, while soil processes or chemistry define other groups. The division of groups into units is based on the presence or absence of diagnostic horizons and properties. One advantage of this scheme is that it is less hierarchical than many other schemes.

The Soil Survey of England and Wales classified soils according to broad differences in the composition or origin of the soil material and the presence or absence of specific diagnostic features. At the highest levels soils are divided into a small number of categories known as major soil groups, the distribution of which is shown in Figure 7.14. Table 7.5

Raw gley soils   Brown soils   Groundwater gley soils
Lithomorphic soils   Podzolic soils   Made ground soils
Pelosols   Surface-water gley soils   Peat soils

Figure 7.14 The distribution of the major soil groups in England and Wales. (Source: Reproduced from National Soil Resources Institute, National Soil Map, 2004. © Cranfield University and for the Controller of HMSO 2004. No part of this publication may be reproduced without the express written permission of Cranfield University)

Table 7.4 Approximate relationship between the USDA soil taxonomy soil orders and FAO-UNESCO major soil groupings

USDA soil taxonomy soil orders	FAO-UNESCO major soil groups
Alfisols	Luvisols
Andisols	Andosols
Aridisols	Calcisols, Gypsisols, Solonchaks, Solonetz
Entisols	Arenosols, Fluvisols, Leptosols, Regosols
Histosols	Histosols
Inceptisols	Cambisols
Mollisols	Chernozems, Greyzems, Kastanozems, Phaeozems
Oxisols	Alsiosl, Ferralsols, Nitosols, Plinthosols
Spodosols	Podzols
Ultisols	Acrisols, Lixisols
Vertisols	Vertisols

BOX 7.1 ➤

Table 7.5 Description of the major soil groups in the UK

Major soil type	Characteristics
Raw gley soils	Soils that occur in mineral material that has remained waterlogged since deposition. They are chiefly confined to intertidal flats or saltings
Lithomorphic soils	Shallow soils usually well drained and formed directly over bedrock in which the only significant soil-forming process has been the formation of an organic or organic-enriched mineral surface
Pelosols	Slowly permeable non-alluvial clayey soils that crack deeply in dry seasons
Brown soils	Generally free-draining brownish or reddish soils overlying permeable materials
Podzolic soils	Soils with dark brown, black or ochreous subsurface layers resulting from the accumulation of iron, aluminium or organic matter leached from the upper layers. They normally develop as a result of acid weathering conditions
Surface water gley soils	Seasonally waterlogged, slowly permeable soils
Groundwater gley soils	Soils with prominently mottled or grey subsoils resulting from periodic waterlogging by a fluctuating groundwater table
Peat soils	Predominantly organic soils derived from partially decomposed plant material that accumulates under waterlogged conditions
Human-made soils	Predominantly organic soils derived from partially decomposed plant material that accumulates under waterlogged conditions

provides a brief description of the main characteristics of these major soil groups. Within each major group, soils are progressively subdivided into soil group, soil subgroup and then into soil series (Avery, 1990). A slightly different system was developed for soils in Scotland (MISR, 1984). However, it is possible to amalgamate the two systems to show the relative occurrence of the major soil groups in the United Kingdom (Table 7.6).

Table 7.6 The occurrence (%) of the major soil groups in the UK

Soil type	England and Wales	Scotland
Lithomorphic soil	7	10
Brown soils (including pelosols)	45	19
Podzols	5	24
Gley soils (including surface water and groundwater)	40	23
Peat soils	3	24

(Source: Avery, 1990; MISR, 1984)

BOX 7.1

7.5 Physical properties of soil

Soil physical properties are those properties of the soil that you can see, feel, taste and smell. By observing **soil colour**, we can estimate organic matter content, iron content, soil drainage and soil aeration. By feeling the soil we can estimate the kinds and amounts of different size particles present. Soil physical properties have a huge influence on how soils function in an ecosystem and how they can be managed.

7.5.1 Soil colour

Soil colour is easy to observe and although it has little effect on the soil, it is possible to use it to determine the nature of soil properties such as organic matter content, aeration and drainage characteristics. Colour also helps us to distinguish the different soil horizons of a soil profile. Soils with a higher amount of organic matter are black or dark brown in colour. Surface horizons are usually darker than subsequent horizons owing to their higher organic matter content (see Figures 7.7, 7.8 and 7.9). Soil colour can be used to determine the drainage characteristics of a soil because of the colour change that takes place when various iron-containing minerals undergo **oxidation** and **reduction**. In well-drained soils iron is oxidized and imparts a reddish or yellowish colour to the soil. In water-logged soils, the iron minerals are reduced, owing to the anaerobic conditions, and impart a grey or blue colour to the soil.

A standard system for soil colour description has been developed using the **Munsell colour chart**. In this system, three measurable variables determine colour. **Hue** is the dominant colour of the pure spectrum (usually redness or yellowness), **value** is the degree of darkness or lightness of the colour (a value of 0 being black) and **chroma** is the purity or strength of the colour (a chroma of 0 being natural grey). By using a standard colour book, the observer can express soil colour as a letter numerical code as well as descriptively. For example, in the case of Black (10Y/R 2/1) the hue is 10Y/R, value 2 and chroma 1.

7.5.2 Soil texture

Mineral particles in the soil vary considerably in size from boulders (greater than 600 mm in diameter) and stones (greater than 2 mm in diameter) down to **sand, silt** and **clay**. The sand-, silt- and clay-sized particles are often referred to

Table 7.7 Soil particle size classification schemes

Size class	Size range (diameter, mm) International or Atterberg System*	USDA-FAO System[†]	SSEW, BS and MIT System[‡]
Clay	<0.002	<0.002	<0.002
Silt	0.002-0.02	0.002-0.05	0.002-0.06
Sand	0.02-2	0.05-2	0.06-2
Gravel	>2	>2	>2

*This system subdivides sand into fine and coarse fractions.
[†]This system subdivides sand into very fine, fine, medium, coarse and very coarse fractions.
[‡]This system is adopted by the Soil Survey for England and Wales, British Standards and the Massachusetts Institute of Technology and subdivides sand into fine, medium and coarse fractions and refers to gravel as stones.

as the fine fraction or **fine earth**, and are usually separated from the larger soil particles by passing through a sieve with 2 mm diameter holes. Within the fine earth fraction, size definitions vary between different systems (Table 7.7). All set the upper limit of clay as 2 μm (2 micrometres; two-millionths of a metre) but differ in the upper limit chosen for silt and the way in which the sand fraction is subdivided (see also Chapter 12, e.g. Figure 12.1).

Soil particles are classified into different size fractions because as particles become smaller they have different properties (see Table 7.8). In particular, as particles become smaller the total surface area of the soil particles in the soil becomes larger and this has a large influence on water holding capacity, **cation exchange capacity (CEC)** (see below) and rate of mineral weathering.

It is extremely rare for soils to be composed of a single particle size class. Thus, soil texture refers to the relative proportions of the sand-, silt- and clay-sized fractions in a soil. The classification of texture in terms of particle size distribution is normally shown as a triangular diagram (Figure 7.15). Combinations of different proportions of sand, silt and clay result in 11 main textural classes although the number of classes may vary between different countries. Triangular diagrams can be used to determine a textural class if the particle size distribution is known, or to determine a range of particle size distributions if a textural class is known. The texture of a soil can be determined by measuring the particle size in the

Table 7.8 Influence of soil separates on some properties and behaviour of soil

Property/behaviour	Sand	Silt	Clay
Water holding capacity	Low	Medium	High
Aeration	Good	Medium	Poor
Drainage rate	High	Slow to medium	Slow
Organic matter decomposition	Rapid	Medium	Slow
Compaction	Resists	Easily compacted	Easily compacted
Susceptibility to water erosion	Low	High	Low
Ability to hold nutrients	Poor	Medium to high	High
Leaching of pollutants	Allows	Moderately retards	Retards

(Source: Brady, Nyle C.; Weil, Ray R., *The Nature and Properties of Soil*, 12th Edition. © 1999. Adapted by permission of Pearson Education, Inc., Upper Saddle River, NJ)

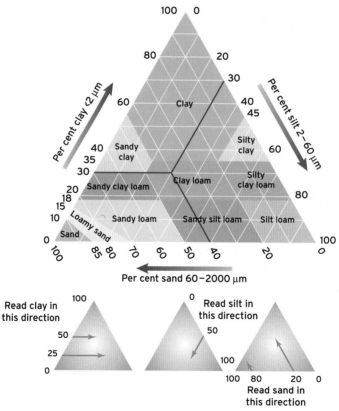

Figure 7.15 The triangular diagram of soil textural classes adopted in England and Wales. For example, a soil with 40% sand, 30% silt and 30% clay is a clay loam, as highlighted in the diagram.

laboratory or in the field by working moist soil between finger and thumb as explained in Box 7.2.

Soil texture is an important property as it greatly influences the soil's ability to absorb and retain water. For example, coarse-textured soils have larger pore spaces because the sand-sized particles do not fit as closely together as the smaller silt- and clay-sized particles. As a result, sandy soils have high percolation rates but lower water retention

DETERMINING SOIL TEXTURE BY FEEL

The simplest and quickest way of assessing soil texture is by working a moist sample of soil in your hand. Begin by removing all stones of more than 2 mm and any large roots, moisten the soil and mould it in your hand for a few minutes, then follow the steps shown in Figure 7.16.

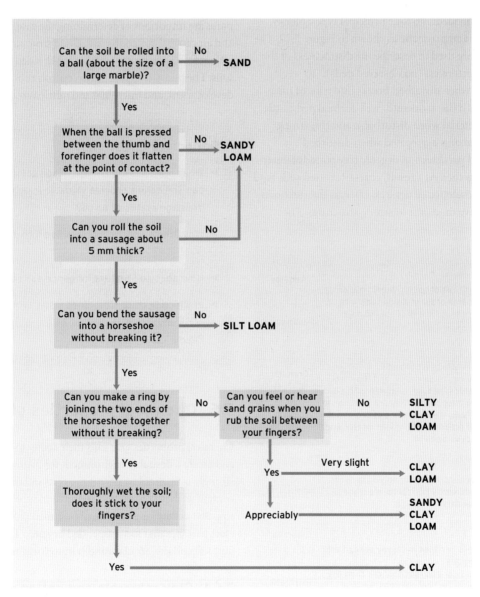

Figure 7.16 Flow chart for determining soil texture by feel.

BOX 7.2

capacities because water passes rapidly through the pores and little sticks to the soil particles. The opposite is observed for fine-textured soils.

7.5.3 Soil structure

Soil particles normally do not remain detached from one another. Instead, soil particles tend to adhere to each other, forming larger groupings called **aggregates** or **peds**. Soil structure is characterized in terms of the shape (or type), size and distinctness (or grade) of these peds. Each ped is separated from another by voids or natural surfaces of weakness. Soil structure is divided into four principal types: blocky, spheroidal, platy and prismatic as shown in Figure 7.17. The following terms are used to describe the distinctness of the structure: (i) structureless (no observed peds); (ii) weak (indistinct peds; when disturbed breaks into a lot of unaggregated material); (iii) moderate (well-formed peds; little unaggregated material when disturbed); and (iv) strong (distinct peds; remains aggregated when disturbed).

The clay-sized particles and organic compounds largely hold the peds together. As a result, coarse-textured soils tend to have weakly developed structures, whereas fine-textured soils generally have moderate to strong structures. The

strength with which the individual peds are held together influences both the soil's resistance to erosion and ease of cultivation. A strong structure holds the soil together and causes it to resist erosion. The same characteristics, however, make it difficult to plough (White, 1997). Ploughing tends to alter and weaken the soil structure, and the passage of farm machinery leads to soil compaction.

Clearly the size, shape and arrangement of the peds determine the pore space or porosity of the soil. A soil with a well-developed structure is typically less compact and has a greater permeability and porosity than does a coarse-grained soil with a poor structure. The size and connectivity of the soil pores are important in determining the ease with which water and air move through the soil. Good structural development is therefore necessary to obtain well-drained and well-aerated soils. The porosity of a soil also controls root movement and development, and microbial and nutrient movement.

Reflective questions

➤ What particles contribute to soil colour and how can soil colour provide valuable insight into the drainage status of a soil?

➤ Why are soil particles classified into different size fractions?

➤ How does soil texture influence soil structure?

➤ In which texture class does a soil with 10% sand, 60% silt and 30% clay fall? (It may help to look at Figure 7.15.)

7.6 Chemical properties of soil

Chemical properties give soils their ability to hold nutrients and create a desirable environment for plant growth. They are strongly influenced by parent material and organic matter content as they control the amount and type of **colloids** in a soil. **Soil colloids** are very small (less than 0.002 mm in diameter) particles that stay suspended in water. The most important soil colloids are clays and humus (organic matter).

7.6.1 Clay minerals and cation exchange

Clay minerals are formed from the weathering products of aluminium and silicate minerals. Box 7.3 explains the structure of clay minerals. Clay minerals are extremely small (<0.002 mm in diameter), and have a large surface area and

Main type	Subtype
Blocky Common in B horizons of soils in humid regions. Peds are of the same order of magnitude with flattened surfaces that form the faces of adjacent peds	Angular Subangular
Spheroidal Characteristic of surface A horizons. Peds are roughly equidimensional and there are two types: *granular* which are relatively non-porous, and *crumb* which are porous. They are common in the A horizon of soils under grass and deciduous forest, where the presence of organic matter and roots help their development	Granular Crumb
Platy Common in E horizons and compacted clay soils. Peds with predominantly horizontal cleavage and short vertical axis	
Prismatic Characteristic of B horizons in clayey soils of arid and semi-arid regions. Peds with the vertical axis much longer than the horizontal one form vertical columns; vertical faces are usually well defined with angular tops. Prisms with rounded tops are referred to as *columnar*	Columnar Prismatic

Figure 7.17 Diagrammatic representation of the main types of soil structure.

FUNDAMENTAL PRINCIPLES

THE STRUCTURE OF CLAY MINERALS

The structure of all clay minerals is based on two types of sheets consisting of repeating units of (i) a silicon (Si) atom surrounded by four oxygen (O) atoms in the form of a tetrahedron as shown in Figure 7.18(a) and (ii) an aluminium (Al) or magnesium (Mg) atom surrounded by six oxygen (O) or hydroxy (OH) atoms in the shape of an octahedron (Figure 7.18b). The individual units are linked together by sharing oxygen atoms to form silicon tetrahedral sheets and aluminium octahedral sheets (Figure 7.19). All clay minerals are built from various combinations of these two sheets.

Alternating sheets of one tetrahedral sheet and one octahedral sheet produce what are known as 1:1 clays. Kaolinite is the commonest 1:1 clay mineral (Figure 7.20). Each pair of sheets is held together by hydrogen ions, making it a relatively rigid and

stable structure. In 2:1 clays, the aluminium octahedral sheet is sandwiched between two silicon tetrahedral sheets. There are many different types of 2:1 clays that are distinguished on the basis of how the unit layers are held together and the spacing between the unit layers. In illite, the unit layers are held together by potassium (K^+) ions, which make it a relatively stable clay (Figure 7.20). In contrast, weak oxygen bonds hold the unit layers in **montmorillonite** together. As a result, water molecules

can penetrate between the layers enabling it to expand and contract on wetting and drying. The structures of some of the different clay minerals are shown in Figure 7.20.

Figure 7.19 The structure of (a) a silicon tetrahedral sheet and (b) an aluminium octahedral sheet.

Figure 7.18 The structure of (a) a silicon tetrahedron and (b) an aluminium octahedron.

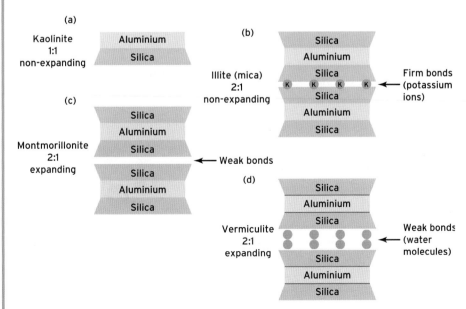

Figure 7.20 The arrangement of silicon and aluminium sheets in (a) kaolinite, (b) illite, (c) montmorillonite and (d) vermiculite. (Source: adapted from Smithson *et al.*, 2002)

BOX 7.3

193

an electrical charge so they are able to attract and hold water and cations. Therefore, they have a fundamental influence on both the physical and chemical properties of the soil. The electrical charge on clay colloids results from **isomorphous substitution** which occurs during the formation of clay minerals. It is when one atom in the crystal lattice is replaced by another atom of similar size without disrupting or changing the crystal structure of the mineral. For example, Al^{3+} may replace Si^{4+} in the tetrahedral sheet (Figure 7.19a) and Mg^{2+} or Fe^{3+} or Fe^{2+} or Ca^{2+} can replace Al^{3+} in the octahedral sheet (Figure 7.19b). The clay mineral structure is electrically neutral, but as the replacing ion generally has a lower positive charge than the ion it replaces, the clay mineral becomes electrically charged. For example, the silicon ion has a positive charge of 4 (i.e. Si^{4+}). If silicon is replaced by aluminium (Al^{3+}), which has a charge of 3, then the clay has lost one positive charge, which leaves one unsatisfied negative charge on the clay.

The net negative charge on clay minerals is balanced by cations (e.g. Ca^{2+}, Mg^{2+}, K^+, Na^+, Al^{3+} and H^+), which are attracted to the surface of the clay minerals and held (**adsorbed**) there by electrostatic attractions. They are referred to as **exchangeable cations** because cations in the soil solution can displace adsorbed cations on the clay surface. Interchange between a cation in solution and another on the surface of a colloid is known as **cation exchange**. All cation exchange reactions must be chemically balanced. For example, if a clay containing Na^+ is washed with a solution of $CaCl_2$, each Ca^{2+} ion will replace two Na^+ ions as Ca^{2+} has double the charge of Na^+, and Na^+ will be washed out in the solution, as shown in Figure 7.21.

Cation exchange reactions are also rapid and reversible. The distribution of the cations between soil and solution depends on their relative concentrations and the force of attraction to the negatively charged surface of the clay mineral. Cations adsorbed on the exchange sites are in equilibrium with cations in the soil solution. For example, if Ca and Mg are the dominant cations in the soil solution they will also dominate the exchange sites. In general, the strength of adsorption increases as the charge of the cation increases and the size of the hydrated cation decreases. The sequence of preferred adsorption is: $Al^{3+} > Ca^{2+} > Mg^{2+} > K^+ > Na^+$ (Cresser *et al.*, 1993).

Cation exchange capacity, commonly abbreviated to CEC, is the ability or capacity of a given quantity of soil to hold cations. This capacity is directly dependent on the overall net negative charge of the colloids present in the soil. It is usually expressed as milliequivalents (meq) per kg of oven-dried soil. The main factors controlling the CEC of a soil are the number of colloids present (soil texture), the type of colloids present and organic matter content. The CEC of a soil is a very important property as it controls both soil fertility and soil acidity. Soils with a high CEC usually have a high capacity to store nutrients and are therefore potentially more fertile than soils with a low CEC.

The soil cations that are readily adsorbed onto soil colloids can be divided into two groups. Firstly there are the base cations, which include the important plant nutrients Ca^{2+}, Mg^{2+}, K^+ and Na^+. Secondly there are acid cations, which include Al^{3+} and H^+. Related to this distinction in cations is the term **base saturation**, which is defined as the

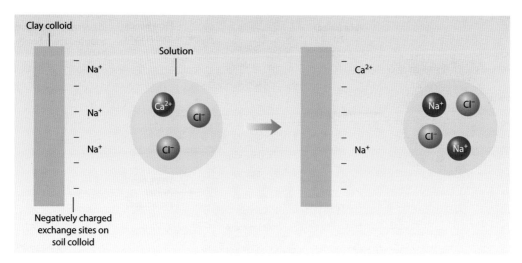

Figure 7.21 Schematic diagram representing cation exchange between calcium ions in solution and sodium ions held on the surface of a negatively charged colloid.

proportion of exchange sites occupied by base cations and is calculated as follows:

Base saturation (%)

$$= \frac{(Ca^{2+} + Mg^{2+} + K^+ + Na^+)}{CEC} \times 100 \qquad (7.1)$$

A soil with a high base saturation (greater than 35%) is more fertile than a soil with a low base saturation.

7.6.2 Soil acidity

The degree of acidity or alkalinity of a soil is an important variable as it affects most soil physical, chemical and biological processes. Whether a soil is acidic, neutral or alkaline is determined by measuring the hydrogen ion concentration in the soil solution. In pure water at 24°C, water ionizes to give equal concentrations of hydrogen (H^+) and hydroxide (OH^-) ions:

$$H_2O \Leftrightarrow H^+ + OH^- \qquad (7.2)$$

The concentration of both H^+ and OH^- ions is 1×10^{-7} (0.000 000 1) **moles** per litre. As it is inconvenient to use these very small numbers to express the concentrations of H^+ ions, a simpler method of using the negative logarithm of the hydrogen concentration was developed, known as pH (see Chapter 15).

As the pH scale is logarithmic, a change of one unit represents a 10-fold change in hydrogen concentration (Table 7.9). For example, a 10-fold increase in H^+ ion concentration from 1×10^{-5} to 1×10^{-4} moles per litre is represented by a one-unit decrease in pH from 5 to 4. Table 7.9 also shows the inverse relationship between the concentrations of H^+ and OH^- ions. As one increases the

other must decrease proportionally as the product of the H^+ and OH^- concentrations must always equal 1×10^{-14}.

Although the pH scale ranges from 1 to 14, most soils have a pH of between 3.5 and 9. Very low values are often associated with soils rich in organic matter, whereas high values usually result from the presence of sodium carbonate. Although the reason a soil becomes acidic is because of excess H^+ ions in the soil solution, it is the presence of aluminium that is largely responsible for producing these H^+ ions as discussed in Box 7.4.

There are a number of natural processes and human-induced changes that result in hydrogen and aluminium becoming the predominant cations in the soil and thus increase its acidity:

- Leaching; percolating water removes base cations (Ca^{2+}, Mg^{2+}, K^+ and Na^+) faster than their rate of release from weathering and therefore cation exchange sites become dominated by aluminium and hydrogen.
- The respiration of roots and microbes, and the decomposition of organic matter, release hydrogen ions.
- The addition of acids such as H_2SO_4 and HNO_3 from the atmosphere in acid rain also increases hydrogen ions in the soil.
- Use of acid-forming fertilizers. For example, the application of ammonium-based fertilizers results in nitrate and H^+ release.
- Harvesting of the crop removes the base cations from the soil as they are not returned to the soil as litter.

Soil pH affects which plants grow well as they vary considerably in their tolerance to soil pH. Soil pH also determines the fate of many pollutants, affecting their

Table 7.9 The pH scale

pH	H$^+$ concentration (moles per litre)	OH$^-$ concentration (moles per litre)	Description
3	0.001	0.000 000 000 01	Excessively acidic
4	0.0001	0.000 000 0001	Strongly acidic
5	0.000 01	0.000 000 001	Moderately acidic
6	0.000 001	0.000 000 01	Slightly acidic
7	0.000 0001	0.000 0001	Neutral
8	0.000 000 01	0.000 001	Alkaline
9	0.000 000 001	0.000 01	Strongly alkaline
0	0.000 000 000 1	0.0001	Excessively alkaline

FUNDAMENTAL PRINCIPLES

HOW ALUMINIUM INFLUENCES SOIL ACIDITY

As clay minerals weather and break down, the aluminium in the octahedral layer is released into the soil solution, where it either reacts with water or is adsorbed onto the exchange sites of negatively charged clay minerals. Al^{3+} ions are adsorbed in preference to all the other major cations. The influence that aluminium has on soil acidity is itself dependent on the acidity of the soil. At pH less than 5, aluminium is soluble and exists as Al^{3+}. When Al^{3+} enters the soil solution it reacts with water (it is hydrolysed) to produce H^+ ions:

$$Al^{3+} + H_2O \Leftrightarrow AlOH^{2+} + H^+ \quad (7.3)$$

Thus the acidity of the soil increases (pH falls). In soils with a pH of between 5 and 6.5, aluminium also contributes H^+ ions to the soil solution but by different mechanisms, as aluminium can no longer exist as Al^{3+} ions but is converted to aluminium hydroxy ions:

$$Al^{3+} + OH^- \Leftrightarrow AlOH^{2+} \quad (7.4)$$
$$AlOH^{2+} + OH^- \Leftrightarrow Al(OH)_2^+ \quad (7.5)$$
$$\text{aluminium hydroxy ions}$$

These hydroxy aluminium ions act as exchangeable cations, just like Al^{3+}, and are adsorbed by the clay minerals. They are in equilibrium with hydroxy aluminium ions in the soil solution, where they produce H^+ ions by the following reactions:

$$AlOH^{2+} + H_2O \Leftrightarrow Al(OH)_2^+ + H^+ \quad (7.6)$$
$$Al(OH)_2^+ + H_2O \Leftrightarrow Al(OH)_3 + H^+ \quad (7.7)$$

In soils where the pH is above 7, Ca^{2+} and Mg^{2+} dominate the exchange sites and most of the hydroxy aluminium ions have been converted to gibbsite ($(AlOH)_3$), which is insoluble and cannot be adsorbed by the negative clay minerals as it has no charge. The general relationship between soil pH and the composition of cations held on the exchange sites of clay minerals is presented in Figure 7.22. In a neutral soil the exchangeable cations that dominate the cation exchange sites are the base cations, whereas in an acidic soil aluminium and hydrogen ions dominate the exchange sites.

Figure 7.22 Relationship between soil pH and the cations held on the exchange sites of colloids, including clay minerals. (Source: Brady, Nyle C. and Weil, Ray R., *The Nature and Properties of Soil*, 13th Edition. © 2002. Adapted (electronically reproduced in case of e-use) by permission of Pearson Education, Inc., Upper Saddle River, NJ)

BOX 7.4

breakdown, solubility and possible movement from the soil to surface waters and groundwaters. For example, many heavy metals become more water soluble under acid conditions and can move down with water through the soil to aquifers and surface waters (see Chapter 15).

There are 16 **essential elements** without which green plants cannot grow normally. The availability of these essential nutrients for plant uptake is greatly influenced by soil pH (Figure 7.23) as are the number, species and activities of soil organisms. All 16 essential elements must be present in the correct proportions as too little or too much of any element will result in symptoms of nutrient deficiency or toxicity. On the basis of their concentration in plants they are divided into **macronutrients** (carbon, oxygen, hydrogen, nitrogen, phosphorus, sulphur, calcium, magnesium, potassium and chloride) and **micronutrients** (iron, manganese, zinc, copper, boron and molybdenum).

Strongly acidic soils (pH 4–5) usually have a low or reduced supply of the macronutrients, particularly calcium, magnesium, potassium, nitrogen, phosphorus and sulphur,

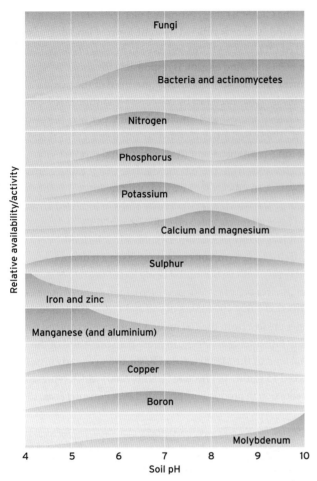

Figure 7.23 The influence of soil pH on the availability of plant nutrients and activity of soil organisms.

7.7 Impact of human activities on soils and soil processes

The soil functions described in this chapter are at risk from human activities. Pressures include agriculture, drainage, extraction, application of wastes and urban development. These pressures can lead to soil degradation such as soil erosion, contamination (by heavy metals, organic contaminants such as pesticides, radionuclides (from nuclear waste) and excessive use of nitrogen and phosphorus fertilizers), acidification, soil compaction, loss of organic matter, salinization and loss of biodiversity. All of these threats lower the current and/or future capacity of the soil to support human life. At a global level, the total area of soil that has been degraded by human activities (almost 20 million km^2) exceeds the total area of farmland (15 million km^2) and the main causes are deforestation, overgrazing and poor agricultural management (Oldeman *et al.*, 1991). In Europe, an estimated 633 million hectares (6.33 million km^2) are affected by some kind of degradation process (Table 7.10). In this section emphasis is placed on the most severe soil degradation problems. For each of these the causes, magnitude, impact on soil function and remedies will be briefly discussed.

7.7.1 Soil erosion

Soil erosion is a two-phase process consisting of the detachment of individual particles from the soil mass and their transport (see Chapter 11). It is a natural process but is accelerated by human activities that expose the soil during times of erosive rainfall or windstorms, or that increase the amount and speed of **overland flow**. Farming practices such

and high, often toxic concentrations of the micronutrients, especially iron, manganese and zinc. While most nutrients are more soluble in acid soils than neutral or slightly basic soils, phosphorus and molybdenum become insoluble at low pH and unavailable to plants. Soil pH also influences plant growth by the influence of pH on activity of beneficial microorganisms and root activity. In acidic soils the number and activity of many soil organisms are reduced, including worms, bacteria that convert ammonium to nitrate, organisms that break down organic matter and nitrogen-fixing bacteria.

In contrast, high soil pH, however, leads to phosphorus and boron becoming insoluble and unavailable to plants and concentrations of the micronutrients, particularly iron, manganese, zinc and copper, becoming so low that plant growth is restrained. Therefore a pH range of 6 to 7 is generally most favourable for plant growth as most plant nutrients are readily available in this range. However, some plants have adapted to a pH above or below this range.

Table 7.10 Estimated areas affected by major soil threats in Europe

Threat	Area affected (million hectares)	Percentage of total European land area
Water erosion	115	12
Wind erosion	42	4
Acidification	85	9
Pesticides	180	19
Over-fertilization	170	18
Soil compaction	33	4
Organic matter loss	3.2	0.3
Salinization	3.8	0.4

(Source: Oldeman *et al.*, 1991)

as overgrazing, removal of vegetation and/or hedgerows, ploughing up and down slopes, abandonment of terraces, compaction by heavy machinery and poor crop management may have these effects.

The GLASOD study estimated that 15% of the Earth's ice-free land surface is afflicted by some form of land degradation (GLASOD, 1990). Of this soil erosion by water is responsible for 56% and wind erosion is responsible for about 28%. During the past 40 years, nearly one-third of the world's arable land has been lost by erosion and continues to be lost at a rate of more than 10 million hectares per year (Pimentel *et al.*, 1995). Soil erosion's most serious impact is its threat to the long-term sustainability of agricultural productivity, which results from the 'on-site' damage that it causes. Erosion by water can quickly remove large volumes of soil as shown in Figure 7.24, which bury or destroy crops in localized areas, and leave channels, **rills** and gullies that in the worst case can inhibit agricultural machinery cultivating the land.

In the long term, soil erosion results in a reduction of soil depth, with fertile topsoil being lost at the rate of several millimetres per year (Morgan, 1986). In the United States, an estimated 3.6×10^9 tonnes of soil and 118×10^9 tonnes of water are lost from the 160 million hectares of cropland each year (Pimentel *et al.*, 1995). In more than one-third of the total land of the Mediterranean Basin, average yearly losses exceed 15 tonnes per hectare (UNEP, 2000) which represents a reduction in productivity of around 8% (Pimentel *et al.*, 1995).

Figure 7.24 Severe gully erosion on farmland in Bolivia. (Source: Ron Giling/STILL Pictures The Whole Earth Photo Library)

In the Russian Federation, it is estimated that the humus content of agricultural soils decreases by about 1% each year (Karavayeva *et al.*, 1991) and that long-term productivity is endangered since the annual erosion rates exceed the rate of humus production. In India, 113 and 38 million hectares are subject to water and wind erosion, respectively, and between 5.37 and 8.4 million tonnes of plant nutrients are lost every year due to soil erosion (Gerrard, 2000).

In addition to 'on-site' effects, the soil that is detached by accelerated water or wind erosion may be transported considerable distances. This gives rise to 'off-site problems', including sediment deposition on roads and in watercourses and reservoirs (see Chapter 12). Another major off-site impact results from agricultural chemicals (fertilizers, pesticides, heavy metals) that often move with eroded sediment. These chemicals can pollute downstream watercourses.

A few studies have attempted to estimate the economic impact of soil erosion. For example, Pimentel *et al.* (1995) estimated that soil erosion in the United States translates into an on-site economic loss of more than $27 billion each year, of which $20 billion is for replacement of nutrients and $7 billion for lost water and soil depth. They also state that the total on- and off-site costs of damage by wind and water erosion and the cost of erosion prevention each year is $44 billion. In 1991, the direct cost impact of soil erosion in Spain was estimated at $212 million per year, including the loss of agricultural production, impairment of water reservoirs and damage due to flooding (ICONA, 1991). In addition, the cost of attempts to fight erosion and restore the soil were estimated at about $3 billion over a period of 15–20 years (ICONA, 1991).

Measures to control erosion include the retention or planting or strips of permanent vegetation (trees or hedges) to form shelter belts to reduce the effects of wind (see Chapter 6), and tillage techniques such as contour ploughing, strip and alley cropping, and use of cover crops to reduce the rate at which water is able to move across the soil surface. Rotation farming, adjusting stocking levels and agro-forestry practices can also be used to help reduce soil erosion. Although severe soil erosion is nearly always irreversible, in less severe cases damage can still be prevented.

7.7.2 Soil acidification

Soil acidification is a natural process that occurs over the long term, but it has recently been enhanced by human action though the emission of sulphur and nitrogen compounds from the combustion of fossil fuels, resulting in acid rain. Humans also influence the acidity of soils by agricultural practices such as the harvesting of crops, the draining of waterlogged soils and the overuse of nitrogen fertilizers. However, in central and western Europe and North America acidic deposition is by far the most important cause of soil acidification.

Soil acidification is not a visually obvious feature although it has a large impact upon the ecosystem. The most sensitive soils to acidification are those derived from base-poor igneous (e.g. granite) and metamorphic rocks. Soils developed on these rocks tend to have low base cation and clay contents. Soils containing carbonates or with higher base cation and clay contents have a greater capacity to buffer acidification. This differential ability of soils to cope with acidification has been examined through use of a 'critical loads' approach as discussed in Box 7.5.

Soil acidification has the effect of increasing the leaching of base cations, such as calcium and magnesium, and depleting the soil's buffering capacity. It also increases the solubility of heavy metals in the soil, such as aluminium, manganese, lead, cadmium and zinc, which can be toxic to plants. This may lead to decreased plant growth or changes in plant communities. For example, forest decline in central Europe is linked with increasing acidity in soils (Figure 7.26). Populations of soil organisms may also change, with a shift towards more acid-tolerant species. As a result, a number of soil processes can slow down. For example, the decomposition of litter becomes slower, leading to surface accumulation. Soil acidification gradually leads to acidification of waters draining from them (see Chapter 15). Acidity and high concentrations of aluminium can lead to deterioration of aquatic life with losses in the diversity and size of invertebrate and fish populations.

Soil acidification can be slowed down by a reduction in acid deposition. Since the 1980s considerable national and international effort has been made to decrease emissions of acidifying pollutants. This is a difficult problem because those countries suffering most from acid rain are not always the main polluters owing to the movement of the pollution in the atmosphere across the planet. In 1979, 34 European and North American countries adopted the Convention on Long-Range Transboundary Air Pollution, which bound them to reducing emissions. The convention agreed firm targets in 1983, when 21 European countries agreed to reduce their sulphur dioxide (SO_2) emissions by 30% from the 1980 levels by 1993. Twelve countries reached this target by 1988. The United Kingdom, however, did not ratify this agreement, but did eventually declare an intention to reduce SO_2 emissions by 30% in the late 1990s. In the United States, an 'Acid Rain Program' commenced in 1995 that aimed to achieve environmental and public health benefits through reductions in emissions of SO_2 and oxides of

HAZARDS

CRITICAL LOADS

Some soils are less able to cope with acidification than others. To help quantify the effects of soil acidification and relate them to the acid deposited, an 'effects-based' approach, known as critical loads, has been developed. A critical load is defined as 'a quantitative estimate of exposure to one or more pollutants below which significant harmful effects on elements of the environment do not occur according to present knowledge' (Nilsson and Grennfelt, 1988). Deposition above that limit may lead to harmful effects on the environment. Maps of critical loads and their exceedances (excess over the critical load) have been used to show the potential for harmful effects to systems at steady state as an aid to developing strategies for reducing pollution. For soil acidification the critical loads are based on the rate of release of base cations from soil minerals by weathering, indicating the capacity of the receiving soil to buffer acid inputs. Figure 7.25 shows critical loads of acidity for soils in the United Kingdom. Critical loads have also been developed for soil–plant systems on the basis of biological indicators that

keq H^+ ha^{-1} yr^{-1}
■ ≤0.2
■ 0.2–0.5
□ 0.5–1.0
■ 1.0–2.0
■ >2.0

Figure 7.25 Empirical critical loads of acidity for soils. (Source: CEH, NSRI, Macaulay Institute and DardNI)

reflect the health of the whole system. For example, the critical molar ratio of calcium (or calcium plus magnesium) to aluminium in soil solution is a commonly used criterion to protect the fine roots of trees and may be used in the calculation of critical loads for forest soils.

BOX 7.5

nitrogen (NO_x). Since the signing of these agreements there has been a substantial reduction in SO_2 emissions in western Europe, although emissions of NO_x continued to increase until the late 1980s before starting to decline.

7.7.3 Soil pollution

A wide range of substances including heavy metals, pesticides and fertilizers can pollute soils. A distinction is often made between soil pollution originating from clearly defined sources (local or point source pollution) and that from

undefined sources (diffuse or non-point sources). The introduction of pollutants to the soil can result in damage to or loss of soil functions and may result in water contamination.

7.7.3.1 Heavy metals

Although heavy metals such as cadmium, copper, chromium, lead, zinc, mercury and arsenic are present naturally in soil, they can originate from a number of other sources including industry (e.g. atmospheric emissions, waste disposal, effluent disposal), agriculture (e.g. application of sewage

Figure 7.26 Forest dieback associated with acid rain in the Czech Republic.

The behaviour of pesticides in soil is influenced by a number of factors including its chemical properties, climate and soil type (particularly soil texture). Many pesticides, particularly the older ones, have a broad activity spectrum. This means they affect organisms that they were not intended to target. Pesticides can affect soil directly by adsorption onto clays and organic matter, by affecting soil microorganisms and plant growth, and by moving through the soil to surface and groundwaters. In Europe, the maximum admissible concentration of pesticides and metabolites in drinking water is set at $0.5 \mu g \, 1^{-1}$ (EEC 80/778). The impact of pesticide use on the soil is very much dependent on the specific pesticide used. However, as there are over 1000 different compounds on the market, all behaving differently in the soil, it is very difficult to identify and evaluate all the threats posed by pesticides.

7.7.3.3 Fertilizer use – nitrogen and phosphorus

Soils used for intensive agriculture require additional nutrients, particularly nitrogen, phosphorus and potassium, to maintain optimum plant productivity. Over the past 50 years the use of inorganic fertilizers has increased by between 5 and 10 times to increase crop yield. Nitrates are not adsorbed on soil particles but remain in solution, from where they may be taken up by the plant, leached out in drainage water or denitrified. In contrast phosphorus is adsorbed strongly on the surface of clay particles and to iron and aluminium oxides. Nitrates are therefore leached out in drainage water whenever there is sufficient excess of rainfall. The concentration of nitrates in drainage water depends on the volume of drainage water and the amount of nitrates available for leaching in the soil. Leaching of nitrates is undesirable because in drinking water they are considered to be a health hazard and in marine waters can cause **eutrophication** (Burt *et al.*, 1993).

Although there is a link between the use of inorganic nitrogen fertilizers and nitrate leaching, it is indirect, as long as the recommended amount is used, it is not applied in the autumn and there is no crop failure. The more fertilizer used, the greater the crop yield and the more nitrogen in the plant residue (straw, stubble, roots). This can lead to an increase in soil organic nitrogen and the potential for nitrate leaching after mineralization of organic nitrogen to nitrate. Mineralization of soil organic nitrogen can also occur when permanent grasslands and woodland are brought into cultivation. Leaching of nitrates is most pronounced on free-draining soils during autumn and winter. The ultimate loss depends on the soil texture, land-use, rainfall pattern,

sludge, farm wastes and fertilizers), waste incineration, combustion of fossil fuels and road traffic. Long-range transport of atmospheric pollutants can also add to the metal load in an area. In general, highest concentrations of heavy metals are associated with local sources of contamination such as mining and industrial facilities both in operation and after closure. Thus the largest and most affected areas are concentrated around heavily industrialized regions of the world, such as north-west Europe. Heavy metals accumulate in the soil as they bind to organic matter and clay minerals and are largely unavailable to plants. However, if soil acidity increases, heavy metals are released into soil solution, where they can be taken up by plant roots and *soil organisms*, or leached into surface and groundwaters, thus polluting the food chain and affecting drinking-water quality.

7.7.3.2 Pesticides and organic solvents

Organic compounds such as pesticides, oils, tars, chlorinated hydrocarbons, PCBs (polychlorinated biphenyls; used in the manufacture of electrical appliances) and dioxins are widely used in industry and agriculture and enter the soil through atmospheric deposition, direct spreading onto land, or contamination by waste waters and waste disposal. Modern agricultural production systems rely on pesticides (mainly fungicides, insecticides and herbicides) for crop protection and disease control purposes. These pesticides are applied either directly or indirectly to the soil. The intensive use of pesticides has occurred since the Second World War. In northern and western Europe pesticide use peaked in the 1980s, whereas in southern Europe its use increased (Stanners and Bourdeau, 1991).

drainage properties, the presence or absence of vegetation, the amount and availability of nitrogen applied, the timing of applications in relation to crop growth and, for grassland, the intensity of grazing by livestock (Burt *et al.*, 1993).

Until recently, the risk of water pollution from phosphorus was believed to be minimal owing to the fact that the majority of phosphorus applied in fertilizer is quickly bound to the soil. However, long-term inputs of phosphorus from fertilizers and manure to intensive crop and livestock agricultural systems have been made at levels that often exceed outputs in crop and animal produce. Calculations of an annual phosphorus balance for European agriculture indicate that Western European countries currently operate an annual phosphorus surplus (Ulén *et al.*, 2007). In the United Kingdom this surplus is 15 kg ha^{-1} yr^{-1} (Withers *et al.*, 2001). Over the past 20 years, fertilizer phosphorus application rates for individual crops have remained relatively constant (Withers *et al.*, 2001). As a result, the total amount and availability of phosphorus in agricultural soils have increased to a point where some soils can be classified as 'over-fertilized'. There is concern that this may lead to increased phosphorus loading to the aquatic environment, through a combination of leaching and eroded soil material. Although phosphorus losses are small in comparison with nitrate, only a small increase in phosphorus concentration is needed to produce a large change in the ecological dynamics of lakes and rivers.

To reduce phosphorus loss from agricultural land the concentration of soil phosphorus and/or the transport of phosphorus from land to water needs to be reduced. As most phosphorus loss is associated with the movement of fine soil particles, the same measures that are used to reduce soil erosion, such as cover crops and vegetated buffer strips between cultivated land and watercourses, would be beneficial.

Nutrient leaching from agricultural land is mainly a problem in areas of intensive agriculture, where fertilizer use is greatest, such as western Europe and North America. In Europe, the 1991 Nitrates Directive aimed to reduce water pollution caused or induced by nitrates from agricultural sources through the designation of Nitrate Vulnerable Zones (NVZs). Within these zones, farmers are required to undertake measures to reduce nitrate leaching. In addition, codes of good agricultural practice aim to narrow the imbalance between fertilizer input and plant uptake through a combination of measures such as: (i) better adjustment of fertilizer application and crop demands by soil testing and taking account of organic manures as sources of nutrients; (ii) application of fertilizers at the most appropriate times (i.e. when the crop most needs it); (iii) improvement of methods of manure application; (iv) minimizing leaching losses from arable land by sowing autumn crops; and (v) less intense use of grasslands. However, nitrate leaching cannot be completely prevented, especially in areas with high precipitation and where high crop yields are achieved on permeable soils (Burt *et al.*, 1993).

7.7.4 Other threats

7.7.4.1 Soil organic matter

As discussed above, soil organic matter is a vital component of productive and stable soils. It also acts as a buffer against many of the threats discussed above. Intensive arable production in Europe and North America has led to a decline in organic matter. This is of particular concern in Mediterranean areas, where 75% of the total area has a low (3.4%) or very low (1.7%) soil organic matter content (COM, 2002). Agronomists consider a soil with less than 1.7% organic matter to be in pre-desertification stage. In England and Wales, the percentage of soils with less than 3.6% organic matter rose from 35% to 42% in the period 1980 to 1995. In the same period, in the Beuce region south of Paris, soil organic matter decreased by half (COM, 2002). The ploughing up of grasslands, the abandonment of crop rotation and the burning of crop residues all reduce the amount of vegetation matter returning to the soil and increase carbon dioxide levels in the atmosphere. Various measures can be taken to increase the organic content of soils. These include the introduction of grass into agricultural rotations, ploughing crop residues back into the soil, and increasing land left with a grass cover.

7.7.4.2 Salinization

Salinization is the accumulation in soils of soluble salts of sodium, magnesium and calcium to the extent that the soil fertility is severely reduced. The process is mainly a problem in arid and semi-arid areas, where there is insufficient rain to leach away soluble salts, and upward soil water movement due to evapotranspiration leads to salt precipitation at or near the soil surface (see Chapters 12, 13 and 16). Irrigation of the land with water of a high salt content dramatically worsens the process. In coastal areas salinization can also be associated with abstraction of groundwater for drinking and industry. This can cause the intrusion of salty seawater as it attempts to replace lost freshwater in the ground. Human-induced salinization is estimated to affect 50% of all irrigated land (Abrol *et al.*, 1988). Salinization is reversible but reclamation of saline soils is expensive, as it requires complex amelioration techniques. It is estimated that:

- in Syria, 45% of the irrigated land is affected (Ilaiwi *et al.*, 1992);
- in Spain, 3% of the 3.5 million hectares of irrigated land is severely affected, significantly reducing its agricultural potential, and another 15% is under serious risk (COM, 2002);
- in the Hungarian plains, irrigation has caused salinization of more than 20% of the region (Stanners and Bourdeau, 1995);
- in the Russian Federation, 7% of agricultural soils are saline, part of them naturally so, such as the Solonchaks (Stanners and Bourdeau, 1995);
- in Romania, 6% of irrigated land became saline between 1970 and 1995 (Stanners and Bourdeau, 1995);
- in Australia, salinization could devastate up to 12 million hectares of land during the next 100 years and already costs at least A$500 million a year in Victoria alone (Gowing, 2003).

7.7.4.3 Soil compaction

Soil compaction occurs when soil is subject to the repetitive and cumulative effect of mechanical pressure through the use of heavy machinery, or to a lesser extent by the trampling of cattle when overstocked, especially in wet soil conditions. Compaction causes reorientation of soil particles and reduces the porosity of the soil. This slows water and air movement and reduces the water holding capacity of the soil. With reduced infiltration rates, compacted soil is likely to produce much greater volumes of overland flow, which increases the risk of soil erosion. Compaction can occur within both the top and subsoil. Soil compaction mainly affects heavily mechanized agricultural land such as central Europe and North America. In Europe about 33 million hectares (4% of land area) is at risk from soil compaction, mainly in the Russian Federation, Poland, Germany, Belgium, the Netherlands and north-west France (Oldeman *et al.*, 1991). The major impact of soil compaction is a decline of agricultural productivity. Surface compaction may reduce yield by up to 13% (average 5%) and subsurface compaction may reduce yields by between 5 and 35% (Stanners and Bourdeau, 1995). Surface soil compaction can be resolved by reworking the soil. However, subsoil compaction is much harder to resolve. Compaction can be prevented by reducing the amount of tillage, reducing traffic over fields, improving land drainage, increasing organic matter and reducing surface pressure of machinery

by increasing the axles and wheels on agricultural machinery, increasing tyre width and reducing tyre pressure.

7.7.4.4 Soil biodiversity

Soil is the habitat for a large variety of living organisms assembled in complex and varied communities. Soil biodiversity reflects the variability among living organisms in the soil. Species range from the myriad of invisible microbes, bacteria and fungi to the more familiar macro-fauna such as earthworms and termites. Soil is by far the most biological diverse part of Earth. Soil organisms are known to play a key role in many of the fundamental functions that occur in soil. However, the roles played by most groups of soil organisms are very poorly known. Soil communities are currently exposed to a wide range of impacts, including soil erosion, agricultural intensification and the deposition of acidic pollutants, with poorly documented effects on diversity of the soil biota, and virtually unknown effects on ecosystem processes.

7.7.5 Policy and legislation

Soil degradation processes are a major worldwide problem with significant environmental, social and economic consequences. Soil is also an important store of global carbon which, if disturbed, can increase greenhouse gas concentrations in the atmosphere. Box 7.6 describes how soils can be managed to store carbon. As the world's population continues to grow there is increased need to protect soil as a vital resource. With the addition of a quarter of a million people each day, the world food demand is increasing at a time when per capita food productivity is beginning to decline (Pimental *et al.*, 1995). Growing awareness of the need for a global response has led to national and international initiatives. In 1972, the Council of Europe's Soil Charter called on states to promote a soil conservation policy. The World's Soil Charter and the World's Soils Policy sought to encourage international co-operation in the rational use of soil resources (COM, 2002). In 1992, at the Rio summit, the participating states adopted a series of declarations of relevance to soil protection. In particular, the concept of sustainable development was agreed and legally binding conventions on climate change, biological diversity and desertification adopted. The aim of the 1994 Convention to Combat Desertification was to prevent and reduce land degradation, rehabilitate partly degraded land and reclaim desertified land. For further information on **desertification** see the 'Ecosystems and Human Well-Being: Desertification Synthesis', a report of the Millennium Ecosystem Assessment (see Box 8.4 in Chapter 8).

NEW DIRECTIONS

MANAGING SOILS TO STORE CARBON

Soil is a major component in the global carbon cycle, containing about 1500 Pg (1 Pg = 1 Gt = 10^{15} g) of organic carbon (Batjes, 1996), which is about three times the amount in vegetation and twice the amount in the atmosphere. Through photosynthesis, plants convert carbon dioxide (CO_2) into organic forms of carbon and return some to the atmosphere through respiration. The carbon that remains in plant tissue is added to the soil through their roots and as litter when plants die and decompose. This carbon is then stored in the soil as soil organic matter. Carbon can remain stored in the soil for millennia, or be quickly released back into the atmosphere as CO_2. Climate, vegetation type, soil texture and drainage all influence the amount and length of time carbon is stored in the soil. Therefore, soils play a major role in maintaining a balanced global carbon cycle. However, the carbon content of soil is smaller today than a few hundred years ago due to the intensification and mechanization of agriculture. Agricultural practices have depleted soil organic carbon pools by two main routes:

1. Reducing the amount of carbon returned to the soil in litter by harvesting and removing the crop.
2. Excessive use of tillage practices which breaks up the soil, increasing the decomposition rate of soil organic matter which leads to an increase in the release of CO_2 from the soil.

Figure 7.27 illustrates the soil carbon changes over time on agricultural land. Estimates of historic soil organic carbon loss range from 40 to 90 Gt (Smith, 2004), of which about one-third is attributed to soil degradation and accelerated erosion and two-thirds to mineralization (Lal, 2004). Conversion of natural ecosystems to fields for crop production and grasslands causes depletion of a soil's carbon content by as much as 75% (Lal, 2004). Severe depletion of the soil organic carbon pool degrades soil quality and leads to a decline in crop production. Currently there is concern that many of the world's agricultural soils are alarmingly depleted of carbon (see Section 7.7.4).

There is, however, a way to reverse the soil carbon release process as research has shown that soils can regain lost carbon by absorbing or 'sequestering' it from the atmosphere. **Carbon sequestration** implies transferring atmospheric CO_2 into long-lived pools and storing it securely so it is not immediately re-emitted to the atmosphere. Thus soil carbon sequestration means increasing soil carbon stocks. This can be best achieved through changes in land-use and management practices in soils that have been depleted in carbon, such as intensively managed agricultural soils and degraded soils. Some of the soil carbon sequestration options available for agricultural land include reduction in tillage, reducing fallow periods, improving efficiency of animal manure use and crop residue use, conversion of arable land to grassland, woodland or bioenergy crops and restoring degraded land. Estimates of the maximum yearly carbon mitigation potential for some of these land management options are shown in Figure 7.28 and compared with the 1990 CO_2 emission from the

Figure 7.27 Changes in soil carbon stocks resulting from changes in agricultural land-use and management.

BOX 7.6 ➤

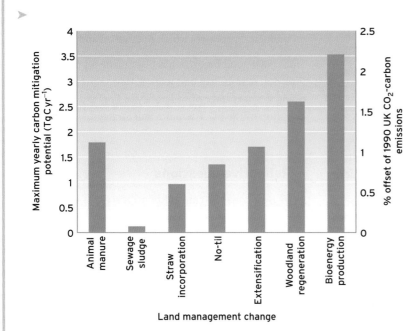

Figure 7.28 Estimates of the maximum yearly carbon mitigation potential, and maximum percentage offset of UK 1990 CO_2 emissions of some different land management changes. (Source: after Smith *et al.*, 2000)

BOX 7.6

United Kingdom. Many of these land management practices also improve soil quality, plant production and water conservation, reduce erosion, and enhance wildlife habitat and species protection, which result in increased biodiversity.

Improvements in measuring, monitoring and verifying changes in carbon stocks in soils are needed for quantitative economic and policy analysis. Currently, world scientists can combine data on soil carbon, land-use and climate to create models that estimate the carbon change related to farm management practices. However, they are continuing to refine measurement methods for greater accuracy.

In 2001, the European Union indicated soil loss and declining soil fertility as a main threat to sustainable development as it erodes the viability of agricultural land (COM, 2002). Although several different EU polices (e.g. on water, waste, pesticides, industrial pollution prevention) contributed to soil protection, they were not sufficient to ensure an adequate level of protection for all soil in Europe. Hence in response to concerns about the degradation of soils in the EU, the European Commission adopted a Communication 'Towards a Thematic Strategy for Soil Protection' in April 2002 and in September 2006 the Commission adopted the Thematic Strategy for Soil Protection. The Strategy's objective is to define a common and comprehensive approach, focusing on the preservation of soil functions, based on the following principles:

1. Preventing further soil degradation and preserving its functions.
2. Restoring degraded soils to a level of functionality consistent at least with current and intended use, thus also considering the cost implications of the restoration of soil.

The strategy outlines why further action is needed to ensure a high level of soil protection across the EU and what kind of measures must be taken. There are currently four main elements to the strategy:

1. Measures to address soil erosion, organic matter decline, compaction, salinization and landslides, obliging Member States to identify risk areas and develop associated programmes of measures and targets.
2. Measures to address soil contamination, including actions such as an inventory of contaminated sites, the production of soil status reports, identification of sites on which potentially polluting activities are taking place or have taken place, and the production of a remediation strategy for such sites.
3. Measures on soil sealing, obliging Member States to take appropriate steps to minimize soil sealing, or mitigate its effects, using construction techniques and products which would allow as many soil functions as possible to be maintained.
4. Awareness raising, reporting and exchange of information.

In May 2004, a Soil Action Plan for England was published, which was a key milestone in soil policy in that it highlighted the first comprehensive statement on the state of soil in England and how Government and other partners were working together to improve soils. The Action Plan addresses issues that are listed under eight headings: (1) protecting soils in the planning system, (2) minimizing contamination of soils, (3) predicting and adapting to the impacts of climate change on soils, (4) soils for agriculture and forestry, (5) interactions between soils, air and water, (6) soils and biodiversity, (7) soils, the landscape and cultural heritage, and (8) soils in mineral extraction, construction and the built environment. The 52 actions proposed in the First Soil Action Plan for England work towards a common vision that recognizes the several vital functions that soils perform for society. In order to achieve this vision, the aims are to ensure:

- soil managers will look after their soils with a view both to their own and society's short-term needs and to the interests of future generations;
- the regulatory, legislative and political framework will provide appropriate protection of soil as an irreplaceable natural resource and empower and encourage people with soil to manage it properly;

- a better understanding of, and access to, information on the state of our soils and the physical, chemical and biological processes which operate on and within them.

Since 2004 good progress has been made on delivery of the First Soil Action Plan and work is now in progress for developing the next stage of soil policy in England given the adoption of the EU Thematic Soil Strategy.

Reflective questions

➤ How does the concept of critical loads work?

➤ What are the main controls on the nature and intensity of soil erosion?

➤ Explain what is meant by the term 'soil compaction' and how the effects of compaction can be reduced in soils.

7.8 Summary

Soil is composed of minerals, organic matter, air, water and living organisms in interactive combinations produced by physical, chemical and biological processes. Soil is an essential component of the terrestrial biosphere and performs a wide range of essential functions that sustain life. It supports plant growth on which humans rely for food, fibre and wood for fuel and building materials. It provides a habitat for large numbers of animals and microorganisms that decompose dead plants and animals into the nutrients needed by all living things. It acts as a reservoir for water, and has a filtering, transforming and buffering role. It also provides raw materials and a physical base for the foundations of buildings and roads.

Soil is made up of mineral and organic materials, and contains pore spaces occupied by water and air. The minerals include residues of the parent material and secondary minerals, which are the product of weathering. The organic matter is composed of readily decomposing plant, microbial and animal products, living organisms and roots and resistant organic matter known as humus. The soil water contains solutes and dissolved gases, and is referred to as the soil solution. The composition of soil air differs from that of atmospheric air in that it generally contains more carbon dioxide and less oxygen owing to the respiration of soil organisms and roots. The soil is an open system, which allows input of materials to the soil, the loss of materials from the soil and internal transfers and reorganization of these materials within the system. It is the processes of additions, removals, mixing, translocations and transformations that are influential in differentiating soil material into a series of horizons that constitutes the soil profile, and in determining the nature and properties of soil.

The soil-forming processes that dominate at a site are controlled by five interacting environmental conditions: parent material, climate, topography, organisms and time. These environmental variables are known as the soil-forming factors and they control the direction and speed of soil formation. Climate and organisms determine the rate at which chemical and biological reactions occur in the soil, while parent material and topography define the initial state for soil development and time measures the extent to which reactions will have proceeded.

There are large differences between soil profiles from place to place throughout the world. To describe soil

➤

profiles in a coherent manner, various classification schemes have been introduced. The physical properties of soil depend largely on the size of the soil particles (soil texture) and on their arrangement (soil structure) into peds or aggregates. Texture and structure influence the distribution and movement of water and air in the soil and thus greatly affect plant growth. Soil colour is used as an indicator of organic matter content, drainage and aeration.

Clay minerals are the product of weathering. The structure of all clay minerals is based on two types of sheets: the tetrahedral sheet, which consists of repeating units of a silicon atom surrounded by four oxygen atoms in the form of a tetrahedron; and the octahedral sheet, which consists of repeating units of an aluminium atom surrounded by six oxygen atoms or hydroxy (OH) groups in the shape of an octahedron. Alternating sheets of one tetrahedral sheet and one octahedral sheet produce what are known as 1 : 1 clays, whereas an aluminium octahedral sheet sandwiched between two silicon tetrahedral sheets produces 2 : 1 clays. Isomorphous substitution (Al^{3+} replaces Si^{4+} or/and Fe^{2+} or Mg^{2+} replaces Al^{3+}) in the crystal lattice of clay minerals results in an overall net negative charge. This charge is balanced by cations, positively charged ions, which are attracted to the surface of the clay minerals and held there by electrostatic attractions. These cations are known as exchangeable cations as they can be displaced by cations in the soil solution in the process of cation exchange. In most agricultural soils, calcium and magnesium are the dominant exchangeable cations. However, as the acidity of the soil increases, aluminium and hydrogen ions dominate the exchange sites.

There are a number of natural and human-induced changes that increase soil acidity. They include leaching of base cations, respiration of roots and organisms, decomposition of organic matter, deposition of acids from the atmosphere, application of nitrogen fertilizers, removal of base cations in crop harvests and draining of waterlogged land. Soil pH determines the fate of many pollutants, affecting their breakdown, solubility and possible movement from the soil to surface and groundwaters. The availability of the essential nutrients for plant uptake is also influenced by soil pH as are the number, species and activities of soil organisms. Low soil pH leads to a decline in the number and activities of many soil organisms, an increase in concentrations of aluminium, iron, manganese and zinc (to the extent of toxicity to plants and other organisms), a decrease in the concentration of the macronutrients (to the extent that plants may show signs of deficiencies) and a reduction in root activity. In contrast, high soil pH results in phosphorus and boron becoming insoluble and unavailable to plants and low concentrations of the micronutrients, particularly iron, manganese, zinc and copper, resulting in restricted plant growth.

Soil, as a resource, is being increasingly exploited throughout the world. The pressures of agriculture, drainage, extraction, application of wastes and urban development have led to soil degradation such as soil erosion, contamination, acidification, compaction, loss of organic matter, salinization and loss of biodiversity. Ensuring that our soil is sustainable, and not damaging to future generations, will involve, as well as regulatory bodies, all sectors of society whose activities and decisions affect soils. This is one of the world's biggest challenges for the future.

Further reading

Brady, N.C. and Weil, R.R. (2002) *The nature and properties of soil*, 13th edition. Prentice Hall, Upper Saddle River, NJ.
This is the latest edition of a comprehensive and popular textbook. Most of the examples are American.

Cresser, M.S., Kilhma, K. and Edwards, A.C. (1993) *Soil chemistry and its applications.* Cambridge University Press, Cambridge.
This textbook demonstrates the role soil chemistry plays with other areas of soil science and environmental science such as water quality and pollution science.

FitzPatrick, E.A. (1983) *Soils: Their formation, classification and distribution.* Longman Scientific and Technical, Harlow.
This textbook includes a comprehensive section on soil classification and a comparison of the many different systems used throughout the world.

Gerrard, J. (2000) *Fundamentals of soils.* Routledge, London.
A textbook designed for undergraduate readership.

Rowell, D.L. (1994) *Soil science: Methods and applications.* Longman, London.
This textbook includes useful examples of practical soil work in the field and laboratory.

Royal Commission on Environmental Pollution (1996) *Nineteenth Report – Sustainable Use of Soil.* HMSO, London.
This report outlines the major uses of soil in the UK and the environmental issues associated with each use.

Stanners, D. and Bourdeau, P. (1995) *Europe's environment: The Dobris assessment.* European Environment Agency, Copenhagen.
Focus on chapter 7 (soil), pages 146–169. This chapter gives a detailed and comprehensive review of the problems and threats facing soils in Europe.

Toy, T.J., Foster, G.R. and Renard, K.G. (2002) *Soil erosion: Processes, prediction, measurement and control.* John Wiley & Sons, Chichester.
Good for case studies of soil erosion and information on practical methods of measuring and preventing erosion.

White, R.E. (1997) *Principles and practice of soil science: The soil as a natural resource,* 3rd edition. Blackwell Science, Oxford.
This textbook covers all aspects of soil science and is divided into three sections: soil habitat, soil environment and soil management.

Wild, A. (1993) *Soils and the environment: An introduction.* Cambridge University Press, Cambridge.
This textbook has two sections: Part A covers soil properties and processes, whilst Part B considers soils in relation to the environment.

Web resources

Department of Environment, Food and Rural Affairs: Soil
http://www.defra.gov.uk/environment/land/soil/index.htm
This page provides links to soil protection, monitoring and research in England and Wales.

Environment Agency: The state of soils in England and Wales
http://www.environment-agency.gov.uk/subjects/landquality/
The state of soils in England and Wales is a report that summarizes our current knowledge about the condition of soils.

European Commission: A Strategy to Keep Europe's Soils Robust and Healthy
http://ec.europa.eu/environment/soil/index.htm
This contains information on the development of the EU Thematic Strategy for Soil Protection and reports on the state of European soils.

Millennium Ecosystem Assessment
http://www.maweb.org/en/index.aspx
The Millennium Ecosystem Assessment considered the consequences of ecosystem change for human well-being.

Soil Association Home Page
http://www.soilassociation.org/
The Soil Association is the UK's leading campaigning and certification organization for organic food and farming. At this site you can find information on 'soil biodiversity' and 'soil – the importance and protection of a living soil' when you go to 'Library' and enter 'soil' in key words.

Soil Compaction: Causes, Effects and Control
http://www.extension.umn.edu/distribution/cropsystems/DC3115.html

This is a 16-page colour brochure that gives examples of soil compaction, focusing on the causes and consequences.

Soil Net
http://www.soil-net.com
This site gives a brief introduction to soils and explains the importance of soil within the environment.

Soil Organic Matter
http://www1.agric.gov.ab.ca/$department/deptdocs.nsf/all/agdex890?opendocument
This online publication deals with the role of organic matter in soil productivity and the effects of various management practices on soil organic matter.

Soil Science Education Home Page
http://soil.gsfc.nasa.gov/
This is a very useful site covering a wide range of interesting topics related to soil and the environment.

Soil Science Society of America: Soil Science Terms Glossary
http://www.soils.org/sssagloss/
This provides a glossary of terms related to soil science, revised in 2001.

United States Department of Agriculture National Resources Conservation Service: Soil
http://soils.usda.gov/
The NRCS soils website covers a wide range of topics related to soil science and includes a good section for teachers and students. The NRSC also supplies a site on soil quality, in which you can find a list of soil quality information sheets covering a wide range of topics including erosion, compaction, salinization, soil biodiversity, pesticides and organic matter:
http://soils.usda.gov/sqi/

Visit our website at **www.pearsoned.co.uk/holden** for further questions, annotated weblinks, interactive models and video-clips showing physical processes in action.

CHAPTER 8

The biosphere

Hilary S.C. Thomas
Department of Humanities and Science, University of Wales, Newport

Learning objectives

After reading this chapter you should be able to:

➤ define and explain the important terms 'biosphere', 'biome' and 'biogeographical realm'

➤ understand the main influences leading to patterns in plant and animal distributions

➤ describe the location and main characteristics of the major global biomes

➤ explain some of the factors leading to change in these regions

8.1 Introduction

The **biosphere** is usually taken as referring to the surface of the Earth, together with those parts above and below it that maintain life. Its field of interest therefore includes parts of the atmosphere as well as the soils and waters of the Earth. Studies of the biosphere are linked with geology, ecology, soils, atmospheric processes and climates, and oceans. The biosphere is dependent on other components of our planet for its functioning as illustrated in Figure 8.1. As geographers, we are aware of the interactions between the atmosphere, biosphere and lithosphere but are fascinated by the patterns of distinctive regions of plants and animals that result on the Earth's surface. We are interested in not only the patterns but also how humans influence the biosphere through a range of deliberate and inadvertent practices. The following quote illustrates *one* viewpoint of the role of humans in the biosphere:

> *The biosphere comprises the natural world in which man has been placed, and which, thanks to his mental capacities, he is able to regard objectively, thus raising himself above it. . . . It is not the sole calling of man to use nature to his own ends. He also bears the responsibility for maintaining the Earth's ecological equilibrium, of tending and preserving it to the best of his ability. If he is to do this and to avoid exploiting the environment in a way which in the long run jeopardises his own existence, he has to recognise the laws of nature and act upon them.*

(Walter, 1985, p. 2)

At the beginning of the twenty-first century, as both the density of human population and its technological abilities increase (see Chapter 21), we are also aware of the extent to which, for good or ill, we are able to alter these patterns. Directing **ecological succession** and the management of energy flows through ecosystems allows commercial harvesting of resources to feed a rapidly growing population. The increase in disposable income and leisure time within richer areas of the world, together with increasing

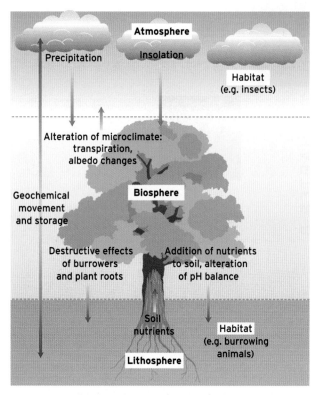

Figure 8.1 Interactions between the atmosphere, biosphere and lithosphere.

appreciation of the aesthetic value of all forms of wildlife, has led to the use of the biosphere for a variety of forms of recreation. This varies from whale-watching, where the activity is dependent on the character of distinct locations, to projects such as those taking place in Cornwall, UK, as part of the Eden Project where tourists are given the opportunity to experience a variety of artificially created 'biomes' (see Box 8.2 below).

Biomes are major (global-scale) zones with characteristic life forms of plants and animals. The alteration and fragmentation of major biomes (such as the evergreen woodlands of the Mediterranean) by non-sustainable practices of exploitation has been a feature of human impact on the environment not just in recent decades, but for thousands of years. This has given rise, in some locations, to new biomes. However, biomes and the biosphere itself are in constant, if usually slow, change in response to fluctuations of climate. Modern geographers have the advantage of technology such as remote sensing and numerical modelling techniques to enable them to track changes in vegetation cover, growth and type (see Chapters 22 and 23). These academic efforts provide data and additionally provide a conceptual framework to aid those working towards conservation and restoration of environments.

This chapter aims to describe the fundamental properties and characteristics of the biosphere. The main factors

leading to both spatial and temporal patterns within the biosphere are outlined, before focusing upon these patterns, at global and continental scales. Finally, ways in which elements of the biosphere may change over time are suggested and consideration is given to the role of the biosphere in human leisure activity. This chapter and the two following it are integrated by the concepts used in common by both biogeographers and ecologists. Cross-referencing is therefore made between them, to indicate where an aspect in a section is examined in more detail within one of the other two chapters. Occasionally Latin names are used for plants and animals during these three chapters. This is the standard practice and in many cases there are no common (or English) names for certain species. However, where common names do exist these are provided. You do not need to remember all the plant and animal names since it is more important to understand the processes and characteristics of the environments discussed. Nevertheless the species names are provided as examples.

8.2 Functions and processes within the biosphere

Although functions and processes within the biosphere are more properly dealt with as part of Chapter 9, it is worth noting here that these stem, for the most part, from the unique relationship between living things and their environment. Both are capable, to a greater or lesser degree, of affecting the other. These interactions can be observed to take place within a generally hierarchical structure, within which flows of energy, nutrients and matter operate at a variety of spatial and temporal scales.

8.2.1 Characteristics of the biosphere

The biosphere, like the atmosphere and oceans, is distinguished by movement and dynamism. For the biosphere this dynamism is of its life forms. The main characteristics of these fluxes are discussed in greater detail in Chapter 10. It is, however, worth noting here that this dynamism may be directional, as in the slow post-glacial return of species to high-latitude forests, or of a more cyclic nature as in regeneration after repeated fire damage within semi-arid zones.

Boundaries within the biosphere may result from a variety of factors, often relating to **environmental gradients** (e.g. climate, altitude, soil type), but increasingly in some areas to intensity of human usage. All geographers should be aware that a boundary depicted as 'a line on the map', especially at an atlas scale, needs considerable caution in interpretation. For example, we must think about how wide an area any line on the map of the tropical biomes (see Figure 8.5 below) covers on the ground, and how constant it is over time and space. Many terrestrial boundaries do not imply a total and sudden change to biomes. The resulting transitional areas are described as **ecotones** and can vary hugely in extent. Ecotones are expected to be early indicators of change in response to climate. Evidence exists to show that this has been the case in the past. For example, the northern Sahara was regarded as a major grain-providing area for the Roman Empire and the shifting patterns of desertification along the southern Sahelian boundary continue to provoke widespread concern.

The Earth's biosphere is not the same throughout, but has developed over time a pattern of distinctive regions at all scales from the global to the very local. These regions have characteristic energy flows, **biomass**, trophic levels and rates and types of nutrient cycling activity, and a number of factors influence these processes.

8.2.2 Major factors producing regions within the biosphere

8.2.2.1 Temperature regime

Both the actual temperature and its seasonal pattern may be of critical importance to plant or animal life. While four seasons are the norm for the middle latitudes, two seasons, or sometimes three, are more frequent on a global scale (see Chapter 5) and are of particular relevance to understanding biomes. The growing season for most plants creates an effective baseline of food for other creatures (see Chapter 10) and is taken to be the length of time when the average monthly temperature is above 10°C. This growing season may vary considerably within a small geographical range. The length of growing season has particular implications for **herbivores** (animals that just eat plants). Herbivores must adapt to the changing availability of food resources through the seasons. This is often done by becoming dormant (e.g. hibernation) for part of the year or by migration, sometimes over very considerable distances. High temperatures may exacerbate the effect of low rainfall. Vegetation change in response to temperature regimes may also affect the capacity for the plant–soil complex to retain moisture.

8.2.2.2 Moisture availability

Moisture availability is related to the local rainfall regime, in particular to the length of the dry season. It is also dependent upon the potential effectiveness of rainfall in comparison with losses of water via evaporation and accessibility of river or groundwater. A range of other factors will act locally to produce diversity within this framework. In particular, the roles of the soil type, geology, slope and altitude are often fundamental in providing zones of increased (e.g. in hillslope hollows) or decreased (e.g. on very coarse substrate) moisture for plants and animals. These major factors, together with others discussed below, can be classed as zonal or azonal in their effects.

8.2.2.3 Zonal factors

The regional macroclimate (equatorial, monsoonal, desert and so on – see Chapter 5) creates characteristically favourable or unfavourable conditions for plants and animals. The level of cold or drought may determine dormancy or deciduousness and can lead to distinctive life forms. It may also affect the relative speed of nutrient cycles. The major world biomes therefore correspond very closely to the major climatic zones (see also Chapter 22, especially Section 22.2). Zonal soil types, such as brown forest soils, have, for example, distinctive mechanical characteristics and types and rates of nutrient supply (see Chapter 7).

8.2.2.4 Azonal factors

Azonal factors have the ability to disrupt the otherwise climatically controlled pattern of biogeographical regions (biomes, ecotones, 'realms' – see below). Geomorphology will affect drainage, provide varying aspect and hence influence the receipt of solar radiation. Geology provides a varying substrate in terms of properties such as pH, nutrient availability and soil texture. The level and frequency of human influences, whether ancient or modern, 'managed' or unintentional, will have effects that are often highly disruptive to the zonal pattern. For example, the Bronze Age forest clearances led to high levels of soil erosion around the Mediterranean, similar to those seen much more recently in parts of the tropics (Evans, 1999).

8.2.3 The major biogeographical realms

The factors discussed above have together resulted in what are described as biogeographical zones or realms, which in turn are divided into biogeographical regions or biomes. The biomes, while often described as major terrestrial vegetation communities, defined by the similarity of the dominant plants (Archibold, 1995), also include characteristic animal communities. Some workers have divided the Earth into realms in relation to vegetation, while others use a

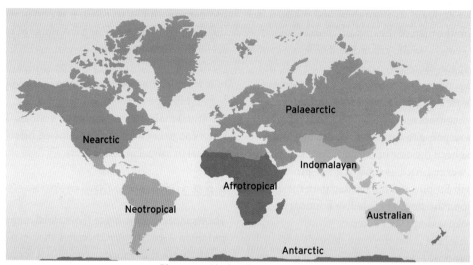

Biogeographical realms

NB Holarctic = Nearctic + Palaearctic

Figure 8.2 A classification of biogeographical realms. The Holarctic consists of the Nearctic and Palaearctic while the Palaeotropic consists of the remaining regions.

faunal basis (Bradbury, 1998). The classification discussed below draws together these two themes but also indicates where there are widely held differences of opinion. It should be noted, however, that although the geographical distribution of any single species within a realm may well coincide with the boundaries of that realm, it is usually the combination of species that will show a strong and characteristic association with it.

Figure 8.2 shows a typical classification of biogeographical realms. These fall into two main regions, which are then subdivided. The first region is the Holarctic region which although based on vegetation also comprises the two faunal realms of the Palaearctic (Eurasia) and the Nearctic (North America and Greenland). This region approximates to the ancient plate tectonic region of **Laurasia**, which was a large continent of the northern hemisphere that existed around 200 million years ago (see Chapter 2). The other major region may be regarded as being formed from **Gondwanaland**, the equally ancient southern continent. These regional similarities therefore indicate the importance of plate tectonics in determining the global distribution of species. The southern region is known as the Palaeotropic region and consists of further subdivisions, although there is often debate as to how these subdivisions should be made. The first subdivision is considered to be the Afrotropical realm (south of the Sahara). Many people think of the Indian (or south-east Asian) realm as a separate region as shown in Figure 8.2. The Asian (or Indomalayan) realm is similar to the African realm on floral grounds but there are distinct faunal differences.

There are also three further southern hemisphere realms that are considered to be distinct. The Neotropical realm refers to South America while the Australian realm sometimes excludes New Zealand on floral grounds (Takhtajan, 1969) but other people often include New Zealand, southern South America and southern South Africa within this realm (Walter, 1985). A further Antarctic realm is often distinguished which again, as in Figure 8.2, is considered to comprise southern South America, southern South Africa and New Zealand, all of which are related to the southern part of the old supercontinent of Gondwanaland. It should be noted that this classification has ignored the ocean realm. The divisions above result from the work of scientists usually interested in terrestrial distributions and as such the full richness of the world's biogeographical regions is sometimes underestimated (see Chapter 3).

Understanding past climatic and tectonic changes is important for understanding the patterns of plant and animal geography that can be identified today. These long-term climate and tectonic changes offer in many cases the clearest explanation for the variety of biodiversity to be found across the planet. The groupings in Figure 8.2 are of great use for purposes such as **taxonomy** (species classification). However, many geographers find that more detailed classification based on climatic controls is of greater usefulness. This is because it allows correlation with factors that have immediate effects on biomass, species dominance and, sometimes, human land-use. Most workers agree that there is a very strong correlation between these biomes, the structural characteristics of their vegetation and the major world climate zones (Grime, 1997). Thus, these biomes form the main structure of the remainder of this chapter.

Since vegetation cover tends, in most cases, to be the determining factor in both the distribution of animals and the potential and actual human activity in the regions involved, the naming of the biomes described in the following sections of this chapter are based on a generally accepted classification based on the predominant vegetation type. As Cox and Moore (1993), who include several marine biomes within their classification, point out, there is no real agreement among geographers about the number of biomes in the world. For convenience, they are discussed in this chapter within three major climatic zones. Remote sensing techniques can be used to distinguish vegetation patterns and change as described in Box 8.1.

Reflective questions

➤ What are the main factors that produce regions of the biosphere and how do they interact?

➤ What is the difference between a biogeographical realm and a biome?

8.3 The tropical biomes

The tropical biomes include the complex and biomass-rich equatorial and tropical rainforests, the tropical woodland–grassland mix known as the savanna and the low-biomass region of the hot desert biome. Table 8.1 and

TECHNIQUES

REMOTE SENSING TO MONITOR LAND COVER CHANGE

Remote sensing is technology that has developed rapidly, from its initial military applications to a tool that enables environmental managers to obtain 'real-time' and 'multi-temporal' images of the globe, focused on specific regions or themes of investigation. Many satellites have instruments on board that enable them to detect different types of land cover. For example, Landsat's Thematic Mapper and Multispectral Scanner and NASA/ National Oceanographic and Atmospheric Administration are frequently used to provide consistent and reliable information. See Chapter 23 for a detailed exposition of remote sensing.

Vegetation change is one of the more easily distinguishable forms of land cover change. The very high reflective characteristics within the near-infrared part of the spectrum are distinctive to vegetation. Remote sensing from space can be used to

'see' these reflective characteristics. Further differentiation is possible using factors such as water content, vegetation density and a range of distinctive vegetation structures, even down to individual plant shapes. These factors combine to give a wide range of reflectance signatures, which can be translated into precise vegetation data. For example, deciduous leaves reflect more strongly than evergreen needles, and both ripening and wilting crops have associated reflectance changes. Each land cover can be assigned specific signatures, allowing mapping of very large areas and of regions not easily accessible. Results can be checked by ground-truth sampling rather than surveying the whole globe, reducing time, expense and hazard for field staff. Data are therefore more readily accessible for some developing countries than traditional field surveys.

An example of a remote sensing classification is shown in Figure 8.3, taken from the Global Land Cover Facility, provided by the University of

Maryland. Although there is loss of information at the scale shown in Figure 8.3, you should be able, in the light of information within the rest of this chapter, to identify many of the major biomes. Comparison of this figure with the other biome maps in this chapter (Figures 8.5, 8.14 and 8.22) demonstrates that the potential natural vegetation has often been supplanted by human land-use.

Monitoring of biomes depends upon reliable, objective base-line data and such data are often heavily reliant upon remotely sensed information. Considerable work is therefore ongoing into the identification of indicators of early change. This type of monitoring can also be used to check compliance with management plans and with regional agricultural and environmental legislation. It can be used to check the relative impacts of human- and climate-induced change (see Box 8.4 below). You may wish to visit NASA's Earth Observatory website which provides many images of geographical interest:
http://earthobservatory.nasa.gov

BOX 8.1 ➤

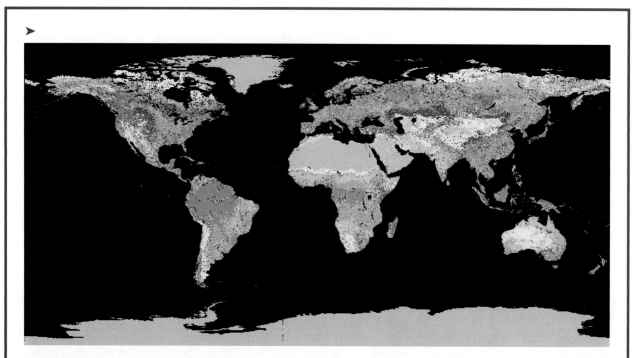

Figure 8.3 Global land cover map based on remote sensing data. (Source: Hansen *et al.*, 1998; the source for this data set was the Global Land Cover Facility, www.landcover.org)

BOX 8.1

Table 8.1 Potential primary production of the Earth by climate zone

Climate zones	Area (km² × 10⁶)	Phytomass Total (t × 10⁹)	Phytomass Average (t h⁻¹)	Primary production Total (t ha⁻¹ × 10⁹)	Primary production Average (t ha⁻¹ yr⁻¹)
Polar	8.05	13.8	17.1	1.33	1.6
Boreal	23.2	439	189	15.2	6.5
Temperate					
Humid	7.39	254	342	9.34	12.6
Semi-arid	8.10	16.8	20.8	6.64	8.2
Arid	7.04	8.24	11.7	1.99	2.8
Subtropical					
Humid	6.24	228	366	15.9	25.5
Semi-arid	8.29	81.9	98.7	11.5	13.8
Arid	9.73	13.6	14.9	7.14	7.3
Tropical					
Humid	26.5	1166	440	77.3	29.2
Semi-arid	16.0	172	107	22.6	14.1
Arid	12.8	9.01	7.0	2.62	2.0
Geo-biosphere					
Land area	133.0	2400	180	172	12.8
Glaciers	13.9	0	0	0	0
Biohydrosphere					
Lakes and rivers	2.0	0.04	0.2	1.0	5.0
Oceans	361.0	0.17	0.005	60.0	1.7

(Source: from Walter, 1985)

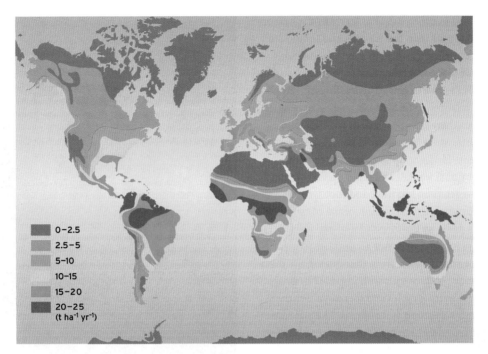

Figure 8.4 Map showing distribution of potential primary production of the Earth. (Source: from Schultz, 1995, after Leith, 1964)

Figure 8.4 show how the potential **primary productivity** (amount of energy fixed by plants during photosynthesis) is distributed across the Earth. Comparison of Figures 8.4 and 8.5 shows how the tropical biomes include regions of both large potential productivity and small potential productivity.

8.3.1 Equatorial and tropical forests

As a general rule tropical forest biomes are located where the climate is hot with mean annual temperatures around 25°C with little seasonal variation, and where there is around 2000 mm of rainfall per year. The rainfall needs to

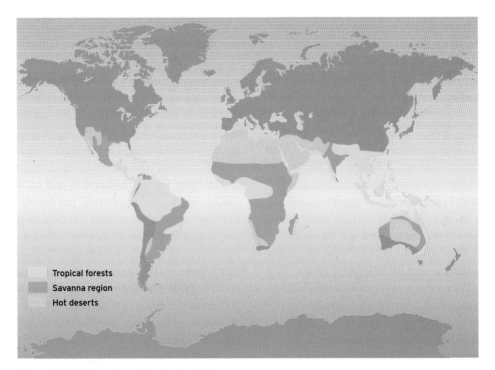

Figure 8.5 The location of the tropical biomes.

be consistent throughout the year, or at least any dry season should not extend longer than five months and should still produce rainfall exceeding 120 mm in every month. Soils reflect high rates of **biogeochemical** activity which occurs at three to four times the rate of other mature forests. This leads to deep but relatively infertile soils, such as **oxisols**. In these environments there may be a lack of recently weathered rock fragments, which would normally release new supplies of nutrients for plant uptake. Fertility therefore depends upon continual leaf fall and rapid cycling of nutrients through the vegetation, leaf litter and root system. Despite the high productivity, as a result of the constant nutrient demands made by the forest's rapid and non-seasonal growth, plant litter does not accumulate to great depths. With increasing distance from the equator, the more seasonal rainfall may produce soils with impeded drainage through drying and wetting cycles and this can affect the distribution of both vegetation and animals. These forests tend to be characterized by high levels of vegetative biomass (Figure 8.4), dependent upon the length of dry season. It is estimated that these forests provide 40% of the world's terrestrial **net primary productivity**. Many of the available nutrients are stored within the biomass and, in normal circumstances, little escapes from the system into rivers.

These tropical forest biomes are often referred to as rainforests. Tall trees are the dominant vegetation. Considerable variation of forest type exists within regions as well as between the major realms. Most tropical forest trees, however, are broad-leaved and evergreen species. Compared with other biomes, there is seldom a good survival reason for an organism to germinate or have its reproductive period at a specific season, since there is little climatic advantage to be gained. This has the effect of producing the characteristic luxuriant and continuous vegetation cover, with high levels of primary productivity and biomass, which benefit lower species throughout the system. Some plants shed all their leaves over a short period, in a similar fashion to those of the deciduous forests in Europe and North America, while others are able to maintain a continuous process of leaf fall and replenishment within the individual plant throughout the year. One effect of this is clearly seen in Figure 8.6 which emphasizes the nature of the closed canopy. However, with increasing distance from the equatorial areas, an increasing seasonality in primary productivity may be observed. Thus the non-seasonal equatorial rainforests can be distinguished from the tropical rainforests, which have one or two short dry seasons, the length of which increases at higher latitudes towards desert biomes.

The main issue for the vegetation is the competition for light. This leads to stratification of canopies, and proliferation

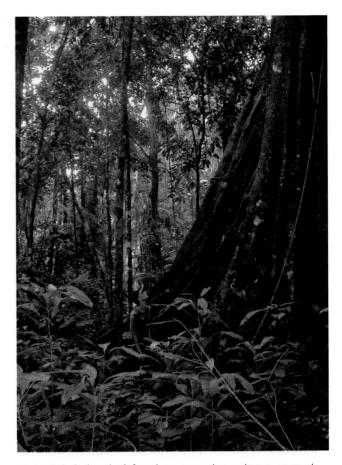

Figure 8.6 A closed rainforest canopy makes a gloomy scene at the foot of this rainforest tree. The struggle for light, often the main limiting factor in this otherwise highly favourable environment for vegetation in particular, can be seen here. Even a small break in the canopy is immediately utilized by climbers and by juvenile trees. (Source: photo courtesy of Nicholas Berry)

of climbers (Figure 8.7) and **epiphytes** (plants which grow above the ground surface using other plants for support and that are not rooted in the soil). There may be discontinuous shrub and herb layers, giving perhaps five layers in all. Species with higher light requirements, such as **lianas** (a type of climbing vine) or epiphytes, display typically adaptive lifestyles. Lianas climb rapidly and will often assume a shrub form in clearings, but will not usually form leaves until sufficient light is available, at the canopy level. Epiphytes attach themselves to the trunks or branches of trees to obtain a similar advantage. The competition for light ensures that the canopy is continuous and dense, even down to ground level at river banks. This gives rise to the illusion of the 'impenetrable rainforest'. The tree layers are distinguished by a huge diversity of animals and birds, reflecting the availability of food sources. Species often restrict their range to a single stratum although they may travel extensively throughout the forest within that layer.

Figure 8.8 A rainforest lizard with suction padded toes. (Source: photo courtesy of Nicholas Berry and Despina Psarra)

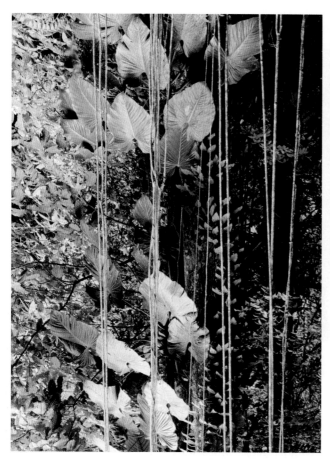

Figure 8.7 Long climbing lianas dangling among large green leaves of epiphytes in a rainforest in Trinidad. (Source: Staffan Widstrand/Nature Picture Library)

Animal adaptations include prehensile tails (tails capable of grasping), grasping feet and clawed or suction-padded toes (Figure 8.8). Birds tend to be fruit rather than seed eaters.

The upper canopy contains the dominant, mature species, often 25–35 m above ground, with still taller 'emergent' individuals that may reach a height of 50 m. These individuals, as seen in Figure 8.9, tend to have few branches below canopy level and fairly slender trunks in relation to their height, but may be supported by buttress roots. The second (~10–15 m) and third (~3–5 m) tree canopies are composed, to a large extent, of immature individuals of the dominant species. Second-canopy individuals are often triangular upward in shape, to take advantage of what light is available, and the second and third canopies are increasingly discontinuous, depending on the pattern of light availability.

The effectiveness of this competition ensures that in undisturbed forest there is relatively little understorey vegetation, since plants must be extremely shade-tolerant to survive. This space near the forest floor with limited vegetation, however, does provide room for large animals, such as wild pigs and jaguars, together with rodents, amphibians and reptiles. Nevertheless, most of the animal biomass is found in the soil fauna, including termites and beetles.

When natural clearings appear in the forest owing to fire, wind or water damage or resulting from the death of large individual trees, the saplings, climbers and other species contained within the lower canopies compete to obtain access to the increased light, fighting for this newly available niche. The regeneration process, if undisturbed, is usually regarded as having three distinct phases: gap, building and mature. Although complete regeneration can take around 250 years, the canopy is often closed once more after 5–20 years. Therefore the incidence of soil erosion and decreases in soil nutrient flows are minimized. Humans have mimicked this type of natural clearance for centuries within tropical forests via low-intensity shifting cultivation (see also Box 6.6 in Chapter 6), otherwise known as slash and burn. When carried out at low intensity, shifting cultivation accommodates this natural regeneration process. However, increasing intensification of clearance, such as that which accompanies large-scale commercial agriculture or forestry operations, does not allow regeneration to take place and coarse grass such as *Imperata cylindrica* replaces the previously complex forest vegetation habitat.

In intact equatorial forests, there is seldom a problem in obtaining sufficient soil water, unlike the competition for light to enable photosynthesis to take place, and evergreen species usually dominate deciduous ones until the length of the dry season becomes appreciable further away from the equator. Some plants display modifications aiding the

Figure 8.9 Rainforest trees. Emergent individuals can be seen towering above other trees with few low branches and slender trunks. Second- and third-canopy trees can be seen in the foreground. (Source: Lloyd Park © Dorling Kindersley)

removal of surplus water, such as the drip-tips of *Ficus* species (e.g. rubber plants). Raunkiaer life forms are classes of plants based on how the new tissue (buds and shoots) develops and grows as shown in Figure 8.10. In terms of

Raunkiaer life forms, the equatorial forest biome is characterized by a high incidence of **phanerophytes** (tall plants, visible all year, with buds high in the air) such as the palm family, together with ephiphytes.

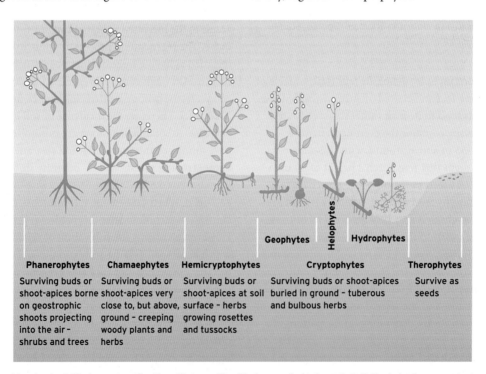

Figure 8.10 Raunkiaer's plant life form classification. (Source: *The life forms of plants and statistical plant geography* by Christen Raunkiaer, (1934))

Although the rainforests contain half of the world's faunal and floral species, stands of vegetation tend to be associations, rather than dominated by a single species. The large number of species is probably due to the complex structure of these forests, together with the stable prevailing climatic conditions and the length of time since radical climatic disturbance. Large islands of forest may have remained relatively undisturbed by global climate changes for hundreds or thousands of years. This equatorial and tropical forest biome is probably the most ancient and stable biome, avoiding the major stresses of the Quaternary glaciations discussed by Dawson (1992) which resulted in a reduction of diversity in the Nearctic and Palaearctic realms.

While the characteristic structure of the equatorial and tropical forest realms tends to remain, the actual species to be found vary between the realms. For example, leguminous trees such as *Dalbergia* species characterize the South American Palaeotropic, while the African region generally has fewer species with leguminous tree species represented by *Brachystegia*. In the Indian region, **dipterocarp** (generally tall and large) trees are more likely to be dominant. However, the conditions and characteristics outlined above may be dramatically altered by local factors. Under specific azonal factors, such as local geology, it may be that drainage and nutrient levels provide a more challenging environment. In the Guiana Highlands of South America, the coarse sandstones have encouraged soil acidity and a form of tropical heathland exists. Under conditions of inundation by water, such as along parts of the Indian and Brazilian coasts, swamp formations such as mangrove forest will develop.

In areas with more than five months of dry season, the tropical forest biome changes in response to the water stresses on plants and animals. First deciduous and then strongly adapted species, such as thorn woodland species, begin to dominate in each realm, such as in the mulga and brigalow scrub of Australia and the acacia thorn scrublands typical of Africa. In all cases, the structure of the forest is of a low, 4–10 m canopy with thin or non-existent shrub and ground layers beneath.

The ecotone representing the forest–savanna boundary has been the object of much research, since forest clearance has been a major signature of human impact on Earth (Atkins *et al.*, 1998). The ecotone is highly variable in width, ranging from 30 m in parts of West Africa to areas of 'savanna-in-forest' and 'forest-in-savanna' mosaic covering hundreds of kilometres elsewhere (Mannion, 2002).

8.3.2 Savanna

The temperature of the savanna biome is very similar to that of the rainforests, but the dry season is sufficiently long to result in seasonal vegetation. Figure 8.5 indicates the typical location of savanna. The structure of this biome is characterized by a more open canopy than the tropical forests and the penetration of light allows the growth of grass and other ground flora. Most areas share the climatic characteristics of a dry season in which rainfall is below 250 mm a month for longer than five months. Where the annual rainfall drops below around 625 mm, thorn scrub begins to take over. A gradient within the biome, as shown in Figure 8.11, can often be recognized, as the increasing length of dry season has more pronounced effects away from the equator moving from savanna woodland, through tree savanna, then sometimes to a thorn/shrub savanna before savanna grassland dominates.

Most workers now assume that the tropical savanna areas are almost always the result of **edaphic** (influenced by the soil) or biotic factors dominating climatic factors. Savanna regions tend to be found in areas of **continental shields** such as the Mato Grosso, Central Africa or central Australia. Here, the effects of resistant, often infertile, substrates are compounded by long-continued human occupancy and usage with increasingly intensive woodland clearance, often by fire, with little chance of recovery. Climatic change during the Quaternary has also altered the tree cover to a very large extent. In general the Australian, *Eucalyptus*-dominated, savanna woodlands are considered the most likely to be related to an earlier warmer climate, while those savannas in other realms may well be anthropogenic in origin and exist as fire climaxes (see Chapter 9). Termites also form an important element of the savanna because of their effects on litter decomposition and soil character.

Savanna areas appear as a mosaic. This is a result of diversity of origin and effectiveness of other factors. Large-scale diversity of structure within the biome is related to differences in the availability of water and soil nutrients. There are usually few trees. Where there are trees this is usually linked either to where the water table is close to the surface at depressions in the ground, or along river valleys, or to where drainage is enhanced within otherwise impermeable areas (e.g. the uplands of Trinidad). The density of tree spacing is dependent upon the level of competition for water. Woody species such as the giant baobab, with its thick bark, short season in leaf and water-retentive trunk, tend to be strongly adapted to withstand fire and drought. Many savanna trees are low with branches forming an umbrella-shaped crown as seen in Figure 8.12. This is in distinct

Figure 8.11 Profile diagrams for four ecoclines: (a) along a gradient of increasing aridity from moist forest in the Appalachians west-wards to desert in the USA; (b) along a gradient of increasing aridity from rainforest to desert in South America; (c) along an elevation gradient up tropical mountains in South America from tropical rainforest to the alpine zone; and (d) along a temperature gradient from tropical seasonal forest northwards in forest climates to the Arctic tundra. (Source: after Whittaker, 1975)

contrast to the form of typical rainforest species (Hoffmann *et al.*, 2003) (see Figure 8.9).

The fruiting period of savanna trees often takes advantage of 'fire seasons' (large fires occur every few years across parts of the savanna) by dropping fruit at the end of the fire into temporarily nutrient-rich soil. Other common adaptations include a deep root system, which allows suckers to respond to the wet season, thorns or spines to reduce water loss and deter grazers, and a general paucity of leaves and branches. Most trees are deciduous as an adaptation to the drought stresses of the dry season, such as *Brachystegia* and *Julbernardia* that dominate the African zone, although some, such as *Acacia faidherbia albida*, keep their leaves. Others, including *Acacia karoo*, produce very large numbers of seeds, since losses of seeds to fire, termites or drought can be great.

The savanna vegetation is often characterized by **xeromorphic** (adapted to dry conditions) grassland species which may produce a canopy several metres high. Elephant grass, for example, can reach 3–4 m in height. These grasses often have **rhizomes** (lateral stems through the ground that send up new shoots) or densely tufted habits to protect

against fire and drought. Grassland stands are often locally dominated by very few species and levels of biomass are highly variable within this biome, related to the amount of trees. Collinson (1997) suggested a biomass range from 150 t ha^{-1} for savanna woodland to 2 t ha^{-1} for sparse grassland without trees and emphasized the direct relationship between productivity and rainfall.

Large animal predators tend to occupy wooded areas that provide shelter, cover and a variety of food. They prey on herbivores such as antelope found in huge numbers within East and Central Africa or deer species and marsupials in South America and Australia respectively. Hyenas exemplify the types of adaptation that lead to success within this environment (they also forage in drier regions), scavenging usually at night, since nocturnalism not only reduces competition for prey, such as from vultures, but also conserves their water supply. Other animal adaptations within this biome include the extensive migration of large grazing mammals and their related predators. Since prey migrate, the Serengeti spotted hyenas form temporary and mobile social groups, rather than permanent groups that would defend a single area (Mills, 1989). They will track

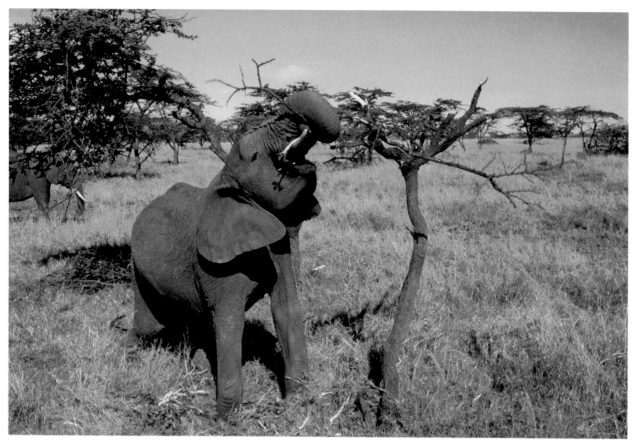

Figure 8.12 The typically sparse ground cover within much of the African savanna biome is shown in this photograph from Kenya. Elephants graze the acacia trees. The landscape is characterized by spiny shrubs and flat-topped trees with browse-lines related to the preferred feeding habits of the local animals. (Source: Stephen J. Krasemann/Photo Researchers Inc.)

seasonal availability of both water and vegetation, which at times conflicts with human administrative boundaries.

8.3.3 Hot deserts

The dominant factor to which the hot desert biome is attuned is the climate, often summarized as hot all year, dry all year. The criterion taken here is of insufficient moisture for complete ground cover, leading to characteristic adaptations of the life forms occupying the biome, or traversing it as part of migratory journeys. Both plants and animals tend therefore to display characteristics controlling evaporation and maximizing the conservation of any water available. Areas such as the Atacama Desert, in South America, include locations that may avoid sporadic rainfall for 30 years, while others may have more regular rainfall but which is insufficient to counteract the locally high levels of evaporation. Fog and dew, however, provide important supplementary water. In addition animals are also challenged by temperature stress. Heat may be extreme, with large diurnal ranges, due to lack of cloud cover (see Chapter 5).

Most desert soils are poorly developed but increased amounts of nutrients and improved structure are usually found around the roots of shrubs. This reflects the increased organic matter provided both by the plant and from those animals using the plant as shelter.

The structure of the vegetation of hot deserts is varied, but low in height and always with very open stands, as shown in the oasis in Figure 8.13. The vegetation density is dependent primarily upon on access to, and competition for, water. This is seen in the dramatic increase in biomass around oases and watercourses. Increased aridity results not only in increasing space between individuals but also in increasing clustering of their distribution. This is more complex than a purely climatic response as it is dependent upon local geology and geomorphology; soil conditions are very poor in hot deserts where the upper soil is of a coarse texture. Where runoff is concentrated, the habitat improves.

Responses to the hot desert's challenging environment often take the form, in plants, of controlling transpiration rates by, for example, only opening their stomata (see

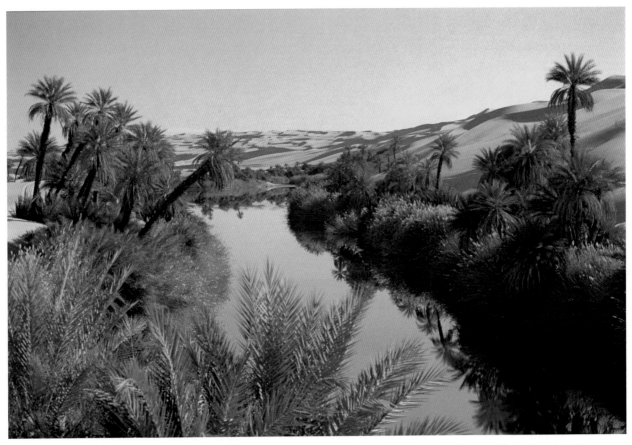

Figure 8.13 An Oasis in the Sahara Desert. When water is available, trees are able to establish in what are otherwise extremely arid areas. (Source: Carsten Peter/National Geographic)

Chapter 4) at night. Cacti such as the Brazilian xique-xique store water within their stems and may display other moisture-conservation features. Leaves are often replaced by thorns (e.g. *Euphorbia ingens*, common in Central and eastern Africa) or, in succulent species, by stems with water-storing cells that also have the ability to photosynthesize (e.g. *Opuntia* species in Mexico and *Aloe* species in Africa). Other adaptations include woodiness, which prevents collapse of plant material during wilting, or a small total leaf area. For grass species, their characteristic short, tufted nature helps protect against drought and heat stress. Photographs of some dryland species are provided in Figures 16.6, 16.7 and 16.8 and some adaptations are provided in Table 16.3 in Chapter 16.

These adaptations may be combined with, or replaced by, a lengthy dormant season. This can involve a total dieback of above-ground tissue, perhaps where a large root system allows access to either deep groundwater or, more commonly where the substrate is coarse, to tap a wide lateral zone. The record for shrub root depth would appear to be held by a tamarisk (a small, narrow-leaved tree) in Suez,

with roots that measured 45.7 m. Seeds may have germination-inhibiting coatings that require significant weakening by moisture before germination. This prevents germination in conditions that will otherwise be too harsh for the young seedling.

Where adaptations do not exist, plants may have very rapid life cycles, responding to the period between rainfall and resumption of drought, thereby avoiding the environmental challenge. They may be 'annual' plants, which means that the plant grows from a seed, flowers, produces its own seed and then dies. The same plant will then not grow again but its offspring will survive where seeds germinate during the next rainfall event. This accounts for the phenomenon of the desert 'blooming' for a few weeks after rainfall. Inevitably, these annuals are usually of small biomass and produce very large quantities of seed.

In the hot desert of Death Valley, USA, instead of grass there is a **xerophytic** broad-leaved ground layer, often called cactus scrub but including mesquite, cottonwood and tumbleweed, with creosote bush and saltgrass nearer saltpans where evaporated water has left concentrated salts behind. The vegetation in equivalent Australian areas is mallee scrub (*Eucalyptus* and *Acacia* species) which can include spinifex grassland but may also develop into mulga shrubland. The caatinga of north-east Brazil is characterized by thorny acacias, together with xerophytic grasses.

Animal adaptations may be either physiological or behavioural. These often slow the loss of body moisture, through excretion by dry faecal pellets or concentrated urine, or through sweat, which is especially dangerous for smaller animals because their surface area in relation to body mass is high. Many animals, like sidewinder rattlesnakes, are without sweat glands. Small animals, such as the jack-rabbit, are nocturnal, hiding in burrows during the day or undergo aestivation which is a state of dormancy during the driest season. Many insects, like Namib beetles, are also nocturnal and **cryptozoic** (shelter-seeking). Namib beetles obtain moisture from dew. Larger animals are relatively unusual within the desert fauna but may, like camels, have hairy coats to enable sweat to evaporate and produce a cooling effect. Camels also have variable body temperature, storing heat (not water!) within the body during the day and releasing it at night. Other species rely upon the moisture released by mist, fog or dew, or scavenge, like the hyena, at night.

It will therefore be apparent that hot desert biomass is mainly underground, with plant life forms mainly geophytes or therophytes (see Figure 8.10) to avoid drought. Biodiversity is low and net primary production strongly related to rainfall.

Reflective questions

➤ What are the main differences between the tropical biomes?

➤ Are there distinct boundaries between biomes?

➤ What are Raunkiaer life forms and which are characteristic of the different tropical biomes?

8.4 The temperate biomes

8.4.1 The Mediterranean/chaparral biome

The location of the temperate biomes can be seen in Figure 8.14. The 'Mediterranean' biome covers a wider range of areas than those surrounding the Mediterranean Sea. The characteristic climate is warm all year but with low rainfall, characterized by summer drought with high evaporation rates (see Chapter 5). This biome forms a transitional zone,

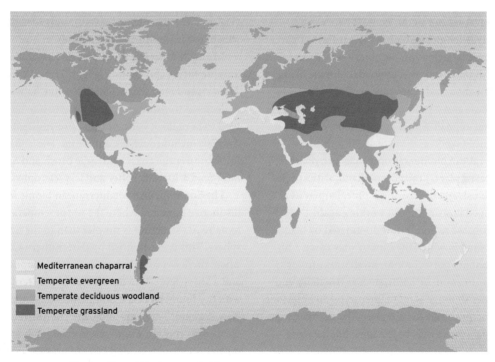

Mediterranean chaparral
Temperate evergreen
Temperate deciduous woodland
Temperate grassland

Figure 8.14 The location of the temperate biomes.

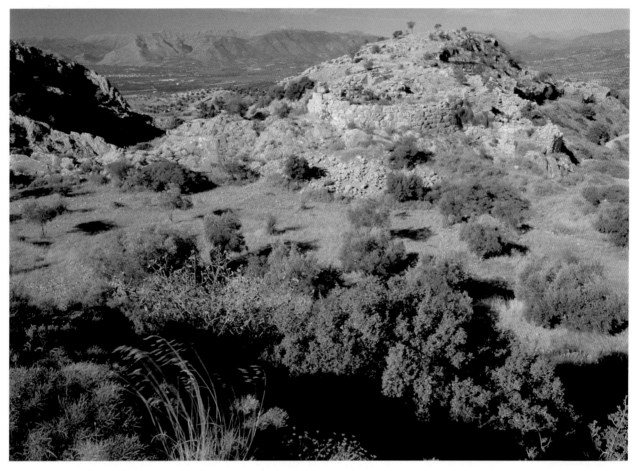

Figure 8.15 Maquis scrubland in Greece. (Source: Joe Cornish/DK Images)

similar to the savanna, but in this context between the desert and the true temperate biomes. Regardless of parent material, the upper soil horizon becomes very dry in summer and soil water is drawn upwards. Climate changes over the past few thousand years have tended to increase the aridity of these areas, producing an environment to which plants and animals would appear to have had to adapt rapidly.

The Mediterranean biome was once **sclerophyllous** (hard, tough-leaved plants) mixed woodland, with species such as the cork oak and the maritime pine. Largely as a result of continued human impact, working against woodland regeneration, the region is now dominated by maquis. Maquis is scrubland in which the canopy may reach 3 m, and typically contains gorse, broom, myrtle, arbutus or olive, together with aromatic herbs, such as rosemary and sage. Where water shortage is more intense, as on limestone substrate, the maquis tends to be replaced by garrigue vegetation. The actual species may be the same but garrigue vegetation tends to have a lower canopy than maquis with wider-spaced individual plants, which are mostly evergreen

(Figure 8.15). There are, however, some deciduous species which lose leaves during dry periods.

In North America, the chaparral has similar, sclerophyllous life forms and canopy height. Typical species include *Eriodictyon tomentosum*, which has fine hairs on the leaves to decrease transpiration, and the creosote bush. Any trees tend to have thick bark, or other fire-resistant qualities. Animals include deer, elk and bears. The 'soft chaparral', found nearer the coast, with sage brush and *Salvia melliflora*, is literally softer and less sclerophyllous in nature.

The South African Mediterranean biome region has a distinctive flora and where the warm temperate forest remains, its structure is simpler than in the northern hemisphere. This is exemplified by the Southern Cape Province, where there is a mainly coniferous *Podocarpus* upper canopy and a lower-level flora dominated by tree ferns, comparable with the southern beech domination of similar regions of New Zealand, Tasmania and Chile. In Australia, mallee scrublands contain *Eucalyptus* species, with an *Acacia* shrub understorey and a ground layer of grasses such as *Trioda*, together with marsupial grazers such as wallabies.

Overall, therefore, there is a tendency towards a xerophytic scrubland, with strongly developed adaptations against the relatively frequent natural fires of this biome. These adaptations include thick, smooth bark and/or deep roots from both of which regeneration may take place. Some seeds open only after exposure to fire and there is a strong representation of cryptophytes (see Figure 8.10), surviving difficult periods as bulbs or rhizomes. As a result, a distinctive post-fire regeneration cycle exists, moving through domination by annuals, then herbaceous species before the scrubland returns. It should be noted, however, that species representing the scrubland are usually present throughout the earlier stages and that the disturbance created by fire therefore increases species diversity within the earlier stages. Animals adapt to, or avoid the stresses of, drought and fire, often by speed (e.g. kangaroos, goats and emus) or by burrowing (e.g. mice).

The Mediterranean biome is one of several biomes that have been recreated as an educational tourist attraction in south-west England called the Eden Project. This attraction is discussed in Box 8.2.

8.4.2 Temperate grasslands

The temperate grassland biome is found in continental interior areas of the Holarctic and within the eastern region of the Neotropical realm. Trees are generally absent, except within the ecotones joining this biome to that of the deciduous forests. The vegetation is dominated by grasses, usually perennial (the same plant surviving for year after year) and often xerophyllous. It has, especially in its most climatically favoured regions, suffered intensive human impacts. Large herds of herbivores such as bison were characteristic, although are now dramatically reduced as a result of human activity. The extinction of the passenger pigeon, *Ectopistes migratorius*, once observed in huge flocks in the temperate eastern grasslands of North America, was less directly but equally certainly due to human influences. Quammen (1996) suggested that its extinction followed hunting that resulted in numbers dropping below that compatible with its social ecology, even though numbers were still very large. The bird depended upon benefits of crowding to identify food sources during foraging flights and to avoid surprise attacks by predators. It is an example of a species where humans grossly underestimated the minimum size of the viable population.

The dry season of temperate grasslands tends to be lengthy, with annual precipitation usually less than 500 mm. Precipitation may take the form of snowfall in winter, while the intense heat of the continental interiors in summer leads to convectional rainfall and high evaporation. As a response to greater continentality, there is a gradual change from the moister, less continental conditions of neighbouring biomes, as shown in Figure 8.19. The geomorphological and glacial heritage of extensive, gently undulating areas with rich soils has led to regions such as the prairies and the steppes becoming important producers of cereal crops. Chernozem molisols together with vertisols and andisols are typical soils over these areas (see Chapter 7). Where it is damper there may be brown earth soils.

Adaptations, such as the height of the grassland, are similar to that of the forest biome and relate to maximizing effectiveness of precipitation and to minimizing damage due to natural fires. These adaptations include the shallow, turf-forming dense roots of many of the grasses typical of the moister areas (e.g. *Agropyron* species) or the tussocks typical in drier areas (e.g. *Poa* and *Festuca* species). These features provide fire resistance and water trapping. Lack of a protective cover from predators has caused animals to develop speed (e.g. antelope and deer), bulk (e.g. elk and bison) or a burrowing habit (e.g. mice and voles). The single vegetation layer results in a relatively limited diversity of birds but plant diversity is often as high as in forest formations. The tendency to summer fires means that systems at or below ground level that allow survival, such as bulbs, rhizomes or tubers (**perennating systems**), are important advantages (see Figure 8.10).

Where protection from fire is combined with increased soil moisture, then scrub and woodland have been seen to develop, notably in the South African veld, in the tussock grassland of New Zealand and in Australia. Sucker-producing trees such as aspen are among the first to appear, along watercourses. This has given rise to some controversy as to the origin of the wetter areas of this biome.

8.4.3 Temperate deciduous forest

Sometimes described as temperate deciduous woodlands, this biome is found only in the northern hemisphere, since in the southern hemisphere the vegetation of the equivalent climatic areas is predominantly evergreen. This is probably a result of the relatively late development of the deciduous habit, after the Laurasia–Gondwanaland split (see above). In all areas, however, there has been very considerable human interference, including transfer of species between regions, sometimes with disastrous effects. The effect of these introductions is discussed within Chapter 9. There is often a fairly sharp boundary between the deciduous forest and the more northerly boreal forest biome. The climate in the deciduous forest biome is moist

THE EDEN PROJECT - BIOGEOGRAPHY AS LEISURE?

The Eden Project in Cornwall, southwest England (Figure 8.16), is a botanic garden focused on education about the world we live in and the ways humans use and abuse the world. Rather than presenting taxonomic collections, the site represents three of the world's biomes, with more planned for the future. The most famous icon of the project is the huge covered tropical biome, capable of showing rainforest trees at a mature size. This contains representations of the flora and land-use of four regions: Guyana, Malaysia, Cameroon and Oceanic Islands. Within this largest 'biome', temperature and humidity are maintained at tropical levels throughout the year and plants are now reaching their full extent (Figure 8.17). Popularity for this venue is now huge and growing. A range of educational activities are focused upon this biome. There is also a covered warm-temperate biome showing South Africa, California and the Mediterranean Basin. The outdoor displays show the temperate zone, covering northern Europe but also including displays that relate to other countries such as northern United States and Chile.

The plant collections are designed to show regional landscapes and

Figure 8.16 The huge scale of the Eden Project can be appreciated in this picture. This is just one of the three triple-domed structures on the site. The walls of the old china-clay quarry can be identified in the background, together with the pre-industrial land surface.

BOX 8.2 ➤

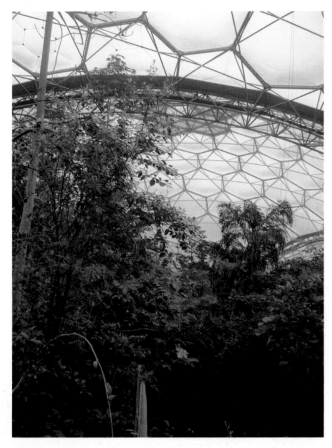

Figure 8.17 Inside part of the tropical dome at the Eden Project.

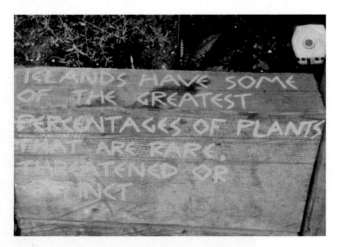

Figure 8.18 Thought-provoking inscriptions are displayed throughout the Eden Project site.

land-use practices, so each biome shows both wild plants and crops appropriate to the region. These cannot be full ecological simulations of the different zones and they will be managed displays rather than living landscapes. For example, although some invertebrates are contained within the displays it is impossible to represent the full range of species and, for many species, processes such as pollination or propagation are carried out by the staff.

Nevertheless, thought has gone into representing the zones as faithfully as possible. For example, the soils of the warm temperate region have been made to be drought prone and nutrient poor to encourage the plants to adopt typical growth forms. The plantings aim for a representation of the typical plant assemblages found in those regions and aim to mimic seasonal displays such as the flush of annuals found in South Africa following rain. An important aim is to infuse the displays with thought-provoking items such as using a log to reflect on island biogeography (Figure 8.18).

The 'outdoor biome' has more opportunity to represent authentic ecosystems and in an area called 'Wild Cornwall' examples of Cornish heaths, Atlantic coastal woodlands and farm and fieldbank communities are being created, using habitat restoration techniques. The educational displays are thus linked to research into restoration methodology that can be fed into real projects both within the region and globally.

BOX 8.2

227

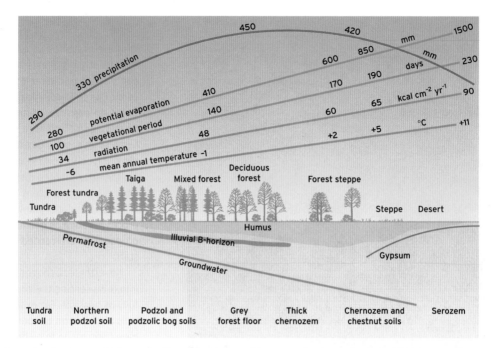

Figure 8.19 Schematic climate, vegetation and soil profile of the east European lowlands from north-west to south-east. See Chapter 7 for discussion of soil properties and Chapter 19 for discussion of permafrost processes. (Source: after Walter, 1990)

but temperate all year, with few months having a mean temperature above 10°C.

Brown earth molisols (see Chapter 7) are typical of many areas, with a rich soil fauna providing mixing of nutrients. Where soils are under conifers, on naturally acidic rocks or have become heavily leached, podzols may be the characteristic soil. In response to this, heathland may develop, where common heather (more correctly called 'ling') and bell heather dominate, with a canopy of less than 60 cm (Figure 8.20). Other components of the formation frequently include grasses and rushes such as *Nardus stricta*, *Molinia caerulea*, *Eriophorum vaginatum* and *Juncus effusus*, which form part of a typical heather cycle, producing a mosaic of age, species composition and structural types within the wider moorland habitat. On lowland heaths, such as Lüneberg Heath, northern Germany, junipers, birch, Scots pine and gorse may also form important elements. Over a lime-rich substrate, however, beech may dominate, producing a thick litter layer with a corresponding decrease in ground flora.

In the most favourable areas, the temperate deciduous biome displays a structure with four main layers. The canopy, where the trees with rounded deep crowns reach 8–30 m in height, is underlain by a shrub layer usually below 5 m in height and a field layer that includes many grass species as well as a ground layer of mosses and

liverworts (Figure 8.21). The extent of the lower layers is dependent upon the nature of the species making up the canopy.

A characteristic of this biome is the very marked seasonality of the vegetation, with a corresponding effect upon wildlife. Since the growing season is restricted, areas may

Figure 8.20 *Calluna vulgaris* (ling heather) in flower across a heathland.

Figure 8.21 The deciduous woodland edge. Photographed in late spring, this picture shows how some ground layer plants are spring flowering, to enable the main part of their reproductive cycle to be complete before the tree canopy closes.

display a succession of dominants. This may begin with corms and other ground flora (e.g. snowdrops and dog's mercury), followed by species such as violets, primroses and bluebells, all of which have a flowering period before the canopy of new leaves closes and limits access to sunlight. To succeed in this environmental niche, they must initiate growth very early in the warmer period, often as soon as daylight hours increase, since their effective growing season is determined by the development of the species forming the upper canopy. As in the rainforest biome, climbers such as traveller's joy, ivy, wild rose and, to a lesser extent, ephiphytes such as mistletoe use the main canopy species for support in their quest for sunlight.

As the name of the biome suggests, trees are typically deciduous, including various oak species, beech and ash. In the northern American regions, maples will often also be important, replaced by hickory in more southern areas. Beech, birch and ash species are also important in the Asian forests. The autumnal leaf fall reduces both water loss and frost damage during the winter. Leaf loss also closes the nutrient cycle, returning absorbed nutrients back to the soil as the leaf litter decomposes. New leaves appear almost as soon as the growing season begins. Flowers appear before the leaves in tree species such as hazel and some willows but also in flowering plants such as coltsfoot and butterbur. This adaptation both maximizes the time available for fruit to ripen before the onset of the next winter and assists wind pollination. Animals may hibernate (e.g. hedgehogs) or burrow (e.g. rabbits) to avoid the challenges of the winter. Deer are found in all regions and there was once an abundance of other large mammals such as bears. Large

mammals are now considerably reduced in both numbers and range in temperate deciduous forests as a result of human impact.

8.4.4 Southern hemisphere, evergreen temperate forest

This is the variant form of the temperate deciduous biome and usually displays a structure of two tree canopies and a shrub layer, having climbers and epiphytes but less frequently a ground flora. Southern beech are an important component, with conifers, of the podocarp family, found towards the boundaries with warmer areas. The New Zealand region has especially rich bird life, including carnivorous (meat-eating) parrots, compensating for a lack of native mammals.

Reflective questions

➤ What are the similarities and differences between the origins of Mediterranean and savanna biomes? How might their effective management differ?

➤ What are the main plant adaptations in temperate biomes?

8.5 The cold biomes

8.5.1 Taiga

The location of the cold biomes, comprising the taiga and the tundra, are shown in Figure 8.22. The lack of any extensive regions with these types of biogeographical character within the southern hemisphere should be noted. The taiga biome is often termed boreal forest. The taiga–tundra boundary often reflects both the post-glacial recovery, as a result of which biomes are continuing their slow poleward migration, and the effects of commercial forestry.

The climate of the taiga is cool all year, with relatively little rainfall, since it is located for the most part within continental interiors (see Chapter 5). There is, therefore, a summer maximum of precipitation, often via convection. The biome is usually taken as extending polewards from where there are less than five months with air temperatures above a mean of 10°C, until only a single month fulfils this criterion. Since most plants will cease to grow when temperatures fall below this value, there is a very limited growing

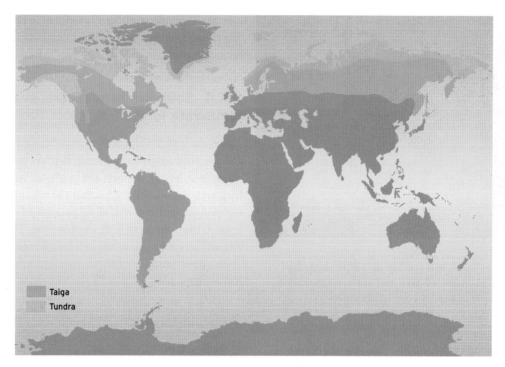

Figure 8.22 The location of the cold biomes.

season which has a major influence upon the faunal (animal) as well as the floral (plant) elements of the biome (Figure 8.23).

Soils tend to be variants of the podzol family (see Chapter 7) and vast areas have been glacially eroded. Characteristics include slow nutrient cycles and the litter layer

Figure 8.23 The taiga at Denali National Park, Alaska, with spruce willows and rugged landscape. The trees have wider branches nearer the foot of the tree tapering to narrower branches at the top. This minimizes potentially damaging snow accumulation. (Source: photo courtesy of Ed Reschke/Peter Arnold Inc./Still Pictures)

may contain three to five times the annual accumulation and strong vertical layering. This is caused, in part, by the acidic nature of the leaf litter but also by the relative lack of soil fauna leading to very slow decomposition, a response to a cold and short growing season and poor drainage. In those areas where glacial effects include impeded drainage, peat bogs may develop rather than forest.

A very wide ecotone exists between the temperate and boreal forests but it tends to be represented by mosaics rather than stands of mixed deciduous trees and conifers. These are often related to local conditions. For example, spruce prefers loamy soils and in the most advantageous areas oaks may survive, while pines are found on sandy outwash plains. In the warmer and wetter taiga areas, forests are characterized by a structure of unbroken stands of tall (24–30 m) conically shaped trees, a discontinuous shrub and herb layer in response to the lack of sunlight penetrating the canopy and a thick resinous leaf litter and well-developed moss ground layer. The true central taiga, however, has lower and more discontinuous vegetation, with greater dominance of ground layer lichens. The trees tend to become more sparse in areas where the growing season diminishes. This ecotone is sometimes referred to as 'sub-arctic parkland' with elements similar to that characterizing the rainforest–savanna boundary.

Figure 8.24 View of migrating reindeer across the tundra landscape of Siberia. The lack of vegetation cover means that vast areas of land are needed to support the deer populations. (Source: Alan & Sandy Carey/Photodisc)

Most Eurasian forests are dominated in the west by Scots pine and Norway spruce giving way to larch, but those in North America are characterized by the lodgepole pine and Alpine fir in the west and white spruce, black spruce and balsam fir further east. Not all trees are coniferous, and larch, birch and alder are frequently of great regional importance. *Larix dahurica* dominates the world's most northerly forest in Siberia. Its deciduous nature enables it to withstand the combination of intense cold and strong winds. Tufted grasses and heathland plants such as crowberry and bilberry are typical of the ground layer. In the colder areas, as tree cover diminishes, the ground is often carpeted with lichens. A 100-year-old tree may be only 1.5 m high, with a 7 cm basal diameter. The deer family is well represented in North America, Europe and Asia, and is usually migrant within the forest (Figure 8.24), although both caribou (North America) and reindeer (Europe) also utilize the tundra (see below). Weasels (*Mustela* species) successfully utilize the more open areas both in the taiga and in neighbouring biomes. The fur of northern species turns white in winter, providing camouflage. The weasels' success is mainly due, however, to the advantages provided by their small but long and thin bodies. These include the ability to follow prey such as mice and rabbits into their burrows. Weasels may also move into these burrows after feeding upon the original owners. This offers protection against the harsh environment and predators.

In the brief spring period available in taiga biomes, evergreen species are able to begin immediate photosynthesis as sunlight and temperature increase. This maximizes the growing season. Where the effective soil layer is shallow as a result of the local geology or permafrost (see Chapter 19), plants may respond with fan-shaped root systems, allowing take-up of water as soon as the spring thaw occurs, achieving a similar result. The typical form of trees in this biome is of a tapering single trunk with the lowest branches being the longest, minimizing potentially damaging snow accumulation (Figure 8.23). Under the closed canopy the lower limbs die and in mature trees live branches often only exist above 6–10 m on the trunk.

Many taiga species display an increased concentration of sugar in their sap during the winter. This offers protection against both cold and lack of water, since plant roots absorb water from the soil far less efficiently than during warmer conditions. A sugar increase, combined with a decrease in water content of the plant cells, means that water absorbed from the soil can enter but not leave the vegetation. Similar adaptations are displayed where transpiration rates are lowered by stomatal closure (see Chapter 4), in the most severe conditions, or by thickened leaves or bark.

8.5.2 Tundra

Tundra regions provide, in many ways, the most challenging environments for plants and animals. Note that the boundary shown in Figure 8.22 within each continental land mass is further north in the west than the east, indicating the warming effect of maritime winds (see Chapter 5). The equatorward limit is taken to be a temperature of 10°C for the warmest month. The 'summer' period, with temperatures rising above freezing, may only last for two months of the year and winter temperatures may plunge to −50°C. Strong, cold, dry winds are generally prevalent, with precipitation most likely during the summer. Precipitation is seldom substantial but snow accumulates from year to year in many sheltered sites. Lowland tundra areas are also characterized by continuous permafrost (see Chapter 19). This results in low levels of soil fauna and limits viable rooting depths.

Tundra soils (cryosols) are characterized by a litter layer of partly decomposed, often highly acidic plant material, commonly up to 10 cm thick but reaching over 1 m in boggy locations. This rests upon a gleyed horizon (see Chapter 7) which in turn lies upon a permanently frozen layer. Geomorphology has a stronger than usual immediate influence on this biome. Glacial erosion and contemporary fluvial activity have left shallow soils at best, but with underlying sands and gravels in valley floors. This provides more opportunities for trees within such valleys, and where there may be a deeper layer that is not permanently frozen. The vegetation generally forms a mosaic closely related to local geomorphological and microclimatic features, in competition for the least stressful sites (Figure 8.25). This has the effect that, although on a microhabitat scale the number of species is often very small, in a regional context there may be very considerable diversity (Matthews, 1992).

In the most favoured localities there may be a characteristic three-fold structure comprising a low shrub layer above tussocky grasses and cushion-form herbs, underlain by a final layer of mosses and lichens. As in other biomes, this structure becomes progressively simpler and lower as conditions become more severe. Vegetation growth throughout this biome is slow. A 400-year-old juniper trunk may measure only 25 mm in diameter. Therefore the colonization of bare ground may take many decades. Similarly, the **metamorphosis** of some tundra insects such as *Gynaephora groenlandica* can occupy several years.

The low productivity of the vegetation means that large areas are required to support the migrating herds of large mammals, such as reindeer. Migratory birds, such as terns, utilize the mosquito and other summer insect populations and are in turn preyed upon by hawks, falcons and owls. Plant-eating lemmings are very important creatures in the tundra zone. They are preyed upon by a variety of species, such as owls, foxes and weasels. Changes in lemming populations therefore result in changes in predator populations.

Figure 8.25 Bones of a musk ox on stony tundra ground. Tundra is the flat, treeless Arctic region that lies between the polar regions of perpetual snow and ice and the northern limit of tree growth. Only the most favourable areas are utilized by plants. These usually occur where protection from the wind is combined with the availability of surface water. In this photograph on Clavering Island, North East Greenland, the nutrients released by the decay of the musk ox carcass have allowed localized plant growth. Tundra vegetation consists mostly of mosses, lichens and small dwarf shrubs. (Source: Simon Fraser/Science Photo Library)

The tundra is limited floristically, probably because throughout the Quaternary this biome was destroyed for long periods during glacial periods as the ice covered the land masses of these regions (see Chapter 20). In the most severely challenging tundra areas, the plant associations are usually composed of sparse mosses, lichens and tufted grasses, all of which may become dormant when necessary for survival. Most plants of the tundra are perennial with perennating tissue at or below the surface. As climatic conditions become increasingly extreme, the importance of various means of vegetative reproduction also increases. Underground runners, rhizomes and bulbs dominate, as seeding becomes less reliable during the short growing season. Some species are self-fertile and do not require a partner for pollination, and others are wind rather than insect pollinated. The dominant plants overall are grasses and sedges although this is highly variable. In the most favoured areas there are patches of dwarf trees, such as the dwarf birch, *Betula nana*, but these form a shrub layer structurally. Rigid sedge and alpine meadow grass together with crowberry, mosses and lichens characterize this variant of tundra. As conditions deteriorate polewards, *Sphagnum* moss is replaced by hardier mosses, such as *Distichium capillaceum.* Flowering plants include members of the buttercup, poppy and saxifrage families, often with bright, insect-attracting flowers.

Heath dominates on the coarser-grained substrates, and is especially important over much of tundra Greenland. Heathland species include members of the *Vaccinium* and *Erica* families such as the arctic blueberry. In marshy areas such as the Mackenzie Delta, grasses, sedges and rushes are more characteristic, with *Sphagnum* moss and willow cotton grass where waterlogging occurs. Dwarf trees such as alder and willow may also exist as a shrub layer. Dwarf forms of vegetation provide several vital benefits arising from the microclimates created. These include a relative calm, since wind speeds are reduced by friction near the ground, which helps reduce evapotranspiration. The insulating effect of a tussock or cushion habit both provides warmer temperatures during the growing season and conserves moisture. Willows may reach only 0.3 m and will occupy the most sheltered areas. A semi-horizontal growth form benefits from the lower wind speeds and also protects shoots to some extent from frost. Many herbs are present here only in cushion or rosette form. Growth can begin from within the cushion or as part of the compressed rosette as soon as the snow melts, giving a slightly longer growing season. The next season's flowering buds are often pre-formed.

Dormancy allows plants to survive cold and drought. Those trees and shrubs able to survive in the most sheltered tundra areas tend to be deciduous where shoots are likely to be killed by frost. This creates a disadvantage, however, since new leaves must form within the short growing season. As in the taiga biome, the high sugar content within sap offers frost protection but also increases nutritional value.

Animal behavioural adaptations include hibernation and migration. Species such as caribou and reindeer, arctic foxes, wolves and lemmings are all known to migrate between the taiga and tundra, although musk oxen and some caribou species will remain within the tundra biome. Burrowing animals include voles and snowshoe hares. Animals and birds may have insulating fur or feathers, such as the ptarmigan, with feathers on the soles of its feet and also white camouflage in winter.

Despite the actual constituents of the plants and animals differing within the tundra biome, their habit and structure in this highly challenging environment are actually very similar throughout the biome, across the realms. This is often referred to as an example of 'convergent evolution' whereby the species in different areas have all evolved in the same way to cope with the conditions.

Reflective question

➤ What are the main differences between tundra and taiga biomes and how do plant adaptations reflect these?

8.6 Mountain biomes

As with climate (see Chapter 6), there are regional modifications to the underlying structure of biomes. For example, topography can play a local role in altering regional patterns. High mountain areas are usually sufficiently unique to be considered as separate biomes to their surrounding regions. They may also form significant barriers to the dispersal of life forms. The effect of aspect on local-scale biodiversity tends to be important within any mountainous area and it can also be influenced by local land-use patterns. Since mountain biomes tend to be unique to the area within which those mountains are found, these mountain biomes will be discussed by use of a case study. The Andes mountain biome is therefore described in Box 8.3.

Reflective question

➤ Why might spatial maps of the biomes be complicated by topography?

8.7 Changing biomes

The biomes described above are depicted as having specific characteristics, either functional or structural, but it is clear that they are all dynamic in nature. With intensification of resource usage and climate change, the ability to regenerate and hence the degree of permanence of any biome cannot be guaranteed. 'Maritime versus continental climates, soil effects and fire effects can shift the balance between woodland, shrubland and grassland types' (Whittaker, 1975). Such natural causes of change have probably always existed, but the effects of human intervention are now equally important. Multinational bodies have been attempting to establish the consequences of biome and ecosystem changes not only for the ecosystem itself but also for human well-being. Box 8.4 provides an example scheme.

CASE STUDIES

THE MOUNTAIN BIOMES – THE ANDES

As you ascend the Andes foothills, precipitation, cloud and fog increase and this is especially important in the arid regions of Peru and Ecuador. In higher zones above significant cloud cover, diurnal temperature ranges can exceed the normal annual ranges of the region especially within tropical areas of the Andes. This results in increased temperature stress for life forms. There are therefore altitudinal differences in mountain biome characteristics. Those for the Andes are described below from the lower slopes to the higher slopes and the zones are indicated in Figure 8.26. The lowest slopes called *Tierra calienta* generally have vegetation characteristics similar to the surrounding lowland biome and so are not discussed here.

Lower slopes: *Tierra templada*

The lower slopes consist of tropical pre-montane forest, mostly within the eastern inland foothills. Two distinctive formations include:

- Lomas: fog/cloud-dependent vegetation within the otherwise hot desert biome of the coastal lowlands. For example, *Tillandsia paleacea* takes in moisture via hair-like structures on leaf surfaces.

- Matorral: evergreen shrubs with an open canopy usually less than 5 m high in the equivalent of the Mediterranean biome, with *Lithaea* and *Acacia* species.

Middle slopes: the upper *Tierra templada* and the lower zones of *Tierra fria*

Temperate rain- or cloud-forests exist in most areas of the middle Andes slopes. This is mixed forest in northern areas which is mostly evergreen, although deciduous beech species become more important with increasing altitude and further south where conditions become cooler. Where locally drier conditions exist, savanna-type structures are found.

Upper slopes: *Tierra fria* and *Tierra helada*

Above the treeline, species are often xerophytic (e.g. *Stipa*) and there may be giant forms of plants such as giant groundsel and giant lobelia. These

Figure 8.26 Schematic representation of the altitudinal belts in Venezuela, with mean annual precipitation and temperature. (Source: after Breckle, 2002)

BOX 8.3 ➤

➤

giant forms may be due to increased ultraviolet light or to lower oxygen levels. Grasses are found mixed with small evergreen shrubs where conditions are sufficiently moist but tend to be sparse and of cushion habit. Llama and alpaca grazing may be important controls on these areas (Figure 8.27).

A 'tropical alpine' zone exists before the true tundra zone, composed of low perennials but with slightly different species from the equivalent tundra genera and with more flowering plants, as opposed to mosses and lichens. Spring flowering

is an important adaptation, to allow a rapid beginning to the growing season, as soon as the snow melts and/or temperatures rise. Adaptations against the cold nights are found, such as within *Espeletia* species, where dead leaves shelter the stem/trunks, or the woolly-leaved *Senecio* species. Tundra-like conditions and life forms occur in the highest zone between the treeline and the snowline. Plant adaptations are similar to those of other tundra biomes, especially a low habit to escape the strong cold and dry winds. Animals at this height, such as

llamas, are adapted to low atmospheric oxygen by modified cardiovascular systems and may migrate to lower areas in winter. However, seasonal differences are not as extreme as those encountered on a diurnal basis.

In the highest regions, expanses of rock face or areas with gravels and **scree** are common. Their rapid drainage can lead effectively to drought conditions and very low biomass is the norm. However, sheltered areas may support lichens, together with some insects and birds.

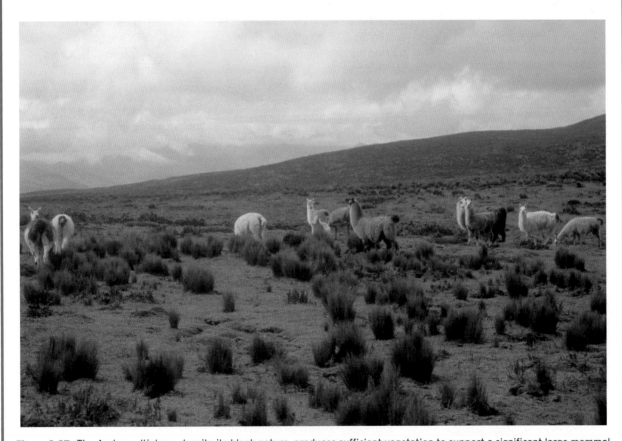

Figure 8.27 The Andean altiplano, despite its bleak nature, produces sufficient vegetation to support a significant large mammal population. Like the tussocky habit of the grassland, these llamas are adapted to withstand the extreme climate conditions of this altitude. (Source: photo courtesy of Wouter Buytaert)

BOX 8.3

NEW DIRECTIONS

MILLENNIUM ECOSYSTEM ASSESSMENT

The Millennium Ecosystem Assessment was conducted under the auspices of the United Nations, running from 2001 to 2005, with contributors from 95 countries. Its aim was 'to assess the consequences of ecosystem change for human well-being and to establish the scientific basis for actions needed to enhance the conservation and sustainable use of ecosystems and their contributions to human well-being' (http://www.maweb.org).

In the context of this chapter, the UN wished to assess the state of the biomes and the consequences of any major changes to them, for whatever reason, for humanity. Those controlling the exercise included non-governmental organizations and indigenous peoples as well as businesses, governments and international institutions. The work makes clear how the health and stability of the biosphere are intimately connected with human social and economic well-being. Changes to the biosphere and to biodiversity were seen as having importance to humankind's economy and well-being and needed to be monitored and, if possible, directed.

Many of the findings are highly significant. Biodiversity is being reduced. This is in part due to intensification of human land management. Often species regarded as without value, such as many insect species, are not protected and where species are considered to be a threat to the local economy, whether elephants or viral infections, they are removed. Increasing globalization has also increased the potential for invasion of

previously isolated ecosystems by alien species (see Chapters 9 and 10), threatening further groups of species. The greatest immediate threat to biodiversity within the next half-century would appear to come not from climate change but from land-use change, although ultimately climate change will become the greater driver of change.

The Millennium Ecosystem Assessment found that although the rate of conversion of 'natural land' to arable or other intensive human use was slowing in biomes such as the temperate woodlands and Mediterranean zones, this was largely due to the fact that, within these regions, land suitable for development was running out. Van Vuuren *et al.* (2006) believed that tropical biomes such as savanna grasslands and shrublands, and tropical woodlands and forests, are the most likely to show marked changes to their vegetation. These will, inevitably, cascade changes throughout the rest of the ecosystems involved.

The Millennium Ecosystem Assessment came up with four potential future scenarios, relating to different types of human development drivers (Figure 8.28):

1. The *techno-garden* form of development, where environmentally sound technology, within a globally connected world, provides highly engineered and monitored ecosystems. Management would constantly be monitoring these ecosystems, enabling them to be pro-active should any problem arise, with the technology and global connectivity available to put strategies into place quickly.

2. The *global orchestration* scenario would focus on high global economic growth, connections and trade.
3. The *adapting mosaic* is where local societies are stronger than global ones and people develop ecosystem management strategies appropriate to their own region, with local monitoring being a strong driver of policy-making. Managers would be very aware of the wider global initiatives but may pre-empt these by local action.
4. The *order from strength* situation is where if a problem is seen to be developing in their region, then institutional and governmental action will certainly be taken, especially if their regional markets are likely to be affected.

It should be emphasized that these four scenarios were not specific states that the UN expected to come into being but were intended to guide discussion by providing a framework of possibilities against which research could be assessed. They are being used (as in the work of van Veeren *et al.*, 2006) to predict, for example, habitat change and species loss. Each scenario has, of course, been the result of scientific model building and discussions and relies upon global development being focused, at least in part, on one of the four main drivers for change. You can obtain more details on each of the scenarios by visiting the Millennium Ecosystem Assessment website:
http://www.maweb.org

BOX 8.4 ➤

Figure 8.28 The four potential future scenarios produced by the Millennium Ecosystem Assessment, relating to different types of human development drivers.

BOX 8.4

Deliberate and accidental introductions of species, which act as weed species without their natural controls of predators, alter local balances. Vast numbers of species were introduced by European colonists into, for example, New Zealand and the United States (see Chapters 9 and 10).

The majority of biomes are identified by the presence of species that share characteristics allowing them to colonize particular geographical and climatic regions. The associations between these 'life zones' and the distribution of species throughout them present a way of monitoring and perhaps ameliorating environmental and climatic change by ecosystem management on a massive scale. Approaches to resource utilization are changing, in the light of changed priorities and values. This is reflected in the popularity of venues such as the Eden Project discussed in Box 8.2.

How resilient is the biosphere in its response to human and natural agents of change? A backward glance in time suggests that the biosphere has the capacity to accommodate vast fluctuations in climatic and environmental conditions and changes in the numbers and density of the species which it supports. However, much of the reassurance gained from this hindsight lies in evidence from the fossil record with, inevitably, gaps in our knowledge. For the biosphere to maintain the capacity to support the current great diversity of life, some form of management is essential. The following two chapters suggest some of the means by which biogeographers and ecologists are attempting to address such issues.

Reflective question

➤ What might cause changes to biome characteristics and what might cause changes to biome location?

8.8 Summary

Within this chapter the main features of the biosphere, especially of the major biomes, have been put forward. Any one of these biomes could form a chapter, or even a book, of its own. The current objective, however, is to draw out common themes to allow geographical comparisons to be made. The themes have been the typical climates, soils and wildlife of all types. Since vegetation tends to underpin most other biogeographical activity, the characteristic vegetation for each biome has been an important part of the descriptions provided for each one. Specific mention has also been made of the plant and animal adaptations to

the challenges of each biome. The factors underlying the geographical patterns may then, in many cases, be reduced to the effects of temperature and available moisture. A useful summary diagram is given in Figure 8.29. However, local human factors can alter biome characteristics. Furthermore, there are often no distinct boundaries between biomes. Instead a biome may slowly grade into another biome; this graded area is often called an ecotone.

The tropical biomes range from high-productivity, high-biomass rainforests to low-productivity, low-biomass deserts. Within each of these, species are adapted to compete for local resources. In tropical forests with fast nutrient cycling, rapid bacterial decomposition and

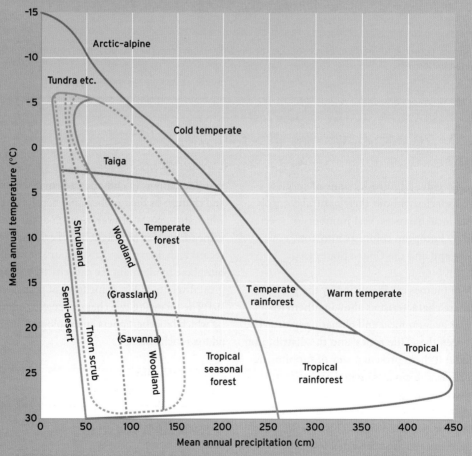

Figure 8.29 The distribution of the major terrestrial biomes with respect to mean annual precipitation and temperature. Within regions delimited by the dashed lines a number of factors, including proximity to oceans, seasonality of drought and human land-use, may affect the biome type that develops. Boundaries are, of course, approximate. (Source: after Whittaker, 1975)

plentiful water, species compete for light resources and at ground level there may be hardly any light penetration. However, in savanna grasslands and hot deserts, species are adapted to compete for water resources. Often this means only being active for short periods when water is available and developing methods for conserving water, including nocturnal activity and daytime dormancy.

The temperate biomes vary from deciduous and evergreen forests to seasonal grasslands with perennial plants and large grazing herbivores. Rainfall, temperature and sunlight are seasonal and Mediterranean and temperate grassland biomes are subject to natural fires to which plants and animals are adapted (e.g. by burrowing or rhizomes). The cold biomes of the tundra and taiga are mainly found in the northern hemisphere and have slow nutrient cycles, a lack of soil fauna and acidic leaf litter. Tree and shrub growth can often be stunted and where water resources are frozen for part of the year, plants may develop wide fanning root systems to take up as much water as possible as soon as it thaws. Evergreen species maximize productivity potential by allowing photosynthesis and growth to take place as soon as spring sunlight becomes available. Hibernation and burrowing allow species to cope with long cold winters. The tundra is dominated by sparse vegetation cover and the low productivity of vegetation means that large areas are required to support migratory mammals such as reindeer. In highly stressful environments only mosses and lichens may be able to survive and the tundra has a very low diversity of plants.

Further reading

Archibold, O.W. (1995) *Ecology of world vegetation.* Chapman and Hall, London.
This has a North American focus and many useful photographs and diagrams. It also provides an unusually large list of references that can be followed up.

Bradbury, I.K. (1998) *The biosphere.* John Wiley & Sons, Chichester.
This is a standard undergraduate text which is good for the general principles.

Cox, C.B. and Moore, P.D. (1993) *Biogeography. An ecological and evolutionary approach.* Blackwell Scientific, Oxford.

This is an excellent and classic undergraduate text which is of relevance to this chapter and for Chapters 9 and 10.

Walter, H. (2002) *Vegetation of the Earth and ecological systems of the geo-biosphere,* translated by Owen Muise. Springer-Verlag, Berlin.
This classic book clearly shows the author's love of his subject and the results of extensive personal research. Therefore there is a slightly patchy coverage.

Whittaker, R.H. (1975) *Communities and ecosystems.* Macmillan, New York.
This is an American classic text, written by the proponent of environmental gradients.

Web resources

Atlas of the Biosphere: Mapping the Biosphere
http://www.sage.wisc.edu/atlas/maps.php
There is a series of extremely good maps here showing the global distribution of factors that characterize and influence the various biomes of the world such as precipitation and primary productivity.

Blue Planet Biomes
http://www.blueplanetbiomes.org
A very good site with extensive information on the various plant and animal species found dominating particular biomes and general biome characteristics. The site is easy to read and navigate around.

CIESIN: Land-use and Land-cover Change: 3.2 The Causes and Processes of Deforestation
http://sedac.ciesin.columbia.edu/tg/guide_frame.jsp?rd=LU&ds=1

Although predominantly a heavy text site, it also provides links to detailed and in-depth discussions into the underlying causes and processes of deforestation and the changing nature of tropical rainforest biomes in an essay format.

Eden Project Home Page

http://www.edenproject.com/
Outlines the aims of the Eden Project in attempting to recreate and monitor processes that occur within different biomes of the world.

Environments of the Andes

http://www.sacha.org/envir/intro.htm
This site covers environments of the Andes (Peru-based) resulting from a major project supported by the *National Geographic*, comparing field records of plants and animals from 70 years ago with present-day observations. There are beautiful images and a range of useful linked sites. It provides handy case study material.

Global Climate Change Research Explorer: The Biosphere

http://www.exploratorium.edu/climate/biosphere/
This site gives news on recent research and case studies exploring the effect of current climate change on various plant and animal communities. Various links to related sites are also provided.

Millennium Ecosystem Assessment

http://www.maweb.org
The Millennium Ecosystem Assessment assessed the consequences of ecosystem change for human well-being (see Box 8.4).

NASA Earth Observatory

http://earthobservatory.nasa.gov
A series of informative and well-explained factsheets on the carbon cycle, biomass burning, tropical deforestation, change in global land surface, global fire monitoring and global warming are provided by NASA.

Nearctica: Major Biomes of the World

http://www.nearctica.com/ecology/habitats/biomes.htm
This site is an excellent online academic journal that supplies links to a variety of sites related to global biodiversity. This particular web page provides links to sites of detailed information and pictures on the characteristics of the major global biomes, their global distribution, structural characteristics, soils, flora, fauna and differences between biomes.

UNESCO Biosphere Reserve Programme Home Page

http://www.unesco.org/mab/wnbrs.shtml
This site provides a description of the function of the programme and the global distribution of UNESCO Biosphere Reserves.

 Visit our website at **www.pearsoned.co.uk/holden** for further questions, annotated weblinks, interactive models and video-clips showing physical processes in action.

CHAPTER 9

Biogeographical concepts

Hilary S.C. Thomas
Department of Humanities and Science, University of Wales, Newport

Learning objectives

After reading this chapter you should be able to:

➤ understand key concepts within biogeography

➤ understand processes leading to changes in biogeographical distributions over time and space

➤ describe the basic principles of landscape ecology

➤ understand the theory of island biography and how it can be used to inform management

➤ show an awareness of current issues in biogeography

9.1 Introduction

Biogeography is an important area of science and informs global environmental policy as well as local land management practice. Biogeography is the study of the distribution and patterns of life on Earth. Biogeography is therefore usually focused upon the study of the current distribution of living things. However, it also recognizes that to understand, predict and manage changes in biogeographical patterns it is necessary to examine the processes that create these distributions. The links with ecology are strong.

Figure 9.1 shows the range of topics shared by both disciplines. Those with the greatest interest for biogeographers are located to the right of the diagram. Most of these biogeographical topics are therefore addressed within this chapter. The various types of spatial biogeographical distributions will be explained, together with factors that may lead to change over time such as climate change or the introduction of 'alien' species. Factors more closely allied to ecology are located to the left of Figure 9.1. Many topics such as island biogeography and landscape ecology are part of both research areas but will form important elements of this chapter. Concepts such as biomes, **succession** and conservation are very important to both and are shown in bold at the centre of Figure 9.1. These topics are dealt with in Chapters 8, 9, 10 and 22. Of course, with biomes being so fundamental to geography there is a separate chapter on this topic (Chapter 8).

Of the many factors and processes that underpin our present understanding of biogeography, certain groups are especially important. These include those relating to the concept of succession, the strategies used by plants, animals, birds and insects as they utilize the biosphere, and the effects of habitat disturbance. Common themes for all these factors include the role of time and how geological history can help explain plant and animal distributions, the relationships between species and their environment and how ecological processes operate to produce biogeographical distributions. These fundamental themes will be explored in this chapter.

9.2 Succession

Classic studies within biogeography have focused on vegetation succession, where the sequential changes to be seen over considerable lengths of time have been used to develop a number of concepts that underlie modern biogeographical thinking. The distribution of living things within the biosphere is rarely constant, at whatever scale it is studied. Birds may migrate vast distances, insects crawl from one blade of grass to another and the seeds of plants may drift with the wind, roll down slopes or be moved through the agency of animals. To begin to understand the processes involved it is worthwhile taking a snapshot view that is revisited over time. When this is done it usually becomes clear that directional change is taking place. Such a change tends to lead towards increased complexity of the **community** structure and increased biodiversity. This is seen, for example, in studies of bare ground such as on a new volcanic island or on areas abandoned from other land-uses (Figure 9.2). In nearly all cases the disturbed or recently exposed ground is colonized by animals and vegetation. This process is described as **primary succession**. Change is fairly slow at first and then becomes very rapid, before becoming much slower again. This final slow phase

generally continues for as long as the area's development is undisturbed.

Disturbance usually causes the whole process to begin once more, shifts the succession back one or more stages, or may cause it to move forward in a different direction. Succession is therefore the result of a range of environmental and ecological processes and has immense importance within biogeography as well as in ecology. The following section discusses the early influential schools of thought on succession while the topic is given further treatment in Chapter 10.

Figure 9.2 Primary succession on Vesuvian lava flows. The lower slopes of Vesuvius show several stages of colonization, with woodland developing on the areas that have been stable for the longest periods since the last eruptions.

9.2.1 The development of succession theory

While Cowles (1899) did pioneering work on dune succession on the shores of Lake Michigan, which demonstrated how plants, soil, water, climate and topography interact, there were two other major contributors to the concept of succession as it is used today: Clements and Gleason. Clements (1916, 1928) based his work around the concept that plant distributions formed recognizable communities that became increasingly complex in their interactions within themselves. He observed that each community operated almost as a single developing organism. As a community develops, the populations of plants and other species change and replace each other and may also alter the environment. This may occur, for example, through a reduction in wind speed and daily temperature ranges as vegetation increases in height. These changes lead to climatically controlled **climax communities.** During the development towards the climax community, the community passes through recognizable **seral stages** such as from an initial cover of lichen to communities dominated by mosses, then by grasses, then by shrubs and finally by forest. At this final stage a dynamic equilibrium (see Chapter 1) would keep the community basically the same despite any temporary local or minor disturbances. A division was seen between primary succession from bare ground and **secondary succession** that would take place after disturbance where a seed-bank, remnant roots and other materials were likely to remain and possibly influence the character of the resulting community. In either case, Clements assumed that six distinctive processes were in operation, controlling the development of the community:

1. *Nudation* – creation of the initial bare surface.
2. *Migration* – seed or vegetative spread.
3. *Ecesis* – establishment, to the extent that a complete life cycle can occur.

4. *Competition* – for space, water and so on.
5. *Reaction* – modification of the habitat by the vegetation.
6. *Stabilization* – usually the end product or climax community related to the regional climate.

A widely accepted amendment to Clements's idea that there would be a single final community in any one area was made by the British ecologist Tansley. He believed that where there were strong local environmental influences, these would control the type of final community. There might be, for example, 'topographic' climax communities, controlled by the local geomorphology, or 'edaphic' climax communities where the major influences were the local soil conditions.

Much modern work, however, from glacier forelands to roadside verges, indicates that communities, rather than becoming locally characteristic over time, frequently become increasingly different. This suggests that neither a **monoclimax** (single final community) nor a set of **polyclimaxes** (several co-existing final communities) may be the end result of succession. Studies continue on this theme because the ability to manipulate succession is of great value, in applications relating to economic usefulness to humans or where maximized conservation value is required.

In what appears, at first, to be the exact opposite of Clements's ideas, Gleason (1926) assumed that communities existed by chance, as a result of their individual response to the environment in which they found themselves. He suggested that the distribution of species would be controlled by factors such as migration opportunities, environmental selection and frequency of disturbance. He also maintained that there was no true directional succession, since change was random and disturbance was so frequent that a stable final assemblage of species was not possible. His ideas, with others, are shown for easy comparison in Table 9.1.

Table 9.1 Some early ideas on succession important to biogeographical prediction of patterns

Author's initial publication	Idea	End result
Clements (1916)	Like a single developing organism	A climatically controlled climax community
Tansley (1920)	Some divergence of outcome	Mosaic of polyclimax communities, reflecting local environmental factors
Gleason (1917)	Individualistic plant responses	No stable, characteristic final assemblage
Whittaker (1953)	Environmental gradients	Gradual changes in response to gradients of environmental change

Whittaker (1953), however, considered that undisturbed communities could be generally observed to change slowly from one type of community to another along habitat gradients such as light or water availability. As a result, he regarded climax communities as a pattern of communities that reflected the pattern of environmental gradients. Within this there would be a central, most extensive community type that would represent the prevailing or climatic climax. Whittaker's succession model is regarded as a climax pattern model, which he considered to be self-maintaining and 'potentially immortal if not disturbed'. He also introduced the term '**ecocline**', for a combination of environmental variants that change together through space. For example, both temperature and exposure change simultaneously as a result of altitude. Matthews (1992) pointed out, however, that despite Whittaker's proposal being more complex and including elements of Gleason's individual responses, it still provided no mechanism. There is therefore a need to move towards more process-based approaches when thinking about succession. As the study of succession moved forward, the emphasis changed towards the search for mechanisms. This helps to increase the practical application of a concept that has become, while still contested, one of the most important elements in the explanation of changes in biogeographical pattern over time. Such successional processes are discussed in Chapter 10.

Reflective questions

➤ How does the concept of succession alter our perception of 'restoration' in degraded or disturbed areas?

➤ Can you summarize succession theories?

9.3 Spatial patterns and processes

A variety of processes lie behind the patterns seen in the distribution of plants and animals. Geographical factors are often very closely allied with those relating to ecology. For example, where the local climate leads to the development of environmental gradients, the result will often determine the range of available **niches** for plants and animals. Biogeographical processes will vary in importance according to the scale or resolution of the region selected by an investigator. It will become apparent that varying scales of study are important for the development of biogeographical understanding.

9.3.1 Global-scale patterns

9.3.1.1 Climate

On a global scale the patterns of species distribution can be seen to reflect those of the major climatic zones (see Chapter 8). The length of the growing season, for example, provides a constraint upon the distribution of species. It would, however, be simplistic to assume that one factor, even at this scale, dominates biogeographical distributions. Whittaker (1975) noted that biomes should be defined by their structure, but in practice they have to be defined by combining this with their environment. The availability of water, for example, is often related to climate, but may also be a response to local geology and topography and hence, perhaps, soil conditions.

9.3.1.2 Geological factors

Tectonic movement, including the break-up of continents, has led to opportunities for species to spread but also to the creation of barriers to spreading, such as mountain ranges or open water. Lengthy geographical isolation, for whatever reason, may produce distinctive communities at a variety of scales. Perhaps the most famous biogeographical concept arising from this is that of 'Wallace's line'. Alfred Russel Wallace was impressed by the sudden difference in bird families (and other fauna and flora) he encountered when he sailed from the island of Bali and landed on Lombok, 30 km to the east, in the 1850s (Figure 9.3). The birds on Bali were clearly related to those of the larger islands of Java and Sumatra and mainland Malaysia. On Lombok, however, the birds were instead related to those of New Guinea and Australia. He marked the channel between Bali and Lombok as the divide between two biogeographical regions, the Oriental and Australian. In his honour this dividing line as shown in Figure 9.3 is still referred to as Wallace's line. This line on the map corresponds roughly to the location of an ocean trench separating global plates. Thus the islands to the north of the line were formed from the Asian continent and the islands to the south were originally part of the Australian continent and they had been separate continents for over 200 million years. Of course it is only since the 1960s that we have known about plate tectonics (see Chapter 2) so Wallace was way ahead of his time.

9.3.1.3 Ecological factors

Ecological factors affecting global-scale biogeographical patterns and distributions include habitat conditions such

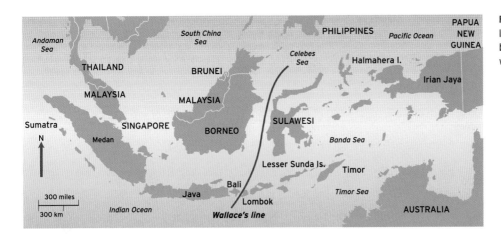

Figure 9.3 Wallace's line. This line marks a distinct change in biogeography and is coincident with an ocean trench.

as the presence or absence of predators and prey, competition for nutrients and living space, and the ability to adapt and migrate (Huggett, 2004). Ecological factors have been the focus of much research and there are a range of terms to describe the geographical patterns of living creatures (Table 9.2). The distribution of some species is regarded as cosmopolitan since they are found worldwide when suitable habitats exist, but this is not the usual situation (Spellerberg and Sawyer, 1999). Geological processes can result in barriers to movement and lead to certain species being **endemic** (confined to a particular area), such as the Hawaiian hawk, *Buteo solitarius*. If species have never been known to occur in other regions their distribution is regarded as being of **primary endemism** such as the Australasian marsupials. However, certain endemic species are only endemic because extinctions of those species have

occurred in the other places where they used to survive (such as the mammals of the West Indies). These species are therefore described as having a **secondary endemic** distribution. If a species is quite widely distributed but with large gaps between regions its biogeographical distribution is described as **disjunct**. The mountain avens, *Dryas octopetala*, for example, is found in the Burren Hills, Ireland, although it is usually considered to be a species of high mountain regions.

The isolation of groups of organisms (such that they do not interbreed and the gene flow from the larger population is suppressed) can lead to the evolution of new species, or **speciation**. This particular mechanism is termed allopatric (or geographic) speciation. A good example of such speciation is the finches from the Galapagos Islands, specimens of which were collected by Charles Darwin and studied back in

Table 9.2 Commonly used terms to describe patterns in the distribution of species and other taxonomic groups

Type of distribution	Definition
Cosmopolitan	A species or taxonomic group that is distributed widely throughout the world
Primary endemic	A species or taxonomic group in a particular region native only to that region
Secondary endemic	A species or taxonomic group in a particular region whose distribution has contracted so that it is now native only to the region in which it is found
Rare	A species or taxonomic group that is restricted geographically or is widespread but never found in abundance
Disjunct	A species that occupies areas that are widely separated and scattered (species with a discontinuous distribution)
Indigenous or native	A species that originates in a particular place or which has arrived there entirely independently of human activity

(Source: from Spellerberg *et al.*, 1999)

Table 9.3 Biogeographical features of three oceanic island systems

	Hawaiian Islands (Pacific)	Galapagos Islands (Pacific)	Mauritius (Indian Ocean)
Vascular plants	900 of which 850 are endemic	540 of which 170 are endemic	878 of which 329 are endemic
Threatened native plants	343	82	269
Introduced plant species	4000	195	–
Animals known to have gone extinct	86	5	41
Extinct endemic plants	108	2	24
Species of land snails	c.1000, all endemic	90 of which 66 are endemic	109 of which 77 are endemic
Extinct snail species	29	1	25
Unusual wildlife	High endemism, superspecies-rich taxa, e.g. fruitflies	High endemism, unusual species, e.g. marine iguanas	High endemism, once home to the dodo

(Source: adapted from Jeffries, 1997)

London by John Gould along with Darwin. The finches probably contain 13 different species, within the subfamily Geospizinae. It is normal for genetic drift to take place within populations such as the gradual change in height or in hair colour over generations, especially if these changes are adaptive to the local environment (e.g. providing better camouflage against predators). In small populations, such as within the Galapagos Islands, the founder population, which may consist of a very few individuals that arrive in a new location (or are left in an almost destroyed old one), may become increasingly diverse over time. The Galapagos finches consist of species with different-shaped beaks (to take advantage of different food sources) and some that are vegetarian and others that are carnivorous. Oceanic islands and other isolated regions are very likely to support such endemic species, as shown by Table 9.3. The table also shows that these islands are also highly susceptible to disturbance and extinction.

Living creatures vary tremendously in their mobility. This is considered to be part of their **vital attributes** and may determine the most likely combinations of species to be found in new or disturbed localities. Most species spread very gradually, as opportunity arises. This movement may take centuries to cover a few tens of kilometres but is still highly effective. It allows species (such as oak trees), but not necessarily individuals, to disperse. For example, species may disperse away from an area that is becoming increasingly cold or nutrient poor. Individuals in cooler margins eventually die but their offspring preferentially succeed in the warmer margins of the local distribution. This may especially be the case where ecological niches are vacated by other species with greater warmth requirements.

A far more rapid change is possible where, for example, plants such as dandelion have highly mobile seeds that drift with even small draughts of air (Figure 9.4). Many species

Figure 9.4 Dandelion seed head. The seeds can travel long distances even in a gentle breeze.

have taken advantage of the increased mobility offered by human forms of transport and are found in clusters around, for example, rural railway stations. Seeds and insects also inadvertently take the train, bus or car from one locality to another and may develop new areas of distribution if the conditions in the new locality are suitable in terms of habitat and a vacant or incompletely filled ecological niche. Often, however, the dispersed individuals are unable to succeed as they cannot overcome competitive exclusion from local resources.

9.3.2 Small-scale patterns

When regions are investigated there are often smaller-scale nested patterns. These may relate to the availability of light, water or nutrients, to the dampness or dryness of the site, or to the intensity of disturbance by humans. Humans may create an environmental gradient that we could call a 'disturbance gradient'. This may range from virtually untouched wilderness to a suburban back garden. Four points along such a gradient have been identified by Westhoff (1983) as shown in Table 9.4. It should be noted, however, that there are very few areas of the world that today would truly belong to the 'natural' end of the gradient, as identified in his classification.

9.3.3 Landscape ecology

In 1939, Troll described **landscape ecology** as 'the study of the entire complex cause–effect network between the living communities and their environmental conditions which prevails in a specific section of the landscape . . . and becomes apparent in a specific landscape pattern'. The living landscape can show clear structure, functions and change and may be studied at a huge variety of scales. At probably the finest scale, a suburban garden can provide evidence of the effects of microclimatic and pedology on the pattern of living things. On a coarser scale (e.g. a national park), a landscape system will display within it a hierarchy of smaller-scale patterns. Too coarse a scale of approach, however, may often lack the detail required for many biogeographical purposes. Equally, too fine a resolution such as concentrating upon a single tree may obscure patterns.

Landscape ecology is a major growth area in biogeography, and in land planning, conservation and ecology. The results of landscape ecological investigations are used to inform planning and management decisions in many countries. Troll developed his ideas when working with air photographs in Africa. Present-day geographers are still likely to use these, but they will also use a **geographical information system** to help map and visualize different effects such as geology, geomorphology, climate, soil, biogeography, economic activity, settlement, culture and social structure. These collectively give a landscape its identity and help determine its biogeographical character. Landscape ecology is of particular value to the biogeographer when defining and helping to explain the distributions seen in groups of species. It helps when identifying species habitats and functions within a landscape and attempting to measure, through analysis of landscape stability, how likely it is that a pattern may change over time. The International Association for Landscape Ecology puts forward several core themes for its activity:

- the spatial pattern or structure of landscapes, ranging from wilderness to cities;
- the relationship between pattern and process in landscapes;
- the relationship of human activity to landscape pattern, process and change;
- the effect of scale and disturbance on the landscape.

Table 9.4 Degrees of naturalness

Natural	a landscape or an ecosystem not influenced by humans
Subnatural	a landscape or an ecosystem influenced to some degree, but still holding to the same formation or vegetation type from which it derived, such as grassland
Semi-natural	a landscape or an ecosystem in which the flora and fauna are largely spontaneous but in which the vegetation structure is altered in such a way that it now constitutes a different type, such as from woodland to heather moorland
Cultural	a landscape or an ecosystem in which the flora and fauna are essentially influenced by humans, such as on arable land or within plantation forests

(Source: after Westhoff, 1983)

Any landscape, regardless of the scale of an investigation, can be analyzed in terms of its landscape ecological structure. The typical **landscape patches, landscape matrices** and **landscape corridors** together define the character of that landscape and hence help define the 'sense of place' vital in preserving the heritage of an area.

9.3.3.1 Landscape patches

Landscape patches are distinctive elements within the wider landscape, such as ponds, woods or towns. These patches may have value put upon them by those living or working in the area. The spread of a town may be described as 'urban sprawl' and be locally unpopular, while the cutting down of part of a woodland might be considered disastrous. The objective analysis of landscape patches usually deals with the patch characteristics of shape, frequency, origin and stability.

The shape of a patch may indicate its vulnerability to change as a result of outside influences. The greater the proportion of 'edge' to 'interior' (Figure 9.5a), the more likely the patch is to be influenced by external factors. For example, a small wooded area will be more exposed to light entering the understorey from surrounding land than a more substantial wooded area, where the central areas may be sheltered from sunlight and wind. This will allow species from outside the smaller patch to colonize more easily. Figure 9.5(a) shows a patch which is elongated and therefore has a greater proportion of edges to interior. The figure also shows a much wider, more circular patch which therefore has a smaller proportion of edges to interior and will be less susceptible to external disturbance. Patches with significant concave boundary sections tend to demonstrate that in those sectors the patch is contracting, as opposed to areas of expansion, indicated by convex sections as shown in Figure 9.5(b).

The origin of a landscape patch may also determine other characteristics (Figure 9.6). Classic geographical sites, such as a wet point (e.g. a spring) or dry point (e.g. a hummock), also produce related landscape patches. In each case, the change in local environmental conditions tends to result in distinctive biogeographical and land-use patterns. These may then become altered or reinforced by other processes such as the bare ground that results from trampling and intensive grazing around waterholes in semi-arid areas. Some introduced patches originate from management, such as the field patterns of arable land, **polders** (land reclaimed from the sea through embankments) or intentionally burned patches of heather moorland or rainforest. Other patches may be inadvertent such as those created by cattle trampling at the approaches to field gateways. Many patches result from natural disturbance such as within a forest where a dominant tree within the canopy dies and falls, creating a new area for regeneration. Of course, most human settlements form patches of varying scale. Within the built-up area of a major city, biogeographically important patches of parks, lakes and gardens exist. Perhaps the most important patches on a global scale in the twenty-first century will be remnants of biomes that have otherwise vanished and thus act as a haven for species that would otherwise be extinct.

9.3.3.2 Landscape matrices

The matrix of the landscape is usually regarded as that element of the landscape that occupies a greater area than any patch type within it. In general, it plays a dominant role in the dynamics and character of the local biogeography, and also in other elements of the local geography, so helping to create a distinctive sense of place. The matrix contains within it the other landscape elements (patches, corridors) and a measure of its stability can be obtained by a study of the extent to which patches appear and develop within it, creating 'porosity'. Slight porosity is usual but if this reaches a very high level, the integrity of the matrix may be threatened. For example, clearance of a forest for agriculture initially produces a forested landscape containing patches of farmland. These patches of farmland may increase in size or number over time to the point at which the landscape is better described as a farmland containing patches of forest. There may be an interim scenario, where two or more matrices contribute to make up a particular landscape. Such a situation may occur

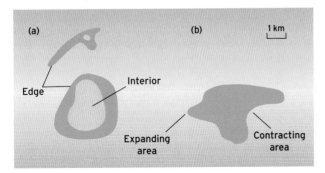

Figure 9.5 Landscape patches: (a) the proportion of edge to interior is an important factor in determining whether a patch will be influenced by external factors; (b) convex edges indicate contracting patches while concave edges indicate an expanding part of a landscape patch.

Figure 9.6 Valley near Killarney in south-west Ireland, where a variety of landscape patches can be identified. Most of these are related to different levels of exposure to wind, to water availability or to geomorphology, as the area has been heavily glaciated with both erosional and depositional features being important biogeographical factors. Humans have also created a settlement patch which can be seen just below centre in the photograph.

across the transition between biomes, seen classically in the forest–savanna boundary of tropical regions, as discussed in Chapter 8.

9.3.3.3 Landscape corridors

Landscape corridors are narrow strips of land that differ from the matrix that exists on either side of the strip. They may be isolated strips but are usually attached in some manner to a patch, often of similar character. For example, hedges may unintentionally link woodland, rivers may run from lakes, and roads join built-up areas. The key characteristics of corridors for the biogeographer relate to their connectivity both with similar corridors and with other landscape features and to the fact that there are often sharp microclimatic and soil gradients from one side of a corridor to another. An example is shown in Figure 9.7 of connected hedgerows acting as

landscape corridors. In this artificial situation natural, parallel corridors such as embankments, verges, carriageways, central reservations and drainage zones, each with their own characteristics, exist within the major landscape feature. Increasingly, corridors such as hedgerows or streams are considered to be of immense potential conservation value, by giving species an escape route from threatened patches, and perhaps providing migration opportunities. Corridors are also geographically important in that they are frequently the element of the landscape that both inhabitants and visitors may regard as typifying the area and giving it its character, such as drystone walls, or small, irregular hedges dividing up the landscape. A new corridor, such as a line of electricity pylons, may attract intense criticism and, despite its small size, be regarded as 'ruining the area'. It may also form a significant barrier or filter to movement for a whole range of species.

Figure 9.7 Hedgerows acting as landscape corridors. The hedges support diverse plants communities and shelter animals.

Reflective questions

➤ What are the major factors influencing global-scale biogeographical patterns?

➤ Can you provide examples of each of the Westhoff landscape types from within your country?

➤ What are the distinctive landscape patches and corridors in your local landscape?

9.4 Temporal patterns and distributions

9.4.1 Geological time

The role of geological processes in biogeography via plate movements and barrier development has already been discussed above and in Chapter 8. However, the geological record also suggests that there have been periods of the past when large-scale mass extinctions occurred. Over the past 600 million years there are believed to have been five mass extinction phases. A good example is the mass extinctions associated with the loss of the dinosaurs 65 million years ago. This extinction resulted in 40% of the terrestrial invertebrate species being wiped out. A range of causes for mass extinctions have been proposed and these include asteroid impact, rising sea levels, prolonged glaciations and other environmental changes (e.g. global atmospheric gas changes following periods of extended volcanism).

9.4.2 Post-glacial change

Over a lesser but still extended period of time, climatic fluctuations over the past 2 million years have been important determinants of biogeographical change. Many regions of the world have experienced major climatic change to which plants and animals have been forced to adjust in order to survive. Many of these changes were relatively slow such that the distribution patterns of most plants were able to respond to these changes. Areas such as southern Europe became refuges for more northern species as conditions became increasingly colder (see Chapter 20). Here they survived during the coldest periods and some species were in a position to recolonize as conditions improved. The British Isles, however, are relatively species-poor for their latitude. The Irish Sea and the English Channel became deeper and wider

following the last Glacial Maximum. As the ice retreated from northern Europe and sea levels rose, those species with dispersal mechanisms unable to operate over extensive water bodies were unsuccessful colonizers. Box 9.1 describes biogeographical change in north-west Europe over the past 10 000 years. It should be clear from this box that such regions have not had natural stable biogeographies and that the landscape is in a constant state of change.

ENVIRONMENTAL CHANGE

POST-GLACIAL BIOGEOGRAPHICAL CHANGE IN NORTH-WEST EUROPE

Around 11 000 to 10 000 years ago north-west Europe was dominated by a tundra biome (see Chapter 8). This consisted of grasses and sedges, often rosette and tussock species with dwarf trees and shrubs such as willow and birch together with alpine species of herbs. Many of these species were survivors from isolated pockets that were maintained during the glacial periods in warmer areas of southern Europe such as around the Bay of Biscay.

By 9000 to 7500 years ago, groves of birch, aspen and juniper and, further south, more substantial blocks of woodland began to develop. As warming progressed, lowlands became increasingly wooded, up to about 1000 m altitude. Species varied with oak, elm, hazel and lime common in the south, with some beech and yew where the land was chalky. Pine, birch and hazel existed further north. The more cold-tolerant plants were gradually pushed into the mountain areas and to the moorlands.

Between about 8000 and 5000 years ago the climate was warm and

wet. The beginnings of peat formation were seen in many regions, together with the inundation of some formerly wooded areas by sea (e.g. south-east England fenlands). The North Sea together with the English Channel rapidly widened as sea levels rose in response to further melting of the world's ice caps. This reduced the chances of species migration into mainland Britain and Ireland. As a result, while Britain at present has around 1500 native species the climatically equivalent French zones contain nearer 6000. The Atlantic period also saw a decline in elm all over north-west Europe. Suggestions for this include a response to climate, the effect of humans, a rise in the occurrence of the weed species *Plantago lanceolata*, or possibly due to disease.

The period around 5000 to 3000 years ago was even warmer and drier and evidence for the beginnings of agriculture can be found in many regions during this period. There is evidence of some decline in woodland, together with an increase in heathland. The period from 3000 to 2000 years ago was generally a wetter and cooler period, with moorland vegetation developing in

many areas. Cereals were introduced to agriculture and there is increasing evidence of use of trees by humans for shelter, cooking and fodder.

Thus we need to think quite carefully about what we mean when we talk about natural landscapes and 'native' species (Figure 9.8). Often in areas disturbed by humans there is a perception that we should restore the landscape to its natural state. However, this natural state is actually one of continuous change. Most species present in north-west Europe today were not present 12 000 years ago. The forests of northern Europe have not been around for thousands of years. If an area has been affected by humans for 2000 years, what should we restore it to? Should we try to restore species that were present 2000 years ago on the site or should we introduce species that we *think* should be there if humans had not affected the land? Historical biogeography can be a very useful tool in allowing us to place contemporary biogeographical distributions into context and in allowing us to understand contemporary patterns.

BOX 9.1 ➤

Figure 9.8 Moorland vegetation in the English Pennines. The vegetation is periodically burnt and grazed to control its growth and yet these landscapes are considered to be ones of 'outstanding *natural* beauty'.

BOX 9.1

9.4.3 Migratory patterns

The migratory movement of species creates biogeographical change over time. This form of change, however, forms a recurring pattern over long periods and may be an important element of a region's biogeography. Species may be present in a region at one time of year and absent at another. Therefore, if a site is being studied at the same time every year (which is often done to record change and avoid the effects of changing seasons) migratory species may be missed from the survey. Figure 9.9 shows that the 'summer' and 'winter' distributions of a migratory species may be not only very distant from each other, rarely just following temperature changes, but also quite different in scale. For Kirtland's warbler the range is restricted to a small area between Lakes Michigan and Huron during the summer, whereas during the winter

the range is much larger covering the entire Bahamas region. Important considerations for migratory species include suitable areas for breeding and the availability of food. Some species travel great distances to enable them to utilize areas for part of the year which for permanent residents would require dormancy, or other adaptations.

9.4.4 Alien introductions

Alien species are any species not part of the 'native' biogeography. Their interest for the biogeographer is the manner in which they are able to arrive and in many cases establish and spread. This colonization can sometimes occur at astounding rates, far outcompeting local species. The aliens may be introduced accidentally or on purpose. Certain

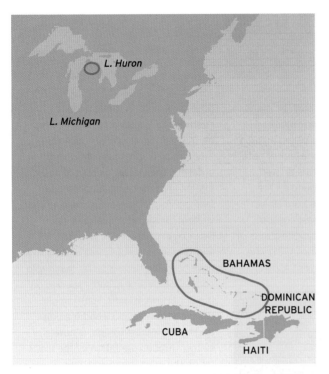

Figure 9.9 The summer and winter range of Kirtland's warbler, *Dendroica kirtlandii*. The bird is restricted to a small area between Lakes Huron and Michigan in the summer while the winter range is much greater with the birds dispersed througout the Bahamas. (Source: Van Tyne, 1951)

communities and regions seem more susceptible to invasion than others. Many continental islands, such as Tasmania, Ireland and mainland Britain, were once connected to a continent by land that, although it once formed a bridge, is now submerged as a result of post-glacial sea-level rises. These islands therefore tend to be relatively species-poor (see above), which makes them more likely to be susceptible to invasions. Box 9.2 describes an example of mammalian invasion of islands and the effect on other species.

Firstly, environmental niches may not be completely filled in these locations so that these areas can actually support a much greater diversity of species than at present. Secondly, in the absence of certain hosts, parasites, predators and diseases may also be absent from such species-poor communities or they may have adapted to utilize different species. Thirdly, certain forms of habitat management, especially those related to farming or to conservation, may also create an environment that is capable of favouring alien species. In such environments it is easy to envisage the easy integration of newly arriving species which are able to occupy some of these biogeographical

gaps. Human colonizers over the years have deliberately brought alien species into areas that were suitable for agriculture or even for nostalgia. In many cases, such as the British rabbits introduced into the New Zealand grasslands, they may now be considered as pests (see Chapter 10). The edible dormouse, *Glis glis*, originally from Hungary, has reached pest status in parts of southern Britain, even though the rate of its spread is slow compared with that of other introductions. Shown in Figure 9.11 is the spread of the American grey squirrel, within mainland Britain, at the expense of the native red squirrel. Although the spread of the grey squirrel has been well documented since the deliberate releases in the late nineteenth and early twentieth centuries, the reasons for its success are still disputed. However, the grey squirrel was adapted to fierce competition in its native North America, so it is not a surprise that it has spread so quickly in the much less hostile British deciduous woodlands (Yalden, 1999). Some species have been targeted for eradication after their populations exploded or became problematic following introduction. These include copyu in Britain and limu seaweed introduced around Hawaii which is threatening native coral by blocking out sunlight. Box 9.3 describes the problem of Japanese knotweed in the United Kingdom.

Although it is often the lack of predators in the new locality that may explain the success of a new species, it should be noted that even should this biogeographically fortunate state exist, it may not be prolonged. The Guernsey fleabane (a type of daisy) has spread since the 1980s throughout London and appears to be a rival for the niches occupied by the butterfly bush species (English Nature, 2002). However, it is now consumed by the insect *Nysius senecionis*, which is another alien species first recorded in Britain in 1992 some nine years after the arrival of the Guernsey fleabane.

Reflective questions

➤ What role have tectonic and glacial processes played on the biographical patterns and processes that we see today?

➤ Why might you have to design a biogeographical research project carefully in terms of its timing during the year?

➤ What factors determine the success or failure of an invading species?

HAZARDS

THE INVADORS OF THE LORD HOWE ISLANDS, AUSTRALIA

The black rat (*Rattus rattus*) is regarded as one of the 'world's worst' invading species by the World Conservation Union (Figure 9.10). Its main adverse effect arises from its opportunistic predation, as opposed to the effects on habitats of other species. The effect of the arrival of rats on isolated communities such as islands has been observed worldwide, although a common problem in quantifying this has been the lack of data on conditions before rats arrived. A typical example is found on the geographically remote Lord Howe Island group, off the east coast of Australia, which has had the status of World Heritage Site since 1982. The black rat reached many of Australia's offshore islands in the nineteenth century and is known to have arrived on the main Lord Howe Island in 1918, after which it spread rapidly. Anecdotal and qualitative records of its effects are now being supplemented by more formal surveys, related to the work of 'Biosecurity Australia' (see Box 10.6 in Chapter 10). The arrival and spread of the black rat is likely to have been the tipping factor leading to population decline in some species on the Lord Howe Islands. For example, the Lord Howe Island wood-feeding cockroach, *Panesthia lata*, is now extinct through predation by the black rat on the main island and only survives in small numbers on the smaller islands, such as Roach Island, where the rats have not yet arrived. The Lord Howe Island gecko, *Christinus guentheri*, has also suffered from the impact of the rats, again through predation, in this case of eggs and young geckos. The geckos are sensitive to such predation because each female lays very few eggs.

Figure 9.10 The Black rat (Source: Klein/Hubert/Bios/STILL Pictures).

BOX 9.2

9.5 Biogeographical modelling

In this section two very different examples are provided. The first, **island biogeography**, contains perhaps the most famous use of the term 'biogeography' and has been the basis of much pure research and applied management techniques. It has also fuelled fierce controversy over the years. Like the ideas of Clements, Gleason and Whittaker discussed above, the concepts arising from the island biogeography model of species change are almost unconsciously used as part of the language of biogeography and ecology. In comparison, the second example of biogeographical climate modelling is far more recent. The ability to predict where and when a species is likely to achieve pest status can be a valuable warning device for a wide range of users including farmers and environmental and economic planners.

9.5.1 Island biogeography

The study of isolated areas such as islands has provided knowledge and understanding of huge importance. Islands often provide, through their clearly defined boundaries and geographical isolation, as near a situation to a scientific laboratory that the biogeographer is likely to encounter. On many islands the biogeography is simplified owing to a lack of external factors, thus allowing us to examine individual processes more clearly.

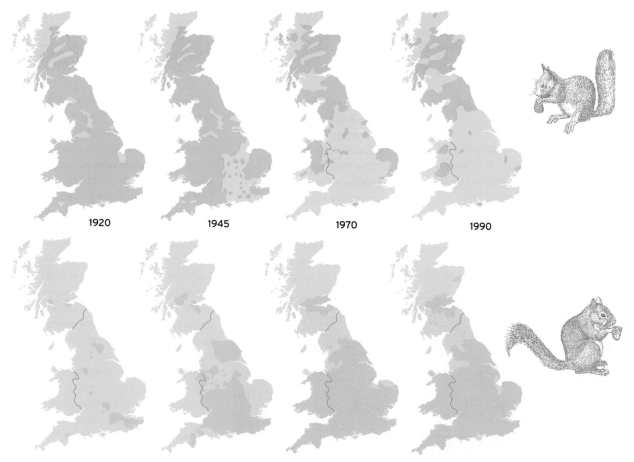

Figure 9.11 The effectiveness of spread among introduced species: the grey squirrel. Since the 1920s the grey squirrel has taken over the habitats of the red squirrel which has consequently been outcompeted and declined. (Source: after Yalden, 1999)

1920   1945   1970   1990

MacArthur and Wilson (1967) considered a balance between the rates of immigration of species to an island and the rates of extinction. Their theory of island biogeography suggested a relationship between the species richness of an island and its size and isolation. The opportunities for species to arrive at a newly created island and be able to sustain themselves depend upon ease of access and the nature of the habitat encountered. Islands close to a mainland would usually be more accessible, while more remote ones would be disadvantaged (Figure 9.13c), but might develop a range of species more slowly, perhaps if they were part of a chain of islands, where species might have the opportunity to cross between islands. If the most favourable habitat was already occupied, later immigrant species might have to adapt to survive. Larger islands or those with a greater variety of habitats might be able to support a greater range of species (Figure 9.13b). Most data sets support the idea that larger islands tend to contain a greater number of species than those that are smaller.

The rate of extinction of species inhabiting the island would initially be low, since competition for resources would be low. As the number of species increased, however, and pressure on resources increased, the rate of extinction would rise. Figure 9.13(a) expresses the curves for extinction and immigration and suggests that there will be an equilibrium number of species when the rate of immigration is matched by that of species extinction. The role of distance from immigration source and size of island is also shown on the figure as it alters the slope of the lines on the graphs (Figure 9.13b and c).

Several important extensions of island biogeography theory are used in biogeography and elsewhere to explain patterns of distribution. Firstly, if an island is created by the loss of a land bridge to the mainland, such as Tasmania, following a sea-level rise, the new 'continental' island might initially be species-rich. Not all species might be able to be supported by the restricted resources of the new island, so

HAZARDS

JAPANESE KNOTWEED IN THE UNITED KINGDOM

Japanese knotweed, *Fallopia japonica* var. japonica (Figure 9.12), is probably the most invasive UK plant species at the beginning of the twenty-first century. Originally used in the United Kingdom as a garden ornamental plant, it was introduced in the nineteenth century and spreads rapidly though its underground rhizomes. It also hybridizes with similar species such as *F. sachalinensis* and *F. japonica* var. compacta. Even small sections of rhizome will form a new plant and the dense clumps formed after two or three years can be 9 m high, substantially altering local habitats and out-competing local species. The spread of this alien is frequently assisted by movement of topsoil by deliberate human action and from material accidentally dropped by construction traffic as well as through natural processes. As a result its distribution can be strongly linear along roads and footpaths, riverbanks, beaches and railway lines, as well as on construction sites.

It is now a criminal offence to grow the species in the wild or to knowingly spread it in the United Kingdom. Japanese knotweed is classed as 'controlled waste' under the UK Environmental Protection Act 1990 and so may only be taken to licensed landfill sites. The Environment Agency's website (**www.environment-agency. gov.uk**) provides advice for those needing to deal with the species and many counties have a 'Knotweed Forum'.

A range of control methods are used against Japanese knotweed. However, it takes several years to destroy a rhizome system using herbicide and removal of the plant requires not just the root system to be taken out but also the surrounding soil to ensure that no fragments of rhizome remain. Cutting above-ground material, with the intention of exhausting the plant, is another long-term technique but this creates the problem of disposal of the cut material. A multi-agency team is working to identify natural predators of the plant that may be suitable for release (after quarantine testing) in the United Kingdom as an aid to elimination or control of Japanese knotweed.

Figure 9.12 Japanese knotweed. (Source: Dorling Kindersley)

BOX 9.3

in this case, an increase in extinction rates and a drop in species richness might reasonably be expected over time. This is opposite to the circumstances governing the developing species richness of a new 'oceanic' island, such as Hawaii.

The second extension of the theory is to encompass within it 'virtual' islands, or landscape patches, such as woodland clearings, ponds or an isolated marshy area. This element of the island biogeography model has had extensive application in determining the most suitable

I : Immigration
E: Extinction
Ŝ: Equilibrium number of species present

Figure 9.13 The MacArthur and Wilson (1967) theory of island biogeography. (a) The island has immigration of new species from the mainland and extinctions of species. The number of species on the island should be in balance since if there are too many then extinctions will be greater than the rate of immigration of new species and if there are too few for the environmental niches available then new immigrants will find it easy to survive and extinction rate will be low. (b) Smaller islands have fewer environmental niches and so can support fewer species than larger Islands. (c) Islands closer to the mainland species' source will have a greater number of species as immigration rates will be greater than for distant islands.

areas for conservation. This has resulted in modelling to determine whether it is better to have a single large patch that can preserve a greater number of species or several small patches. Several smaller areas when added together may have the same total area as one large patch. However, these two situations (lots of small patches and one large patch) do not produce the same diversity even though they have the same total area (May, 1975). The 'single large' school of thought is supported by Wilson (1994) and by Diamond (1975), Diamond and May (1976) and Diamond and Veitch (1981). Diamond supports the use of a few, large reserves, such as Yellowstone National Park, especially if the conservation of large mammals or those requiring migration routes is involved. Studies monitoring species loss have shown that single large reserves such as Yellowstone and similar very large 'islands' have lost fewer species than the

smaller American national parks. It is also contended that neither initial species diversity nor latitudinal range (both of which are used as gauges of habitat diversity) are as important as the size of the reserve in predicting species diversity. A major political disadvantage, however, is that agencies have used this school of thought to downgrade the protection of certain reserves on the grounds that they must be too small to be viable.

Others such as Simberloff (1983) and van der Maarel (1997) regard habitat diversity as the most important factor influencing the number of species existing in a given reserve. In this light, it can be argued that 'small' is not only acceptable at times but can also provide insurance against site loss by providing replication. In other words, having species in lots of small sites means that if species were lost from one site there would still be others preserving them. However, if we only protected one large site and a species was lost from that, then the species would be lost completely. In many countries there is simply insufficient land or too great a demand for land to allow the creation of large nature reserves.

9.5.2 Species distribution modelling

There is often an economic as well as a scientific reason for learning about the preferred 'geography' of a species. This may be in order to increase the efficiency of its production, such as breeding varieties of sheep able to cope well in specific conditions. It may also be to control the spread of a pest species. Using climate change models to predict how species distributions might change is becoming very popular today. The range of many insects will expand or change, and new combinations of pests and diseases may emerge as natural systems respond to altered temperature and precipitation profiles. It will be important to predict such problems in advance in order to prepare for and mitigate against them.

Any species will have, for each environmental variable, not only a preferred ecological niche where conditions are ideal, but also a less favourable wider area within which it is still able to survive and reproduce. Outside this area conditions are so stressful that the species is unlikely to be found. Such areas can be measured indirectly by observation of population abundance. These areas can then be related to the local climatic conditions. It should then be possible to predict where species might be distributed for a given climate regime. This can first be tested by applying the predictions to other areas where the population has been observed but not measured.

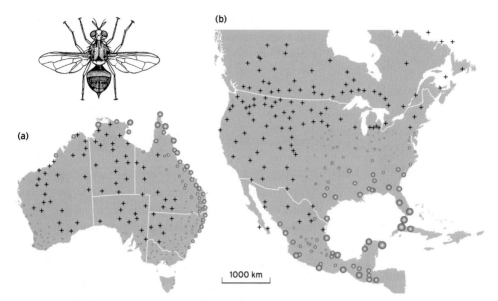

Figure 9.14 Climatic model-ling of the Queensland fruit fly for (a) Australia and (b) North America. The area of each circle is proportional to the Ecoclimatic Index of the fly; crosses indicate that the fly could not permanently colonize that area. (Source: courtesy of CSIRO Australia, © CSIRO Australia 1991)

For example, in many parts of Australia, fruit growing has become important as the basis for agriculture, both for home use and for export. A major pest is the Queensland fruit fly, *Bactrocera tryoni.* Sutherst and Maywald (1985) developed a model allowing the potential distribution of this fruit fly to be predicted. They developed an 'Ecoclimatic Index' to allow precise evaluation of areas suited to the Queensland fruit fly to be carried out. The evaluation is based upon factors such as the stress created for the species as a result of localities where temperatures or moisture levels are either too high or too low. The resulting model can predict distribution of the fly, based upon the Ecoclimatic Index, not just for Australia but for other areas where there is the potential for inad-vertent introduction of the species by accident (e.g. poor hygiene). The Ecoclimatic Index is shown for Australia in Figure 9.14(a) and shows those areas most favourable to colonization. The potential of the fly as a pest for North America has also been predicted using the model and predictions are shown in Figure 9.14(b). This has been done by matching the climate of North America with that of its native range. The figure shows that acci-dental transport of the fly to North America could lead to large areas of potential colonization. Most of Mexico and the east coast of the United States are vulnerable. Canada and western United States, however, are unlikely to support a permanent colonization. Thus it is possible to calculate which regions of a continent are at risk and therefore which regions should have their imports more carefully checked.

Reflective questions

➤ What is the theory of island biogeography?

➤ How is island biogeography theory relevant to places that are not oceanic islands?

➤ What benefits are there in modelling species habitats in a world of changing climate and mass transportation of foodstuffs?

9.6 Biogeography and environmental management

Many aspects of biogeography have great relevance and impact today. It has enabled the development of techniques for managing a range of environmental factors within an increasingly technological environment, where the need for land makes multiple land-use demands common (Figure 9.15). The management of wild or semi-natural animals and plants may have a variety of objectives, methods and levels of intensity. Four important manage-ment aspects of biogeography are those of:

1. agriculture – sometimes taken as being a form of 'directed' succession;
2. conservation – either of a particular species or group of species, or as a means of increasing biodiversity;

Figure 9.15 A picnic area in the upper Taff Valley, Wales, managed as an amenity area while controlling access to a reservoir and helping to minimize sedimentation flows into the water.

3. recreation/amenity environments;
4. environmental tools – plants or animals may be a means to an end rather than being present for their intrinsic value. This can include the planting of grasses to increase the stability of new slopes or the keeping of fierce guard dogs.

The management aims of agricultural systems are focused upon optimizing conditions for one particular species of plant or animal, often through the control and direction of the normal successional processes. This may involve the reduction of competition by removing unwanted 'weed' species. Other methods include increasing nutrient supplies in combinations particularly appropriate to the selected species (e.g. via fertilizers). The reduction or removal of predators, pests and parasites is also a feature of this type of biogeographical management. These unwanted species, however, may be regarded by conservationists as endangered species, or by agriculturists elsewhere as valued potential crops. A weed is often merely a plant in the wrong place at the wrong time. The same is true of many faunal 'pests'. A further complication arises when, for example, a generally unwelcome species, such as

bracken, provides a habitat for species with a high conservation value, such as the high brown fritillary butterfly. Some management practices may seem severe, such as the burning of heather moorland, and often they are carried out as a tradition rather than being based on best possible practice (Holden *et al.*, 2007). Often the biogeographical research has not yet been done to establish best practice or the most efficient techniques. It is also important to understand the resistance or **resilience** (ability to 'bounce back') of biogeographical systems to make management practices efficient (Box 9.4).

Often a wide range of biogeographical techniques have to be adopted which are directed towards economic goals such as tourism and sporting activities. These may be combined with those of conservation and the preservation of cultural values. Frequently there are multiple land-use requirements made upon management (see Chapter 24) and priorities change as seasonal demands on the site alter. As a result, development of multifaceted plans for parks and reserves is increasingly common, often including biodiversity action plans. Amenity management can include extreme forms of management. The creation of appropriate turf conditions

NEW DIRECTIONS

RESILIENCE AND PANARCHY

The biosphere is dynamic and often has the ability to 'bounce back' or to recover from stresses applied to it. This capacity to return to its previous state is described as resilience. In Figure 9.16, the area of grassland on the headland has remained for many years, despite the stressors of the coastal environment and from trampling by visitors along the coastal path. Unlike straightforward physical resilience, such as within a rubber band, ecological resilience is highly complex and may involve a range of short- and longer-term adaptations. It

is problematic to measure or to predict the ecological limits of any one locality. With increasingly intensive human land-use, coupled with climate and other environmental changes, it is important to be able to measure ecological resilience and to find the tipping point. In other words it is important to determine how much stress or disruption can be accommodated before a change to a new state is inevitable. This new state may result in biogeographical boundary changes and the processes involved in the adaptive capacity of ecosystems are therefore important.

The term 'resilience', in this context, was first used by C.S. Holling in 1973 and the measurement of resilience is becoming an important tool in ecosystem management. The concept should be treated with care, however, as it can be problematic if taken too literally. It is sometimes equated, if on a simplistic level, with 'sustainability' or with successful environmental management. Ecological succession (see Chapter 10) can be regarded as a series of stable states, separated by dynamic, transitional conditions. Some workers, who regard ecological resilience as both a functional and

Figure 9.16 Tintagel, Cornwall, England. An example of an area where habitats are displaying their adaptive capacity to both environmental and human-related stresses.

BOX 9.4 ➤

> philosophical tool, consider it to be the determinant of movement between stable states and that a system's adaptive capacity can act as a buffer against movement from one state to another (grassland and desert, for example) but that adaptive capacity may also be the medium of change form one state to another.
>
> The natural environment, however, is 'noisy' and the complexity of relationships within and between, for example, trophic levels is becoming increasingly recognized, their measurement and analysis appearing to go beyond the traditional hierarchical structures used in ecology and biogeography. This complexity has given rise to a concept which is old in philosophical terms (1860), but relatively new in terms of its ecological meaning. This is the concept of 'panarchy' which is upheld by many of the workers supporting resilience theory. It claims that, to fully understand environmental change, it is necessary to consider long- and short-term interactions. These interactions might occur both across and within trophic levels adapting to change and creating resilience. Panarchy equates these interactions with the evolutionary development of ecosystems but in a more random way than is traditionally accepted. Instead of ecological hierarchies, there are panarchies that incorporate adaptive cycles, signifying their dynamism. With this concept there is an emphasis on the connectivity between panarchy levels.
>
> BOX 9.4

for international standard soccer pitches may involve massive inputs of energy and biogeographical and ecological expertise at a microscale. Even civic amenities such as parks and playing fields require the control of vegetation succession.

Biogeographical management may involve land restoration or land reclamation. Land reclamation suggests that the land can be used again whereas land restoration is about returning the site to its former state. Large areas of many countries may have initially unpromising environments that make either of these proposals seem very difficult. Figure 9.17 lists major characteristics of industrial wastes that limit vegetation development. However, an understanding of biogeography suggests that some of these harsh areas may in fact act as sanctuaries where species unable to tolerate the competition found in more usual environments may be able to flourish. Thus the harsh conditions shown in Figure 9.17 may not be considered as limiting but actually as enabling. For example, it has been possible to locate areas such as those with high levels of

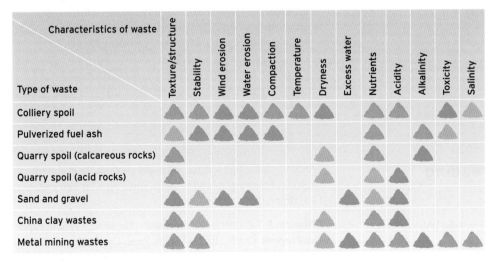

Type of waste \\ Characteristics of waste	Texture/structure	Stability	Wind erosion	Water erosion	Compaction	Temperature	Dryness	Excess water	Nutrients	Acidity	Alkalinity	Toxicity	Salinity
Colliery spoil	▲	▲	▲	▲	▲	▲	▲		▲	▲		▲	▲
Pulverized fuel ash	▲	▲	▲	▲	▲				▲		▲	▲	
Quarry spoil (calcareous rocks)	▲						▲		▲		▲		
Quarry spoil (acid rocks)	▲						▲		▲	▲			
Sand and gravel	▲	▲	▲	▲				▲	▲	▲			
China clay wastes	▲	▲					▲		▲	▲			
Metal mining wastes	▲	▲					▲	▲	▲	▲	▲	▲	▲

Figure 9.17 Major characteristics of industrial wastes which limit soil and vegetation development. (Source: after Wheater, 1999)

Characteristic may:

▲ be present and/or cause some problems

▲ cause moderate problems

▲ cause major problems

lead in the soil by their distinctive vegetation. Thus, where expenditure for restoration is prohibitive or the land has been affected too severely, other uses for that land may still provide environmental opportunities. Flooded gravel pits may provide valuable wetland nature reserves, for example.

9.7 Summary

This chapter has shown the importance of a range of processes in determining biogeographical patterns in time and space. Large-scale factors such as climate and plate tectonics may play a role in the distribution of species and species groups. Ecological factors such as succession, evolution, extinction, species mobility and the immigration of alien species are also vital in determining the biogeographical patterns we see today. On a smaller scale local topography or pedology may control the spatial distribution of species. Landscape ecology provides a useful concept for exploring regional and small-scale biogeographical distributions. This theory suggests that landscapes are made up of a matrix which contains patches and corridors. The patches provide oases surrounded by the more usual distributions and the corridors provide pathways for species dispersal and movement between patches.

Overriding most of this chapter is the element of change in the distribution of living things. Current biogeography can produce only a snapshot of a constantly changing situation. Habitats develop and change in response to gradual changes, such as in a regional climate, or more dramatic disturbance through fire or the agency of humans. In turn the community occupying the habitat may respond by losing or gaining species. The theory of island biogeography suggests that islands (or landscape patches) are more likely to have a greater number of species if they are larger because the habitats within them are more likely to be diverse. This is because any given area will be subject to immigration of new species over time and extinctions of species as the increased number of species results in increased competition. There is thus an equilibrium point for the number of species any given island area can hold. This sort of theory, while it has its critics, also has important practical applications in the management of the living elements of the physical environment, whether for economic gain or for conservation. Biogeography has a wide range of applications from land reclamation on industrial sites to predicting the influence of climate change on the spatial distribution of crop growth potential and crop pests.

Further reading

Bradbury, I.K. (1998) *The biosphere.* John Wiley & Sons, Chichester.
This is a standard undergraduate text which is good for the general principles and discusses evolutionary aspects of biographical pattern.

Cox, B. and Moore, P. (1993) *Biogeography: An ecological and evolutionary approach.* Blackwell Scientific, Oxford.

Another classic text which is especially good on the evolutionary aspects of biogeographical patterns, going back far into geological time.

Forman, R.T.T. (1995) *Land mosaics – The ecology of landscapes and regions.* Cambridge University Press, Cambridge.
Forman is not a supporter of island biogeography but this is still one of the best texts on landscape ecology and is written in a very accessible style.

Global Ecology and Biogeography, Volume 9 (2000).
This particular volume of the journal includes several papers discussing the need for 'a new paradigm of island biogeography'.

Huggett, R.J. (2004) *Fundamentals of biogeography.* Routledge, London.
Huggett has written several textbooks in this area, all worth exploring. This one is full of illustrative examples.

Jeffries, M.J. (1997) *Biodiversity and conservation.* Routledge, London.
This book is very much a product of the Rio Earth summit but somewhat less pessimistic than most and it provides a very wide range of examples and ideas.

Kent, M. and Coker, P. (1992) *Vegetation description and analysis – A practical approach.* Belhaven Press, Chichester.
This is a good hands-on guide to many of the techniques needed in fieldwork design and the analysis of data collected. It provides a range of useful case studies.

Matthews, J.A. (1992) *The ecology of recently deglaciated terrain: A geoecological approach to glacier forelands and primary succession.* Cambridge University Press, Cambridge.
This excellent text is aimed at a higher level and will be suitable for those interested in either succession or harsh upland or semi-arctic environments.

Quammen, D. (1996) *The song of the dodo: Island biogeography in an age of extinctions.* Hutchinson, London.
An excellent example of narrative non-fiction is provided by Quammen. It works better for the fast reader since it is a chunky volume, but you might want to dip into it for fascinating accounts of interviews with a range of well-known scientists.

Spellerberg, I.F. and Sawyer, J.W.D. (1999) *An introduction to applied biogeography.* Cambridge University Press, Cambridge.
This book provides a southern hemisphere view and covers far more than the title suggests. It includes an overview of the discipline and an extensive section on the application of landscape ecology techniques.

Whittaker, R.J. and Fernández-Palacios, J.M. (2007) Island biogeography: Ecology, evolution, and conservation. Oxford University Press, Oxford.
A detailed overview of the theory of island biogeography and the role of islands as ecological hotspots, the threats to species and measures to protect species.

Web resources

Ecological Succession and its Application to Forestry
http://www.jimswan.com/111/succession/succession.htm
Here there is an in-depth discussion of the processes of ecological succession in forests and the effects of forestry methods upon natural succession.

Ecology and Society
http://www.ecologyandsociety.org/
At this site there is an online peer-reviewed journal on ecological conservation and sustainability that provides an easy-to-use format for finding related articles.

The International Association for Landscape Ecology
http://www.landscape-ecology.org/
The home page of the association, it provides an explanation of the science behind landscape ecology, news of current situations across the globe and some related links.

Island Biogeography and Evolution: Solving a Phylogenetic Puzzle using Molecular Genetics

http://www.ucmp.berkeley.edu/fosrec/Filson.html
There is an exercise here designed by the Museum of Paleontology at the University of California, Berkeley, to demonstrate the complexity of evolutionary problems and the way geographical and geological data are used to solve such problems. It is an interesting site and useful in emphasizing the interdisciplinary nature of the study of biogeographical patterns.

Lord Howe Island
http://www.lordhoweisland.info/
Provides more information about this remote area discussed in Box 9.2, albeit initially from a tourism context.

Nearctica: Ecology Main Page
http://www.nearctica.com/ecology/ecology.htm
This site is an excellent online academic journal that supplies links to sites on global biodiversity, conservation, succession and island biogeography, some of which are very useful.

Resilience Alliance
http://www.resalliance.org
A group who do research on resilience as a basis for sustainability.

 Visit our website at **www.pearsoned.co.uk/holden** for further questions, annotated weblinks, interactive models and video-clips showing physical processes in action.

CHAPTER 10

Ecological processes

Hilary S.C. Thomas

Department of Humanities and Science, University of Wales, Newport

Learning objectives

After reading this chapter you should be able to:

➤ **understand important ecological characteristics**

➤ **describe ecosystem processes of energy, nutrient and material flows**

➤ **describe spatial and temporal ecological processes**

➤ **understand how humans influence spatial and temporal ecological processes and how they can create unique ecological communities**

10.1 Introduction

The study of ecology is focused upon the interactions of living things with their environment. The study of these interrelationships has led to the development of ideas that are seen as increasingly important in the twenty-first century. In 1989, the British Ecological Society, as part of its 75th anniversary celebrations, published the results of a poll of its members. The aims had been to determine the key ideas in ecology and those areas of the subject that were considered to have contributed most to our understanding of the natural world. The ideas that most ecologists agreed upon included: the ecosystem, **succession**, energy flow, conservation of resources, competition, niche, materials recycling, the community, life-history strategies and ecosystem fragility, along with a range of other themes that will be discussed in this chapter. Cherrett (1989) statistically analysed this survey to see if there were clusters of responses that might indicate groups of ecologists with similar priorities. The analysis resulted in Table 10.1 which indicates how certain important ecological concepts formed four distinct groups. There has been little change to these priorities since the original survey and so the concepts addressed in this chapter will be similar in character to the four groups shown in Table 10.1.

The characteristic features underpinning any understanding of modern ecology, such as the nature and role of ecosystems, habitats, communities and the environmental niche, will be discussed in this chapter. The chapter will consider spatial patterns created via ecological processes at a range of scales including the scale of an individual tree. It will then move on to consider temporal patterns and distributions, such as the effects of climate change. In particular, the variety of cycles and flows of nutrients, energy and matter that underlie these patterns and distributions are considered, since the study of these interrelationships is the defining feature of ecology. The checks and balances operating within these patterns lead to a

Table 10.1 Important areas of ecological study, based upon Cherrett's (1989) analyses

The ecosystem
  Energy flows
  Geochemical cycles
  Succession
	Studies in 'holistic' theories

Ecosystem fragility
  Conservation of resources
  Habitat restoration
  Sustainability
	Practical/applied studies

Life-history strategies
  Predator-prey relationships
  Ecological adaptations
  Plant-herbivore interactions
	Studies of biotic influences

Environmental heterogeneity
  Stochastic processes
  Natural disturbance
	Studies related to external influences

Figure 10.1 A simple ecosystem. The photosynthesizers use the energy of the Sun, and nutrients and water within the soil, to help produce organic matter. This plant matter is then eaten by herbivores (plant eaters) which in turn are eaten by carnivores (meat eaters). Some of these carnivores may be preyed upon by other carnivores. During this process waste is produced. Some of this waste may be recycled back into the soil to be taken up by plants at a later stage, whereas other wastes may be lost from the ecosystem by wind or water movement.

consideration of the drivers, both natural and anthropogenic, of ecological change. Some of the resulting issues, relating to the conservation and rehabilitation of ecologically sensitive areas and to the effects of climate change, are discussed towards the end of the chapter. They are indicators of areas where ecology has great relevance and impact today.

10.2 The functions and characteristics of ecology

The following section briefly introduces major themes in ecology that will be elaborated on throughout the rest of this chapter.

10.2.1 The ecosystem

Huggett (2004) suggested that 'ecosystems are communities together with the physical environment that sustains them'. An ecosystem may be a pond, a catchment basin, a biome or the Earth's biosphere. Selecting the appropriate boundary is dependent on the problem and on the timescale (Waring, 1989). However, more important than scale issues is the idea that ecosystems are about the characteristics of and interrelationships between the properties of the environment and

its living inhabitants, permanent or temporary, and of the flows of a variety of materials between these elements. This characteristic of interconnection means that if one part of an ecosystem changes then this will affect other parts of the ecosystem too.

Most ecosystems are taken to be 'open systems' where there is an element of input to and/or output from the system, rather than complete self-containment. At its simplest level as shown in Figure 10.1 the major input is usually taken to be energy, in the form of solar radiation. Outputs include energy loss at a variety of levels. In the simplified structure of Figure 10.1 sunlight, rainfall and soil are used to indicate typical ecosystem inputs, and various types of **detritus** (waste) typify ecosystem outputs. Within it, an ecosystem can be divided into several nutrient levels. In reality, as indicated by Figures 10.6 and 10.7 below, both inputs and outputs may be far more complex, making the realistic modelling of many ecosystems and ecological processes problematic (Murray, 1968).

10.2.2 The habitat

The physical environment provides the distinctive habitat or habitats of the ecosystem. This is the living area for a species or group of species and may be very small, such as a single branch of a tree for a lichen, or very large, such as the areas needed by predators of grazing animals in tropical grassland biomes (see Chapter 8). Sometimes the terms habitat and ecosystem are used incorrectly, as if they were the same, but the term habitat does not include the large-scale interrelationships that are an essential part of the description of an ecosystem. An ecosystem may contain a range of habitats within it. For example, a sand-dune ecosystem with dunes and slacks provides a wide range of habitats, at a variety of scales.

10.2.3 Populations

Populations of species vary over time often in response to changing environmental conditions (Beeby and Brennan, 1997). Chain reactions are the norm such as a decrease in predators leading to an increase in prey species and vice versa. Sometimes more indirect changes may occur owing to the removal of one important species upon which lots of other species and processes were dependent. Limits in population size arise from the limited resources available within the ecological niche and from the balance between births and deaths within the population as a whole. Some of the commonly used models of population growth and decline are shown in Figure 10.2. If there were no limits then population growth would occur as indicated in Figure 10.2(a). Population curves are produced for situations where limiting factors include finite amounts of nutrients or space (Figure 10.2b), or where altitude limits population growth owing to more severe climate conditions (Figure 10.2c). Population curves can also be produced for a range of other scenarios including those where disease in one species allows increased survival of another species, or of younger generations of the same species. Disturbance of a species may impact on the population, and models such as that shown in Figure 10.2(d) help indicate the typical recovery time (or relaxation time – see Chapter 1) and what may happen if disturbances occurred more frequently than the full time needed for total population recovery. Therefore these population models are of enormous benefit for management planning.

10.2.4 Ecological communities

A community is made up of groups of individuals that may be of any living organisms, although most familiar are often those defined by plants or animals, which occupy a specific area. From the geographer's viewpoint, the community is therefore a mappable entity. Communities operate at a variety of scales (e.g. Figure 10.3) with boundaries that may be sharp or very gradual but their character is determined by the 'core' of the community, which may be some definite species, such as the Tasmanian blue gum which often co-exists with other species. Within the community, there may well be competition and change over time, so that an early community may evolve into a very different later one. This may be through effecting changes within the habitat or by responding to changes imposed upon it from outside.

Communities are often referred to by the Latin classification term for their dominant species. A *Callunetum* is therefore a community dominated by *Calluna vulgaris* or

Figure 10.2 Some typical simple population curves: (a) rapid growth in the absence of limiting factors; (b) growth of population after a certain size (*x*) comes under increasing environmental friction until carrying capacity (*y*) of that location is reached. Population will fluctuate around the carrying capacity. (c) A typical environmental limiting factor. The number of species declines with increasing elevation. (d) Effects of repeated disturbance. Note the importance of sufficient recovery time between events.

Figure 10.3 An early simple ecological community on a boulder.

the common heather. Where other species are also present in certain subtypes of that community, they are described as **differentials**. These terms provide a useful way of describing the likely features and processes operating

within an area. The ecological community is also characterized by its interactions. For example, the heather in a *Callunetum* provides shelter, food and access to water for a range of creatures including birds, insects and mammals.

10.2.5 Ecological functions

Within ecology, the study of the global functions of the biosphere has particular relevance for humans. A classification of those functions purely in terms of their utility for humans (van der Maarel, 2000) is suggested below:

- A carrier – where the local ecology provides space and surface for human activities.
- Production – as a supplier of matter and energy from natural and agricultural resources (Figure 10.4).
- Information – for orientation, aesthetic appreciation, philosophical identification, research, education, recreation and, through monitoring, to detect changes (Figure 10.5).
- Regulation – based on ecosystem stabilization, for example via chemical cycles and damping of climatic change.

Figure 10.4 One of the functions ecosystems serve from a human perspective is to supply food.

Figure 10.5 Ecosystems serve an aesthetic and recreational function.

The environmental function of production is fundamental to human existence. Humans survive by eating foods and certain crops can also be used for other purposes. These include building materials, fuel and clothing. Ecosystems also provide a regulatory function that is fundamental for ecology as a modern science and one of great modern relevance. They regulate atmospheric carbon dioxide concentrations, for example by promoting photosynthesis, absorbing carbon dioxide and releasing oxygen back into the atmosphere. However, the human alteration of ecosystems is of great concern as this may result in change to these regulatory systems. Regional alterations could have global implications. The inevitable interdependence and interaction between individuals, species and their environments provides ecosystems and communities with both strength and potential fragility. This is drawn in part from their habitats but also from their living components.

The functions and characteristics described above form important components of ecological research and will be elaborated on within the following sections.

Reflective questions

➤ What is the difference between an ecosystem and a habitat?

➤ What are the main ecological functions from a human perspective?

10.3 Ecological processes

10.3.1 Energy and nutrient flows

Most plants and some other fairly simple life forms such as certain bacteria and algae take energy from sunlight. Through the process of photosynthesis they convert this to sugars, which then become available for animals, birds or insects to use for nutrition. The life forms involved in this initial process within the ecosystem are described as producers or **autotrophs** and they form the lowest level of the food chain and the first trophic level (or nourishment level). As shown in Figure 10.6, primary producers can include a range of vegetation types. The purpose of eating is to generate energy in order to live and to collect nutrients in order to grow. The consumers or **heterotrophs** feeding on the autotrophs in the example shown in Figure 10.6 are grasshoppers, rabbits and mice. These creatures, usually herbivores, feed directly on the producers and are described as first-order consumers and form the next trophic level. The energy gained by consuming the autotrophs is used by these herbivore consumers for a range of functions such as digestion, movement, reproduction and respiration or is lost as heat. As a result, there is relatively little energy passed on, perhaps only 10%, from one trophic level to the next. Therefore a consumer requires a large amount of biomass from lower trophic levels. For example, rabbits must spend a considerable amount of their time grazing. Herbivores are consumed by the **carnivores** (such as the snake) and **omnivores** of the next trophic level. These, in turn, are likely to form the food and therefore energy base of higher-level carnivores or omnivores such as the hawk shown in Figure 10.6. When they die, their remains will probably form the diet of the decomposer **saprovores**, such as maggots, becoming part of a detritus-based, rather than vegetation-based, food chain. This final stage releases the last of the energy as heat, and is also important in the recycling of inorganic nutrients that

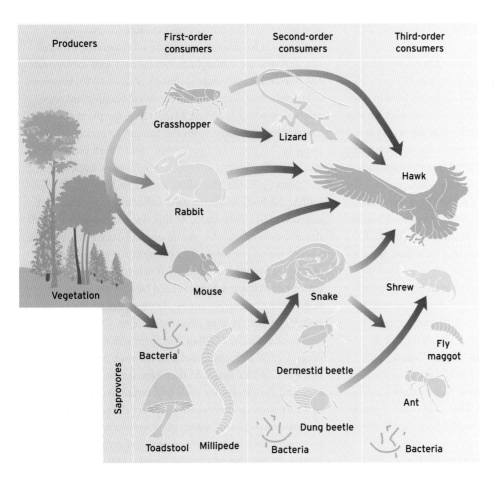

Figure 10.6 A simplified example of nutrient flows within a temperate ecosystem, indicating trophic levels. (Source: after Murray, 1968)

have been passed through the food chain. These include materials such as nitrogen and phosphorus, which entered the system as nutrition for those original producers, perhaps taken up from the soil water (see Chapter 7).

Food chains are not always simple and, as Figure 10.7 shows, many animals are able to operate at a variety of

levels within what is better described as a food web. This is less confusing if translated into human dietary terms. The wealthy human who is primarily and perhaps preferentially a carnivore will still, in most cases, add vegetables to their diet (e.g. steak and chips/French fries). The less wealthy may subsist for the most part on

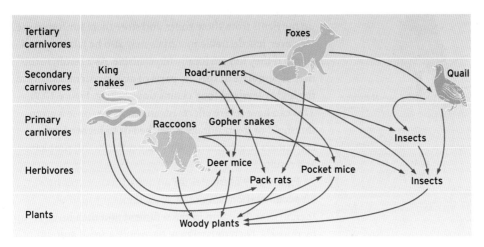

Figure 10.7 The complexity of a typical food web. This example is from the chaparral scrub of California. The various trophic levels indicate the animals of different groups and it is noticeable that not all animals are confined to one trophic level. Raccoons, for example, are both herbivores and carnivores. However, they are ultimately dependent on the plants.

vegetables such as beans and rice but will augment this with meat or fish if possible.

Since not all of the biomass in any one trophic level is eaten, terrestrial food webs are often shown, as in Figure 10.1, as biomass 'pyramids'. The lower trophic levels tend to have the greatest bulk, not all of which will be consumed. This is because there are always some animals that die of old age rather than being caught and eaten by predators. However, in aquatic systems the trophic pyramid is not always accurate. Some forms of algae, for example, are both rapid reproducers and highly nutritious, so they may form the diet of a larger biomass than exists within their own trophic level. This is like a hospital patient attached to a 'drip' of such highly nutritious material that very little is required to keep the patient alive.

In addition to energy transfers, material is transferred through an ecosystem. The material within ecosystems follows complex paths that often include temporary storage for varying amounts of time. The difference in time for chemicals held within the Atlantic Ocean compared with those held within a single potato is enormous. **Biogeochemical** cycles involve links with the local variations in geology and climate. There are many types of such cycles but for simplicity the main components of the phosphate form of the phosphorus cycle are shown in Figure 10.8 in order to illustrate the biogeochemical cycling process.

Many rocks contain phosphates. When weathering takes place, the water-soluble phosphate goes into solution, often as part of the soil water or in water bodies such as lakes

and streams. Here it becomes available for producers within local food webs, becoming incorporated into cell membranes. Subsequent consumers may also incorporate the phosphate into their skeletons (Figure 10.8). However, a subcycle may occur if a **mycorrhizal** fungi–plant root relationship is able to take place. Here, the plant roots extract the phosphate from the fungus, so making the plant in this instance a consumer rather than producer. In exchange, the fungus is a supplier of sugars. Back within the main cycle, the phosphate moves on from its storage within the body when released either as a component of defecation or upon death. At this stage saprovores will aid decomposition and the phosphate will once more become part of the soil or water. The phosphorus cycle can have almost endless repetition until it is leached out of a soil into a watercourse where it could be used for an aquatic ecosystem. On a larger timescale phosphorus can be incorporated as part of deep ocean sediments. Over time the sediment could be **lithified** and form sedimentary rock. As geological processes continue the rock may eventually reach the surface again. At this stage it may become weathered and the phosphate cycle will continue. Other chemicals may undergo very similar or very different pathways through geology, soil, water, plant and animal transfer processes. Thus ecosystems consist not only of complex food webs but also of complex energy and material cycling processes.

10.3.2 Bioaccumulation

Through the biogeochemical cycling process there may be locations where chemicals build up or become concentrated. Some chemicals have far greater potential than others to accumulate in zones where they become bioavailable (e.g. within soil water). Producers and consumers take up nutrients from these zones and store them for future use. However, less desirable materials may also be taken up and stored if they too are locally bioavailable. As a result, toxins may accumulate in specific parts of an ecosystem, usually in the higher levels of food chains or webs (e.g. Figure 10.9). This can have severe effects on an ecosystem once a given threshold of toxin storage is reached or to higher-level consumers (e.g. humans) who may suffer health problems as a result.

Figure 10.9 illustrates for an aquatic ecosystem how **bioaccumulation** can occur. A contaminant such as mercury can build up in the sediment and then be taken up by mussels. Each mussel may have only a small amount of mercury. However, as small fish eat a lot of these mussels the mercury becomes more concentrated inside the fish.

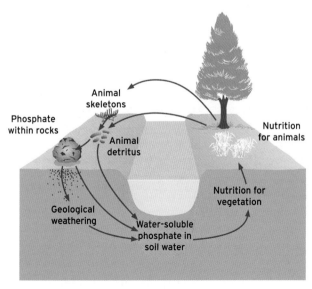

Figure 10.8 An example of a biogeochemical cycle.

3 Predatory species continue to concentrate toxins by eating the small fish. These larger fish are then consumed by humans

2 Mussels concentrate toxins and are then eaten by small fish

1 Contaminants (e.g. mercury) are washed into water and build up in sediment

∴ = toxins

Figure 10.9 An example of bioaccumulation in an aquatic ecosystem. The toxin is found in low concentrations in mussels but as these are consumed in large quantities by a fish then the toxin concentrations are much greater in that fish. The toxin is concentrated even further higher up the food chain.

Then, as other predators eat lots of these fish the mercury can accumulate to high levels within them. This can have toxic consequences for humans who may eat the larger predatory fish.

To be able to 'bioaccumulate', materials must not merely be stored in the consumer, but must be stored in those parts of the individual that will be consumed. Therefore, those materials stored in 'meat' are far more likely to become bioaccumulated than those stored in bone. Water-soluble toxins are less likely to produce this type of problem, since they tend to be lost as part of the more general water loss of the consumer and, unless the toxin is very long-lived, may not be taken up again. In the immediate period following the Chernobyl nuclear explosion, soil and water pollution occurred in northern Europe as the local winds spread the radioactive materials over large areas. This resulted in long-term bioaccumulation in vegetation. These toxins were then consumed by grazing animals in northern Europe during the following year, spreading the negative effects to the wider ecosystem. **Bioconcentration** levels, which are the extent to which a material is found in tissue compared with background levels, are generally site specific. Therefore the effects of the Chernobyl fallout were not found in Australia.

Reflective questions

➤ Why do herbivores such as rabbits or sheep need to spend a significant amount of their time grazing?

➤ Where is energy gained and lost in an ecosystem?

➤ What is bioaccumulation?

➤ Why may small, seemingly harmless, concentrations of a chemical found at the lower end of the food chain eventually prove harmful to the ecosystem?

10.4 Spatial patterns and distributions in ecology

10.4.1 The ecological niche

As with so much in ecology, understanding the spatial patterns of living things comes from studying not just the individual requirements of a species but their interrelationship with both their physical environment and the other species occupying it. In any locality, the status and character of populations are connected, in particular, with the level of competition for the ecological space available: the environmental niches. Competition is related to the type and amount of resources available at any one location and the ecological niche is the combination of those resources required by the individual organism. The niche is not the same as the habitat, since the ecological niche relates more to the distribution of the species within the wider community and is explained further below.

The resource requirements for most living things usually centre upon provision of appropriate climate, shelter, food and water. The ecological niche, in the forms described below, is the basis of most ecological patterns. In the simplest of situations, where there are no competitors for any of the resources required by an individual or species, the **fundamental niche** can be occupied, which contains ideal conditions. This is rather like your being able to use your educational institution's library resources as the only client for the duration of your course. Most species, however, have to utilize a **realized niche** that is the result of competitive interaction between several species attracted to the resources. This is like taking the only textbook left on the library shelf as other books are not available because the keenest students took the best books out of the library at the beginning of term. You therefore have to hope that at least one of these other books becomes available before the examination. This competition is usually strongest between similar species, since their ecological niches are likely to overlap. The species able to survive on the lowest amount of the limiting resource will be advantaged.

There are many instances of niches that are used by different groups at different points in time. For example, daytime and night-time predation or the use of scarce water resources may allow resources to become partitioned. Different parts of a resource, such as a tree, are often used by different species. Observations in a British wood showed a characteristic tree-use pattern by blue, great and marsh tit birds (Gibb, 1954) which was later confirmed by ornithologists elsewhere. Blue tits were far more likely to spend feeding time among the leaves during the summer and on the twigs and buds in the winter, than elsewhere on the tree. Great tits obtained their prey from the ground beneath the trees, in winter especially, although they also used the leaves in the summer. Marsh tits used a far greater proportion of the trees throughout the year.

10.4.2 Competition

Both interspecific and intraspecific **competition** are important factors determining ecological distributions. Interspecific competition arises in situations such as at an African waterhole in an otherwise arid area. To be able to drink the water, herbivores and other vulnerable species may need to wait until carnivores have drunk their fill and moved away. The disadvantage of the long wait is compensated by the knowledge that eventually the herbivores too will be able to drink. The species in this case share the same spatial distribution, but not at the same time. Intraspecific competition can be seen where a number of squirrels need to collect acorns from a small area within a city park. Intraspecific competition may lead to the exclusion of weaker individuals and explain the patterns of territories which control both feeding and reproduction opportunities for the squirrels.

10.4.3 Life strategies

A further factor in the patterns and distributions found in ecology relates to the life strategies of species and individuals (Drury and Nisbet, 1973). These are usually related to elements of the life cycle of a species, particularly its means of dispersal. Understanding a species' life strategy

Table 10.2 Characteristics of 'r' and 'K' life strategists

'r'-selected	'K'-selected
Rapid development	Slow development
Short-lived	Long-lived
Small body size	Larger body size
Have many offspring – tend to overproduce	Have few offspring
Single reproduction	Repeated reproduction
Do not care much for individual youngsters	Care for their offspring
Population not regulated by density – boom and bust	Population stabilizes near a capacity level
Opportunistic – will invade new territory	Maintain numbers in stable ecosystems
Lax competition	Keen competition

may help to predict or to explain its ecological distribution (Table 10.2). There are two main types of strategies, known as **'r' and 'K' selection**. Both are likely to occur within the wider landscape but the 'r' strategists are more likely to be found in new or disturbed sites, since they have good colonizing ability. They include many weed species and are often described as **ruderal**. The second type of strategist is the 'K' strategist. These species are more likely to do well in a less disturbed environment. Table 10.2 illustrates the typical characteristics of 'r' and 'K' strategists.

Both stress and disturbance may alter a community. Stress on species, perhaps as a result of recurrent drought, affects productivity. Disturbance, such as a forest fire, destroys biomass. The 'r' strategists are likely to succeed in conditions of high disturbance but low stress. 'K' strategists are more likely to outcompete other species in conditions of low disturbance and low stress. This is an area of ecological research which is very important given potential climate change and human impacts on stress and disturbance.

10.4.4 Biodiversity: patterns of species richness

Figure 10.10 illustrates the links to biodiversity in terms of its definition, why we should be bothered about it, the potential threats and solutions. In recent decades there

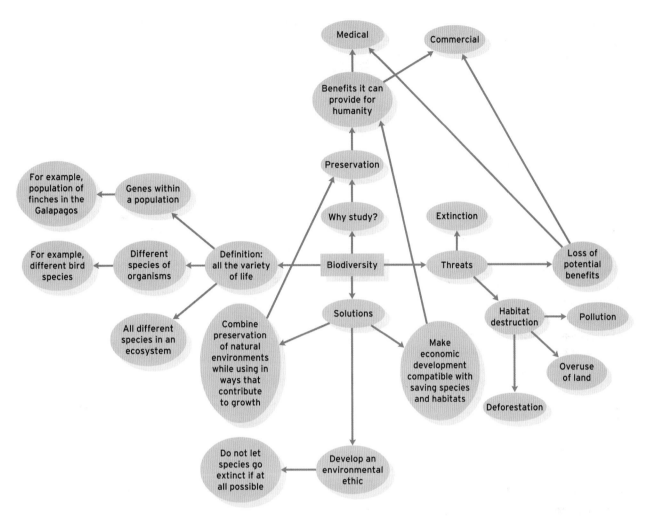

Figure 10.10 A biodiversity concept map. (Source: 'A biodiversity concept map', Michael J. Robinson, from the website of the 1999 Summer Biology Institute on Diversity, produced by the Leadership Programme for Teachers at the Woodrow Wilson National Fellowship Foundation (Princeton, NJ, USA: Woodrow Wilson National Fellowship Foundation 1999). **http://woodrow.org/teachers/bi/1999/projects/group9/Robinson/conceptmap.html**. Used by permission of the Woodrow Wilson National Fellowship Foundation, 2004)

have been increasing interest in and concern about 'biodiversity'. This is a measure of species richness. It is the variation among living organisms and ecosystems of which they are part. There are a number of diversity indices and some are described in Box 10.1. Some areas may have many more species than others. For example, the tropical forests are believed to provide habitats for 40% of the species on Earth. There have therefore been increasing attempts both to identify biodiversity 'hot spots' and to conserve biodiverse areas. Concerns exist because of increased extinction in areas of altered land-use such as tropical forest regions. In these areas, species may become extinct before their existence (and potential human function, e.g. for medicine) is known. Therefore it is important to recognize factors that may lead to enhanced species

richness. Global-scale areas of species richness are usually the result of factors including lack of disturbance and lack of isolation.

Tropical areas have the greatest species diversity (Wilson, 1992). This is, to a great extent, a result of the relatively minor impact (lack of disturbance) of the Quaternary glaciations on tropical areas. Where a region's ecology has been uninterrupted, in terms of major climatic or other disturbance for lengthy periods, species are less likely to have become locally extinct. These areas may also benefit if species from disturbed areas arrive as immigrants. Barriers such as high mountains or oceans will lessen immigration opportunities. While many isolated areas are species-poor, they may contain species now unique to the area that were originally more widespread. Regional and local patterns of

TECHNIQUES

DIVERSITY INDICES

It is often important to measure the species richness of a site or habitat. Measurement might be very general: concerned with numerous vegetative, insect, animal or bird species; or quite specific: confined, for example, to rodents, or mosses, or a particular family of birds. Quantifying the abundance and diversity of species in any taxonomic group can be done using a diversity index.

Diversity indices are a useful way of objectively measuring species abundance or diversity in almost all terrestrial and aquatic habitats. They can be used to describe spatial changes (differences in species between different habitats or between different areas of the same site), or to establish temporal change (the impact of pollution or management by repeated sampling of the same site periodically over time). One of the most common, and most flexible, diversity indices is the Shannon–Wiener Diversity Index (also known as the Shannon index or the Shannon–Weaver index). This is often used to compare two or more situations as the index is of most use as a comparative measure. It is important to compare 'like with like'. In other words, comparing the diversity of wetland birds between two areas or the diversity of grasses in a field at specific time intervals is fine, but comparing the diversity of moths at one site with the diversity of fish at another is meaningless. In addition, care must be taken to sample a similar area between sites or monitor the site for a similar length of time in order that the index is not biased.

The Shannon-Wiener index has a complicated-looking formula but is quite simple to work out by hand or on a computer if you take things one stage at a time:

$$H = -\sum^s[p_i \times \ln(p_i)] \quad (10.1)$$

where H is the Shannon–Wiener Index, S is the total number of species at a site, and p_i is the proportion of a particular species. Σ indicates the sum of, while ln is the natural logarithm, which can be calculated on most scientific calculators or in most computer spreadsheet packages. For ease, this can be expanded to:

$$H = -\sum[N_i/N_{tot} \times \ln(N_i/N_{tot})] \quad (10.2)$$

where N_i is the abundance of an individual species and N_{tot} is the total abundance. A worked example is provided below.

Table 10.3 provides data on wetland bird species from two marshes. In this case, five species were found at Marsh 1 and only three species were found at Marsh 2, but the number of individuals found at each site was the same (70). This is

for ease in this example and need not be the case. The first step is to calculate p_i for each species at each site, then calculate $\ln(p_i)$, and $p_i \times \ln(p_i)$. Thus for Marsh 1 the values are given in Table 10.4 with values for Marsh 2 shown in Table 10.5. In this example the calculations have been done to two decimal places. Once this has been completed you should then sum all the individual results (one for each species) from the formula $p_i \times (\ln p_i)$. For Marsh 1 this gives the result −1.54. Finally, take the absolute of this value to give the Shannon-Wiener score: $H = -(-1.54) = 1.54$. For Marsh 2, $H = 0.61$.

The most important thing to remember when interpreting values for the Shannon-Wiener index is that they are relative values. This means that if one area has an index value of 3 this is not double (twice as species-rich) as another area with an index of 1.5. In this instance all that can be said is the first site has a higher species richness or is more species diverse than the second. Shannon-Wiener values typically range from 1.5 to 3.5 but (as here) can range from 0 to over

Table 10.3 Data on wetland bird species from two marshes

Marsh 1	Marsh 2
Species 1 = 20	Species 1 = 4
Species 2 = 5	Species 2 = 9
Species 3 = 14	Species 3 = 57
Species 4 = 15	
Species 5 = 16	

BOX 10.1 ➤

➤

4.5 depending on the number of species involved. In the above example, Marsh 1 has a higher (but not double) diversity of wetland bird species than does Marsh 2. This might be as a result of location, greater habitat diversity, better management or many other site-specific factors. In terms of setting conservation priorities, Marsh 1 has a greater species richness and might be considered as being of higher conservation importance than Marsh 2. Conversely, it might be considered that more conservation strategies are needed at Marsh 2 to increase its species diversity.

Source: Material courtesy of Anne Goodenough, University of Gloucestershire.

Table 10.4 Calculations to help work out the Shannon-Wiener index for Marsh 1

Formula	Species 1	Species 2	Species 3	Species 4	Species 5
$N_i \div N_{tot}$ (gives p_i)	$20 \div 70 = 0.29$	$5 \div 70 = 0.07$	$14 \div 70 = 0.20$	$15 \div 70 = 0.21$	$16 \div 70 = 0.23$
$\ln(p_i)$	-1.24	-2.66	-1.61	-1.56	-1.47
$p_i \times \ln(p_i)$	-0.36	-0.19	-0.32	-0.33	-0.34

Table 10.5 Calculations to help work out the Shannon-Wiener index for Marsh 2

Formula	Species 1	Species 2	Species 3
$N_i \div N_{tot}$ (gives p_i)	$4 \div 70 = 0.06$	$9 \div 70 = 0.13$	$57 \div 70 = 0.81$
$\ln(p_i)$	-2.81	-2.04	-0.21
$p_i \times \ln(p_i)$	-0.17	-0.27	-0.17

BOX 10.1

species richness may result from short-term disturbances or habitat diversity. Short-term disturbances such as natural fires are likely to encourage high levels of species richness, as in Mediterranean regions (see Chapter 8). A mosaic of patches of differing age-since-disturbance often develops, each with a distinctive community. This gives the area as a whole a high species richness. Some environments are able to provide a range of habitats within a local area. Woodland tends to have a range of light and shelter conditions, providing a diversity of habitats within it as well as those created by the woodland edges, where species from both the woodland and the surrounding area may partition the resources available.

Reflective questions

➤ What is the difference between a fundamental niche and a realized niche?

➤ What is the difference between intraspecific and interspecific competition?

➤ Think about endangered species that you have heard of: are they mostly 'r' strategists or 'K' strategists? What type of strategists are humans?

10.5 Temporal change in ecological patterns and distributions

10.5.1 Succession

Many ecologists have noted that from the community to the ecosystem scale, the composition, complexity and biomass of an ecological unit changes over time (e.g. Connell and Slatyer, 1977; Tilman, 1985; Wilson, 1992). Most of these observers have concluded that such change is seldom completely random in nature and is more often directional, moving towards or away from a particular state or character.

This form of change is known as succession. Important schools of thought have developed concerning succession and these are discussed in Chapter 9. However, more process-based assessments have also been developed and these are discussed in Box 10.2.

10.5.2 Human influence

A wide range of ecological stressors may initiate changes to ecological systems or to habitats, which may alter the population and community flows and possibly lead to the extinction of certain species. Such changes have always

ENVIRONMENTAL CHANGE

ECOLOGICAL OLYMPICS – THE SUCCESSION RACES

For many years, ecologists have accepted succession as a module to work from. The changes over time could be measured but research continued in order to determine why these changes occurred. Table 10.6 provides a list of ideas that attempt to describe the processes involved in succession. Egler (1954), in searching for a mechanism for succession, studied individual species. He looked at initial invasion of a site through to species disappearance from the local community. From this analysis he developed the concept of **relay floristics**. This suggested that as one group of plant species establishes it is replaced by another and then they are replaced until a stable state is achieved. Many ecologists have studied the importance of evolutionary 'strategies' such as life-cycle or physiological strategies as a succession mechanism. Noble and Slatyer (1980) developed the concept of vital attributes. These could be characteristics such as the speed of

reaching reproductive state and would determine the place of that species in the vegetation replacement series.

Three different models of succession mechanisms were compared by Connell and Slatyer (1977) and each achieved wide ecological acceptance under different situations. These are facilitation (the relay floristics model), tolerance and inhibition and are shown in Figure 10.11. The facilitation model suggests that a species may colonize and then change the

local environment in terms of pH, moisture availability, structure and shading, for example. This may mean that the conditions are facilitated for another species. Therefore when another species comes along and finds the conditions more to its liking, it can out-compete the first species.

The tolerance model proposes that most species are present from the start. The fastest growers are the early dominants. These have little or no effect on slow growers or later

Table 10.6 Ideas on the processes involved in succession

Authors	Ideas
Egler (1954)	Relay floristics
Noble and Slatyer (1980)	Vital attribute
Drury and Nisbet (1973), Grime (1997)	Life strategies
Connell and Slatyer (1977)	Facilitation, tolerance and inhibition
Tilman (1985)	Resource ratio/allocation

BOX 10.2 ➤

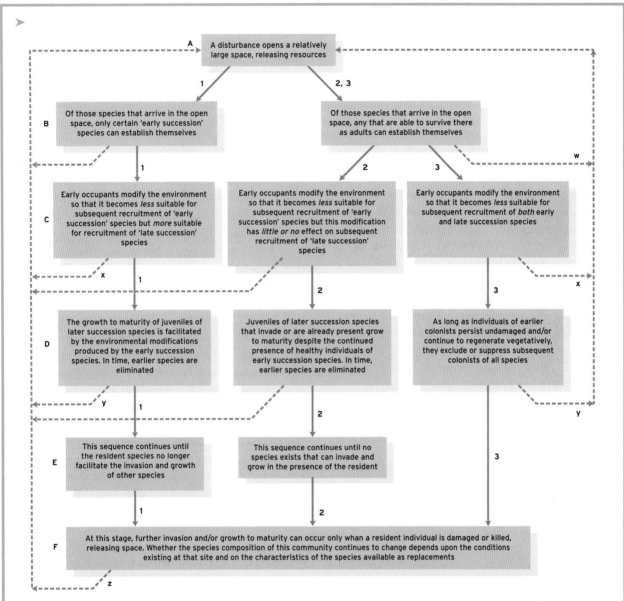

Figure 10.11 Three models (1, 2, 3) of the mechanisms producing the sequence (A, B, C, D, E, F) of species in succession. The dashed lines represent interruptions to the process, in decreasing frequency in the order w, x, y and z. (Source: after Connell, J.H. and Slatyer, R.O., 1977, Mechanisms of succession in natural communities and their role in community stability and organisation, *American Naturalist*, **111**, p. 1121, Fig. 1, © University of Chicago Press. Reprinted by permission of the University of Chicago Press)

colonizers, though late successional species will suppress the early species. There is thus a competitive hierarchy. Several attempts have been made to distinguish between species on the grounds of their competitive abilities under different circumstances. As such, lists similar to league tables have been produced for a range of species. Tilman (1985) linked this competitive hierarchy to the resources of soil and light and how these might change during succession. Where there is a single limiting nutrient for survival or reproduction, the species that can be successful at the lowest availability level of that resource will win.

Finally the inhibition model proposes that colonists capture space and inhibit colonization of later species. It therefore requires disturbance of pioneers before later species can establish.

BOX 10.2

277

been part of the natural processes operating within the environment. However, humans have been responsible for a variety of deliberate alterations. Species-selective agricultural production or clearance of an area followed by urban construction is usually intentional. However, the after-effects may well be unintentional and uncontrolled.

10.5.2.1 Agriculture and its effects upon ecosystems

Agriculture provides the most obvious example of intentional human manipulation of natural ecological patterns. Its effects cover huge areas of the Earth's surface. It is estimated that as much as 70% of the developed areas of the world might already be in a state of an agroecosystem, which is where the agriculture is the dominant part of the system. In the developing world agroecosystems are a major driver of landscape change. Many of the more successful agroecosystems mimic elements of the original ecosystem such as the replacement of prairie grasslands by large-scale cereal cropping. However, our increasing technical abilities can produce far more original landscapes. In Figure 10.12 commercial forestry has replaced the moorland and clear-felled areas are clearly visible on the hillsides, vulnerable to erosion until new planting takes place.

Agricultural systems can be regarded as examples of ecosystems with managed inputs and outputs of energy and nutrients with controlled species diversity. The unintended results may, however, act as a trigger to a variety of

unwelcome patterns and distributions. Intentional nutritional increases to crops may be leached (see Chapter 7) into areas downslope of the original application, into water systems or to the sea. This may cause eutrophication, giving rise to secondary problems such as a phytoplankton bloom (NEGTAP, 2001). The main effects of agricultural expansion may be seen as an increase in cropland, but other known effects include an increase in savanna woodland as a result of the overgrazing of grasses.

Slash and burn or shifting agriculture allows the 'useless' but difficult-to-clear species to remain and also allows those species regarded locally as useful to remain. These species therefore remain on site while crops are planted around them. The land will remain in agricultural use until either weed growth or dropping yields become unacceptable and then a move to a new plot is made. In the Apo Kayan area of Borneo, for example, temporary long houses and grain stores are located in the centres of areas of shifting cultivation. Intermixed with areas in current use are unused plots of varying ages, which slowly revert to secondary forest after about 20 years. The old plot is therefore traditionally allowed to regenerate. However, too often population or other pressures mean that secondary regrowth, with the accompanying return of soil nutrient levels, will not be complete before the plot is cleared once more. Over time this process creates a situation where the seed bank is not large enough to allow regeneration of those intermediate species lost during each clearing phase. An altered pattern of species and competition therefore arises over time.

More intensive agriculture, both before and after sowing, may remove unwanted species either through selective herbicide, reinforced by the use of selective fertilizing, or, where organic means are preferred, through manual or mechanized weeding. The end result, of decreased biodiversity, remains the same.

10.5.2.2 Urban ecology

The growth of urban areas is often regarded as ecologically destructive; 50% of the world's population lives in urban areas. It is accompanied by domestic animals, pests, parasites and other species that are able to compete successfully in these relatively new ecological environments. The resulting urban ecosystems, while varying in response to wider-scale geographical patterns such as latitude, are characterized by: **encapsulated countryside** (e.g. Figure 10.13) (either ancient habitats or previously managed land – both likely to suffer ecosystem degradation unless management is provided); highly managed places such as gardens; and abandoned land including post-industrial habitats that

Figure 10.12 Softwood plantations at varying stages of development below upland heather moorland and blanket bog on the Berwyn Mountains, mid-Wales. Newly planted trees can be seen in the foreground.

Figure 10.13 An example of encapsulated countryside where a rural area has been incorporated within an urban setting.

may now be sustaining unique communities that are a mixture of native and alien species.

The distinctive character of urban microclimates includes, in most instances, increased temperature. Local air flows may be modified by the physical structure of the city surface and the differences in temperature regime with the surrounding rural areas (see Chapter 6). Air pollution is often a problem, with high levels of particulates at certain times of day. Absolute humidity and the incidence of fog, including photochemical smog, are also likely to be increased (Oke, 1987). Water quality in urban areas may be affected, depending upon constructional and meteorological conditions (see Chapter 15). A general decrease in vegetation cover, coupled with an increase in impermeable surfaces, creates increased likelihood of flooding coupled with lower soil moisture and groundwater levels. Increased runoff may stress local aquatic ecosystems in terms of water flow, sedimentation and pollutants, which may include road salts during winter periods (see Chapter 15). Land quality is generally regarded as being diminished after urbanization with problems relating to exhausted soils, altered slopes and the concentration of wastes such as at landfill sites.

Probably the greatest interest in urban ecology is the 'urban wildlife'. Certain species have found the new combination of resources advantageous and have made significant dietary and/or habitat changes to enable them to fill vacant urban niches. The cliff-ledge habits of the rock dove have transferred easily to the window-sills utilized by feral pigeons. The wide nutritional range provided by urban waste dumps attracts a range of birds and other species. House mice and house spiders, together with the brown rat, have long been the prey of domesticated cats and dogs. More exotic pets intentionally or accidentally released are often able to survive, if at the edge of their ecological niche, such as terrapins which are now regarded with affection as long-term residents in an urban lake in Cardiff, Wales. Inevitably, the existence of these smaller creatures has allowed higher trophic-level predators to move into towns and cities, such as birds of prey. Foxes moved into many British urban areas during the early part of the twentieth century when improved transport systems led to the building of suburban housing in once rural areas. Rural foxes quickly urbanized, taking advantage of the food and shelter provided in gardens from compost heaps, bird tables and garden buildings. Following the destruction of hedgerows and woods, food and shelter resources may now actually be better for some species within towns and cities. Garden escapes of a range of plants have also increased local biodiversity and competition (Figure 10.14). Urbanization of species may create problems for humans, such as the urban raccoons of Washington, DC, and New York that are believed to have been responsible for a rabies outbreak (Chang *et al.,* 2002).

10.5.2.3 Recombinant communities

Recombinant communities are working communities of plants and animals that are a mixture of native and alien species. The arrival and integration of introduced species into the specific recombinant community may be relatively recent or may stem from very ancient introductions, such as the sycamore into the British countryside. Such communities are not restricted to urban or to industrial areas, although this is often their most common occurrence. It has been suggested that, in European cities, some 20–35% of the flora may be composed of recently established species and discussion is under way about how we categorize such systems.

10.5.3 Ecosystem fragility

The fragile nature of many ecosystems subject to exacerbated change is the object of much current research and debate (e.g. Solé and Montoya, 2001; see also Box 9.4 in Chapter 9). The complexity of many ecosystems makes them robust under a range of conditions, such as the

Figure 10.14 The Hottentot fig, originally from southern Africa. This photo shows a garden escape now dominating native species along the cliffs at Lizard Point, Cornwall, England.

random removal of species. However, it also makes them vulnerable when highly connected **keystone species** are removed. The nodes in Figure 10.15 represent individual species within a specific food web. The responses to species removal include both secondary extinction of other species and the fragmentation of originally complex food webs that were always assumed to provide effective buffering, into discontinuous smaller webs. Those systems where the network is somewhere between a regular and a totally random distribution of connections are able to withstand perturbations relatively well and to respond rapidly to change. However, the remaining ecosystems are much more fragile. Decay of the original web into disconnected species may result in ecosystem collapse. Species that are highly connected with the rest of the web are those whose elimination is likely to be most destructive. These are known as keystone species. Keystone species may be damaged if bioaccumulation results in toxic conditions for those species. These species may also be those posing a perceived

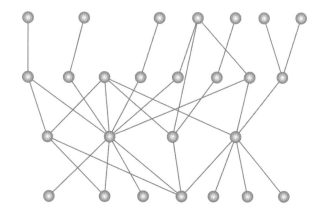

Figure 10.15 Schematic representation of ecological relationships between species. The links may not just include predator-prey links but many other links including seed dispersal and shelter. Few species have many links and many have just one or two links. If one of the species is removed then because of the many links with the other species connected to it they will also be affected. Those species that are highly connected to other species are known as keystone species. (Source: after Solé and Montoya, 2001)

threat to humans. If such a species is removed then the resulting chain of effects might include a population explosion of previous prey, or the extinction of species dependent upon the removed species for nutrition or seed dispersal. Organisms other than the top trophic-level predators may be keystone species. In many food webs, species may appear at several trophic levels as shown in Figure 10.7.

The keystone relationship, often vital to successful conservation planning, may be derived from a variety of characteristics seen in the examples below:

- Elephants are generally regarded as grassland keystone species in tropical Africa. Their browsing deters succession from changing these habitats into shrubland or forest.
- A subspecies, the forest elephant, is key in maintaining some West African forests through its role in differential seed dispersal.
- In temperate grasslands, prairie dogs are a major keystone species as a result of their high level of linkages. They appear as grazers, predators, prey or, through their burrows, as providers of microhabitats for other species.
- The starfish *Pisaster ochraceus* underpins the stability of many rocky shoreline ecosystems on the North American coast, through its preferred diet of Californian mussels, which are otherwise the competitive dominants of the community.

Reflective questions

➤ What are the main process-based models of succession?

➤ How do humans influence ecological patterns?

➤ What are recombinant communities?

➤ What are keystone species and why are they important for ecosystem management?

10.6 Ecological processes and environmental management

10.6.1 Conservation and sustainability

Conservation is almost universally regarded as a good thing but definitions as to what conservation means are varied. Conservation always implies intervention and value-judgements have to be made as to what should be conserved and what should be ignored. This forms an uneasy alliance with the objectivity implicit in ecological science. Protected conservation areas are intended to maintain functioning ecosystems, minimizing known vulnerabilities and maximizing biodiversity. Often management is required to increase local diversity or protect what is considered to be a rare and important species (Box 10.3). For example, the Kenfig National Nature Reserve, South Wales, consists of a dune system which provides a habitat for rare plants such as the fen orchid. The main problem at the site is that a lack of fresh sandy sediment supply has led to an increased rate of succession. During soil development, increasing amounts of organic matter and nutrients and decreasing pH have created opportunities for rapid-growing species to replace species characteristic of the dune habitat. Thus a programme of removal of litter was introduced to remove organic matter from the system. Artificial bare areas (blow-outs) were also created in the dunes to promote plant communities of earlier stages as shown in Figure 10.18 and sheep and rabbit grazing is actively encouraged. The results were that levels of diversity increased dramatically.

However, there are questions about whether we should strive to maintain ecological communities that are now no longer self-sufficient (see Chapter 22). The Kenfig system will need continuous management intervention if biodiversity is to be maintained at a high level. There may also be more interesting and distinctive recombinant communities which are sometimes the only refuge for species with unusual niche requirements. The idea of sustainability is related not only to the ecosystem we are trying to manage but also to how we behave in the world as a whole and how sustainable our use of biosphere resources actually is. Calculations of sustainability (and unsustainability) are discussed in Box 10.4.

Shaw *et al.* (1998) showed for urban landscapes in the Tucson metropolitan area that among the 33 land cover categories, golf courses and neighbourhood parks had the highest vegetative cover (93% and 77% respectively). Schools (45%) and moderate density housing (2–8 houses ha^{-1}) had a higher vegetative cover (44%) than natural open space in this semi-arid region which had only 22% cover. The study concluded that the single most important strategy for integrating conservation into planning for Tucson's growth would be to protect an interconnected matrix of habitats. Networks of reserves with corridors maintained between them are often regarded as a potential conservation solution in areas of fragmented landscapes.

CASE STUDIES

CONSERVATION OF THE HAWAIIAN GOOSE

Ecology and historic decline

The Hawaiian goose (*Branta sandvicensis*), or nene (pronounced nay-nay), is a small brown goose endemic to the Islands of Hawaii and has the smallest biogeographical range of any living goose (Figure 10.16). The state emblem of Hawaii, the nene was a common species on many of the larger islands with an estimated historic population of 25 000 birds. However, after colonization of the islands by Europeans in 1778, the population began to decline dramatically. This decline was fuelled by direct threats including hunting and egg-collecting, and also by the indirect threat posed by habitat loss as the landscape became increasingly farmed. The single most serious threat, however, was the introduction of non-native predators to the islands. Most devastating was the predation of adults, chicks and eggs by the small

Indian mongoose (*Herpestes auropunctatus*) which was introduced to the islands in 1883. The combination of these human-induced problems of hunting, habitat change, and introduction of alien predators resulted in massive population decline: a demonstration of the interactions between a species and its biotic and abiotic environment. In 1907, the species was listed as a protected species but this had little effect, and by the late 1940s, only about 20 to 30 wild birds remained in Hawaii.

The nene's decline demonstrates the vulnerability of island populations to environmental change. In the case of the nene, the situation was made all the more serious by its endemic status: extinction in Hawaii meant extinction from the biosphere. Moreover, the nene is a keystone species, dispersing seeds and playing an important role in vegetation development, particularly in early successional communities on lava slopes. This illustrates how the decline of one

species can have consequences for many more.

Conservation: captive breeding and reintroduction

Because of the critical nature of the species' population, several birds were captured during 1949 and 1950 and a captive breeding programme was started with two main bases, one on Hawaii and one in England at the Wildfowl and Wetland Trust at Slimbridge, Gloucestershire. Fortunately, the species adapted well to captive conditions and this conservation strategy probably saved the species from extinction.

A reintroduction programme of captive-bred individuals was started on three Hawaiian Islands (Hawaii, Kauai and Maui) in 1960 and over 2300 birds have now been released. The wild population has increased significantly as a result of this reintroduction programme and there are now about 1000 birds in eight populations. However, at least five of these populations are not self-sustaining, still requiring continued release of captive-bred birds.

Although it is probably only the captive breeding and subsequent reintroduction programme that have saved the species from extinction, there are still problems with these conservation approaches. Recent DNA analysis of the captive breeding nene populations suggests captive populations show low genetic variation. This is not surprising given that the population went through such a bottleneck (i.e. the number of individuals fell so low) meaning that all birds alive today come from a very few

Figure 10.16 The Hawaiian goose, ringed for identification purposes. (Source: Photo courtesy of Anne Goodenough)

BOX 10.3 ➤

breeding individuals. Such chronic inbreeding may have led to inbreeding depression in the reintroduced population, although this is very difficult to test scientifically. There are also concerns that reintroduced birds are showing altered patterns of behaviour and a lack of adaptability to change as a result of being captive-bred. These problems demonstrate the value of *in situ* conservation approaches whenever possible, such as suitable habitat management and control of alien species, rather than captive breeding and reintroduction. Of course, this can only be done well before the situation becomes critical.

The current situation

Because of its small and non self-sufficient population, the nene is still listed on the IUCN Red List as 'vulnerable' and is federally listed as 'endangered'. The current population dynamics of, and conservation strategies for, the nene in Hawaii are summarized in Figure 10.17. Many of the current threats are the same as those which caused the historic decline, but the difference is that most of these are being addressed through habitat management and other conservation strategies. In addition, environmental education programmes seek to inform people about the nene and reduce illegal poaching, and extensive research programmes to establish and inform future conservation are continuing in the hope that the wild populations of nene can one day become self-sustaining.

Material courtesy of Anne Goodenough, University of Gloucestershire.

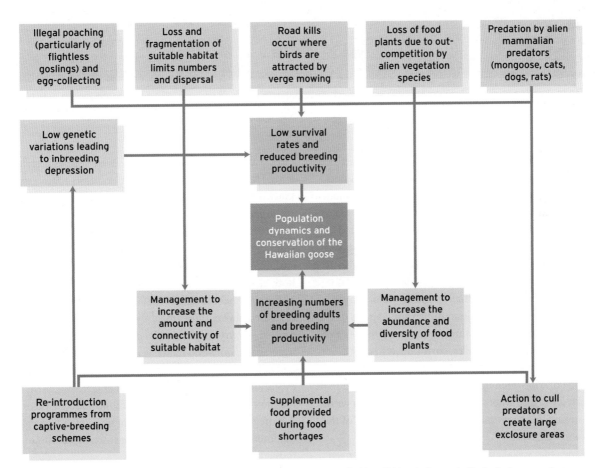

Figure 10.17 The current population dynamics of wild Hawaiian geese in Hawaii. Purple boxes indicate factors causing population decline; green boxes show management and conservation strategies. (Source: courtesy of Anne Goodenough)

BOX 10.3

Figure 10.18 Kenfig National Nature Reserve There are a variety of habitats within the dune system with shrubby vegetation in some of the drier slacks. This photo shows ponding in an artificially created dune slack which has been created to try to establish earlier succession species (Source: Kenfig National Nature Reserve)

UNESCO Biosphere Reserves (such as the one described in Box 10.5) are areas of terrestrial and coastal ecosystems promoting solutions to reconcile the conservation of biodiversity with sustainable use. They are internationally recognized sites for testing out and demonstrating integrated management of land, water and biodiversity. The emphasis is on *conservation* rather than *preservation*. There is also an importance placed upon the use of the land by the local human population and potentially a much wider population. The emphasis on the human dimension appeals to most geographers and is a distinctive feature of the reserves. Each reserve contains a core area, surrounded first by a buffer zone and then by a zone of transition. The transition zone is the one in which the human impact is intended to be the greatest but also where there are likely to be areas within which sustainable economic and social development of the region will take place.

The idea that social and economic considerations need to be taken into account when thinking about ecological management suggests that multidisciplinary concepts are required (see also Figure 10.10). **Social ecology**, for example, is having increased influence on ecological thought. This is where conflicts between social values and ecological sustainability may be addressed. For Brunckhorst

(1995) all environmental problems were seen as social problems. He put forward as a potential solution a framework known as cultural **bioregional theory**. This is based on landscape ecology (see Chapter 9) and integrates the arts and culture. Landscape and conservation values usually develop from the cultural history of the locality. In the belief that such a bioregional framework for planning should include both economic and ecological issues, it is similar to the philosophy behind the UNESCO Biosphere Reserves discussed above. Bioregional theory suggests that there are unified ecosystems defined by geophysical indicators (local watersheds, flora and fauna, geology and weather patterns) and cultural indicators (history of local human habitation, evolution as a community, presence and arrangement of material constructions and socio-economic structures). A bioregional approach attempts to understand and value the interconnected and sustainable workings of these individual components.

10.6.2 Climate change

Root *et al.* (2003) considered data sets extracted from 143 earlier ecological or biogeographical studies, where temperature information was available as well as species occurrence

NEW DIRECTIONS

THE ECOLOGICAL FOOTPRINT

The ecological footprint is a recently developed sustainability index. It can provide estimates of the amount of ecological resources in hectares per person that are used by individuals, companies, specific activities and so on. The footprint is a measure of the amount of biologically productive area needed to both produce resources used and to absorb waste created. The measure reflects the intensity of land-use and population densities as well as levels of resource use. When the index is used some startling results can be seen.

Table 10.7 shows how, of the countries listed, only Australia and Canada are below capacity. All of the other countries use more resources than they can sustain. It has been estimated that at current rates the annual consumption of resources by the world's population requires 14 months to be renewed and that many countries are running at an ecological deficit. The footprint can therefore be a potential measure of sustainability. Goals can be set for reducing local or national footprints by decreasing consumption or increasing productivity. Technology can be utilized to improve current productivity such as the use of roof gardens and solar panels, creating multiple use of space in urban areas. Farming methods such as slope terracing or multiple harvesting from individual fields can also be adopted. Consumption can be reduced through the use of recycled materials and using public rather than private transport and energy-efficient materials and products. You can calculate your own ecological footprint on the web; see the list of web resources at the end of the chapter.

Table 10.7 Examples of ecological footprints calculated for a range of developed countries at the end of the twentieth century

	Population (millions)	Footprint[*]	Current capacity[*]	±[*]
World	6210.1	2.3	1.9	−0.4
Australia	19.7	6.9	14.7	7.8
Canada	31.2	6.9	15.9	9.0
Japan	127.2	4.6	0.8	−3.9
Netherlands	16.1	5.7	0.8	−4.9
South Africa	44.2	4.0	2.6	−1.3
USA	288.3	9.6	6.0	−3.6
UK	60.2	5.5	1.7	−3.8

[*]Global hectares per capita.

(Source: from the Redefining Progress Ecological Footprint Accounts for 1999)

BOX 10.4

CASE STUDIES

THE PÁLAVA UNESCO BIOSPHERE RESERVE, CZECH REPUBLIC

The Pálava UNESCO Biosphere Reserve area, in the south of the Czech Republic near Brno, is one of the driest parts of the Czech Republic and is dominated by Jurassic limestone hills (Figure 10.19). The main range, the Pavlov Hills, give their name to the Biosphere Reserve set up in 1986 which contains a range of nature reserves and other protected areas, both ecological and cultural. There is evidence of settlement in the area from the Stone Age with pasture and coppicing as land-uses. The historic small town of Mikulov, from where the Biosphere Reserve is administered, is the focus of vine-growing villages and is also of tourism importance (Figure 10.19).

The Biosphere Reserve is administered by the Czech Nature Conservation Authority, which works with other institutions, especially the Masaryk and Mendel Universities in the nearest city, Brno. Like most biosphere reserves, it is a working landscape with emphasis on management to maintain the current balance between humans and nature, to protect the distinctive landscape and to facilitate environmental research and education.

The area is at a biogeographical crossroads, containing species from alpine as well as northern and southern continental zones, and contains many endangered species, such as the nationally scarce Old World swallowtail (*Papilio machaon*).

Although the potential natural vegetation is mixed oak forest, the type of oak forest alters with height and includes species-rich shrub and herb layers. Dry grassland areas are also important elements of the landscape. Some of these are naturally occurring 'karstic forest steppes' but others reflect the intensive grazing of the area in the past. Although grazing has a long history in the region, in Communist times the State ran game reserves in the area, with mouflon and goats stocked at higher rates than are regarded as acceptable today. Other problems stem from mining, with some hills almost destroyed, and from soil erosion. Stock management in parts of the reserve has changed, especially from 1996, but intensive grazing persists in some areas.

Figure 10.19 The Pavlov Hills and the small town of Mikulov which is the focus of vine-growing and tourism. The hills have a thin cover of soil and vegetation over the limestone.

BOX 10.5

and location. The intention was to identify spatial trends over time. Some **phenological** data (timing of events such as leaf fall and buds appearing) were also incorporated into the study. Overall, their results identified a consistent temperature-related shift. Where mobile, many species in the regions studied have moved either polewards or to higher altitudes over the past century. There has been a discernible impact of recent global warming on animals and plants. Walker and Steffan (1997), in their overview of the implications of global change for natural and managed terrestrial ecosystems, consider changes in the species composition and structure of ecosystems and how these might affect the system's functions. They predict that the biosphere will be generally weedier and structurally simpler, with fewer areas in an ecologically complex old-growth state. Biomes will not shift as intact entities and the terrestrial biosphere may become a carbon source rather than a sink. Future research

should consider the interactive effects of changes in carbon dioxide, nitrogen and temperature rather than discrete elements of change. The effects of changes in land-use will probably continue to reduce species richness and an increasing number of 'natural' ecosystems will move towards production systems for human use.

10.6.3 Biosecurity

Isolated regions such as Australia and New Zealand have always been vulnerable to the arrival and establishment of alien species. For example, Figure 10.20 was produced in 1950 and shows the alien birds and mammals that had successfully established in New Zealand at the time. Therefore concern over the introduction of species is not a new phenomenon. Biosecurity is one response to this, as shown in Box 10.6. New Zealand, for example, has

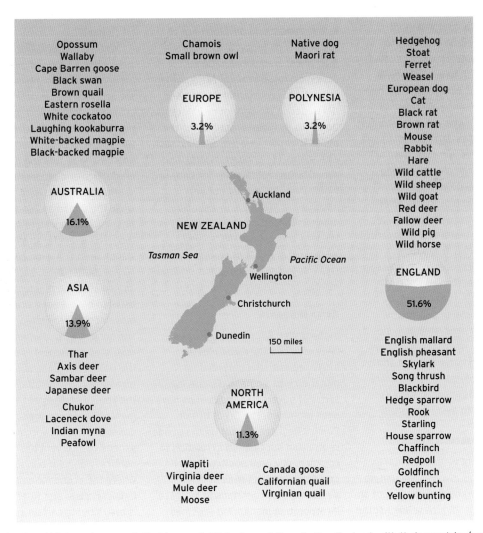

Figure 10.20 Introduced birds and mammals that have established populations in New Zealand, with their countries/continents of origin. The percentage from each origin location is shown. (Source: after Wodzicki, 1950)

BIOSECURITY IN THE TWENTY-FIRST CENTURY

As human use of the natural environment has intensified and interactions between regions and continents become easier, so the development of factors liable to interrupt the stability (and especially the economic stability) of ecological units has increased. Biosecurity measures often originate as a result of the threat of economic loss or damage to human health. Recent global and regional biosecurity threats include avian influenza, swine fever and 'foot and mouth' disease. Biosecurity attempts to secure the boundaries of an ecological unit against generally known threats. Just as ecosystems operate at a very wide range of scales, so biosecurity can operate in the context of a pond, field or farm breeding unit, a nature reserve, a nation state or globally (for example, quarantine measures in relation to space research). While biosecurity often relates to an 'invading species', a distinguishing feature of biosecurity is that it is generally taken to be a set of human control processes. It is associated with the perceived value of species or habitats and their use by humans. The level of potential damage to the ecosystem may be minor, such as where fencing stops sheep straying into woodland and destroying saplings, or have potentially disastrous impacts, such as those relating to avian influenza.

How does biosecurity work?

Risk assessment is fundamental to biosecurity. This involves calculating the probability and likely impact of an event and balancing these against the costs of implementing measures to counteract that event or to minimize its impact. Most states now have some from of legislation against the importation of potential or proven pest species but these evolve rapidly as new hazards arise. In Australia legislation is in place in relation to 'prevention of entry, surveillance, emergency response and eradication, and containment and control'. This uses a database divided into permitted species, prohibited imports and species where the import risk analysis process is still ongoing. The establishment and development of the powers of 'Biosecurity Australia' is a typical example of national ecological hazard minimization, where regulation is constantly revised and developed in the light of new information. This includes a 'hotline' for the public to report invasive species, such as fire ants (*Solenopsis invicta*), and surveillance programmes for known highly invasive species not yet observed in Australia. In a similar manner, 'Biosecurity New Zealand' is the lead agency in New Zealand, working in a 'whole of system' leadership role including economic, environmental, social and cultural issues.

Avian influenza

In the United Kingdom, as the possibility of a 'bird flu' epidemic has increased, with the potential for mutation into a human disease, a series of government guidelines have been produced. The Great Britain Poultry Register was set up to provide centralized information in relation to commercially bred poultry. The growth of various forms of 'animal passports' also continues to increase, incorporating monitoring schemes and, as required, restrictions on animal (and animal product) movement. These schemes are also intended to bolster consumer confidence in agricultural products. As a result of the threat of avian flu, surveillance of wild birds has been added to the existing monitoring of poultry welfare by government agencies. A general recommendation is that poultry be kept under cover to minimize contact with wild birds. However, this can raise problems with small-scale and free-range production units, where existing housing may not be suitable.

When an outbreak occurs or is suspected, a range of buffer zones around potentially infected sites are set up. Once the disease is confirmed, culling of infected and other individuals is usually required. In 2007, turkey chicks being reared in certain poultry sheds, forming part of a major production and processing unit in southern England, were exterminated, while the Canadian Food Inspection Agency, in response to an outbreak of the H7N3 avian flu strain in 2004, slaughtered 17 million birds within a 70 km^2 buffer zone. Beyond the immediate zone of infection, it is usual to have a zone where poultry movement is prohibited, in turn surrounded by one where only restricted movement is permitted. In the United Kingdom, should these measures prove insufficient to curb the spread of disease, the Government has also acquired millions of doses of avian flu vaccine. Such expensive interventions are often not politically popular but, especially where the economic and human impact is clear and where the threat is restricted to certain species, it may

BOX 10.6 ➤

➤ be the only means to allow trade to continue. Doses of vaccine would also be available, from a separate stockpile, for the preventive vaccination of zoo birds, in the interests of conservation.

Community awareness and involvement

Costs and benefits of biosecurity are not always quantifiable and measures may be seen as heavy-handed by the public or organizations affected. In such situations, public education and awareness are vital elements of hazard control. A successful biosecurity campaign with high levels of community awareness and involvement is the 'didymo' campaign in New Zealand. *Didymosphenia geminata* is a non-native freshwater alga that, by attaching itself to the beds of streams, rivers and lakes, can form a thick brown mat on rocks, plants and other materials, destroying habitats. *D. geminata* was first identified during a routine survey of algal growth and has since been found in many rivers in New Zealand's South Island. Since the algae are most likely to be spread by humans moving items between waterways, people using rivers and lakes must comply with control measures or face penalties of up to five year's imprisonment or a heavy fine. These measures focus upon a regime of 'check, clean, dry', especially when moving between waterways. Posters, websites and other public information attempts to re-enforce the message.

BOX 10.6

had a Minister of Biosecurity since 1997 and New Zealand's systems for protecting its borders are acknowledged to be among the best in the world. The most obvious measures to those arriving from abroad are those taken by the Quarantine Service. Ships, aircraft and mail are searched and sanitized as a matter of routine. Stopping alien species on arrival or dealing with their impact once recognized within the New Zealand environment are reactive measures. However, more proactive measures include measures such as requiring ships travelling between Australia and New Zealand to empty their water-ballast tanks mid-voyage. This reduces the risk of importing known invasive species such as the Northern Pacific starfish.

Reflective question

➤ Think of your local environment: what conservation projects are taking place and to what extent does social ecology play a role in those projects?

10.7 Summary

This chapter has presented a range of ecological concepts and processes. While an ecosystem may appear to be static, closer observation shows that important links between plants, animals and soil are related to processes involving movement of energy, organic and inorganic materials through the system, so that ecosystems are in a constant state of change. There are a range of trophic levels which result in inefficient energy transfers between a hierarchy of species. In most situations, rather than a simple linear or circular route for processes there are more likely to be a range of options at various stages. Hence food chains are often better described as food webs with the same species appearing in a range of stages. Biogeochemical cycles may incorporate reservoirs for matter or nutrients. These reservoirs may form a vital buffer against change resulting from pulses of increased flow or the removal of a source of material.

Often, if a substance is bioavailable it can be taken up and stored for future use. Sometimes toxic substances are also bioavailable and these can accumulate and become concentrated within certain parts of an ecosystem. This may result in a range of problems for the system as a

➤

whole if the bioaccumulation results in severe toxicity to a species feeding on it. This may be very problematic if the species being adversely affected is a keystone species. These keystone species are very important to the whole ecosystem because they are so well connected to many other parts of the system. Identifying such keystone species is an important goal for environmental managers. Losing keystone species may result in ecosystem collapse and loss of biodiversity. Biodiversity is positively related to the number of habitats within an ecosystem. Some ecosystems may be more fragile than others, depending on the nature of the linkages within the ecosystem. The role of environmental niches, competition and a range of life strategies all play a role in biodiversity and the succession of an ecosystem.

Human disturbance of ecosystems through agriculture and urbanization provides important areas of research. While urbanization may seem a negative factor for ecosystems it can result in new recombinant communities. Environmental management requires consideration not only of ecosystem processes such as population dynamics and succession but also of human needs for the ecosystem. In other words, ecosystems can be considered to serve a range of functions. Ecological thought must therefore also consider socio-economic, value-laden ideas within the area of management.

Further reading

Beeby, A.H. and Brennan, A.M. (1997) *First ecology.* Chapman and Hall, London.
This is a relatively easy read with some useful boxed studies, self-assessment exercises (with answers) and a nice glossary section.

Bradbury, I.K. (1998) *The biosphere.* John Wiley & Sons, Chichester.
This book provides a good overview of many of the ecological concepts discussed in this chapter and provides detail regarding taxonomy and the functions of ecosystems.

Ganderton, P. and Coker, P. (2005) *Environmental Biogeography.* Prentice Hall, Harlow.
Strong, accessible introduction to the area.

Quammen, D. (1996) *The song of the dodo: Island biogeography in an age of extinctions.* Hutchinson, London.
An excellent example of narrative non-fiction is provided by Quammen. It works better for the fast reader since it is a chunky volume, but you might want to dip into it for fascinating accounts of interviews with a range of well-known scientists.

van der Maarel, E. (1997) *Biodiversity: From Babel to biosphere management.* Opulus Press, Uppsala.
This is, essentially, the inaugural lecture at Groningen of one of Europe's most distinguished vegetation scientists and is written in an entertaining, accessible and at times intentionally provocative style.

Web resources

Best Foot Forward Home Page
http://www.bestfootforward.com/
This is a site on urban sustainability which contains links to a range of related sites.

Community Ecology
http://www.bio.bris.ac.uk/research/community/index.html
The University of Bristol site on community ecology contains useful information on food web ecology and invasion ecology, including case study examples.

Conservation International Home Page
http://www.conservation.org/English/Pages/Default.aspx?nointro=true
Directing through the site allows information to be obtained on the threats to biodiversity of certain ecosystems, conservation projects currently under way and links to a range of related sites.

Ecology and Society
http://www.ecologyandsociety.org/
At this site there is an online peer-reviewed journal on ecological conservation and sustainability that provides an easy-to-use format for finding related articles. A collection of articles with responses to them (by readers of the academic journal

Conservation Ecology) on ecosystem integrity and keystone species can be found on this site at:

http://www.ecologyandsociety.org/news/

Extension Toxicology Network: Bioaccumulation

http://extoxnet.orst.edu/tibs/bioaccum.htm

This is a detailed text site on bioaccumulation describing what it is, the processes of bioaccumulation (uptake, storage and elimination) and the factors affecting bioaccumulation.

Franklin Institute Online: Ecosystems, Biomes and Habitats

http://www.fi.edu/tfi/units/life/habitat/habitat.html

This site provides a good overview of ecosystem characteristics with links to numerous related sites.

Hawaiian Ecosystems at Risk Project

http://www.hear.org

This is the home page of the US Geological Survey 'Hawaiian Ecosystems at Risk' Project, including a discussion on the threats from new invading species, a regional climate model and lots of interesting links.

IUCN – The World Conservation Union Home Page

http://www.iucn.org/

This describes what IUCN does, conservation news and information on current conservation programmes.

The Large Scale Biosphere–Atmosphere Experiment in Amazonia (LBA)

http://lba.cptec.inpe.br/lba/index.php?lg=eng

You can see what is happening in a large-scale biosphere–atmosphere experiment in the Amazon region, researching a wide array of ecological processes on this site.

r–K selection: the development–reproduction trade-off

http://pespmc1.vub.ac.be/RKSELECT.html

This site provides a discussion on the differences between 'r' and 'K' selection.

UNEP World Conservation Monitoring Centre

http://www.unep-wcmc.org/

Here there are details on various global conservation issues and programmes. There are extensive lists of links to sites of biodiversity and conservation organizations and general ecological organizations. Information on the threats of climate change on biological communities and the possible ecological responses can be found at:

http://www.unep-wcmc.org/climate/index.html

Urban Ecology Research Laboratory

http://www.urbaneco.washington.edu/

The University of Washington Urban Ecology Laboratory home page describes the programmes undertaken and a discussion of some of the underpinning issues.

 Visit our website at **www.pearsoned.co.uk/holden** for further questions, annotated weblinks, interactive models and video-clips showing physical processes in action.

PART V

Geomorphology and hydrology

Figure V.1 Mountains and moorlands of northern Scotland with streams eroding into the hillsides.

Part contents

Scope

This part of the textbook deals with the nature of water (especially Chapter 13), sediment (especially Chapters 11, 12, 14, 16, 17, 18 and 19), solute (especially Chapter 15) and ice movements (especially Chapters 18 and 19) in a range of environments including drylands (Chapter 16), spectacular glacial landscapes (Chapter 18), periglacial environments (Chapter 19) and coastal environments (Chapter 17). It describes such motions of energy and mass on the hillslope, in the floodplain, within channels dominated by water and ice, and in coastal zones. The first five chapters of this part detail processes that can occur on any landscape on Earth whereas the remaining four chapters describe the processes and nature of particular Earth environments.

Studying the form of the Earth and its various characteristics seems an obvious facet of geography and is commonly perceived as the entirety of the work physical geographers do. However, this refers only to geomorphology. Geomorphologists are interested in trying to ascertain the processes and mechanisms that interact with, and create, the form of the landscapes we study. Often, while most processes may be operating on all land environments, the dominating processes depend on the regional climate characteristics or tectonic context. Chapter 11 highlights that it is often the relative abundance of the various processes, and the balance between the process rates, that determines the nature of the landforms in different environments.

Chapter 11 touches on issues of form–process relationships in relation to hillslope profiles, erosion, weathering and landscape evolution. However, the landform itself is not simply a product of geomorphological or hydrological processes: the processes themselves will be influenced by the landform. In addition, landscapes may incorporate a legacy from the operation of geomorphological processes under past climatic regimes such as dryland and post-glacial environments. Therefore contemporary processes and landscapes are likely to be linked to historic processes and landscapes and there may be time lags in landscape response.

On a rudimentary level, it is the Earth's attempt to create some kind of balance of inputs and outputs of energy that leads to the internal reorganization of sediment and materials, resulting in the changing shape of landscapes. It is through uplift, erosion and sedimentation that landscape forms are maintained. Erosion and sedimentation are dependent upon the physical agent exchanging the energy in the system, which is most commonly a transporting fluid (water, wind or ice), although gravity also plays a significant role. Although erosion and sedimentation are contrasting processes, they are inextricably linked and the same sediments can be reworked many times. Chapters 11 and 12 delve into issues of weathering, erosion, sediments and sedimentation to provide a basis for the following chapters. Sediments (Chapter 12) and solutes (Chapter 15) are transported from hillslopes (Chapter 11) to fluvial environments (Chapter 14), most commonly by hydrological processes (Chapter 13), which also interact with glacial sediment and solute systems (Chapter 18), finally ending at the coastal zone (Chapter 17). This input–output budgeting concept is most commonly used in catchment hydrology and fluvial geomorphology as well as within glaciology where the mass balance (the budget) of glacier ice is studied with respect to its influence on glacier movement and hydrology. Coastal studies are another classic example in which the sediment budget is used as a framework for studying dynamic system adjustments to wave, tide and current processes.

Within every chapter in this part, consideration is given to the influence of humans on landscape processes and how understanding of the processes can result in improved management strategies.

Erosion, weathering and landform evolution

Mike Kirkby
School of Geography, University of Leeds

Learning objectives

After reading this chapter you should be able to:

➤ understand the important role of hillslopes in water-eroded landscapes

➤ describe the key processes of weathering and sediment and solute transport

➤ explain how the balance and rate of processes influences the evolving form of hillslope profiles

➤ use slope profiles to interpret the processes shaping them

11.1 Introduction

Over 90% of landscapes that are currently not glaciated consist of hillslopes, and the remainder consist of river channels and their floodplains. Although hillslopes are not generally the most active part of the landscape, they provide almost all of the material which eventually leaves a river catchment through the more active channelways. The processes by which material is weathered and transported

to the streams are therefore vital to an understanding of how the catchment works as a geomorphological machine. The weathered debris on hillslopes (the **regolith**) is also the raw material from which soils are developed (see Chapter 7). Land management such as agricultural practice strongly affects the rate and types of hillslope processes. The way in which farmland is managed can dramatically influence whether soil erosion remains at an acceptable level, or is increased to a rate which leads to long-term and often irreversible degradation of the soil.

Terrestrial landscapes are dominated by erosion, and the material removed is ultimately transported to the oceans where it takes part in the continual slow recycling of the Earth's crust as tectonic plates spread apart and collide to form new mountains (Chapter 2). Geomorphological processes form an essential part of this crustal recycling which periodically renews the surface of our planet in episodes of orogeny, erosion and isostatic response. Water helps to break up rocks as part of the process of weathering, and drives sediment transport processes that carry soil materials down to the ocean, progressively eroding the land. This chapter reviews the various hillslope sediment processes, the factors which influence their rates and the ways in which the processes in an area influence the form of the hillslopes. The balance between process rates has a very strong influence on the form of both the landscape and its soils, and plays a large part in the distinctive appearance of landscapes in different climatic regions of the Earth. Weathering and erosion provide the raw material which rivers transport through their valleys to the oceans. However, changes in hill-slope erosion rate may not be matched by similar changes in river transporting capacity, resulting in either floodplain **aggradation** and widening, or valley incision.

11.2 Slope profiles

The appearance of hillslopes around the world varies considerably (Allen, 1997). Some of these differences are related to the vegetation and soils on the hillslopes, but there are also important differences in the form of hillslope profiles. Most profiles are convex near the divide (top of the slope) and concave near their base (Figure 11.1). The most significant differences between them are in their total length, gradient and convexity.

11.2.1 Slope length

Slope profiles are generally surveyed along a straight line (on a map) from the divide, following a line of steepest descent, to the nearest point at their base (Figure 11.1), which is usually in a river or floodplain, or, for coastal slope profiles, directly into the sea. The length of inland slope profiles is linked to the **drainage density** of channels (see Chapter 14), which varies, depending on climate and rock type. In temperate areas such as Britain, slope profiles are almost a kilometre in length, whereas some semi-arid areas

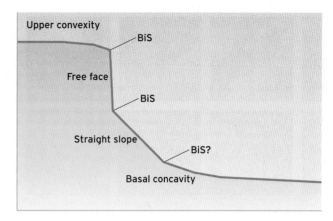

Figure 11.2 Elements of a slope profile: upper convexity, free face, straight slope and basal concavity. BiS is break in slope.

have slopes that are less than 10 m long. Drainage density generally tends to be low (1–5 km km^{-2}) in humid areas such as north-west Europe, and higher (10–500 km km^{-2}) in semi-arid areas, particularly where the soils are developed from impermeable rocks such as shales, clays or marls, on which intensely dissected **badlands** may occur.

11.2.2 Slope steepness

Slope profiles also differ in their average gradients, from steep cliffs to gentle slopes with almost imperceptible gradients. Figure 11.2 shows the four possible elements of a profile, consisting of an upper convexity, a free face, a straight slope of almost constant gradient and a basal concavity. Not all of these features are found in every slope profile, and some may be repeated more than once within a profile. For example, many profiles do not have a free face at all, but grade smoothly from convexity to concavity, as in Figure 11.1, whereas many steep profiles have a number of free faces, each with a straight slope and/or basal concavity below it.

- Upper convexity: from the divide, there is generally a convex slope of increasing gradient and with slopes from level (0°) to a maximum of up to 35° (although usually less).
- Free face: below the upper convexity there may be a cliff of free face, usually consisting of bedrock at a slope of up to 70° (although exceptionally steeper and/or locally overhanging). The free face may be somewhat stepped if there are rock layers of different resistances.
- Straight slope: there may also be a straight section of almost uniform gradient. Below a free face this usually consists of a scree or **talus** slope, at 30° to 40°, with a surface of loose stones. In some cases the straight slope is cut into bedrock, which is usually visible in patches

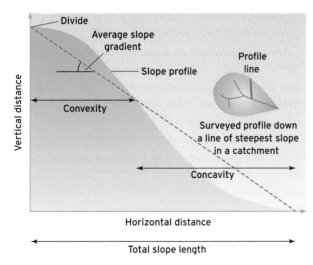

Figure 11.1 A typical hillslope profile.

beneath a thin layer of loose stones, and is called a boulder-controlled slope. Where the rock is layered, the sequence of free face and talus slope may be repeated several times. Straight slopes at 10° to 25°, in fine materials, are also commonly found in sands and clays, although not associated with a free face, but merging directly into the upper convexity.

- Basal concavity: at the foot of the slope there is usually a basal concavity which leads down towards the valley-bottom river and floodplain. This is usually in sand or finer materials.

Between these slope elements, there may be breaks in slope (BiS in Figure 11.2), within which the slope gradient changes relatively abruptly. There are commonly breaks in slope at the top of a free face, and between the free face and the straight slope below it. In arid and semi-arid areas, there is usually another break in slope between the straight slope and the basal concavity, although this is not normally found in humid areas, and is associated with the different balance of processes. Comparing hillslopes of different overall steepness, the proportion of different slope elements changes, and the free face and/or straight slope, are commonly absent on lower-gradient slopes. These differences are sketched in Figure 11.3, and this represents one of the very many possible evolutionary sequences over time. On even steeper slopes, the basal concavity and/or the straight slope may also disappear, so that the cliff plunges directly into a river as a gorge.

The slope steepness reflects the relationships between the hillslope and conditions at the slope base. Where there is a river that is cutting rapidly downwards at the slope base, then slopes are inevitably steeper than where the river is stable. The most rapid downcutting is usually associated with tectonic uplift, since the river is generally able to cut down almost as fast as the land is uplifted, while the upper part of the hill-

slopes is initially little affected. Steep slopes are also associated with coastal areas, owing to undercutting by wave action (Chapter 17). Coastlines are some of the commonest locations for good free face development. Cliffs are also common in formerly glaciated areas, where slopes have been steepened by glacial erosion, and post-glacial processes have not yet (after 10 000–15 000 years) had time to eliminate them.

Over time, and in tectonically stable areas, slope processes progressively erode the landscape to produce more and more gentle slopes. The gentlest slopes are therefore found in stable shield areas, such as West Africa, central Australia and northern Canada and Eurasia. Where rocks are more easily eroded, slopes flatten more quickly, so that, in any area, the gentlest slopes are usually found on clays and shales. However, there are exceptions which include badland areas where the dissection is severe, promoting steep gully sides.

11.2.3 Slope convexity

The amount of the hillside that is convex in profile is usually expressed as a proportion of the total slope length. Hillslopes vary from almost complete convexity (100% convex) to a narrow convexity and a much broader concavity, although most slopes are mainly convex. Where slopes appear to have long concavities, as in Figure 11.4, the profiles seen from a distance are not the lines of steepest slope, but a line or a series of lines along the divides between channels. These divides follow the concave profiles of the streams between them, whereas the true slope profiles, which run from the divides into the nearest stream, are convex, or convex and straight.

The proportion of convexity is related to both the balance between slope processes and the relationship between the slope and the river at its base. Where the river is cutting down rapidly, or undercutting the base of the hillside, then slopes are not only steeper but also tend to be more convex. Similarly, where the streams are aggrading, more of the slope profile tends to become concave.

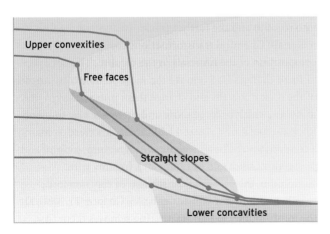

Figure 11.3 Slope elements on steep and gentle gradient hillslopes.

Reflective question

➤ Compare slope profiles you have seen or studied in different areas: how do they differ in the proportions of each slope component (Figures 11.2 and 11.3) – or are they all the same? You could look for pictures on the Internet to address this question (e.g. http://www.astronomynotes.com/nature/beauty.htm).

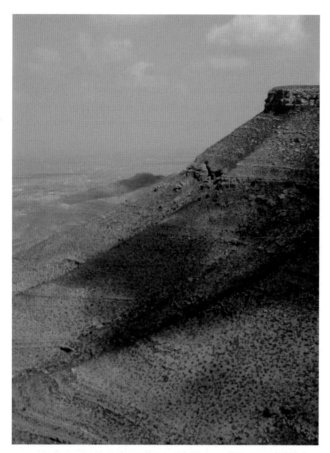

Figure 11.4 Cliffs and gullies south of Tatouin, Tunisia. The gullies are in sandstone cliffs and across the basal concavity in shales. Visually the skyline slope profile follows the divides between the concave gully thalwegs. However, true slope profiles are from the divides into the nearest channel thalwegs.

Table 11.1 Classification of the most important hillslope processes

	Weathering processes	Transport processes	Type (S/T)
Chemical	Mineral weathering	Leaching	S
		Ionic diffusion	T
Physical	Freeze-thaw	Mass movements:	
	Salt weathering	Landslides	S
	Thermal shattering	Debris avalanches	S
		Debris flows	S
		Soil creep	T
		Gelifluction	T
		Particle movements:	
		Rockfall	S
		Through-wash	T
		Rainsplash	T
		Rainflow	T
		Rillwash	T
Biological	Animal digestion	Biological mixing	T
	Root growth	(often included within soil creep)	

Types: T = transport-limited; S = supply-limited removal (see text).

11.3 Weathering

Slope processes are of two very broad types: (i) weathering and (ii) transport of the regolith. Within each of these types, there are a number of separate processes, which may be classified by their particular mechanisms into groups (Table 11.1), although many of these processes occur in combination. Most slope processes are greatly assisted by the presence of water, which helps chemical reactions, makes masses slide more easily, carries debris as it flows and supports the growth of plants and animals. For both weathering and transport, the processes can conveniently be distinguished as chemical, physical and biological.

Landscapes evolve over time in response to the internal redistribution of sediment, usually with some net removal of material to rivers or the ocean. The way in which landscapes and slopes evolve depends on their initial form, the slope processes operating and the boundary conditions which determine where and how much sediment is removed (Allen, 1997). These relationships between process and form are discussed later in this chapter. The following section deals with weathering and the subsequent section deals with transport processes.

11.3.1 Weathering processes

Weathering is the *in situ* transformation of parent materials into regolith and the further transformation of regolith materials. Chemical transformations change the chemical composition of the minerals in the regolith. The net effect of these changes is to remove the more soluble constituents of the rock minerals, and change them into a series of new minerals, which become more and more like clay minerals. Rock composition (Figure 11.5) can be described as mainly a combination of bases (calcium, magnesium, sodium and potassium), silica and **sesquioxides** (mainly with aluminium, the principal constituent of clay minerals, together with minor constituents; see Chapter 7). Bases are the most soluble, so that weathering removes them first. Silica, although at least 10 times less soluble than bases, is itself at least 10 times more soluble than the sesquioxides

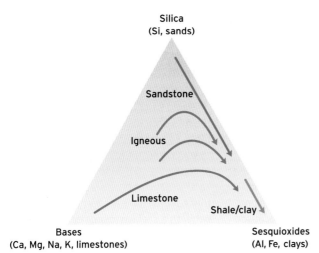

Figure 11.5 Weathering of rocks, represented as a mixture of bases, silica and sesquioxides. The arrows show the course of weathering for typical rock types. The lengths of the arrows indicate relative rates of change.

(Selby, 1993). The changes due to chemical weathering can be shown on a triangular diagram (Figure 11.5), which shows the proportional composition in terms of these three components. Weathering first reduces the proportion of bases, and then the proportion of silica, so that all parent materials eventually end up in the sesquioxide corner, essentially as clays. These changes are inevitable, since most rock minerals are created at great depths and high temperatures. When they are brought close to the surface, they evolve towards an equilibrium with the much lower pressures and temperatures in the regolith, although this may take a very long time. In general, minerals that are formed close to the surface, such as the clays, are less liable to change in this way, and so undergo less chemical weathering than, for example, igneous rock minerals, which were formed at high temperatures and pressures. Chemical weathering usually produces new minerals, which are physically weaker than before, and so makes the material more easily transported (Yatsu, 1988).

Coming towards a new equilibrium with near-surface conditions, minerals come into contact with the atmosphere, so that many of the chemical changes are in the direction of oxidation (adding oxygen), carbonation (adding carbon dioxide), hydration and hydrolysis (adding water with or without CO_2). Oxidation is a process akin to rusting, and is most visible in the soil through oxidation of iron compounds, giving reddish coloration to the soil as iron is weathered from reduced (greenish blue) to oxidized (reddish) forms. Carbonation is the process involved in

solution of limestones, in which:

$$\begin{aligned} \text{Limestone} + \text{carbonic acid} \\ \rightarrow \text{ soluble calcium}(\text{Ca}^+)\text{ and} \\ \text{bicarbonate }(\text{HCO}_3^-)\text{ ions} \end{aligned} \quad (11.1)$$

Hydration is the conversion of mineral crystals to forms with higher water (H^+) and OH contents. Hydrolysis commonly involves both water and CO_2, in reactions of the form:

$$\begin{aligned} \text{Feldspar} + \text{carbonic acid} + \text{water} \\ \rightleftharpoons \text{clay mineral} + \text{silicic acid in solution} \\ + \text{potassium and bicarbonate ions in solution} \end{aligned} \quad (11.2)$$

In all of these normal weathering reactions, the more soluble bases are being removed in solution as ions, while the new minerals tend to be higher in aluminium and silicate, the key constituents of clay minerals.

In addition to these relatively simple chemical reactions, the presence of complex humic acids in soil organic matter provides additional potential for weathering which can increase the acidity (reduced pH) of the soil and create soluble complexes that are able to mobilize even the normally insoluble aluminium. In base-poor environments, for example in sandy soils, this additional acidity can leach aluminium and iron from the surface soil, redepositing them as hardpans at depths of 30–100 cm, where the soil pH rises due to reduced concentrations of soil organic matter.

As the initial chemical weathering of bedrock goes on, most silicate and sesquioxide minerals show little change in volume, but lose some of their mass, which is transported away in solution – one of the most important processes of mass removal. However, this loss of material is not observed as lowering of the surface until the weathered material is removed by the physical processes of erosion. Limestone, which is rich in bases, forms a major exception to the general rule that there is little change in bulk and, for the purest limestones, there can be a loss of over 90% of its original volume as it dissolves, leaving an insoluble clay, rich in iron and aluminium. This can often produce dramatic landforms and caves known as karst landforms. Details of karst landforms are discussed in Box 11.1.

Physical weathering transforms rock and regolith materials by mechanically breaking them into smaller fragments. In some cases chemical and biological processes help this mechanical breakdown. The most important processes of physical weathering are freeze–thaw, salt weathering and thermal cracking. One of the most powerful and widespread physical weathering processes is freeze–thaw action, by which water freezing in small cracks expands as it freezes by 10% and is able to open the cracks sufficiently to split the

CASE STUDIES

KARST LANDFORMS

Where rocks are highly soluble, so that most of their mass can be removed in solution, a set of distinctive, or 'karst', landforms may be formed. Karst occurs most commonly in limestone, but may also occur in other soluble rocks, such as gypsum. Some comparable forms can also be generated through melting of ice to form 'thermokarst'. Where the rocks are strong, water flow and solution are concentrated along joints and other porous areas, enlarging them in a complex network of passages, many of them underground, although generally with some connections with the surface. These passages may evolve over very long periods, and some become enlarged into caves, some of which can be hundreds of metres in extent and form networks over many kilometres.

Joint passages evolve most rapidly where there is most water flow, so that they tend to be formed mainly above the water table, although this can alter over time through climate or base-level change. This also means that caves frequently develop along the lines of stream valleys, because the streams have the largest and most continuous water flow. Streams, in this way, can progressively bring about their own demise, as more and more of the water sinks into the bed and follows underground passages, which may only emerge at the boundaries of the soluble rock mass.

Other factors that increase the rate of solution are the presence of soil organic matter and cold temperatures, because CO_2 is more soluble in the cold. The combination of these factors with the flow means that karst development is most rapid in humid tropical areas, and slowest in hot deserts.

Rapid karst development lowers the hills until they intersect the water table, so that they eventually form towers rising out of an almost level plain (Figure 11.6). In temperate areas, karst usually develops more slowly, so that cave systems are formed beneath a surface topography of hills and valleys, as occurs on the Carboniferous limestone in New Zealand and other temperate areas (Figure 11.7). Where the landscape has been glaciated, as in northern Britain, the rock is, in many places, initially smoothed by glacial abrasion. The action of solution and joint enlargement only gradually destroys the initial smooth surface, leaving a very characteristic surface of clints and grikes (Figure 11.8). In unglaciated landscapes, no visible trace of the original surface is left, and a landscape of pinnacles, bounded by joint planes, is equally characteristic.

Figure 11.6 Towering karst pinnacles above flat lying paddy fields south of Guilin, Yangshuo, Guangxi, China. (Source: Tony Waltham/Robert Harding World Imagery)

BOX 11.1 ➤

Figure 11.7 Limestone sink holes overlying a cave system west of Timaru, New Zealand. (Source: G.R. Dick Roberts/Natural Science Photos)

(a)

(b)

Figure 11.8 Clints (higher rock) and grikes (gaps in between) formed in limestone in Wharfedale, northern England; (b) shows a close up of one of the grikes which contain a habitat for plants and animals.

BOX 11.1

rock. Salts released by chemical weathering and brought towards the surface may also be recrystallized in small cracks under desert conditions. The formation of these crystals can also split and flake rock and regolith materials in the process of salt weathering (see Chapter 16). During fires, started either naturally by lightning strikes or by human action, the temperature range (>300°C) is so great that the differential expansion of the rock is sufficient to crack and shatter it. In combination with weakening by chemical weathering, the much smaller diurnal range of temperatures in desert areas (~50°C) may also be sufficient (see Chapter 16) but this process is generally less important than frost weathering in cold areas, and salt weathering in deserts.

Biological processes are effective through a combination of biochemical and mechanical methods. The main types of weathering are by root action and faunal digestion. Large and small roots put organic acids into the soil which help them to extract the nutrients needed by the plant, and in doing so, they help chemically to weather the soil. As roots grow they also expand, opening small cracks until the chemically weakened rock is physically broken apart. Many of the small animals in the soil pass soil material through their bodies, altering it both biochemically and mechanically as they extract nutrients from it. In temperate climates, earthworms are generally the most important fauna in this process.

> ### Reflective questions
>
> ➤ What are the main weathering processes?
>
> ➤ How would you expect the main weathering processes to vary with climate?
>
> ➤ How active are weathering profiles in different soil horizons?

11.4 Transport processes

Where there is a plentiful supply of material, and the process which moves it can only move a limited amount for a short distance, the rate of transport is limited by the **transporting capacity** of the process, which is defined as the maximum amount of material which the process can carry (Kirkby, 1971). A transport process like this, such as **rainsplash** (see below), is described as **transport limited**. Some other processes are limited, not by their capacity to transport, but by the supply of suitable material to transport, and are described as **supply limited**. For example,

rockfall (see below) from a cliff has a very large potential capacity to carry material, but is limited, fortunately, by a shortage in supply of freshly weathered material.

There is not always a clear distinction between transport- and supply-limited processes, but it is an important distinction which has a substantial impact on the way in which hillslopes evolve over geological time periods. Landscapes that are dominated by transport-limited removal are generally covered by a good layer of soil and vegetation, and slope gradients tend to decline through time. Landscapes where removal of material is mainly supply limited, however, tend to have sparse vegetation, thin soils and steep slopes which tend to remain steep throughout most of their evolution (Carson and Kirkby, 1972).

11.4.1 Chemical transport processes (solution)

The process of solution is closely linked to chemical weathering, as rock and water interact. The chemical reactions that are altering rock material in place are, at the same time, releasing the lost material dissolved in the water. Rocks that weather rapidly therefore lose material in solution rapidly too. In Figure 11.9, rain falls on the soil, where it picks up solutes from the regolith, in proportion to the concentration of each constituent in the regolith, and its solubility. Some water is lost by percolation, containing solutes carried, for example, into limestone cave systems. Some water is lost to evapotranspiration, and this carries little or no solutes, so that the remaining overland flow and subsurface flow runoff have an increased concentration of solutes. This concentration effect is only marked in relatively arid climates, where the evapotranspiration is high.

In extreme cases, some of the soluble material reaches its maximum saturated concentration, and any further concentration leads to redeposition of the dissolved material near the surface. This occurs most commonly for calcium, which

Figure 11.9 Cycling of solutes in soil water and runoff.

is often found to form crusts of calcrete near the surface in arid and semi-arid areas. The concentration of solutes is therefore generally highest in dry climates, but the total amounts removed in solution are much less than in more humid areas. Where a flow of water contains dissolved material, the rate at which the solutes are carried away or advected is determined by their concentration in the runoff water. This is called **advective solution** or leaching, and is very effective at removing solutes from regolith near to the surface, both in runoff and in percolating waters. Once material is leached out, it generally travels far downstream, and its rate is supply limited. Leaching is not very effective, however, in removing material from the bedrock–regolith boundary because little water usually flows across this boundary. However, there is a rapid change in solute concentration at this boundary as water remains in contact with material of different composition (see Chapter 15). Close to the bedrock, there is a high concentration of solute ions in the water. Further above the bedrock, there is a lower concentration in the slightly more weathered regolith.

This difference in concentration results in a net upward movement of ions, which means that there will be a movement of solutes away from the bedrock towards the regolith, even though no water is moving. This is because of **ionic diffusion**. Ions move about randomly over short distances and Figure 11.10 compares the number of downward movements from the upper area of low concentration and the number of upward movements from the lower area of high concentration. Even though the movements are random, Figure 11.10 illustrates how random movements in all

directions cause a net diffusion of material from areas of higher ion concentration to areas of lower concentration. Around the regolith–bedrock interface the concentration of ions is higher near the unweathered bedrock than in the partially weathered regolith, driving a net upward movement of ions, carrying solutes away from the bedrock, even though no water is moving. In this case the solute load depends not on the flow of water but on the differences in concentration between the layers. This movement of solutes by ionic diffusion is not as fast as by leaching where there is appreciable water moving, but can be very important in the early stages of rock weathering, when little water is able to flow through the almost intact rock (Yatsu, 1988). Because material moves only a short distance by ionic diffusion, and is not limited by the supply of suitable material, solution by ionic diffusion is a transport-limited process.

Leaching is generally the most important process in carrying solutes down the slope and into the rivers (see Chapters 7 and 15). Both leaching and ionic diffusion, however, play an important role in moving solutes vertically. One particularly significant role of vertical leaching is in carrying plant nutrients down to the roots from decaying leaf litter deposited on the surface, and so completing the nutrient cycle of the vegetation (see Chapter 10).

11.4.2 Physical transport processes

When material is physically transported down a hillslope, it may travel as a mass or as independent particles. In a **mass movement**, a block of rock or soil moves as a single unit, although there may be some relative movement within the block. The movement of the block is mainly determined by the forces on the block as a whole, and the individual rock or soil fragments within the block are in close contact, so that they are moved together, almost irrespective of the properties of the individual constituent grains. The alternative to a mass movement is a **particle movement**, in which grains move one, or a few, at a time, and do not significantly interact with one another as they move. For a particle movement, forces act on each particle separately, and they move selectively, mainly depending on their sizes, but also on other factors such as shape and density (Selby, 1993). Some processes can behave in either way, according to the size of an individual event. For example, in small rockfalls there is little interaction between the few blocks coming down the cliff face, but larger blocks may break up into fragments which interact as they fall, giving them some of the characteristics of a mass movement.

Both mass movements and particle movements can occur at a range of rates. In general, however, movements

Figure 11.10 Net transport resulting from random diffusion. Because the concentration is greater nearer the bottom of this figure than at the top then there will be more upward movements than downward movements and so the net transport will be upwards.

driven by large flows of water tend to be more rapid than drier movements. The more rapid movements also tend to carry material farther, and so tend to be supply limited, whereas slower movements tend to be transport limited.

11.4.2.1 Force and resistance

Movement of material is decided by a balance of forces, some of which promote movement and some of which resist movement. For mass movements, the forces act on the block of material which is about to move, and for particle movement on each individual particle. The main forces promoting movement are those of gravity and water detachment. On a slope, there is always a component of the weight of the material that tends to pull it downslope, and this applies equally to particle and mass movements. Flowing water can detach fragments of rock or soil if it passes over them rapidly. It can do this in three ways: (i) flow of water over the surface picks up material from the surface, (ii) detachment by raindrop impact occurring at up to 10 m s^{-1} which splashes back at an almost equal speed, and

can detach soil grains up to 10 mm in diameter; and (iii) in a deposit which has enough fine-grained material in it, water can also permeate the entire deposit, and convert it into a mixture of water and sediment which moves as a thick slurry. This process is called **fluidization**, and is able to carry large masses of material in **debris flows** (see below). Friction and **cohesion** provide resistance to movement. Box 11.2 provides details of such resistance processes and the balance of forces operating on hillslopes. If you are not so mathematically minded then you can just read on without delving into the box. However, if you wish to understand more about forces and resistance then please do read Box 11.2. In simple terms, material begins to move when the forces promoting its movement become larger than the resistances holding it back. The ratio of these forces is known as the **safety factor**:

Safety factor (SF)

$$= \frac{\text{sum of forces resisting movement}}{\text{sum of forces promoting movement}} \quad (11.3)$$

FUNDAMENTAL PRINCIPLES

RESISTANCE AND THE BALANCE OF FORCES

Resistance to movement is mainly due to friction and, to a lesser extent, cohesion. When a particle, or a block of material, rests on another, a component of its weight (together with the weight of any other material on top of it) provides the 'into-slope' force shown in Figure 11.11. For given materials in contact, the frictional resistance that can be exerted is a fixed proportion of the into-slope force. This is called the **coefficient of friction**, μ. Another way to express the coefficient of friction is as the angle of friction, ϕ. For a dry slope, the angle of friction is related to the coefficient of friction by the equation:

$$\mu = \tan \phi \quad (11.4)$$

In Figure 11.11, where there is no water entrainment ($E = 0$), the downslope

component of the weight ($mg \sin \beta$) exactly balances the frictional resistance ($F = N \tan \phi = mg \cos \beta \tan \phi$) when the slope angle, β, reaches the angle of friction, ϕ. The angle of

friction is therefore very easy to measure experimentally (Figure 11.12), and, for coarse material, is approximately equal to the angle of repose found in natural scree slopes, of 30–35°.

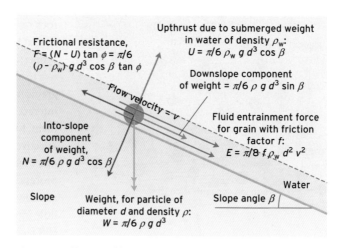

Figure 11.11 Forces acting on a particle resting on a slope in overland flow.

BOX 11.2 ➤

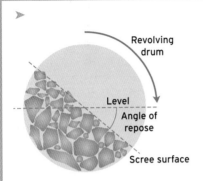

Figure 11.12 A transparent drum is slowly turned so that the angle of repose of a scree material can be measured.

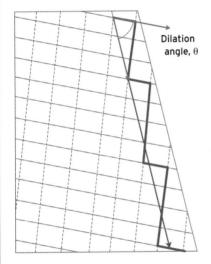

Figure 11.13 The stable slope angle for a cliff with interlocking joint blocks and/or bedding planes. Blocks are only released from the slope when they are able to slide out at an angle to the slope surface, the 'dilation angle' shown, which may be 40° or more.

In many kinds of partly weathered bedrock, the material consists of roughly rectangular blocks, separated by joints or bedding planes. The possibility of sliding parallel to the surface along the zig-zag line indicated in Figure 11.13 is very much hindered by the interlocking of the two surfaces, and can begin only when the surfaces

are lifted apart, or dilated in the direction shown by the arrow, at an angle θ to the slide direction, so that free sliding only takes place along the surface after the sliding mass has lifted clear of the main cliff. Under these conditions, which are typical of most cliffs, the effective angle of friction, Φ, is increased by the dilation angle, so that:

$$\Phi = \phi + \theta \qquad (11.5)$$

It is for this reason that cliffs are commonly able to stand at angles of 75° or more, made up of an angle of friction of 35° plus a dilation angle of 40° or more.

The frictional resistance to sliding of a block of material is very strongly affected by the water within the material. If the material is saturated with water, then part of its weight is carried by the water, following **Archimedes' principle** that the upthrust is equal to the weight of water displaced. On a hillslope, the surface of the flowing water is parallel to the surface, and so this upthrust, or weight relief, affects only the into-slope component of the weight and not the downslope component. The effect of saturating the material is therefore to reduce the frictional resistance, which is proportional to the into-slope force. This is a very important factor, and is able to reduce the frictional force to roughly half its value under dry conditions. This means that a slope that becomes saturated from time to time can be stable only at slopes of about half of the angle of friction.

In some materials, especially unweathered rocks and some clays, there is still some resistance to movement even when there is little or no overburden weight. This residual

resistance is called the cohesion of the material, and the total cohesive force is equal to the cohesion value for the material multiplied by the area of effective contact. Cohesion is thought to develop in materials that have been consolidated at depth. Hence clays formed close to the surface, such as some tills, have little cohesion, while older clays, particularly those consolidated over geological time periods, have substantial cohesion. When these compressed clays are brought up to the surface, however, weathering along fissures in the clay gradually reduces the cohesion, so that, over a period of 50–100 years, the cohesion becomes very small again.

Material strength is made up of friction and cohesion. Both of these are usually expressed as a **stress**, or force per unit area, and measured in megapascals (equal to millions of newtons per square metre) (see also Box 18.3 in Chapter 18). The total strength, s, exerted to prevent sliding is expressed as a stress, or force per unit area:

$$s = c + \sigma \tan \phi \qquad (11.6)$$

where c is the cohesion, σ is the normal force (N) per unit area, and ϕ is the angle of friction. The angle of friction and the cohesion are essentially properties of the material. Some typical values are summarized in Table 11.2. The general pattern in this table is that materials with a lot of clay minerals (clay, till and shale) have lower angles of friction than others, and heavily consolidated materials (limestone, granite, sandstone and shale) have higher cohesion than unconsolidated materials which have been formed closer to the surface (sands, gravels, clay, till and chalk).

BOX 11.2 ➤

305

Table 11.2 Typical values for soil strength parameters in soil and rock materials

	Sand/gravel	Clay	Glacial till	Chalk	Limestone	Granite	Sandstone	Shale
Cohesion (MPa–million N m^{-2})	0	0-10	0-10	10-20	100-400	100-400	10-30	10-40
Angle of friction (degrees)	30-40	5-20	15-35	30-35	30-35	30-35	30-35	10-25

BOX 11.2

If SF > 1 then movement will not occur; if SF ≤ 1 then movement will begin. As soon as movement begins, the resisting forces on the material usually decrease, as the moving material detaches itself from the bonds which originally held it in place. The moving material therefore accelerates at first. Material slows down again only when the promoting forces also decrease. This usually happens where the material comes down to lower gradients where the downslope component of the gravity force becomes less, or where the water flow carrying the material spreads out and moves more slowly, or seeps out laterally or into the ground. The critical value of the safety factor is also influenced by the effect of turbulence in the flow, so that the critical value for entrainment is not 1.0, but lies between 10 and 20 in experimental practice.

11.4.2.2 Rapid mass movements

There are many names for different types of rapid and slow mass movements. In rapid mass movements, the crucial distinction is between **slides**, in which the moving mass essentially moves as a block, and **flows**, in which different parts of the mass move over each other with differential movement or **shear**. Figure 11.14 shows the difference between velocity profiles for a slide and flow. It is usually found that flows occur in masses with more water mixed into the moving mass, in proportion to the amount of regolith or rock material. In a slide, water is often very important in reducing the frictional resistance and allowing movement to begin, but there is little water within the moving mass. In a flow, there is usually almost at least as much water as solids, and sometimes many times more. Water and regolith materials can be mixed together in almost any proportions if they are moving fast enough, although coarse materials (sand, gravel and boulders) can

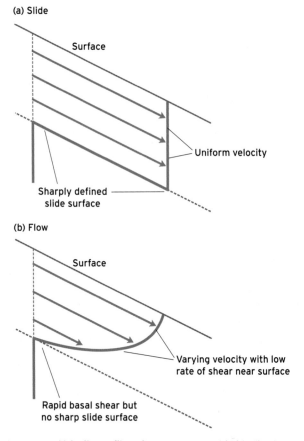

Figure 11.14 Velocity profiles of mass movement in idealized slides and flows. Shear is the differential movement between layers. In a slide, shear is concentrated at the slide surface. In a flow, shear is spread throughout the moving mass.

only remain suspended in the mixture in the fastest and largest flows. Table 11.3 shows a classification of mass movements based on their water content, and the resulting type of flow.

Table 11.3 Rapid mass movements classified by water content

Water content		Density (kg m^{-3})	Types of flow	Sediment forms
More solids than water	↑	2600	Slides	Back-scar and toe
		1900	Debris flows	Thixotropic forms
		1700	Debris avalanche	Marginal levees and lobes
More water then solids	↓	1000	Fluvial sediment transport	Mid-channel bars

In a slide, the form of the original block can usually still be seen, particularly at the upslope end, where the **back-scar** is usually clear. The slide mass may show multiple scars and cracks where it has moved. The downstream end, or toe, shows much more severe deformation producing a hummocky topography where the mass has advanced over the previous surface. Slides may be more or less planar when there are lines of weakness which follow geological structures or are parallel to the ground surface. Cliffs with lines of weakness parallel to the face often fail in this way, creating slab failures in which a flake of rock collapses completely or partially (Figure 11.15a). Sometimes the flake only partly separates and begins to lean progressively outwards until it fails by toppling. Many low-angle (10–20°) clay slopes also fail in planar slides, along surfaces near the base of the weathered regolith (Figure 11.15b). Planar slides are long (in the downslope direction) relative to their depth (measured into the slope), with length : depth ratios of 10 : 1 to 20 : 1 (Skempton and DeLory, 1957).

Slides also occur deep within the mass of a slope. In tills or consolidated clays, these rotational slides (Figure 11.15c) move on surfaces 5–10 m deep, but in strong rock, the slides may be at depths of 50–250 m, in proportion to the much greater cohesive strength of the rock compared to the clay (Table 11.3). Their length : depth ratios are typically 3 : 1 to 6 : 1. The largest slides, in rock, may therefore move entire mountain sides, and may be very destructive, such as the Hope and Turtle Mountain slides in the Canadian Rockies and the Saidmarreh slide in south-west Iran. Where the slope material moves in a wetter mixture, the mass movement becomes more like a flow and **debris avalanches** may occur. These are discussed in Box 11.3.

Where there are sufficient fine-grained materials (silt and clay) in the mixture, then a true debris flow can occur, with a still higher ratio of solids to water. Drainage at the edges of the flow is less evident, and grains are supported because they sink only slowly in the mixture, which is both dense and viscous (sticky), so that the flow is much less

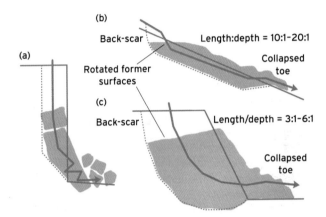

Figure 11.15 Types of rapid mass movement: (a) slab failure on a steep gradient; (b) low-gradient planar slide, length : depth = 10 : 1 to 20 : 1; (c) steep rotational failure, length : depth = 3 : 1 to 6 : 1. Red lines show general path of moving mass.

turbulent and even large rocks sink very slowly. The flow may still move at dangerously high speeds, but also tends to stop suddenly, as drainage lowers the water content to a critical level, and the whole mass suddenly sets into a rigid mass. As water drains out, the viscosity rises sharply as grains are no longer separated by flowing water, but collide with one another and coagulate. This behaviour, in which the viscosity decreases as the rate of shear (relative movement) increases, is called **thixotropic**. This creates more and more rapidly moving masses once the movement has been triggered, and very rapid solidification as the movement slows down, through drainage and/or on lower gradients. Debris flows can be triggered by landslides during intense storms, particularly in semi-arid areas where there is plenty of loose material and in steep mountain areas worldwide. After volcanic eruptions, freshly deposited ash commonly provides a plentiful supply of loose, low-density material, which is particularly prone to debris flows, known as **lahars**. Flows may also be triggered by subsurface seepage of water into the bottom of an unstable human-made deposit, as occurred in the 1966 Aberfan slide in South Wales, where springs saturated a low-density mine

HAZARDS

DEBRIS AVALANCHES

In a pure flow, such as a river, debris is lifted into the flow by turbulent eddies and settles to the bed again under its own submerged weight. Material is deposited as a series of within-channel bars. As the concentration of solids is increased, many of the moving grains strike other grains before reaching the bottom, and are supported on a bouncing layer of grain-to-grain collisions. The lift provided by these collisions, called the **dispersive grain stress**, becomes more important than turbulence in a **hyper-concentrated flow**, and is maintained by the power of the flow. Along the edges of the flow, the water drains out sideways, leaving a levee of material, while the centre of the flow continues downslope. When the flow eventually comes to a stop on gentler slopes, the levees are joined by a loop of debris around the front edge of the flow. This kind of flow is relatively common on steep mountain slopes, and is generally known as a debris avalanche. It can be a major hazard to walkers and livestock. In many formerly glaciated valleys, debris avalanches come down the steep side slopes, and their lobes spread over the gentler foot slopes, each flow following the lowest path available over the lobes of previous flows (Figure 11.16a). Creation of the mixture which behaves as a hyper-concentrated flow is thought to occur through small slides on gully sides which fall into a rapidly flowing stream of water down the centre of the gully. These flows can occur at a wide range of scales some of which can damage buildings and roads whereas others are on a small scale as shown in Figure 11.16(b).

(a)

(b)

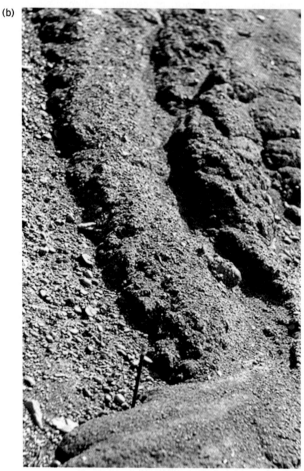

Figure 11.16 Form of central track and outline of levee around debris avalanches: (a) Franconia Notch, New Hampshire – track at left, levees at right; (b) miniature flow (pencil for scale) – terminal lobe in foreground.

BOX 11.3

spoil heap and coal waste flowed into a school and houses, killing 144 people of whom 116 were children.

11.4.2.3 Slow mass movements

The essential characteristic of slow mass movements is that they do not involve movement bounded by a discrete slip surface. Failures occur between individual soil aggregates, and not over the whole of an area. Movements are usually driven either by 'heaves' of expansion or contraction, or by apparently haphazard movements between aggregates. Heaves are usually caused by freezing and thawing of soil water, or by wetting and drying of the soil. Haphazard or apparently random movements are usually caused by biological activity, which mixes the soil in all directions. In all cases, these movements do not cause any net movement when they occur on level ground but on a slope, the steady action of gravity causes more downhill than uphill movement, and there is a gradual transport of regolith material, at a rate that increases with gradient (Figure 11.17). One particular form of slow mass movement, which is intermediate between slow and fast behaviour and is driven by freeze–thaw activity, is **gelifluction**, which occurs in periglacial areas (see Chapter 19).

Wetting or freezing of the regolith has been shown experimentally to produce an expansion that is almost perpendicular to the soil surface. As the soil then dries (after wetting) or thaws (after freezing), it sinks back closer to the vertical, under the influence of gravity on the more open texture of the expanded soil (Figure 11.17). The rate of movement is therefore thought to be roughly proportional to the slope gradient. At different times of year, the expansion penetrates to different depths, so that, totalling over the year, the lateral movement of material is greatest close to the surface, and dies away gradually with depth.

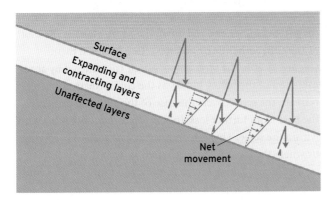

Figure 11.17 Soil creep due to expansion and contraction in a sloping regolith. The net movement is downslope.

Biological organisms move material more or less randomly in all directions but gravity again provides a bias which leads to net downhill movement on a slope at a rate roughly proportional to gradient.

The three main drivers for soil creep, wetting–drying, freeze–thaw and biological mixing, can all be of similar magnitudes, although one or other dominates in any particular site (Selby, 1993). For example, in temperate deciduous forests, biological mixing may dominate, whereas in upland areas, freeze–thaw is probably the most important. Most measurements suggest that rates of near-surface movement by soil creep are typically 1–5 mm per year, dying away to nothing at depths of 300 mm. The total sediment transport capacity is usually estimated, from these measurements, as about:

$$C = 10 \times \text{(tangent slope gradient)} \, \text{cm}^3 \, \text{cm}^{-1} \, \text{yr}^{-1} \quad (11.7)$$

Because soil creep moves the regolith material, and only operates when there is an ample soil cover, the process is considered to be transport limited, so that transport is always at the transporting capacity.

An important anthropogenic process that behaves like an accelerated soil creep is **tillage erosion**, which is the result of ploughing, either up- and downslope or along the contour. Each time the soil is turned over, there is a substantial movement of soil. Up- and downhill ploughing produces a direct downhill component of movement as the turned soil settles back. Contour ploughing can move material either up or down, according to the direction in which the plough turns the soil. Contour ploughing in which the soil is turned downhill moves approximately 1000 times as much material as soil creep. Contour ploughing in both directions (soil turned uphill and then downhill or vice versa) or ploughing up- or downhill produces a smaller net movement, but the overall rate is still about 100 times greater than natural soil creep. Sediment transport is more rapid using modern heavy machinery than with primitive ploughs, but it is clear that tillage erosion may have been responsible for more soil movement in some areas during the past few centuries than natural soil creep during the whole of the last 10 000 years. The accumulated effect is often apparent from the build-up of soil behind old field boundaries and from the infilling of hollows within arable fields.

11.4.2.4 Continuous creep

For several days or weeks, both before and after landslides (Figure 11.18), the soil commonly moves slowly along the slide surface in a process which is intermediate between a landslide and soil creep. The rates may be characteristic of

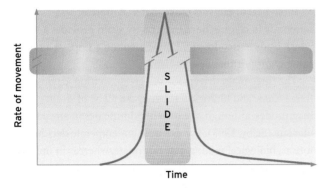

Figure 11.18 Continuous soil creep before and after a slide.

soil creep, but the movement is concentrated along a well-defined surface. In this type of movement, the main movement is a downslope glide rather than an expansion and contraction, but some sites show a combination of normal soil creep and continuous creep, so that many combinations of heave and glide are possible.

11.4.3 Biological mixing

Plant roots and burrowing by creatures of all sizes produce an overall biological mixing. Because there is more material in dense soil than in uncompacted soil, there is a net diffusion of material from denser to looser soil (Figure 11.19). Small organisms such as bacteria are very abundant, but generally move little material, as they are much smaller than the soil aggregates. Larger organisms such as rabbits may move large quantities of material, but are much less abundant than earthworms. The greatest total effect is generally

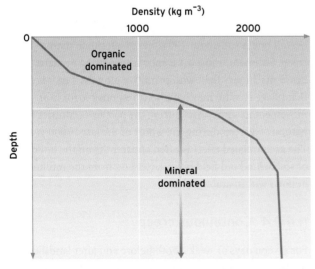

Figure 11.19 A bulk density profile, showing the combined effects of increasing pore space and increasing organic content towards the surface.

due to creatures of moderate size, such as earthworms, termites and ants, which are just large enough to move the aggregates. The rates of mixing are much lower when the soil environment is not suitable for them due to waterlogging, cold, lack of nutrients and/or lack of air (usually at depth). However, in some areas, larger rodents, such as gophers and mountain beavers, have the greatest impact.

The regolith is normally loosely packed near the surface, and denser at depth, and this **bulk density** profile is a result of two processes, both strongly driven by biological mixing: (i) the balance between a net upward diffusion of mineral material by biological mixing, and its settlement under the action of gravity, at a rate which is greatest where the bulk density is least; (ii) the balance between the net downward mixing of low-density organic litter and its decomposition to carbon dioxide. The combination of these two processes gives a bulk density which is very low at the surface, and gradually increases with depth to a constant value at 0.2–2.0 m according to conditions, where the soil is deep enough (Figure 11.19). Within this layer of biological mixing, the mechanical mixing processes are generally much more rapid than any chemical processes, so that the mineral soil shows only very small chemical differences. If weathering goes to depths beyond the reach of biological mixing, there is a clear distinction between the homogenized mixed upper layers and an undisturbed **saprolite** in which the detailed rock structures are preserved intact within the weathered regolith.

11.4.4 Particle movements

11.4.4.1 Rockfall and screes

Although cliffs may lose material in large slab failures, they more commonly lose smaller blocks, which fall as they are released from the cliff face by weathering along the joints and/or bedding planes around them (Terzaghi, 1962). These blocks often break into smaller fragments on impact, and the pieces bounce and roll down the **scree slope** at the foot of the cliff, with little or no interaction between them (Figure 11.20). The scree slope is itself constructed by accumulation of the falling blocks as the cliff retreats. As it retreats, the scree covers the base of the cliff, and protects the base from further loss. In this way a rock core is built up within the scree. This can occasionally be seen in road-cuts, or where the loose scree is quarried away.

The blocks falling onto a scree have a range of sizes, and the scree slope acts as a dynamic sieve which sorts the stones as they bounce, roll and slide down its length. Each time a block makes contact with the scree surface, it may come to a

11.6 Summary

Hillslope processes transform and transport parent materials to river channels which eventually deliver almost all sediment and solutes to the sea. Weathering consists of largely *in situ* transformation to regolith; and erosion consists of the net removal of material. Solution dissolves rock and soil materials, progressively leaving behind the less soluble minerals in the regolith. Physical and biological expansion processes break the regolith into finer grains which can be transported more easily. Mechanical processes, usually aided by water, transport material downhill, in mass movements and surface wash. Although the different processes are all driven by water and gradient, they produce distinctive small-scale features in the landscape. Because processes depend differently on flow and gradient, they also create different and distinctive hillslope forms, which can be analysed through models. Both small- and large-scale forms can be used to infer the processes acting in the landscape.

Further reading

Allen, P.A. (1997) *Earth surface processes*. Blackwell Science, Oxford.
This is probably the best up-to-date book on hillslope processes. It is moderately demanding, and written from an earth science perspective.

Anderson, M.G. (ed.) (1988) *Modelling geomorphological systems*. John Wiley & Sons, Chichester.
This is a good survey of different approaches to modelling in landscape systems.

Carson, M.A. and Kirkby, M.J. (1972) *Hillslope form and process*. Cambridge University Press, Cambridge.
This is a classic covering the topics of this chapter in more detail. However, it is now out of print and so available only in libraries.

Kirkby, M.J., Naden, P.S., Burt, T.P. and Butcher, D.P (1993) *Computer simulation in physical geography*, 2nd edition. John Wiley & Sons, Chichester.
This is a guide to programming some simple models in geomorphology and hydrology.

Middleton, G.V. and Wilcock, P.R. (1994) *Mechanics in the Earth and environmental sciences*. Cambridge University Press, Cambridge.
This book is a state-of-the-art guide to relevant continuum mechanics for the mathematically able and committed student.

Selby, M.J. (1993) *Hillslope materials and processes*, 2nd edition. Oxford University Press, Oxford.
An excellent introductory text.

Web resources

Best Management Practices for Erosion
http://cobweb.ecn.purdue.edu/~epados/erosbmp/src/title.htm
This is an online book with introductory information regarding different kinds of hillslope soil erosion.

British Society for Geomorphology
http://www.geomorphology.org.uk

The BSG provides services to those involved in teaching and research in geomorphology, both in the UK and internationally.

GOLEM
http://www.colorado.edu/geolsci/gtucker/Software/Golem/GolemIntro.html
GOLEM is a numerical model that simulates the evolution of topography over geological timescales. This site provides an overview of GOLEM and links to a number of online journal articles involving the application of GOLEM.

International Association of Geomorphologists

http://www.geomorph.org/

The International Association of Geomorphologists is a scientific, non-governmental organization that aims to develop and promote geomorphology as a science through dissemination of knowledge on geomorphology. It supports a number of working research groups on most geomorphological topics. If you put 'hilllslope erosion' or 'landscape evolution' into the search engine, resources and links can be found. An image gallery, glossary and list of weblinks are also available.

TOPMODEL webpage

http://www.es.lancs.ac.uk/hfdg/freeware/hfdg_freeware_top.htm

TOPMODEL is a rainfall–runoff model that bases its distributed predictions on an analysis of catchment topography; TOPMODEL can be downloaded from this site.

University of Virginia: Geomorphology Home Page

http://erode.evsc.virginia.edu/

A set of resources and lecture material (including detailed PowerPoint presentations) on geomorphological topics such as landform processes and models and drainage basin simulation, and modelling landform evolution on Mars.

 Visit our website at **www.pearsoned.co.uk/holden** for further questions, annotated weblinks, interactive models and video-clips showing physical processes in action.

CHAPTER 12

Sediments and sedimentation

Kevin G. Taylor
Department of Environmental and Geographical Sciences, Manchester Metropolitan University

Learning objectives

After reading this chapter you should be able to:

➤ understand the origin, composition and classification
  of sediments

➤ appreciate the products and patterns of sedimentation
  in surface environments

➤ understand the response of sedimentation to
  environmental change

➤ be aware of the role and impact of sedimentation in
  natural and anthropogenic catchments and basins

12.1 Introduction

Sediments are an important component of the Earth's
surface and their movement and deposition are key
processes in environmental systems. In conjunction with
weathering and erosion (Chapter 11), sedimentation shapes
the face of the Earth. Sedimentation creates floodplains,
builds coastlines and fills the ocean basins. Sediments have
a major impact on natural and engineered systems. In this
chapter, we will introduce the concepts of sediments and
sedimentation, describe the features that sedimentation

produces, and assess the impact that sedimentation has on
the environment.

Sediments are collections of grains of pre-existing rocks,
fragments of dead organisms or minerals precipitated
directly from water at Earth surface temperatures. In Earth
surface environments these grains accumulate (through the
process of sedimentation) to form sediment packages and
sedimentary successions. Sediment is a term restricted to
loose, unconsolidated material (e.g. sand). If the sediment
is buried and hardened, through the action of heat and
pressure, to form a rock (a process that can take from tens
to millions of years) it is then termed a sedimentary rock,
and the suffix-stone is added (e.g. sandstone). Sediment
classification is based on a combination of the sediment's
constituents and the mode of origin of the sediment. The
three major types of sediment are terrigenous **clastic
sediments** (simply termed clastic sediments throughout
this chapter), **biological sediments** and **chemical
sediments**.

Clastic sediments are composed of particles (clasts)
derived from pre-existing rocks (which may be igneous,
metamorphic or sedimentary in nature). These clasts are
derived from the weathering and erosion of bedrock and
soil material and comprise both primary and secondary
minerals. The material that is formed as a result of this
weathering is transported away from the site by water,

wind or ice and will ultimately settle out and accumulate in a range of continental or marine environments. Biological sediments are derived from organic materials and can be either the remains of dead organisms (e.g. shells, plants) or build-ups of framework-building organisms (e.g. coral reefs). Shell material and other skeletal fragments are commonly composed of calcium carbonate ($CaCO_3$), in the form of either calcite or aragonite. Biological sediment composed of such material is often referred to as carbonate sediment. Sediments in which organic material (e.g. plant remains) is a significant component are commonly termed organic-rich biological sediments. For example, peat is an organic-rich biological sediment composed of dead plant material accumulated in wet environments. Chemical sediments are those that are produced by chemical processes and are formed predominantly as a result of precipitation of minerals directly from a water body. A good example of this is the precipitation of salt from evaporating seawater or an inland sea, as is happening in the Dead Sea today.

Sedimentation is the process by which sediment is deposited, leading to its accumulation. The most common cause of deposition is the settling out of sediment from a transporting fluid (water, wind or ice). As such, sedimentation is the opposite of erosion and transportation. While erosion removes material from a location, sedimentation leaves material behind. Although they are contrasting processes, sedimentation and transportation are closely interconnected processes. Many sediments can be sedimented and then reworked and transported many times prior to the material permanently accumulating in sediment packages. Sedimentation is therefore a dynamic process, which cannot be considered in isolation from erosion and transportation.

12.2 Clastic sediments

As stated above, clastic sediments are composed of grains of rock, weathered and eroded from pre-existing bedrock material. The dominant grains present within clastic sediments are quartz, feldspar, mica, clay minerals and iron oxides (Nichols, 1999). In sediments those minerals which are most resistant to weathering are concentrated relative to those that are less resistant. This is because more easily weathered material will be more likely to be dissolved into solution (see Chapters 11 and 15), remain in suspension or be transported well away from the source area. In general, sediment that accumulates near its bedrock source bears a greater resemblance, compositionally, to the bedrock material than does sediment deposited a long way from its source.

Quartz dominates the composition of most clastic sediments as this mineral is most resistant to weathering and transport.

12.2.1 Classification of clastic sediments

The most widely used scheme for classifying clastic sediments is based on the size of the clasts, or grains, within the sediment. Three major grain size classes can be recognized. Gravel refers to grains greater than 2 mm in size. Sand refers to grains less than 2 mm but greater than 63 μm (1/16th of a millimetre) in size. Mud refers to grains less than 63 μm in size. These three major classes can be subdivided further using the Udden–Wentworth grain size scale (Figure 12.1). The Udden–Wentworth grain size scale is the scheme most widely used to classify the grain size of sediments. Each grain size class within the Udden–Wentworth scale is a factor of two larger than the previous one, and is therefore a logarithmic scale (logarithmic to base 2). As can be seen from Figure 12.1, each of the three major classes is subdivided into smaller classes (e.g. very fine sand, fine sand, medium sand, coarse sand, very coarse sand). The Φ (pronounced 'phi') scale is a numerical representation of the Udden–Wentworth scale, based on the logarithmic nature of the scale. The Φ value $= -\log_2$ grain size (in mm). This scale is mathematically more convenient to use than fractions of millimetres. Note that a larger value of Φ represents a smaller grain size.

mm	Φ	Class terms	
		Boulders	
256	−8		
128	−7	Cobbles	
64	−6		
32	−5		
16	−4	Pebbles	
8	−3		
4	−2		
2	−1	Granules	
1	0		Very coarse
0.5	1		Coarse
0.25	2	Sand	Medium
0.125	3		Fine
0.0625	4		Very fine
0.0312	5		Coarse
0.0156	6	Silt	Medium
0.0078	7		Fine
0.0039	8		Very fine
		Clay	

Figure 12.1 The Udden-Wentworth grain size classification scheme for sediment grains. The Φ value $= -\log_2$ grain size (in mm).

Grain size on unconsolidated sediment is readily estimated by observing the sediment through a hand lens or a binocular microscope. A more accurate determination of grain size can be made by sieving. Sediment is shaken through a stack of sieves of reducing mesh size and the percentage mass of sediment trapped at each sieve is weighed and calculated. Determination of grain size by sieving has the benefit of allowing statistical calculations to be made on the sediment (e.g. the mean and modal grain size). In consolidated sedimentary rocks grain size is commonly determined qualitatively by microscope examination.

12.2.2 Clastic sediment grain shape and texture

A great deal of information about the origin, history, source and environment of deposition of a sediment can be inferred by studying textural properties of the grains within a sediment. The **sorting** of a sediment is a measure of the degree to which the grains in the sediment are clustered around one grain size (in other words, a measure of the spread, or standard deviation, of grain sizes in a sediment). Although this can be determined statistically by sieving analysis, sorting is more usually estimated visually through a microscope by comparison with sorting charts. In such a case, the terms poorly sorted, moderately sorted, well sorted and very well sorted are used. Although a number of factors control sorting within a sediment, in general the further a sediment has been transported from its source, the better the sorting of the grains.

Grain roundness is another textural property that contains environmental information. Grains can be very angular, angular, subangular, sub-rounded, rounded and well rounded (Figure 12.2). Grains within sediments become rounded by continual abrasion during transport as a result of the impact of the sediment grains with each

other. Well-rounded grains indicate that the sediment has undergone extended transport prior to deposition. Wind-blown dune sands in deserts are commonly very well rounded as a result of continual grain impacts during wind transport. In contrast, sediment grains in **scree slopes** (see Chapter 11 and below) are commonly very angular as they have undergone only limited transport.

12.2.3 Sediment transport and sedimentation

With the exception of *in situ* organic build-ups, such as reefs (see Chapter 17), and chemical precipitation of minerals directly from water, virtually all sediments are deposited after some element of transport. This is particularly true for clastic sediments. Sediment transport can take place as a result of gravity, but more commonly transport is by water, wind or ice. The density of the transport medium has a major control on the ability of the medium to carry sediment. The higher the density of the medium then the larger the grains that can be transported.

Transport of sediment by gravity alone is only important on steep slopes and can be thought of as the first stage of erosion and transport of weathered material. Material may move down a slope, under the action of gravity, by a number of mechanisms, depending on the grain size and cohesiveness of the material, and the slope angle (see Chapter 11). Major mechanisms are **rockfalls, landslides, soil creep** and **slumping** (see Chapter 11). In rockfalls, consolidated material falls and breaks up into a jumble of material at the base of a cliff or steep slope. In contrast, a landslide is where a large coherent mass of material moves down a slope undeformed. Slumping is similar to landsliding, but contains saturated slope material (pore spaces are full of water) which deforms upon movement. Rockfalls, landslides and slumps are rapid events. Soil creep is the very slow, imperceptible, movement of material down a slope (see Chapter 11). Screes are accumulations of sediments that build up adjacent to mountain fronts, developed through the collection of loose sediment material removed from the mountain by gravity-driven sediment transport. Sediments fall onto the surface of the scree and move down the scree surface. They come to rest at the base of the scree where the slope shallows out. Scree slopes are composed mainly of poorly sorted, angular gravels, and exhibit only crude layering.

Another form of subaerial gravity flow is a debris flow. A debris flow is a slurry-like flow containing both solid material and water but with a high ratio of solids to water (see Chapter 11). Debris flows are highly destructive and are common in all climate regimes. They are often started after

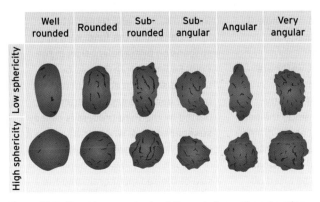

Well rounded	Rounded	Sub-rounded	Sub-angular	Angular	Very angular

Figure 12.2 Roundness and sphericity scale for sediment grains. (Source: after Pettijohn *et al.*, 1987)

heavy rainfall on debris-laden mountain slopes, but can also be initiated by earthquakes, volcanic eruptions and even forest fires. The sediments deposited by debris flows are generally poorly sorted with little, or no, internal stratification.

Water is by far the most common medium for sediment transport. Water moves as a result of flow in channels or as a result of currents generated by wind and tides. If water movement is fast enough it may carry sediment, and in many cases this transport can be for hundreds of kilometres before the sediment grains are deposited. Box 12.1 describes important principles associated with sediment transport and water. Transport of sediment by air (wind) can also be an important mechanism, but its effectiveness is limited by the low density of air and as a result only small grains can be transported by the wind. Wind-blown sediments are important indicators of climate change in the past and may

FUNDAMENTAL PRINCIPLES

SEDIMENT TRANSPORT BY WATER

To understand transport of sediment by water we need to have a basic understanding of the nature of flowing water. Water can flow in one of two ways: **laminar flow** and **turbulent flow**. In laminar flow the water molecules all flow in the same direction, parallel to each other (Figure 12.3). As a result, almost no mixing of water takes place during laminar flow. Laminar flow is uncommon in surface waters, being restricted to low flow velocities and very shallow water. In turbulent flow, water molecules move in many directions, with an overall net flow in one direction. As a result water undergoing turbulent flow is well mixed.

The Reynolds number (Re) is a dimensionless (it has no units) quantity which indicates the extent to which a flowing fluid is laminar or turbulent. The Reynolds number relates the velocity of a flow (u), the ratio between the density and viscosity of the fluid (v, the fluid **kinematic viscosity**) and the length of the pipe or channel through which the fluid is flowing (l). The Reynolds number (Re) = ul/v. It has been experimentally determined that when the Reynolds number is low (less than 500) laminar flow dominates, and when the Reynolds number is high (greater than 2000) turbulent flow dominates. The Reynolds number is applicable to both water and air, but the lower viscosity of air results in turbulent flow dominating at lower flow velocities than water.

Sediment grains in water are transported by one of three processes (Figure 12.4). Firstly, grains can be moved along the bed surface by rolling. Secondly, grains may bounce along the bed surface; this process is termed **saltation**. Thirdly, material may be lifted off

(a) Rolling

(b) Saltation

(c) Suspension

Figure 12.4 The three mechanisms of sediment transport in flowing water: (a) rolling, (b) saltation and (c) suspension.

Laminar flow

Flow ➡

Turbulent flow

Flow ➡

Figure 12.3 Laminar and turbulent flow in water.

the bed surface and transported in suspension in the fluid, kept in suspension by turbulent flow in the fluid. Sediment transported by the first and second mechanism is termed **bed load** and that transported by the third process is called **suspended load**. Whether a sediment grain will be transported as bed load or suspended load will depend on the size of the sediment grain and the

BOX 12.1 ➤

velocity of the fluid flow. At low current velocities only small sediment grains (clays) will be transported in suspension. At higher velocities larger grains may be transported in suspension, but it is rare for grains

larger than sand sized to be transported as suspended load.

The **Hjulström curve** (Figure 12.5) shows the nature of flow velocity required to move sediment of different grain sizes in water. This

graph shows both the velocity of water required to keep a sediment clast in transport and the velocity required to move a stationary clast. Therefore, as a consequence it also shows the velocity below which sediment of a specific grain size will be sedimented. As can be clearly seen in Figure 12.5, as sediment grain size increases, a higher velocity is required both to keep the grains in transport and to move stationary grains. For fine sediment grain sizes only low flow velocities are required to keep sediment in transport. A consequence of this is that fine-grained sediments will accumulate only under very quiet water conditions. In contrast to this, fine-grained sediments require relatively more energy to be moved from a stationary position. This is due to the fact that clay grains, which dominate fine-grained sediments, are cohesive in nature and clump together. As a result of this, it takes greater flow velocity to transport stationary clay material than sand-sized particles. This means that, although clay-sized particles are deposited only when current velocity effectively falls to zero, once deposited, clay size material is not easily eroded and retransported.

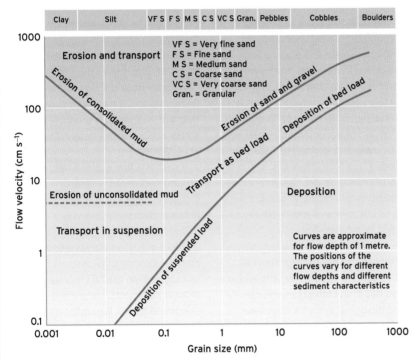

Figure 12.5 The Hjulström diagram, illustrating the relationship between grain size and current velocity for sediment grain transport. The two curves show the energy required to keep sediment in transport (lower curve) and the energy required to transport grains from a stationary position. (Source: after Press and Siever, 1986)

BOX 12.1

also have important impacts on climate itself (Lowe and Walker, 1997). Such issues are addressed in Box 12.2. Although ice is a solid material, it moves and deforms slowly in the form of glaciers and ice sheets and can therefore be thought of as a fluid. As a result it can transport large amounts of sediments slowly over relatively short distances (see Chapter 18; Leeder, 1999).

12.2.4 Products of sedimentation – bedforms

A **bedform** is a morphological feature formed when sediment and flow interact on the sediment bed surface.

Ripples on a sandy beach and sand dunes in a desert are both examples of sediment bedforms. As sediment is moved along the bed surface by the current, small irregularities on the bed surface influence the manner in which sediment is transported and deposited. Bedforms deposited under **unidirectional flow** (currents that are flowing in dominantly one direction such as river currents and wind) are different from those formed under **oscillatory flow** (currents that oscillate backwards and forwards such as wave currents).

Current ripples form under unidirectional flow and are small bedforms (up to 5 cm in height and 30 cm in wavelength) and form predominantly in sand-sized

ENVIRONMENTAL CHANGE

WIND-BLOWN TRANSPORT OF SEDIMENT

Wind-blown sediment transport is called aeolian transport. Although such transport can take place in many environments, it dominates in arid and semi-arid environments with little water. As air has a lower viscosity than water, higher flow velocities are needed to move sediment grains. At typical wind speeds, medium sand grains (up to 0.5 mm) are the largest grains that can be transported (this contrasts with the pebbles and boulders that can be transported by water). As a result the deposits built up from wind-blown sediment grains are generally fine grained in nature. Transport distances for such material can be vast, with dust deposits transported thousands of kilometres from the Sahara of North Africa to the Atlantic Ocean (Figures 12.6 and 12.7). This material settles out on the sea

floor, contributing to sedimentation in the oceanic environment. The amount of material removed is huge and for the Sahara approximately 2.6×10^8 tonnes per year. The extent of this dust transport has changed through time, a fact documented through climate and dust records preserved in ice cores and deep-sea sediments. Dust transport from land surfaces was greater during the last glacial period than today. Thompson *et al.* (1995) documented increased dust transport during the Late Glacial Stage in ice cores from glaciers in the high Peruvian Andes. They concluded that atmospheric dust contents were up to 200 times as high as today as a result of increased aridity.

As well as responding to changes in global climate, the transport of mineral dust through the atmosphere also has direct impacts upon climate and biological systems. Very fine-grained mineral dust is highly effec-

tive at scattering light and, therefore, may have a cooling effect on climate (see Chapter 2). It has also been proposed that dust transported from the Sahara has had a detrimental impact on the development of Caribbean corals. It has been recognized that since the late 1970s, fluxes of wind-blown dust to the Caribbean from the Sahara have increased and it has been suggested that this has led to environmental stress on Caribbean corals (Shinn *et al.*, 2000).

Loess is a fine-grained (less than 50 μm) sedimentary deposit composed of grains of quartz, feldspar, carbonate and clay minerals transported by wind from arid land surfaces and deposited elsewhere, often thousands of kilometres from its source (Pye, 1987). Thick deposits of loess sediments are present in the Czech and Slovak Republics of central Europe, and in central China, in an

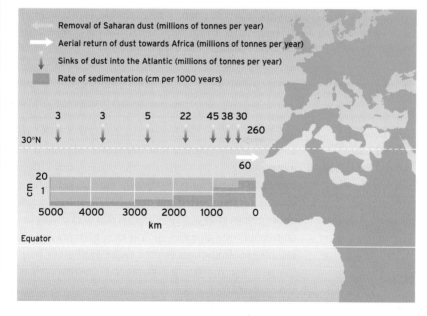

Figure 12.6 Mineral dust transport from the Sahara into the Atlantic Ocean. (Source: adapted from Leeder, 1999)

BOX 12.2 ➤

area known as the Loess Plateau. It is believed that these loess deposits were formed during full glacial conditions. The loess was derived from winds blowing across arid glacial outwash plains. These loess deposits also contain soil layers which are interpreted to reflect times of wetter climatic conditions, with negligible loess sedimentation (Lowe and Walker, 1997).

Figure 12.7 Dust transport by wind in the northern Sahara, Morocco. (Source: photo courtesy of Andrew Thomas)

BOX 12.2

sediment. Current ripples are not symmetric but have a shallow stoss side and a steep lee side as shown in Figure 12.8(a). Sediment is transported up the stoss side and avalanches down the lee side. As a result the ripple migrates in a downstream direction. In plan view (looking from above) the shape of current ripples can vary from straight to sinuous to linguoid (Figure 12.8b). This variation in ripple shape can be in response to water flow and water depth.

Dune bedforms are larger than current ripples (up to 10 m high) but have similar cross-sections and form in a similar manner. Ripples are more predominant in silt to medium sand-sized sediments, whereas dunes are more predominant in medium to coarse sands (Figure 12.9). A clear relationship has been documented between flow velocity and bedforms (King, 1991) (Figure 12.9). At low flow velocities, bedforms do not form. At greater flow velocities current ripples form within fine to medium sand-sized sediment, dune bedforms forming at higher flow velocity. At very high velocities bedforms do not form owing to the speed of sediment transport. This is known as the upper-plane bed stage.

Wind-generated waves in water bodies (predominantly shallow marine settings, but sometimes also present in lakes) produce circular, oscillatory water motion (see Chapter 17). Beneath the surface of the water body, at the sediment surface, this motion is translated into horizontal oscillatory current movement (Figure 12.10). This motion sweeps grains away from a central zone and deposits particles as symmetrical ripples on the sediment bed. In

cross-section and plan aspect, wave-formed ripples are symmetrical in shape and as such can easily be distinguished from current ripples formed under unidirectional flow conditions (Nichols, 1999).

Both sand ripples and sand dunes may be formed by wind-transported sediment. The most predominant environment for wind-produced bedform formation is that of the arid desert environments (Chapter 16), but localized sand dunes may also form in coastal environments as coastal sand dunes (Chapter 17). Arid zone aeolian (wind-blown) bedforms can be highly variable in size, ranging up to 600 m in wavelength and 100 m in height. When sediments are deposited from wind, the most common result is sand dunes (Figure 12.11). Under simple conditions of unidirectional wind patterns, simple dunes may form, with stoss and lee slopes. However, in many desert regions wind directions can change seasonally, which leads to more complicated dune bedforms. Unidirectional dunes can take two forms. **Transverse dunes** are linear features, with a shallow windward side and a steep lee slope. The internal structure is similar to dunes and ripples formed below water. **Barchan dunes** also have a shallow windward and steep lee side, but these are isolated dunes, with a characteristic crescent shape. These dunes are most commonly formed as sediment is moved across a hard substrate, such as a dried-up lake bed. **Stellate** (star-shaped) **dunes** form under conditions in which wind directions are variable, and with no particular direction prevailing. These dunes do not migrate and may be initiated at irregularities in the ground surface. **Seif** (or linear) **dunes** form where two distinct wind

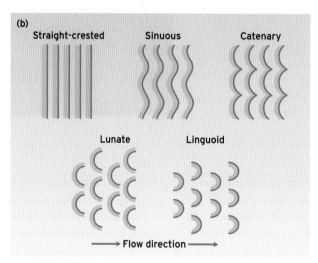

Figure 12.8 (a) Schematic diagram to illustrate the formation of a current ripple under unidirectional current flow. (b) Shape of current ripples in plan view. The change from straight-crested to linguoid is governed by current strength and water depth. (Source: after Tucker, 1981)

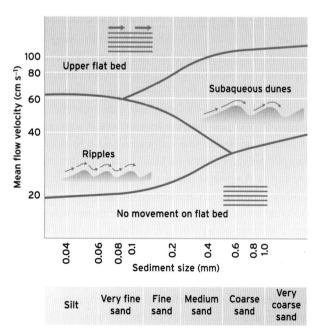

Figure 12.9 A bedform–flow diagram to illustrate the grain size and current velocity regimes under which sediment bedforms are present. (Source: after Nichols, 1999, and King, 1991)

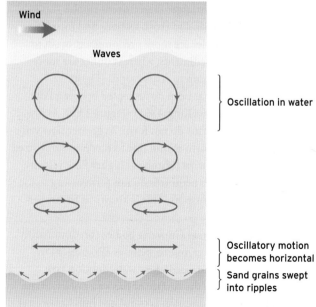

Figure 12.10 The formation of symmetrical wave ripples by oscillation of a water body. (Source: after Nichols, 1999)

Figure 12.11 Aeolian bedforms in arid environments. (Source: after Nichols, 1999)

directions are present at approximately right angles to each other. Wind-formed ripples are commonly present on the surface of sand dunes. Chapter 16 provides further details on sand-dune forms and processes.

Reflective questions

➤ What grain roundness would you expect for a beach sand?

➤ By reference to the Hjulström curve, what can you say about the transport of boulders within rivers and streams?

➤ How would you use ripples to distinguish between sediment deposited in a river and that deposited in a shallow marine wave-dominated setting?

12.3 Biological sediments

In areas where there is a minimal supply of clastic sediments, other components form the major contribution to sediments. The main component is the accumulation of dead organisms or the build-up of framework structures (**bioherms** such as coral reefs). Two main types of biological sediments can be recognized, carbonate sediments and organic-rich sediments. The major components in carbonate sediments are skeletal fragments of marine or freshwater organisms (e.g. brachiopods, molluscs, echinoids,

corals, foraminifera, calcareous algae), bioherms (e.g. corals) and non-biological components (e.g. ooids, peloids) (Figures 12.12, 12.13 and 12.14). Reefs are the result of framework building by organisms composed of calcium carbonate material (Masselink and Hughes, 2003). Organisms most characteristic of reef build-up are corals, but bryzoa, coralline algae and brachiopods have also produced bioherms in the geological past. Non-biological components form as the result of direct precipitation of calcium carbonate in the form of grains, although some of these grains may form via microbiological processes. Ooids and pisoids are concentrically coated grains whereas peloids are grains with little internal structure (Figure 12.13).

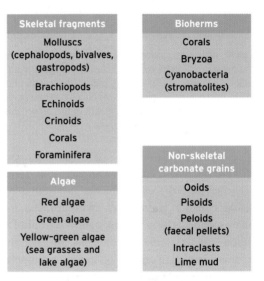

Figure 12.12 Components of carbonate sediments.

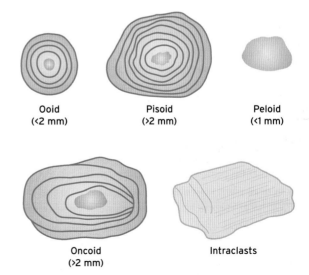

Figure 12.13 Non-skeletal components of carbonate sediments. (Source: after Nichols, 1999)

Figure 12.14 A tropical carbonate sedimentary environment. Note the presence of large coral colonies, short-upstanding *Halimeda* (blue-green algae) and patches of bare carbonate sand. All material in such environments becomes transported and deposited as sediment upon the death of the organisms. (Source: photo courtesy of Chris Perry)

(a)

(b)

Figure 12.15 Peat formation occurs in waterlogged areas such as in (a) and can build up thick deposits as shown in (b).

Carbonate sediments are most abundant within warm tropical waters, as carbonate production is favoured in strong solar radiation, warm-water environments. Carbonate sediments form predominantly in tropical and subtropical shallow marine environments, but calcium carbonate accumulation can also occur within temperate shallow marine environments and lakes. Sediments composed of shells of single-celled organisms (e.g. diatoms and foraminifera) also accumulate in deep-sea environments. Such sediment is commonly termed calcareous ooze, or siliceous ooze, depending on the composition of the shell material making up the sediment. When buried and lithified, carbonate sediments are known as limestones. Chalk, a pure white limestone formed throughout north-west Europe 80–65 million years ago, is an example of a sedimentary rock composed of the remains of shells of a calcareous blue–green alga (Tucker and Wright, 1990).

The most widely distributed type of organic-rich biological sediment is peat, which forms through the accumulation of dead plant material in waterlogged swamp and bog environments (Figure 12.15). The accumulation of peat is favoured in areas where rates of plant breakdown are low, which is most common in waterlogged stagnant conditions. Under such conditions, low oxygen, coupled with low pH, inhibits the breakdown of organic matter by bacteria and fungus. This leads to the accumulation of organic material.

Peat can be cut and dried for use as a burning fuel. If peat layers are buried beneath further sediment layers, water is squeezed from the material and volatiles (water vapour and carbon dioxide) are lost. This process leads to the formation of lignites and coals. Many of the large coal deposits of north-west Europe were deposited in freshwater swamps during the Carboniferous era, 350 to 300 million years ago.

Reflective question

➤ What types of landform are composed largely of biological sediments?

Reflective question

➤ In which environments are chemical sediments more likely to be found?

12.4 Chemical sediments

The most important chemical sediment is that termed evaporative sediment. These sediments are formed as a result of minerals precipitated out of lake or seawater as waters are concentrated by evaporation. As seawater is evaporated the least soluble mineral precipitates out first. This is usually calcite (calcium carbonate, $CaCO_3$) followed by gypsum (calcium sulphate, $CaSO_4$) and halite (sodium chloride, $NaCl$) as the water becomes more saturated. If water becomes very concentrated a number of salts of potassium become precipitated (bittern salts). The most commonly encountered evaporite mineral is **gypsum**. Water concentrated to 19% of its original volume will precipitate gypsum. Halite is only deposited once the water has been reduced to less than 10% of its original volume and, therefore, is less common than gypsum. Many water bodies undergoing evaporation are periodically recharged by addition of water, either by rainfall or river water input, making it rare for such concentrated evaporation. The high solubility of $NaCl$ also means that it is readily redissolved on exposure to water. Bittern salts are very rare and form only after complete evaporation of a standing body of water.

Evaporative sediments are common in arid climates (see also Chapter 16) and may form in coastal settings or in standing bodies of water. In the case of coastal settings, the best documented examples are **sabkhas**. These are low-angle tidal flat environments (the coasts of the Persian Gulf being a classic example). Evaporation of groundwater draws in seawater, which upon evaporation precipitates gypsum. However, in these settings the water rarely becomes concentrated enough to precipitate halite. Thick deposits of halite have been formed in the past, such as those extracted in the Cheshire region of northern England. However, such thick deposits are not forming in the present day. To produce such thick accumulations of halite complete evaporation of large water bodies has been invoked, such as the Mediterranean Sea. However, this may become a phenomenon of the future if climate and land management change impact on evaporative processes.

12.5 Sedimentation in Earth surface environments

Sedimentation takes place in a wide range of Earth surface environments. In general, sedimentary environments can be thought of as a traverse from upland mountainous environments, through lower-lying continental environments, shoreline and shallow marine settings, eventually to oceanic environments, and this is the structure adopted in the following section (Figure 12.16). This passage is particularly true in the case of clastic sediments where the transport of sediment by water, wind or ice moves sediments from continental environments to offshore environments. Given the scope of this chapter, only a brief summary can be given here of the characteristics of sedimentation in Earth surface environments. Other chapters of this book provide further details within specific environments.

12.5.1 Continental environments

12.5.1.1 River environments

The term fluvial is generally used to describe river environments and processes. There are three major types of fluvial

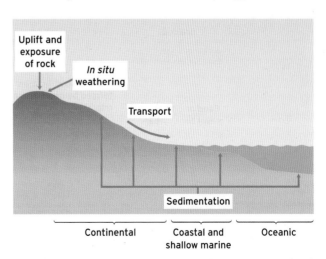

Figure 12.16 Schematic diagram to illustrate the pathways and major sites of sedimentation of clastic sediments in surface environments. (Source: after Nichols, 1999)

environment that can be recognized: braided (where the river consists of several shifting channels separated by islands, or bars); meandering (where singular sinuous channels are surrounded by low-lying floodplains); and anastomosing (where the river consists of stable, splitting and rejoining channels, thereby possessing elements of both braided and meandering form). For a detailed discussion of fluvial geomorphology see Chapter 14.

Braided channels form most commonly where water flows over loose sand or gravel, often in mountainous or upland areas. A characteristic feature of braided rivers is the mobile nature of the channels and the intervening bars (Figure 12.17; see also Figure 14.4 in Chapter 14). The sediments deposited in braided rivers are most commonly composed of gravels and coarse sands. Fine-grained sediments may be deposited on bars, but they are not common. The predominant sedimentary bedforms present within braided river sediments are dune bedforms, formed as bars accrete and move forward within and between the channels.

In contrast to braided rivers, meandering rivers form in low-lying areas, with a predominance of finer sediment. Meandering is a term used to describe the sinuous nature of the channels (Figure 12.17). Sediments deposited in meandering river environments are quite different from braided rivers, being predominantly finer grained than braided river sediments. Sediments deposited in the channels are coarser than those deposited in floodplains, as the water flowing within the channels transports the finer materials downstream. During periods of high stage (floods) water may overflow the channels and deposit suspended sediment upon the floodplain. As the sediment carried in suspension

is silt and clay sized, the resulting sediment on floodplains is fine grained. Such fine-grained sediments are high in nutrients and as a result floodplains are fertile areas. Indeed, it is the flooding and deposition of fresh sediment that keeps the land fertile, which is why flood management and dam building can have a major negative impact on soil fertility in river areas. In addition to the floodplains, sediments accumulate on the inside of meanders, which leads to the formation of point bars (Figure 12.17). Current ripples, especially on the surface of point bars, are the most common sediment bedform in meandering river environments (Tucker, 1981).

12.5.1.2 Arid environments

Desert environments are those regions of the Earth where potential evaporation exceeds rainfall (often the definition that potential evapotranspiration is more than twice the precipitation is used; see Chapter 16) and in these environments standing bodies of water are rare. The major process of sediment transport and deposition is by wind action, although flood events can also lead to fluvial conditions prevailing under wet seasons. Three major forms of sedimentation take place in arid environments: **sand seas**, **alluvial fans** and **playa** lakes.

Sand seas (also known as **ergs**) are areas of sand accumulations, and large sand seas are present in the Sahara, Namibia, south-western North America and western central Australia. Sand seas are not the same as deserts. Deserts are simply dryland environments and may or may not contain sand seas. These sand seas are composed of dune bedforms deposited from wind-blown sediments (see above). Sediment

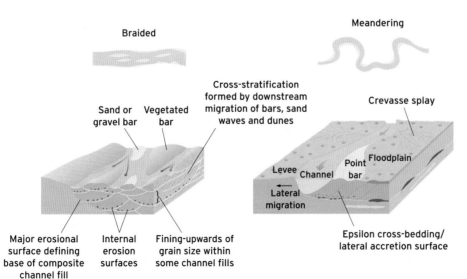

Figure 12.17 Sedimentation associated with braided and meandering river systems. (Source: after Tucker, 1981)

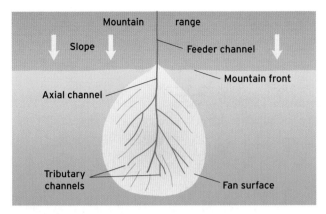

Figure 12.18 Typical features of an alluvial fan. (Source: after Harvey, 1997)

grains in these environments are typically well rounded as a result of grain–grain collisions during transport, and composed almost entirely of quartz. The red colour of desert sands is the result of a thin coating of iron oxide as a result of the oxidizing conditions in arid environments (see Chapter 16).

Alluvial fans are cones of sediment that accumulate at mountain fronts. The major agent of sediment transport on alluvial fans is flowing water (which is generally present only during wet seasons). This flowing water spreads out and deposits sediment as it slows down in distant parts of the alluvial fan surface. The result is a semicircular fan (Figure 12.18). Sediment deposits on alluvial fans are a mixture of gravels and coarse sands, with a general decrease in grain size away from the mountain front. Alluvial fans are particularly common in arid mountainous settings, with well-developed fans present in Death Valley (e.g. Figure 12.19), California, and the Atlas Mountains of Morocco (Harvey, 1997).

Figure 12.19 An alluvial fan in Death Valley, California. (Source: Marli Bryant Miller, Visuals Unlimited)

Playa lakes are **ephemeral** (seasonal) bodies of water which accumulate during rainfall events. Water flowing into these lakes deposits a layer of silt and clay. As the lake dries up, evaporite minerals are precipitated (gypsum and halite), and desiccation cracks form within the sediment surface. If the water body dries up completely, a salt crust known as a salt pan is produced. At the next wet period further sediment is deposited followed by additional formation of evaporite minerals. This alternation of muds and evaporite minerals is characteristic of playa lakes.

12.5.1.3 Glacial environments

Ice plays a major role in sediment transport and sedimentation and gives rise to a number of characteristic landforms and sediment deposits. However, erosion and sedimentation through glacial activity are significantly different from those resulting from water and air. Detailed information on sediments in glacial systems can be found in Chapter 18.

12.5.2 Coastal and marine environments

12.5.2.1 Delta environments

There have been many definitions for deltas but a broad definition can be given as a discrete shoreline protuberance formed at a point where a river enters an ocean or other body of water. Deltas are sites where sediment supplied by the river is accumulating faster than it is being redistributed in the ocean by waves and tides. These environments vary depending on whether a delta is dominated by river processes (e.g. the Mississippi Delta), wave processes (e.g. the Rhone Delta) or tidal processes (e.g. the Ganges Delta). Further details of delta environments are provided in Chapter 17.

12.5.2.2 Estuaries and salt marshes

An estuary is a semi-enclosed coastal water body where there is a mixture of river and seawater and where there is a mixture of fluvial and marine processes. At the present time estuaries are common as a result of the post-glacial rise in sea level drowning the mouths of rivers. Two major morphological elements are present in estuaries: tidal channels (Figure 12.20) and tidal mudflats. Tidal channels are major sites of sediment transport and hence consist of

Figure 12.20 An estuarine environment in a tropical setting, Queensland, Australia. Note the presence of sinuous tidal channels, with areas of vegetated fine-grained sediments stabilised by mangroves.

sand-sized sediment. Within tidal channels subaqueous dune bedforms and ripples commonly form. In estuaries both flood (incoming tide) and ebb (outflowing tide) currents can be present, but in general either the ebb or the flood current is strongest and this is reflected in the nature of the bedforms. Tidal mudflats are regions away from strong flood or ebb currents. Suspended sediment (silt and mud) is carried over the mudflats during high tide and deposited upon the mudflat as the tide turns and water velocity falls. The frequency of flooding depends on the height of the tide and the elevation of the mudflat. Mudflats gradually build upwards and therefore become flooded less frequently over time. If flood frequency is low, salt-tolerant plant species will colonize the mudflat, and these systems are termed salt marshes. In tropical environments, mangroves form in such environments as a result of similar processes (Figure 12.21). Chapter 17 provides more information on estuary and salt marsh systems.

12.5.2.3 Beaches, barriers and lagoons

Between deltas and estuaries coastlines may be sites of sediment erosion or sediment deposition (see Chapter 17). Along coastlines that are sites of deposition, sedimentation may take place on beaches, lagoons or barriers. A beach is

Figure 12.21 Details of a tropical coastal mangrove environment in Queensland, Australia. Note the presence of tree roots trapping and stabilizing sediment.

an area that is continuously impacted by waves. Sediment accumulating on beaches (which may be supplied by cliff erosion or by **longshore drift**) is continuously reworked and is characteristically well sorted and well rounded. On very shallow-sloped beaches, waves and wind may form ripples, but on steeper-sloped beaches low-angle sediment accumulation may be present, especially on wave-dominated beaches. A barrier island is a beach detached from the main coast to form a ridge of sediment parallel to the coast (see Chapter 17). Such islands are most common along shorelines with a low tidal range and high wave energy. Sediment accumulates along the front of the island in a beach environment, whereas landward of the island, quiet conditions allow the accumulation of fine-grained material, either in salt marshes or lagoons. Lagoons are areas of low energy and are normally formed behind a barrier such as a barrier island. Along clastic sediment shorelines, muds and salt marshes develop, whereas in carbonate-dominated shorelines, fine-grained carbonate mud accumulates.

Within tropical coastal settings in which clastic sediment input by rivers is minimal, the deposition of carbonate biological sediments can dominate. In such environments, sediment may be formed by the build-up of reef-building organisms, the accumulation of skeletal material and the inorganic precipitation of calcium carbonate. Reefs composed of corals accumulate in shallow, high-energy conditions and commonly act as barriers to shallow lagoon environments behind. Within these shallow lagoons, fine-grained sediment composed of precipitated calcium carbonate (lime mud) accumulates.

12.5.2.4 Shallow marine environments

The nature of sediment deposited in shallow marine environments is governed by the strength of currents produced by tides and storms. Sands are the predominant sediment deposited, commonly as sand dunes up to 10 m in height. In wave- and storm-dominated environments (micro-tidal) sands are deposited as wave-rippled and symmetrical hummocks. In tidal-dominated shallow marine environments (macro-tidal) sand waves and sand ribbons form. Sand waves form under lower tidal flow and are aligned perpendicular to the direction of tidal flow. Sand ribbons form under higher tidal flows and are aligned parallel to tidal flow. Fine-grained sediment tends to be deposited in waters deep enough not to be affected by major storms and tides. At the present time, many shallow marine environments (e.g. the North Sea) are covered with a layer of coarse sand and pebbles. These deposits are relict from the time of much lower sea level during the last Ice Age. Finer-grained sediment is currently being trapped in estuaries.

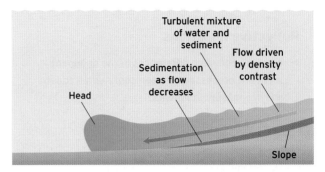

Figure 12.22 Schematic drawing of a turbidity current. (Source: after Nichols, 1999)

12.5.2.5 Oceanic environments

Oceanic environments include those environments from the edge of the continental shelf into the abyssal plains (see Chapter 3), and span a water depth from 100 m or less to over 8000 m in some of the deep-sea trenches. Sedimentation in oceanic environments takes place via two major processes: **turbidite currents** and **pelagic sedimentation**. Turbidite currents are mixtures of sediment and water which, because of their increased density relative to seawater, flow down and along the bottom surface of the oceans (Figure 12.22). In this process, they transport sand and clay-sized sediment from shelf slopes to deeper oceanic environments, depositing sediment as a thin bed widely across the sea floor. Turbidite flows are commonly triggered by earthquake events, with one of the best documented examples being in the Grand Banks area of Newfoundland in 1929. The resulting turbidite flows broke transatlantic telephone cables on the seabed.

Pelagic sedimentation is the slow background sedimentation of fine-grained material falling through the water column to the seabed. The best developed pelagic sediments accumulate in the deep sea where clastic sediment input from continents is minimal. Three types of pelagic sediments have been documented from deep-sea environments: brown clay, carbonaceous ooze and siliceous ooze. Brown clay is a sediment accumulation of fine-grained clay grains and glass fragments. The sediment is derived predominantly from wind-blown continental material, volcanic material and micrometeoric grains. Carbonaceous ooze is an accumulation of calcite tests of microscopic organisms living in the water column (e.g. foraminifera and coccoliths), and is widely distributed on the ocean floor. Siliceous ooze is composed of the tests of microscopic organisms made of silica (e.g. radiolaria and diatoms) and has a localized distribution on the ocean floor. See Figure 3.10 in Chapter 3 for further information.

12.6 Response of sedimentation to environmental change

Both natural and anthropogenic activities can have a major impact on the style and rate of sedimentation in surface environments. Therefore it is important that we can measure the rates of sedimentation on different environments and Box 12.3 provides some examples of how this can be done. The natural changes that impact most upon sedimentation are climate change and sea-level change. Climate change leads to changes in rainfall and vegetation

TECHNIQUES

MEASUREMENT OF SEDIMENTATION RATES

The rate of sedimentation is a measure of the thickness of sediment (normally measured in centimetres) that accumulates at a specific location over a specific amount of time. Generally, accumulation rates are quoted in centimetres or millimetres per year, but may also sometimes be quoted as grams per cm^2 of sediment. However, in many environments sedimentation rate is generally very low, and so rates in centimetres per hundred years, or even per thousand years, are commonly quoted.

Sedimentation rates can be measured for environmental systems using a range of techniques. Short-term measurements can be made by collecting sediment that accumulates in a sediment trap and measuring the amount of sediment deposited over a month or a year. Alternatively, in salt marsh or floodplain environments short-term sediment accumulation rates have been measured by laying down grass mats on the marsh and measuring sedimentation upon the

mats over durations of 1-10 years. Longer-term measurements of sedimentation rates can be made using radionuclide determination. Caesium-137 (^{137}Cs) is a radioactive element that was released into the

atmosphere by atomic weapons testing in the 1950s and by the Chernobyl nuclear power station incident in 1986 (Figure 12.23). By measuring the vertical location of these two concentration peaks of

Figure 12.23 The inputs of radioactive ^{137}Cs from the atmosphere to the northern hemisphere since 1950. The peaks in the 1950s and 1960s as a result of atomic weapons testing, and in 1986 as a result of the Chernobyl incident, allow for estimates of sedimentation rate to be made for sediments over the past 50 years, assuming constant sedimentation rates. (Source: after Owens *et al.*, 1996)

BOX 12.3 ➤

➤
^{137}Cs in the sediment, an estimate
of annual sedimentation rate
can be made. For longer-term
sedimentation rate estimations,
archaeological artefacts or
^{14}C dating can be used to estimate
sediment accumulation rates over
hundreds to thousands of years
(see Chapter 20). Under special
circumstances, yearly layers of
sediment (commonly called **varves**
in lake sediments) can be recognized, allowing for accurate estimates of sedimentation rate in
such cases (Figure 12.24).

Typical sedimentation rates in
natural systems display a wide range
of values. In general, sedimentation
rates are low, being of the order
of less than 1 cm yr^{-1}. Sedimentation
rates are very low in oceanic environments (commonly less than 1 cm
per 1000 years), as a result of
their great distance from sediment
sources. Sedimentation rates in
lakes, floodplains and coastal
settings can be much higher (in
the range of 0.1-10 cm yr^{-1}). It
should be remembered, however,
that sedimentation is a dynamic
process and rates of sediment
accumulation may vary over timescales from daily to yearly. Longer-
term changes in sedimentation
rate will also result from natural
and anthropogenic changes to
the system.

Figure 12.24 Annual sediment layers (varves) deposited in Quaternary Glacial Lake Riada, central Ireland. Core is approximately 25 cm in length. (Source: photo courtesy of Cathy Delaney)

BOX 12.3

which have major impacts upon sediment supply in the catchment, and thereby sedimentation in associated receiving water bodies. Sea-level change results in changes in the base level of sedimentary systems. The result is commonly either the marine inundation of coastal and continental environments, or the exposure of shallow coastal shelves. In general, these natural changes are slow and gradual, although there are numerous examples in the geological past where such changes have produced marked

changes in sedimentation style. At the present time sea-level rise, associated with global climate warming, is having marked impacts on low-lying coastal systems (see Chapter 17).

Of greater short-term impact and concern are the effects of anthropogenic activities on sedimentation. Such activities can be either direct, through the engineering of water bodies (e.g. dams, reservoirs), or indirect, through changes in catchment characteristics (e.g. mining, urbanization).

12.6.1 Dams and reservoirs

Dams and reservoirs have been constructed since early human history for regulation of water, but in the last 50 years the construction of major dams for water supply and hydroelectric generation has increased markedly. As well as having marked impacts upon water flow within catchments downstream of the dam, they also have a marked impact on sedimentation and sediment transport throughout the catchment. The two most significant impacts are the trapping of sediment behind the dam (see below) and the reduction in the sediment load of the river downstream. Sedimentation in lakes behind dams leads to less floodplain sedimentation downstream, which reduces nutrient supply and increases the potential for erosion (clear-water erosion). Before 1930 the Colorado River, USA, carried up to 150 million tonnes of suspended sediment annually to its head in the Gulf of California. Since that time, a number of dams have been built on the Colorado. The Glen Canyon Dam, for example, was constructed as a sediment trap to prolong the life of Lake Mead behind the Hoover Dam which was completed in the 1930s. Sediment is currently being deposited and trapped behind these structures (e.g. in Lake Mead and Lake Powell) and is no longer discharged into the sea. Indeed, as a result of dams and water abstraction in southern California, water no longer enters the sea from the Colorado River.

The largest dam-building project in the world is being undertaken on the Yangtze River in China. The Three Gorges Dam building scheme will result in a dam 2 km long and 100 m high; it will become fully operational in 2011. The resultant reservoir will stretch for 600 km upstream. As well as for hydroelectric generation, the Three Gorges Dam is designed to protect 10 million people downstream from devastating floods that have killed up to 300 000 people in the past 100 years. However, there is concern as to how long the reservoir will last, given that silt carried down by the Yangtze will sediment behind the dam, perhaps eventually filling the reservoir. On the Yellow River, also in China, a reservoir behind the Sanmenxia Dam filled with silt within four years of construction and had to be emptied, dredged and rebuilt. In the year 2000 the reservoir had less than half its original capacity (Chengrui and Dregne, 2001). The Yangtze carries 530 million tonnes of silt through the Three Gorges area each year. A similar, smaller dam on the Yangtze (the Gezhouba Dam), a test run for the Three Gorges, lost more than a third of its capacity within seven years of opening. To minimize this loss of capacity in the Three Gorges area, two approaches have been recommended.

Figure 12.25 Peat erosion in Upper Teesdale, northern England. The sediment contributes to the infilling of Cow Green Reservoir downstream.

The first is to keep the reservoir levels low during high-flow seasons to allow more transported sediment into the lower reaches of the river. The second is to increase tree cover in the catchment in order to decrease erosion of silt into the river.

The construction of water reservoirs in upland environments, especially in north-west Europe, has been a common practice. Soil erosion in such catchments is high, leading to significant sediment supply within the catchment (Figure 12.25). This erosion has been exacerbated by human land-use practices and pollution. Sediment within upland catchments has major impacts on the water reservoirs within them. The most important is that sedimentation within the reservoir (Figure 12.26) reduces the effective volume of water that can be held by that reservoir. Another impact is that the sediment has a deteriorating effect on the colour of the water as a result of the high organic content of the sediment (mainly peats). Although sedimentation impacts vary between catchments, it has been estimated, for example, that a typical water reservoir in upland United Kingdom loses 10% of its volume as a result of sedimentation over a 100 year lifetime.

12.6.2 Mining

The activity of mining economic deposits from the Earth can have major impacts on sedimentation in both river and coastal systems (Box 12.4). Mining activity can take the

Figure 12.26 An upland water reservoir after drought conditions showing the layers of sediment that have built up. Such sedimentation reduces the capacity of these reservoirs.

form of subsurface mining for metals, or the open-cast mining of metals, coals and aggregate material. In all cases, the mining activity exposes large amounts of material to subaerial and fluvial erosion through the production of piles of spoil and waste material. This increase in erodibility leads to increased sediment loads in rivers and increased deposition of material in downstream environments. In addition to large **sediment yields**, the sediment that is deposited in downstream settings is commonly highly contaminated with metals, which has an impact on organisms living within those environments. In mining activities that are close to the coast, estuaries will experience increased sedimentation. In Cornwall, UK, historical mining of metals, especially tin, released large volumes of particulate mine waste into river systems (Pirrie *et al.*, 2002). This sediment was deposited in the coastal zone, leading to the rapid siltation of small estuaries and the loss of ports.

HAZARDS

TOXIC SEDIMENT DISCHARGES AS A RESULT OF MINING ACTIVITIES

Potentially toxic elements (e.g. arsenic, mercury, zinc, lead, cadmium) are a major concern in the environment, and in many cases these are associated with sediment in aquatic systems as a result of the affinity of metals for the particulate fraction, and their low solubility in water. Therefore, the dispersion of contaminated sediment in aquatic systems can often have major negative impacts upon ecosystems. The dispersion of this contaminated sediment can be a consequence of the disposal of overburden or low-grade ore material as unconfined spoil heaps (Figure 12.27) from mines such as the copper mine shown in Figure 12.28, or the release of

fine-grained processed mineral waste (tailings) from an impounded tailings reservoir. In the former case, metal contamination of river sediments is as a result of long-term hydrological remobilization of sediment from the spoil heaps. In the case of the latter, contamination is generally the result of an acute, short-lived event resulting from engineering failure.

An example of a severe pollution event linked to the dispersion of metals associated with a tailings dam spill is the Aznalcóllar copper–silver–lead–zinc mine in Spain, 45 km west of Seville. On 25 April 1998 the tailings dam, which held the fine-grained metal-rich tailings waste from the mining activities, failed, releasing metal-rich sediment into the Agrio and Guadiamar Rivers. Approximately 4600 ha of floodplain land was flooded with an estimated 2 million m^3

of metal-rich tailings, as well as over 5 million m^3 of metal-rich acidic water. This was the worst recorded pollution event in Spanish history. The contaminated sediment was deposited along a length of river 40 km downstream from the mine, and 800 m wide.

The clean-up of the affected rivers took the form of mechanical excavation, and manual clearing of the deposited layer of tailings sediment. Recent research has shown that there are long-term impacts upon the river systems as a result of the deposition of these tailings' sediment on the floodplain and the channel bed (Hudson-Edwards *et al.*, 2005; Kraus and Wiegand, 2006). Similar toxic sediment inputs to river systems as a result of mining have happened recently in Bolivia (Rio Pilcomayo; Hudson-Edwards *et al.*, 2001) and Romania (Tisa Basin; Macklin *et al.*, 2003).

BOX 12.4 ➤

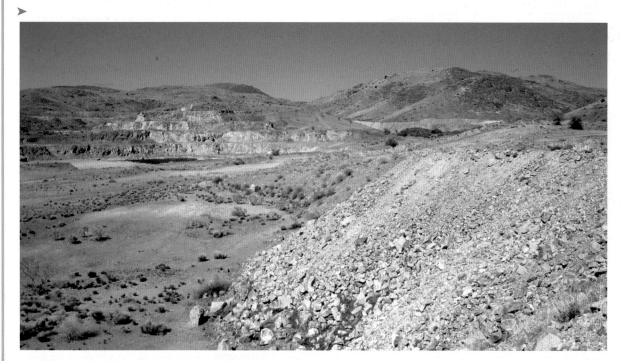

Figure 12.27 Unconfined spoil heaps of metal-rich material, Copper Flats copper mine, New Mexico, USA.

Figure 12.28 Santa Rita open-cast copper mine, New Mexico, USA.

BOX 12.4

12.6.3 Urbanization

Urbanization has a marked effect on sediment sourcing and sedimentation, mainly through the anthropogenic nature of sedimentary material and the engineering of the land surfaces. There are two main types of urban sediment: aquatic sediments in urban water bodies (e.g. canals, docks), and as street sediments on road surfaces ('street dust'). Urbanization of sediment catchments has a number of effects on sedimentation in water bodies. Watercourses become engineered, commonly by channelization and culverting, and land surfaces are paved over. All this has the effect of increasing the rate and extent of sediment supply to receiving water bodies, while vegetation loss reduces the sediment storage capacity of the system. In addition to sediment yields, sediment quality also markedly decreases as a result of urbanization. Sediment can become contaminated by sewage, industrial pollution and vehicular pollution. As a result of this, sediment pollution is a problem within urban water bodies, and chemical reactions in the sediment can lead to the remobilization of this pollution and the generation of noxious methane gas (Taylor *et al.*, 2003). Sediment accumulates on street surfaces as a result of industrial, vehicular and building activities and these sediments commonly contain high concentrations of lead and other metals (Robertson *et al.*, 2003). These sediments have been implicated in respiratory diseases, as the fine-grained fraction of this sediment can be resuspended in the atmosphere and inhaled by humans living and working in urban environments.

12.6.4 Sediment management

The field of sediment management has rapidly expanded in response to the environment pressures exerted upon sediment systems. As Box 12.5 indicates, many systems suffer from issues of sediment quality (e.g. pollution) and sediment quantity (e.g. sediment requiring dredging). Management of sediment in such cases may take the form of source control by cleaning up contaminated sediment sources such as mine sites, industry and waste water treatment works, for example. Site-specific remediation, or clean-up, can take the form of dredging, sediment nourishment, sediment capping (to isolate contaminated sediments from overlying water) or *in situ* chemical or biological treatment.

Reflective questions

➤ Thinking of dams, mining and urbanization, which of one or two of these activities are most likely to (a) increase rates of sedimentation; (b) change the type (or quality) of the sediment; and (c) decrease rates of sedimentation?

➤ Can you describe some methods of measuring sedimentation rates?

NEW DIRECTIONS

CONTAMINATED SEDIMENT MANAGEMENT AND TREATMENT, PORT OF HAMBURG, GERMANY

The contamination of sediments with potentially toxic elements (e.g. metals, persistent organic pollutants) is increasingly being recognized as a major problem. In many cases, the presence of this contamination needs management and remediation. Such an example is the Port of Hamburg,

Germany, which is a major economic shipping port on the River Elbe, approximately 100 km from the North Sea. Sediment is delivered to the port from upstream sources on the River Elbe and from tidal transport of sediment from the North Sea. Sedimentation rates are high, and sediment accumulation in the port significantly reduces water depth. As a result a dredging programme is needed to maintain water depths in the port to allow continued shipping access

(Figure 12.29). Approximately 3 to 4 million m^3 of sediment are dredged each year. Once the sediment has been dredged from the port, it cannot be dumped at sea as the sediment is contaminated and must be treated as a controlled waste. The River Elbe upstream of Hamburg flows through industrial and mining areas in Germany and the Czech Republic and for this reason the sediment in Hamburg Port is contaminated (Netzband *et al.*, 2002), containing

BOX 12.5 ➤

Figure 12.29 Dredging operations in Port of Hamburg. (Source: photo courtesy of Philip N. Owens)

high levels of arsenic, mercury, chromium, lead and organic pollutants (polychlorinated biphenyls, PCBs, and polyaromatic hydrocarbons, PAHs).

The option taken to deal with this dredged contaminated sediment is to treat it in a specially built sediment treatment plant. The dredged material is passed through rotary screens and sieves to remove the coarser fraction of the sediment, leaving behind the fine sediment fraction ($< 63 \mu m$). This fine sediment fraction contains the majority of the contaminants as a result of the high capacity of silt and clay minerals to adsorb contaminants. The resulting coarser sediment fraction is lower in contamination and can be used for building material. The fine sediment fraction is dewatered and disposed of to a specially built landfill facility. By removing the coarse fraction from the dredged material the volume of material which requires landfill disposal is significantly reduced. This dredging of sediment from the port also acts as a pollutant filter to the North Sea, annually removing approximately 30% of metal contaminants from the River Elbe that would otherwise be discharged to the North Sea.

This management of sediment represents a site-specific approach. A more sustainable approach for contaminated sediment management is one that considers sediment management on the river basin scale, identifying and minimizing contaminant inputs at source. It is increasingly being recognized that river basin scale approaches to sediment management are the most effective, from both economic and ecological viewpoints (Owens, 2005).

BOX 12.5

12.7 Summary

Sedimentation is a major process acting to shape the Earth's surface. Sediments are derived from fragments of pre-existing rocks (clastic sediments), the remains of organisms (biological sediments) and the direct precipitation of minerals from seawater (chemical sediments). Sediments are deposited in all surface environments, from continental settings to oceanic settings. Within these environments distinct grain composition, grain shape and sediment bedforms are preserved. This allows the reconstruction of past sedimentary environments through the recent and geological past. Both natural and anthropogenically induced changes in environmental conditions impact upon sedimentation processes. Of these, anthropogenic impacts (e.g. engineered structures and urbanization) have the most marked and rapid effects. It should be clear that a thorough understanding of sediments and sedimentation processes is required for physical geographers and environmental scientists to interpret Earth surface geomorphology.

Further reading

Leeder, M.R. (1999) *Sedimentology and sedimentary basins.* Blackwell Science, Oxford.
This textbook is aimed at higher-level undergraduate students in the earth science and geographical disciplines. Although much of the detail is in depth, the student will find further information on aspects of sedimentology not covered in this chapter. Subjects covered well include the physical transport of sediment by water, wind and ice, the role of tectonics on sedimentation and large-scale sedimentary processes on the Earth's surface. Readers interested in the more physical and mathematical aspects of sediment transport and deposition will be well served by this book.

Lowe, J.J. and Walker, M.J.C. (1997) *Reconstructing Quaternary environments.* Longman, Harlow.
This deals extensively with methods and examples of how Quaternary environments can be reconstructed from the rock record. As such, it has a large amount of information upon sedimentary successions and would give the student a good background to ancient sedimentary deposits. It also provides useful information on loess and wind-blown deposits.

McManus, J. and Duck, R. W. (eds) (1993) *Geomorphology and sedimentology of lakes and reservoirs.* John Wiley & Sons, Chichester.
This edited volume contains a series of papers dealing with the topic of sedimentation within lakes and reservoirs. Of particular interest are a set of papers on the impact of sedimentation in engineered reservoirs. This book takes a more geographical approach than most textbooks on sedimentology.

Nichols, G. (1999) *Sedimentology and stratigraphy.* Blackwell Science, Oxford.
This book is designed for undergraduates studying sedimentology in the earth sciences. Although latter parts of this text are based on interpreting sedimentary environments from the geological record, the first half of the book provides a good, clear overview of the major processes operating on sediments, the formation of bedforms and the sediments deposited in the full range of Earth surface environments.

Perry, C.T. and Taylor, K.G. (eds) (2007) *Environmental sedimentology.* Blackwell Publishing, Oxford.
This book provides an extensive introduction to the sedimentology of contemporary Earth surface environments and the impacts of climatic and environmental change upon these environments. A large range of terrestrial, coastal and marine environments are covered, and aspects of the biology, physics and chemistry of sediments are included, as well as good case examples of sediment pollution, management and remediation.

Reading, H. (1998) *Sedimentary environments: Processes, facies and stratigraphy,* 3rd edition. Blackwell Scientific, Oxford.
This is an updated and revised version of a classic text. The book covers sedimentation in detail within the full range of Earth surface environments. Each chapter deals with specific sedimentary environments.

Tucker, M.E. and Wright, V.P. (1990) *Carbonate sedimentology.* Blackwell Scientific Publications, Oxford.
This book gives full coverage of the composition, origin and depositional environments of carbonate sediments. There are very good chapters on the constituents of carbonate sediments and a wide range of carbonate depositional environments are covered. There are also chapters on the geochemistry of carbonate sediments and on carbonate sediments in the geological record.

Web resources

European Sediment Research Network – SedNet
http://www.sednet.org
This site contains a monthly newsletter and links to further resources on sediment within river basins throughout Europe.

International Association of Sedimentologists
http://www.iasnet.org
This site presents articles and news on research into sedimentology and papers published in the journal *Sedimentology.*

Research Centre Ocean Margins: Sedimentation Processes
http://www.rcom-marum.de.html
Information on the research into ocean sediments and sedimentation processes undertaken by the Research Centre Ocean Margins (RCOM) in Bremen is provided. The site includes background information to the scientific questions it is most concerned with.

Society for Sedimentary Geology
http://www.sepm.org/index.htm
This site presents news of recent research in the field and papers published in the *Journal of Sedimentary Research.*

Three Gorges Dam Case Study
http://www.arch.mcgill.ca/prof/sijpkes/arch374/winter2001/
dbiggs/three.html
Here you will find information on the Three Gorges Dam
Project, the controversy over its existence and links to other sites
about the dam and the river.

US Environmental Protection Agency (EPA) contaminated
sediments
http://www.epa.gov/waterscience/cs/
This site contains information upon the assessment and management of contaminated sediment in the United States.

USGS: Types of Sand Dunes
http://pubs.usgs.gov/gip/deserts/dunes/
This site provides photos and descriptions of the major different
forms of sand dunes.

 Visit our website at **www.pearsoned.co.uk/holden** for further questions, annotated weblinks, interactive models
and video-clips showing physical processes in action.

CHAPTER 13

Catchment hydrology

Joseph Holden
School of Geography, University of Leeds

Learning objectives

After reading this chapter you should be able to:

➤ describe and critically evaluate the measurements involved in producing a catchment water budget

➤ understand the main hillslope hydrological processes that lead to river flow

➤ explain spatial and temporal changes in runoff generation

➤ evaluate the form of river hydrographs and describe the likely processes leading to their form

➤ understand how land-use and climate change may affect catchment hydrological processes, and how these might alter runoff and storm hydrographs

13.1 Introduction

A **catchment** is defined as an area of land in which water flowing across the surface drains into a particular stream or river. The **drainage basin** or catchment has been a fundamental unit of study in hydrology and geomorphology since the time of Robert E. Horton in the 1930s. The catchment is a convenient unit because it is normally well-defined topographically, it can be studied as a series of nested units of increasing size (so that larger catchments are made of many smaller subcatchments), and it is an open system for which inputs and outputs of mass and energy can be defined and measured. Catchments are delineated on the basis of land surface topography. The boundary of a catchment is called a **drainage divide**. Catchments are made of hillslopes and channels. The proportion of hillslope area to channel density or total stream channel length may determine how efficiently water can be removed from a catchment since water in channels tends to move much more quickly than water across and through hillslopes.

Most water reaches river channels (and lakes) via hillslopes. Various hillslope flow pathways exist and these delay flow to different extents. The flow paths also determine water quality; water that has been in contact with soil or rock for long periods, for example, often has a very different chemistry from **precipitation** water (see Chapter 15). The dominance of different types of runoff process is controlled by local climate and catchment features such as geology, topography, soils and vegetation. Knowledge of the relevant mechanisms and their controls is important for determining catchment hydrological response to a precipitation event. The amount of precipitation that reaches the river channel can be very great (almost 100% in some urban areas) or very low (less than 5%), depending on soil or rock water storage and evaporation. Changes in catchment management (or land-use) such as urbanization, ploughing, afforestation, deforestation and artificial soil drainage can result in changes to the flow paths for water across hillslopes and therefore to changes in the timing, volumes and quality of water reaching the river channel. Thus, flood risk and water supply can be altered if there are changes in the relative dominance of runoff flow paths occurring within the catchment.

The movement of water on hillslopes occurs in a number of forms including overland flow (of different types), **subsurface flow** involving **micropores**, **macropores** and natural **soil pipes**, to **displacement flow** and **groundwater** discharge. The dominance of these processes varies with climate, topography, soil character, vegetation cover and land-use, but may vary at one location (e.g. seasonally) with soil moisture conditions and with storm intensity and duration. The spectrum of runoff processes is related to both the type and intensity of erosion process (both in particulate form and dissolved or solutional form) and to the resulting mode of hillslope and landform evolution (see Chapters 11, 12 and 15). Therefore, while the landscape plays a role in determining runoff processes, the runoff processes themselves help to shape the landscape.

Historically many catchment studies have adopted a **water balance** approach:

$$P = Q + E + I + \Delta(M, G, S) \qquad (13.1)$$

where P is precipitation, Q river discharge, E evapotranspiration, I interception, M soil water storage, G groundwater storage and S channel and surface storage (Ward and Robinson, 2000). Seasonal variations in the balance (especially between precipitation and evaporation) help to control the seasonal patterns of river flow. In arid zones such as central Australia or south-east Spain high rates of evapotranspiration may mean that river flows are small (many are completely dry) even after a rainfall event. Traditionally, physical geographers adopted the water balance approach to catchment problems with interest centred on the relationship between the inputs and outputs. The problem with these approaches is that they hide the true complexities within the system. The various processes of runoff act on different spatial and temporal scales such that a ratio of inputs to outputs tells us little about *why* something happens in a drainage basin.

This chapter therefore aims, in addition to investigating the main components of a catchment water balance, to examine runoff production processes and to evaluate the role of these processes in generating river flow. This will help explain why there is spatial and temporal variability in

the dominance of particular processes and therefore such variability in river flow. The chapter will then be able to demonstrate how environmental change may lead to both simple and complex (non-linear) changes to hillslope runoff processes and river flow.

13.2 Measuring the main components of catchment hydrology

Figure 13.1 shows the main components of a traditional catchment water budget. It does not include the transfer processes that will be discussed later. Measuring the main components shown in Figure 13.1 might at first seem straightforward but there are many problems associated with the techniques employed. The following sections discuss each component of the water budget and associated measurement strategies.

13.2.1 Precipitation

The main hydrological inputs to any catchment system are in the various forms of precipitation (e.g. hail, mist, dew, rain, sleet or snow). Precipitation is usually expressed in units of length such as mm or cm (or inches in the United States). This is because it is assumed that the precipitation has fallen uniformly over a given area and so the volume of water is divided by the surface area of the catchment (or gauging instrument) to give a depth of water. Rain gauges are traditionally used to collect inputs of precipitation to the

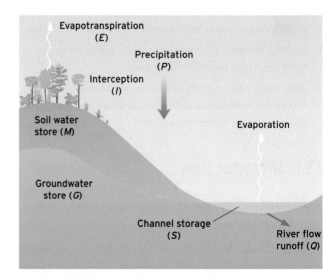

Figure 13.1 Main measurement components of a catchment water budget. The letters used form part of equation (13.1).

(a)

(b)

Figure 13.2 Rain gauges. (a) A rain gauge at ground level surrounded by a metal grid. The grid reduces potential splash and turbulence effects which might influence the amount of precipitation that enters the gauge. (b) A tipping bucket rain gauge with the internal mechanism visible. The water drips through a funnel at the top into the sea-saw buckets. Each time the bucket tips this is recorded by a small datalogger.

ground surface (Figure 13.2). These gauges funnel water into a collecting device. Many gauges can automatically record the volume of water received (usually by a tipping bucket or siphon mechanism) and if gauges are remote or flood warnings need to be provided, data can be collected by **telemetry** (e.g. mobile phone). There are two sets of problems with these devices related to areal coverage and the form of precipitation.

13.2.1.1 Areal coverage

Rain gauges provide point measurement of precipitation. Often large areas of catchments are only gauged by one or two rain gauges. Yet it is well known that there can be great local differences in rainfall received. Topography, aspect and the localized nature of many storm events may mean that readings from one rain gauge cannot be applied to the whole catchment. If there is a network of gauges then spatial averaging can be done using arithmetic means, **Thiessen polygons** (Figure 13.3) or **isohyets** (contours of equal rainfall; Figure 13.4). Often topography complicates spatial interpolation between gauges since slight changes in altitude can result in large changes in precipitation. It is sometimes difficult to install and maintain a sufficiently dense network of gauges and therefore remote sensing technology is becoming widely used in precipitation measurement (see Chapter 23). Radar techniques allow rainfall totals to be estimated across catchments much more accurately through time and space (Campos *et al.*, 2007). Radar can remotely pick up changes in raindrop density across a catchment allowing improved

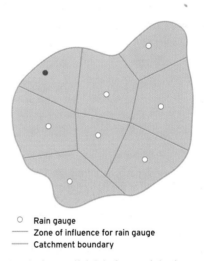

○ Rain gauge
— Zone of influence for rain gauge
— Catchment boundary

Figure 13.3 Producing spatial data from point rain gauges using Thiessen polygons; polygons are constructed around a set of points in such a way that the polygon boundaries are equidistant from the neighbouring points.

flood forecasts and catchment water budgets to be developed. Nevertheless it is often necessary to check radar data using ground rain gauges. Satellite imagery can also be used to estimate snow cover and how much water might be stored in snowpacks.

13.2.1.2 Form of precipitation and turbulence effects

Precipitation in forms other than rainfall, and wind turbulence around gauges, both cause problems for precipitation measurement. Careful siting of a rain gauge, usually at

Figure 13.4 Producing spatial data from rain gauges using isohyets joining points of equal precipitation around Hong Kong on 6 August 2002. (Source: Courtesy of the Hong Kong Observatory of Hong Kong Special Administrative Region)

ground level, is needed so that turbulence does not affect precipitation entering the gauge orifice (Figure 13.2a). In temperate and polar regions snowfall is not immediately recorded on most rain gauges since water is not freely available to trickle into the collecting device. Even those gauges that are heated suffer from the problem that snow that has already fallen elsewhere can blow into the gauge, causing precipitation totals to be overestimated. Many mists and dews can precipitate water onto vegetation or the ground surface. Rain gauge surfaces have different textures from surrounding vegetation and may not accurately record these inputs. For example, in Newfoundland, Price (1992) used wire netting to catch fog precipitation and found that 50% of summer precipitation to the vegetation and ground surface came in the form of fog that was not correctly measured by rain gauges. Therefore it is often necessary to have separate recording devices for different types of precipitation (e.g. using automatic snow depth, pressure or light reflectance recorders).

13.2.2 River flow

River flow is the only phase of the hydrological cycle in which water is confined into well-defined channels allowing 'accurate' measurements to be made of the quantities involved. Good water management is founded on reliable river flow information. Many river flow gauges exist around the world and generally these take the form of a water-level recorder (normally a pressure sensor or a float) housed in a protective well which is fixed into the river at a suitable location. In order to get good control over water levels and to allow for small changes to result in measurable changes in water level around the recording instrument, many gauging sites take the form of weirs or flumes (Figure 13.5). These large in-stream engineering features, however, are not always practical on large rivers.

If measurement of water volumes is required in order to produce a water budget then recorded water levels must be converted into a discharge reading for the river. This

(a)

(b)

(c)

Figure 13.5 Gauging stations on the Tees catchment, northern England: (a) compound Crump weir, (b) V-notch weir, (c) flume-type weir.

requires a **rating equation** to be determined for each river flow gauging site. Sometimes this is well known for a particular shape of weir but generally field calibration is required. Box 13.1 shows how stream discharge can be measured and how the data can then be used to derive a rating equation. There are inevitable errors involved with using rating equations for river flow, not least those associated with the techniques of measuring discharge itself (e.g. inaccurate measurement of stream cross-section, effect of instruments on local water velocity, inadequate mixing of tracers, vertical changes in water velocity not accounted for by a measuring instrument), the difficulties with obtaining reliable discharge data at high flows and the fact that many river channel cross-sections change shape over time (see Chapter 14). Ultrasonic discharge gauges are now becoming more common in order to get around some of these problems. They work by recording the time for a beam of acoustic pulses to cross a river at different depths. The main benefit of ultrasonic gauges is that they can be used in conditions where no stable rating equation can be developed and where other conventional methods are not applicable. However, they cannot be used well where there is dense aquatic vegetation or where high suspended sediment loads are likely.

Often the volume of water that has been discharged by a river is divided by the catchment area in order to allow runoff volumes to be compared with precipitation depths, **infiltration rates** or evaporation losses which are also commonly expressed in units of length. By working out how much water entered the catchment as precipitation and measuring how much came out (by measuring stream discharge) we can determine the 'efficiency' of the catchment. For example:

- Precipitation = 40 mm in 7 days.
- River discharge = 120 000 m^3 of water in 7 days.
- The catchment area is 15 km^2 (this catchment area is equivalent to 15 × 1000 × 1000 m^2 (since there are 1000 × 1000 m^2 in a km^2) = 15 000 000 m^2).
- If we were to spread the discharge evenly over the surface of the entire catchment this would give us a depth of water of 120 000 m^3/15 000 000 m^2 = 0.008 m = 8 mm of water.
- So when compared with the depth of rainfall that fell over the entire catchment, which was 40 mm, a discharge of 8 mm is equivalent to 20% of the precipitation.
- Hence the catchment efficiency was 20%.
- If there was no change in catchment storage then we can say that 80% of the water was lost through evapotranspiration.

TECHNIQUES

RIVER DISCHARGE MEASUREMENT

Measuring discharge directly

Several methods can be adopted to measure river discharge. Two main methods are illustrated below.

Velocity–area method

This method requires estimation of the water velocity to be multiplied by an estimation of the cross-sectional area of water. The river cross-section is surveyed and water and bed level are plotted. Water velocity can be measured using a variety of instruments (see Herschy, 1999). Figure 13.6 shows a river survey with the water velocity being measured using an impeller meter. These are commonly used and work by counting the number of revolutions of an impeller as the water flows past it. A close-up view of the impeller can be seen in Figure 1.7 in Chapter 1. It is usual to place the impeller at three-fifths of the water depth to give an estimate of mean velocity in the water column. This is because water will move at different velocities at different heights and it is assumed that three-fifths of the depth provides the best average. Measurements of velocity can be done at several points across the river. Each separate velocity measurement can then be multiplied by a separate area value associated with that reading as indicated in Figure 13.7. The separate discharge values for each segment can then be summed to give the discharge for the whole cross-section.

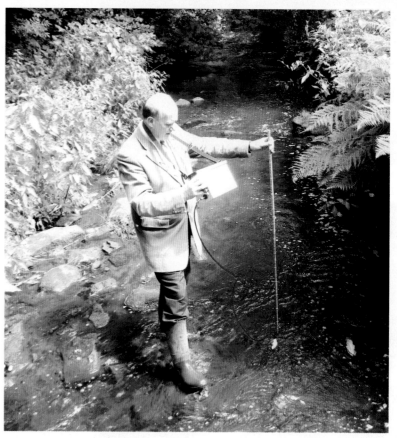

Figure 13.6 Measuring water velocity using an impeller meter.

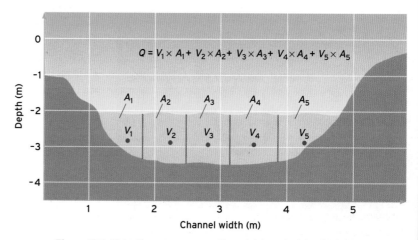

$$Q = V_1 \times A_1 + V_2 \times A_2 + V_3 \times A_3 + V_4 \times A_4 + V_5 \times A_5$$

Figure 13.7 Velocity–area cross-section plot to calculate discharge.

BOX 13.1 ➤

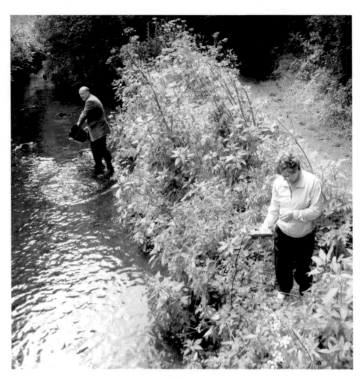

Figure 13.8 Measuring the concentration of sodium chloride in river water after it has been added upstream. The man in the background is adding a gulp injection from a bucket and the woman in the foreground is measuring the conductivity of the stream water using a conductivity probe and has a stopwatch to record the time. However, for the method to work effectively the two people should be much further apart for this size of river because the sodium chloride should be fully mixed with all of the stream water by the time it reaches the conductivity probe.

Figure 13.9 Dilution curve measured in the river shown in Figure 13.8 as the tracer flows past the sampling point.

Dilution gauging

An alternative method is to calculate discharge based on dilution. This involves adding a known concentration of chemical (e.g. sodium chloride) to the river and measuring the amount of dilution in the river water (Figure 13.8). A graph of concentration against time can then be plotted (Figure 13.9). Using the following equation the discharge of the river can then be established:

$$Q = \frac{(C_i - C_b)V}{\int (C_d - C_b)\mathrm{d}t} \quad (13.2)$$

where Q = discharge; C_i = concentration in added water; C_b = background river water concentration; V = volume of water added to river; C_d = measured downstream concentration after water is added; and t = time of measurement. The equation simplifies to:

$$Q = \frac{(C_i - C_b)V}{\text{shaded area under graph shown in Figure 13.9}} \quad (13.3)$$

Further details on how this technique can be used are given in Burt (1988).

Measuring the discharge of a river using a rating curve

If water levels are being measured at a point on a river using an automatic gauge these levels can be converted

BOX 13.1 ➤

> into discharge values. This is done
using a rating equation. It is necessary to measure (e.g. by using one of
the techniques above) the discharge
of the river when the water level is at
different heights. The values can then
be plotted as shown in Figure 13.10
and the equation for the line of best
fit is calculated. This can then be
used to infer the discharge from
any water height at that point on the
river. Care must be taken when inferring discharge values that are
beyond the range of water-level
values that have been calibrated.
This is because the curve can change
shape when the river floods or for
very low flows. It is also very difficult
to measure river discharge using
velocity–area techniques or dilution
gauging when a river is in flood or
for very large rivers.

Figure 13.10 Example of a stage-discharge rating curve and equation.

BOX 13.1

In many catchments river flow is not gauged but it is sometimes estimated using the water balance equation (13.1) when the other variables (e.g. precipitation, evapotranspiration) are known. On an annual basis, or over longer periods, it is often assumed that storage of soil and groundwater is zero and thus $Q = P - E$.

13.2.3 Evapotranspiration

Evapotranspiration is evaporation plus transpiration (the biological process by which water is lost from a plant though its leaves). Evapotranspiration is difficult to measure and is affected by: solar radiation (providing **latent heat**; see Chapters 4 and 5); temperature of the air and the evapo-ration surface (these influence the vapour capacity of the air to hold moisture and the rate at which evapotranspiration can occur); wind speed (removing saturated air); humidity; turbulence (and hence surface roughness as determined by topography or vegetation); plant biology (transpiration processes); and availability of water. As for precipitation, evapotranspiration can vary widely over short distances and

there are often mosaics of values coinciding with changes in vegetation cover.

Evaporation can be measured using an evaporation pan, which is simply a container of open water in which the water depth is measured. Evaporation is calculated as precipitation minus the fall in water level. However, these measurements tend to be unreliable because of the effect of direct sunlight and heating of the pan material. **Atmometers** can give direct readings of evaporation; a water supply is connected to a porous surface and the amount of evaporation over a given time is measured by the change in water stored. However, measurement of evapotranspiration rather than evaporation is necessary since most catchments of interest are not made up entirely of open water. **Lysimeters** are often used to measure evapotranspiration but they can be very difficult to install and maintain. Essentially, a lysimeter (Figure 13.11) isolates a block of soil (with its vegetation cover) from its surroundings so its water balance can be measured. The weight of the block is measured to determine how much water has been lost. If the soil in the lysimeter is kept

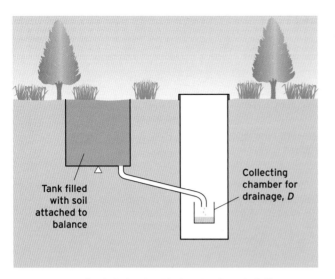

Figure 13.11 A simple lysimeter. The inputs of precipitation are measured as are the outputs from soil drainage out of the block. The block is weighed and evapotranspiration is estimated by determining the change in weight over time once drainage and precipitation are taken into account.

moist by the addition of water, and well covered by vegetation (a grass sward is ideal), evapotranspiration is controlled by the weather and is largely independent of the amount (biomass) of the vegetation. This allows measurement of **potential evapotranspiration** (PE) which is the evapotranspiration from a vegetated surface with unlimited water supply. Because of difficulties in directly measuring PE it is often estimated based on equations that take account of solar radiation, air temperature, dew point temperature, wind speed and the vegetation roughness properties (e.g. vegetation height and density). However, as with precipitation and lysimetry measurements, there are problems with spatial variability of evapotranspiration and the applicability of local calculations to entire catchments.

13.2.4 Soil water

The ability of a soil to absorb and retain moisture is crucial to the hydrology of an area. The soil water store will vary seasonally and will depend on soil properties, soil depth and precipitation. Deep permeable soils can store large quantities of water, providing a moisture reserve through times of drought and helping to sustain river flow during dry periods. There are a wide range of techniques for the measurement of soil water content. These are listed in Table 13.1. The most common method is the gravimetric technique whereby soil samples are extracted, weighed

and then heated in an oven for 24 h at 105°C to remove the moisture and then reweighed. However, obtaining acceptable point measurements and reliable areal measurements is still difficult. Remote sensing of soil moisture characteristics is possible in some environments either using airborne or satellite instruments or using ground-based geophysical equipment such as **ground-penetrating radar** (see Chapter 23). Airborne or satellite techniques are preferable as they can provide wide spatial coverage in a short period of time and can often illustrate the nature of medium-scale variability in soil moisture (e.g. comparing hillslope to hillslope or hillslope top to hillslope bottom). However, most of these Earth observational techniques are limited to examining soil moisture in the upper 5 cm of the soil profile (Schmugge and Jackson, 1996) which may not be representative of the entire soil profile. A range of techniques exist for examining soil water movement and these will be discussed later in this chapter.

13.2.5 Groundwater

Groundwater is water in the **saturated zone** below the land surface. Therefore groundwater can be held within soils or in bedrock. Nearly all rocks are porous and can hold water within them. The depth at which a soil or rock is fully saturated (all the pore spaces are full of water) is called the **water table** (Figure 13.12). If we dig a hole into the ground (such as a well) then water will flow from saturated rock into the well. The water level within the well will rise to a constant level which demarcates the water table. Therefore measuring the water table is relatively simple and requires an instrument to record the water level in a well (similar to those used in river-level gauging). The water table is often not flat depending on topography and geology. Sometimes the water table will be at the surface, particularly after heavy rain in hillslope hollows.

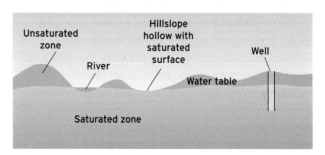

Figure 13.12 Saturated and unsaturated zones and the water table.

Table 13.1 Some common methods for determination of soil water content

Technique	Method	Main advantages	Main disadvantages
Feel	Estimate by touch	Very quick, cheap	Not quantitative and requires vast experience
Gravimetric	Soil samples weighed and heated in oven at 105°C for 24 h and then reweighed to determine moisture loss by mass	Accurate, little equipment required	Slow, destructive, requires many samples
Neutron probe	Radioactive source lowered into hole and fast neutrons released – these are impeded by hydrogen nuclei in soil water. A detector senses scattering of impeded neutrons	Repeatable at any given site, rapid measurement	Must be calibrated for each new test site, radiation hazard, high cost
Time domain reflectometry	Waves of electromagnetic energy sent into ground and reflections of these waves from subsurface features are collected. The presence of water affects the speed of the wave	Rapid collection of spatial data	Must be precisely calibrated. Lots of noise on the signals
Capacitance sensors	Uses dielectric constant of soil as a measure of water content	Small-scale differences between soil layers can be measured as well as values close to the soil surface	Difficult to install and calibrate as air gaps and local inhomogeneities affect readings
Heat dissipation sensors	The temperature in a porous block is measured before and after a small heat pulse is applied to it. The amount of heat flow is proportional to the amount of water contained within the soil	Variation in the conductivity of water is not a problem, easy to automate, non-destructive	Large power requirement, high cost
Carbide method	CO_2 gas is emitted when water is added to calcium carbide. A small wet weight soil sample is mixed with calcium carbide in a sealed container and the pressure of the produced gas is measured	Relatively inexpensive, rapid in field measurements	Destructive, specialized equipment and reagents needed
Remote sensing	Satellite/airborne imaging	Large spatial coverage, fast	Difficult to calibrate the images, coarse scale

The groundwater store is important because in some areas it is the only source of water for humans and it represents virtually all (97%) of the Earth's non-saline and non-frozen water resources (Price, 2002). However, measuring exactly how much water is in any groundwater store is very difficult and requires an estimation of the porosity of the rock or soil that is holding the water. If the bedrock contains only a few small pores then it cannot hold much water, yet if all the pores are full then it is still saturated. Water is not just stored as groundwater but moves through saturated soil and rock. This water movement will be discussed later in this chapter. One assumption that is often made in water balance studies is that catchments are watertight. However, bedrock geology may allow water to flow out of a catchment below the surface without reaching the local river system. Often surface topography does not tell us exactly which direction water may be flowing below the surface and so it is sometimes difficult to determine groundwater losses and gains from a catchment.

compared with clay soils that have low hydraulic conductivities. Therefore water will drain through sand more quickly than clay. Lateral throughflow through the matrix will occur in any soil in which the hydraulic conductivity declines with depth. If both soil and bedrock remain permeable at depth, however, then percolation remains vertical and little lateral flow can occur; infiltrating water will serve only to recharge groundwater storage (Burt, 1996).

Water also moves through soils on hillslopes that are unsaturated. Even after a long drought most soils contain some water. This suggests that gravitational drainage and evapotranspiration are not the only forces at work in moving water within soils and that other forces involved must be very strong. Chapter 7 discusses the processes by which water is held within soil and it was shown that water remains in the soil after gravitational drainage because of the combined attraction of the water molecules to each other and the water to the soil particles. This water that is held in soil against the force of gravity is known as capillary water and it will move within the soil from wet areas to dry areas. It is more difficult to get water out of small soil pores (spaces between the solid particles) than it is to obtain water from larger soil pores. This is why sandy soils are much dryer than clay soils. The attractive forces holding the water to the soil particles and between the thin layers of water are much greater in small pores. This means that large soil pores lose their water before small pores. However, the forces exerted by small empty soil pores mean that when water is added to a soil it will fill small pores first; thus the 'suction' or 'soil water tension' exerted by small pores is greater than that exerted by larger pores. When all of the soil pores are full of water the soil is saturated and there are no suction forces. Instead there will be forces associated with gravity and the pressure of water above a given point. There will be a positive **pore water pressure** caused by the pressure of water from above. When the soil is unsaturated there will be forces associated with gravity and negative forces associated with the suction effect. It is therefore possible to measure these suction or positive pore water forces and to determine in which direction water is likely to move through a hillslope.

13.3.4.2 Macropore flow

Macropores are pores larger than 0.1 mm in diameter and can promote rapid, preferential transport of water and chemicals through the soil, not only because of their size but also because they are connected and continuous over sufficient distances to bypass agriculturally and environmentally important soil layers (Beven and Germann, 1982). Therefore if a field has many macropores, surface fertilizer applications may get washed through the macropore channels and may

not enter the main part of the soil. Fertilizer applications could therefore be transported out of the hillslope before they can be taken up by plants, potentially resulting in downstream water quality problems. Macropores can be formed by soil fauna, plant roots and cracking (Figure 13.19). They

Figure 13.19 Macropores identified using dye staining. The blue dye is added to the soil surface. It is possible to see areas in the soil where water has quickly percolated down the macropores as these are stained blue. These macropores have mainly been created by crane fly larvae and result in much higher infiltration and percolation rates. In fact two of the larvae can be seen in this photograph. (Source: photo courtesy of Katy Gell)

NEUTRON IMAGING OF WATER FLOWPATHS

Neutron radiography provides images similar to X-ray images but with the advantage that organic materials or water are clearly visible in neutron radiographs because of their high hydrogen content, while many structural materials such as aluminium or steel are nearly transparent. Neutron imaging experiments have been carried out on soil using the neutron imaging station, Neutrograph, at the Institut Laue–Langevin (France). This is the world's most intense beam of its kind. The samples are placed approximately 11 m from the neutron source (the nuclear fission reactor), and are irradiated with neutrons. As the neutron beam passes through the sample, it is attenuated (the intensity is reduced) depending on the length of the neutrons' path through the sample (the sample's thickness) and the material present along that path. The attenuated beam then passes into a lead-shielded, light-tight box, where a screen converts the neutrons to

photons and a special camera detects their presence.

Projections with an exposure time of between 50 and 150 ms were taken every 0.225° as the sample was rotated through 180°, resulting in 800 unique images. These images were then computationally reconstructed to produce cross-sectional slices, which are perpendicular to the beam and have an in-slice resolution of approximately 160 μm. Figure 13.20(a) shows a neutron image of a whole soil

core when all the slices have been put together while Figure 13.20(b) shows one of the slices. A small macropore can be seen near the bottom and top of the core. It is also possible to see that the macropore near the top has rough walls and may have only recently formed whereas the lower macropore has smooth walls and may have water flow through it to smooth its sides.

Part of the method description in this box was kindly provided by Martin Dawson, University of Leeds.

(a)   (b)

Figure 13.20 Neutron images of a soil core. (a) Macropores near the top and bottom of the core. The core is only 8 cm long and 4 cm in diameter. (b) A thin horizontal slice through part of the core with black areas within the core indicating macropores. (Source: Photos courtesy of Martin Dawson)

BOX 13.3

have been identified by using microscope and visual observation techniques (e.g. dyes) and even using neutron imaging (Box 13.3). However, the occurrence of a macropore does not necessarily mean that there will be preferential flow of water through the channel. The process of water flow in macropores has three components: water is delivered to macropores, the available water then flows some distance into the macropores, and finally the water may be absorbed through the walls or the base of the macropores. A macropore must be sufficiently connected to a supply of water in order for there to be flow.

Macropores may not take up much space in the soil and if they are open at the soil surface they will often only take up a tiny proportion of the soil surface. Despite their small

spatial role, macropores can still have a high impact on runoff and play a large role in throughflow as water can preferentially flow through them. A study in Niger on a crusted sandy soil showed that 50% of infiltrated water moved through macropores (Leonard *et al.*, 2001). Some studies in upland peat catchments have indicated that 30% of throughflow moves through macropores (Holden *et al.*, 2001).

13.3.4.3 Pipeflow

Natural soil pipes are subsurface cavities of diameter greater than 1 mm that are continuous in length such that they can transmit water, sediment and solutes through the soil and

(a)

(b)

(c)

Figure 13.21 Natural soil pipes: (a) large pipe outlet with coarse sediment delivered from its base; (b) pipe on a peatland river bank; (c) inside the mouth of a peatland pipe.

bypass the soil matrix. Soil pipes are larger versions of soil macropores. Soil pipes are created by a wide range of processes including faunal activity (animal burrows) and by turbulent flow through desiccation (shrinkage) cracks, biotic (e.g. roots) and mass movement cracks enhancing macropores into pipe networks. Climate, biota, human activity, soil chemistry, soil texture, erodibility, soil structure, hydraulic conductivity, clay minerals, cracking potential and dispersivity are all important controls on soil piping (Jones, 1981). Pipes can be up to several metres in diameter and several hundred metres in length (e.g. Holden and Burt, 2002a) and occur in a broad range of environments. Some typical pipe outlets are shown in Figure 13.21. Pipes may transmit a large proportion of water to the stream in some catchments. Holden and Burt (2002a) found 10% of river discharge in peat catchments had moved through the pipe network on the hillslopes while 43% of discharge moved

through pipe networks in semi-arid loess soils of China (Zhu, 1997). Figure 15.12 in Chapter 15 shows a small soil pipe releasing large amounts of water.

Piping is common in arid and semi-arid areas such as south-east Spain and Arizona where shrinking and desiccation cracking are common. Often soil crusting in these environments can result in infiltration-excess overland flow. If a few cracks at the surface exist then infiltration will be concentrated at those points. Turbulent flow may then enlarge cracks to form soil pipes. In these areas pipes can grow so large that they eventually collapse, forming large gullies, and so they are important in shaping the landscape of several regions. Often the networks of pipes can be quite complex and may look something like that shown in Figure 13.22. Pipes can therefore provide rapid connectivity of water, sediment and solutes throughout the soil profile and form complex meandering and branching networks.

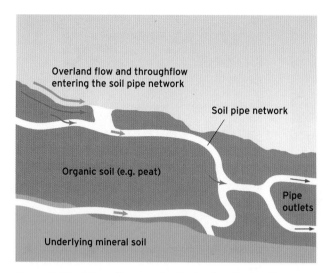

Figure 13.22 Schematic example of a short vertical section through a soil pipe network. The pipe allows water to move quickly through different layers of the soil.

Figure 13.23 A confined aquifer. Between X and Z the aquifer is unconfined and has a water table; to the right of Z the aquifer is confined and has a potentiometric surface. Wells A, B, D and E enter permeable material and strike water. Wells C and F are in impermeable material which will yield water only very slowly. (Source: after Price, 2002)

13.3.4.4 Groundwater flow

Groundwater is water held below the water table both in soils and rock. Therefore groundwater flow has, to some extent, already been discussed in the sections on matrix flow, macropore flow and pipeflow. However, further treatment of groundwater flow as a separate component is required because of its worldwide importance. In many catchments water is supplied to the stream from groundwater in the bedrock. This is water that has percolated down through the overlying soil and entered the bedrock. Rock has small pores, fractures and fissures. Therefore it is possible to use Darcy's law (see Box 13.2) to investigate flow rates through bedrock where fissures are at a minimum. Where there are large fractures such as in cavernous limestone areas (e.g. Cutta Cutta caves area of northern Australia) then it may not be so useful.

Groundwater may be a large store of water, but in order for it to be available to supply river flow the holding material (rock or soil) needs to be not just porous but permeable. That is to say, a rock (or soil) may be porous but relatively impermeable either because the pores are not connected or because they are so small that water can be forced through them only with difficulty. Conversely a rock that has no voids except one or two large cracks will have a low porosity and therefore a poor store of water. Nevertheless because water will be able to pass easily through the cracks the permeability will be high (Burt, 1996). Layers of rock sufficiently porous to store water and permeable enough to allow water to flow through them in economic quantities are called **aquifers**. Sometimes aquifers can be confined between impermeable

rock layers (**aquitards**) and are open only for recharge and discharge at certain locations (Figure 13.23). Sometimes aquifers are recharged from stream and lake bed seepage rather than supplying river flow itself. However, groundwater is generally an important supplier of river baseflow (although glacier ice melt is another source). Depending on the nature of the aquifer, baseflow discharge may be uniform throughout the year or peak discharge may lag significantly behind precipitation inputs. The next section of this chapter details how variations in runoff production processes such as groundwater flow can cause changes in river discharge.

Reflective questions

➤ Why are the source areas for infiltration-excess overland flow and saturation-excess overland flow likely to be different in any given catchment?

➤ Why can Darcy's law not be applied accurately to most hillslopes?

➤ Why might fertilizers and pesticides applied to some fields move through the soil very quickly and bypass the plant roots leading to river pollution?

13.4 River discharge, hydrographs and runoff production processes

Seasonal variation in river flow which tends to be repeated each year is known as the **regime** of the river. Often this is expressed as monthly discharge. Chapter 14 provides some examples of river regimes in different climates. Rivers in equatorial areas tend to have a fairly regular regime, tropical rivers show a marked contrast between discharge in dry and wet seasons, while in other climatic areas snow may complicate the regime as it does not contribute to runoff until melting occurs. Thus regimes can often be complex and despite important climatological controls may also depend on the catchment geology and soils that control the relative role of hillslope runoff production processes.

There are several important components of river flow that need to be assessed in order to determine flood risk and water availability for reservoirs and abstraction for drinking water and industry. The most important of these is river discharge and its variability over time. A graph of the way discharge varies over time is called a **hydrograph**. Hydrographs can be analysed for long periods or for short storm events. Long records of river discharge are useful because they can tell us about the likely frequency of particular high- and low-flow events and the seasonality and overall variability in flows. Unfortunately long-term records of river flow are not widely available.

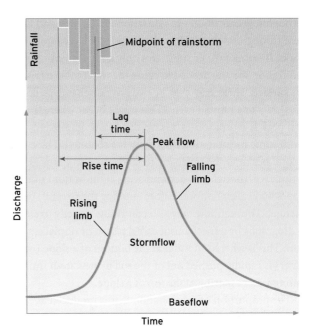

Figure 13.24 Components of a storm hydrograph.

13.4.1 Stormflow

It is often useful to analyse the characteristics of individual storm hydrographs. This is because it gives us information on how quickly rainwater moves from the hillslope to the stream channel. Figure 13.24 identifies the main components of a storm hydrograph. The separation on the hydrograph of baseflow and stormflow is often arbitrary and a variety of techniques have been adopted; these are discussed in Ward and Robinson (2000). The baseflow is the amount of water in the river channel that is derived from groundwater sources and can be considered to be a 'slowflow' component of the hydrograph. In the case of Figure 13.24, baseflow rises only slowly and some time after the rain has fallen. This is because water in the groundwater zone typically travels much more slowly and by longer, more tortuous flow paths than by other processes. The amount of baseflow

in a river depends on seasonal variations in precipitation, evapotranspiration and vegetation. However, there is a wide range of baseflow runoff responses. In cavernous limestones water will very quickly move through cracks and fissures providing rapid, peaked groundwater hydrographs. However, it is unusual for aquifers to provide a major contribution to storm hydrographs (Price, 2002).

The hydrograph in Figure 13.24 shows that there is a lag time between the precipitation and the peak discharge of the river. This lag time is affected by the hillslope runoff processes discussed above. Where infiltration-excess overland flow dominates the hillslope runoff response then the hydrograph is likely to have a short lag time and high peak flow (Figure 13.25a). If throughflow in the small soil pores (matrix flow) dominates runoff response on the hillslopes then the hydrograph may look something like that in Figure 13.25(b). The lag time is long and peak flow low. However, since

Figure 13.25 Example hydrographs for a single storm: (a) for an infiltration-excess overland flow dominated catchment; (b) for a throughflow dominated catchment.

throughflow contributes to saturation-excess overland flow then throughflow can still lead to rapid and large flood peaks. In some soils or substrates only a small amount of infiltration may be needed to cause the water table to rise to the surface. There may even be two river discharge peaks caused by one rainfall event. This might occur where the first peak is saturation-excess overland flow dominated (with some precipitation directly in the channel). The second peak may be much longer and larger and caused by subsurface throughflow accumulating at the bottom of hillslopes and valley bottoms before entering the stream channel. Throughflow may also contribute directly to storm hydrographs by a mechanism called piston or displacement flow. This is where soil water at the bottom of a slope (old water) is rapidly pushed out of the soil by new fresh infiltrating water entering at the top of a slope.

The proportion of precipitation that is produced as stormflow in a river may vary from storm to storm. During a wet season saturation-excess overland flow may be more common since the area of the catchment over which it is generated is greater. As a result river discharge peaks will be greater and lag times shorter than during drier antecedent conditions.

The occurrence of hillslope flow processes in the catchment and their relative dominance affect the speed at which water is delivered to the stream (Table 13.2). Catchments dominated by infiltration-excess overland flow have the shortest lag times and the highest peak flows. This is why urbanization can lead to increased flood risk downstream. Dekker and Ritsema (1996) showed for a clay soil in the Netherlands that flow through shrinkage cracks when the soil was relatively dry produced higher peak flows than when the soil was wet and lacked macropore channels.

Table 13.2 Relative flow speeds for runoff processes

Pathway	Relative speed of delivery
Infiltration-excess overland flow	Very fast
Saturation-excess overland flow	Fast
Throughflow:	
Matrix flow	Slow
Macropore flow	Medium
Pipeflow	Fast

13.4.2 Flow frequency

In very large catchments it is often difficult to spot individual storm hydrographs on the river at the lower end of the catchment. This is because the length of time for flood peaks to travel down the various tributaries and through the system may be many days or weeks. Therefore longer-term analysis of daily or weekly flows is required. Furthermore, over long periods it is possible to identify important differences in catchment response through seasons. Figure 13.26 shows hydrographs for two adjacent catchments in Dorset, England. There is a great difference in hydrograph response which is not related to different climates. The Sydling Water catchment is dominated by baseflow and lies on permeable chalk limestone with high soil infiltration capacities. Therefore little storm runoff is generated within the catchment and groundwater flow dominates river flow. For the River Asker, however, there are many events when stormflow is produced and the catchment is more responsive to rainfall. There are many storm hydrographs during the year and it is likely that overland flow or rapid throughflow mechanisms (through macropores and soil pipes) generate most of the runoff in the Asker catchment. The infiltration capacities of the soils are quite high and so infiltration-excess overland flow is unlikely to be a frequent occurrence. However, the substrate is impermeable clay in the Asker catchment and so there is little chance for percolating water to penetrate deeper subsurface layers. Therefore the soils above the bedrock are more easily saturated and can rapidly generate saturation-excess overland flow. The soils are also quite thin and there are steep slopes.

When analysing long-term discharge records it is sometimes useful to compare **flow duration curves**. Hourly or daily flows are grouped into discharge classes and the percentage of time that any particular flow is equalled or exceeded is plotted. Often these curves are plotted on a probability scale (this is the scale on which a normal distribution plots as a straight line) as shown in Figure 13.27. This allows the lower and higher ends of the curves to be examined and compared with other curves in more detail. Generally, if rivers are being compared it is usual to plot the other axis as the discharge divided by mean discharge to allow fair comparison between large and small catchments. Examples of flow duration curves are plotted in Figure 13.27. Flow duration curves that plot steeply throughout, such as the Tees and the Tamar (UK), denote highly variable flows with a large stormflow component, whereas gently sloping curves such as the River Ver (UK) indicate a large baseflow component. The slope at the lower end of the flow duration curve characterizes the perennial storage in the catchment so that a flat lower end indicates a large amount of storage.

Figure 13.26 River discharge on two Dorset catchments during 1998. (Source: data courtesy of National River Flow Archive, CEH Wallingford, and the Environment Agency, South Wessex area)

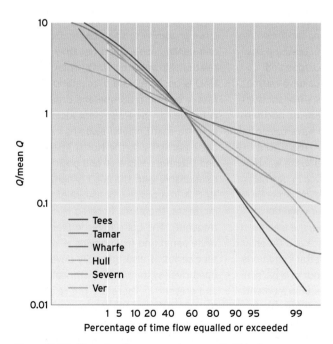

Figure 13.27 Flow duration curves for some British rivers. (Source: after Ward, R.C. and Robinson, M., *Principle of hydrology*, 4th edition, 2000, McGraw-Hill, Fig. 7.14. Reproduced with the kind permission of the McGraw-Hill Publishing Company)

Flow duration curves and regime analysis tend to suggest that river flows are stable. Unfortunately long-term flow records are not always available, but where they are available, it is possible to identify variations in river flow that are more longer term. Figure 13.28, for example, plots mean annual flow for the Willamette River at Albany, north-west Oregon. There is great variability of flow from year to year. There does appear to be some cyclicity to the long-term trends, however, in that dry years tend be grouped together and wet years tend to be grouped together. Research is currently attempting to link trends in annual river flow with climatic factors such as El Niño events or changes in the North Atlantic Oscillation (NAO). Chapter 4 provides further detail on El Niño and the NAO and Box 13.4 illustrates how the NAO may impact river flows.

The severity of any flooding will depend on the response of hillslope runoff production processes to heavy or prolonged rainfall or snowmelt and to the nature of the area being flooded. Often flood frequency is analysed by examining flow duration curves. However, it is more common to analyse the flood record for the number of times water level in a river peaked above a given (often critical) level. Sometimes this water-level value is not the

Figure 13.28 Annual mean discharge for the Willamette River at Albany, north-west Oregon, showing how dry years and wet years appear to be grouped together. A water year is one year starting from the end of the driest season of the year. It is not the same as a calendar year. For example, in Oregon (and in the United Kingdom) a water year runs from I October to 30 September. (Source: after http://water.usgs.gov/pubs/circ/cirell61/tn-nawqa91-15.gif)

THE IMPACT OF THE NORTH ATLANTIC OSCILLATION ON RIVER DISCHARGE

The North Atlantic Oscillation (NAO) is a fluctuation of atmospheric pressure that see-saws between the subtropical and polar regions of the North Atlantic Ocean. The term 'North Atlantic Oscillation' is the oldest known atmospheric phenomenon with written evidence dating back to the diaries of Hans Egede Saabye, a Danish missionary who was writing on the topic between 1770 and 1778. The most commonly used method of measuring the NAO is using mean atmospheric pressure at sea-level gauge stations based in the Azores, Gibraltar or Portugal and comparing those with measurements from gauge stations in Iceland, usually located in Reykjavik.

Renewed interest in the NAO has come about since the 1990s due to a sustained positive phase which has significant implications for the entire North Atlantic region. During a positive phase northern Europe and eastern North America have warmer and wetter conditions while southern Europe, the Mediterranean, northern Africa and Greenland have cooler but drier conditions. When the NAO is in a negative phase the opposite weather patterns occur, resulting in northern Europe and eastern North America having cooler but drier conditions while southern Europe, the Mediterranean, northern Africa and Greenland having warmer but wetter conditions. Therefore the behaviour of a particular NAO phase can result in a potential increase in frequency of flooding in certain areas and drought in others.

Figure 13.29 highlights the opposing relationships that exist

between river flow and the NAO. The River Agueda in Portugal and River Guadalquivir in Spain both lie in close proximity to the Atlantic Ocean and hence have a good negative relationship with the NAO (when the NAO is positive, little precipitation falls in southern Europe and drought is frequent). The River Kent located in north-east England also lies in close proximity to the North Atlantic and therefore has a strong positive relationship with the NAO (when the NAO is in a positive phase, rainfall increases, therefore flooding is more frequent). The Loire River in France, while benefiting from having a long time series, is also a heavily developed river system both historically and presently. Therefore the relationship between the NAO and river discharge is somewhat masked. However, a negative relationship is evident. Current research interest

BOX 13.4 ➤

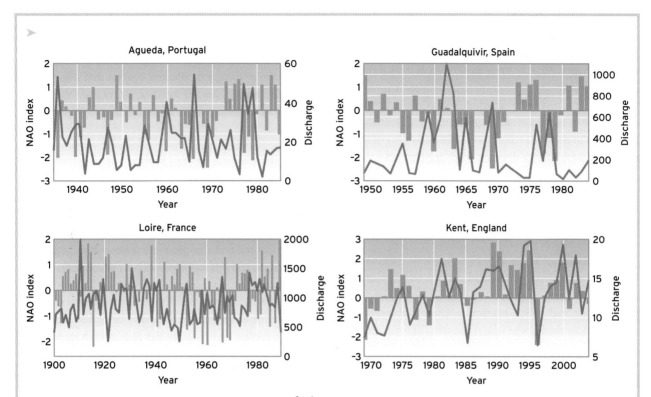

Figure 13.29 Mean annual discharge (blue, measured in $m^3\,s^{-1}$) of four rivers in Europe and corresponding NAO index (red). The strength of the NAO phase, and therefore the associated impacts, is shown by their distance from the zero axes. Note that the length of the available data record for each river is different. (Source: Figure based upon river data kindly supplied from Global Runoff Data Centre (GRDC). Courtesy of Sara Alexander, University of Leeds)

is focused on trying to locate where the boundaries of the different relationships between river discharge and the NAO lie. If northern Europe behaves a certain way to a certain phase, and southern Europe behaves in opposition, where does the hydrological boundary between northern and southern Europe lie? Furthering our understanding of this complex atmospheric phenomenon and its impacts on river flow will result in better prediction and mitigation of flooding and drought. This is increasingly important as global climate change is expected to generate more extreme weather behaviour.

The text in this box was kindly provided by Sara Alexander, University of Leeds.

BOX 13.4

same as a discharge value because the lower parts of many catchments are affected by tides. If high storm discharge from upstream coincides with a spring high tide or a sea-level surge then flooding can be exacerbated. If long-term river-level records are available then simple return period calculations are possible. For example, if a water level greater than 20 m occurred 10 times in 10 years we would say that the return frequency of the 20 m flood at a given point is on average once per year. This helps us determine how often certain flood events might take place. Of course we could get a 20 m flood occurring three times in one year and the return frequency is just an average value. However, inferring what might be expected in the future from flood events that have happened in the past may not be reliable given that land management change and climate change might impact hydrological processes operating within a given catchment.

13.4.3 River flow in drylands and glacial regions

Arid zones, particularly those in subtropical drylands, are characterized by rare but intense rainfall events. Such high rainfall intensities coupled with sparse vegetation cover tend to result in infiltration-excess overland flow, rapid runoff and high flood peaks. However, many dryland soils are coarse and sandy in nature and thus have high infiltration capacities. In these areas, such as the Kalahari in Africa, overland flow is a rare occurrence (see Chapter 16). There can thus be a wide variation in response depending on the type of substrate and vegetation cover. Typically river flows in drylands will cease within a few days of the rainstorm. Much water does not reach the channel system because evaporation rates are high and soil water storage capacity is great since the soils were dry prior to the rainfall. Water is also often lost via seepage into river beds. Chapter 16 provides more information on dryland hydrology.

River discharge in glacial regions tends be dominated by two features. Firstly a seasonal control, which means that river discharge is greatest in early summer when snow and ice melt is at a maximum. Discharge can be extremely low during the winter months. The glacier ice acts as a long-term water store so that annual precipitation inputs to the catchment do not necessarily match the outputs. Changes in the size of the glacier will be a main determinant of catchment water balance. Secondly there is a diurnal control so that night-time discharge tends to be much lower than that of the mid-afternoon. This reflects melting cycles of the glacier ice during daylight hours. Typically peak discharge lags a few hours behind peak temperature as it takes time for the energy to melt the ice and then for the meltwater to be routed through the river network. Further detail on glacial and periglacial hydrological processes is provided in Chapters 18 and 19.

Reflective questions

➤ What are the main ways stormflow can be generated?

➤ Why might the dominance of particular runoff processes vary over time and space?

➤ What shape do you think the flow duration curves of Sydling Water and the River Asker should be? (Both catchments are discussed above and their annual hydrographs are shown in Figure 13.26.)

13.5 Impacts of environmental change on runoff production

Future climate change predictions suggest that some parts of the Earth's land surface will become wetter and others drier. Changes in seasonality will also bring changes in precipitation delivery, the nature of soil saturation and therefore the dominance of particular runoff processes. Changes in precipitation regimes and local temperatures may result in migration of vegetation communities and to associated changes in soil structure and soil formation processes. At the same time changing land management such as deforestation or grazing may result in reduction of the infiltration capacity of a soil and changes in the water budget (e.g. evapotranspiration and interception suddenly decrease on deforestation; afforestation has the reverse affect and will therefore reduce river flow). Population increases, shifts in their location and climate change may bring about great strain on surface and groundwater resources in some parts of the world over the coming decades. Impoundment and river water abstraction alter river regimes and over-abstraction of groundwater can lead to the source of water becoming deeper and deeper as the water table lowers. In some areas the water table may take many decades, or millennia, to recover even if abstraction has ceased. This is because some groundwater is 'fossil' water, formed in earlier wetter periods (such as the large aquifers beneath the Sahara). Because of the links between groundwater and baseflow, rivers can dry up if groundwater is abstracted. Box 13.5 provides an example of the problems associated with such groundwater overuse in Australia.

In some locations flooding has increased in recent years due to both changing precipitation regimes (e.g. positive NAO in northern Europe) and land management activity which results in enhanced production of stormflow. One of the problems with building larger and better flood defences around our towns and cities in order to counter greater river discharges is that the flood water has to go somewhere. Many solutions to flooding have involved either building taller levees or embankments next to rivers or straightening the river channel and clearing out the sediment and vegetation to allow faster, more uniform river flows. However, there are many examples from around the world where these techniques have led to worse flooding downstream (e.g. the Mississippi had its worse (and disastrous) flooding in 1993 after many years of river engineering works). This is because sending the water more quickly through one part of the river system simply reduces the lag time downstream and increases the overall flood peak. Floodplains normally

CASE STUDIES

GROUNDWATER ABSTRACTION IN AUSTRALIA

As a vital resource groundwater supplies over 65% of irrigation water for farming in South Australia, Victoria and New South Wales. Western Australia uses groundwater to supply 72% of its urban and industrial demand. Up to 4 million people in Australia rely totally or partially on groundwater for domestic supply. Such demand puts pressure on groundwater resources. This is because if the inflow into the ground is much smaller than the outflow from the aquifers then the groundwater resource will decrease. While there are large groundwater reserves in Australia, many are in very remote areas or where the water is difficult to access. Therefore the more accessible groundwater reserves may become depleted. In fact 30% of Australia's groundwater management units are either almost overused or certainly overused.

Additionally there is the problem of salinization. Salinization is the build-up of salt within the soil (Figure 13.30). This causes severe damage to farmland as the land is no longer able to support crops. Water from deep below the surface contains salts. Normally these salts are held deep below the surface where they do not affect plants. However, when the deep groundwater is brought to the surface and evaporated in hot conditions this leaves behind salts as a deposit, killing plants and soil organisms. There are two main ways excess salts are brought to the surface. The first source is from irrigation water which is abstracted from deep sources. The second source is when

Figure 13.30 Salinization of soils in New South Wales, Australia. (Source: Mark Edwards/STILL Pictures The Whole Earth Photo Library)

BOX 13.5 ➤

land-use is changed so that trees are removed. The trees may keep the water table fairly deep by using a lot of water. However, when they are removed, the water table may quickly rise to the surface and mobilize the salts that are stored in Australia's soils. Approximately 5.7 million hectares of Australia are within regions at risk or suffering from dryland salinity. In 50 years this may increase to 17 million hectares (ANRA, 2001). There are a series of knock-on effects of soil salinization. For example, the high concentrations of salt at the soil surface can also pollute streams when runoff occurs and the native vegetation also becomes damaged. Damage to river ecology due to salt pollution is also associated with loss of streambank vegetation, which can then also lead to exacerbated bank erosion.

BOX 13.5

have a function: they act as a temporary store of water, which means that the flood peak downstream is not as great as it would otherwise be. However, the demand for building or farming on the flat fertile floodplains of the world means that fewer and fewer of them are freely available for a river to store its water. Thus flooding in lowland areas is becoming an ever-increasing problem as the extra water is brought downstream more quickly and in greater quantities than ever before.

Changes in flood risk caused by land management change are not always as simple as would be first expected. For example, it is not possible to say that if you cover 20% of a catchment with trees to take up more water that you will reduce the flood risk by 20%. Indeed, such an activity that reduces flooding in a small area might actually lead to more flooding downstream. Figure 13.31 shows a catchment with hydrographs at different points down the system. Within this larger catchment a small tributary catchment is highlighted. If land management were to change in this catchment that resulted in a lower flood peak and delayed lag times (e.g. dense trees were planted) it would be expected that this would be beneficial to those downstream. However, the figure shows that because the peak from the small tributary has been delayed, it now peaks at the same time as the peak in the main river channel. Therefore it is contributing most discharge at the same time as the main river. The effect of this synchronous peaking is an increase in the flood peak in the main river channel. This will therefore have a negative impact on those living downstream and increase flood risk. This example highlights the importance of considering whole catchments and not just single parts of catchments. Flood risk mitigation strategies must examine whole catchments and models are needed that allow examination of the timing and volume of water moving through the

Rainfall

30 min
60 min

—— Before tree planting in subcatchment
—— After tree planting in subcatchment

Figure 13.31 The importance of flood wave synchronicity when considering impacts of land management change on flood risk. Despite the tree planting activity causing the small tributary catchment to have a lower flood peak, this has still resulted in a higher overall flood peak in the main channel. This is because the timing of the flood peak from the tributary has been delayed so that it now matches that in the main channel and so it makes the overall flood larger.

drainage network across whole catchments. Some parts of a catchment might be more sensitive to change than others and the same change in one part of a catchment (e.g. tree planting) might not have the same impact if that change had occurred in another part of the catchment. Box 13.6 provides more information on the flooding problem.

HAZARDS

FLOODS

Floods are natural hazards and should be expected to occur on any river. Each year flooding causes hundreds of deaths around the world. Floods damage buildings and infrastructure, drown humans and livestock, spread disease, contaminate water supplies and damage crops. Recent major floods include those in Jakarta (January 2007) which killed 80 people, the Malaysia Peninsula (December 2006) which resulted in the evacuation of 100 000 people, Mumbai (July 2005) which killed 700 people and the southern Alberta floods in Canada (June 2005) which killed 72 people. Other large floods include the Mozambique floods of 2000 which killed thousands and devastated the country and the 1998 Yangtze flood which left a staggering 14 million people homeless. Maps of recent and current flooding can be found on the website of the Dartmouth Flood Observatory (**www.dartmouth.edu/ ~floods/**). If you visit the website you will see that there is always a flood occurring somewhere in the world.

The magnitude of floods has been both increased and decreased by human action in different locations. Additionally humans have decided to live in low-lying areas subject to flooding. These areas tend to be where there are fertile soils (often made more fertile by regular flooding) suitable for crops and where navigation of rivers by boats allows transport of goods and people. So floods should be expected if people live on floodplains and it is possible to produce maps of areas prone from flooding. Nevertheless people still choose to develop land and live in zones where flooding is likely.

Humans have modified the landscape of many regions to change both deliberately and inadvertently the size and frequency of floods. Covering more of the landscape with concrete and tarmac, which are impermeable, and then channelling flow into drains that feed streams, will inevitably lead to increased flood risk. Large-scale change across entire catchments could change flood risk too, such as deforestation or overgrazing which compacts the soil, leading to reduced infiltration.

Additionally it is possible that human modification of the climate may also lead to changes in flood frequency and flood magnitude. For example, it is forecast for some areas that a warmer climate could mean more intense rainfall events. This could lead to more floods. However, predicting flood risk is very difficult, especially in a changing climate.

If a flood of a particular size has occurred twice in the past hundred years it might be called the 1 in 50 year flood. However, that does not mean it will occur only twice in the next 100 years. With changing climate (or land management) the same size flood could occur 20 times (1 in 5 year flood). The difficulties of predicting the risk from flooding are of major importance to humans because often designs of flood defence are made based on flood return periods (e.g. to protect against the 1 in 100 year flood).

Flooding from extreme events is the subject of a scientific research initiative funded by the UK's Natural Environment Research Council called 'Flood Risk From Extreme Events'. The research being conducted aims to (1) improve the estimation and prediction of flood risk from extreme events through consideration of the processes involved as part of an integrated system; (2) seek ways to reduce uncertainty and improve the quantification of flood risk; and (3) identify and articulate critical guidance on how flood risk is changing over a range of timescales. Much of this work must be done by modelling to predict how floods will occur under different rainfall scenarios or different land management scenarios. For further details of this project please see **www.nerc.ac.uk**

BOX 13.6

Because the hydrograph shape and relative size of the peak flows are often determined by hillslope runoff processes, human activities can affect the form of a storm hydrograph. Soil erosion, deforestation, afforestation, grazing intensity (compacting the soil surface) and other agricultural activities (such as ploughing, irrigation, crop growth) or the development of towns and drain systems are ways in which the hydrograph shape may be affected. Building dams and changing or diverting stream courses are other ways the hydrograph shape could be altered. If the soil surface or the subsurface structure of the soil is altered in any way then it is likely that the flow paths for water from the hillslope will alter and the shape of the hydrograph will change. Many of these changes may not be simply reversible. For example, many organic soils such as peats change their structure when dried. If they are rewet, their structure does not revert to its original form and is permanently damaged. Therefore in some cases, changing the land management and causing damage to soils might lead to permanent change in flowpaths, which could lead to enhanced flooding. If land management was altered back again to its original use, this may not reduce flood risk because the damage to the soil has already occurred.

Reflective questions

➤ Based on climate change predictions for your area how might your local river flow change over the next 50 years?

➤ For a river near your home find out what land management changes have taken place within its catchment: what are the likely impacts of these land-use changes on the hydrology of the catchment in terms of water balance, runoff production processes, shape of the storm hydrograph and long-term river flow?

13.6 Summary

This chapter illustrates the main components of catchment water budgets and shows how each of these can be measured. The main problems with measurement techniques are related to extrapolating point data over an entire catchment and errors associated with the measurement techniques themselves. Advances are being made on collecting spatial data such as precipitation radar and remote sensing of soil moisture but these still need to be calibrated using point data.

Water reaches the river channel in a range of ways, some by direct precipitation on the channel but mostly by movement through soil or bedrock. Water that runs over the land surface as overland flow is often water that is being returned to the surface from the soil except where infiltration-excess overland flow occurs or where fresh rainwater meets an already saturated land surface and mixes with return flow. Throughflow occurs in a variety of forms, each of which affects residence time of water in the soil and contact with soil or bedrock

constituents. Throughflow can contribute to river stormflow and to river baseflow. The dominance of particular runoff processes in contributing to river flow both changes over time (e.g. during a storm or over seasons) and varies spatially across a catchment.

The nature of river discharge is controlled by the water balance and by the ways in which water reaches the river in any given catchment. Some catchments are dominated by flashy stormflow with minimal baseflow (sometimes none) whereas others with deep permeable soils and rocks are dominated by baseflow with limited stormflow runoff. There is a spectrum of responses in between. Environmental change can lead to changes in the dominance of particular runoff production processes within any catchment or to timing of delivery of water to the river channel and through river channel systems. This leads to changes in catchment response to precipitation in both the short and long term. Some land management or climate-induced changes to runoff production processes may not be simply reversible.

Further reading

Anderson, M.G. and Burt, T.P. (eds) (1985) *Hydrological forecasting.* John Wiley & Sons, Chichester.
This text contains a wide range of important chapters covering hillslope hydrology; overland flow, subsurface flow, infiltration and channel routing are all covered very well.

Arnell, N. (2002) *Hydrology and global environmental change.* Pearson Education, Harlow.
This is an excellent companion to this chapter. The book is full of great examples and explains terms and processes clearly. There are many useful comments and diagrams about how environmental change and hydrological processes interact.

Brutsaert, W. (2005) *Hydrology: An introduction.* Cambridge University Press, Cambridge.
More advanced principles are delivered in this book for those who wish to follow up particular topics.

Herschy, R.W. (ed.) (1999) *Hydrometry, principles and practice.* John Wiley & Sons, Chichester.
This book contains an overview of river gauging techniques and instruments. It is a little limited on some aspects of the water budget, but written very clearly with lots of useful photographs.

Jones, J.A.A. (1997) *Global hydrology: Processes, resources and environmental management.* Pearson Education, Harlow.
This book provides a larger-scale perspective on hydrology than many other books (as well as containing material on catchment processes). There is some informative material on human water demand, water management and human impacts on hydrological processes and river regimes as well as a good section on modelling runoff.

Lehr, J.H. (ed.) (2005) *Water encyclopedia.* John Wiley & Sons, New York (5-volume set).
This is a comprehensive reference work with excellent articles on all aspects of hydrology. I contributed eight articles to this encyclopedia.

Petts, G.E. and Calow, P. (eds) (1996) *River flows and channel forms.* Blackwell Scientific. Oxford.
There are lots of good examples of runoff production processes and catchment management in this book.

Price, M. (2002) *Introducing groundwater.* Nelson Thornes, Cheltenham.
This is a clear, well-illustrated and nicely written book on groundwater. The best introduction to groundwater you can read. It also contains material on analysing river flows.

Shaw, E.M. (1994) *Hydrology in practice.* Stanley Thornes, Cheltenham.
This is a detailed (and sometimes technical) overview of hydrology.

Ward, R.C. and Robinson, M. (2000) *Principles of hydrology.* McGraw-Hill, London.
These authors provide one of the best general textbooks on hydrology.

Web resources

Australian Centre for Groundwater Studies Home Page
http://www.groundwater.com.au/
This site provides information on various groundwater issues and research, an extensive glossary and links.

Australian National Resources Atlas
http://audit.ea.gov.au/ANRA/
This website provides information and case studies on several hydrological topics including salinization problems as discussed in Box 13.5.

Centre for Ecology and Hydrology (CEH)
http://www.nerc-wallingford.ac.uk/ih/
CEH is a UK institute (formerly the Institute of Hydrology) that undertakes research into the effects of land-use, climate, topography and geology on the volume and character of surface water resources. Information on hydrological topics, research and links to similar sites is provided.

Climate Change 2001, Impacts, Adaptation and Vulnerability: Hydrology and Water Resources
http://www.grida.no/climate/ipcc_tar/wg2/159.htm
This site provides extensive discussion on the observed and predicted impacts of recent and predicted climate change on the hydrological cycle at all scales throughout the world.

Cooperative Research Centre (CRC) for Catchment Hydrology Home Page

http://www.catchment.crc.org.au/

The CRC provides resources for assessing the hydrological impact of land-use and water management decisions at a catchment scale. This site provides information on current research and projects in catchment hydrology and catchment modelling.

Dartmouth Flood Observatory

http://www.dartmouth.edu/~floods/

An excellent site for seeing where floods are occurring today (the site is updated every day) and where they have occurred in the past.

Groundwater – Nature's Hidden Treasure

http://www.ec.gc.ca/water/en/info/pubs/FS/e_FSA5.htm

Provided by the Canadian Department of the Environment, this site has an informative discussion on groundwater, its various sources, processes and contamination.

Hyperlinks in Hydrology for Europe and the Wider World

http://www.nerc-wallingford.ac.uk/ih/devel/wmo/

This site provides an extensive set of links to websites concerned with various different aspects of catchment hydrology at a regional, national and international level. The site is provided by the World Meteorological Organization.

International Hydrological Programme

http://typo38.unesco.org/index.php?id=240

UNESCO's intergovernmental scientific cooperative programme in water resources to better manage global water resources is

presented in this website. The site offers numerous case studies on world water issues and conflicts, worldwide water events, and various useful and interesting links to related hydrology sites by theme and by region.

TOPMODEL web page

http://www.es.lancs.ac.uk/hfdg/freeware/hfdg_freeware_top.htm

TOPMODEL is a rainfall-runoff model that bases its distributed predictions on an analysis of catchment topography; a version of TOPMODEL can be downloaded from this site.

US Geological Survey Website

Information on current flooding, drought and general hydrological issues can be found at the USGS Surface Water Information Page:

http://water.usgs.gov/osw/

An informative online article/booklet can be found at the 'Groundwater and Surface Water, a single resource' page. It discusses the processes of interaction between groundwater and surface waters and the human influence upon this system:

http://pubs.usgs.gov/circ/circ1139/

Finally, the 'sustainability of groundwater resources' site provided by the USGS gives a helpful online booklet on groundwater, its uses and its sustainability; it is clear to read and provides helpful diagrams:

http://pubs.usgs.gov/circ/circ1186/

 Visit our website at **www.pearsoned.co.uk/holden** for further questions, annotated weblinks, interactive models and video-clips showing physical processes in action.

CHAPTER 14

Fluvial geomorphology and river management

David J. Gilvear

School of Biological and Environmental Science, University of Stirling

Learning objectives

After reading this chapter you should be able to:

➤ understand basic controls on the processes of flood generation and sediment delivery to rivers

➤ define and measure the size and shape of river channels and drainage networks

➤ explain how water and sediment move in river channels

➤ describe the environmental controls on the size and shape of river channels and rates of channel change

➤ show an awareness of how an understanding of river channel behaviour can be used to manage rivers in a more sustainable and environmentally sensitive way

14.1 Introduction

Rivers vary greatly in appearance with changes both from source to mouth and between individual rivers. It is this morphological diversity that is the fascination for many people. In the case of Europe, contrast the quietly flowing small chalk-fed streams shaded by overhanging willows of southern England with the turbulent milky-coloured torrents draining glacial regions of the Alps. Similarly, contrast the raging torrents in the Alps that form the headwaters of the River Rhine with the more tranquil character of the same river as its passes through the 'polder' landscape of the Netherlands and discharges into the North Sea. At a local scale, salmon fishermen identify reaches and 'pools' of unique character to improve their chances of a successful catch. The size, shape and location of a river can also be transformed overnight by a single large flood, by depositing sediment in some areas and reactivating other reaches by erosion. Few people in the United Kingdom will, for example, have escaped seeing the horrifying images of the destructive forces of the flood that hit the small Cornish village of Boscastle, England in the summer of 2004, the January 2005 flooding of Carlisle, northern England, or the July 2007 floods across Yorkshire and Worcestershire. Meanwhile in 2005, floods in China directly affected 210 million people, killed approximately 1500 people and damage was estimated at $19.3 billion. At the same time in Western India 1000 lives were lost with, for example, 940 mm of rain in 24 h falling on Mumbai; this storm immediately following the Boxing Day Indian Ocean tsunami (see Chapter 2) was described by the media as a 'tsunami from the sky'. Floods and the associated fluvial processes of sediment erosion and deposition are a major threat to humankind and one likely to get worse under global warming scenarios. Of lesser human consequence, the salmon fishermen may return following a flood to find that the location and nature of their favourite pools have changed. Indeed, on the River Tay in Scotland, salmon fishermen pay large amounts of money to obtain lifetime fishing rights to a particular 'pool' only to return following a very large flood to find it infilled with sediment.

Rivers are dynamic landscape features that adjust their morphology, in both time and space. They are a significant hazard to humans and yet have also been vital to human civilization. They provide landscapes that have a very high nature conservation value via the habitats they provide for plants and animals. Describing and explaining variability in space and time of in-channel sediment mobilization, transport and deposition processes together with the landforms they create are encompassed within the scientific discipline called fluvial geomorphology. The movement of sediment from catchment-wide sources via hillslope runoff, together with subsequent deposition on floodplains, also forms part of the subject of fluvial geomorphology but is the focus of other chapters within this book (Chapters 11 and 12). This chapter will examine the geomorphology of river channels.

Fluvial geomorphology, until relatively recently, was a subject practised by a few academics in their quest to understand the evolution of river landscapes. Modern fluvial geomorphology involves detailed field study of processes using precise measurements, often involving sophisticated equipment, statistical analysis of the field data obtained, mapping and monitoring change on rivers using airborne and spaceborne remote sensing and modelling using large and powerful computers and numerical models. Fluvial geomorphologists now work alongside civil engineers, water resource managers, planners and ecologists in a common quest to utilize rivers for water supply, power and navigation, to alleviate flooding and destruction of houses and transport networks, and yet to maintain the nature conservation value of rivers. Fluvial geomorphologists are also leading the way with regard to river restoration, which focuses on reversing the historical legacy of environmental degradation of rivers caused by inappropriate **channelization**, bank protection and river flow regulation methods.

14.2 Catchment processes: energy and materials for rivers

14.2.1 Runoff, river regimes and floods

Catchment runoff is controlled by regional climate and catchment characteristics such as topography, geology, soils, vegetation and land-use. The percentage of rainfall that reaches the channel may vary from more than 90% to less than 10% depending on the water balance (see Chapter 13). Seasonal variability in the water balance, together with a number of other variables, controls the pattern of stream-flow throughout the year. Such patterns are known as runoff regimes and may be revealed simply by plotting daily flow against time, but more often are shown using monthly flow data, sometimes expressed for each month as a ratio to mean annual flow. At the global scale four major river regimes can be identified (Figure 14.1). The first type of runoff regime is one dominated by snow and ice melt which produce a major peak of streamflow during the late spring in the case of snow melt or early summer with glacial **ablation** (see Chapter 18). In temperate, oceanic areas precipitation occurs all year with a winter or autumn maximum, but the runoff regime is more the reflection of the marked peak of evapotranspiration during the summer months. Tropical, non-equatorial, river systems receive high precipitation during the summer but experience a marked dry season during the winter. Evapotranspiration is high at all times so that the streamflow mirrors the seasonal pattern

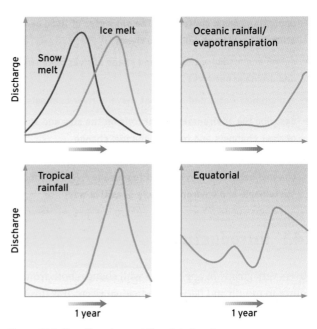

Figure 14.1 Runoff regimes at the global scale.

of rainfall. Finally equatorial rivers have more complex regimes because precipitation has a bimodal distribution with two clear maxima.

At the regional scale, although river regimes may show broad similarity, closer inspection of flow fluctuations reveals distinct differences. In steep, impermeable mountainous catchments with intense precipitation episodes, regimes with rapid rates of water-level change and large flood peaks are apparent. These are known as **flashy regimes**. In contrast, lowland catchments with permeable geologies have slower rates of water-level change and smaller flood peaks. These are often termed 'subdued' river regimes. Such differences, particularly in the size of flood peaks, have important implications for river channel morphology. Many river regimes have been highly modified by human river flow regulation; in most cases this results in a reduction in the size and frequency of floods and an increase in low flows.

14.2.2 Sediment sources and delivery

Sediment is transferred to river systems from a variety of catchment sources, including surface erosion on hillslopes, by rill and gully erosion and landsliding (see Chapters 7, 11 and 12), usually during or following intense precipitation and runoff. Sediment delivery to river systems is therefore often highly pulsed (the sediment arrives in distinct phases or pulses, rather than continuously). Delivery will often be seasonal in nature, most notably in glacial rivers where summer glacial meltwater flushes out **rock flour** produced by glacial abrasion (see Chapter 18). In addition sediment can enter the river, owing to river banks eroding floodplain materials or adjacent valley sides. The bulk of a river's sediment comes from the headwater located within mountainous areas. For example, over 80% of the Amazon's sediment load comes from 12% of the catchment covering the Andes Mountains. It is also important to realize that not all sediment eroded in the catchment reaches the stream. It may be temporarily stored in locations such as at the base of slopes or floodplain margins.

As with river regimes, there are marked variations in the volume of sediment eroded at the global and local scales. At the global scale the highest natural **sediment yields** are observed for catchments receiving between 250 and 350 mm precipitation annually with a sparse vegetation cover. With increasing precipitation, vegetation cover increases and sediment release from the catchment markedly declines (Langbein and Schumm, 1958). Sediment yield thus tends to reach a minimum with a forest cover at 750 mm precipitation

Table 14.1 Average sediment yield ($t\,km^{-2}\,yr^{-1}$) for a variety of rivers

Kosi, India	3130
Brahmaputra, Bangladesh	1370
Colorado, USA	424
River Ystwyth, UK	164
Orinoco, Venezuela	90
Zambesi, Mozambique	75
Amazon, Brazil	67
Danube, Russia	27
River Exe, UK	24
Volga, Russia	19

annually. Further increases in annual precipitation appear to offset any further vegetation biomass increase and slightly elevate sediment yields beyond the 750 mm maximum (Douglas, 1967). Given that precipitation varies widely at global and regional scales and the Earth's surface varies from horizontal plains to steep, sometimes near vertical, mountainous terrain the average sediment yield of rivers varies widely (Table 14.1). The sediment yield per unit area is highest for small rivers although obviously the total load of large rivers usually exceeds that of smaller ones. In the case of the United Kingdom, the average value lies somewhere in the range $30–40\,t\,km^{-2}\,yr^{-1}$ but values range from less than 1.0 to $500\,t\,km^{-2}\,yr^{-1}$. It is important to realize that human activities can also release vast amounts of sediment to river systems and modify the natural sediment yield. Construction activity and deforestation are two activities that are known to inject large pulses of sediment to river systems. Poor agricultural practices, such as ploughing downslope, lead to sheet wash, rilling and gullying and can also result in elevated sediment input to rivers.

Apart from the total load the **calibre** of sediment also varies widely and this has important implications for channel morphology. Tropical weathering tends to result in the majority of sediment delivered to river systems being silts and clays. In temperate areas silts and clays are also prevalent but in mountainous terrain, sands and gravel and even boulders enter the stream system. In Arctic and alpine areas coarse sediments are well represented but glacial erosion also produces vast amounts of silt-sized material. Knowledge of river sediment loads and the calibre of the material is important in river engineering. It can determine the size of sediment traps at water intakes and the lifetime of a flood storage reservoir, for example.

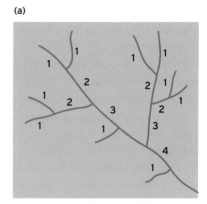

Reflective questions

➤ Apart from water what else does a river transport in the natural environment?

➤ What are the main controls on how much sediment reaches the mouth of a river?

14.3 River channel morphology: measuring rivers

Critical to understanding the geomorphic behaviour of river systems is the ability to measure their morphological attributes precisely (Kondolf and Piegey, 2003). Stream channels consist first and foremost of a drainage network whereby each time two river networks join, the downstream flow channel dimensions nearly always increase. Discharge can diminish downstream in arid areas owing to evapotranspiration, such as in the Okavango River in southern Africa. For a given river channel reach there are five basic variables or **degrees of freedom**: slope (which is also influenced by the nature of channel network), channel width, depth, **channel planform** and bed-roughness. Reliable measurement of these variables needs to take into account the fact that their values change not only at the regional scale but also locally. Furthermore, precise definition of the terms in a field situation can be complex.

14.3.1 Channel networks and slope

Despite the variable size and nature of stream patterns, each network has been found to have an ordered internal composition. Horton (1945) used this observation to promote the **stream order** as a way of describing drainage networks in a numerical manner. In his ordering system each source stream was designated as a first-order stream; two first-order channels meet to generate a second-order stream, and so on. He then drew a line from the highest-order stream to the headwaters along the channel, which involved least deviation from the line of the main stem of the channel network. He then related this line to changes in altitude. Therefore it is along this line that channel slope changes from source to mouth are measured. Measurement is undertaken either by using contour levels and distance measurements from a map or by field survey. When the values are

Figure 14.2 Stream networks: (a) stream ordering according to the systems of Horton and Strahler; (b) measuring mainstream length (MSL) and total stream length (TSL) and drainage area (DA).

plotted against distance from the stream source this is known as the channel **long profile**.

Ordering methods have been refined subsequent to the work of Horton. Strahler's (1957) modification of Horton's ordering system has been most widely adopted (Figure 14.2a). Strahler also designated headwater tributaries as first-order streams with their meeting producing a second-order stream and so on, but he omitted the second reordering procedure so that all headwater tributaries are designated first order and the main channel is not redesignated along its length. Other variables used to quantify the stream network include: **main stream length** (MSL, km) which is the distance of the main river channel from source to mouth and equates with the length over which the long profile is measured (Figure 14.2b); **total stream length** (TSL, km) which is the combined length of all components of the channel network; and **drainage density** (DD, km) which is the drainage area (DA, km^2) divided by the TSL (Figure 14.2b). These variables are important because they reflect the combined effects of topographical, geological, pedological and vegetation controls on catchment hydrology. Indeed, these variables have been utilized as a surrogate for river flow in the analysis of channel morphology and provide clues as to the nature of the flood hydrology. High drainage densities, for example, often mean flashy flow regimes.

14.3.2 Channel cross-section: width, depth

River channels are three-dimensional linear features. To describe their reach morphology, in terms of size and shape, their form is broken down into channel cross-section and planform. Channel size and shape vary locally within a reach. In order for reliable measurements to be made that can be compared with measurements elsewhere on the river, the cross-section should be measured on a straight reach and located consistently with reference to river bed landforms (see section below on channel bed morphology). Channel size and shape are usually quantified by measuring **bank-full** channel dimensions in cross-section. Bank-full conditions are usually defined by when water rapidly spreads out onto a floodplain (the river channel is full of water). The significance of the bank-full water level is that flow resistance (via friction of the bed and banks) reaches a minimum at bank-full stage and so the conveyance of water at this level is most efficient. Demarcation of the position of bank-full stage in the field, however, can be problematic in many cases. A simple uniform channel cross-section with two straight banks that intersect a floodplain at a sharp angle is extremely rare in nature. The valley floor on either side of the channel may be at different elevations, the banks have irregular sides that gradually merge with the floodplain and vegetation may overhang and colonize banks masking their presence.

Several techniques have been used to provide objectivity in defining bank-full stage, including various morphological criteria, and vegetation and lichen limits, but all have some inherent subjectivity. Measurement of the channel cross-section can be used to determine channel bank-full width (W), channel depth (d), length of the **wetted perimeter** (length of the two river banks plus channel bed) at bank-full (P) and channel cross-sectional area (A) (Figure 14.3). These variables can also be used to calculate two other key variables: the width–depth ratio (W/d) and the **hydraulic radius** (R) which is A divided by P.

14.3.3 Channel planform

The planform, or pattern, of a channel can be divided into three broad types, namely braided (Figure 14.4), meandering (Figure 14.5) and straight. In reality many channels

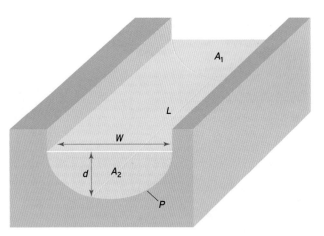

Figure 14.3 Morphological measurements of stream channel cross-section: channel bank-full width (W), channel depth (d), length of the wetted perimeter (length of the two river banks plus channel bed) at bank full (P) and channel cross-sectional area (A). L is the distance from one cross-section to the next where the area, depth, width, etc., may be different.

are intermediate types and in addition to these three broad types, **anastomosing** and wandering planforms have been identified. Variables quantifying the form of undivided or single channels are numerous. The most common is **channel sinuosity** (Figure 14.6). This is the ratio of the length of river between two points to the length of the valley between these two points. Rivers display a continuum of sinuosities from a value of 1 to more than 5. Straight channels are defined as having a sinuosity of less than 1.5. A meandering channel refers to a single channel with a number of bends which result in a channel sinuosity in excess of 1.5. However, other variables can be used to describe meandering channels further. These include the meander width, meander wavelength and radius of curvature. In the case of divided channels, channel multiplicity can be calculated by measuring the total length of the perimeters of sand and gravel islands (known as **bars**) and vegetated islands in the river and dividing by the length of the channel reach over which this process was undertaken. One problem with this approach, however, is that the value obtained is not a constant and varies with the water level at the time of survey. Bars can become 'drowned' during high water levels and emerge during low flows.

14.3.4 Channel boundary materials

Channel boundaries can consist of cohesive sediments, sands and gravels, bedrock or vegetation. Bedrock or vegetation boundary types need only be recorded as such, but as a basis for analysis of sediment transport, detailed field

Figure 14.4 The braided channel of the Chandra River, Tibet. (Source: photo © Dominic Habron)

Figure 14.5 The meandering River Tollenese, Germany. (Source: Bernard Edmaier/Science Photo Library Ltd)

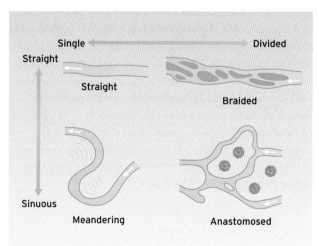

Figure 14.6 Planform of the four main river planform types defined according to channel sinuosity and channel multiplicity: braided, meandering, anastomosing and wandering.

sampling and subsequent analysis of bed sediments are required. For sand and silts, or mixed sands and gravels, a bulk sample or grab sample is abstracted and analysed for particle size. Sediment mineralogy and roundness can be measured using a hand lens and used to help determine sediment sources. For coarse channel bed sediments, which typically have a great range of sizes, often 100 particles are selected randomly and their size measured individually with callipers or a pebble sizing plate. If all the particles are roughly of the same size the sediment is referred to as well sorted, whereas if they are highly variable in nature they are poorly sorted. In general, riverine sediments tend to be well sorted but can show a bimodal distribution. Roundness is judged visually. **Imbrication** of particles, whereby their long axes tend to be aligned along a common vector due to flow direction, can also be noted and used to understand flow paths during floods.

Movement of stream bed material can be monitored using tracers. Traditionally bed material was painted and the loss and occasionally subsequent discovery of the pebbles could be used in the first case to identify movement and in the second track movement distances (Kondolf and Piegey, 2003). More recently use has been made of magnetized pebbles, which enables relocation using a metal detector, or pebbles with a radio-tracking device inserted. Sediment load can also be calculated by recording the amounts of material in sediment traps and stored in reservoirs (see Chapter 12).

Reflective questions

➤ Imagine yourself by your local river. Try to decide where you would define the bank-full limit to be on the channel.

➤ How do you overcome the problem of measuring sediment size when the river bed comprises pebbles with a wide range of sizes?

14.4 River channel processes: understanding water and sediment movement

14.4.1 Water flow and flow hydraulics

Understanding what governs the velocity of water flow and amount of flow that can be accommodated by a channel of a given size, shape, **boundary roughness** and bed slope is central to fluvial geomorphology. Water within a channel is subject to two principal forces: the force of gravity, which induces water flow, and frictional forces. Hence, a steep-gradient smooth channel in which gravitational forces are high and channel boundary roughness is low will have a high average water velocity. A similar channel, however, with a channel boundary consisting of boulders, and thus high frictional forces, will have a slower average water velocity, even though the turbulence of the water may give the impression of fast-moving water. Mean water velocity in an open channel can be estimated using the Manning equation:

$$V\,(\mathrm{m\,s^{-1}}) = \frac{R^{0.66}\,S^{0.5}}{n} \tag{14.1}$$

where V is mean water velocity, S channel bed slope (expressed as a gradient in m per m such as 0.005), R hydraulic radius (see above) and Manning's 'n' is a measure of channel roughness. Indices of bed grain size are usually used to quantify channel roughness although in-channel vegetation can be an important component of channel roughness. Defining a representative value of bed roughness for heterogeneous channel beds is problematic, particularly since larger particles have a greater effect on roughness than small particles. Values will thus only be an approximation of roughness.

For a given cross-section the way in which water velocity, depth and width increase with a rise in water level has been termed hydraulic geometry. Water velocity, width and depth express themselves as power functions of discharge:

$$w = aQ^b \tag{14.2}$$

$$d = cQ^f \tag{14.3}$$

$$v = kQ^m \tag{14.4}$$

where Q is stream discharge in $\mathrm{m^3\,s^{-1}}$, w is water width in metres, d is water depth in metres and v is water velocity in $\mathrm{m\,s^{-1}}$. Since wdv is equal to Q it can be established that the sum of the exponents b, f and m is 1 and the sum of the intercept values a, c and k is equal to 1. In most river channels, velocity increases more rapidly than water depth and width. Typical values for b, f and m are 0.1, 0.4 and 0.5. Stream power is another useful measure of river flow hydraulics and is described in Box 14.1.

Within a river channel cross-section, considerable variation in water velocity and flow characteristics occurs. The velocity distribution in cross-section can be shown by **isovels**, lines of constant velocity. Generally water velocities are greatest in mid-channel and much lower at channel margins and close to the stream bed (Figure 14.7). At river bends centrifugal forces give rise to an excess fluid pressure on the outer bank and a deficit on the inner bank. This

STREAM POWER

One of the most important expressions of the hydraulics of channel flow is **stream power** which is the work done or energy loss of the river. Stream power is therefore a key parameter in determining rates of erosion, sediment transport and instability. Differing types of channel morphology and levels of channel instability have thus been related to stream power:

$$\text{Stream power}\,(\text{W}\,\text{m}^{-1}) = \rho QS \quad (14.5)$$

Here Q is stream discharge in $\text{m}^3\,\text{s}^{-1}$, S channel slope and ρ water density. Stream power at **bank-full discharge** can vary over a 1000-fold range according to channel type and is proportional to the cube of velocity. Thus slight changes in velocity can significantly affect potential stream power. Unit stream power (P) is an important dimensionless index for comparative studies:

$$P(\times\,10^{-5}\,\text{N}\,\text{s}^{-1}) = \frac{\rho QS}{w} \quad (14.6)$$

where w is width in metres and the other variables are as in equation (14.5). The index expresses the amount of energy loss per unit area of channel bed. Stream power is a useful parameter for determining the sensitivity of alluvial channels to activities such as channel straightening, gravel removal or scenarios of climate-change-induced flooding.

BOX 14.1

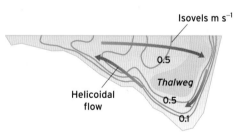

Figure 14.7 Patterns of velocity distribution within river channels shown using isovels, which are lines of equal velocity (m s⁻¹): (a) broad shallow and rectangular channels; (b) meandering channels with a helicoidal flow and the thalweg indicated. The helicoidal flow is superimposed on the downstream movement of water with a downward flow of water near the outer bank of the meander bend and an upward flow near the inner bank of the bend.

induces a surface water gradient between the inner and outer bank and a circulatory pattern of flow, termed **helicoidal flow,** is superimposed on the downstream movement of the water. In fairly uniform medium-sinuosity channels this creates outward flow close to the water surface,

downward flow close to the outer bank and inward flow towards the inner bank on the apex of the meander bend. The area of maximum water velocity, known as the **thalweg**, moves towards the outer bank in response (Figure 14.7). Convergence of flow by channel narrowing, funnelling of flow between boulders and the directing of water towards the outer bank of meander bends increase the velocity of water and can result in bed scour and erosion, while divergent flow often leads to cessation of bed load transport and depositional landforms.

14.4.2 Sediment movement

Sediment in rivers may be classified in terms of its origin. The bed material refers to sediment of mixed size found in the stream bed, which can be carried as bed load or suspended load. It is important to be aware that the bed material may have once originated from the river bank and the channel bed upstream or from catchment sources such as landslides. Transport of fine sediments, sourced from soil erosion and river bank erosion, is termed washload and is predominantly transported in suspension. Sediment mobilization and transport in rivers is inextricably linked with flow hydraulics which governs the nature of water flow in channels. For a particle to be 'lifted' from the stream bed or 'sheared' from the river bank a threshold has to be passed whereby a critical velocity or **shear stress** exceeds the frictional forces that resist erosion.

Bed load transport is almost entirely a function of flow volume, velocity and turbulence. Particles roll, slide or saltate (hop) along the bed in a shallow zone only a few grain diameters thick (see Chapter 12). Given no change in flow, a particle will come to rest only if it becomes lodged against an obstruction or falls into an area sheltered from the main force of the water by a larger particle. With further increases in the strength of flow, the smaller particles may be carried upwards into the main body of water and transported in suspension. Particle mobilization (otherwise known as **entrainment**) is dependent upon channel slope, particle size and shape, and immersed weight in relation to the bed shear stress and fluid kinematic viscosity. Deposition and cessation of bed load movement for an individual particle occur when velocity falls below the critical conditions. **Hydraulic sorting** is an inevitable consequence of selective transport whereby, in non-cohesive sediments, finer particles are preferentially moved downstream. Another useful concept is that of **stream competence**. This is the largest size of particle that a stream can carry as bed load at a single time or position. Stream capacity refers to the maximum volume of debris that a stream can carry.

In contrast to bed load, suspended load transport is determined not only by the discharge of the river and nature of flow, but also by its rate of supply from the drainage basin. In many cases the suspended load is therefore 'supply limited'. This means that the sediment transport capacity of the river exceeds the rate of supply of sediment from the catchment. Supply limitation occurs most of the time in rivers because bank and hillslope sediment sources are activated only during precipitation events and high flows. However, many channels have flow conditions that are capable of transporting the washload over a wide range of flows. Thus, although there is a broad correlation between discharge and suspended sediment transport, a plot of the two can exhibit a wide scatter of points which relate to temporal variations in hillslope controls on sediment mobilization. Indeed sediment movement, in both suspended and bed load form, is highly pulsed, with 'waves' of sediment moving through the river system. As a result sediment transport in rivers has been likened to a jerky conveyor belt (Ferguson, 1981).

Within a given river channel reach, if erosion exceeds deposition, a lowering of the river bed will occur (degradation). If net erosion and deposition are equal, vertical stability will be maintained. Channel and floodplain landforms and their stability, in this context, can also be viewed in terms of the balance between sediment input and output. Where the influx of sediment is less than removal there will be channel deepening and/or widening. Stable channels are

where there is no net loss of sediment, although short-term variations may occur during floods. Depositing channels occur when the input exceeds output, resulting in aggradation of the bed or lateral deposition along channel margins. Identifying such situations is of critical concern to river management. Eroding channels may undermine structures such as bridge piers while depositing channels may submerge engineering works. Stable channels, especially bedrock channels, are less likely to be a problem although short-term fluctuations in channel form caused by flooding or sediment pulses moving down the river can bring about erosional or depositional hazards. The role of floods and discharge of varying magnitude in controlling sediment transport is discussed in Box 14.2.

> ### Reflective questions
>
> ➤ Do you think that water moves faster in upland boulder-bed rivers or gentle lowland rivers?
>
> ➤ Why do river channels sometimes usually have clear water during low flows but are turbid at high flows?
>
> ➤ Do you think that river channels adjust their size and shape to optimize their carrying of water or sediment or both?

14.5 River channels: linking channel processes and morphology

Linking channel size and shape to water and sediment transport processes would be relatively easy if channels responded proportionally and instantaneously to the size of floods. This does not occur, because different elements of a channel's morphology have different susceptibilities to change and they may change over different timescales (Figure 14.9). Rivers have been described as having historical hangovers because of the lag between process change and landform response.

Time is a continuous variable but representative periods can nevertheless be defined. These include:

- instantaneous time (10^0–10^1 years) (1–10 years);
- short timescales (10^1 and 10^2 years) (10–100 years);
- medium timescales (10^3 and 10^4 years) (1000–10 000 years);
- long timescales ($>10^5$ years) ($>$100 000 years).

FUNDAMENTAL PRINCIPLES

THE DOMINANT DISCHARGE CONCEPT

Rivers erode their beds and banks and receive inputs of sediment from hillslope sources during times of flood. In general, floods with higher peak flows have greater potential to erode and transport sediment per unit time. Medium-sized floods, however, although transporting less sediment per unit time, occur more frequently and can thus transport greater sediment loads in the long term. In the long term the total amount of geomorphic work is the product of the sediment transported during those size floods and their frequency (Figure 14.8). The flood discharge that achieves the greatest

total geomorphic work is referred to as the dominant discharge and may equate with bank-full discharge, although this is not necessarily the case (Emmett and Wolman, 2001). Small floods are generally ineffective in mobilizing coarse river beds and causing bank erosion but can transport washload intro-duced to the river by hillslope runoff. In many large temperate rivers the majority of sediment transport is accomplished by flood events which occur between twice each year and once in every five years. However, while this concept is useful, it is also overly simplistic because morpholog-ical change usually lags behind hydro-logical change. In addition, in highly active river channels morphological

change (e.g. bar evolution on **braided channels**) can occur under normal (non-flood) conditions.

Figure 14.8 Illustration of the derivation of dominant discharge.

BOX 14.2

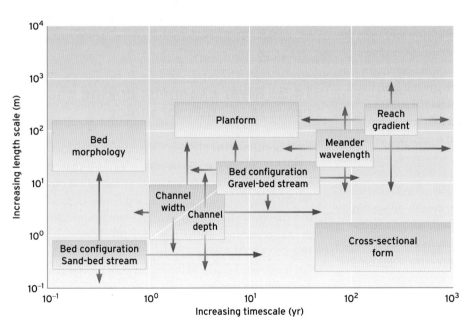

Figure 14.9 Schematic diagram of the timescale of adjustment of various channel form components with given length dimensions for a hypothetical temperate river (10^{-1} is 0.1 years, 10^0 is 1 year and 10^1 is 10 years, while 10^2 is 100 years and 10^3 is 1000 years).

Discharge and sediment loads will vary in instantan-eous time as a result of individual flood events. Cross-sectional form parameters adjust over instantaneous and short timescales, planform and local-scale profile change over short and medium timescales, and the

overall longitudinal river profile only changes over the medium and long term. Figure 14.9 illustrates this diagrammatically by showing how different features of varying size (or length scale) change over different timescales.

Figure 14.10 Gabion protection being used to stabilize a river bank adjacent to a new housing development aptly named 'Riverside' in Bridge of Allan on the Allan Water, Scotland.

The concept of equilibrium is central to linking channel processes and morphology. True equilibrium could occur only if all morphological variables responded in instantaneous time but as stated above this is rarely the case. Negative feedback is an important component of equilibrium between form and process. For example, if a channel is too narrow for a given flow and sediment concentration during an exceptional flood event, bank erosion may enlarge the channel and reduce flow velocity. Channel form thus oscillates around an average form and this is referred to as dynamic equilibrium (see Chapter 1, especially Figure 1.4).

Of course no adjustment will occur if the channel boundary prevents erosion because it is resistant bedrock or strengthened by a vegetation root mat, concrete or artificially placed bank protection such as **gabions** (wire cages filled with rocks; Figure 14.10). In these cases the discharge will be accommodated in the existing channel or by inundation of the valley floor. Valley floor inundation occurs during a flood even if channel adjustment is occurring, because the rate of change of discharge is more rapid than the rate at which the channel can adjust its dimensions and bed-roughness. If the river returns to a period with smaller floods, flow velocities will be reduced in the enlarged channel, resulting in sediment deposition and a return towards the morphology that existed before the exceptionally large flood event. However, if there is no sediment delivery from upstream for deposition then no adjustment will occur in channel dimensions following a flood. The time taken for a channel to return to its original form after a flood-induced change is termed the relaxation time or recovery period. In temperate rivers this is usually fast in relation to the return periods of floods. Rapid revegetation is an important component of the recovery. Human activity, such as channelization or river flow regulation by altering fluvial processes, may result in a river's morphology being knocked out of equilibrium. The river will then strive to create a new and undesirable equilibrium form by either bed incision, bed deposition or bank erosion. An understanding of equilibrium is therefore important for river management.

14.5.1 Long profile

The long profile (slope of a river from its source to mouth) is typically concave with progressively lower gradients downstream. The degree of concavity, however, varies among rivers according to a host of factors. These include inherited landscape form, geology, tectonics and variability in runoff. Indeed many rivers that drain passive continental margins (see Chapter 2) have a significant convexity. Abnormal profiles also occur owing principally to interruption by lakes or resistant rock bands. Where particularly resistant rock bands exist, waterfalls or rapids result (Figure 14.11). At the reach scale the long profile may locally steepen or be reduced in gradient owing to localized aggradation and degradation processes. At the larger scale such aggradation or degradation can be caused by a rise or fall in sea level.

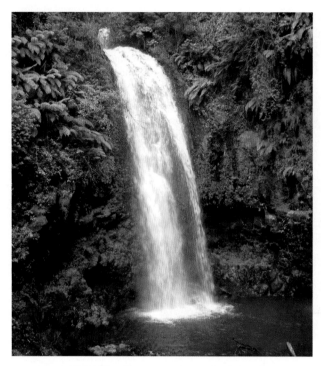

Figure 14.11 Waterfalls often occur at a particularly resistant rock band. This waterfall is in Montagne d'Ambre National Park, Madagascar. (Source: photo courtesy of Despina Psarra)

14.5.2 River channel cross-sections

Channel cross-sections adjust to accommodate the discharge and sediment load from the drainage basin, as permitted by the constraints of relative and absolute bed and bank erodibility values and channel slope. This can most easily be demonstrated with reference to how the channel bank-full cross-section varies along a river from source to mouth. Generally, change in channel form can be summarized by stating that with distance downstream there is an increase in bank-full cross-sectional area. However, there is not a 1 : 1 relationship between increase in discharge and channel cross-sectional area. This is firstly because larger channels are hydraulically more efficient (as indexed by the hydraulic radius). Thus a given increase in channel cross-sectional area will result in a proportionally greater increase in bank-full discharge capacity. Secondly, boundary roughness generally decreases downstream, promoting faster flow for an otherwise unchanged channel form. This reduced roughness is able to offset the decrease in channel slope that often declines downstream also resulting in a hydraulically more efficient channel. The relationship between downstream increase in discharge and bank-full indices of channel morphology is known as 'downstream

hydraulic geometry' (Leopold and Maddock, 1953) and average values are:

$$w^b = aQ^{0.5} \tag{14.7}$$

$$d^b = cQ^{0.4} \tag{14.8}$$

$$v^b = kQ^{0.1} \tag{14.9}$$

where w^b, d^b and v^b are channel bank-full cross-sectional width, depth and velocity values, respectively. Such relations imply an adjustment of channel shape whereby channels become broader and shallower in proportional terms while velocity only increases slowly downstream. These equations would describe the change in channel form downstream in uniform sediments. However, bed and bank erodibility and sediment load are important in controlling channel cross-sectional geometry. Channels with a high percentage of silt/clay in their banks, and rivers transporting much of the sediment load in suspension, tend to be narrower and deeper than sand and gravel-bed rivers. This fact can be related to flow hydraulics. Channels carrying material in suspension will be most effective if the movement of water is optimized. Semicircular channels are hydraulically most efficient. This is why guttering tends to be semicircular and engineers build channels of similar shape. Bed load transport, however, is optimized where bed shear is greatest close to the bed and there is a wide channel over which the material can be mobilized and transported. This demonstrates how water and sediment transport processes together with the three-dimensional nature of channel morphology are intertwined. Vegetation can also be important in controlling cross-sectional form by influencing bank resistance. In general root systems provide increased strength to bank sediments and as a consequence vegetation-lined channels tend to be narrower. This fact has not always been appreciated by river managers and farmers, who have often removed bankside vegetation. This deliberate or inadvertent removal has often resulted in bank erosion which, once initiated, is difficult to stop. In many cases rubble has been used to armour the bank but during large floods this may be mobilized and then redeposited downstream. This deposition reduces channel capacity and exacerbates the problem even further. There is therefore a need to understand geomorphic processes and the controls on channel morphology if sensible river management decisions are to be made.

14.5.3 Channel planform

When viewed from above, channels vary greatly in appearance. They range from those with tortuous bends that snake through the landscape to straight and multiple-channel rivers. The processes controlling the development of these channel patterns through time have long been the focus of fluvial geomorphology (e.g. Leopold and Wolman, 1957; Kellerhals and Church, 1989; Van der Berg, 1995). A conceptual model of morphological types in relation to some controlling process variables is shown in Figure 14.12. Studies have consistently demonstrated that channel slope and discharge are important controlling variables. Early work by Leopold and Wolman (1957) based on numerous field studies suggested that for any given discharge there is

a threshold slope above which channels will meander. They also found another higher threshold above which they will braid. They also found that these critical slope thresholds decreased with increasing discharge (Figure 14.13a). Thus, braided planforms are found on large rivers, or alternatively on small rivers with steep slopes (where stream power is high). Braided streams have several channels which except at high flood are divided by active coarse-grained bars. Each channel may be sinuous but overall a fairly straight planform exists and the width–depth ratio is large. The term anastomosing is usually confined to rivers with semi-permanent sinuous channels with cohesive banks, although a variety of forms have been identified in temperate, tropical and Arctic environments. The relationship between planform and the primary control of stream power is complex owing to other influencing variables; Simpson and Smith (2001) working on the Milk River in North America found that it

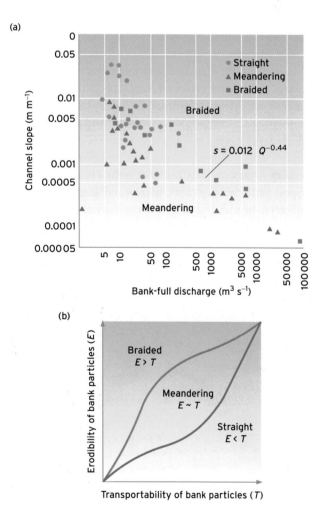

Figure 14.12 Conceptual model of morphological types of channels indicating the conditions (sediment size, channel gradient, ratio of bed load to suspended load) under which river channels will be straight, meandering or braided.

Figure 14.13 (a) The relationship between braided and meandering planforms and the controlling variable of slope and discharge. (b) Straight, meandering and braided patterns defined in terms of erodibility and transportability of bank particles.

failed the Leopold and Wolman (1957) slope–discharge test in that it was braided when its discharge and slope values suggested it should be meandering. They suggested that this was because the river had a sandy bed whereas Leopold and Wolman had worked mainly on gravel-bed rivers in the United States. Clearly bed sediment calibre is important in controlling river planform.

Where the bed material is coarse, the bank material highly erodible, or the dominant mode of sediment transport is by bed load movement there will be a tendency towards wandering and braided rivers. River channel planforms have, for example, been explained by the ratio of the transportability of bank particles to the erodibility of bank particles (Figure 14.13b). Braided and **wandering gravel-bed rivers** tend to have uncohesive coarse floodplain sediments that maintain high rates of bank erosion and bank collapse and yet only travel short distances during floods. In areas where bank erosion cannot occur, such as in bedrock-lined rivers, channels tend to be straight or follow the pattern of geological structure (e.g. faults). Elsewhere rivers are likely to meander.

It is important to understand the potential response of channel planform when river management alters a process control. If the activity takes a stream across a threshold then the river can change its form and level of instability. Furthermore, if channels are constructed and a planform is adopted which is not in equilibrium with the processes it is unlikely to be a success for long. Bulldozing the Arve River between Argentiere and Chamonix in the French Alps (Figure 14.14) into a straight, narrow and deep channel is unlikely to be a long-term solution to channel instability

Figure 14.14 The Arve River between Argentiere and Chamonix in the French Alps following bulldozing and straightening.

and flooding. The Arve River's natural tendency is towards a braided pattern and so it will try to return to this form following the channelization, causing major local instability.

14.5.4 Channel bed morphology

The stream bed changes downstream. Often there are bedrock channels in the upper section of the long profile with an exponential decline of bed material particle size downstream. This downstream fining is due to particle sorting and abrasion. Particle sorting occurs because smaller particles are preferentially entrained and transported further downstream. Abrasion, involving grinding and chipping, causes the bed material to become more rounded downstream. Perturbations to this general pattern are caused by injections of sediment from tributaries or bank collapse, lithological changes along the long profile and reach variability in stream power. Lakes will also interrupt the pattern of sediment transport.

Superimposed on the long profile and cross-sectional geometry are a variety of erosional and depositional bedforms. The type of forms depends upon whether the bed is composed of predominantly sands, gravels or bedrock and upon channel gradient. Within gravel-bed rivers the most common bedforms are pool–riffle sequences. Pools are characterized during low flow by relatively slow flowing water with fine bed material. **Riffles** are formed by accumulation of relatively coarse material and are characterized by shallow, more rapidly flowing water (Figure 14.15). The spacing of pools and riffles is often five to seven times the channel width but a variety of spacings have been found in natural channels. In mountain streams where the size of bed materials is large relative to the size of the channel and/or the channel slope is steep, step–pool systems sometimes replace pools and riffles.

The channel bed of sand-dominated reaches is commonly composed of ripples (less than 40 mm in height and 600 mm in wavelength) and larger **dunes** forming traverse sand bars (see Chapter 12). During periods of high discharge sand dunes can be formed, as seen in Figure 14.16. The size and shape of these bedforms change with time since their form is directly related to flow velocity. Downstream migration of dunes and ripples occurs as material is carried up the stoss side and avalanches over the crest and down the lee side (see Chapter 12). At very high flow velocities a flat river bed can be formed or **anti-dunes** created. Anti-dunes migrate upstream as erosion from the downstream side of the anti-dune throws material into saltation and suspension more rapidly than it can be replenished from upstream.

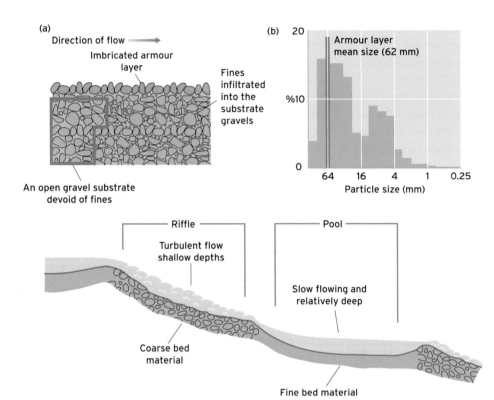

(a)

Direction of flow →

Imbricated armour layer

Fines infiltrated into the substrate gravels

An open gravel substrate devoid of fines

(b)

Armour layer mean size (62 mm)

%10

Particle size (mm)

Riffle

Turbulent flow shallow depths

Pool

Slow flowing and relatively deep

Coarse bed material

Fine bed material

Figure 14.15 A pool-riffle sequence showing the main morphological attributes of pools and riffles (viewed from the side) including (a) a schematic diagram of stream bed armouring with and without inflitrated fine sediment and (b) a particle size distribution of the bed sediments on the gravel-bed River Tryweryn, Wales.

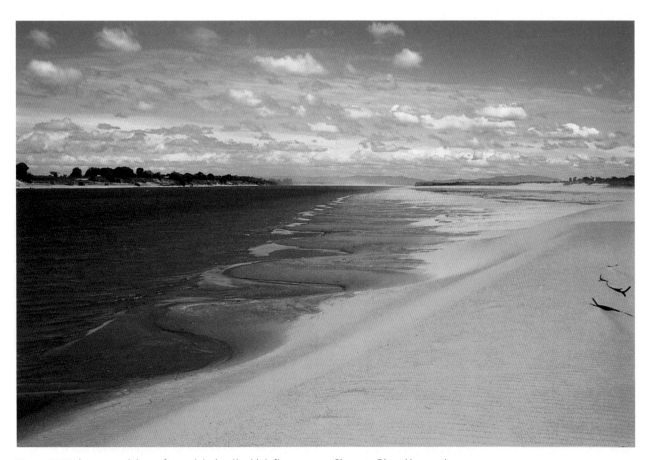

Figure 14.16 Large sand dunes formed during the high flow season, Cinaruco River, Venezuela.

Such sedimentary structures can be related directly to stream power and the diameter of the bed material. In bedrock channels a variety of sculptured forms can be found, in part controlled by the rock type and structure. The most well-known feature is the **pothole** caused by **corrasion** (mechanical wearing and grinding), **cavitation** (pressure changes due to bubble collapse in turbulent flow) and **corrosion** (chemical weathering). In some bedrock rivers rhythmical forms similar to pools and riffles have also been identified. Thus, the rapids within the Grand Canyon of the Colorado River, which are well known to white-water rafters, are spaced at intervals of 2.6 km with deeper sections in between.

Sizeable accumulations of bed load material are referred to as bars (Figure 14.17). For example, riffles are **lobate** gravel bars but are usually studied independently of other bar forms owing to their regular spacing. Bars may consist of coarse or fine material, be relatively stable or highly mobile, and be attached or detached from the river banks. In general, mid-channel bars are found on braided rivers, lateral and diagonal bars on wandering gravel-bed rivers, whereas point bars form on the inside of bends within **meandering rivers** (Figure 14.17). Most bar forms consist of an upstream portion of sediment known as the bar head, with a tail of finer material. Associated with bar forms and channel bed topography is variability in grain size. Grain size tends to be greatest in the centre of channels and finer at the channel margins (relating to the normal pattern of flow velocity across the channel). Within meandering channels, because water velocity is greatest close to the outer bank, grain size tends to be at a maximum here and slowly decreases towards the inner bank. Exceptions to these generalizations will reflect complex flow hydraulics. In natural channels with non-uniform channel morphology, patterns of flow velocity during floods are highly variable and complex. The vertical structure of bed forms in gravel-bed rivers typically consists of a surface layer of stones with finer material beneath. The coarser layer is referred to as an **armoured layer** as it protects the finer material from being transported (Figure 14.18).

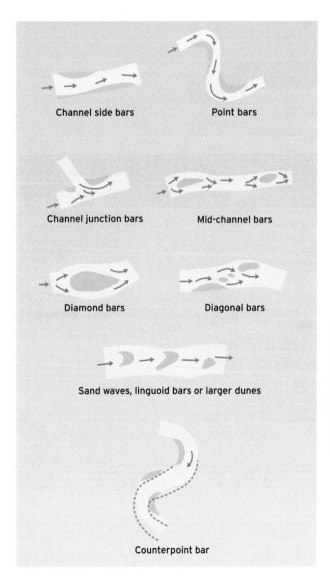

Figure 14.17 A classification of simple bar forms and flow patterns.

Channel side bars

Point bars

Channel junction bars

Mid-channel bars

Diamond bars

Diagonal bars

Sand waves, linguoid bars or larger dunes

Counterpoint bar

Figure 14.18 A vertical 'freeze core' of a gravel substrate. The core has been collected by inserting a tube into the sediment and then filling the tube with liquid nitrogen. This freezes the water around the tube and the nearby sediment thereby allowing it to be pulled out with the sediment intact for investigation. The core shows mixed grain sizes below the coarser 'armour' layer that is characteristically found at the surface.

14.6 River channel changes: rates and types of channel adjustment

River channels adjust their channel slope, cross-section, planform, bed morphology or stream network over a range of timescales in relation to natural and anthropogenic changes (Figure 14.19). This chapter is focusing on short timescale changes because of their direct relevance to river management. Some changes take place slowly and gradually while others can occur almost instantaneously in response to individual flood events. Channel change can also be distinguished according to whether it is **autogenic** or **allogenic**. Autogenic change refers to fluctuations of channel form about an equilibrium condition. Allogenic change refers to adjustments of channel form in response to a change through time in the sediment and water regime of the river. Box 14.3 provides a simple way of determining how channel shape might change if discharge or sediment input changes.

14.6.1 Cross-sectional change

The extent of bank or bed erosion will depend upon the resistance of the channel bed and banks. Sand-bed rivers with highly mobile beds, such as the Fraser River in Canada, can experience changes of 5 m in bed elevation during a flood season. On such rivers even greater bed lowering can occur around bridge piers where scour is exacerbated. Knowledge of such bed-level change during floods is critical in creating sound foundations for river structures such as

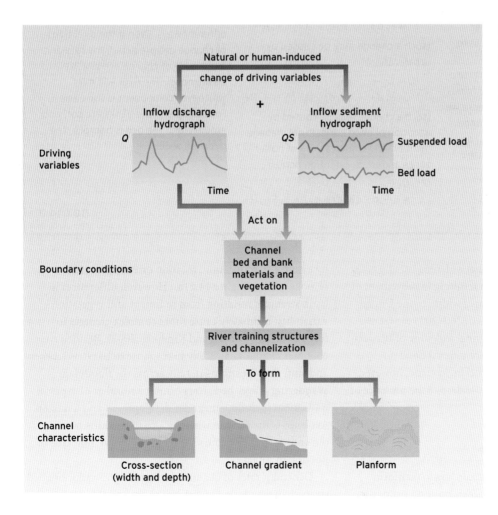

Figure 14.19 Principles of channel form and change. Alteration of the driving variables and/or boundary conditions causes changes in channel characteristics.

TECHNIQUES

PREDICTING CHANNEL RESPONSE TO CHANGES IN DISCHARGE AND SEDIMENT YIELD

Predicting how channels might adjust to changes in sediment yield, sediment type and discharge is complex. Rivers vary in their sensitivity to discharge or sediment load. Similarly their sensitivity in terms of morphological adjustment to direct human intervention such as channelization, river margin vegetation removal or gravel extraction, and indirectly via land-use change, varies greatly (Kondolf *et al.*, 2002). The highly eminent American fluvial geomorphologist Stan Schumm, however, produced a set of relationships as to how channels would respond to change in discharge or sediment yield (Schumm, 1977). Using a plus or minus sign to denote an increase or decrease respectively, the effects of a change in discharge (Q) or bed load (Q_b) on channel forms can be hypothesized:

$Q+ \quad w+, d+, (w/d)+, \lambda+, S-$

$Q- \quad w-, d-, (w/d)-, \lambda-, S+$

$Q_b+ \quad w+, d-, (w/d)+, \lambda+, S+, P-$

$Q_b- \quad w-, d+, (w/d)-, \lambda-, S-, P+$

where w is width, d mean depth, λ meander wavelength, S channel gradient and P channel sinuosity.

The algorithms were based on Schumm's observations of predominantly sand-bed channels in semi-arid and sub-humid regions. Therefore their wider applicability has yet to be fully tested. Furthermore, changes in discharge or sediment load rarely occur alone because of their interdependence on climatic and catchment variables. Four combinations can be postulated:

$Q+, Q_b+$

$\quad w+, d+ -, (w/d)+, \lambda+, S+, P-$

(such a change may be caused by urbanization)

$Q-, Q_b-$

$\quad w-, d+ -, (w/d)-, \lambda-, S+ -, P+$

(such a change may be caused by adoption of good agricultural management practices such as use of buffer zones and water retention ponds)

$Q+, Q_b-$

$\quad w+ -, d+, (w/d)+ -, \lambda+ -, S-, P+$

(such a situation may occur when a river is in receipt of water from a 'donor' stream)

$Q-, Q_b+$

$\quad w+ -, d-, (w/d)+ -, \lambda+ -, S+, P-$

(such a situation may occur where construction or mining activity takes place in a catchment).

These simple algorithms can be useful in assessing how a channel might respond to climate change or human activity and they may predict the direction of change. Such simplistic models, however, should be used with caution. Channels with highly variable ratios of bed to bank resistance and complex modes of sediment transport can respond in different ways. Even if the directions of change are predicted, the rate and magnitude of change are still likely to be unknown. The river will alter whichever component is the easiest. For example, if the banks are soft and erodible and the bed is bedrock, the river will widen.

BOX 14.3

---

that shown in Figure 14.20. Channel widening usually takes place during extreme floods often resulting in movement of a river by more than a whole channel width. In the United States there is evidence that floods of 100–200 year recurrence interval have resulted in channel widening of between 60 and 600%. A 100 year recurrence interval flood on natural reaches of the River Tummel in Scotland, however, resulted in channel widening of 10% despite the presence of uncohesive sand and gravels.

14.6.2 Planform change

Changes in channel pattern vary according to planform geometry. Meandering rivers shift their position primarily by extension, translation, rotation or enlargement (Hooke, 1980) (Figure 14.21a). However, differences in bank strength, resulting from sediment and vegetation variability, ultimately cause more complex changes in form to occur (Micheli and Kirchner, 2002). Braided river channels change chaotically in response to bar development and shift their position laterally across the floodplain. Wandering gravel-bed rivers exhibit a number of types of movement including meander development and **avulsion**. Avulsion is the process whereby a channel shifts from an old course to a new course, leaving an intervening area of floodplain intact.

Generally, rates of channel shifting are greater for braided rivers than wandering gravel-bed rivers and slowest for

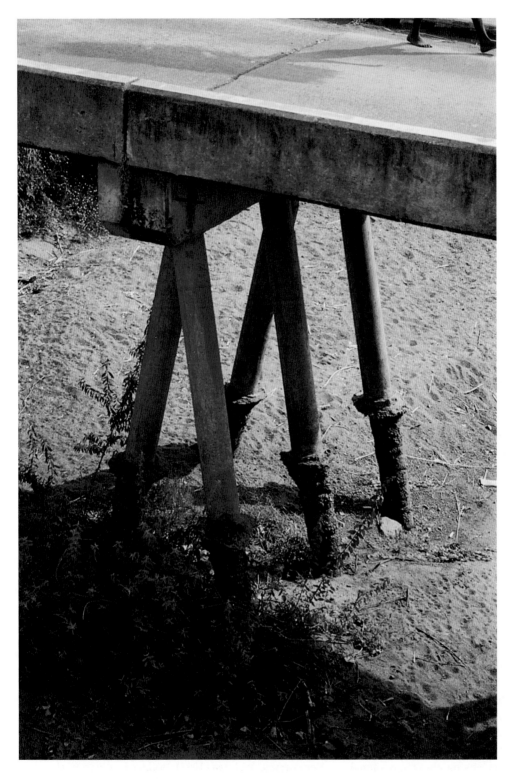

Figure 14.20 Exposure of bridge piers on a tributary of the Luangwa River, Zambia. The original bed level can be seen by the 'ragged' concrete on the piers.

meandering rivers. Locally, however, bank erosion on the outside of meander bends can be quite rapid and ultimately result in meander cutoffs (Figure 14.21b). In the United Kingdom, medium-sized meandering rivers have been reported to have migrated across more than 50% of the floodplain in less than 200 years. For small and medium-sized meandering rivers, average bank erosion rates are often less than 0.1 m yr^{-1}, usually less than 0.5 m yr^{-1}, but may reach values of 5 m yr^{-1} or more. In contrast, some meandering channels appear static with little or no change

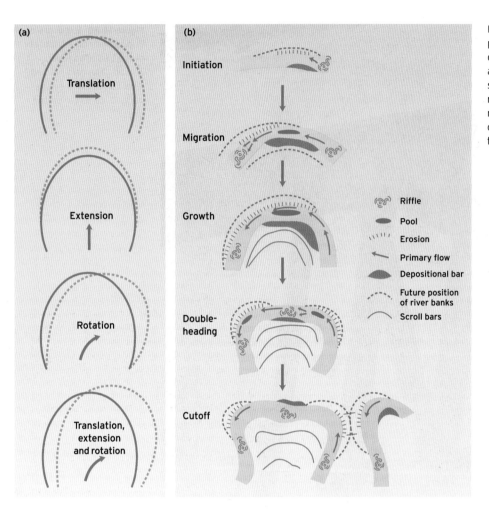

Figure 14.21 Modes of channel planform change on a meandering river: (a) types of change and (b) the development of a straight reach into a meandering reach and then into a straight reach as the meander bend is cut off leaving 'oxbow' lakes on the floodplain.

in planform over a few hundred years or more. Mobility appears to relate to the degree of incision and stream power. The meandering Luangwa River in Zambia is an example of a highly mobile meandering river. On the Luangwa River average annual bank erosion rates have varied between 1 and 20 m over the 40 years from 1956 to 1997 (Gilvear *et al.*, 2000). Figure 14.22 shows a reach of the Luangwa in 1956 and 1988 where channel migration and meander cutoff have occurred. Such vast changes in the channel have resulted in safari lodges being swept into the river. The safari lodges are located on the outside of the meanders, despite the erosion threat, because animals come down to the river edge to drink. They do so via the point bars on the opposite side of the river from the safari huts and thus many animals can be easily spotted. Prediction of the rate and direction of meander migration is problematic on meandering rivers. The rate of bank erosion varies around the meander bend and as the meander develops and alters its sinuosity. The rate of erosion will also be dependent upon the size and timing of floods over proceeding years and on the resistance

of river bank sediments encountered as the river migrates across the floodplain.

14.6.3 Human-induced change

Over the last few thousand years human activity has modified river channels. This modification accelerated over the past 200–300 years through channelization, straightening and embanking. In addition there have been changes resulting from human activity altering river flows and sediment yields. Such activities include reservoir construction, urbanization, building construction, mining, land drainage and vegetation change such as afforestation and deforestation. For example, Wolman (1967) produced a model of the response of channels to land-use change in the Piedmont region of the United States and this model is shown in Figure 14.23. His model of land-use change over the past 200 years deals with the conversion from forest to urban land with interim stages of arable agriculture, reversion to woods and grazing and construction activity. The model suggests

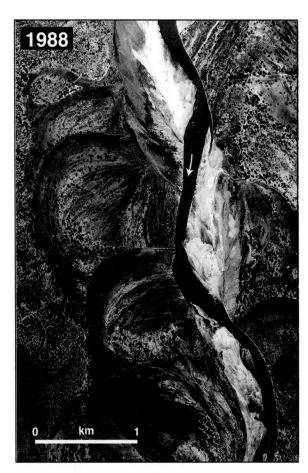

Figure 14.22 Channel planform changes on the meandering Luangwa River, Zambia, over a 32 year period.

Figure 14.23 A model of variations in sediment yield and channel response over time, Piedmont region, USA.

episodes of aggradation, scour, stability and bank erosion within the affected river channels. For example, channels below urban areas in the United States have been shown to have channel capacities of up to six times the size of their rural counterparts. In the United Kingdom values up to 150% are more common (Gregory *et al.*, 1992). The nature of channel change in detail is complex and a good example is the case of rivers regulated by dams as discussed in Box 14.4.

ENVIRONMENTAL CHANGE

RIVER CHANNELS BELOW DAMS

Reservoir construction often results in a sizeable reduction in the magnitude of flood peaks downstream. It also reduces sediment discharge, especially in those reaches immediately below the dam (see Chapter 12). A common response, except in bedrock channels, to the release of sediment-free water below dams is degradation of the channel bed. This occurs, despite a reduction in flood peaks, because of the greater erosive potential of 'clear water' but more importantly is due to any output of sediment from the reach not being matched by an input from upstream because of the interruption of sediment transfer by the dam. In time

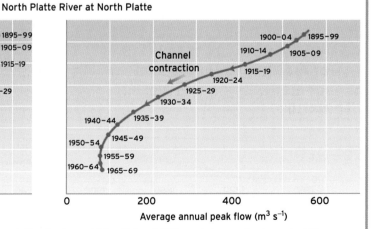

Figure 14.24 Channel changes resulting from river regulation by dams. (a) Spatial variability and changes in channel width (1945-1989) downstream of the Spey Dam, Scotland. (b) Changes in channel width with mean annual and annual peak flows averaged over five years for the North Platte River, Nebraska. (Source: after Williams and Wolman, 1983)

BOX 14.4 ➤

degradation progressively moves downstream. This process often continues unless bedrock is met, the channel becomes armoured to the effects of clear water erosion, or an unregulated tributary injects sediment to the regulated river. In the absence of large injections of sediment, degradation and erosion can be prevalent along the whole course of the river downstream of the dam. Increased coastal erosion 965 km downstream of the High Aswan Dam on the Nile has even been blamed on the impoundment.

The primary response of the regulated river to the change in flood regime, however, usually seems to be a decrease in channel capacity. This usually takes the form of a reduction in channel width which can often be over 50%. Width reduction is largely achieved through the formation of depositional bars and **berms** and subsequent vegetation colonization. Shrinkage of over 50% in the width of the River Spey in Scotland, downstream of the Spey Dam constructed in 1938, occurred over 50 years (Figure 14.24a; Gilvear, 2003).

Work on the North Platte River in Nebraska has demonstrated a progressive reduction in channel width through time in response to annual flow and peak flow reduction (Figure 14.24b) following impoundment upstream (Williams and Wolman, 1983). However, channel reduction in some reaches was shown to lag well behind flow reduction because the redistribution of large quantities of sediments may be involved and flows able to transport such sediment are rare. Petts (1984) has speculated that a century or more may be required before the full effects of dams occur downstream. This demonstrates how river management needs to consider impacts over a variety of timescales. More about this subject can be found in Petts (1979, 1984).

BOX 14.4

Reflective questions

➤ Why do rivers adjust their form and how rapidly do they change width and depth?

➤ Do you think that geomorphologists now understand enough about rivers to predict how they will change in the future, particularly where human interference is present?

14.7 Fluvial geomorphology and environmentally sound river management: living and working with nature

The above sections have demonstrated that river channels are not static features in the landscape but have high levels of dynamism and are very sensitive to changes in flow or sediment yield or changes in bed and bank resistance. Such temporal variability has to be incorporated into river management if environmentally sound and sustainable solutions are to be reached. This is the focus of this section in the chapter. Rivers can be classified into different types and this often provides a simple way for managers to determine which rivers might be more dynamic than others in particular environmental settings. Box 14.5 provides a description of how this is being used to inform river management.

14.7.1 River management and the engineering tradition

Management must be based on the best available science. The science of fluvial geomorphology until recently was rarely used in river management and river engineering. As a consequence civil engineers, the people who traditionally dealt with river management problems, have witnessed dramatic failures of structures such as river bank revetments and bridges. Other costly structures have been lost to many a river by bank erosion. In addition, there are many cases where channels have been widened, deepened and straightened only for the rivers to move back to their natural dimensions and planform over proceeding years. A notable example of this is the Mississippi River which regained much of its sinuosity following straightening in the early twentieth century. This is because rivers are naturally in equilibrium with their discharge, in-channel flow hydraulics and sediment transport. Moreover, many rivers are naturally inherently dynamic and adjust their bed and banks in time and space. Thus, only where channels have been totally lined with concrete have they tended to remain stable.

Moreover, numbers of riverine animals and plants have declined substantially, almost to extinction, in many river systems owing to engineering approaches to river

TECHNIQUES

RIVER TYPOLOGIES

The link between fluvial processes and channel morphology types has now been recognized in relation to the concept of river styles or river typing. Individual river types such as lowland active meandering or step-pool rivers are thought to maintain particular physical habitat characteristics and relate to a certain set of fluvial controls determining sediment supply and sediment transport. This latter fact results in individual river types having differing geomorphic sensitivities to river engineering. It is also likely that differing river types also exert a strong control on

the biota present. The Scottish Environmental Protection Agency, prompted by the stimulus of a new piece of European legislation, the Water Framework Directive, is currently in the process of automatically typing (categorizing) all their river channels within a geographical information system using independent controls such as slope, valley confinement, distance downstream (as a surrogate for discharge) and geology. The typology being used is a modified version of the Montgomery and Buffington (1997) approach developed in the western United States. Channel types include cascade,

step-pool, pool-riffle, braided and meandering. Once river typing has been accomplished for all rivers across Scotland it will be used in two major ways. Firstly it will be used to aid assessment of whether applications to undertake river engineering activity should be approved or not, dependent upon the geomorphic sensitivity of the river type at the specified location. Secondly where the present-day channel morphology deviates massively from that which should naturally occur due to the effect of historic river engineering the reaches will be seen as potential targets for river restoration schemes (see Section 14.7.5).

BOX 14.5

---

management destroying the fluvial habitats and processes to which they are adapted. River engineering typically results in uniform flows through the channel whereas a range of flows are required in order to increase the number of available habitats.

14.7.2 Living with rivers

As a result of the failure of a number of river management and engineering schemes there has been a radical shift in the nature of river management (Leeks *et al.*, 1988; see Chapter 24). This shift is towards working with, rather than against, natural processes and accepting the dynamic nature of river channels. This is because the long-term viability of many engineering structures cannot be assured given the highly mobile nature of rivers. Flood embankments can be lost to bank erosion, and current deflectors meant to deepen the river and curtail its lateral movement may well be largely ineffective in a river with high stream power. In this context, geomorphologists have a role in: (i) deciding on what type of rivers are activities such as channel straightening or floodplain development permissible; (ii) deciding where and how far from rivers structures can be built; and (iii) designing long-term and environmentally sensitive bank protection and engineering

solutions to river management problems. Such an approach places the emphasis on living with rivers rather than fighting against the forces of nature. The role of the geomorphologist in the three situations above is illustrated here with reference to a number of research projects.

In a study of the success of channelization schemes Brookes (1985) examined channel response in England, Wales and Denmark. The schemes varied in their type, age and extent and were on a variety of river types. Channelization schemes were found to be successful on low-energy streams but on streams with stream powers above 35 W m^{-2} stream channels reverted back towards their natural morphology (Figure 14.26). Such a finding is obviously of direct relevance to assessing where channelization schemes may be an appropriate solution to an environmental management problem (Brookes, 1985).

The role of the geomorphologist in assessing where structures should or should not be built is illustrated by the River Tay in Scotland. Here reconstruction of channel planform evolution over the past 250 years and scrutiny of the present-day morphology revealed that flood embankment failures along the river occurred more often on the outside of meander bends and where embankments had been constructed over old courses of the river (Gilvear and Black, 1999)

HAZARDS

FLOODING AND BANK EROSION ON THE BRAHMAPUTRA RIVER

The Brahmaputra River drains a catchment of 580 000 km², with more than half the area lying within China and Tibet, the remainder being composed of parts of Bhutan, India and Bangladesh. In total the river is 2906 km long with its source being the Kangklung Kang Glacier in Tibet at an altitude of 4877 m. On leaving the Himalayas it flows for 640 km within the state of Assam, before entering Bangladesh and flowing into the Bay of Bengal. At its mouth the mean flow of the Brahmaputra River is 19 830 m³ s⁻¹, making it the fourth largest river in the world in terms of discharge. The Brahmaputra also has one of the highest sediment loads of any river in the world with an estimated annual load of 402 million tonnes.

Once the river meets the alluvial plains of India and Bangladesh it forms a large, wide, braided river system up to 20 km across and has a valley floor width of 70-80 km. Each year in the wet season widespread inundation of the valley floor brings misery to millions of people. To combat this flooding thousands of kilometres of embankments have been built to try to limit the extent of flooding. However, the Brahmaputra, like all braided rivers, shifts its course from year to year, threatening cities and the stability of embankments and destroying agricultural land. It is estimated that 868 km² of land in India was lost by erosion in the twentieth century. The river is also reported to have widened and become shallower with bank erosion rates almost quadrupling in the twentieth century and bed aggradation rates of 16.8 cm yr⁻¹ having been reported. This increase in erosion and sedimentation has been linked to increased sediment delivery to the river, resulting from deforestation and an upward trend in flood magnitudes possibly linked to climate change.

Understanding how the Brahmaputra River will alter its morphology

Figure 14.25 A satellite image of the Brahmaputra River, India, classified into major land cover types. The image shows a reach with a large mid-channel island and braided channel patterns. Note the scale.

Water deep
Water medium
Water shallow
Sand
Vegetation dense
Vegetation medium
Vegetation low
Rough pasture/soil
Vegetation grassland

0 _____ 10 Kilometers

BOX 14.6 ➤

405

and course in the future is of crucial importance. However, on such a large river, traditional field-based approaches to collecting geomorphological data have severe limitations. High spatial and spectral resolution satellites, not to mention Google

Earth, however, can now provide important information on channel planform and channel instability. These remote sensing technologies allow the mapping of water surfaces and sand bars. Images can be compared over time to determine

channel planform change and the shifting position of the river (Figure 14.25). Wet season images also allow the extent of inundation to be mapped. Such hazard mapping is important in assessing risk and developing flood mitigation strategies.

BOX 14.6

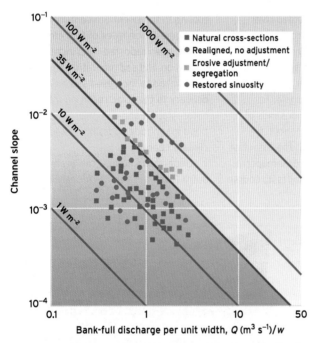

Figure 14.26 The response of river channels to channelization in relation to stream power. (Source: after Brookes, 1985)

Table 14.2 Percentage of flood embankment failures on the River Tay in comparison with their geomorphic setting (1993 flood event)[*]

Overlying relic channels less than 250 years old	14
Overlying relic channels older than 250 years	5
Outside bend of meanders	28
Overlying relic channels and outside of a bend	8
Perpendicular to valley sides	13
Others	32

[*]Subsequent to 1993, 70% of the locations have experienced failure at least once more during floods in 1998, 2003, 2005 and 2007.

The structures are located in order to generate a secondary flow cell. This secondary cell has opposite polarity to the main flow cell that results from helicoidal flow in the meander bend. Convergence of faster water occurs in the centre of the channel and causes bed scouring. Eroded material is then deposited towards the outer bank. As shear stresses against the outer bank, toe region and bank height are reduced, bank retreat is prevented. Installation of such structures has been successfully utilized on the River Roding in England (Hey, 1996).

14.7.3 River maintenance

River maintenance is needed where human use of rivers does not allow natural fluvial processes to be fully maintained or where the channel dynamics are in conflict with human use. For example, river flow regulation by dams can cause sedimentation and siltation of gravels below tributary junctions due to the reduced frequency and magnitude of floods. This, in turn, can have adverse impacts on aquatic organisms such as salmon, which require uncovered gravels to spawn.

(Table 14.2). This suggests that, in the case of the River Tay, and probably more generally, it is unwise to build flood embankments over areas of relatively recent fluvial activity on high-energy and thus inherently mobile river systems.

Traditional engineering approaches to bank protection on meandering channels have starved the river of sediment, accelerated water velocities towards the outer bank and as such in some cases accelerated erosion even where rock-filled gabions and concrete have been used to protect the river bank. An example of an alternative geomorphic approach to stemming bank erosion is the use of submerged vanes or hydrofoils. These work on the basis of modifying the flow adjacent to the bank such that the processes of erosion are reduced and bank failure prevented.

In recent years there has therefore been a move towards releases of reservoir water down salmon rivers to remove unwanted sediment. These pulses are termed either flushing flows or channel maintenance flows. The most well-known, and probably largest, example of a channel maintenance flow was the huge one from the Glen Canyon Dam on the Colorado River in 1997 (Kaplan, 2002). However, to a water resource company this water may be seen as a loss that affects its profit margin. The company may wish to know the effective discharge and duration for the release to meet its objectives. This is an area where fluvial geomorphologists can provide advice. For channel maintenance, a general rule of thumb is that sediment within gravel-bed channels begins to be mobilized when flow depth is greater than 80% of channel bank-full depth. Flushing flow recommendations in the United States have been either hydrologically determined and based on a percentage of the mean annual flow or geomorphologically determined. The latter involves either direct field measurement of the threshold discharge for sediment transport or modelling of sediment transport to predict the discharge that causes incipient motion of particles. Some geomorphologists argue that the flood that occurs once every 1.5 years is theoretically the most suitable discharge given that, on average, river channels have a channel capacity equivalent to this flow level. However, in reality the dominant discharge in terms of sediment transport is highly variable between reaches and rivers.

Another area where a geomorphological understanding of river systems has led to a change in river maintenance is in the field of dam sedimentation and clear water erosion problems downstream. In many places sands and gravels are now lifted over dams and put into the river downstream. The major hydroelectric company in Scotland, for example, now undertakes this practice after 50 years of removing sediments from behind its diversion dams and stockpiling the material. Alternatively the Grande Dixence hydroelectric power company in Switzerland has traps that automatically purge material downstream when they become full of sediment. This can occur up to 50 times per year owing to the high bed load yield of alpine streams. However, such river maintenance alone may not successfully impact on the downstream river unless there are peak discharges able to redistribute the material along the course of the regulated river. Indeed such schemes may exacerbate sedimentation problems that are a feature of regulated rivers. This sedimentation can actually increase flooding by reducing the channel capacity (via infilling), despite discharges being lower. It is thus apparent that a good understanding of the relationship between flow processes, erosion, sediment transport and depositional processes is necessary for improved river management.

14.7.4 Building new river channels

Fluvial geomorphologists are now being asked to help engineers build new river channels. This is required when the natural channel is in the way of urban development or mining on the valley floor (Gilvear and Bradley, 1997). Formerly the channel would have been diverted along straight concrete-lined channels. From aesthetic and environmental viewpoints these traditional techniques are no longer acceptable. A 'natural' channel morphology is preferable. It ought to be in equilibrium with the processes of water movement and transfer and create habitat conditions that conserve in-stream biota. At its simplest the channel can 'mirror' the former natural course but in other situations design criteria need to be based on geomorphic principles and processes while incorporating some hard engineering. For example, a 2.7 km sinuous gravel-bed river diversion was built on the River Nith, Scotland, in 2005 (Figure 14.27). This partially mirrors the old channel and was constructed with reference to geomorphic principles. There was some hard engineering but only at critical locations where prevention of erosion was of paramount importance. The old natural river has previously been lost to valley floor mining for coal. However, before this river was obliterated the fish were rescued and introduced to the new channel. The author is currently monitoring the stability of the diversion channel to learn more about how rivers behave in such circumstances so that future rivers can be constructed even better.

Figure 14.27 The River Nith diversion, south-west Scotland, showing its engineered morphology. The point bar features visible on inside of the meander bends were not engineered but developed over the first few flood events to flow through the diversion. The spoil behind is from a coal mine which lies beyond.

(a)

(b)

Figure 14.28 The re-meandered Sinderland Brook near Manchester, England, on the day after completion (a) and three months later (b). (Source: photo (b) courtesy of Charles Perfect)

14.7.5 River restoration

River restoration is often undertaken where past human activity has resulted in ecological devastation of river systems. Unsympathetic engineering treatments reduce channel morphological diversity and it has long been recognized that morphological diversity benefits the ecological carrying capacity of a river and provides suitable habitats for the various life stages of fauna. River restoration also occurs in places where river channels have previously been straightened. For example, to reduce future flooding, the Swiss Government in 2004 announced plans to reinstate meanders on over 100 rivers at a cost of over £80 million. Re-meandering should reduce water velocities and create flood wave attenuation leading to smaller flood peaks. The process often includes reinstatement of meanders and creation of pools and riffles. This may be as simple as mirroring an upstream reach or producing an exact copy of what once existed at that location. Relocation to the original form may not be the best solution, however, since under natural conditions the original river form would have changed. In many cases the nature of the channel that existed before the straightening occurred is not known. Therefore geomorphologists have to design channels in equilibrium with river regime, sediment load and calibre and must consider what the natural channel pattern ought to be.

A good example of a river restoration project involving re-meandering is the case of Sinderland Brook near Manchester, England. Land was sold off for housing development. A condition of the purchase was that the straightened and highly channelized stream that runs

across the land was re-meandered. Figure 14.28 shows the re-meandered morphology immediately following construction and one year later after vegetation colonization of the riparian zone had occurred. The scheme has led to increased in-stream species diversity and reduced flood risk to an existing housing estate. A different type of restoration project was undertaken on the Highland Water in the New Forest, England. This river once had pools and riffles providing excellent fish habitat. However, dredging followed by headward incision had led to an over-deepened channel without any gravel. Figure 14.29 shows the river

Figure 14.29 The Highland Water in the New Forest, England, following introduction of bed gravels. Prior to substrate replenishment, dredging and subsequent incision had created a channel devoid of gravel and hydraulic features such as pools and riffles.

following reintroduction of gravel during a process known as substrate replenishment. The success of the project relied upon determining the correct bed material particle size distribution so that the material did not undergo large-scale removal during the first large flood but still remained sufficiently mobile so that pools and riffles would be created. Another concern was that the permeability of the gravels did not result in the complete disappearance of water during low flows; to combat leakage the gravel was laid down in layers, consolidated and some finer material added to create less permeable layers. From these two examples of river restoration the direct and paramount importance of an understanding of fluvial geomorphology is evident.

Reflective questions

➤ Based on your knowledge of fluvial geomorphology, do you think you could design a river to be in equilibrium with a given flood regime and sediment loading?

➤ What are the main problems associated with traditional engineering approaches to river management?

➤ Why is morphological diversity in river channels important to the flora and fauna?

14.8 Summary

This chapter has focused on fluvial processes, the linkage between processes and landforms, and natural and human-induced channel change. It has also examined the relevance of fluvial geomorphology to the needs of river management in the twenty-first century.

Catchments vary in the amount of water and sediment they carry. This variation is a function not simply of catchment size but also of local topography, land management, geology, soils, vegetation, and so on. Catchments with high drainage densities are likely to have high peak flows and a flashy regime. River channel networks can be described and classified in a number of ways including the stream order systems of Horton and Strahler. The size and shape of river channels can be described in terms of channel cross-section and planform which may vary over short distances. River channel planform ranges from braided and anastomosing to meandering and straight river channels. River cross-sections are not uniform and water and sediment flows through a cross-section vary. At river bends a circulatory pattern of flow (helicoidal flow) is superimposed on the downstream movement of the water.

Water within a channel is subject to gravity and frictional forces. Steep, smooth channels will have a high average water velocity. A similar channel with a channel boundary consisting of boulders, and thus high frictional forces, will have a slower average water velocity, even though the turbulence of the water may give the impression of fast-moving water. Mean water velocity in an open channel can be estimated using the Manning equation which accounts for channel slope, hydraulic radius and channel roughness. For a given cross-section the way in which water velocity, depth and width increase with a rise in water level is known as hydraulic geometry. Stream power is a key parameter in determining rates of erosion, sediment transport and instability. Slight changes in velocity can significantly affect potential stream power.

For a particle to be entrained from the stream bed or bank, a threshold has to be passed whereby a critical velocity or shear stress exceeds the frictional forces that resist erosion. This is dependent upon channel slope, particle size and shape, and immersed weight in relation to the bed shear stress and fluid kinematic viscosity. However, processes such as imbrication may also play a role. Deposition and cessation of bed load movement for an individual particle occur when velocity falls below critical conditions. Hydraulic sorting occurs under these conditions. Sediment can be carried as bed load or as suspended load and bed load transport is almost entirely a function of flow volume, velocity and turbulence. Particles roll, slide or saltate along the bed.

Bed morphology can vary depending on bed material while the nature of a channel's boundary materials may also play a role in channel stability. In coarse sediments pool-riffle sequences may dominate, whereas in fine sandy channels dune structures may be found. Bedrock channels, however, may be subject to cavitation and corrosion processes. River channel change can occur very quickly during a flood event and channels can avulse from one site to another. Channels may take a long time to recover from a large event in terms of adjusting their size and shape back to suit lower flows but this relaxation time varies from river to river.

Many channelized rivers have altered their course or shape following engineering works and this has caused on-site and upstream and downstream problems. Rivers are naturally dynamic and yet humans often require stable river channels. Thus fluvial geomorphology has a very important role to play in modern-day river management. It can identify the causes of management problems, at both reach and catchment scales, and predict the impacts of human intervention on rivers with mobile bed sediments and erodible banks. River restoration, design of new river channels and river maintenance all require fluvial geomorphological insight. However, much remains to be understood about the behaviour of river channels.

Further reading

Brookes, A. (1985) River channelisation, traditional engineering methods, physical consequences and alternative practices. *Progress in Physical Geography*, 9, 44–73.
This is a review article that shows how traditional engineering approaches to channelization can lead to geomorphic problems and environmental degradation. It also explains how geomorphic knowledge can be used to design better channelization schemes. Professor Andrew Brookes is well known for his work on channelized rivers.

Gilvear, D.J. and Black, A.R. (1999) Flood induced embankment failures on the River Tay: implications of climatically induced hydrological change in Scotland. *Hydrological Sciences Journal*, 44, 345–362.
This is a paper that notes the importance of past channel changes to flood embankment stability and how future changes in flood hydrology that may occur owing to climate change could heighten levels of flood embankment instability.

Gilvear, D.J. and Bradley, S. (1997) Geomorphic adjustment of a newly constructed ecologically sound river diversion on an upland gravel bed river, Evan Water, Scotland. *Regulated Rivers*, 13, 1–13.
This is a short case study paper examining how a newly constructed river diversion responded to a large flood just a few months after its completion.

Gupta, A. (2007) *Large rivers: Geomorphology and management*. John Wiley & Sons, Chichester.
This book has many very useful case studies and illustrates techniques for managing very large rivers.

Gurnell, A.M. *et al.* (2006) Initial adjustments within a new river channel: interactions between fluvial processes, colonising vegetation and bank profile development. *Environmental Management*, 38, 580–596.
This paper illustrates the role of geomorphology in river engineering and how fluvial processes shape physical habitat and biota.

Kondolf, M. and Piegey, H. (2003) *Tools in fluvial geomorphology*. John Wiley & Sons, Chichester.
This is a volume aimed at providing comprehensive details of the techniques required to map, monitor and investigate the geomorphology of rivers.

Montgomery, D.R. and Buffington, J.M. (1997) Channel-reach morphology in mountain drainage basins. *Geological Society of America Bulletin*, 109, 596–611.
This paper classifies river types as found in Oregon and Washington and relates the types to fluvial processes and topographic and geological controls – relevant to Box 14.4.

Petts, G.E. (1984) Sedimentation within a regulated river. *Earth Surface Processes and Landforms*, 9, 125–134.
This is an examination of how river regulation caused pronounced sedimentation downstream – relevant to Box 14.5.

Petts, G.E. and Calow, P. (eds) (1996) *River restoration*. Blackwell Science, Oxford.
This is a series of very useful chapters on river engineering and restoration. The chapter by Hey is particularly relevant.

Thorne, C.R., Hey, R.D. and Newson, M.D. (1997) *Applied fluvial geomorphology for river engineering and management*. John Wiley & Sons, Chichester.
This is a volume aimed at providing an overview of fluvial geomorphology as a basis for effective management and engineering within rivers.

Web resources

Exploring Earth: Surface Waters

http://www.classzone.com/books/earth_science/terc/navigation/chapter13.cfm

An excellent site is presented here containing various animations of the processes that are involved in fluvial geomorphology, pictures, questions and answers and general information on river systems.

Google Earth

http://earth.google.com/

Google Earth can be used to examine the planform morphology of many of the world's great rivers. Try to find rivers such as the Thames, Rhine, Danube, Murray–Darling, Mississippi, Yukon, Orinoco, Congo and Brahmaputra. Analyse their morphology and also consider whether there is any evidence to suggest in the case of the first five rivers where their planfrom has been altered by river engineering.

International Rivers Network

http://www.irn.org/index.html

This is the home page of the IRN, an organization promoting equitable and sustainable river management strategies for meeting human needs for water, energy and flood management.

NASA Exploring the Environment Program: Stream Ordering

http://www.cotf.edu/ete/modules/waterq/wqphysmethods.html

NASA provide a definition of stream ordering and how it works, including pictures of representative streams.

River Landscapes: Restoring Rivers and Riparian Lands

http://www.rivers.gov.au/

This is a site provided by the Australian Government Department on Land and Water. The site offers a number of factsheets covering issues that are faced when attempting to manage river environments, including definitions and basic information.

Visit our website at **www.pearsoned.co.uk/holden** for further questions, annotated weblinks, interactive models and video-clips showing physical processes in action.

Solutes

Kate V. Heal

School of GeoSciences, The University of Edinburgh

Learning objectives

After reading this chapter you should be able to:

➤ **understand how solutes are affected by hydrological processes operating within catchments**

➤ **appreciate the interaction of geology, climate, soil, biotic and human factors that produce the solute characteristics of a catchment**

➤ **understand how spatial and temporal patterns in solutes occur**

➤ **explain why changes in climate and land-use affect solutes**

➤ **understand the difference between solute flux and solute concentration**

15.1 Introduction

Water is essential for life, and in the natural world it consists of more than pure hydrogen and oxygen. It contains many other substances in dissolved or solid forms. This chapter deals with dissolved solids known as solutes. Solutes are naturally occurring. However, human activities can discharge solutes into the environment and also alter the processes affecting both the **concentrations** and fluxes of naturally occurring solutes. A contaminant is a chemical compound present in the environment, at a level significantly greater than its natural abundance, owing to anthropogenic activities. An understanding of solute sources and transport processes underpins many topics in physical geography and environmental science including mineral weathering, soil development, plant and animal nutrition, aquatic ecology, and the quality of water resources for human use.

This chapter will examine solute processes within a catchment framework since these processes are intimately linked with the hydrological processes discussed in Chapter 13. A range of chemical and physical factors affect solutes, and understanding solute transport processes is important because of the implications for devising effective strategies for managing water quality in catchments. The factors that determine spatial and temporal patterns of solutes can be examined at a number of scales and these will be outlined within this chapter. Knowledge of these factors is required in order to predict the impacts of environmental change on solutes in catchments. Such environmental change may include pollution incidents, land-use change or changes in atmospheric chemistry.

15.2 Solute chemistry: some key points

It is vital to appreciate some of the chemical and physical factors affecting solutes before solute processes are explored within catchments. Solute concentrations at a given point and the spatial distribution of solutes are constantly changing as the result of physical, chemical and biological reactions. This section briefly discusses the aspects of solute chemistry relevant for understanding solute processes in catchments. For further information on solute chemistry see Langmuir (1997).

15.2.1 Solute form

Solutes exist in a number of different forms in the environment. Many substances, particularly metals, are transported in water chiefly in the form of **complexes** with other compounds rather than as a free ion (see Chapter 7). It is important to know the form of a solute as well as the

concentration since different forms of solutes have differing toxicities and bioavailabilities (see Chapter 10) to living organisms. For example, the free aluminium ion (Al^{3+}) is much more toxic to fish than aluminium that is complexed to **organic** compounds. In addition to the formation of complexes, many elements can occur in more than one **oxidation state** in the environment. For example, iron may be present as iron +2 (ferrous, Fe^{2+}) or iron +3 (ferric, Fe^{3+}). The oxidation state of an element exerts an important control on its solubility and availability for transport by water in the environment. The oxidation state of an element at a particular point in the environment is mainly governed by the surrounding pH and **redox potential** conditions.

15.2.2 pH and redox potential

The influence of pH and redox potential on the form and environmental impact of solutes is illustrated by the occurrence of iron in freshwater. The Eh–pH stability field (see Box 15.1 for explanation) for iron shows that Fe^{2+} occurs in

FUNDAMENTAL PRINCIPLES

pH AND REDOX POTENTIAL

pH is a measure of acidity from the activity of hydrogen ions in a system. The pH scale ranges from 0 (extremely acidic) to 7 (neutral) to 14 (extremely alkaline). Chemically, pH is the negative logarithm to the base 10 of the concentration of hydrogen ions in solution (see Chapter 7):

$$pH = -\log_{10}[H^+] \qquad (15.1)$$

Redox potential is a measure of how oxidizing or reducing the environment is in terms of the occurrence of electrons (e^-). Redox potential is normally measured as Eh, in units of volts (V). From electrochemical theory, **Eh–pH stability fields** can be plotted for different elements that show the forms expected under particular Eh and pH conditions, if chemical equilibrium is assumed. The Eh–pH stability field for iron is shown in Figure 15.1.

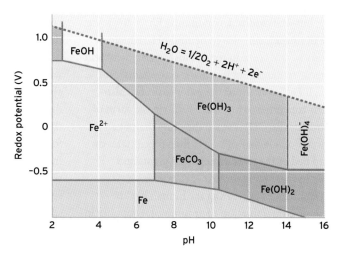

Figure 15.1 Eh-pH stability field for aqueous iron at atmospheric pressure and 25ºC. If the water redox potential (Eh) and pH are known the figure shows the form in which iron will occur. For example, at pH 4 and Eh 0, iron would be present as Fe^{2+}. (Source: after Morgan and Stumm, 1965)

BOX 15.1

HAZARDS

ARSENIC IN GROUNDWATER SUPPLIES IN BANGLADESH AND WEST BENGAL

The world's largest outbreak of arsenic poisoning has been occurring in Bangladesh and West Bengal, India, since the mid-1990s. The initial symptoms of arsenic poisoning are skin darkening and lesions, followed by swelling of the liver and spleen. Eventually cancers of the intestine may develop. An estimated 30 million people in Bangladesh and 2 million people in West Bengal are at risk of cancers caused by arsenic poisoning. The outbreak is attributed to the consumption of groundwater containing high concentrations of arsenic.

Until the 1960s, most of the population in the affected area obtained water supplies from surface waters which were contaminated with sewage, resulting in widespread gastrointestinal disease. From the 1960s thousands of low-cost wells were dug to access groundwater supplies with a better microbiological quality. Many of the new wells were sunk into aquifers with a naturally high arsenic concentration, from arsenic bound to iron minerals in delta sediments which were deposited 25 000–80 000 years ago. Decomposition of vegetable matter in the delta sediments consumed all the available oxygen, creating reducing conditions that converted insoluble Fe^{3+} to the soluble Fe^{2+} state. The solubilization of iron also mobilized arsenic bound to the iron minerals so that arsenic concentrations of 0.05–1 mg l^{-1} have been measured in wells, compared with the new World Health Organization limit for drinking water of 0.01 mg l^{-1}. As well as the total arsenic concentration, the chemical speciation of arsenic is important in assessing the impacts on public health. Arsenic has two oxidation states in the environment. Arsenic +3 is the most toxic but is difficult to remove from water. In comparison arsenic +5 is less toxic and easier to remove from water. Geologists, chemists and engineers are working together to identify alternative water supplies for the affected population and, in the short term, to develop technologies to reduce arsenic concentrations in the well waters.

BOX 15.2

highly acidic, but well oxidized, conditions (Eh > 0 V) (e.g. in acid mine drainage), and also in neutral pH, reducing conditions (Eh < 0 V) (e.g. in fens). This is partly why acid mine drainage is so damaging to aquatic ecosystems. Not only is the high Fe^{2+} content of the water toxic, but also soluble Fe^{2+} is converted to insoluble Fe^{3+} (in the form of iron +3 hydroxide, $Fe(OH)_3$) when the mine drainage mixes with water of a higher pH. Fe^{3+} then precipitates on the banks and bed of the river, smothering plants and animals living in these habitats. Box 15.2 explains how redox potential is responsible for creating the human health hazard of arsenic contamination of groundwater supplies in Bangladesh and West Bengal.

15.2.3 Temperature and pressure

Increasing temperature increases the solubility of some solids in water and speeds up the rate of chemical reactions, but reduces the solubility of most gases. The last is particularly significant for the dissolved oxygen content of rivers and lakes that is necessary for organisms, such as fish, to survive. In contrast, at high pressure, the solubility of gases in water increases. This can influence the pH of water in different parts of the catchment hydrological system as follows. In groundwater, the solubility of carbon dioxide in water increases because the pressure is higher than atmospheric. Carbon dioxide dissolves in water to form carbonic acid, a weak acid, causing a decrease in pH in groundwater. When the groundwater comes into contact with the surface environment again in, for example, a spring, the reverse effect occurs. Carbon dioxide in the spring water can come out of solution, because of the reduced pressure, and the pH of the water increases as it now contains less carbonic acid.

15.2.4 The role of particulates

The movement of particulate material in catchments can affect the distribution of solutes within the catchment hydrological system. For example, eroded soil from mixing model calculations can be performed for a number of points in time to separate storm hydrographs for a river into different flow components on the basis of solute concentrations.

Using solutes as tracers has increased the understanding of hydrological processes in catchments. The advantage of using solutes as chemical tracers is that interference with the catchment hydrological system is minimal, whereas direct measurements of hydrological pathways by physical methods can alter the flows being measured. In particular, solute tracer studies in catchments have shown that, in temperate latitudes, groundwater frequently makes a significant contribution to storm runoff (Pearce *et al.*, 1986). This challenged traditional hydrological theory which assumed that groundwater only constituted a small proportion of storm runoff because slow groundwater flow rates meant that it could not respond rapidly to rainfall. As a result, new hydrological mechanisms were proposed to account for the rapid response of groundwater to rainfall events (Sklash *et al.*, 1986).

Knowledge of hydrology and hydrological pathways in a catchment is important in explaining and modelling solute concentrations in rivers and also in understanding temporal and spatial patterns of solutes in catchments. Conversely, solute concentrations and fluxes can be used as tools in understanding catchment hydrology. Box 15.3 illustrates how hydrological pathways are important in devising measures to reduce diffuse phosphorus pollution from agricultural land.

Figure 15.13 Mean monthly nitrate concentrations in river water from 13 upland catchments with semi-natural vegetation in northern Scotland which were sampled fortnightly from June 1996 to May 1997. (Source: adapted from Clark *et al.*, 2004)

Reflective questions

➤ Which combination of hydrological pathways would make a catchment most vulnerable to spillage of a polluting chemical?

➤ Which flow path would you expect to produce the greatest concentration of solutes: infiltration-excess overland flow, return flow or macropore flow? Why?

15.5 Temporal patterns of solutes

Solute concentrations in rivers vary over time because the hydrological processes that generate runoff within catchments are dynamic. Four main factors control temporal patterns of solute concentrations in catchments. Firstly, the quantity and mobility of the solute supply determine its availability for transport to the river channel. For example, nitrate concentrations in river water display a typical seasonal pattern as shown in Figure 15.13 for catchments in northern Scotland with semi-natural vegetation and no arable land. Nitrate concentrations are lowest in spring and summer due to plant and microbial uptake and highest in the winter months when cooler temperatures mean that there is less uptake and therefore more nitrate available for leaching from the soil.

Secondly, the composition of precipitation inputs to the catchment affects solute concentrations in river water. For example, precipitation in coastal regions is often enriched by sea salt during stormy weather conditions when droplets of seawater are incorporated into precipitation. These **sea-salt events** can have a dramatic effect on river water composition since chloride ions cause an increase in **specific conductance**. Sodium ions in the precipitation exchange with hydrogen ions adsorbed onto soil cation exchange sites, resulting in a flush of acidity.

Thirdly, catchment size and heterogeneity affect solute concentrations. In large catchments, solute concentrations in the main river channel represent the sum of solute processes operating in the individual tributary catchments, each of which may differ considerably in their individual concentrations.

Finally, changes in hydrological pathways in catchments over time affect solute concentrations in river water as different pathways have different solute signatures. Groundwater and throughflow deep within the soil are the major source of base cations to river water from mineral weathering. Therefore base cation concentrations in river water are normally higher in drier conditions when these hydrological pathways are the main source of river water.

Base cation concentrations decline in wetter conditions since more runoff from macropore flow and overland flow is generated, with lower base cation concentrations, thereby diluting the groundwater inputs.

Temporal patterns in solute concentrations in river water are studied at three different timescales: from short-term changes of the order of a few hours during storm events, to annual patterns occurring over days and months, and long-term changes occurring over decades. Some of the patterns of variation and the processes responsible are examined below for each timescale.

15.5.1 Patterns of solutes in storm events: short-term changes

Considerable research effort in solutes has focused on storm events. Dramatic changes in solute concentrations occur and a significant proportion of solute loadings is transported during storm events. The relationship between solute concentrations and river discharge is of particular interest for incorporation into hydrological models. Studies of solute patterns in river water during storm events have found significant variations between solutes, and also between individual storm events and between catchments for the same solute. The response of individual solutes during storm events varies because of the different magnitude and location of solute stores within a catchment and the extent to which they are accessed by the hydrological pathways that generate storm runoff.

Nevertheless, three generic relationships have been identified between solute concentrations and discharge: positive, negative and **hysteresis**. Positive relationships between solute concentrations and discharge occur when solutes are washed rapidly into the river channel by overland flow and shallow throughflow. In Figure 15.14 dissolved organic carbon (DOC) concentrations increase with discharge owing to the flushing of organic matter from more organic-rich surface soil horizons in the catchment. Negative relationships between solute concentrations and discharge occur for solutes that are mainly derived from mineral weathering and transported to the river in deep soil throughflow and groundwater flow. During storm events, inputs from these hydrological pathways are diluted by runoff generated from near-surface inputs. This effect is evident in Figure 15.14 where calcium concentrations decrease as discharge increases owing to dilution by near-surface inputs with lower calcium concentrations.

More complex hysteresis relationships occur between solute concentrations and river discharge during storm events when very different solute concentrations are

Figure 15.14 Solute concentrations and discharge in a storm event in the Catamaran Brook catchment, New Brunswick, Canada. (Source: adapted from Caissie et al., 1996)

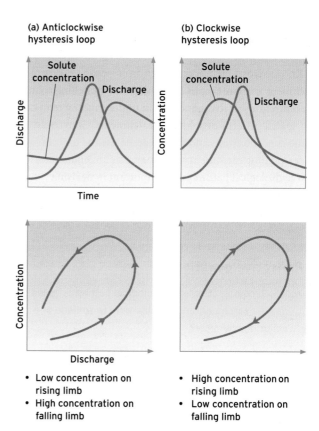

(a) Anticlockwise hysteresis loop

(b) Clockwise hysteresis loop

- Low concentration on rising limb
- High concentration on falling limb

- High concentration on rising limb
- Low concentration on falling limb

Figure 15.15 Schematic diagram of the formation of (a) anticlockwise and (b) clockwise hysteresis loops for solute concentrations in storm events.

measured at the same discharge on the **rising** and **falling limbs** of the storm hydrograph. This results in hysteresis loops when solute concentration is plotted against discharge. Figure 15.15 illustrates how anticlockwise and clockwise hysteresis loops can develop for solute

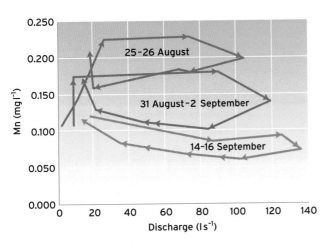

Figure 15.16 Hysteresis relationship between manganese concentrations and discharge during storm events in the Upper Nidderdale catchment, North Yorkshire, England, in 1994.

Figure 15.17 In climates where a snowpack accumulates in winter, such as at Lake Agnes in the Rocky Mountains, Alberta, Canada, solutes are released into river water when the spring snowmelt occurs.

concentrations and river discharge. Anticlockwise loops occur when solute concentration is higher on the falling limb of the storm hydrograph compared with the rising limb. This could be caused by displacement into the river channel of soil water containing high solute concentrations by rainfall percolating through the soil. Clockwise hysteresis loops are formed when a higher solute concentration occurs on the rising limb of the storm hydrograph compared with the falling limb. This can arise when solutes are flushed from ground and vegetation surfaces and surface soil horizons at the start of storm events. Clockwise hysteresis loops for manganese concentrations and discharge in the River Nidd are shown in Figure 15.16 and are caused by flushing of manganese from surface soil horizons on the rising limb of the storm hydrograph. Depletion of manganese soil stores during consecutive storm events results in changes to the shape of the hysteresis loops.

15.5.2 Annual patterns of solute concentrations

Annual patterns of solutes in catchments occur because of systematic seasonal changes in hydrological pathways, weather, and biological and human activities. In temperate latitudes, with drier summers and wetter winters, solutes that occur mainly in deep soil throughflow and groundwater hydrological pathways tend to result in high river water concentrations in summer. During the winter such solutes are diluted as a greater proportion of the runoff occurs as near-surface flow. Evaporation can also cause greater summer solute concentrations than in winter. In contrast, solutes that accumulate on ground and vegetation surfaces

from dry deposition during drier summer conditions exhibit maximum annual concentrations in river water in the autumn when they are flushed into the river by increasing rainfall.

Annual cycles of biological activity also affect concentrations of nutrients such as nitrogen and phosphorus in river water. The lowest nitrate concentrations in river water occur in the summer growing season because uptake from soil water by plants and microorganisms is at a maximum. Maximum stream nitrate concentrations occur in the autumn as the result of reduced demand from plants and also increased production of nitrate by **nitrifying bacteria** in the soil, stimulated by wetter soil conditions.

In climates where a snowpack accumulates in winter (Figure 15.17), the melting of the snowpack in spring causes extremely high solute concentrations in river water. This is caused by the release of dry deposited solutes, accumulated on the surface of the snow during the winter, and also by flushing of solutes from the soil that have remained in a frozen state during the winter. High levels of aluminium and acidity, released into rivers during snowpack melting, can be particularly damaging to fish populations (Davies *et al.*, 1992). Seasonal variations in the composition of precipitation inputs to catchments can also cause annual patterns of solute concentrations in river water. For example, the sea-salt content of precipitation in coastal areas at temperate latitudes is higher in the winter months as stronger winds incorporate more seawater droplets into precipitation (Figure 15.18). The increased

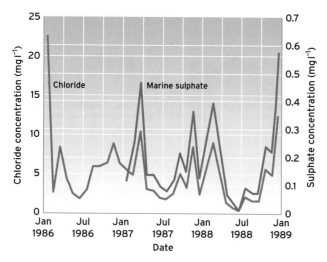

Figure 15.18 Monthly mean concentrations of wet deposited chloride and marine sulphate at Beinn Eighe, north-west Scotland, 1986-1988.

Figure 15.19 Total ammonia concentrations in the River Ouseburn, upstream and downstream of Newcastle Airport, and minimum grass temperatures. When the temperature drops below 5°C de-icer is used at the airport and consequently the ammonia concentrations increase in the rivers downstream when the de-icer is washed off the runways. (Source: after Turnbull and Bevan, 1995)

sea-salt input can result in elevated levels of chloride, specific conductance and acidity in river water in winter.

Seasonal variations in human activity such as winter ploughing of fields, spring application of fertilizer and salting of road surfaces in winter may all cause distinct annual patterns of solute concentrations in river water. The effect of applying glycol and urea de-icers at Newcastle International Airport, northern England, on solute concentrations in river water is shown in Figure 15.19. Total ammonia concentrations increase rapidly downstream of the airport when ground temperatures fall below 5°C (Turnbull and Bevan, 1995).

15.5.3 Long-term patterns of solute concentrations

Changes in solute concentrations in river water over decades are difficult to evaluate because of the frequent unavailability of records of sufficient quality and length. Figure 15.20

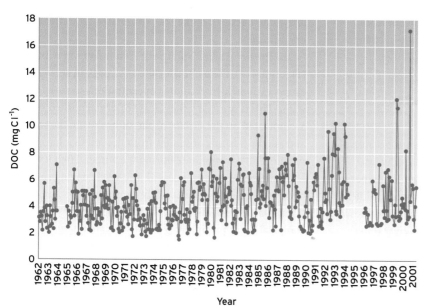

Figure 15.20 Long-term trends in dissolved organic carbon concentrations in the River Coquet, Warkworth, northern England, 1962-2001. (Source: after Worrall *et al.*, 2004)

illustrates the difficulty of identifying a long-term trend in river water dissolved organic carbon (DOC) concentrations hidden amongst considerable storm event and annual variability. Graphical analysis techniques show an upward trend in DOC concentrations the cause of which is hotly debated (see Box 15.4).

The main causes of long-term patterns of solute concentrations in river water are climatic variability and climate change, land management change and changes in human polluting activities (e.g. atmospheric pollution). Indirect evidence from lake sediment cores links acidification of freshwaters in north-west Europe and North America with increasing atmospheric pollution. Measurements of the abundance of acid-tolerant **diatoms** in lake sediment cores indicate that water pH declined from 5.5 to 4.3 from 1700 to 1970 in many areas such as that shown in Figure 15.21 (Battarbee *et al.*, 1985). The concentration of heavy metals, associated with atmospheric pollution

by fossil fuel combustion for domestic, industrial and transport purposes, also increased in lake sediment over the same time period.

The precise causes of long-term patterns of solute concentrations in river water are difficult to identify because of the interaction of causal factors. For example, climate change often results in changes to land-use and management. For example, measurements of **biochemical oxygen demand (BOD)** in the River Clyde, Scotland, shown in Figure 15.22(a), indicate that river water quality improved in the lower reaches of the river during the 1970s (Curran and Robertson, 1991). One explanation for this trend is the development and enforcement of stricter environmental laws concerning water pollution in the catchment. However, the improvement in water quality could also be caused by greater dilution by rainfall and river discharge which increased in the catchment over the same time period (Figure 15.22b).

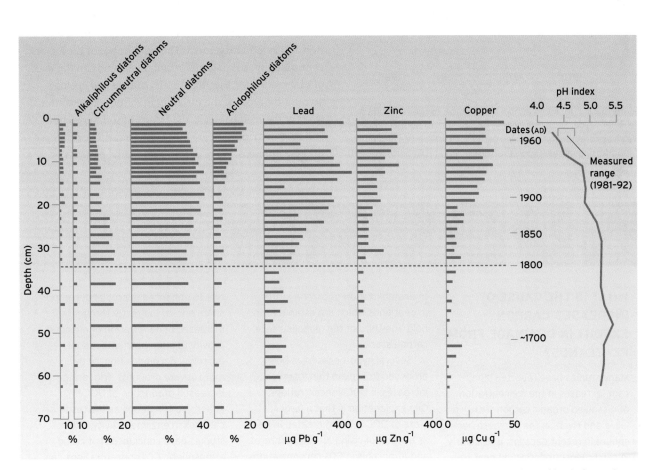

Figure 15.21 Abundance of diatoms with different water pH requirements, reconstructed pH and concentrations of lead, zinc and copper in a sediment core from Loch Enoch, south-west Scotland. (Source: adapted and reprinted with permission from *Nature*, Battarbee *et al.*, 1985, Lake acidification in Galloway: a palaeoecological test of competing hypotheses, *Nature*, **314**, Figs 1 and 3. Copyright 1985 Macmillan Magazines Ltd)

(a)

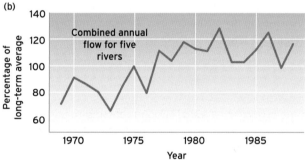

(b)

Figure 15.22 (a) Mean annual biochemical oxygen demand in the River Clyde, Scotland, 1970–1988. (b) Sum of annual flows for five rivers draining to the Clyde estuary, expressed as percentage of long-term average. (Source: after Curran and Robertson, 1991)

Reflective questions

➤ Thinking about dilution and other processes, do solute concentrations always decline as discharge increases? What are the reasons for your answer?

➤ What effects could climate change have on solute patterns at the storm event, annual and long-term timescales and how would you explain the processes which cause the effects you have identified?

15.6 Spatial patterns of solutes

Solute concentrations and fluxes (concentration × discharge) in rivers and lakes vary over space as well as over time owing to differences in climate, geology, topography, soils, vegetation and land management. The significance of the factors controlling spatial patterns of solutes depends on the scale of observation. At the global scale, climate and geology account for most of the variation in solute concentrations, while at the regional scale, the factors of soil, vegetation and land management are more important. The influence of these different sets of controls on spatial patterns of

ENVIRONMENTAL CHANGE

WHAT IS THE CAUSE OF INCREASED CARBON EXPORT IN DRAINAGE FROM PEATLANDS?

Many studies have reported significant increases in the concentration of dissolved organic carbon (DOC) in lakes and rivers in the northern hemisphere in recent decades, particularly in catchments underlain by peat soils (Figure 15.23). These observations have been interpreted as evidence of destabilization of carbon stores in peatlands which are estimated to hold one-third of the global surface carbon stock.

Several hypotheses have been proposed to explain the observed increases in DOC concentrations. One explanation is that a larger store of DOC is being created in peat as a result of rising air temperatures and atmospheric CO_2 concentrations. Rising air temperatures could create a larger store of DOC, either directly due to enhanced microbial decomposition and/or indirectly through increased mineralization of nitrogen, providing more nutrients to fuel microbial decomposition, or by more severe droughts, triggering increased production of DOC in anaerobic conditions as the result of changes in enzyme activity. Other studies have concluded that rising atmospheric CO_2 concentrations can stimulate plant growth activity, resulting in increased production of

BOX 15.4 ➤

Figure 15.23 DOC concentrations have increased in many rivers and lakes in the northern hemisphere in recent decades, particularly in catchments underlain by peat soils as in the uplands of south-west Scotland. (Source: photo courtesy of A.M. Hardie)

compounds exuded from plant roots that are rich in DOC (Freeman *et al.*, 2004).

Another explanation is that more intensive land management activities, such as land drainage, forest planting and harvesting, burning, and livestock grazing in the UK uplands, have resulted in greater DOC mobilization. However, although changing land-use and management may have a local effect, it is not thought to be the main cause of the observed increase in DOC. Yet another hypothesis is that reductions since the 1980s in the deposition of acidic atmospheric pollutants, particularly sulphur, have altered soil water chemistry, resulting in increased

production and mobility of DOC in peat soils. A further explanation focuses on catchment hydrology in which either decreases in water flow could result in increased measured DOC concentrations or changes in hydrological flow paths could mean that richer sources of DOC in peatlands are accessed.

There is much debate and ongoing research concerning the explanation and interpretation of the observed increase in DOC concentrations in rivers and lakes and it has been suggested that attention should focus on DOC flux (DOC concentration multiplied by flow) rather than concentration. An investigation of DOC fluxes found fewer apparent

long-term increases in total annual flux in rivers in Great Britain from 1975 to 2003, in contrast to the marked increases in DOC concentrations. The flux trends are consistent with increases in air temperature and atmospheric CO_2 over the same time period but the spatial pattern of fluxes does not correspond with changes in atmospheric deposition. The main control on DOC fluxes was concluded to be the throughput of water through catchment soils, with DOC flux increasing as runoff increased, though there was also a small effect of previous rainfall and runoff conditions (Worrall and Burt, 2007).

BOX 15.4

solutes will be examined at the global and regional scales in the following sections.

15.6.1 Global patterns of solutes

Global controls on solute concentrations and loads in river water have been analysed by Walling and Webb (1986). Solute concentrations in rivers often increase with mean annual temperature because of greater evaporative loss from

river systems, thereby increasing river solute concentrations. The total ionic load, however, increases with mean annual runoff and slope steepness in rivers. This is because it is the rate of water movement at the weathering front that largely determines the rate of solute release. There is also a marked difference between rock types (Figure 15.24). Ionic loads in rivers draining catchments developed on different geologies are in the order: sedimentary > igneous extrusive > igneous intrusive and metamorphic.

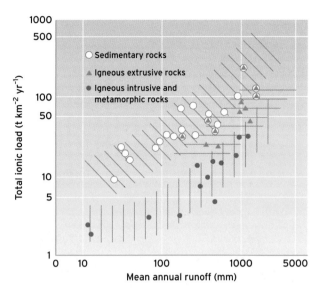

Figure 15.24 Relationship between total ionic load and annual runoff for major world rivers draining catchments developed on different geologies. (Source: adapted from Meybeck, 1980)

15.6.2 Regional patterns of solutes

At the regional scale, land-use is the dominant control on solute concentrations in river waters. Land-use affects solute concentrations by altering the magnitude and mobility of solute sources in the catchment and also by altering the hydrological pathways that transport solutes to the river channel. The effects of conifer plantations, agricultural and urban land uses on solute patterns are discussed below as examples.

15.6.2.1 Conifer plantations

The alteration of solute concentrations in river water by conifer plantations depends on the stage of the forestry cycle. The most significant effect of conifer plantations on nutrient concentrations is an increase in nitrate concentrations if the trees are harvested by clear-cutting. This effect is caused by creation of an improved microclimate for microbial decomposition in forest floor soils, resulting in increased rates of production of nitrate. Elevated nitrate concentrations in river water persist until new vegetation establishes and takes up nitrogen from the soil.

Conifer plantations in catchments developed on base-poor geologies contribute to high levels of acidity and aluminium in river waters. These effects are most marked in mature plantations (Figure 15.4) and are caused by several processes as summarized in Figure 15.25. The major cause of acidification is the interception of acidic atmospheric pollutants by the forest canopy. Conifer trees are more efficient scavengers of atmospheric pollutants than broad-leaved trees and grassland vegetation because the

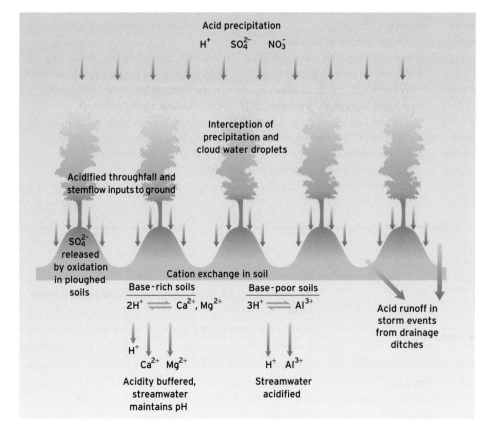

Figure 15.25 Processes of acidification of river water under conifer plantations.

nature of the canopy surface and the large surface area of pine needles encourages the condensation of acidic cloud droplets and deposition of acidic gases and particles. Consequently, high inputs of acidity (hydrogen ions) occur to soils underlying conifer plantations. Acid inputs to forest soils are initially neutralized by the exchange of hydrogen ions with calcium and magnesium ions at cation exchange sites. However, if acid inputs continue faster than base cations are released into the soil from mineral weathering, the soil cation exchange sites become dominated by hydrogen ions and aluminium. Thus, levels of acidity and aluminium increase in river water. The neutralizing capacity of the soil can be depleted further by the uptake of base cations by the tree crop. Another factor that contributes to river water acidification in conifer plantations is the preparation of ground for planting. Drainage ditches are often created by ploughing to improve the soil moisture conditions for the trees. The ditches alter hydrological pathways in the catchment by forming a more efficient drainage system, transferring runoff rapidly to river channels during storm events (Holden *et al.*, 2007). As a result, runoff only passes through the acidic surface soil horizon and has limited opportunity to percolate to less acidic deep soil horizons where neutralization of acidity can occur. Drainage ditch construction also increases the acidity of river water through allowing soil oxidation to occur. This results in production of the acidifying anion, sulphate.

15.6.2.2 Agricultural land management

The main effects of agricultural land management on solute concentrations in river water are an increase in concentrations of nutrients and pesticides. Increased concentrations of these may arise from point sources such as leakage of pesticides, silage, slurry and milk wastes from storage facilities, or from diffuse pollution, such as from farmyards (Figure 15.26). Sources of diffuse pollution are numerous and are summarized in Figure 15.27. Where fertilizer or manure applications exceed crop nutrient requirements, or are applied just before

Figure 15.26 One source of diffuse pollution from agriculture is runoff from farmyards and roofs when faecal material from animals and birds, sediment and spillages of oil and hydrocarbons are washed off by rainfall.

heavy rainfall, leaching of soluble nitrogen and phosphorus occurs from the soil. Artificial land drainage can increase the percolation of water, water residence time and also improve soil aeration. These factors increase nitrate leaching from the soil because there is more opportunity for leaching to occur and conditions are more favourable for nitrate production by microbial activity. Ploughing permanent grassland for conversion to crops also increases nitrate leaching from the soil for similar reasons. Decomposition of crop residues results in the release of nitrogen and phosphorus in the soil, which may be transported into river courses. Finally, soil erosion transports nutrients attached to soil particles from agricultural land to river channels. This transport pathway is particularly important for phosphorus (see Box 15.3).

15.6.2.3 Urbanization

Urbanization typically results in increases in concentrations of metals, nutrients, pesticides and organic matter in river water. These elevated concentrations arise from an

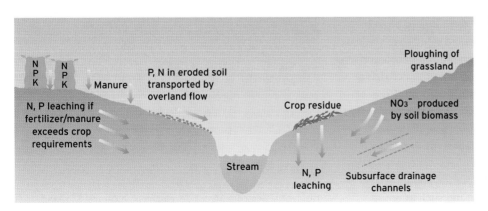

N
P
K

N
P
K

Manure

P, N in eroded soil transported by overland flow

N, P leaching if fertilizer/manure exceeds crop requirements

Crop residue

Ploughing of grassland

NO₃⁻ produced by soil biomass

Stream

N, P leaching

Subsurface drainage channels

Figure 15.27 Diffuse sources and transport of nitrogen and phosphorus from agricultural land to river channels.

Table 15.3 Sources of selected solutes in urban areas

	Vehicle use	Pesticide use	Industrial/other use
Copper	Metal corrosion	Algicide	Paint
			Wood preservative
			Electroplating
Lead	Fuel	-	Paint
	Batteries		Lead pipe
Zinc	Metal corrosion	Wood preservative	Paint
	Tyres		Metal corrosion
	Road salt		
Chromium	Metal corrosion	-	Paint
			Metal corrosion
			Electroplating
Nitrogen	Combustion of fuel	Fertilizers	Cleaning operations
			Animal excrement
Chloroform	Formed from salt	Insecticide	Solvent
PCBs (polychlorinated biphenyls)	-	-	Electrical insulation

(Source: after Novotny and Harvey, 1994, and Makepeace *et al.*, 1995)

increased magnitude of solute sources such as metals deposited from tyre breakdown and corrosion of vehicle brake linings, and intensive pesticide use on gardens, public parks and road verges (Table 15.3). They also result from the dramatic changes in catchment hydrological pathways brought about by urbanization. Urban drainage engineering aims to remove runoff as rapidly as possible from urban surfaces to prevent inundation. Traditionally this aim was achieved through the use of nearly impervious surfaces for car parks, highways and roofs, from which water drained rapidly into a waste-water treatment works or a nearby river. However, this causes flushes of solutes into urban rivers during storm events, often in toxic concentrations, as dry deposited solutes are washed from impervious surfaces. Solute pollution in urban rivers is worse during medium-sized storm events, occurring two or three times a year, since sufficient runoff is generated to mobilize dry deposited solutes from urban surfaces but there is not enough water to dilute solute concentrations. Urban **Best Management Practices (BMPs)** or Sustainable Urban Drainage Systems (SUDSs) are now being introduced to reduce the pollution of urban rivers. These include structures such as ponds, wetlands and porous paving that can increase the storage of runoff in the catchment and to allow water quality improvement by physical, chemical and biological processes (Figure 15.28). A new and increasing water

Figure 15.28 Example of a SUDS: a wetland and retention basin with a permanent water pool treating diffuse pollution from a residential and commercial development in central Scotland.

HAZARDS

NEW SOLUTES OF CONCERN IN THE TWENTY-FIRST CENTURY: ORGANIC MICROPOLLUTANTS

In the past few decades there has been increasing production of numerous synthetic organic chemicals, such as pesticides, pharmaceuticals, microbial disinfectants and personal care products, for use in industry, agriculture, medical treatment and household products. While many of these chemicals have undoubted benefits, their release into the water environment as organic micropollutants may have potentially damaging impacts on ecology and human health, including increased incidences of cancer and impairment of reproduction. Organic micropollutants can enter the water environment from both point sources, such as discharge from sewage treatment works containing organic micropollutants excreted by humans, and diffuse sources such as runoff from animal feeding operations that use pharmaceuticals. Since organic micropollutants typically occur at very low concentrations in the environment, it is only recently that analytical methods have been developed that are capable of detecting them. Consequently the extent of occurrence of organic micropollutants is still largely unknown. The first US-wide survey of the occurrence of 95 organic micropollutants in 139 streams downstream of areas of intense urbanization or livestock production found that they were prevalent in 80% of the streams sampled. The most frequently detected organic micropollutants by type of product are shown in Figure 15.29 and include steroids, non-prescription drugs, insect repellent, detergent by-products and disinfectants. Although the concentrations detected in the study were generally low and rarely exceeded existing standards, for many organic micropollutants no standards have yet been set and the potential combined effects of different organic micropollutants is not yet known (Kolpin et al., 2002).

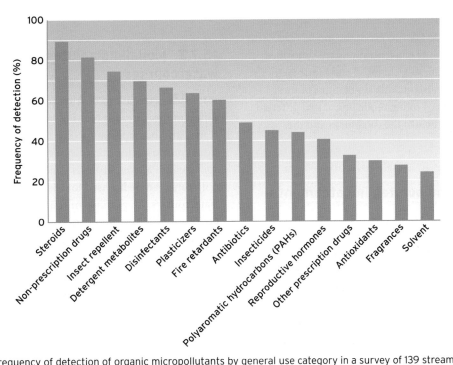

Figure 15.29 Frequency of detection of organic micropollutants by general use category in a survey of 139 streams in the USA, 1999-2000. (Source: adapted from Kolpin et al., 2002)

BOX 15.5

quality concern associated with urban runoff is the presence and effect of organic micropollutants and these are examined in Box 15.5.

Reflective questions

➤ Would water draining a catchment underlain by a metamorphic rock type tend to have a high or low solute concentration compared with other bedrock types?

➤ How do the types, sources and dominant transport processes of solutes from urban and agricultural land compare and contrast?

15.7 Modelling solutes

Models are widely used to predict solute concentrations in catchments, rivers and lakes for catchment management and planning. This may include simulating the effects of pollution incidents on river water quality, for example. As with all models, solute models require field or laboratory measurements for model calibration and validation to evaluate the success of the model (see Chapter 1). Since solute concentrations are controlled by many hydrological, soil, chemical and biotic factors, solute models must also include simulations of these processes. Examples of solute modelling at two different scales are discussed below: the catchment and the watercourse.

15.7.1 Modelling solutes in catchments

Many catchment solute models have been developed. For example, it may be necessary to predict the effect of land-use change or atmospheric pollution on solute concentrations in river water. An example of this type of model is the MAGIC (Modelling Acidification of Groundwater In Catchments) model which was developed to predict the response of solutes in surface waters to acid deposition (Cosby *et al.*, 1985). The model assumes that the concentration of solutes in surface waters is governed by atmospheric deposition, soil weathering reactions, uptake of solutes by biomass and runoff. A

series of chemical equations simulates the reactions of solutes in the soil, including cation exchange and adsorption. The model is a **lumped model** which assumes that the catchment is homogeneous, with uniform soil properties and precipitation inputs.

Application of the MAGIC model to a catchment requires data for soil physical and chemical properties, rainfall/runoff characteristics, precipitation chemistry and base cation weathering rates. The model uses a **stochastic** procedure in which hundreds of simulations are performed, each with slightly different input data. Instead of producing one result, the model produces a **frequency distribution** of outputs. The MAGIC model was applied to predict the effect of reductions in acid deposition on the **alkalinity** of lakes in southern Norway (Hornberger *et al.*, 1989). Figure 15.30 shows the distribution of simulated lake alkalinity values in 1974 compared with predictions of alkalinity in 2020 for different scenarios of acid deposition reduction. The peak of the frequency distributions is a measure of the average alkalinity of lake waters. The model results suggest that the average alkalinity of lakes in southern Norway will not increase significantly by 2020 unless the most radical reductions of 50 or 70% in acid deposition compared with 1980 levels are achieved.

15.7.2 Modelling solutes in watercourses

It is often necessary to develop models to examine different measures for treating instream pollution. It may also be important to use such models for operational purposes, for example to respond to a pollution spillage into a river, or to set concentration limits for polluters that are licensed to discharge pollution into watercourses. An example of this type of model is the QUASAR (QUAlity Simulation Along Rivers) model which was developed to predict the effect of pollution on solute concentrations in rivers (Whitehead *et al.*, 1981). The model performs a mass balance of flow and solute concentrations in a river system and includes simulations of biological and chemical decay processes operating in the river channel. The model requires input data for solute concentrations and flows entering the **river reach** that is to be modelled. The QUASAR model was applied to the upper reaches of the River Pelenna, South Wales, from which salmon have

5% deposition reduction

30% deposition reduction

50% deposition reduction

70% deposition reduction

Figure 15.30 Frequency distributions of simulated alkalinity for 1974 and 2020 in lakes in southern Norway under different deposition reduction scenarios, modelled with the MAGIC model. The frequency distributions represent the range of results predicted by the model as it was run repeatedly. The most common result is therefore represented by the value with the maximum frequency. (Source: after Hornberger *et al.*, 1989)

Figure 15.31 Iron staining in a stream channel from a mine-water discharge.

disappeared because of water pollution from discharges from abandoned coal mines with low pH and high iron concentrations (Figure 15.31). In order to restore salmon populations a water pH of at least 6.0 and iron concentrations below 1 mg l^{-1} are required. QUASAR was used in a stochastic modelling procedure to assess the effectiveness of different treatments of the mine-water discharges in improving water quality in the river. The results identified the four key mine-water discharges for treatment and specified the level of treatment required (Whitehead *et al.*, 1995). A very different type of model that has been applied to model solute concentrations in watercourses is **artificial neural networks (ANNs)** and these are described in Box 15.6.

Reflective question

➤ A model is being developed to predict the effect of urbanization on solute concentrations in the river draining a catchment. What do you think are the key processes and solutes to include in the model, and why?

NEW DIRECTIONS

MODELLING WITH ARTIFICIAL NEURAL NETWORKS

Artificial neural networks (ANNs) have been inspired by the functioning of the brain. They process information in parallel, compared with other types of solute models which process information sequentially. The equations in ANNs that transform input data to model outputs are not pre-programmed, as in other forms of modelling. Instead ANNs formulate their own equations by finding relationships between historical input and output data sets for a catchment. If it is assumed that the learned relationships are applicable to the future, then ANNs can apply them to make predictions. ANNs have been used to provide advance forecasts of salinity in the River Murray, south

Australia (Maier and Dandy, 1996). High salinity levels in water from the River Murray cause problems for domestic, industrial and agricultural water users. For example, saline waters require increased amounts of soap and detergents and also cause soil salinization. Advanced forecasts of high levels of salinity in the River Murray upstream of the city of Adelaide allow pumping schedules to be adjusted to reduce salinity in the public water supply. ANNs were trained with data on travel time, salinity and groundwater level from the catchment for one year and then used to predict salinity levels for another year. Comparison of actual and predicted salinities shows that ANNs are a useful tool for providing advance forecasts of high salinity levels in the River Murray (Figure 15.32).

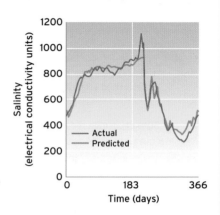

Figure 15.32 Comparison of actual salinity with 14 day forecast at Murray Bridge, River Murray, for 1991. Predicted values were calculated using artificial neural network modelling. (Source: after Maier and Dandy, 1996)

BOX 15.6

15.8 Summary

This chapter demonstrates the importance and relevance of solute hydrology to physical geography and environmental science. Chemical and physical factors affecting solute behaviour include solute form, pH, redox potential, temperature and pressure, and particulates. The behaviour of solutes in all components of the catchment hydrological system is important in determining the spatial and temporal variation of solute flux and solute concentration. Solute flux is the concentration of a solute multiplied by the discharge. Precipitation, evaporation/evapotranspiration, interception, soil water, groundwater, and river and lake processes are important in determining both solute concentrations and fluxes. In addition the interaction between solutes and hydrological pathways controls the downstream solute patterns. Different solute concentrations are contained and modified by different hydrological pathways. Slow-moving water that has a long soil or substrate residence time will tend to have very different solute contents and concentrations from water that has moved more quickly over or through the catchment. This faster-moving water may have solute concentrations that closely match that of the input precipitation. It is because of this that solutes are often used as flow tracers in catchments in order to identify the pathways through which the water has travelled.

Spatial patterns of solutes at the global and regional scales are caused by differences in climate, topography, geology, soil, vegetation and land-use and management. Temporal patterns of solutes at the storm event, annual and long-term timescales are the result of changes in the relative dominance of runoff production processes within the region or catchment of study and by changes to inputs or stores of solutes within the system. Antecedent conditions (e.g. soil moisture content) and land management or environmental change may affect the way in which water moves across the catchment as well as the volume of water moving across the catchment. In addition inputs of fertilizer or changes in forestry practice may change the amount of certain solutes available for movement and removal. Modelling solutes within catchments and individual water bodies is required for management, planning and operational purposes. While many models help us with our predictions, a complete understanding of solute and contaminant behaviour cannot be attained without appreciating the interaction of geology, climate, soil, biotic and human activities with catchment hydrology. Such an understanding is necessary to advance our understanding of how the world works and to predict the effects of land management and climate change on solute and contaminant hydrology. Furthermore, adverse and unexpected consequences for human health, environmental resources and terrestrial and aquatic ecosystems can occur when our well-intentioned management interventions attempt to manipulate solute behaviour without taking account of these interactions.

Further reading

Chapman, D.V. (ed.) (1996) *Water quality assessments: A guide to the use of biota, sediments and water in environmental monitoring*, 2nd edition. Chapman and Hall, London.
This book contains clearly written and practical advice on water quality sampling programmes and methods. It is essential reading if you ever carry out a water quality survey.

D'Arcy, B.J., Ellis, J.B., Ferrier, R.C., Jenkins, A. and Dils, R. (eds) (2000) *Diffuse pollution impacts: The environmental and economic impacts of diffuse pollution in the UK.* Terence Dalton Publishers, London.
A good introduction to the sources and impacts of different diffuse pollutants affecting waters.

Langmuir, D. (1997) *Aqueous environmental geochemistry.* Prentice Hall, Englewood Cliffs, NJ.
This text provides detailed coverage of the chemical processes that affect solute concentrations in catchments.

Likens, G.E. and Bormann, F.H. (1995) *Biogeochemistry of a forested ecosystem*, 2nd edition. Springer-Verlag, New York.
This book provides an excellent example of solute processes and measurement and solute changes over space and time in a long-term catchment experiment in north-east United States.

Novotny, V. and Harvey, O. (1994) *Water quality: Prevention, identification and management of diffuse pollution.* Van Nostrand Reinhold, London.
This is a comprehensive tome on diffuse pollution; a book to dip into, rather than to read from cover to cover.

Soulsby, C. (1997) Hydrochemical processes. In: Wilby, R.L. (ed.), *Contemporary hydrology.* John Wiley & Sons, Chichester, pp. 59–106.
A book chapter that describes fundamental solute processes and contains a range of useful examples. Chapters 1 to 8 of this book provide very detailed coverage of natural solute processes in catchments.

Web resources

The Hubbard Brook Ecosystem Study

http://www.hubbardbrook.org/

This is the website of the foremost long-term catchment experiment in the United States which included watershed manipulations to examine the effects of forestry practice on water resources and quality.

UK Acid Waters Monitoring Network

http://www.ukawmn.ucl.ac.uk/

These web pages contain information on the current programme and sites and access to water quality data.

UK Centre for Ecology and Hydrology (CEH) publications

http://www.ceh.ac.uk/products/publications/CEHPublications-H-All.html

Many CEH publications are available to download from here, including authoritative reports on the effects of forestry on water quality.

UK Foundation for Water Research

http://www.fwr.org/

Look for the Eutrophication Forum home page which has access to eutrophication case studies and current research into this topic.

UNESCO Water Portal

http://www.unesco.org/water/

The UNESCO Water Portal web pages aim to enhance access to information related to freshwater and provide links to the current UNESCO and UNESCO-led programmes on freshwater.

US Environmental Protection Agency: Water

http://www.epa.gov/water/

US Environmental Protection Agency 'water' web pages provide access to information on water quality standards for many different uses, such as drinking water, groundwater, bathing water and ecological quality.

 Visit our website at **www.pearsoned.co.uk/holden** for further questions, annotated weblinks, interactive models and video-clips showing physical processes in action.

CHAPTER 16

Dryland processes and environments

David S.G. Thomas

School of Geography, Oxford University Centre for the Environment, University of Oxford

Learning objectives

After reading this chapter you should be able to:

➤ understand what aridity is and what causes drylands and deserts

➤ describe where dryland areas are located today and the diversity of environments that drylands represent

➤ understand the main processes that have shaped drylands and operate within them

➤ understand why drylands are important regions for people and how their importance may change in the future

16.1 Introduction

Deserts frequently feature in the media, because of their spectacular landscapes. They are a popular setting for films and television programmes and for magazine advertisements, especially for cars and perfumes. In many of these instances deserts, or drylands, are portrayed as barren, plant-less regions that are dominated by ochrous sand dunes. The image of deserts as sand-dune landscapes is, however, only part of the story of drylands. Dryland environments are diverse (Figure 16.1), reflecting variations in moisture availability, structural settings, parent rocks, ecological characteristics and local geomorphic conditions. Nor is it the case that deserts and drylands are devoid of human interest and occupation (Figure 16.2). Over 5000 years ago, early Egyptian civilizations developed and flourished in a desert setting, as did those of Mesopotamia. In both cases, major rivers, bringing water from wetter regions, were key to people's ability to develop in otherwise hostile places. More generally, the ability to find strategies to cope with seasonal and longer moisture deficits contributed to the successful development of migratory pastoral lifestyles throughout North Africa, the Middle East and central Asia and, in the waterless Kalahari of southern Africa, to the evolution of a hunter–gatherer lifestyle by the San, or bushmen.

Explorers and travellers from western Europe began increasingly to visit such regions from the eighteenth century onwards. Dryland areas were frequently viewed with disdain and fear, and sometimes as barriers in the quest for more hospitable environments, as illustrated in the writings of Thomas Mitchell, following his travels in central Australia in the mid-1830s: 'After surmounting the barriers of parched deserts and hostile barbarians, I had at last the satisfaction of overlooking from a pyramid of granite a much better country' (Mitchell, 1837). In contrast to well-watered and well-vegetated temperate areas, the

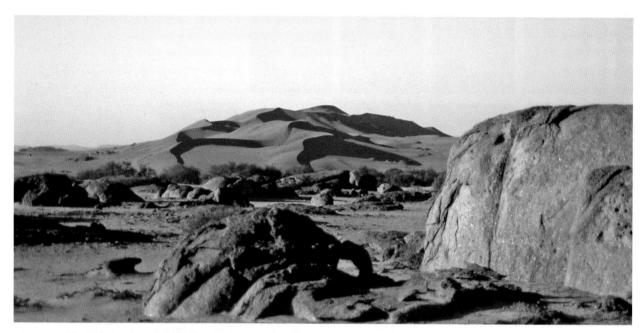

Figure 16.1 Three different dryland landscapes in the space of a few hundred metres, Namib Desert, looking due south. At the northern edge of the Namib sand sea, which contains mega-dunes that can reach heights of over 300 m (background), there is an abrupt transformation to a rocky and stony plain, where processes such as salt weathering operate on the largely bare rock surfaces (foreground). This sharp transition occurs because of the east–west-orientated Kuiseb Valley. This valley, which has flowing water in it for no more than a few days a year, is a linear oasis fed by rains that fall in the highlands to the east of the desert. The small annual flow of the river is sufficient to inhibit the northward extension of the sand dunes, and therefore to preserve a distinction between the southerly sand desert and northerly stoney desert. There is enough seepage water in the valley sediments to support plants all year round.

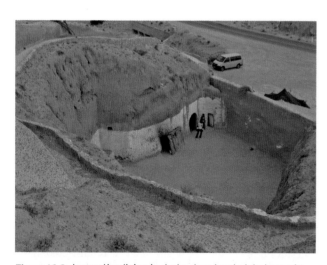

Figure 16.2 Innovative living in drylands: a troglodyte house in the Matmata loess plateau, Tunisia. The even-grained but massive wind-blown dust deposits of loess have provided an ideal medium in which to excavate underground houses. The rooms that lead off this central courtyard maintain cool, even temperatures even at the hottest times of the year. There are many examples of human ingenuity being used over the millennia to cope with the harsh environmental conditions that can exist in dryland regions.

deficit of water often proved a major obstacle to movement, while the paucity of plant cover presented desert landscapes as spectacular and bizarre. This contributed to reports about deserts that emphasized the unusual aspects, which continue to fuel the images portrayed in the media today.

During the twentieth century, technologies evolved so that today, in the twenty-first century, the world's drylands support over 1 billion people. Drylands possess many cities with populations in excess of 2 million, including the extensive urban areas of such notable metropolises as Los Angeles and Beijing. Half of Africa's total population lives in drylands, while dryland countries and regions are amongst those with the highest annual population growth rates today. For example, the population growth rate is 4.77% in Afghanistan and 1.45% in Mongolia, compared with a global average of 1.15% per annum.

There is clearly an important social dimension to understanding how dryland environmental systems function. The widespread availability of satellite imagery from the 1970s onwards (Figure 16.3) together with a marked increase in systematic environmental research in drylands, have

Figure 16.3 Satellite image of part of the Namib Desert. In the southern part of the image, the Namib sand sea is abruptly truncated at its northern margin by the Kuiseb Valley, north of which lies the stoney-gravelly Namib. Figure 16.1 shows conditions in this boundary zone from the ground. (Source: courtesy of NASA)

contributed both to a more comprehensive view of the diversity of deserts and drylands on the one hand, and, on the other hand, to a recognition that the processes that occur in these areas are not unique, but simply differ from those in other environments in the frequency and magnitude of occurrence. This chapter explores dryland characteristics, landforms and dominant processes.

16.2 Aridity

The main characteristic of deserts and drylands is clearly a lack of available moisture. This does not mean that precipitation does not occur. Rather, it is erratic: it can be spread unevenly through the year, contributing to a long 'dry season', can be patchy in where it occurs in a given region, and can have a high propensity not to fall in any given year, contributing to uncertainty. Deserts and drylands are not necessarily hot (another common image), though some certainly are. Winters can be cool, even cold, and for example in parts of interior North America and Asia, winter snowfall can be an important precipitation source. Clear cloudless skies can lead to high diurnal temperature ranges of several tens of degrees Celsius. Of all the climatic characteristics of these regions, it is the annual overall net negative moisture balance, or **aridity**, that is the defining characteristic of deserts and drylands.

A number of methods exist for measuring the annual moisture balance at any location, all of which require meteorological data, preferably collected over a number of

years so that mean values and trends can be determined. Data on monthly precipitation (P) and monthly potential evapotranspiration (PET) are required for the moisture balance to be calculated. Direct measurements of PET are not widely available, so that methods for its determination from other climate data have been used, even data from weather stations that collect only rudimentary information such as mean monthly temperature and the number of daylight hours.

Moisture balance values of less than 1.0 indicate an annual excess of PET over P, a moisture deficit. In practice areas where values fall below 0.5 are regarded as drylands. Since the 1950s and the work of Perivail Meigs (1953), areas that are too cold for seasonal crop growth have been excluded from the classification of drylands. This convention was introduced by UNESCO in its arid lands programme, which was concerned with areas that support human populations, however sparsely. This means that tundra and extremely low-latitude areas, where actual precipitation amounts are very low, are not included in the classification. For example, the Dry Valleys area of Antarctica is extremely arid but temperatures are so low that this area does not meet the crop growth criterion mentioned above. Where P/PET values are below 0.5, a number of subdivisions have been introduced since the late 1970s by organizations such as the United Nations in its Environment Programme (UNEP), so that regions with different degrees of aridity can be distinguished.

16.2.1 Drylands

The driest, or **hyper-arid** areas, which can be regarded as the 'true deserts', have P/PET values of less than 0.05. These areas, which also have periods in excess of 12 months without rainfall being recorded, include large tracts of the central Sahara, the Arabian Peninsula and the coastal Atacama and Namib Deserts (Figure 16.4). Arid areas have P/PET values between 0.05 and 0.2 and include areas of central Australia and the northern and southern fringes of the Sahara. Annual rainfall may be up to 200 mm where it occurs predominantly in winter months and up to 300 mm where it is principally a summer occurrence. Much of the Sahel belt of Africa, large areas of the western interior of North America, north-east Brazil and areas of Mediterranean Europe are semi-arid. P/PET values fall between 0.2 and 0.5, and rainfall may be up to 500 mm a year in winter rainfall areas and up to 800 mm in summer rainfall areas. Whether hyper-arid, arid or semi-arid, dryland regions are all susceptible to large year-to-year variations in precipitation, and are susceptible to drought events. As a rule of

Figure 16.4 Barren rocky landscape in part of the coastal Atacama Desert, Chile. This is one of the driest places on Earth, where the most effective geomorphic agent is neither the work of water nor the wind, but rather it is the effect of the rapid uplift of the Andes region due to tectonic processes. This induces some slope instability, with Earth movement and gravity being the most effective geomorphic agents operating to shape the long-term development of the landscape.

thumb, the human occupation and agricultural use of drylands increases from hyper-arid to semi-arid, just as do natural vegetation amounts (see Chapter 8). In the early 1990s UNEP added dry–subhumid areas (P/PET = 0.5 − 0.65) to the areas that it regarded as drylands, because such regions experience similar problems, including susceptibilities to drought, as semi-arid areas (see Middleton and Thomas, 1997). These areas include the Canadian Prairies, parts of Kenya and southern Russia.

Considered together, the different types of dryland cover over 37%, or 47% if dry–subhumid areas are included, of the Earth's land surface (Figure 16.5). As well as aridity, one of the principal characteristics of drylands is uncertainty about whether precipitation will occur. High inter-annual rainfall variation is therefore common. Since convection is a major cause of dryland rainfall, this variability can occur at a very local scale so that annual rainfall totals can vary dramatically over distances of a few kilometres. Perhaps the only certain thing about drylands is that all regions are drought-susceptible and any dryland area has the potential to experience hyper-arid conditions in any year. Dryland environments represent a significant, but difficult to habitate, part of the globe. Droughts in Africa, and the resultant social consequences, are frequently reported in the media, for example through the 1970s and 1980s in the Sahel belt of Africa and in 2006 from Dafur Province, Sudan. When precipitation fails, consequences can be dramatic even in drylands in the developed world. For example, the Province of Saskatchewan, in the Canadian Prairies, experienced in 2002 its driest recorded year ever, with both winter snowfall and summer rains at very low levels. This had drastic effects on this agriculturally important region, both for crop production and for farming communities. Understanding the causes of aridity, and the processes operating in drylands, is therefore more than of academic interest.

16.2.2 Causes of aridity

The distribution of drylands shown in Figure 16.5 yields some clues as to why dryland environments occur. At the

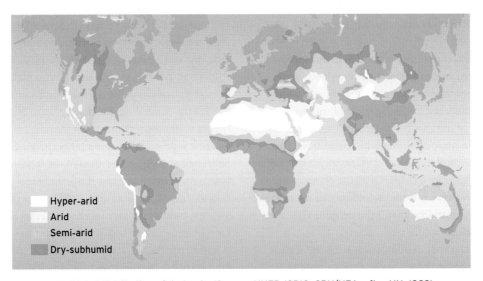

Figure 16.5 Global distribution of drylands. (Source: UNEP, ISRIC, CRU/UEA, after UN, 1992)

16.4 Geomorphological processes in drylands

16.4.1 Dryland landscapes

Dryland landscapes are as diverse as those found in any other global climatic zone. This reflects the range of structural (tectonic and geological) settings found in the regions that experience dryland climatic conditions. It also reflects the interactions between climate and landforms, which determine both the specific geomorphic processes that occur and their rates and effectiveness of operation. Some researchers have attempted to quantify the main landform types present in different areas, with the results providing useful illustrations of the diversity of dryland landscapes and the relative importance of different geomorphic processes (see chapter 1 in Thomas, 1997c). Three contrasting dryland regions are illustrated in Table 16.4.

The south-west United States, known as the basin and range region, is a high relative relief environment, which is reflected in the large extent of mountainous areas and intervening desert flats. The relatively sparsely vegetated mountain surfaces can generate high runoff rates when rain falls. Therefore alluvial fans are an important feature at the junction between mountains and flats. The Australian dryland interior has altogether much flatter terrain, and in contrast to the south-west United States is a tectonically 'quiet' region. Ephemeral river systems, which often have their headwaters in wetter areas next to the central deserts, are common. The abundant availability of sediment for transport by wind has generated extensive sand-dune systems.

Table 16.4 Percentage estimates of different landscapes within three dryland areas

	SW USA	Australia	Sahara
Mountains	38	16	43
Alluvial fans	31	0	1
River plains and channel systems	5	13	3
Playas/dry lakes	1	1	1
Sand dunes	1	40	28
Undifferentiated flats	21	18	10
Other	3	12	14

(Source: adapted from Thomas, 1997a)

The Sahara is an extensive area with a diversity of landscapes. These include recently tectonically active areas that have contributed to the development of mountains, and extremely dry continental basins that have accumulated eroded sediments that have been reworked under extremely arid conditions into sand dunes.

The landforms present in any particular dryland region may also incorporate a legacy from the operation of geomorphic processes under past climatic regimes during the Quaternary and even Tertiary periods (see Chapter 20). Thus in the south-west United States, dry lake beds, or playas (see Chapter 12), represent the sites of previously permanent lakes that existed during wetter climatic regimes in the late Quaternary. Even though the vegetation cover of the Australian dune systems is often relatively sparse, the dunes owe much of their construction to periods when conditions were even drier and windier than at present.

The examples in Table 16.4 show that dryland landscapes are composed of features that result from the effective operation of a range of geomorphic processes. These processes are not in themselves unique to dryland regions. However, the manner in which rock weathering, slope and channel processes, and aeolian (wind-blown) processes operate is greatly influenced by dryland climatic conditions, both directly and through their effect upon plant cover. We will now consider these three sets of geomorphic processes, with particular emphasis on the elements important to drylands.

16.4.2 Rock weathering in drylands

The presence of bare or relatively bare rock surfaces in dryland hills, mountains and plains (Figure 16.1), the high diurnal temperature ranges that drylands can experience, and the excess of evapotranspiration over precipitation, which can lead to salts becoming concentrated in surface locations, have all been considered as significant factors controlling dryland rock weathering regimes. Flaking rock surfaces are widely seen in drylands and have long been regarded as ample evidence of the favourable conditions that drylands present for rock weathering. However, it must not be forgotten that different rock types and structures vary in their weathering potential and susceptibility. In recent years field measurements and experimental laboratory simulations have been enhancing our understanding of how these factors may influence weathering processes.

Insolation, or thermal weathering, has been widely cited as a major form of desert rock weathering, occurring effectively under the influence of high rock surface diurnal temperature ranges, which may exceed 50°C. This is thought to cause the expansion and contraction of rock particles and

1 Formation of surface salt blisters

Void   Mud

Cracks extending to
subsurface mud

**2 Formation of polygonal salt plates by
crystallization from subsurface brines**

Salt
polygon

Broken
blister

**3 Lateral growth of polygons and vertical
salt pinnacles**   ← Dry wind

Point of intense
evaporation and
continuous salt
crystal growth

Forms plate margin
ramparts. May develop
into salt pinnacles
as preferential
crystallization site

Plates about
1 cm thick

4 Thickening and marginal trimming

Periodic
rain

Plate margins
trimmed by
solution

Brine on plate
crystallizes causing
thickening

5 Development of thrust polygons

Plates now up
to 30 cm thick

Thermal
expansion
of mud

Flow of
plastic
wet mud

6 Extrusion of mud pinnacles

Mud squeezed
through void by
loading of
thick plates

Figure 16.10 Salt crust development sequences on a playa surface. (Source: after Shaw and Thomas, 1997, based on a diagram originally in Krinsley, 1970)

minerals, with different minerals experiencing different expansions and contractions, creating stresses that weaken rocks and loosen particles. The role of temperature change alone as the principal cause of this type of weathering, which often takes the form of 'granular disintegration', has been questioned. This is because the presence of moisture, even in extremely small quantities, may be vital in contributing to rock breakdown through its influence upon the movement and crystallization of salts and in extreme cases on the occurrence of interparticle freeze–thaw processes. Even in the extremely dry Namib Desert, fog (which contributes more than rainfall to total precipitation in parts of the Namib) and dew fall are common and can provide sufficient moisture to effect such processes.

Salt weathering is also viewed as a significant process in drylands, since salt accumulation in surface sediments and rocks is favoured by both high evaporation rates and limited leaching opportunities, as well as the addition of wind-borne salt by aeolian processes (Figure 16.10). Salt weathering is a mechanical process that occurs principally through three mechanisms: crystallization when temperature increases lead to the growth of salt crystals, hydration when moisture inputs cause the salt volume to increase, and thermal expansion when salts increase in volume on heating. Different salts have different susceptibilities to these processes. Sodium sulphate is a good example of a salt that is both common and has a solubility that declines rapidly in

line with falling temperatures such that crystallization occurs. It also experiences a substantial increase in volume upon wetting (Goudie, 1997). The overall impacts of salt weathering and the occurrence and importance of different salt weathering processes have been the subject of much experimentation and debate. What is certain is that temperature and moisture changes are closely interlinked with salt weathering, and that insolation weathering and salt weathering effects are not necessarily discrete processes.

As well as contributing to the overall weathering of rock surfaces, these processes can be important in the development of particular weathering landforms. Cavernous and tafoni (honeycomb) weathering features have often been ascribed to salty weathering agencies. However, these landforms tend to be restricted to relatively even-grained rock types that permit the movement of salt-laden waters, or to the zone of capillary rise. Salt precipitation on flat, drying surfaces such as the floors of ephemeral dryland lakes and pools can contribute, through crystallization, to surface disruption and cracking that persists and develops until the next rainfall and flooding event (Figure 16.11).

16.4.3 Hillslope and channel processes

Although rainfall totals are, by definition, low in drylands, two factors mean that water can be a very effective medium for downslope sediment transport in drylands. Firstly,

Figure 16.11 The saline surface of the Chott el Jerid, a playa lake in southern Tunisia. The extensive flat floor of this salt lake may experience shallow inundation by water during the winter months when rain occurs. Evaporation during spring and summer soon dries the basin out, with the result that salt crusting occurs on the surface of the silt and clay floor. Palaeoenvironmental investigations show that there was an extensive water body during the late last glacial at this location when climate was both wetter and cooler in the region.

rainfall events are frequently intense. For example, rainfall in subtropical drylands is often associated with high rates of convection and thunderstorms. Secondly, vegetation cover is often only partial and at the end of the dry season when rains begin, biomass and ground cover are usually at their lowest levels. Runoff is therefore rapidly generated in many

dryland environments (Figure 16.12). In some areas, however, it should be noted that infiltration capacities are high, even when plant cover is limited, because of the dominance of sandy sediments. In these locations overland flow is rare, such as in the Kalahari of southern Africa (see Thomas and Shaw, 1991) and large areas of interior Australia.

In the areas where runoff is readily generated, it is usually in the form of infiltration-excess overland flow (see Chapter 13). At one extreme this occurs when the ground surface is bare rock. However, the effect also applies to soft unconsolidated sediments where raindrop impact can further enhance the effect by promoting surface sealing and crusting. Sealing and crusting usually result from the breakdown of soil aggregates under raindrop impact, and the washing of the finer particles into the voids between larger ones (see Chapter 11). Overall infiltration rates tend to be higher where larger plants are present, not only because leaves and leaf litter on the ground act to intercept raindrops and thereby prevent sealing, but also because roots increase the presence of macropores in soils and sediments.

Although overland flow may be readily generated owing to high rainfall intensities, rainfall totals are usually low in individual storm events. Furthermore, evaporation rates are high and surface vegetation cover is patchy. Therefore during many storm events overland flow is discontinuous. Often only in the case of higher-magnitude or longer-duration rainfall events does overland flow actually connect to channel systems. Dryland soils are often poorly developed

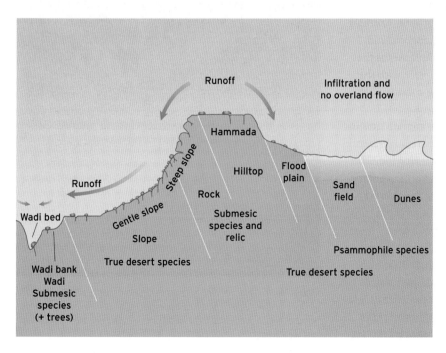

Figure 16.12 Elements of the hydrological cycle in drylands and the ability of different sub-environments to generate runoff. A hammada is a flat rocky area blown free of sand by the wind. (Source: after Shimda *et al.*, 1986)

Figure 16.13 Gullied and eroded loess deposits near Matmata, central Tunisia. The fine silty sediments are susceptible to flocculation (forming woolly, cloud-like aggregations) when wetted during rainfall events, which favours rapid dispersal and removal.

Figure 16.14 The Langbein and Schumm (1958) curve, relating mean sediment yield to mean effective rainfall (effective rainfall is that which produces runoff and is not lost to evapotranspiration or the soil moisture store). Drylands generate the highest sediment yields according to this model, which was based on the average values of measurements of sediment yields in different environments. (Source: after Langbein and Schumm, 1958)

and stony. Slopes with almost non-existent soils are commonly mantled by coarse rock debris. Recent research has shown that the macro-roughness of slopes, which includes both the distribution of plants and that of stones, is a critical determinant of the continuity and nature of overland flow. Even when flow does not reach channel systems, rill development is a common feature of dryland hillslopes (see Chapter 11). In sediments and rock types that are susceptible to very efficient overland flow generation, gullies and even extensive badland systems can develop such as that shown in Figure 16.13.

The relative bareness of hillslopes means that in global terms sediment yields in drylands are the greatest of all environments. This was illustrated by the ground-breaking work of the American geomorphologists Langbein and Schumm in 1958 (Figure 16.14). Rapid runoff means that flood hydrographs for dryland rivers peak early, with the highest-magnitude events capable of doing considerable structural damage (Figure 16.15).

The key features of dryland channel systems are summarized in Table 16.5. With the exception of perennial rivers in drylands that have their sources in neighbouring humid areas, channel flow is usually of short duration, irregular and frequently does not result in flow from the channel head to the end of the system. Two factors account for this. Firstly, the localized nature of dryland rainfall events means that a flow event may only receive a contribution from a small part of the catchment. Secondly, high infiltration losses through the bed may occur owing to unsaturated conditions prior to flooding. Owing to the presence of relatively bare slopes in dryland catchments, sediment transport, unlike in humid regions, is usually **transport limited**

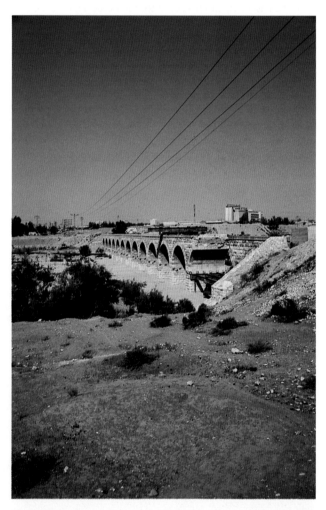

Figure 16.15 This ancient arched bridge crosses a wadi in the city of Beer Sheva, Israel, and has been partially destroyed by a large flood event. A wadi is a river bed that remains dry except during wet weather.

Table 16.5 Some major characteristics of ephemeral dryland channel systems

Flow features	Geomorphic and ecological implications
Irregular and unpredictable	Wide channels, short-distance changes in channel form
High event-to-event variability	Channel systems may change rapidly and channel systems in a region may vary markedly
High water losses to evaporation and through the stream bed	Flow-limited transport rates
	Flow may not travel far down the channel system
	Subsurface moisture may be important for subsequent plant growth
Peaked hydrograph	Rapid flow onset during rainfall events. Flow may also end quickly
High velocities (>4 m s^{-1}) common	High sediment loads (bed load and/or suspended load)

rather than sediment **supply limited**. There is, however, a major downside to this efficiency: dams and other impoundment structures can have relatively short lifespans compared with their counterparts in wetter regions.

Owing to the irregular nature of dryland channel flow and high seepage losses, many such river systems do not terminate at the coastline. Inland-draining river systems are called **endoreic**. Some of these systems may end at inland salt pans, or playas, for example at Lake Eyre in Australia (Croke, 1997) and the Makgadikgadi pans in the Kalahari (Shaw, 1997). Although surface water may fill these basins after flow events or at the end of the wet season, evaporative losses eventually result in drying out and the concentration of salts at the surface.

16.4.4 Aeolian processes and forms

The wind can be an effective agent of sediment transport, provided that sediment suitable for aeolian entrainment is available. The limited vegetation cover of drylands and the propensity for weathered sediment to accumulate in desert basins, because of the disjointed and ephemeral nature of flow in fluvial systems, explain why aeolian processes can be effective in these areas. The modern scientific investigation of aeolian processes and sand-dune development has its roots in the work of Brigadier Ralph Bagnold. He began observations in the North African deserts in the 1930s and 1940s. His seminal work on the physics of blown sand and desert dunes is still regarded as the baseline against which modern investigations are compared (Bagnold, 1941).

Only about 20% of drylands are covered by aeolian sand deposits and dunes. Many dunes occur in more vegetated semi-arid and dry–subhumid areas and are often relict from past drier climatic conditions at those sites. Silt, which has a mean particle diameter of 0.002–0.063 mm, is finer than sand, which has a mean particle diameter of 0.063–2.0 mm. Being much smaller and lighter than sand, aeolian silt (or dust) is often lifted several kilometres up into the atmosphere (Figure 16.16) and is therefore frequently transported beyond dryland areas and ultimately deposited in the oceans or other land areas. Aeolian dust presents a range of hazards

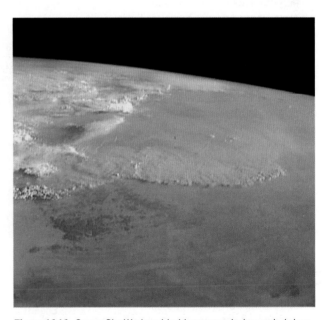

Figure 16.16 Space Shuttle hand-held camera photograph, taken in 1992, of a major dust storm front in the Sahara, near the border between Algeria and Niger. (Source: courtesy of the NASA JSC Digital Image Collection)

within drylands, but is now also seen to be a critical part of a number of major global processes (see Box 16.1). In some regions thick deposits of wind-blown silt, known as **loess**, have accumulated, often on the margins of dryland areas and frequently banked up against hills. Somewhat paradoxi-cally, however, silt can be less readily entrained by the wind than larger sand particles, principally because the small grains pack down better on the ground surface, thus offering more resistance to the forces of entrainment. Sand transport processes are described in Box 16.2.

HAZARDS

DESERT DUST: A BENEFICIAL HAZARD

Silt-sized particles, transported in the atmosphere by suspension as dust, are a significant component of dryland geomorphic systems. Locally in drylands dust is widely documented as a hazard, and dust storms are known to disrupt transport by road and air, through reducing visibility and clogging mechanical parts, as well as being a health hazard. The highest recorded annual frequencies of dust-storm days are recorded in parts of central Asia, where over 100 dust-days a year have been recorded in parts of Iran and Uzbekistan.

Dry lake beds are a particularly important source of dust to the atmosphere, because these are fluvial system end points where fine suspended sediment has accumu-lated. Lake beds may be exposed to the atmosphere because of climate change. For example, major lakes are known to have existed in parts of the Sahara, Kalahari and the south-west United States during the past 20 000 years. Lake beds may also be exposed to atmosphere because of human activity. In recent decades, the drying-out of the Aral Sea due to over-exploitation of its river-inflows for irrigation is a notable example of human actions that have had major negative human consequences. The shrinking of the sea to less than 30% of its former area has exposed 36 000 km^2 of former lake sediments, heavily polluted with chemical fertilizer residues, to the wind. Dust storm frequencies are high, with high rates of lung disease and cancer incidences in Uzbekistan at least partially attributable to the 'Aral Sea dust effect' (O'Hara et al., 2001).

The ability to monitor the move-ment of dust in the atmosphere by TOMS (Total Ozone Mapping Spec-trometer) satellite data has, since the 1990s, revolutionized the under-standing of long-distance dust move-ment from drylands. It has been possible to identify major persistent transport pathways, and the Earth's major dust sources in drylands. The Sahara produces approximately 66% of all the dust present in the Earth's atmosphere, with the Bodele Depres-sion, in the dry Lake Chad Depression in the central Sahara, responsible for about half of this (Figure 16.17). The Bodele is particularly favoured for dust generation because of its topographic setting, whereby strong dust-generating winds are focused onto the dry, erodible, former lake bed.

Recently, the export of dust from the Sahara has been attributed a major role in the complex functioning of the Earth's environmental system. Using satellite data, Koren et al. (2006) have calculated that 40 million tons of dust is transported by the wind from the Sahara to the Amazon Basin each year, with this dust being a critical source of nutri-ents for the fertilization of the Amazon rainforests.

Figure 16.17 Total Ozone Mapping Spectrometer annual mean aerosol index data for 1980 to 1992. Red areas are of greatest dust concentrations, with the red area in the central Sahara being the Bodele Depression (Source: Image courtesy of Sebastian Engelstaedter; Background image courtesy of NASA from Blue Marble: Land, Surface, Shallow Water, and Shaded Topography http://visibleearth.nasa.gov/view_rec.php?id=2433).

BOX 16.1

FUNDAMENTAL PRINCIPLES

SAND TRANSPORT

Sand is transported primarily by saltation (a hopping motion) and creep (a rolling motion) (Figure 16.18). A process called **reptation** is also sometimes referred to (e.g. Anderson and Haff, 1988), which represents grains set into a low hopping motion due to the high-velocity impact of a descending saltating grain. The nature of sand movement is affected by the particle size (or mass), the wind strength and any forces (including gravity and pore space moisture) that resist entrainment (see Chapters 11 and 12). Additional to these forces are ground surface conditions. However dry and sandy a surface is, aeolian entrainment will not take place even under high wind velocities if the surface is protected (e.g. by vegetation).

However, the relationship between surface cover and entrainment is complex, as noted by Wiggs et al. (1993) and others. This is because air moving over a surface develops a velocity profile whereby the surface imparts a frictional drag on air immediately adjacent to the surface, including a very thin (often less than 1 mm thick) zone of zero wind velocity, known confusingly as the aerodynamic roughness length or z_0. It is necessary for these frictional effects to be overcome for particles to be mobilized, picked up by the wind, and moved vertically into faster-moving air away from the immediate surface. This process may be enhanced when air flow is not laminar but is turbulent (see Figure 12.3 for a diagrammatic explanation of laminar and turbulent flow).

Turbulence is now seen as a vital part of the entrainment process, with small variations in surface topography, introduced by factors including variations in particle size and the occurrence of a partial vegetation cover, and variable surface heating by the Sun, contributing to the development of a turbulent velocity profile that enables particle entrainment to be effected. For 2 mm diameter sand particles on a bare, dry surface, a wind velocity of about 5 m s^{-1} is sufficient to generate shear velocities (usually labelled as $u*$ and a term used to describe the gradient of the velocity profile) that enable entrainment to occur. As shear velocities increase, the sand transport rate rises exponentially. Once entrainment has commenced, the impact of descending saltating grains contributes further to particle movement on the surface. Sandy surfaces are usually rippled during aeolian transport events, which is a reflection of the impact of saltating grains on the ground surface. Ripples are usually transient features, with their size and spacing changing frequently and rapidly in response to gustiness and subtle directional changes in the wind.

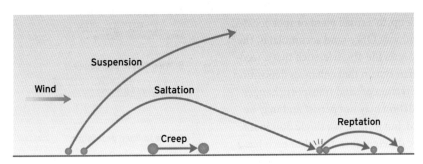

Figure 16.18 The processes of sediment movement by the wind.

BOX 16.2

While sand deposits, usually in the form of dunes, can occur as individual features, the vast majority of aeolian sand deposits are found in spatially extensive (greater than 2 km^2) accumulations known as sand seas (or ergs) (Figure 16.19). These occur at the end of regional-scale sand transport pathways. This reflects the resultant direction of sand-transporting winds, and topographic factors. Although the Sahara Desert may be the largest dryland area on Earth, its topography means that there are a number of discrete sand seas within it. The relatively low relief of much of the

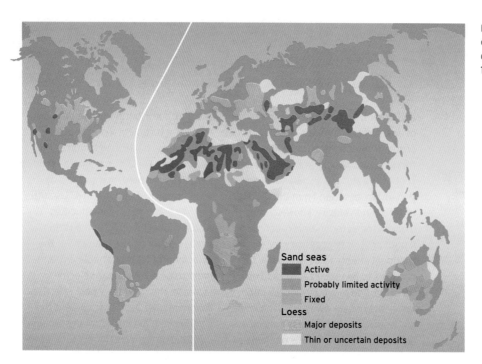

Figure 16.19 Global distribution of sand seas and other aeolian deposits. (Source: after Snead, 1972, and Thomas, 1997b)

Sand seas
Active
Probably limited activity
Fixed
Loess
Major deposits
Thin or uncertain deposits

Arabian Peninsula and the southern African and Australian interiors results in larger overall sand seas in these areas, with the Kalahari sand sea extending to almost 2.5 million km^2.

A sand dune forms when, at a specific location, the rate of arrival of sand in entrainment by the wind exceeds the rate of loss. This can occur when moving sand meets an obstacle that disrupts wind flow or reduces wind velocities, such that the capacity to transport sand is reduced. Obstacles that cause this to occur can range from hills, against which a dune may bank up, to a small plant or rock located within the transport pathway. Once sand accumulates, the accumulation itself becomes the obstacle, such that a feedback occurs in the aeolian system that enhances deposition (Figure 16.20). If the direction of sand transport changes frequently, the accumulation may not persist and may be destroyed and reworked, but if there is a single dominant direction, or if there are different directions but each has a sufficiently long duration, the accumulation will grow. As this occurs, the feature will begin to intrude into the lower atmosphere, which can lead to modification of air flow patterns and strengths. Of particular note is that wind velocities often increase (through compression of flow lines) as dune intrusion occurs. This allows sediment to be transported up the windward dune slope but eventually can limit the vertical accretion of the dune. Wind velocities can eventually create sufficient shear to move all of the sediment arriving at the dune crest through the system. As this occurs, steepening of the leeside of the dune can lead to air flow

being separated from the dune surface, the development of a cell of reversed air flow, and the formation of a steep slip face (Figure 16.21), at the angle of repose for dry sand, as shown in Figure 16.20.

Dunes can form rapidly, in a matter of hours, or slowly over days, weeks or years. The rate depends on the sand

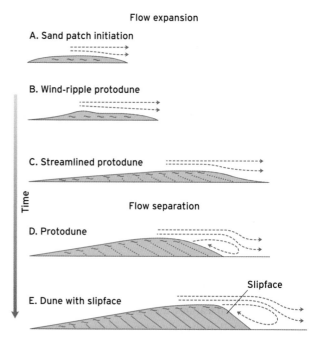

Figure 16.20 The stages of dune initiation and development. (Source: after Kocurek *et al.*, 1990)

Figure 16.21 The slip face of a transverse dune in the United Arab Emirates.

Figure 16.22 Barchan dunes, approximately 4 m high, formed in the interdune area between major linear dune ridges, in the Central Namib sand sea.

Figure 16.23 Giant linear dune ridges, with seif-like crests rising over 100 m above the interdune areas, in the Central Namib sand sea.

transport capacity of the wind and on sediment availability. Dune size is partially dependent on sediment availability and is also a function of dune type and the duration of accumulation (Livingstone and Warren, 1996). Sand dunes occur in a variety of forms, and have a variety of names that reflect both scientific factors and local knowledge. For example, *barchan* is Turkish for 'active dune', and is widely applied to mobile crescent-shaped transverse dunes (Figure 16.22); *seif* is Arabic for a 'curved sword', and is sometimes applied to linear dunes with sinuous crests (Figure 16.23). In some environments sand transport occurs from a single dominant direction, in others, seasonal changes occur. This is a principal factor in determining the type of dune that develops. Chapter 12 details the main dune types (see for example Figure 12.11). Often dunes are found in extensive dune fields within sand seas (Figure 16.24). This is probably due to the area of dune development possessing more than a single obstacle that encourages sand deposition. It is also because a single dune modifies the wind environment, creating downwind flow perturbations that encourage further deposition.

Of the main dune types identified in Figure 12.11 and discussed in Chapter 12, star dunes usually attain the greatest size (Figure 16.25). This is because they develop in the depositional centres of sand seas, where net sediment accumulation and sand-transporting wind directional variability is greatest. Linear dunes in the Namib Desert may exceed 100 m in height but in other regions such as the Kalahari and central Australia, they are more usually 10–15 m high. Sediment availability is an important factor in determining the size of these features, which can in some situations extend uninterrupted for tens, and sometimes hundreds, of kilometres. Transverse dunes are more mobile than other dune types, since they form in more unidirectional wind environments and are therefore able to migrate forward more readily. Sand supply is again an important factor in respect of the maximum size attained by dunes in

Figure 16.24 High dune ridges forming part of the Namib sand sea. The ridges, though orientated into quasi-linear patterns in this eastern part of the dune field, also have star-like peaks that rise over 300 m above the interdunes. The complexity of pattern reflects the seasonal changes in the wind regime in the area, and shows that, in reality, sand dunes often do not present themselves as neat 'textbook' examples. At the bottom of the image the dunes extend to the floor of Sossus Vlei, which occasionally experiences flowing water emanating from rains in the neighbouring Naukluft Mountains. (Source: image courtesy NASA/GSFC/MITI/ERSDAC/ JAROS, and US/Japan ASTER Science Team, 2002)

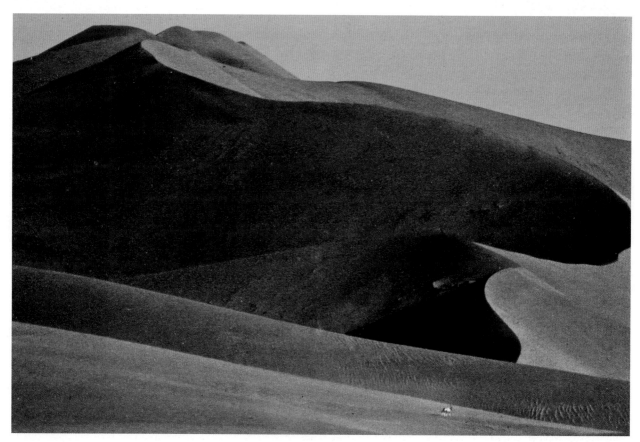

Figure 16.25 Star dune, Sossus Vlei, Namib sand sea.

any area, and this in turn plays a major role in the migration rate. All other factors being equal, a larger dune will take longer to 'roll' forward. Net migration rates that have been recorded include over 60 m yr^{-1} for a 3 m high barchan dune in Mauritania (Sarnthein and Walger, 1974), 18 m yr^{-1} for a 17 m high dune in the same environment, and 20 m yr^{-1} for a 6 m high dune in the Algodones dune field, California (Norris, 1966). Movement may be highly seasonal and dominated by changes over a few days or weeks of the year.

Not all sand sea surfaces are devoid of vegetation. A partial vegetation cover may even enhance sand transport since the turbulence of lower wind layers is enhanced over rough surfaces, and turbulent winds can be more effective in initiating sediment entrainment. Various attempts have been made to establish the vegetation cover limits on sand movement. The issue has relevance not only to understanding dune dynamics but also to understanding the potential risks of wind erosion from agricultural fields. Factors including the spacing of individual plants, plant height and structure all influence vegetation–wind–entrainment

interactions, as well as ambient wind strengths. Grasses, and plant litter, are more effective than shrubs and bushes in limiting sand transport. About 90% of sand movement occurs in the lowest 50 cm of the atmosphere. Given the complex array of plant and surface variables that can affect sand transport, a single simple threshold percentage plant cover separating entrainment-susceptible surfaces from stable surfaces does not exist. However, sand transport (assuming winds exceed threshold velocities) tends to increase rapidly once plant cover falls below around 15%, but can still take place to a limited degree with covers of about 30%.

Sand and dust particles in entrainment can act as an effective abrasional agent creating wind erosion features when they come into contact with immovable objects that are less resistant than the moving particles themselves. When abrasion is persistent, wind-sculpted landforms can result (Figure 16.26). These may range from smoothed surfaces of individual stones (termed **ventifacts**) to the smoothing and rounding of whole hills. Since sand particles

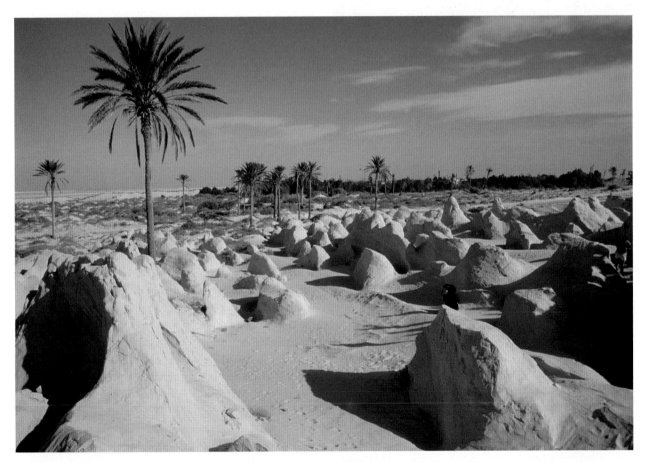

Figure 16.26 Wind-eroded soft sediments on the northern edge of Chott El Jerid, Tunisia. These features are commonly called yardangs, and are formed by the abrasive effects of saltating sand and suspended dust.

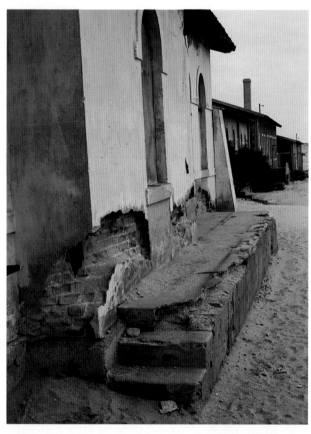

Figure 16.27 Wind erosion of a building. The height of the erosion notch represents the height of grain saltation, Kolmanskop, Namibia.

largely move in saltation, the abrasive effect is limited to the maximum height of saltation and is usually concentrated in the lower 50–100 cm of rock surfaces. Buildings located in sand transport pathways are especially susceptible to wind blasting, such as that shown in Figure 16.27. Dust particles are therefore largely responsible for the wholesale smoothing of hills into features that are known as **yardangs**. As dust can be transported over long distances, extensive yardang fields, streamlined in the direction of sediment transport, can result, and have been observed in satellite imagery both in the Lut Desert of Iran and on the Tebesti Plateau in the Sahara Desert.

Reflective questions

➤ What influences the large-scale nature of dryland landscapes?

➤ How significant is insolation weathering in drylands?

➤ If drylands are moisture deficient, why do we need to consider the work of water in sediment transport?

➤ Why can the wind be an effective geomorphological agent in drylands?

16.5 Environmental change in drylands

Drylands are diverse, dynamic environments. Seasonal climatic variability and the occurrence of drought events are normal elements of these regions. Longer, more persistent changes have also affected these regions as a result of the major climatic shifts the Earth has experienced during the Quaternary period (see Chapter 20). Drylands have been both more extensive and more contracted in the past (Figure 16.28). For example, during the last glaciation many dryland regions were more extensive for periods lasting several hundreds to thousands of years. A colder climate meant a more arid climate. Presently inactive, and in many cases well-vegetated, sand seas in the northern Kalahari, central Australia and in the Sahel were once dynamic features. Dust emissions from the Sahara were greater than at present, as evidenced by significant aeolian dust concentrations in ocean sediment cores from the mid-Atlantic Ocean and the Caribbean. Not all drylands were more extensive at this time, since the global circulation changes associated with major climatic shifts led, for example, to more effective precipitation falling over the presently arid south-western United States and in the currently hyper-arid Atacama Desert of Chile. During the Holocene (the last 10 000 years) evidence from the American mid-west, which is presently highly productive agriculturally, shows that arid episodes occurred on a number of occasions that were of sufficient duration to permit marked sand-dune development in some localities. Box 16.3 describes the types of evidence for expanding and contracting dryland conditions.

Today drylands are susceptible to dual agencies of environmental change: human activities and climatic (both natural and anthropogenically enhanced) change. Human activities, particularly in the face of growing dryland populations, attempt to increase environmental utility and manage uncertainty. Both these affect the operation of environmental processes. For example, expansions of agriculture in drylands often rely on the extraction of groundwater,

Figure 16.28 The maximum extent of major sand seas during the Late Quaternary period. When this map was first published in 1978 it was assumed that all these sand seas attained their maximum extent at the last Glacial Maximum. However, better dating control today indicates a more complex picture of the timing of the maximum activity of some of these sand seas. (Source: reprinted with permission from *Nature*, Sarnthein, M., 1978, Sand deserts during the last glacial maximum and climatic optimum, *Nature*, **272**, Fig. 1b. Copyright 1978 Macmillan Magazines Ltd)

Dune areas
Ice caps
2000 km

TECHNIQUES

INDICATORS OF LONG-TERM ENVIRONMENTAL CHANGE IN DRYLANDS

Past environmental changes, over hundreds and thousands of years, can be determined from a range of available data sources. These include the characteristics of preserved sediments and the occurrence of landforms that today are either inactive or out of equilibrium with present environmental processes. The timing of changes can be established either in a relative sense ('younger than' or 'earlier than') or numerically through the application of a radiometric dating technique. These include radiocarbon dating, luminescence dating and uranium–thorium dating (see Chapter 20 for further information

about these techniques). Different methods are applicable over different time periods. For example, radiocarbon dating can be applied to organic remains spanning the past 30–40 000 years, while luminescence dating can be applied to quartz sands over a time range in excess of 200 000 years and possibly up to 1 million years old. The following are examples of data sources that indicate (a) the greater extent of dryland conditions in the past and (b) the expansion of more humid conditions in the past into presently arid areas:

(a) Evidence of expanded dryland conditions:

- Stabilized, vegetated or degraded desert sand dunes.
- Dunes crossing river valleys,

now breached by more recent fluvial action.

- Wind-blown dust accumulations in offshore ocean sediment cores.
- Wind-eroded hills in presently humid environments.
- Evaporite deposits in lake sediment sequences.

(b) Evidence of contracted dryland conditions:

- Dry valley networks in present drylands.
- Valleys crossed by active sand dunes.
- Flow stone development in caves in drylands.
- Humid plant species' pollen preserved in dryland sediments.

BOX 16.3

which is then used to support livestock or in irrigation systems that support crop growth. This attempts to overcome the temporal variability in rainfall but, as well as positive effects, also results in a lowering of groundwater tables and a concentration of salts (through evaporative effects) in the soil. An outcome is often that such agricultural systems are unsustainable; water resources become depleted and the soil unsuitable for crop growth. This is a particular problem in a range of dryland areas such as south-east Spain, where irrigation schemes implemented in this semi-arid area in the 1970s and 1980s are now proving unproductive. These sorts of problem can lead to enhanced **desertification**. Desertification is described in Box 16.4.

Although most dryland rivers flow only ephemerally, they are increasingly subject to human management. This management may have a number of motivations including

ENVIRONMENTAL CHANGE

DESERTIFICATION

Dryland environments are increasingly subject to human pressures while at the same time being vulnerable to the impacts of naturally varying climatic conditions, especially drought. Indigenous plant and animal communities are well adapted to cope with the highly seasonal climates of drylands and all but the most extreme cases of drought, as are traditional agricultural and pastoral methods. The expansion and intensification of agricultural systems in drylands, particularly during the twentieth century, have led to a marked increase in environmental pressures and stresses, notably during periods of drought. The result of these pressures can be desertification. This is an often misunderstood and abused term, but is used to describe land degradation in drylands. Misunderstanding often arises because natural environmental (especially plant system) responses to drought, from which recovery usually occurs, have been confused with longer-term degradation.

Desertification has had many definitions, one of which is now widely used, and which was endorsed at the 1992 Rio Earth Summit. This is: 'land degradation in arid, semi-arid and dry-subhumid areas [the "susceptible drylands"] resulting from various factors, including climatic variations and human actions'. This definition recognized the natural environments in which desertification occurs, its human agency, and that the propensity for human actions to cause degradation is often enhanced during droughts when stresses are at their greatest. This is illustrated by the fact that desertification received wide global attention during the late 1970s when human pressures in the environment had been exacerbated in the Sahel region by a decade of drought.

Desertification is closely associated with unsustainable land management practices which affect both the soil and vegetation. Vegetation degradation includes, for example, the loss of natural plant cover through the lowering of water tables, and the replacement of palatable grasses used by livestock with unpalatable weeds or bushes, owing to excessive grazing levels. Distinguishing natural dryland ecosystem variability from longer-term changes can, however, prove difficult. This is particularly the case in the absence of environmental monitoring programmes and also because some land-use-related changes in plant communities can recover if usage pressures are removed or if a natural drought event reduces livestock numbers, creating a window in which natural system dynamics are restored.

Soil degradation in drylands takes on two main forms: erosion and internal changes. Wind and water erosion can both be enhanced by land-use practices, particularly if natural vegetation systems are disturbed, leaving slopes vulnerable to water erosion during storm events or sandy sediments vulnerable to wind action. Internal degradation embraces physical and chemical changes. The former includes soil crusting and compaction, which can again result from vegetation removal that increases the effect of raindrop impact. The latter includes the processes of nutrient depletion. This is a particular problem in developing world drylands where chemical fertilizers are expensive and often beyond the means of subsistence farmers. Salinization also affects irrigated lands whereby the irrigation waters evaporate, leaving salts behind which in turn render the soil intolerant to plants.

Establishing the extent of desertification has proved problematic,

BOX 16.4 ➤

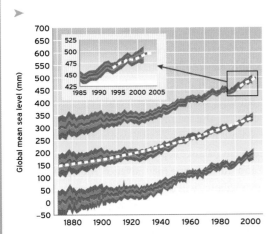

Figure 17.2 Reconstructed global mean sea level from 1870 to 2004. The three lines shown in the diagram are offset by 150 mm and represent, from bottom to top: (1) the monthly global average; (2) the yearly average with the quadratic fit; and (3) the yearly averages with satellite altimeter data superimposed. The inset shows a zoomed-in section of the most recent data with the satellite data represented by the thick black-white dashed line. The dark and light shading represent, respectively, the one and two standard deviation error estimates. (Source: after Church and White, 2006)

Table 17.1 Observed rate of sea-level rise and estimated contributions from different sources. The sea-level data prior to 1993 are from tide gauges and after 1993 from satellite altimetry

Source of sea-level rise	Rate of sea-level rise (mm yr^{-1})	
	1961-2003	1993-2003
Thermal expansion	0.42 ± 0.12	1.6 ± 0.5
Glaciers and ice caps	0.50 ± 0.18	0.77 ± 0.22
Greenland ice sheets	0.5 ± 1.2	2.1 ± 0.7
Antarctic ice sheets	1.4 ± 4.1	2.1 ± 3.5
Sum of individual climate contributions	1.1 ± 0.5	2.8 ± 0.7
Observed sea-level rise	1.8 ± 0.5	3.1 ± 0.7
Difference (observed minus sum of estimated climate contributions)	0.7 ± 0.7	0.3 ± 1

(Source: IPCC, 2007a)

0.3 mm yr^{-1} unexplained. The reduction in the ice volume of these ice sheets is believed to be the result of increased flow speeds of outlet glaciers (especially for the Antarctic ice sheet) and because losses due to melting have exceeded accumulation due to snowfall. The uncertainty surrounding the contribution to sea-level rise of the melting of ice sheets is, however, very considerable.

Using sophisticated computer models, the effect of different emission scenarios on climate change and sea-level rise by 2100 has been predicted (Table 17.2). The rise in global sea level by 2100 will be in the range from 18-38 to 26-59 cm depending on the emissions scenario. However, the predictions do not take into account any future rapid changes in ice flow, which are unpredictable and can potentially have very dramatic impacts. The actual sea-level rise may therefore be larger than predicted by the IPCC.

Assessing the impacts of sea-level rise on our society, and formulating sustainable strategies to manage these, are obviously of great importance. An initial, and admittedly rather crude, approach is to quantify the extent of the land inundated and the number of people affected by rising sea level. Using a Digital Elevation Model (DEM) of the Earth's surface, Rowley *et al.* (2007) have determined that for a sea-level rise of 1 m, an area of 1.1 million m^2 will be inundated, affecting 108 million people. For a sea-level rise of 2 m, these numbers increase to 1.3 million m^2 and 175 million people.

Table 17.2 Projected globally averaged surface warming and sea-level rise by 2100 for different emission scenarios. The scenarios listed are described in the IPCC report (see also Section 21.2.3 in Chapter 21)

Case	Temperature change by 2100		Sea-level rise by 2100
	Best estimate (°C)	Likely range (°C)	Model-based range (m)
B1 scenario	1.8	1.1-2.9	0.18-0.38
A1T scenario	2.4	1.4-3.8	0.20-0.45
B2 scenario	2.4	1.4-3.8	0.20-0.43
A1B scenario	2.8	1.7-4.4	0.21-0.48
A2 scenario	3.4	2.0-5.4	0.23-0.51
A2F1 scenario	4.0	2.4-6.4	0.26-0.59

(Source: IPCC, 2007b)

BOX 17.1

limits to which coastal processes have extended during the Quaternary geological period. During this period, which lasted from approximately 2.4 million years ago until present, sea level fluctuated over 100 m vertically due to expansion and contraction of ice sheets (see Chapter 20). The landward limit of the coastal system therefore includes the coastal depositional landforms and the marine erosion surfaces formed when the sea level was high (slightly above present-day sea level) during warm interglacial periods. During cold glacial periods sea level was low and so coastal processes were close to the edge of the continental shelf (see Chapter 4). The seaward limit of the coastal system is therefore defined by the edge of the **continental shelf**, which typically occurs in water depths of approximately 150 m.

17.2 Coastal morphodynamics

Coastal landforms and processes can be considered over a variety of temporal and spatial scales (time and space), ranging from the response of wave ripples to large wave groups on a timescale of minutes, to the infilling of estuaries following the drowning of river valleys due to sea-level rise over millennia. Regardless of the scale involved, a vital element in the coastal response is the presence of strong feedback between form and process. The morphodynamic approach, introduced by Wright and Thom (1977), formalizes this feedback by considering a coastal morphodynamic system comprising three linked elements (Figure 17.3):

1. *Processes* – this component includes all coastal processes that affect sediment movement. Hydrodynamic (waves, tides and currents) and aerodynamic (wind) processes are important. Weathering contributes significantly to sediment transport along rocky coasts, either directly through solution of minerals, or indirectly by weakening the rock surface to facilitate further sediment movement. Biological, biophysical and biochemical processes are

important in coral reef, salt marsh and mangrove environments (Masselink and Hughes, 2003).

2. *Sediment transport* – a moving fluid imparts a stress on the bed, referred to as 'bed shear stress', and if the bed is mobile this may result in the entrainment ('picking up') and subsequent transport of sediment (see Chapter 12). The resulting pattern of erosion and deposition can be assessed using the sediment budget (see Box 17.2). If more sediment enters a coastal region than leaves it, then the sediment balance is positive and deposition will occur and the coastline may advance. A negative sediment balance will occur if more sediment leaves a coastal region than enters it and net erosion will ensue, with possible coastline retreat.

3. *Morphology* – the surface of a landform or assemblage of landforms such as coastal dunes, **deltas**, **estuaries**, beaches, coral reefs and **shore platforms** is referred to as the morphology. Changes in the morphology are brought about by erosion and deposition.

As the coastal system evolves over time, its evolution is recorded in the sediments (clay, silt, sand and gravel) in the form of the **stratigraphy**. It is important to realize that stratigraphic sequences are a record of the depositional history and that erosional events are represented only by gaps in the stratigraphic record.

Coastal systems exhibit a certain degree of autonomy in their behaviour, but they are ultimately driven and controlled by environmental factors, often referred to as 'boundary conditions'. The three most important boundary conditions are geology, sediments and external forcing (wind, waves, storms and tides), with sea level serving as a meta-control by determining where coastal processes operate. When contemporary coastal systems and processes are considered, human activity should also be taken into account. In fact, along many of our coastlines, human activities, such as beach nourishment, construction of coastal defences and land reclamation, are far more important in

Figure 17.3 Coastal morphodynamic system with its energy input and boundary conditions.

Figure 17.4 The coastal frontage of Treport in north France is an excellent example of a coastline much affected by human activities.

driving and controlling coastal dynamics than the natural boundary conditions, and cannot be ignored (Figure 17.4). It can even be considered that, through climate change, humans are altering the boundary conditions themselves!

A characteristic of coastal morphodynamic systems is the presence of strong links between form and process (Cowell and Thom, 1994). The coupling mechanism between processes and morphology is provided by sediment transport and is relatively easy to comprehend. There is, however, also a link between morphology and processes to complete the morphodynamic feedback loop. For example, under calm wave conditions sand is transported on a beach in the onshore direction, resulting in beach accretion. As the beach builds up, its seaward slope progressively steepens and this has a profound effect on the wave breaking processes and sediment transport. At some stage during beach steepening, the hydro-dynamic conditions may be sufficiently altered to stop further onshore sediment transport. Owing to the close coupling between process and form in morphodynamic systems it is often not clear whether the morphology is the result of the hydrodynamic processes, or vice versa. This makes it very difficult to predict coastal development over long timescales.

The feedback between morphology and processes is fundamental to coastal morphodynamics, and can be nega-tive or positive. **Negative feedback** acts to oppose changes in morphology. For example, beach erosion during storms generally results in the development of an offshore bar. Wave breaking on the bar significantly reduces the amount of wave energy reaching the shoreline, thereby limiting further beach erosion. However, many coastal systems do not always behave that predictably and often the feedback between morphology and processes is positive, rather than negative. **Positive feedback** pushes a system away from equilibrium by modifying the morphology such that it is even less compatible with the processes to which it is exposed. A morphodynamic system driven by positive feedback seems to have a 'mind of its own' and exhibits self-forcing behaviour. An example of positive feedback is the infilling of deep estuaries by marine sediments due to uneven flows between an incoming tide and an outgoing tide. In a deep estuary, flood currents are stronger than ebb currents (currents running back out to sea) and this tidal 'asymmetry' results in a net influx of sediment and infilling of the estuary. As the estuary is being infilled, the tidal asymmetry increases even more as friction effects are enhanced by the reduced water depths. In turn, the increase in tidal asymmetry speeds up the rate of estuarine infilling. This constitutes positive feedback between the estuarine morphology and the tidal processes, resulting in rapid infilling of the estuary. Eventually, tidal flats and salt marshes start developing in the estuary and this marks a reversal in feedback. As the intertidal areas become more extensive, the flood asymmetry of the tide progressively decreases so that the estuarine morphology approaches steady state as sedi-ment imports and exports balance each other.

Adjustment of coastal morphology to changing conditions involves a redistribution of sediment which requires time. The time required for the adjustment to occur is known as the **relaxation time.** The relaxation time strongly depends on the volume of sediment involved in the adjustment of morphology and is related to the size of the landform. Large coastal landforms, such as coastal barriers, have longer relax-ation times than small morphological features, such as **beach cusps**. Generally, the relaxation time exceeds the time between changes in environmental conditions. It is therefore unlikely that a 'steady-state equilibrium' (see Chapter 1) is ever reached, particularly for large coastal landforms.

Reflective questions

➤ How do negative and positive feedback affect our ability to make long-term predictions of coastal development?

➤ What does a morphodynamic approach to coasts involve?

17.3 Coastal processes: waves

Ocean waves are the principal agents for shaping the coast and driving **nearshore** sediment transport processes. Of course, wind and tides are also significant contributors, and are indeed dominant in coastal dune and estuarine

FUNDAMENTAL PRINCIPLES

SEDIMENT BUDGETS

Morphological change directly results from sediment transport processes. Sediment budgets help us to understand the different sediment inputs (sources) and outputs (sinks) involved (Figure 17.5). A sediment budget involves accounting for the sediment volumes (m^3) rather like you would account for money. Key components of the sediment budget are the sediment fluxes, which represent the direction and amount of sediment transport by certain processes and which are expressed as the quantity of sediment moved per unit of time ($kg\ s^{-1}$, $m^3\ yr^{-1}$). If the sediment fluxes are known, sediment budgets can be used to predict how the morphology changes through time in a quantitative fashion.

For example, assume there was an estuary with a surface area of 1 km^2 (this is the same as 1 000 000 m^2) that receives an annual input of sediment from marine and fluvial sources of 100 000 m^3 per year. If it is further assumed that this sediment is evenly spread over the estuary floor, then the depth of the estuary will decrease by:

$$\frac{\text{sediment input}}{\text{surface area}} = \frac{100\ 000\ m^3\ yr^{-1}}{1\ 000\ 000\ m^2}$$
$$= 0.1\ m\ yr^{-1}$$

If the average depth of the estuary is 10 m, then the estuary will be infilled in:

$$\frac{\text{depth}}{\text{accretion rate}} = \frac{10\ m}{0.1\ m\ yr^{-1}}$$
$$= 100\ \text{years}$$

Of course, this simple illustration assumes that the amount of sediment entering the estuary does not change while the estuary is infilling. It therefore ignores feedback between morphology and process which is one of the main principles of morphodynamic systems. Nevertheless, such simple calculations can still tell us a lot about environmental change in coastal zones.

Figure 17.5 Qualitative examples of coastal sediment budgets. (a) Riverine sediment may be deposited into an estuarine embayment, which may also receive sediment from the seaward direction through tidal processes. (b) On a deltaic coast, riverine sediment contributes to the coastal sediment budget, and is moved up the drift-aligned coast, giving rise to a beach-ridge plain. (Source: adapted from Masselink and Hughes, 2003)

BOX 17.2

environments, respectively, but the action of waves is dominant in most settings. Waves are generated by wind and the stronger the wind, the larger the waves. Not surprisingly, therefore, there is a strong climatic control on the global distribution of wave heights, with the largest waves found at the stormiest latitudes around the 'roaring forties' (40° north and south) (Figure 17.6).

It is important at the outset to make the distinction between regular and irregular (or random) waves. The motion of regular waves is periodic, that is the motion is

Figure 17.6 Global values for significant wave height which is exceeded 10% of the time. (Source: after Young and Holland, 1996)

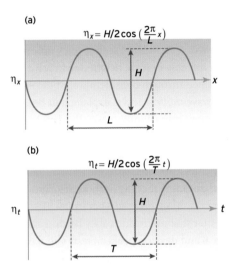

(a)

$$\eta_x = H/2\cos\left(\frac{2\pi}{L}x\right)$$

(b)

$$\eta_t = H/2\cos\left(\frac{2\pi}{T}t\right)$$

Figure 17.7 Schematic showing a regular wave train. (a) The spatial variation in water level η_x is measured at a single moment in time along the direction of wave travel. From such data the wavelength L can be derived. (b) The temporal variation in water level η_t is measured at a single location in space over a representative time period. Such data enable the determination of the wave period T. The wave height H can be derived from both types of wave data. (Source: Masselink and Hughes, 2003)

repetitive over space (Figure 17.7a) or through fixed periods of time (Figure 17.7b). Regular waves can be described in terms of a single representative **wave height** H, **wavelength** L and **wave period** T. The wave height is the difference in elevation between the **wave crest** and the **wave trough**, the wavelength is the distance between successive crests (or troughs) and the wave period is the time it takes for the wave to travel a distance equal to its wavelength. Of these three parameters, the wave period is the easiest to determine in the field: simply count and time the passage of a large number of waves (at least ten) past a fixed point and divide the time by the number of waves. The **wave steepness** is also an important parameter and given by the ratio of wave height to wavelength H/L.

Natural waves are, however, highly irregular, and a range of wave heights and periods are present (Figure 17.8). To properly describe the wave conditions of irregular waves in quantitative terms, statistical techniques are required. For example, a widely used measure of the wave height is the **significant wave height** H_s, which is defined as the average of the highest one-third of the waves. The significant wave

Figure 17.8 Example 2 minute time series of water depth showing 'real' waves measured just outside the surf zone on a sandy beach. These highly irregular and asymmetric waves are characterized by a significant wave height H_s of 0.41 m and a period of 4-5 s.

height can be obtained by recording a time series of water depth and multiplying the associated **standard deviation** by four.

17.3.1 Linear wave theory

The behaviour of ocean waves can be estimated using **linear wave theory**. The equations associated with this theory are widely used (Komar, 1998). Based on the ratio of water depth h to deep-water wavelength L_o, we can identify three different wave regions, each characterized by different water particle motions under the waves (Figure 17.9):

1. *Deep water* ($h/L_o > 0.5$) – as waves travel across the sea surface, the water particles beneath undergo an almost closed circular path. The particles move forward under the crest of the wave and move seaward under the trough of the wave. The diameter of the orbits decreases with increasing depth until at some distance below the water surface, referred to as the **wave base**, the wave motion ceases. The wave base is thus defined as the depth below which wave motion cannot stir bed sediment. In deep water, the wavelength L is given by $gT^2/2\pi$ and the wave speed or celerity C is given by $gT/2\pi$, where g is the gravitational acceleration.

2. *Intermediate water* ($0.5 < h/L_o < 0.05$) – with decreasing water depth, the wave motion extends to the seabed and the surface waves are much affected by the presence of the seabed. As a result, the water particles now follow an elliptical path with the ellipses becoming flatter and smaller as the seabed is approached. At the seabed, the water particles merely undergo a horizontal to-and-fro motion. In intermediate water depths, L and C are not easily computed and fall between the values obtained by using the deep- and shallow-water equations.

3. *Shallow water* ($h/L_o < 0.05$) – very close to the **shore**, all water motion consists of horizontal movements to-and-fro which are uniform with depth. In shallow water, L and C are given by $T\sqrt{(gh)}$ and $\sqrt{(gh)}$, respectively.

The deep- and shallow-water wave equations are straightforward to apply. For example, think about a wave with a period of 10 s. In deep water, the wavelength and wave velocity are only dependent on the wave period and application of the relevant equations gives $L = 156$ m and $C = 15.6$ m s^{-1}. In shallow water, the water depth also needs to be considered. For a water depth of 1 m and a wave period of 10 s the relevant equations yield $L = 31$ m and $C = 3.1$ m s^{-1}. The example highlights a fundamental characteristic of waves: as waves propagate from deep to shallow water, they get closer together and slow down.

17.3.2 Wave processes in intermediate water

Wave **shoaling** is the process whereby the waves change in height as they travel into shallower water and start feeling the seabed. This process is particularly pronounced just before wave breaking at the seaward edge of the **surf zone**, where the waves are suddenly seen to 'pick up' and become steeper. The increase in wave height due to wave shoaling can be explained through consideration of the energy associated with wave motion and is explained in Box 17.3.

Perhaps more importantly, waves also change their shape during shoaling. In deep water, waves are characterized by a **sinusoidal** shape (as shown in Figure 17.7) and the water particle velocities associated with the wave motion are symmetrical, meaning that the onshore velocities are of equal strength and duration as the **offshore** velocities. However, as the waves enter shallower water, they become increasingly asymmetrical and develop peaked crests and flat troughs (Figure 17.10; see also Figure 17.8). The associated flow

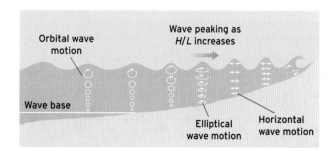

Figure 17.9 Motion of water particles under waves according to linear wave theory.

Figure 17.10 Asymmetric waves in shallow water characterized by peaked wave crests and broad wave troughs. The waves peak up more just prior to breaking.

FUNDAMENTAL PRINCIPLES

WAVE ENERGY FLUX AND SHOALING

The rate at which wave energy is carried along by moving waves is known as the **wave energy flux** and is the product of the amount of energy associated with the waves and the speed at which this energy travels. The total wave energy E is given by:

$$E = \frac{1}{8}\rho g H^2 \qquad (17.1)$$

where ρ is the density of water, g is gravity and H is the wave height. The wave energy is expressed as the amount of energy per unit area ($N\,m^{-2}$) and is more appropriately referred to as the wave energy density. Wave energy depends on the square of the wave height. A doubling of the wave height therefore results in a fourfold increase in wave energy. The speed by which this energy is carried along is not the same as the speed of the individual waves, but is given by Cn, where C is the wave speed and n is given by:

$$n = \frac{1}{2}\left[1 + \frac{2kh}{\sinh(2kh)}\right] \qquad (17.2)$$

where sinh is sine hyperbolic. The parameter n increases from 0.5 to 1 from deep to shallow water. This means that deep-water waves travel at twice the speed of the energy ($n = 0.5$), whereas shallow-water waves travel at the same speed as the energy ($n = 1$). The wave energy flux P is then given by:

$$P = ECn \qquad (17.3)$$

As waves travel from deep to shallow water, the change in the wave height due to shoaling can be calculated by examining the wave energy flux P. Assuming that energy losses due to bed friction can be ignored, the wave energy flux remains constant during wave propagation. This can be expressed as:

$$P = (ECn)_1 = (ECn)_2 = \text{constant}$$

where the subscripts 1 and 2 indicate two different locations along the path

of wave travel. Substituting the wave energy $E = (1/8)\rho g H^2$ in this equation and rearranging the result produces equation (17.4):

$$H_2 = \left(\frac{C_1 n_1}{C_2 n_2}\right)^{1/2} H_1 \qquad (17.4)$$

If we would like to compute the wave height in shallow water, where $n = 1$, from the wave height in deep water, where $n = 0.5$, the equation can be simplified to:

$$H_2 = \left(0.5\frac{C_1}{C_2}\right)^{1/2} H_1 \qquad (17.5)$$

After inserting the appropriate deep- and shallow-water wave speed, the equation becomes:

$$H_2 = \left(0.5\frac{(gT/2\pi)}{\sqrt{gh}}\right)^{1/2} H_1 \qquad (17.6)$$

Assuming a deep-water wave height of 1 m and a period of 12 s, the wave height in 2 m water depth becomes 1.45 m, representing a significant increase.

BOX 17.3

velocities also become asymmetric with the onshore phase of the wave being stronger, but of shorter duration than the offshore phase of the wave. The development of **wave asymmetry** is very important from a sediment transport point of view because it promotes the onshore transport of sediment particles. In fact, without the development of wave asymmetry, beaches would not exist because there would be no mechanism to transport sand and gravel to the shore.

Wave **refraction** is another important process that takes place during shoaling. It occurs because the part of the wave that is in shallower water travels slower than the part of the wave in deeper water. This results in a rotation of the wave crest with respect to the bottom contours, or, in other words, a bending of the wave rays (Figure 17.11). The resulting change in wave direction causes the wave crests to be more aligned parallel to the bottom contours, reducing

the wave angle. Irregular seabed topography can cause waves to be refracted in complex ways and produce significant variations in wave height and energy along the coast. Spreading of the wave rays, referred to as **wave divergence**, occurs when waves propagate over a localized area of relatively deep water (e.g. depression in the sea floor) and this causes a reduction in the wave energy and wave height. Focusing of the wave rays is known as **wave convergence** and occurs when waves travel over a localized area of relatively shallow water such as a shoal on the sea floor and causes an increase in the wave energy and wave height.

17.3.3 Wave processes in shallow water

When the water depth becomes too shallow for a stable wave to exist, the wave breaks. **Wave breaking** occurs when a

Figure 17.11 Wave refraction across a bay.

wave becomes too steep and the horizontal velocities of the water particles in the wave crest exceed the velocity of the wave and the wave disintegrates into bubbles and foam. Waves break in a depth slightly larger than their height and the region on the beach where waves break is known as the surf zone. Wave breaking is an important process, because when waves break, their energy is released and is used for generating nearshore currents and sediment transport. A continuum of breaker shapes occurs in nature; however, three main breaker types are commonly recognized (Figure 17.12):

1. *Spilling breakers* are associated with gentle beach gradients and steep incident waves. A gradual peaking of the wave occurs until the crest becomes unstable, resulting in a gentle forward spilling of the crest (Figure 17.12a and b).
2. *Plunging breakers* occur on steeper beaches than spilling breakers, with waves of intermediate steepness. The shoreward face of the wave becomes vertical, curling

over, and plunging forward and downward as an intact mass of water. Plunging breakers are the most desirable type of breaker for surfers, because they offer the fastest rides and as they plunge over they produce 'tubes' (Figure 17.12a and c).
3. *Surging breakers* are found on steep beaches with low-steepness waves. The front face and crest of surging breakers remain relatively smooth and the wave slides directly up the beach without breaking (Figure 17.12a and d).

Breaker type can be predicted based on a consideration of the wave characteristics and the gradient of the beach using the Iribarren number also known as the surf similarity parameter, ξ (Battjes, 1974):

$$\xi = \frac{\tan \beta}{\sqrt{H_b/L_o}} \tag{17.7}$$

where $\tan \beta$ is the gradient of the beach and the subscripts 'b' and 'o' indicate breaker and deep-water conditions, respectively. Small values for ξ are attained when the beach has a gentle gradient and the incident wave field is characterized by a large wave height and a short wavelength (or a short wave period). Large values of ξ are found when the beach is steep and the incident wave field is characterized by a small wave height and a long wavelength (or a long wave period). Spilling breakers occur for $\xi < 0.4$; plunging breakers for $\xi = 0.4$ to 1; and surging breakers for $\xi > 1$.

For large values of the Iribarren number ($\xi > 1$), the incident wave energy is not dissipated by breaking, but is reflected at the beach, very much like light is reflected off a mirror. The proportion of reflected energy increases with ξ and is generally modest on beaches, unless they are very steep. When waves encounter the vertical face of a seawall,

Figure 17.12 The three main types of breakers: (a, b) spilling, (c, d) plunging and (e, f) surging. (Source: (a), (c), (e) Briggs *et al.* 1997, photos courtesy Rob Brander)

Figure 17.13 Reflection of breaking waves at the vertical face of a seawall giving rise to a criss-cross wave pattern.

however, wave **reflection** approaches 100%. This may give rise to standing wave motion in front of the seawall and/or complicated criss-cross wave patterns (Figure 17.13).

When waves break, they produce **wave set-up**, which is a rise in the mean water level above the still water elevation of the sea (Figure 17.14). In popular terms, wave set-up is conceived as a piling up of water against the shoreline due to the waves and is caused by the breaking waves driving water shoreward. As a general rule of thumb, the set-up at the shoreline is 20% of the offshore significant wave height. At the shoreline, surf zone waves propagate onto the 'dry' beach in the form of **swash**. Swash motion consists of an onshore phase with decelerating flow velocities (uprush) and an offshore phase characterized by accelerating flow velocities (**backwash**). For a number of reasons such as

infiltration effects, **advection** of sediment and turbulence from the surf zone into the **swash zone**, the uprush is a more efficient transporter of sediment than the backwash, and this onshore swash asymmetry is responsible for maintaining the beach gradient.

17.3.4 Nearshore currents

In the surf zone, incident wave energy is lost, or dissipated, due to wave breaking. Much of this energy is used for generating nearshore currents and sediment transport, ultimately resulting in the formation of distinct coastal morphology. Nearshore currents derive their energy from wave breaking and the intensity of these currents increases with increasing incident wave energy level. The strongest currents are therefore encountered during storms. There are three types of wave-induced currents and these systems dominate the net water movement in the nearshore (Figure 17.15):

1. **Longshore currents** are shore-parallel flows within the surf zone. They are driven by waves entering the surf zone with their crests aligned at oblique angles to the shoreline. Longshore currents increase with the incident wave energy level and the angle of wave approach, and may reach velocities in excess of $1 \, \mathrm{m \, s^{-1}}$. They are also affected by alongshore winds and can be particularly strong when strong winds are blowing in the same direction as the longshore current.

2. The bed return flow, often somewhat misleadingly referred to as undertow, is an average flow near the bed flowing offshore. The current is part of a circulation of water characterized by onshore flow in the upper part of

Figure 17.14 Wave set-up and set-down measured in the laboratory. Measurements were obtained using a wave height of 6.45 cm, a wave period of 1.14 s and a beach gradient of 0.082. SWL refers to still water level. (Source: Masselink and Hughes, 2003)

(a)

(b)

(c)

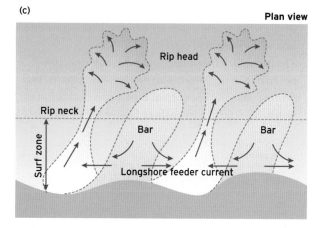

Figure 17.15 Quasi-steady currents in the surf zone: (a) shore-parallel longshore currents due to obliquely incident waves, (b) vertically segregated bed return flow or undertow and (c) horizontally segregated rip currents as part of the nearshore cell circulation system. (Source: after Masselink and Hughes, 2003)

(a)

(b)

Figure 17.16 Rip currents: (a) a single rip current flowing out between sand bars and (b) two rip currents present at either side of an intertidal rock outcrop. (Source: photo (a) courtesy Rob Brander)

the water column and seaward flow near the bottom (Figure 17.15b). Measured bed return velocities are typically 0.1–0.3 m s^{-1}, but under extreme wave conditions may reach values of up to 0.5 m s^{-1}.

3. **Rip currents** are strong, narrow currents that flow seaward through the surf zone in channels and present a significant hazard to swimmers (Figures 17.15c and 17.16).

They consist of onshore transport of water between rip currents, longshore feeder currents, and offshore transport of water in the rip itself. Maximum current velocities associated with circulation of water in rip currents may reach up to 2 m s^{-1} under extreme storm conditions when so-called 'mega-rips' form. Typical rip current velocities are 0.5–1 m s^{-1} and flows are generally stronger during low tide than during high tide.

Nearshore currents are capable of transporting large quantities of sediment. This is partly due to their often significant flow velocities, but also because sediment entrainment is considerably enhanced by the stirring motion of the breaking waves. The amount of sediment transported by longshore currents is known as the **littoral drift** and is generally of the order of 100 000 m³ yr^{-1}. Such large transport rates have a major effect on coastal morphology and shoreline change.

17.4 Coastal processes: storm surge, tides and tsunami

Although ocean waves are the principal agents for shaping the coast and driving nearshore sediment transport, the contributions of storm surge, tides and **tsunami** are also significant and require consideration. These three processes represent changes in the water level of the order of metres, but their timescales vary considerably. Storm surges generally last several days, tides occur daily or twice-daily and tsunami represent water-level fluctuations of around 10 minutes.

17.4.1 Storm surge

During severe storms, the water level near the shore can be significantly elevated compared with tidal predictions, and the difference between the measured and the predicted water level is the storm surge (Figure 17.17). Storm surge

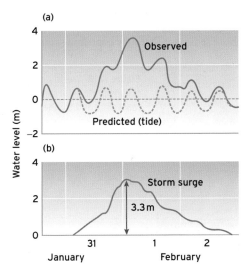

Figure 17.17 (a) Graph showing the predicted (dashed line) tidal water level and observed (solid line) water level at the Hook of Holland, January to February, 1953. (b) Graph showing the residual tide, which is attributed to storm surge, reached 3.3 m. (Source: Masselink and Hughes, 2003)

levels can be in excess of 5 m under extreme conditions, such as during hurricanes or cyclones, leading to extensive coastal flooding and erosion. The amount of storm surge at the coast depends on three main factors:

1. *Low pressure* – sea level will rise approximately 1 cm for every 1 millibar fall in air pressure. Storms are always characterized by low pressure and hence the water level under a storm is always raised.
2. *Onshore wind* – if the wind is directed shoreward it can pond water against the coast, causing an increase in the water level.
3. *Coastal topography* – the effect of the storm surge on the coast depends greatly on coastal configuration. Relatively low-gradient, funnel-shaped coastal settings are particularly prone to extreme surges. Examples of these are the Bay of Bengal or the North Sea. Straight coastlines and promontories are generally less sensitive to storm surge.

17.4.2 Tides

The tidal rise and fall of the ocean surface are barely noticeable in the deep ocean, but on shallow continental shelves, along coastlines, and within estuaries, tidal processes can be the dominant morphological agent. The two driving forces for ocean tides are the gravitational attraction of the Earth–Moon system and the Earth–Sun system, with the latter being almost half that of the former. The theory of tides is rather complicated and it is more useful to describe how tides are manifested along our coasts, than concern ourselves about how they are generated. In fact, the 'proper' theoretical explanation of tides involving bulges of water at either side of the Earth, known as the equilibrium tide theory, does not provide much insight into actual characteristcs of ocean tides other than their dominant periods (12.5 h for the lunar component and 12 h for the solar component).

According to the more practical explanation of tides, the dynamic tide theory, the global tidal water motion is broken up into a large number of tidal systems constrained by the coastal topography. In these systems, known as **amphidromes**, the tide travels around the centre of the amphidrome as a wave. Owing to the Coriolis force (see Chapter 4), the tidal wave travels clockwise in the southern hemisphere and anticlockwise in the northern hemisphere. The difference in water level between high and low water is the **tidal range**, and is practically zero at the centre of the amphidrome and maximum at the edge. Figure 17.18 shows the three amphidromes in the North Sea. At a particular location, the observed tidal range depends on which is the influential amphidrome and how far it is away from the

Figure 17.18 Map showing amphidromes in the North Sea. The dashed lines are lines of equal tidal range, whereas the solid lines show how the tidal wave propagates in an anticlockwise direction around the centre of the amphidrome. The numbers indicate where the tide is at the same stage in the tidal cycle (in hours). Along a line labelled '1', for example, the tide is one hour into the tidal cycle. (Source: Masselink and Hughes, 2003)

centre. It should be noted that even the situation shown in Figure 17.18 is a simplification, because interaction between the tidal wave and the local topography can significantly modify the tidal water motion.

Along most coasts, tides cause a twice-daily rise and fall in the water level (some coasts have mainly once-daily tides, but these are relatively rare). The tidal range varies greatly around the world and it is common to distinguish between micro- ($<2\,$m), meso- (2–$4\,$m) and macro-tidal ($>4\,$m) ranges (Figure 17.19). The largest tidal ranges are generally associated with rather complex coastal configurations, such as semi-enclosed basins like the North Sea and Irish Sea, funnel-shaped bays like the Bay of Fundy, Canada, and regions with wide continental shelves and island groups such as around north-west Australia.

The tidal range also varies over time, mainly due to the interaction between the tidal forces of the Earth–Moon system and the Earth–Sun system. When the Earth, Moon and Sun are all aligned, during either a full or new moon, the tidal forces of the Moon and the Sun are combined, resulting in extra-large tides, known as **spring tides**. When the Moon is at a right angle to the Earth with respect to the Sun, however, the tidal forces of the Moon and the Sun are competing, resulting in extra-small tides, known as **neap tides**. The Moon revolves around the Earth in 28 days; therefore, a spring-to-spring tidal cycle (or neap-to-neap tidal cycle) takes 14 days to complete (Figure 17.20).

Associated with the rising and falling tides are **tidal currents** whose strength increases with the tidal range. At any one place, ebb- and flood-tidal currents vary in strength and duration. This is especially the case for estuaries (see Section 17.7.3), but also along open coasts, in deep water and even on beaches. Therefore, over time there exists a tidal current (the tidal movements do not balance out over time) and this is known as the **residual tidal current**. These currents are important because they induce a net movement of nearshore sediment that may contribute significantly to

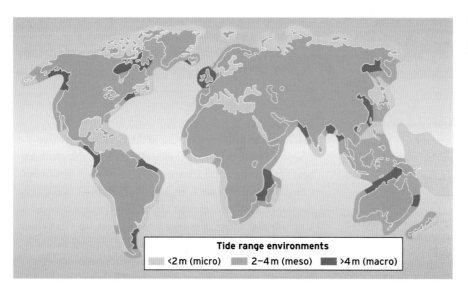

Figure 17.19 World distribution of mean spring tidal range. (Source: Masselink and Hughes, 2003)

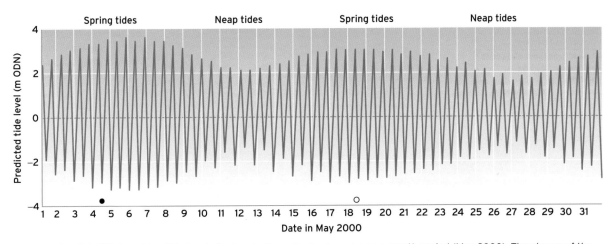

Figure 17.20 Predicted high and low tide levels for Immingham, England, over a one-month period (May 2000). The phases of the Moon are indicated by the solid (New Moon) and open circle (Full Moon). Spring tides occur around the New and Full Moon, whereas neap tides occur halfway between these Moon phases. ODN refers to Ordnance Datum Newlyn, which is approximately mean sea level in the United Kingdom.

coastal morphological development. An example of the residual tidal current pattern around the coast of the United Kingdom is shown in Figure 17.21. Areas of current convergence (where currents meet) and divergence (where currents part) are clearly discernible and it is at these locations where sediment accumulation and erosion, respectively, are likely to occur.

17.4.3 Tsunami

Often a tsunami is referred to as a 'tidal wave'. However, this term is inappropriate because tides and tsunami differ from each other in many respects, particularly in the way that they are generated. Tides are generated by astronomical forces involving the Earth, Moon and Sun, whereas a tsunami is generated as a result of the displacement of a large water mass by any of the following three mechanisms (Masselink and Hughes, 2003):

1. A displacement of the seabed by a submarine earthquake. The 2004 Sumatra earthquake produced a tsunami that killed more than 100 000 people (see Chapter 2).
2. A large landslide into the ocean, perhaps during a volcanic eruption. The 1883 Krakatau eruption produced a tsunami that killed more than 36 000 people.
3. An impulse generated by a meteorite impact striking the ocean. This generation mechanism has not occurred in recent historical time, but mega-tsunami events recorded in the geological record have been attributed to meteorites (Bryant, 2001).

In the open ocean tsunami typically have a wavelength of a few hundred kilometres and a height less than a metre. They travel at the shallow-water wave speed given by $\sqrt{(gh)}$, which for a typical water depth of 5 km implies a speed of around 700 km h^{-1}. When they cross the edge of a continental shelf, however, they rapidly begin to shoal, and become shorter in wavelength and larger in height. This process is identical to the shoaling process of wind waves (see Box 17.3). By the time tsunami reach the coastline, they can be several tens of metres high and in a period of only a couple of hours they can cause enormous property damage and loss of life. Chapter 3 provides more detail on tsunami processes.

Reflective questions

➤ How are neap and spring tides generated?

➤ What factors control the magnitude of a storm surge?

17.5 Coastal classification

Coastal landforms have often been classified to provide useful ways to help assess the different forcing factors and controls such as sea-level history, geology, climate, waves and tides that lead to the great variety of coastal landforms we encounter (Bird, 2000). Most early classification schemes were based on the realization that coastal landforms are largely the product of sea-level variations. Such classifications

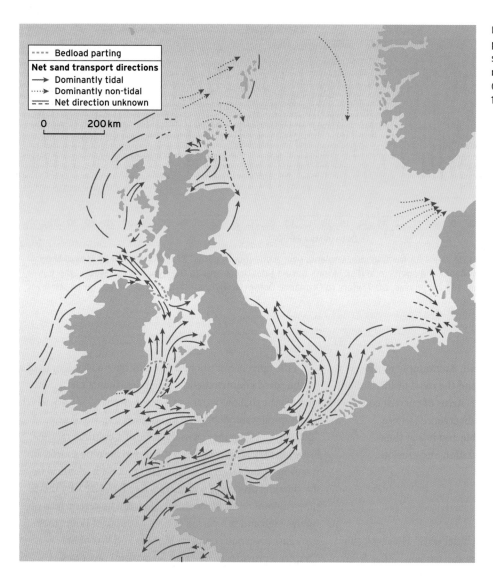

Figure 17.21 Net sediment transport paths on the continental shelf around the British Isles mainly driven by tidal currents. (Source: after Open University, 1989)

distinguish between submerged and emerged coasts. Typical submerged coasts are drowned river and glacial valleys, often referred to as rias and fjords, respectively. Coastal plains are characteristic of emerged coasts. Another type of classification distinguishes between primary and secondary coasts. Primary coasts have a configuration resulting mainly from non-marine processes and include drowned river valleys and deltaic coasts. Secondary coasts, on the other hand, are coasts that have a configuration resulting mainly from marine processes or marine organisms. Examples of such coasts are barrier coasts, coral reefs and mangrove coasts.

The main shortcoming of these early classifications is that the emphasis on geological inheritance and sea-level history leaves only limited concern for the hydrodynamic processes. The morphology of depositional coastal environments (those consisting of mud, sand and gravel, rather

than eroding rocky shores) responds to the relative dominance of river, wave and tidal factors (Boyd *et al.*, 1992). A diagram can be constructed that expresses the relative importance of river outflow, waves and tidal currents (Figure 17.22). In this diagram, deltas are positioned at the fluvial point of the triangle because a fluvial sediment source dominates, while prograding, non-deltaic coasts are located on the opposite wave-tide side of the triangle, because sediment is moved onshore by waves and tides. Estuaries occupy an intermediate position, because they have a mixed sediment source and are affected by river, wave and tidal factors. Different coastal typologies can be identified, reflecting the degree of fluvial, wave or tide dominance, and it is this classification that will be used for the remainder of this chapter. Specifically, the next three sections will discuss the dynamics of wave-, tide- and fluvial-dominated coasts, followed by rocky coastlines.

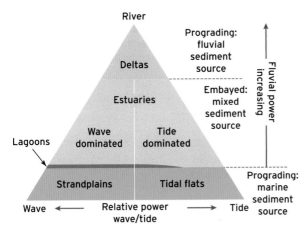

Figure 17.22 Process-based coastal classification based on the relative importance of wave, tide and river processes. (Source: after Boyd *et al.*, 1992)

Reflective question

➤ What is the point of classifying coastlines in distinct types?

17.6 Wave-dominated coastal environments

The most easily recognized landform of wave-dominated coasts is the beach. However, beaches are only one component of wave-dominated coasts. The underwater slope that lies seaward of the beach, known as the shoreface, is also dominated by wave processes and actually occupies a much larger area than the beach. Coastal dunes behind beaches are also common elements of wave-dominated coastal environments. Dunes, beaches and shorefaces are strongly linked by sediment transport pathways and collectively they make up 'coastal barriers', which are considered the basic depositional elements of wave-dominated coasts (Roy *et al.*, 1994). We will first discuss the dynamics of barriers and then move on to beaches and dunes.

17.6.1 Barriers

A large variety of barrier types exist, including **barrier islands** separated by tidal inlets and fronting wide shallow lagoons as along the coast of the Netherlands (Figure 17.23),

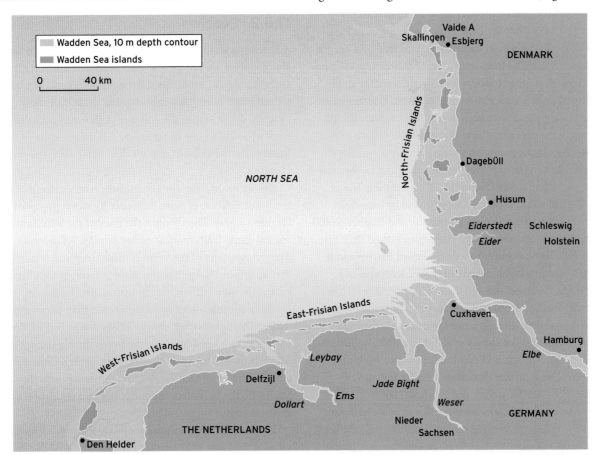

Figure 17.23 Barrier islands of the Wadden Sea in north-western Europe. (Source: after Hofstede, J., 2005, Danish-German-Dutch Wadden Environments, in E.A. Koster (editor). *The Physical Geography of Western Europe*. Oxford University Press, Oxford, 185–205. Copyright Oxford University Press. Reprinted by permission.)

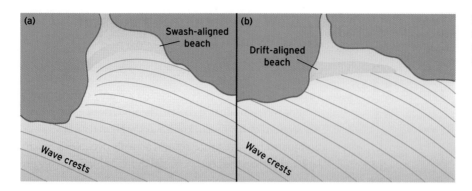

Figure 17.24 Two main types of coastal alignment: (a) swash-alignment and (b) drift-alignment. (Source: adapted from Davies, 1980)

continuous barrier systems that are backed by a coastal plain, **lagoon** or estuary, and mainland beaches on steep coasts with very little additional barrier morphology. Most barrier systems are made up of sand, but gravel barriers are also frequently found, especially at higher latitudes where glacial processes over the last few hundred thousand years have produced vast quantities of gravel-size material.

It is useful to distinguish between two fundamentally different types of barrier coasts: swash-aligned barriers and drift-aligned barriers. Swash-aligned barriers are oriented parallel to the crests of the prevailing incident waves (Figure 17.24a). They are closed systems in terms of long-shore sediment transport and are characterized by a curved planform shape. The long-term average shoreline configuration of swash-aligned barriers is relatively constant since the net littoral drift is zero. However, short-term changes in incident wave conditions such as seasonal or inter-annual changes in the prevailing wave direction will induce minor adjustments in the planform shape.

Drift-aligned barriers are oriented obliquely to the crest of the prevailing waves (Figure 17.24b). The shoreline of drift-aligned coasts is primarily controlled by longshore sediment transport processes (Masselink and Hughes, 2003) and such coasts are sensitive to changes in littoral drift rates. A **spit** is a classic example of a drift-aligned barrier and is a narrow accumulation of sand or gravel, with one end attached to the mainland and the other projecting into the sea or across the mouth of an estuary or a bay. Spits grow in the littoral drift direction and can exist only through a continuous longshore supply of sediment. If this sediment supply ceases, the spit will eventually subsume itself and disappear. For a further discussion of spits and also of barrier origin see Masselink and Hughes (2003).

Transgressive barriers are those that migrate landward under the influence of rising sea level and/or a negative sediment budget (Figure 17.25a). Transgressive barriers consist mainly of tidal delta and/or washover deposits, and are underlain by back-barrier estuarine or lagoonal deposits.

Sediments deposited in seaward environments end up on top of sediment that originated in more landward environments. **Regressive barriers** or strandplains are those that develop under the influence of a falling sea level and/or a positive sediment budget (Figure 17.25b). Here landward sediments are deposited on top of more seaward ones. The barrier is generally overlain by windblown sand, below which there is beach and nearshore sand underlain by silt and clay that had been deposited on what was formerly the continental shelf.

17.6.2 Beaches

A beach is a wave-deposited sand or gravel landform found along marine, lacustrine and estuarine shorelines. It represents the upper part of the shoreface and is generally characterized by an overall concave-upward shape. On most beaches, deviations to the concave profile may occur in the form of smaller-scale features and it is the presence of these features that gives beaches their distinctive morphology (Figure 17.26).

Berms result from the accumulation of sediment at the landward extreme of wave influence by swash processes (Figure 17.27a). They protect the back of the beach and coastal dunes from erosion under extreme wave conditions. The seaward part of the berm is often steep and is termed the beachface. Its equilibrium slope is thought to reflect the balance between differences in uprush and backwash. Onshore uprush forces are often greater than backwash forces. This results from energy losses due to bed friction and infiltration of water into the beach during the uprush, which is promoted by coarse and permeable sediments. The equilibrium beachface gradient is therefore positively correlated with the beachface sediment size and the steepest gradients occur on gravel beaches.

Beach cusps are rhythmic shoreline features formed by swash action and may develop on sand or gravel beaches. The spacing of the cusps is related to the horizontal extent of the swash motion and may range from about 10 cm on lake

(a)

Flood tide delta

Coastal plain

Lagoon

Barrier shoreface

Ravinement surface

Tidal inlet deposits

Relict estuarine muds

Sea level rising

(b)

Abandoned highstand estuary

Coastal plain

Strandplain

Erosion surface

Regressive beach and shoreface sand

Former high sea level

Sea level falling

Figure 17.25 (a) Model of transgressive barrier during sea-level rise. Transgressive barriers are almost entirely composed of tidal delta and washover deposits. The barrier migrates into estuarine and lagoonal environments as sea level rises. (b) Model of regressive barrier during sea-level fall. The surface of the regressive barrier forms a wide strandplain, generally without estuaries. (Source: after Roy *et al.*, 1994)

shores to 50 m on exposed ocean beaches. Beach cusps are considered self-organizing features, resulting from feedback between beachface morphology and swash processes as described in Box 17.4.

The surf zone on energetic beaches is often characterized by a nearshore bar morphology as shown in Figure 17.27(b) which can be rhythmic in form. The development of rhythmic bar morphology has also been attributed to self-organization (Box 17.4). Significant progress in our understanding of nearshore bar dynamics has been obtained over the past two decades through the use of video-monitoring as described in Box 17.5. On beaches subjected to large tidal ranges, the intertidal zone is often flat and featureless. But when the upper part of the intertidal zone is relatively steep, because it is composed of coarse sediments for example, a distinct break in slope is generally found separating the steep upper part from the low-gradient low-tide terrace.

Beaches respond to changing wave conditions and of greatest significance is the exchange of sediment between the upper beach and the surf zone, and the development of berm and bar profiles (Figure 17.29). Under calm conditions, sediment transport in the nearshore zone tends to be in the onshore direction, resulting in a steepening of the

beach profile. If bars are present, these tend to migrate onshore and become part of the beach, resulting in the development of a steep beach with a pronounced berm. Such beaches are referred to as reflective beaches, because a significant part of the incoming wave energy is reflected back from the shoreline. In contrast, energetic wave conditions induce offshore sediment transport with prolonged high-wave conditions, resulting in the destruction of the berm and the formation of a flat beach with subdued bar morphology. The surf zone is likely to be wide with multiple lines of spilling breakers. The majority of the incoming wave energy is dissipated during the wave-breaking process and these beaches are known as dissipative beaches. Most beaches fall within these two extremes and are characterized by nearshore bar morphology over which a significant amount of wave energy is being dissipated due to wave breaking. The upper part of intermediate beaches is, however, rather steep and reflective. Therefore these beaches are referred to as intermediate beaches.

Depending on the wave conditions, beaches tend to move from one beach type to the other (Wright and Short, 1984). Along some coastlines, stormy conditions in the

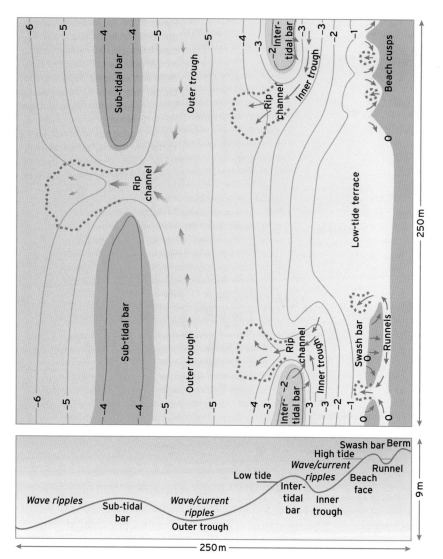

Figure 17.26 Schematic showing dominant morphological features on a beach. The top panel represents a contour plot of the beach morphology, whereas the bottom panel shows a typical cross-shore beach profile. High-tide level on this beach is at an elevation of 0 m and the tidal range is 2 m. (Source: after Masselink and Hughes, 2003)

Figure 17.27 Features of a beach: (a) beach cusp morphology with erosion scarp; (b) intertidal bar exposed at low tide.

NEW DIRECTIONS

SELF-ORGANIZATION

Morphodynamic systems often display a sequence of positive feedback driving the system towards a new state, followed by negative feedback, which stabilizes the system, resulting in equilibrium. This is referred to as self-organization. The result of this process is a rather orderly arrangement of sediments and landforms.

An example of self-organization is the formation of beach cusp morphology. Beach cusps are rhythmic shoreline features formed by swash action and are characterized by steep-gradient, seaward-pointing cusp horns and gentle-gradient, seaward-facing cusp embayments (Figure 17.28). The self-organization theory of beach cusp formation considers beach cusps to be the result of feedback between morphology and swash flow (Werner and Fink, 1993). Positive feedback causes small topographic depressions on the beach face to be amplified by attracting and accelerating water flow, thereby promoting erosion. At the same time, small positive relief features are enhanced by repelling and decelerating water flow, thereby promoting accretion. The sequence of positive feedback is followed by negative feedback processes, which inhibit erosion and accretion on well-developed cusps, and maintain equilibrium. The important feature is that the morphological regularity arises from the internal dynamics of the system.

Self-organization is a relatively new concept in geomorphology, but is increasingly employed to explain and describe a wide range of geomorphic systems (Werner, 1999). The concept has been applied to analyse, describe and explain a large number of coastal features, including gravel barriers, estuaries, beach cusps, coastal dunes and nearshore bars.

Figure 17.28 Beach cusps on a gravel beach. The longshore spacing of the cusps is approximately 3 m.

BOX 17.4

Figure 17.29 Idealized beach profiles with and without bars. Storm conditions induce offshore transport, beach erosion and the formation of a nearshore bar. Calm wave conditions result in onshore sediment transport, beach accretion and the formation of a berm. (Source: Aggard and Masselink, 1999)

TECHNIQUES

VIDEO-MONITORING

Coastal processes and morphological change occur over a range of timescales, varying from seconds to millennia. Each timescale of interest has its own set of equipment and research methodologies to collect and interpret relevant data to advance our understanding at that timescale. During the past two decades, a significant increase in our understanding of the daily-to-yearly behaviour of intertidal and subtidal bar systems has been obtained using video imagery.

Currently over 20 video-monitoring stations have been installed across the world as part of the ARGUS network. At each of these stations, several video cameras are used to cover coastal areas of up to several kilometres. Three types of image are routinely collected every hour: a snap shot, a 10 min average image (referred to as the 'timex' image) and the variance associated with the 10 min image. The timex image is particularly useful in picking up wave breaking patterns, and since waves usually break over bars, the timex provides information on the position and configuration of

the bars. By using points in the images with known coordinates, the oblique pictures can be geo-referenced and merged to produce an accurate map of the nearshore bars (Figure 17.30). The videos yield data on day-by-day changes to bar morphology over years. In fact, the oldest ARGUS

station, deployed on the east coast of the United States, has been collecting data since 1980. By combining video data with hydrodynamic information collected using wave buoys, the processes that drive bar dynamics (migration, formation and destruction) can be investigated in detail.

(a)

(b)

Figure 17.30 Result of ARGUS video-monitoring on Egmond Beach in the Netherlands (12/May/1998 GMT 12 h): (a) oblique image obtained by merging the timex images of five video cameras; and (b) rectified image. The images show four lines of breakers representing in the seaward direction (up): (1) the swash zone; (2) an intertidal bar; (3) a subtidal bar; (4) an underwater beach nourishment deposit. (Source: photo courtesy Aart Kroon)

BOX 17.5

winter and calm conditions during summer give rise to a seasonal cycle of beach change comprising a winter profile with a nearshore bar and a summer profile with a berm, although this will vary in any given year. Depending on the wave conditions, beaches tend to move from one beach type to the other (Wright and Short, 1984). The occurrence of different types of beach morphology can be parameterized by the dimensionless fall velocity Ω given by:

$$\Omega = \frac{H}{w_s T} \tag{17.8}$$

where H is the wave height, w_s is the sediment fall velocity (the speed at which a sediment particle falls through still water) and T is the wave period. Reflective beaches tend to develop when $\Omega < 1.5$, intermediate beaches are characterized by $\Omega = 1.5-5.5$, and dissipative beaches form when $\Omega > 5.5$.

17.6.3 Coastal dunes

Coastal dunes are common features in wave-dominated coastal environments and their dynamics are closely linked to that of the beach (Sherman and Bauer, 1993). Their formation requires an energetic wind climate and a suitably large supply of sand. Onshore winds capable of inducing sediment transport must occur for a significant amount of time. Dunes protect the coast from erosion by providing a buffer to extreme waves and winds. Extreme storm activity inevitably results in elevated water levels and beach erosion, and may lead to coastal flooding. However, well-developed dune systems dissipate the energy of storm waves through dune erosion. The sand eroded from the dune system will be transported offshore, but will eventually return to the beach under fair weather conditions. As the sediment is returned to the beach, wind processes may result in renewed dune development. Maintenance of coastal dune systems is thus an important component of coastal protection and management.

Coastal dunes generally begin to develop around the drift line above the spring high-tide line. Here, tidal litter (seaweed, driftwood) represents an obstacle to the wind, promoting the formation of **shadow dunes** with tails stretching out downwind (Figure 17.31a). Shadow dunes cannot reach elevations higher than that of the obstacle, but ongoing accumulation of sediment can occur following the establishment of pioneer plant species. Pioneer plants are all characterized by a high tolerance to salt, elaborate root systems that can reach down to the freshwater table and rhizomes that grow parallel to the upper dune surface. The sand-trapping ability of the pioneer plants enables the shadow dune to grow upward and outward into incipient

(a)

(b)

(c)

Figure 17.31 Different stages in the formation of foredunes: (a) isolated shadow dunes associated with clumps of vegetation; (b) incipient foredune in front of older dune cliff; and (c) terrace-like foredune ridge backed by older dunes.

foredunes, which are 1–2 m high vegetated mounds of sand (Figure 17.31b). Given suitable conditions (onshore winds and adequate sand supply) and sufficient time, the incipient foredunes will coalesce, forming a **foredune ridge** (Figure 17.31c). Foredunes can grow quickly reaching a height of several metres over a period of 5–10 years.

Reflective questions

➤ Use the Google Map search engine to compare the barrier islands of the Dutch Wadden Sea in the Netherlands with those of the New Jersey coast in the United States. Why are the Dutch barrier islands shorter?

➤ What is the difference in stratigraphy between transgressive and regressive barrier systems?

➤ Given the large wave energy level and the smoothing action of tides that beaches are exposed to, why are most beaches not flat and featureless?

➤ In what way are coastal dunes different from terrestrial desert dunes described in Chapters 12 and 16?

17.7 Tide-dominated coastal environments

Estuaries are the main tide-dominated coastal landform and represent zones of mixing between fluvial and marine processes. The development of present-day estuaries started

when coastal river valleys were flooded as sea level rose following ice melt at the end of the last glacial period. Following stabilization of the sea level around 6000 years ago, infilling of the estuaries occurred as a result of the influx of sediments from both marine and terrestrial sources. Most estuaries can be divided into three zones: the inner zone, central zone and outer zone (Figure 17.32). These three zones are unique with respect to their energy regime, sediment type and morphology (Masselink and Hughes, 2003). River processes dominate at the head of the estuary and their influence decreases towards the mouth of the estuary. Marine processes are most important at the mouth and their role decreases towards the head. The energy regime in the inner zone is therefore river-dominated, in the outer zone it is marine-dominated (waves and tides) and in the central zone it is mixed (tide and river processes). There are many types of estuaries and a popular distinction is that between wave- and tide-dominated estuaries (Dalrymple et al., 1992).

17.7.1 Wave- and tide-dominated estuaries

Wave-dominated estuaries are found in coastal regions subjected to relatively high levels of wave energy (Figure 17.33). The outer zone consists of a barrier system and a tidal inlet, often with an ebb-tide delta and a flood-tide delta at the landward and seaward side, respectively. The outer

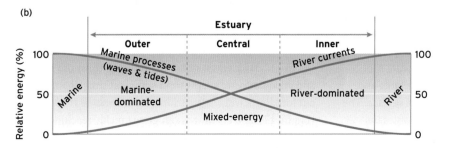

Figure 17.32 (a) Plan view of an estuary showing sediment and hydraulic boundaries. (b) Chart showing the changing mix of wave, tide and river processes along the estuary axis. (Source: after Dalrymple et al., 1992, SEPM (Society for Sedimentary Geology))

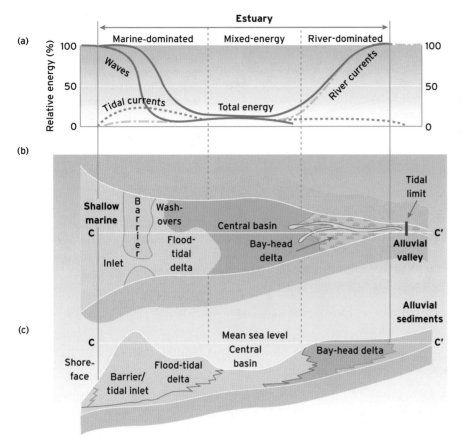

Figure 17.33 (a) Chart showing the change in energy regime along the axis of a wave-dominated estuary. (b) Plan view of the estuary showing positions of principal morphological features. (c) Section view along the estuary axis showing stratigraphy. (Source: after Dalrymple *et al.*, 1992, SEPM (Society for Sedimentary Geology))

zone is dominated by wave processes, but their effects decline rapidly with distance from the inlet due to wave breaking over both the barrier and tidal deltas. The tidal energy also decreases away from the coast, because the narrow tidal inlet restricts the tidal water motion between estuary and sea. A distinguishing characteristic of wave-dominated estuaries is the very low energy level in the central zone. If the estuary is relatively young, a deep central mud basin accumulates the finest sediments; if the estuary is mature, then the central zone is infilled and dominated by salt marshes or mangrove flats also composed of predominantly muddy sediments (Masselink and Hughes, 2003). If the river entering the estuary at the head carries significant amounts of sediment, a **bay-head delta** can be found in the inner zone. Wave-dominated estuaries infill through seaward progression of this bay-head delta and landward extension of the flood-tide delta. Eventually, fluvial and marine sands bury the central basin muds.

Tide-dominated estuaries are found in coastal regions experiencing relatively large tidal ranges and therefore strong tidal currents (Figure 17.34). The scouring action of the tidal currents keeps the entrance of the estuary relatively

open and gives tide-dominated estuaries their typical funnel shape. The strong tidal currents in the outer zone shape the sediments into linear sand bars separated by tidal channels. Waves are of secondary importance, but their influence can sometimes extend further into a tide-dominated estuary due to its funnel shape. Wave energy in the central basin is insignificant, whereas tidal energy is still relatively high. Therefore, the central zone of tide-dominated estuaries is more energetic than that in wave-dominated estuaries (Masselink and Hughes, 2003). A single meandering channel is commonly found in the central zone and this channel is tide-dominated for most of the time, but is significantly influenced by river processes during times of high discharge. The central zone is again a sink for fine sediment and includes extensive intertidal morphology, such as tidal flats and salt marshes. Because tidal currents remain strong, even up to the head of the estuary, fluvial sediments become progressively mixed with estuarine sediments in the inner zone and there is no discrete fluvial delta. The infilling of tide-dominated estuaries is often rapid and occurs through a steady seaward migration of the inner, central and outer zones along the drowned valley.

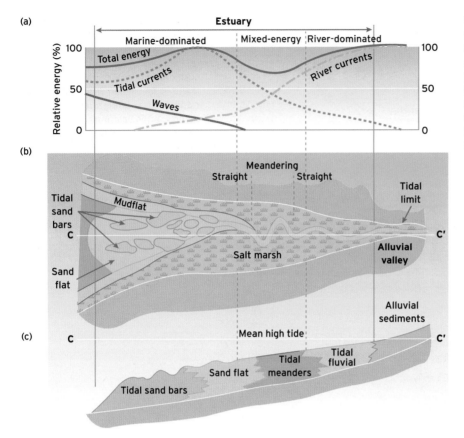

Figure 17.34 (a) Chart showing the change in energy regime along the axis of a tide-dominated estuary. (b) Plan view of the estuary showing positions of principal morphological features. (c) Section view along the estuary axis showing stratigraphy. (Source: after Dalrymple et al., 1992, SEPM (Society for Sedimentary Geology))

17.7.2 Estuarine mixing

An important process occurring in estuaries is the mixing of salt- and freshwater masses that are delivered to the estuary by tide and river flows, respectively. The mixing is accomplished by the turbulence associated with river and tidal flows and is opposed by the density difference between the two water masses. On the basis of the mixing process, we can identify three types of estuaries (Dyer, 1998) as described below.

17.7.2.1 Stratified estuaries

Stratified estuaries tend to have fresh river water sitting on top of saline seawater, separated by a sharp interface, known as the halocline (Figure 17.35). The saltwater near the bed is in the form of a salt wedge, which becomes thinner towards the landward end of the estuary. Stratified estuaries commonly occur along micro-tidal coasts with low to intermediate river discharge. The river discharge should be sufficient to develop a fresh surface water mass, but should not be so large as to expel the saltwater from the estuary. If the tidal or riverine flows become too dynamic, turbulence is generated and the stratification breaks down.

17.7.2.2 Partially mixed estuaries

In **partially mixed estuaries** there is a more gradual salinity gradient, but still with freshwater at the surface and saltwater near the bed. However, there tends to be a more mixed layer in the central part of the water column. The salinity decreases towards the landward end of the estuary both at the surface and at depth. Partially mixed estuaries develop if the tidal energy is sufficient to cause mixing of the fresh- and saltwater, and are typical of meso- to macro-tidal coasts.

17.7.2.3 Well-mixed estuaries

In **well-mixed estuaries** the mixing is so effective that the salinity gradient in the vertical direction vanishes entirely. If such mixed estuaries are sufficiently wide, the Coriolis force (see Chapter 4 for an explanation of this process) starts to play a significant role in affecting the water circulation. Its effect on outflowing river water is to push the flow to the margin of the estuary (Figure 17.35c) and may result in a horizontal separation of river water and seawater. This segregation is best developed on the flooding tide when the fresh- and saltwater masses are opposed and the effect of the Coriolis force is to separate the two flows.

Figure 17.35 Diagram illustrating three main types of estuaries based on density stratification: (a) stratified; (b) partially mixed; and (c) well-mixed estuary. Water masses are indicated by shading on the front face of each block. Salinity contours (arbitrary scale) are indicated by thin lines on the side face of each block in (a) and (b) and on the top face of the block in (c). In (a) and (b) vertical mixing is indicated by thin arrows and non-tidal currents are indicated by thick arrows. (Source: after Pethick, 1984)

Wave-dominated estuaries are usually stratified, whereas tide-dominated estuaries are either partially mixed or well mixed.

17.7.3 Ebb- and flood-dominance

The total volume of water entering an estuary on the flooding tide is called the **tidal prism** and can be estimated by multiplying the tidal rise in water level in the estuary during high tide by the surface area of the estuary. Ignoring the contribution of the freshwater discharge and evaporation, this same volume of water leaves the estuary during the falling tide. The duration and strength of the flood and ebb flow, however, are

significantly different in most estuarine systems. Estuaries or channel sections that display a flooding tide that is larger in velocity magnitude and shorter in duration than the ebbing tide are said to be flood-dominant, whereas those that display an ebbing tide that is largest in magnitude and shortest in duration are said to be ebb-dominant.

Flood- or ebb-dominance often translates directly to net landward or seaward sediment transport, respectively (Friedrichs and Aubrey, 1988). Even a small difference in the velocity magnitude between the flood and ebb tide can lead to a large difference in the total amount of sediment transported by each, and therefore a net sediment transport. In general, flood-dominant estuaries tend to infill their entrance channels by continually pushing coastal sediment landward and as a result are often intermittently closed, whereas ebb-dominant estuaries tend to flush sediment seaward from their entrance channels and as a result are often stable (Masselink and Hughes, 2003). There are a number of mechanisms that can generate tidal flow asymmetry, and hence ebb- or flood-dominance. The two most important are tidal distortion (which is the change of the tidal wave shape due to shoaling, typical of long estuaries and leads to flood-dominance) and a high proportion of intertidal areas such as salt marshes and tidal flats (which is more typical of small, wave-dominated estuaries and leads to ebb-dominance).

17.7.4 Salt marsh and mangroves

The lower intertidal zone in most estuaries is devoid of vegetation due to excessive bed shear stress preventing seedlings taking anchor in the sediment. The upper intertidal zone is less energetic, however, and in temperate environments is colonized by salt-tolerant grasses and reeds known collectively as salt marsh, and in (sub)tropical environments by mangroves (see Figure 12.21 in Chapter 12). Both salt marshes and mangroves exhibit a distinct zonation across the upper intertidal zone with the most salt-tolerant species (the 'pioneers') lowest in the tidal frame, and the least salt-hardy species near the Highest Astronomical Tide (HAT) level.

Despite the obvious difference between these two types of ecosystems, their functioning from an estuarine evolutionary point of view is very similar: both significantly affect the tidal flows and through a variety of mechanisms enhance sediment deposition of clastic material, mainly silts and muds. In addition, in both systems there is a steady supply of organic detritus (roots, stems, leaves, branches) which also contributes to sedimentation. Vertical accretion rates are highly variable, both spatially and temporally, and are of the order of millimetres per year. Generally, sedimentation rates exceed current and even projected rates of sea-level rise, enabling these intertidal environments to keep up with rising sea level. Salt marshes and mangroves thus provide a good example of natural resilience to sea-level rise.

Reflective questions

➤ How do the morphodynamics of wave- and tide-dominated estuaries differ?

➤ To what extent do you agree with the statement that estuaries are rather short-lived features on a geological timescale?

17.8 Fluvial-dominated coastal environments

Coastal deltas are accumulations of sediment deposited where rivers enter into the sea. River sediments may also accumulate at the head of coastal embayments if the coastline is drowned (bay-head delta), but these deposits are also controlled by estuarine processes and were briefly discussed in the previous section. In deltas, the amount of sediment delivered into the coastal margin by a river system outpaces the ability of the marine processes (waves and tides) to remove these sediments, causing the coastline to advance seaward. Since deltas are mainly associated with large river systems, they occupy only a relatively small proportion of the world's coastline. In Europe, for example, there are only six delta systems of significance (Danube Delta in Romania, Ebro Delta in Spain, Po Delta in Italy, Rhine/Meuse Delta of the Netherlands, Rhone Delta in France, Volga Delta in Russia). Nevertheless, deltas are very important from a societal point of view because they are characterized by relatively large population densities.

Although the detailed morphology of deltas varies from one example to the next, depending on the delta regime, there are three morphological units common to almost all deltas (Figure 17.36). The **delta plain** is the sedimentary platform that covers recent seaward advance, the **delta front** represents the seaward front of the delta that is located in relatively shallow water and is being reworked by wave and tidal processes, and the **pro-delta** is situated at the toe of the delta front in relatively deep water and is generally out of reach of wave processes (Masselink and Hughes, 2003). The delta builds out because the river continuously delivers sediment to the coastline. The coarsest sediments are deposited close to the river mouth and the finer sediments settle out further seaward. As the delta front progrades horizontally, the delta plain aggrades vertically. The resulting sediment

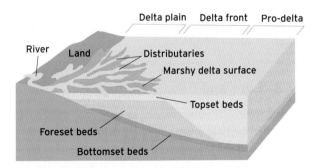

Figure 17.36 The three generic morphostratigraphic units found in all fine-grained deltas: delta plain (**topset beds**), delta front (**foreset beds**) and pro-delta (**bottomset beds**). (Source: after Haslett, 2000)

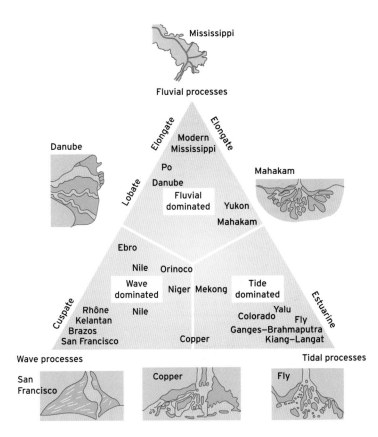

Figure 17.37 Classification of fine-grained deltas proposed by Galloway (1975). The end-member types are fluvial-, tide- and wave-dominated. (Source: From Briggs *et al.*, 1997)

distribution through the delta wedge is then a general fining in the seaward direction and a general coarsening upwards in the vertical direction (Reading and Collinson, 1996). Based on the relative magnitude of river, wave and tide power, Galloway (1975) proposed a classification scheme of deltas with three end-members (Figure 17.37).

Fluvial-dominated deltas are associated with large catchments, river discharge into protected seas with minimal nearshore wave energy, and a small tidal prism. The freshwater river effluent is generally less dense than the saltwater in the receiving basin, and the river water flows out on top of the receiving water. If the outflowing river water is denser than the water in the receiving basin, for example due to extremely high suspended sediment concentrations, the river water will flow out along the seabed. Buoyancy limits the mixing between river and basin waters, allowing the river sediment to be transported further into the receiving basin before settling to the bed. As a result, the morphology of river-dominated deltas is characterized by pronounced seaward protrusions, a classic example of which is the so-called 'bird foot' of the Mississippi Delta (Figure 17.38).

Wave-dominated deltas are typically found in open-coast settings with a steep shoreface gradient, causing the deltaic coastline to be exposed to energetic waves. The action of

Figure 17.38 The Mississippi Delta is shaped like a 'bird's foot' and is the classic example of a river-dominated delta. The scene shown in this false-colour image covers an area of 54 km by 57 km and represents the currently active delta front of the Mississippi known as the Balize.

Figure 17.39 Outlines of the five delta lobes that make up the Mississippi Delta. The currently active delta lobe, the Balize, is shown in Figure 17.38. The youngest delta lobe is the Atchafalaya, which is not yet fully occupied by the river. (Source: adapted from Pilkey, 2003)

Delta lobe	Age
Maringouin	7500–5000 years BP
Teche	5500–3800 years BP
Saint Bernard	4000–2000 years BP
LaFourche	2800–800 years BP
Balize	1000 years BP to present
Atchafalaya	present to–

waves at the river mouth induces strong mixing between the river flow and the receiving water. There are therefore limited buoyancy effects and the sediment transporting capacity decreases rapidly away from the mouth of the river, resulting in the formation of a distributary mouth bar across the river mouth. When waves approach with their crest parallel to the coastline they are refracted symmetrically around the distributary mouth bar. When they approach obliquely, however, they cause longshore transport and spit growth, with the river entrance constantly migrating downdrift. Regardless, the characteristic feature of wave-dominated deltas is a relatively straight, only weakly protruding coastline.

Tide-dominated deltas develop when the tidal prism is larger than the fluvial discharge and are commonly found along meso- and macro-tidal coastlines. In such settings, strong tidal currents flow in and out of the river mouth, also causing strong mixing between the river effluent and the receiving water. Elongated bars in the mouth of the delta often result, separated by tidal channels. A fundamental difference between tide-dominated deltas and tide-dominated estuaries is that in the former case the bottom contours at the mouth of the river bend out (indicating advance of the landform out to sea), whereas in the latter case they bend in (indicating retreat of the landform inland).

Delta development relies heavily on an active sediment supply by the river. If this supply is less than the sediment removal and dispersal by wave and tidal processes, the deltaic shoreline will erode. A natural cause for this to

happen is through **delta switching**, which occurs when the active region of coastal accumulation switches from one location on the delta to another. The interval between switching varies from hundreds to thousands of years, depending on the size of the delta, and many contemporary deltas have gone through several stages of delta switching during the past few thousand years. For example, the Mississippi Delta has had four major delta lobe switches over the past 5000 years, the last of which occurred approximately 800 years ago (Figure 17.39). Delta switching occurs due to an overextension of the presently active **distributaries**, which reduces their competency to convey sediment across the delta plain. At some stage, the existing channel (or channel network) is abandoned in favour of another more competent channel (or channel network) cut through a shorter, steeper section of the delta plain. When this happens, the former area of active delta formation becomes completely inactive with respect to fluvial processes and may undergo subsequent erosion by waves and tides.

Humans can modify the sediment supply in a river through actions on land such as building of dams, sediment abstraction and water abstraction for irrigation. This can lead to delta erosion since these interventions all reduce the sediment supply to the delta and may cause the deltaic shoreline to erode. Erosion problems in deltas are exacerbated by the fact that delta systems are generally subsiding due to the weight of the deltaic deposits on the Earth's crust and hence experience a **relative sea-level** rise. Not

(a)

(b)

(c)

(d)

Figure 17.40 Examples of rocky coastlines: (a) plunging limestone cliffs at the Bill of Portland, Dorset, UK; (b) rugged coastline around Tongue, northern Scotland, UK; (c) cliffed coastline with embayed beaches, south Devon, UK; and (d) sloping platform near Minehead, Somerset, UK.

surprisingly, most deltaic shorelines are currently displaying large erosion rates, with the Nile Delta being the best-documented example (Stanley and Warne, 1998).

Reflective questions

➤ What is the difference between a tide-dominated delta and a tide-dominated estuary?

➤ What are the distinguishing morphological features of fluvial-, wave- and tide-dominated deltas?

17.9 Erosive coasts

Rocky coasts are continually being cut back by the sea and are characterized by erosional features. The erosive nature of rocky coasts results in often stunning coastal scenery

(Figure 17.40), but makes it difficult to deduce their evolutionary history, because the different evolutionary stages are not preserved in the stratigraphy. Along depositional coasts we can usually observe relatively quick morphological changes using measurements, maps and aerial photographs. However, in rocky coastal settings there tends to be a very slow rate of change. This slow rate of change averaged over long periods does not mean that changes cannot be dramatic and sudden.

Sunamura (1992) categorized rocky coast morphology into three main types: sloping shore platform, sub-horizontal shore platform and plunging cliff as shown in Figure 17.41.

17.9.1 Rocky coast processes

Rocky coast erosion is accomplished by a wide range of processes, often working together (see Trenhaile, 1987, for a comprehensive overview of these processes). In terms of

Figure 17.41 Three major morphologies on rocky coasts with their characteristic erosional features: (a) sloping shore platform, (b) sub-horizontal shore platform and (c) plunging cliff. (Source: after Sunamura, 1992)

their function in controlling rocky coast morphology, these processes can be grouped into three main types: mass movements, rock-breakdown processes and marine rock-removal processes.

Mass movements are common along rocky coasts due to the prevailing steep, and therefore unstable, slopes. A spectrum of mass movements can occur on rocky coasts, depending primarily on the properties of the rock (lithology and structure). They include rockfalls which are characteristic of hard rocks, landslides which typically occur in thick, fairly homogeneous deposits of clay, shale or marl, and flows which are mass movements that involve movement of material with a high liquid content. All types of mass movements are episodic and occur more commonly in winter than in summer due to increased rainfall and undercutting of the base of the rocky slopes by wave processes. The principal roles of mass movements are the downwearing of cliffs and the introduction of cliff material into the nearshore zone.

There are a host of physical, chemical and biological processes that weaken and loosen rock material, which then becomes available for removal by marine processes. Their relative importance depends principally on wave energy level, climate and rock type. Mechanical wave action (abrasion and hydraulic action) is the main erosional agent in most swell and storm-wave environments. In sheltered areas and on particularly susceptible rocks, weathering (see Chapter 11) is probably the major erosive mechanism along rocky coasts. Physical weathering breaks down rock through the formation and subsequent widening of cracks in the rock. These can occur due to frost action, alternating cycles of wetting and drying and the growth of salt crystals. Chemical weathering of rocks is most significant in hot, wet climates. Finally, bio-erosion is the removal of rock by organisms and is most important in tropical regions due to the enormously varied marine biota and the abundance of calcareous substrates. Chapter 11 provides more detail on slope stability and weathering processes.

Mass movements and weathering weaken the rock and produce loose rock material that becomes available to marine processes for removal. Without the export of this material, rocky coasts stop evolving and simply develop into weathered terrestrial slopes. The larger the waves, the greater the efficiency of the cross-shore and longshore sediment transport processes to remove loose material and keep rocky coasts 'fresh'.

17.9.2 Coastal cliffs

Coastal cliffs can be defined as 'steep slopes that border ocean coasts' and occur along approximately 80% of the world's coastline. A bewildering variety of cliff profiles are found in nature and this reflects the large number of factors involved in the development of coastal cliffs. In addition to rock type and sea-level history, the relative roles of marine or land processes are crucial in determining cliff morphology. If the ability of marine processes to remove the debris exceeds the supply of material by mass-wasting and weathering processes, sediment will not accumulate at the base of the cliff. In this case, the angle of the cliff profile depends mainly on the structure and lithology of the rock. If, on the other hand, the supply of debris exceeds the capacity of removal at the base of the cliff, the material accumulates into a talus slope (see Chapter 11). The resulting angle of the cliff profile is the **angle of repose** (see Chapter 11) of the debris. Between these two extremes lies an infinite range of slope forms, each of which depends on the relative rates of sediment supply and removal at the shoreline (Pethick, 1984).

The main factor that controls cliff erosion is the hardness of rock. A rock's resistance to erosion is mainly determined by its lithology, although factors such as wave energy and cliff height are also considered significant. As an example, typical cliff erosion rates for granite, shale and glacially deposited materials (glacial till) are <0.001 m yr^{-1}, $0.01–0.1$ m yr^{-1} and $1–10$ m yr^{-1}, respectively (Sunamura, 1992). It is telling to translate these recession rates to predictions of cliff retreat over the next 100 years: granitic cliffs can expect to retreat by less than 0.1 m, while glacial till cliffs are likely to be cut back by 100 to 1000 m. The number for the glacial till cliffs may appear excessive. However, the glacial till cliffs along the Holderness coast in eastern England have retreated by almost 3 km since Roman times, representing a cliff retreat rate of 1.5 m yr^{-1}. It is further noted that cliff retreat is a highly episodic process and that cliff recession rates represent long-term averages. In other words, cliffs do not erode at a uniform rate, but can go through phases of little erosion followed by a sudden phase of erosion.

17.9.3 Shore platforms

Shore platforms are erosional features that develop when erosion of a rocky coast and the subsequent removal of the debris by waves and currents leave behind an erosional surface, forming a horizontal or gently sloping rock surface in the intertidal zone (Figure 17.40d). They are often referred to as 'wave-cut platforms', but this term should not be used because it assumes that shore platforms result from wave action, which is not always true because weathering processes also significantly contribute to shore platform formation (Stephenson, 2000). The rate of platform lowering depends mainly on the hardness of the rock and averages between 0.1 and 2 mm yr^{-1} (Trenhaile, 1987). Shore platform formation is intrinsically linked with cliff erosion, and the rate of vertical lowering of the cliff base (top of the shore platform) is 2–5% of the horizontal cliff recession. The junction between the shore platform and the cliff is usually close to the high-tide level and sometimes a high-tide beach is present at this location. Shore platforms also form along cohesive coasts where they front clay cliffs. These shore platforms are generally covered by a thin layer of sand/gravel material, but become exposed when severe storms remove this veneer.

Reflective question

➤ What are the main factors controlling coastal cliff erosion and shore platform lowering?

17.10 Coastal zone management

Coastlines are the most important and intensely used of all areas settled by humans for a number of reasons, including historical settlement, trading or political linkages, climate, availability of fertile alluvial soils, proximity to fish stocks and, more recently, aesthetic and recreational reasons (Carter, 1988). From a human point of view, the coastal zone is a resource to be used and exploited, whereas from an environmental perspective, the coastal zone is an environment often adversely affected by human activities (French, 1997). The coastal zone is used for various activities, ranging from nature conservation to waste disposal, with most coasts supporting multiple activities. Interactions inevitably occur between two or more coastal uses, and management is required to plan and coordinate the different uses of the coastal zone to avoid conflicts (see also Chapter 24). In the past, coastal management was mainly concerned with single issues that could be dealt with by a single authority. This is no longer the case. The increased complexity of coastal management issues, and the varying spatial and temporal scales at which they operate, bring in many different organizations with an interest in the management of a coastline. These organizations typically include administrative authorities (councils, government agencies, environmental organizations), industry and other interest groups (residents, tourism). For effective management of the coast, an integrated approach should be adopted and the term 'integrated coastal zone management' (ICZM) is used to indicate this approach.

Any ICZM initiative requires sustainability so that human activities should be non-destructive and the resources we exploit should be renewable. It is clear therefore that many coastal practices are not sustainable. However, Kay and Alder (1999) noted that sustainability is not a set of prescriptive actions but a 'way of thinking' about our use of the coastal zone and the resulting impacts. Generally the concept of sustainability to coastal management has resulted in a management approach with a longer-term view and more holistic perspective.

The output of ICZM consists of both legal policies and advisory initiatives. The latter are generally in the form of coastal management plans or shoreline management plans, which chart out a course for the future development of a stretch of coast and/or assist in resolving current management problems. Legally binding initiatives are a very powerful means to direct practices in the coastal zone. For example, the Dynamic Preservation Strategy adopted by the Dutch national government in 1991 (Koster and Hillen, 1995) included a legal provision that prescribed that the Dutch coastline be maintained at its 1990 position, irrespective of uncertain future developments. In other words, land losses due to coastal erosion are considered unlawful and have to be compensated for by beach nourishment. On a local level, councils can use by-laws to control activities in the coastal zone.

From a geographical point of view, the main issue associated with ICZM is to protect the coast from erosion and flooding. Both these aspects are particularly relevant at the moment, because 70% of our sandy coastlines are eroding (Bird, 1985) and sea level is rising at an increasing rate. There are four principal management options available to cope with coastal erosion and flooding due to sea-level rise:

1. *No active intervention* – this option is viable only if the coastline under question is undeveloped and nothing is at stake by giving up the land to coastal erosion.
2. *Managed realignment* – this option involves the relocation of coastal communities and industry, with a prohibition on further development. In this strategy, risks are minimized and costs of protection are avoided. However, social and economic costs associated with relocation and compensation are potentially high. The retreat option requires a strong governmental role with supportive legislation.
3. *Accommodation* – this allows continued occupancy and use of vulnerable coastal areas by adapting to, rather than protecting fully against, adverse impacts. It means learning to live with the sea-level rise and coastal flooding. Accommodation options include elevating buildings, enhancing storm and flood warning systems, and modifying drainage. The accommodation option can also involve changing activities, such as changing farming practices to suit the new environment, or simply accepting the risks of inundation and increasing insurance premiums. The accommodation option requires high levels of organization and community participation.
4. *Hold the line* – this option involves physically protecting the coast through **hard engineering** with structures such as seawalls and **groynes** or **soft engineering** through beach nourishment, for example. A summary of coastal protection measures and the problems associated with their implementation is given in Table 17.3. Protection has clear social, economic and political advantages, because assets and investments are safeguarded while economic activity can continue largely unhindered. Protection is the most expensive option to implement and maintain, and is only economically justifiable if the land to be protected is of great value.

The first three strategies are based on the premise that increased land losses and coastal flooding will be allowed to

Table 17.3 A selection of coastal engineering works

Management issue	Engineered solution(s)	Types	Description	Problems	Illustration
Cliff erosion	Sea walls	Vertical wall	A wall constructed out of rock blocks, or bulkheads of wood or steel, or simply semi-vertical mounds of rubble in front of a cliff	Rock walls are highly reflective, bulkheads less so. Loose rubble, however, absorbs wave energy	
		Curved wall	A concrete constructed concave wall	Quite reflective, but the concave structure introduces a dissipative element	
		Stepped	A rectilinear stepped hard structure, as gently sloping as possible, often with a curved wave-return wall at the top	The scarps of the steps are reflective, but overall the structure is quite dissipative	
		Revetment	A sloping rectilinear armoured structure constructed with less reflective material, such as interlocking blocks (tetrapods), rock-filled gabions and asphalt	The slope and loose material ensure maximum dissipation of wave energy	
Coastal inundation	Sea walls	Earth banks	A free-standing bank of earth and loose material, often at the landward edge of coastal wetlands	May be susceptible to erosion, and overtopped during extreme high-water events	
	Tidal barriers		Barriers built across estuaries with sluice gates that may be closed when threatened by storm surge	Extremely costly, and relies on reliable storm surge warning system (e.g. Thames Barrier)	
Beach stabilization	Groynes		Shore-normal walls of mainly wood, built across beaches to trap drifting sediment	Starve downdrift beaches of sediment	
	Beach nourishment		Adding sediment to a beach to maintain beach levels and dimensions	Sediment is often rapidly removed through erosion and needs regular replenishing; often sourced by dredging coastal waters	
Offshore protection	Breakwaters		Structures situated offshore that intercept waves before they reach the shore. Constructed with concrete and/or rubble	Very costly and often suffer damage during storms	
Tidal inlet management	Jetties		Walls built to line the banks of tidal inlets or river outlets in order to stabilize the waterway for navigation	The jetties protrude into the sea and promote sediment deposition on the updrift side, but also sediment stavation and erosion on the downdrift side	

(Source: after Haslett, 2000)

Chapter 17 Coasts

occur and that some coastal functions and values will be changed or lost. On the other hand, these strategies help to maintain the dynamic nature of the coast and allow it to adjust to rising sea levels naturally. It is beneficial to allow as many coastal regions as possible to retreat naturally, because erosion of these natural areas will liberate sediments, which may lessen the impact of sea-level rise on those areas that are not allowed to retreat naturally. The overall outcome is an increase in the resilience of the coastal system to sea-level rise. Hence, the first three options are most sustainable from a geomorphological point of view (although not necessarily from a socio-economic perspective). Certainly in developed countries there seems to be an increased push by national governments to pursue these more sustainable coastal protection strategies (see Box 17.6).

Notwithstanding the desire to maintain the dynamic nature of coasts, there will always remain a large role for coastal protection measures for the simple reason that many coastal areas are too valuable to be given up. When properly designed and constructed, hard engineering structures do serve an important purpose: storm surge barriers, such as constructed across the Thames and in the south-west of the Netherlands, have prevented serious flooding on several occasions; **sea walls** and **breakwaters** protect coastal development from damage during extreme wave events; and groynes are successful to some extent in trapping sediments and maintaining a beach (Figure 17.42). The 'side effects' of

(a)

(b)

(c)

(d)

Figure 17.42 Examples of hard coastal engineering structures: (a) vertical seawall with curved top; (b) wooden groynes deployed on Blyth Beach, Northumberland, UK. It can be seen that the left hand side has a raised beach where it has trapped longshore drift compared to the right hand side which has a shortened beach with water much closer to the camera. (c) **Jetties** are designed to prevent the silting up of tidal inlets and harbour entrance channels by blocking the littoral drift. In the photo, the littoral drift direction is from right to left. (d) Clay embankments or sea dikes installed as part of land reclamation efforts near the spring high-tide level seem inoffensive enough. However, the land they are protecting from flooding is at the same time deprived of the influx of sediment that would otherwise enable it to keep up with rising sea level. In this photo, the land to the left of the embankment which is flooded during spring high tide is at least 0.5 m higher than the land to the right. (Source for (c): Photo courtesy of Aart Kroon)

CASE STUDIES

NORTH NORFOLK COAST

It is being increasingly appreciated that natural coastal systems have a greater capacity to respond to disturbance than managed coastal systems. The key to such resilience is the ability for sediment to move freely through the system and this requires a non-fixed coastline. This realization has brought about a shift in coastal management ethos from working against nature, to working with nature, and the framing of shoreline management scenarios which

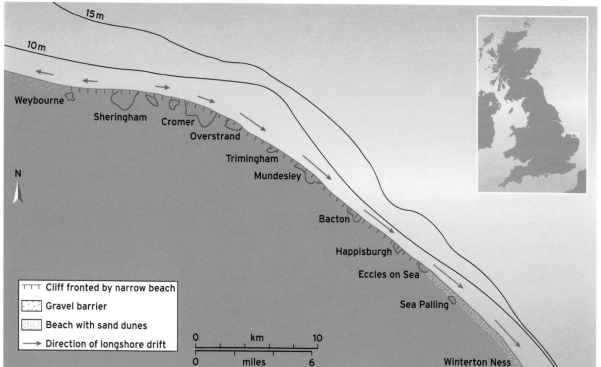

Figure 17.43 The north Norfolk coast is characterized by large littoral drift rates (up to 400 000 m³ yr⁻¹) and rapidly eroding soft cliffs (1-2 m yr⁻¹). Holding the line is not sustainable and the preferred management option for most of this coastline is managed realignment.

BOX 17.6 ➤

➤

move towards a more dynamic and functional coast. The UK Government's Department for Environment, Food and Rural Affairs (DEFRA) has also recognized that future shoreline management policies that work with natural processes, rather than against them, are likely to be more sustainable over the long term and have recently formulated a new government strategy for flood and coastal erosion risk management appropriately named 'Making Space for Water'. This new strategy is far-reaching and proposes that solutions for flood management and coastal erosion that work with natural processes to make more space for water should be identified and pursued wherever possible, including proposals for managed realignment

Table 17.4 Preferred management policies for Kelling to Lowestoft, Norfolk, UK, according to the draft Shoreline Management Plan

Policy unit	Recommended management policy		
	0-20 years	20-50 years	50-100 years
Kelling Hard to Sheringham	N	N	N
Sheringham	H	H	H
Sheringham to Cromer	N	N	N
Cromer	H	H	H
Cromer to Overstrand	N	N	N
Overstrand	H	M	M
Overstrand to Mundesley	N	N	N
Mundesley	H	H	M
Mundesley to Bacton Gas Terminal	N	N	N
Bacton Gas Terminal	H	H	M
Bacton, Walcott and Ostend	H	M	M
Ostend to Eccles	N	N	N
Eccles to Winterton Beach Road	H	H	M
Winterton to Scratby	N	N	N
California to Caister-on-Sea	H	M	M
Caister-on-Sea	H	H	M
Great Yarmouth	H	H	H
Gorleston	H	H	H
Gorleston to Hopton	N	N	N
Hopton	H	N	N
Hopton to Corton	N	N	N
Corton	H	M	M
Corton to Lowestoft	N	N	N
Lowestoft North (to Ness Point)	H	H	H

H = hold the line; M = managed realignment; N = no active intervention

BOX 17.6 ➤

> of coasts and rivers corridors where appropriate.

In line with these principles, the draft second-generation Shoreline Management Plan (SMP) for the north Norfolk coast from Kelling to Lowestoft Ness (Figure 17.43) is promoting more extensive policies of realignment and no active intervention in order to accommodate, rather than restrict, future coastal evolution (Table 17.4). The north Norfolk coast is character-ized by large littoral drift rates (up to 400 000 m^3 yr^{-1}) and rapidly eroding soft cliffs (1-2 m yr^{-1}). The coastal erosion is largely a natural process, unrelated to sea-level rise or coastal management interventions. Rather, due to its geologically very young age (less than 7500 years old), the north Norfolk coast is terribly out of equilibrium with the prevailing wave and tide conditions and is rapidly being modified in an attempt to assume a more stable configuration.

The draft SMP has far-reaching implications if it is implemented. For example, the anticipated loss of housing is 80, 80-450 and 450-1300 houses by 2025, 2055 and 2105, respectively. Not surprisingly, the draft SMP attracted a large amount of local interest and the public consulta-tion attracted more than 2500 objec-tions. It will be interesting to see whether the management policy recommendations of the SMP will be accepted, or whether the coastal planners will have to 'dilute' the recommendations in response to objections raised by local communi-ties, developers and politicians.

BOX 17.6

hard engineering are considerable, however, and it is well established that, following the construction of hard coastal structures, erosion problems on the downdrift unprotected coastline are often exacerbated (or even created). Soft engineering practices, in the form of beach nourishment or beach recharge, largely circumvent the main problem associated with hard engineering. The arti-ficial placement of a large amount of sediment, either on the underwater slope or on the beach itself, protects not only the recharged coast, but also the neighbouring coast-line, because sediment transport processes will redistribute the nourished sediment. This redistribution also represents a major downside of beach nourishment and treatment will have to be repeated at regular intervals. To reduce sediment losses following beach nourishment, groynes may be placed at the boundaries of the nourished area. On the whole, beach nourishment is more aligned with sustainable coastal management and is now very widely used (Bird, 1996).

Reflective questions

➤ What are the main factors involved in deciding on the best coastal management strategy for a particular stretch of coast?

➤ What are the problems associated with using hard engineering structures for coastal protection?

17.11 Summary

Coastal environments are arguably the most important and intensely used of all areas settled by humans. At the same time, they are currently at great risk from coastal erosion and flooding due to climate-induced sea-level rise. Over the last few decades, global sea level has been rising at a rate of 3 mm yr^{-1} and over the next 100 years this rate may increase to 6 mm yr^{-1}. Coastal morphologies are controlled and driven by a set of environmental boundary conditions, including sea-level change, geology, sediment supply and external forcing (wind, waves and tides), and changes in these boundary conditions will modify coastal processes and morphology. The current paradigm of coastal research is the so-called 'morphodynamic approach', which considers coastal systems and their dynamics to be mutually linked by negative and positive feedback. The presence of positive feedback between coastal form and process makes it difficult to predict coastal evolution, especially over longer timescales.

At the most basic level, coastal environments can be divided into clastic (comprising mud, sand and gravel) and rocky coasts. Clastic coastal environments are depositional and their morphology responds to the relative dominance of wave, tidal and fluvial factors. Barriers are the basic depositional elements of wave-dominated coasts and comprise the underwater shoreface and the above-water beach and dunes. They can be swash-aligned or drift-aligned, and respond strongly to changes in the sediment supply and sea level. Estuaries are the dominant tide-dominated coastal landform and represent zones of mixing between fluvial and marine (wave and tidal) processes. On a geological timescale, estuaries are rather short-lived features because they are characterized by relatively rapid infilling by fluvial and marine sediments. Deltas are accumulations of sediment deposited where rivers enter into the sea. Waves and tides significantly affect river outflow processes and play a significant role in determining the delta morphology. Rocky coasts are eroding coasts and the main factor controlling the erosion rate is the strength of the rocks relative to that of the eroding waves. The two dominant rocky coast landforms are (co-evolving) cliffs and shore platforms.

Effective management of the coast requires an integrated approach that considers all coastal users and stakeholders, at the same taking a long-term view to come up with sustainable solutions and policies. The term 'integrated coastal zone management' (ICZM) is used to indicate this approach. An important remit of ICZM is to address problems associated with coastal erosion and flooding, and there are two fundamentally different types of management approaches available. On the one hand, there are strategies that allow some loss of coastal land, functions and values, but help maintain the dynamic nature of the coast. On the other hand, there are strategies that protect the coastline using hard and soft engineering techniques. Allowing coastal regions to retreat naturally increases the resilience of the coast to sea-level rise and is preferable. However, in many instances the coastline under threat is simply too valuable to be sacrificed and coastal protection must be sought.

Further reading

Carter, R.W.G. (1988) *Coastal environments.* Academic Press, London.
Despite being somewhat outdated, this remains a classic text with comprehensive treatment of coastal environments and their management.

Dyer, K.R. (1998) *Estuaries: A physical introduction.* John Wiley & Sons, Chichester.
This text provides an in-depth, but accessible, introduction to tidal dynamics, stratification and mixing in estuaries.

French, P.W. (2001) *Coastal defences – Processes, problems and solutions.* Routledge, London.
A reasonably up-to-date and comprehensive overview of the different types of coastal defences.

Masselink, G. and Hughes, M.G. (2003) *Introduction to coastal processes and geomorphology.* Edward Arnold, London.
This text is co-written by the author of this chapter and provides an excellent continuation for further studies into coastal processes and morphology.

Pilkey, O.H. (2003) *A celebration of the world's barrier islands.* Columbia University Press, New York.
This book may be difficult to get hold of, but is a true gem for anyone interested in barrier islands. This is the only coastal coffee table book available, owing to the inclusion of many original batiks by the American artist Mary Edna Fraser, written at an advanced academic level.

Short, A.D. (ed.) (1999) *Handbook of beach and shoreface morphodynamics.* John Wiley & Sons, Chichester.
This is an advanced text with reviews on shoreface, beach, dune and barrier morphodynamics.

Trenhaile, A.S. (1987) *The geomorphology of rock coasts.* Oxford University Press, Oxford.
A must for anyone interested in learning more about rocky coasts.

Viles, H. and Spencer, T. (1995) *Coastal problems.* Arnold, London.
This is an excellent text dealing specifically with environmental problems in the coastal environment.

Woodroffe, C.D. (2003) *Coasts, form, process and evolution.* Cambridge University Press, Cambridge.
This is an advanced-level text on coastal dynamics largely written from a geological perspective.

Web resources

ARGUS Data Collection
http://argus-data.wldelft.nl/index.html
This material is relevant to Box 17.5.

Google Map Search Engine
http://maps.google.com/
A great tool for looking at coastal morphology – anywhere in the world.

Intergovernmental Panel on Climate Change (IPCC)
http://www.ipcc.ch/
Up-to-date information on climate change and sea-level rise. The impact of these on coastal environments and communities can be downloaded from this site.

Making Space for Water
http://www.defra.gov.uk/environ/fcd/policy/strategy.htm
A new government strategy for flood and coastal erosion risk management proposed by the UK's Department for Environment, Food and Rural Affairs (DEFRA).

NetCoast: A Guide to Integrated Coastal Zone Management
http://www.netcoast.nl/
This site provides the latest relevant information, knowledge, documents, publications and links to other sites related to coastal management. There is a large list of links to websites of coastal zone management programmes throughout the world.

OzEstuaries: Information about Australia's Estuaries and Coasts
http://www.ozcoasts.org.au/
This site provides comprehensive information about Australian estuaries and coastal waterways. It includes in-depth descriptions of the different types of coastal morphologies, and satellite images of all Australian estuaries and deltas can be viewed.

Permanent Service for Mean Sea Level (PSMSL)
http://www.pol.ac.uk/psmsl/
This website is hosted by the Proudman Oceanographic Laboratory (POL) and is the global data bank for long-term sea-level change information from tide gauges. Sea-level data from hundreds of locations around the world can be downloaded for free.

US Geological Survey: Coasts in Crisis
http://pubs.usgs.gov/circ/c1075/
This is a detailed online booklet covering different types of coasts (with pictures), natural and human-induced coastal change and coastal management.

US National Oceanic and Atmospheric Administration (NOAA): National Ocean Service
http://www.oceanservice.noaa.gov/
A very comprehensive website containing a wealth of information on coastal ecosystem science, coastal management and oceanography.

Visit our website at **www.pearsoned.co.uk/holden** for further questions, annotated weblinks, interactive models and video-clips showing physical processes in action.

CHAPTER 18

Glaciers and ice sheets

Tavi Murray

Department of Geography, Swansea University

Learning objectives

After reading this chapter you should be able to:

➤ describe different types of ice masses and their characteristics

➤ explain the concepts of glacier mass balance

➤ explain how glaciers move downslope, and understand glacier thermal regimes and their effect on glacier dynamics and glacier water systems

➤ describe how water moves through a glacier and understand why the water system is important in glacier dynamics

➤ explain how glaciers and ice sheets are reacting to climate change

➤ explain how glaciers erode their beds and modify sediments, and understand the differences in the sediment deposited via a range of mechanisms

➤ describe typical landscapes of glacial erosion and deposition

18.1 Introduction

Over the past 2.4 million years there has been repeated expansion and contraction of glaciers and continental-scale ice sheets in the middle latitudes. Today glaciers cover about 10% of the Earth's land surface. These ice masses store large volumes of water on land, thereby reducing the amount in the oceans. During phases of maximum ice advance, global sea levels fell by 130 m. The amount of water stored in glacial ice today is equivalent to approximately 75 m of global sea-level rise. Most of the water is stored in the great ice sheets of Antarctica and Greenland. For example, the West Antarctic and Greenland ice sheets individually contain the equivalent of about 5 and 7 m of sea-level rise. During the twentieth century global sea level rose by 12–22 cm, and the rate of sea-level rise accelerated: it is currently approximately 0.31 cm per year (IPCC, 2007b). Much of the sea-level rise is due to simple expansion of the water contained in the seas and oceans as it has warmed, but about 40% is attributed to the melting of glaciers and ice sheets.

Glaciers can be used as valuable resources. For example, glacial runoff is used to generate electricity in a number of countries such as Norway, Canada and Switzerland and in many regions it is an important water source for agriculture. In the Karakoram Himalayas, glacial meltwater makes a major contribution to the Indus and Yarkand Rivers and hence to the livelihood of 130 million people (Hewitt, 1998). An advantage of glacially fed water systems is that the water supply actually increases during hot summers, which are generally times of increased need. In addition skiing is a popular pastime, and in many areas this occurs on the surface of glaciers. The engineering of ski lifts and safe operation of these ski areas requires knowledge of glacier dynamics. Increasingly oil, gas and mineral extraction is taking place in glaciated environments. Furthermore icebergs are calved from glaciers, meaning that world shipping is affected by changes in ice dynamics at major outlet glaciers.

Changes in glacial extent have also affected the Earth's climate, for example by altering the Earth's albedo, and glaciers have made major modifications to much of the Earth's landscape by eroding, transporting and depositing sediment. These processes have helped shape the landscape over much of the middle latitudes in places where today there are no glaciers or ice sheets. Where they do occur, these landscape change processes operate at a variety of rates and over a range of scales. This chapter discusses glacial systems, glacial processes and the effects of glaciation on the landscape.

18.2 Glaciology

The aim of this section is to develop an understanding of glaciers and ice sheets, the controls on glacier extent and the effect of climate change on them, the evacuation of water through glacier hydrological systems and the processes and rates of ice flow.

18.2.1 Types of ice mass

There are a wide variety of ice masses including ice sheets, ice caps and valley glaciers. The largest ice masses are known as ice sheets, which cover continental-size regions

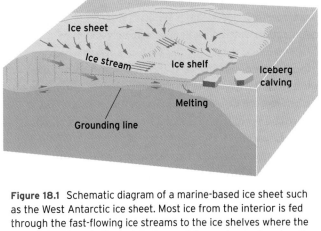

Figure 18.1 Schematic diagram of a marine-based ice sheet such as the West Antarctic ice sheet. Most ice from the interior is fed through the fast-flowing ice streams to the ice shelves where the mass is lost largely through iceberg calving and some basal melt beneath the ice shelves. (Source: after Alley, 1991)

(Figures 18.1 and 18.2). Typically an ice sheet is 1–3 km thick. The Antarctic ice sheets cover approximately 11.97×10^6 km² and the Greenland ice sheet covers around 1.74×10^6 km². The bases of the Greenland and East Antarctic ice sheets are above sea level and these are known as terrestrial ice sheets, whereas a marine-based ice sheet has its base below sea level such as the West Antarctic ice sheet (Figure 18.1).

Ice sheets flow relatively slowly and most of the mass reaches the ice sheet margins through smaller, fast-flowing **ice streams** or outlet glaciers as shown in Figure 18.1. An ice stream is a fast-flowing 'river' of ice within more slowly moving ice sheet walls whereas an outlet glacier is

Figure 18.2 Antarctic and Greenland ice sheet balance velocities. The balance velocity of an ice sheet is the velocity required to evacuate the snow accumulated and maintain the current geometry. Calculating the balance velocity requires knowledge of the surface accumulation and surface slope as well as the bed geometry. (Source: courtesy of Jon Bamber/Adrian Luckman)

a fast-moving section of ice surrounded by rock. An example of an ice stream is Whillans ice stream, West Antarctica, which flows at approximately 850 m yr^{-1}. In Antarctica, the ice streams feed **ice shelves** which are formations of floating ice as shown in Figure 18.1. Ice is lost from the ice shelves by melting into the ocean below or when **iceberg calving** occurs (Figure 18.1). This is when large chunks of ice are released from the ice shelf and float into the ocean.

Ice caps are smaller than ice sheets, but still cover considerable areas. For example, Vatnajökull, situated in south-east Iceland, covers some 8100 km^2. Many mountain regions are glaciated with valley glaciers, which are much smaller in area. Different regions and different types of ice mass have characteristic flow rates, thermal and hydrological systems.

18.2.2 Where do glaciers occur?

Glaciers are formed wherever the snow that falls in winter does not melt over the subsequent summer. This suggests that there are two fundamental requirements for the formation of glacier ice: precipitation in the form of snow (the percentage of precipitation in an area falling as snow is known as the **nivometric coefficient**) and low temperatures. Neither is sufficient on its own. For example, 23 000 years ago the majority of Canada and the northern United States was glaciated (see Chapter 20), but Alaska remained largely free of glacier ice because the precipitation in this area was very low (a cold arid region). Further factors that control the distribution of snow and ice include latitude, altitude, relief, aspect and continentality or the distance from the ocean. Glaciers differ greatly in locations where different factors are dominant. For example, the ice masses of Antarctica are characterized by very low surface temperatures and very low precipitation. These are examples of polar glaciers, which occur at high latitudes. The glaciers of the European Alps occur in a relatively warm climate characterized by very high precipitation rates. These are examples of alpine glaciers, which exist because of their high altitude. Table 18.1 gives a summary of the current glacial coverage.

18.2.3 Glacier mass balance

Examination of glaciers in late summer generally shows that the **snowline** (where fresh snow still lies) has retreated up the glacier and that the lower region of the glacier is bare ice. An example can be seen in Figure 18.3. This lower region is known as the **ablation zone** and despite its becoming snow

Table 18.1 Present-day glacial coverage. There are glaciers on every continent on Earth. Total world coverage is approximately 14 953 945 km^2

	Area	
	km^2	%
South polar region	12 588 000	84.2
North polar region	2 137 274	14.2
North America	76 880	0.5
South America	26 500	0.2
Europe	7 410	< 0.1
Asia	116 854	0.8
Africa	12	< 0.1
Pacific/Australasia	1 015	< 0.1

(Source: after Sugden and John, 1976, updated from Williams and Ferrigno, 1999)

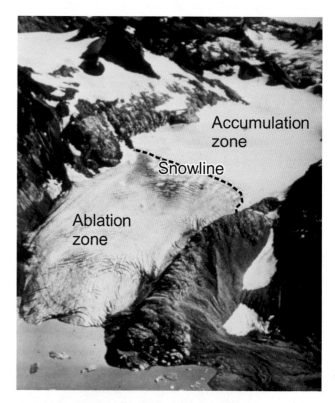

Figure 18.3 Photograph of a typical alpine glacier at the end of summer. The snowline has retreated up the glacier leaving exposed ice close to the glacier margin. The lower part of the glacier, the ablation zone, is melting, and has lost mass over the year. Mass is also lost from this glacier by iceberg calving into a lake. The upper part of the glacier, the accumulation zone, has retained snow and has gained mass over the year. (Source: after Post and LaChapelle, 2000)

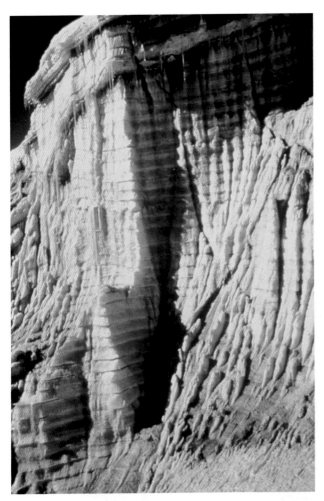

Figure 18.4 Ice cliff (~55 m high) showing annual layers of snow and ice in the accumulation zone of the Quelccaya ice cap in Peru. The layers are approximately 0.75 m thick. Each layer is demarked by a dirty layer of snow from the dry season. Typically the winter layers of ice are light in colour while summer layers are darker because the partial melting and lower summer accumulation rate produces higher concentrations of impurities. The age of each layer can be calculated by counting from the glacier surface. By identifying such layers in the accumulation zone the mass input to the glacier can be estimated, and such layers are used to date ice cores. (Source: photo courtesy of Lonnie G. Thompson, The Ohio State University)

covered in winter it experiences net mass loss during the year (see also Figure 18.8 below). Ice may be lost by melting at the glacier surface and by iceberg calving. Most ice loss in Antarctica occurs through iceberg calving. The region of the glacier that experiences net mass gain through the year is known as the **accumulation zone** (Figures 18.3 and 18.4). In this zone some or all of the snow that falls in winter survives the summer season to form **firn**, which is wetted and compacted snow more than a year old. Slowly this firn becomes denser and forms ice when all the connecting air passages are closed

off. Often annual layers of ice can be detected as a result of the differences in ice formation during the winter snow season and summer dry season as shown in Figure 18.4.

The division between the accumulation and ablation zones occurs at the **equilibrium line** (see Figure 18.8 below). The mass of water that is added to or lost from a glacier over a year is known as the **mass balance**. A glacier is said to be in positive mass balance if it is gaining in mass over a year and in negative mass balance if it is losing mass. For a typical alpine glacier, mass balance is primarily a function of winter precipitation and temperature, and summer solar radiation and temperature.

When calculating mass balance it is better to use the mass of water gained or lost rather than the volume of change because of the extreme differences between the densities of snow, firn and ice (Table 18.2). Despite this, several techniques of estimating the mass balance actually measure the volume balance, because the mass change of a glacier is very difficult to measure (see Box 18.1). In many regions, such as Svalbard in the Norwegian High Arctic, small mountain glaciers are losing mass to the ocean. For two glaciers at Svalbard there has been a net balance of −0.25 m of water equivalent lost per year since the 1960s (Dyurgerov and Meier, 2000). Over 40 years this represents a mass loss of 10 m of ice. The values are given as depths because the volume loss is averaged over the glacier's area. In other words, the glaciers are losing 0.25 m depth of ice per year averaged over their whole area.

The accumulation area of a glacier can be broken down into a number of zones although not all of these zones are

Table 18.2 Typical densities of snow and ice. Snow is largely unaltered since deposition. Firn is wetted snow that has survived at least one summer. Firn becomes ice when there are no longer any interconnecting air passages

Material	Density (kg m^{-3})
New snow	50-70
Damp or settled snow	100-300
Depth hoar	100-300
Wind-packed snow	350-400
Firn	400-830
Very wet snow or firn	700-800
Glacier ice	830-910
Water	1000

(Source: from Paterson, 1994)

NEW DIRECTIONS

WEIGHING THE ICE SHEETS FROM SPACE

Measuring the mass balance of the major ice sheets is a difficult task, mainly because of their large size (Figure 18.5). The Gravity Recovery and Climate Experiment (GRACE),

funded by NASA and the German Aerospace Center, is an exciting new method for monitoring ice sheet mass balance from space. GRACE consists of two satellites orbiting the Earth (Figure 18.5a), which are separated by a distance of around 220 km. The distance between them varies

slightly as the satellites pass over anomalies in the Earth's gravity field (Figure 18.5a). The gravity field allows calculation of changes in ice sheet mass (Figure 18.5b) once correction for glacial isostatic rebound and other factors is made.

(a)

(b)

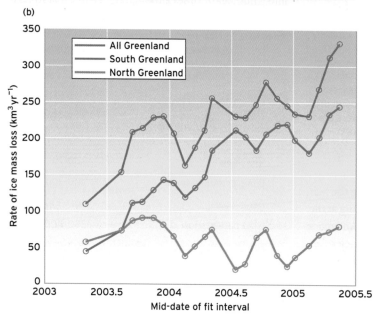

Figure 18.5 (a) Schematic diagram of the GRACE (Gravity Recovery and Climate Experiment) satellites. The two satellites orbit the Earth and variations in the gravity field cause small changes in their distance apart. This allows the production of a gravity field anomaly map shown on the globe. (b) GRACE measurements of the mass of the Greenland ice sheet. (Source: (a) University of Texas Center for Space Research/NASA; (b) Velicogna and Wahr, 2006, reprinted by permission from MacMillan Publishers Ltd.: NATURE, Isabella Velicogna and John Wahr, Acceleration of Greenland ice mass loss in spring 2004, vol. 443 (7109): 329-331: Copyright 2006

BOX 18.1

present on all glaciers. The **dry-snow zone** is an area where there is no surface melt, even in summer. Most alpine glaciers do not have a dry-snow zone as some melting occurs, but much of Antarctica and the central region of the Greenland ice sheet are characterized by no melting. The region where the entire snowpack is saturated at the end of the summer is known as the **wet-snow zone**. Water that refreezes at the base of the snowpack is known as **superimposed ice**. Because this ice has formed by refreezing rather than compression, it has slightly different chemical and physical properties than glacier ice. Superimposed ice formation is an important component of glacier mass balance in High Arctic glaciers such as those in Svalbard and the Canadian Arctic.

18.2.4 Transformation of snow into ice

Glacier ice is formed from compacted snow. The processes that result in this transformation depend on whether there is water present in the snowpack. Changes are slow in the absence of water and result from packing changes and settling, changes in the ice crystal size and shape resulting from sublimation, and deformation of the crystals. The presence of water results in much faster changes, because melting, percolation and refreezing occur in the snowpack. The transformation of snow to ice may take only a few years where the snowpack becomes saturated (e.g. three to five years at the Upper Seward Glacier, Yukon Territory), but more than 100 years in Antarctica or Greenland. See Paterson (1994) for further details of the transformation processes.

18.2.5 Glacier thermal regime

There is an important distinction between cold and warm ice. The melting point of ice reduces with increasing pressure. At atmospheric pressure the melting point is 0°C, whereas beneath 1 km of ice the melting point is about −0.7°C. Therefore, the greater the thickness of an ice mass, the more likely it is to be at the pressure melting point. Warm ice is at the pressure melting point and contains water; cold ice is at temperatures below the pressure melting point and does not. The 'thermal regime' of a glacier exerts a fundamental control over the water system and over the range of processes that can operate at the bed of the glacier. Furthermore, the properties of warm and cold ice vary strongly. For example, the deformation rate of ice increases by about five times between −25 and −10°C. Many other properties of the ice are temperature or water-content dependent, such as the speed and attenuation of acoustic or electrical waves through the ice. This allows us to detect

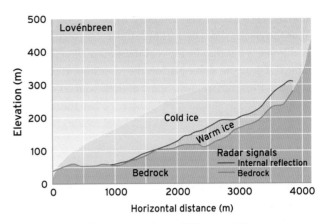

Figure 18.6 The internal structure of Midre Lovénbreen, a glacier in Svalbard, based on radar. Warm ice near the glacier bed contains water and scatters the radar producing a reflection of the radar waves (indicated as an internal reflection on the figure). Note that in this glacier the downslope end is made of cold ice throughout. Because this glacier has both warm and cold ice, it is called a polythermal glacier. (Source: after Bjørnsson *et al.*, 1996)

the thermal structure within glaciers using radar (see Chapter 23). An example of the thermal structure of a glacier is shown in Figure 18.6.

Generally three types of thermal regime are recognized. The first is a **temperate glacier**, which consists of warm ice throughout except for a layer, between 10 and 15 m in thickness, that is seasonally warmed and cooled by temperature variations at the surface (this layer is analogous to the active layer in permafrost, see Chapter 19). Typical examples of temperate glaciers are those in the European Alps, the south island of New Zealand and the Canadian Rockies. The second thermal regime is a **cold glacier**, which consists entirely of cold ice. A typical example might be Meserve Glacier in Antarctica. The final type is a **polythermal glacier**, which comprises both warm and cold ice; an example is given in Figure 18.6.

It is the temperature of the bed that is critical. If the bed of a glacier is cold then there will not be a basal water system and the occurrence of sliding and deformation of basal sediments will be greatly reduced. **Geothermal heat** (from inside the Earth) and heat generated by basal friction may warm the bed of a glacier, producing an active basal water system even in regions where the surface temperature is very cold such as beneath the Whillans ice stream, West Antarctica.

18.2.6 Glacier water systems

Water is produced at the surface of many glaciers in summer from the melt of both snow and ice. Small amounts of water are also produced by geothermal heating or as the glacier slides over its bed or deforms sediments. Additional water may be input at the surface directly due to rainfall. For many alpine glaciers the most important water source is from surface melt. Some of these sources are seasonal (such as surface melt), whereas other sources will persist throughout the year (such as basal melting).

Unless it refreezes, **supraglacial** (surface) meltwater will percolate downwards through any snow or firn and will flow at the surface of the glacier ice. The rate of water flow through snow or firn is fairly slow. Once the ice surface is reached, water will flow downslope along the ice surface and will emerge at the snowline as a series of supraglacial channels. These channels (Figure 18.7a) transport water

(a)

(b)

Figure 18.7 (a) Deeply incised supraglacial stream at the surface of a Svalbard glacier, Norway. (b) A moulin on an Icelandic glacier. Supraglacial water flows into the body of the glacier down moulins that lead to englacial channels. Eventually the water may reach the bed. Arrow indicates two people for scale. (c) The water outlet at many alpine glaciers is a single or a small number of channels. This suggests that the lower region of this type of glacier is often drained by a channel-ized drainage system. (Source: photo (c) courtesy of Mike Crabtree)

(c)

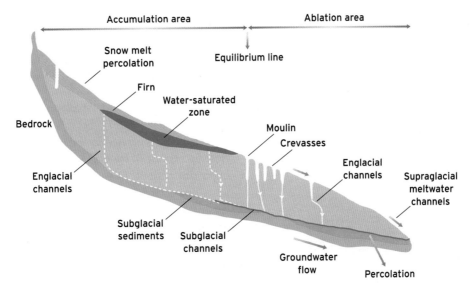

Figure 18.8 Schematic showing a typical temperate glacier hydrological system.

efficiently owing to their smooth ice walls. The water may continue to be routed to the glacier front across the glacier surface, which is common for High Arctic glaciers, or may enter the glacier to flow **englacially** (within the glacier) (Figure 18.8).

Although a small amount of water may percolate directly through the ice in small veins along the boundaries of individual ice crystals, the rate of this percolation is negligible. Most water flowing through a glacier does so through a system of englacial channels as shown in Figure 18.8. Water typically enters the glacier through **moulins** (Figure 18.7b) that tend to form where crevasses (fracture cracks in the ice) cut the drainage path of supraglacial water. Moulins usually descend rapidly to depth as semi-vertical shafts linked by short horizontal channels.

Englacial channels may intersect the glacier bed and flow along the floor of the glacier. Direct studies beneath glaciers are difficult and so ideas about the configuration of **subglacial** (beneath the glacier) water systems are derived from largely theoretical considerations. Four main morphologies have been proposed, namely: (i) flow in a thin sheet; (ii) flow through a network of channels; (iii) flow through linked cavities; or (iv) flow through a braided system of 'canals' at the glacier bed (Figure 18.9). Study of subglacial drainage is an important branch of glaciology, because glacial meltwater is used to generate electricity and for agriculture, and because the water system has a profound effect on glacier dynamics.

18.2.6.1 Sheet flow

One evacuation route for basal water is thought to be through a thin film or sheet, just a micrometre (1000th of a millimetre) to a millimetre in thickness (Figure 18.9a). The thickness of the sheet is thought to vary, being thinnest over bumps in the bed because of the higher pressure at their upstream side. Thicker films cannot develop because enhanced melting would result in the formation of subglacial channels. Flow through sheet flow is typified by very high water pressures and research on bedrock at Blackfoot Glacier, USA, showed that sheet flow occurred over about 80% of the glacier bed (Walder and Hallet, 1979). The role of sheet flow is thought to be small for most alpine glaciers, perhaps transporting water into a channelized system. Sheet flow is thought to be dominant only beneath ice masses where the majority of basal water is derived from subglacial melting due to geothermal heating, sliding or bed deformation rather than surface meltwater.

18.2.6.2 Channelized flow

Water that descends from the glacier surface in englacial conduits will arrive at the bed at discrete locations and is unlikely to become dispersed into a thin sheet. This water will tend to flow at the bed in conduits or channels (Figure 18.9b). Evidence for channelized systems comes from the rapid rate of transfer of dye from moulins to the outlet stream at some glaciers and from the

Figure 18.9 Conceptual forms of the basal water system. Water may flow at the bed of a glacier (a) through a sheet of water; (b) through a channel system eroded either down into the bed (Nye channels) or up into the ice (Röthlisberger channels); (c) through a tortuous series of linked cavities; or (d) through canals, which are thought to be braided in their planform.

observation that many alpine glaciers have one or a few major outlet channels emerging at their margins (Figure 18.7c). Glacier outburst floods from subglacial sources (see Box 18.2) occur through efficient channelized water systems.

Two types of channels are thought to exist beneath glaciers, namely Nye (N) channels and Röthlisberger (R) channels (Figure 18.9b). N-channels are incised into the bedrock, and R-channels are melted upwards into the ice. N-channels are fixed in space and can be preserved on exposed bedrock surfaces. However, they may be destroyed

by subglacial erosion due to ice sliding and may be slow to form because water erodes rock inefficiently. When they are seen they tend to form short channels approximately parallel to ice flow, rather than exposing entire water systems.

R-channels are melted upwards into the glacier ice, which can be much more rapidly thermally eroded than the bedrock. R-channels form because the ice melts owing to frictional heat generated by water flow. These channels can close up because of **ice creep**, especially in winter. An important attribute of R-channels is that an increase in water flux decreases the water pressure in steady-state conditions. This inverse relationship between pressure and flux suggests that large R-channels will 'capture' small R-channels. Thus, water in R-channels will be concentrated in a small number of large channels and hence there is a tendency to form a tree-shaped channel network (Figure 18.9b).

Variations in surface melt drive variations in basal water pressure on both diurnal and longer timescales (Figure 18.10). Diurnally, water pressures are typically highest in the afternoon and lowest in the early morning as a result of melt cycles. In particular, abundant surface melt is routed to the glacier bed during spring, resulting in high water pressures until the water system evolves sufficiently to evacuate the water. Conversely, the beginning of winter is characterized by low basal water pressures. These characteristics result in an increase of sliding rate in spring and early summer at most alpine glaciers and lowest sliding rates during the winter.

18.2.6.3 Linked cavities

An alternative model of water flow across a hard bed is through linked cavities (Figure 18.9c), formed downstream of bedrock bumps as the glacier slides over its bed. These consist

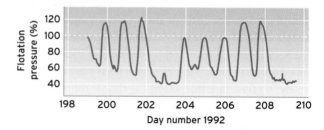

Figure 18.10 Water pressure variations measured by a basal water sensor during summer at Trapridge Glacier. Figure shows the percentage of the flotation pressure (which is the pressure at which the ice will float on its own water). Day 198 is 17 July. Note the diurnal pressure peaks corresponding to high-water input during the late afternoon.

of a tortuous system of basal cavities, typically less than 1 m high and from less than 1 m to 10 m in length, linked by narrow connections less than 0.1 m in height. While the cavities have a large volume, water flow through the system is at high pressure and is very slow, so thermal erosion is inefficient. This means that the water system cannot increase its discharge rapidly in response to increased input.

18.2.6.4 Canals

The models of a glacier hydrological system described so far have assumed that there is hard and impermeable bedrock beneath the glacier. However, as most glaciers retreat they expose extensive glacial sediments, suggesting that many glaciers may be underlain by soft and potentially deformable beds (e.g. Figure 18.12). The wide coverage of glacial till in the middle latitudes suggests that the beds of many major ice sheets in the past were soft. Furthermore, soft beds have been identified beneath a number of contemporary glaciers, including Breiðamerkerjökull, south-east Iceland, Storglaciären, Sweden, and Trapridge Glacier, Yukon Territory.

Sediments are not just deformable, many also have greater permeability than rocks. This means that water can drain through the bed. However, for most ice masses the sediment is too fine to drain the volume of water produced,

HAZARDS

GLACIER LAKE OUTBURST FLOODS

Lakes are common at the margins of retreating temperate glaciers. Lakes can be dammed by glacier ice, by glacial moraines, or the water can be created underneath the glacier ice by a volcanic eruption or geothermal heating. These lakes tend to drain rapidly and cause a glacier lake outburst flood. A typical glacial lake outburst flood has a total discharge of water and debris up to 50 million m³ and a floodwave that may be up to 10 m high. These floods may cause widespread damage and destruction downstream, in some cases for hundreds of kilometres.

Lakes dammed by glacier ice can form when a glacier surges (see Section 18.2.7.5) or advances. One such flood occurred when Hubbard Glacier advanced and blocked Russell Fjord in Alaska in 2002. This formed Russell Lake, which rose 24 m over the summer before the moraine dam finally

failed in August 2002, causing the second largest glacial lake outburst flood in historical times (Figure 18.11a).

Many rapidly retreating mountain glaciers are forming lakes behind moraine dams, threatening villages downstream. An example of a glacier lake outburst flood occurred in Bhutan in the Himalayas where there are more than 2000 glacial lakes, many of which are increasing in volume as glaciers melt. In 1994, the Lugge Lake in northern Bhutan drained, releasing 18 million m³ of water. The floods killed 17 villagers and affected 91 households downstream.

Glacial outburst floods from a subglacial lake occur regularly from beneath Vatnajökull in south-east Iceland. Vatnajökull is a temperate ice cap of about 8100 km², and the volcanic fissure system of the Mid-Atlantic Ridge (see Chapters 2 and 3) lies beneath the western portion of the ice cap where it causes substantial subglacial melt. One of the largest floods from

Vatnajökull occurred in 1996, after an earthquake of magnitude 5 on the Richter scale. Subsidence bowls 100 m deep and cracks in the glacier surface showed that extensive melting was occurring at the base of the glacier along a fissure 5–6 km long and an eruption cloud emerged a few days after that extending to a height of 3 km (Figure 18.11b). Meltwater from the eruption flowed into the 10 km diameter Grímsvotn subglacial lake at a rate of 5000 m³ s⁻¹.

The water stored in Grímsvotn produced an outburst flood, with a 4 m high wave crossing the floodplain. During the course of the flood two bridges were destroyed or badly damaged as were phone and power lines. Large chunks of ice were broken from the glacier margin and carried across the floodplain. The water peak flow was estimated to be 45000 m³ s⁻¹ and the estimated damage was $15 million to roads, bridges and other infrastructure.

BOX 18.2 ➤

(a)

(b)

Figure 18.11 Outburst floods: (a) Outburst flood from Russell Lake, formed by the advance of the Hubbard Glacier to block the Russell Fjord; (b) Vatnajökull October 1996, where a fissure and subsidence bowl formed at the glacier surface followed by an eruption cloud 3 km high. (Source: (a) USGS; (b) Mats Wibe Lund/epa/Corbis

BOX 18.2

of shallow, wide channels distributed more or less evenly across the bed and connected in a braided pattern. Canals are thought to be most likely to form where low-slope glaciers overlie soft sediment.

18.2.7 Glacier dynamics

In order to transfer ice formed from snow that accumulates in the upper reaches of all glaciers to the ablation zone it must flow downslope. To maintain constant glacier geometry, the mass transferred down the glacier should equal the mass lost by melting or calving. The **balance velocity** is the velocity required to maintain the ice mass in equilibrium (Figure 18.2). Those glaciers where the actual and balance velocities are similar are said to be in balance with the current climate, and are unlikely to experience major changes in flow rate unless the climate changes.

There are three mechanisms by which glaciers flow. These are internal deformation or creep, sliding and bed deformation. The driving force for glacial flow is gravity, which is resisted by frictional forces at the sides and bed of the glacier. The gravitational driving force exerts stress on the ice, which can be divided into two components, the normal stress and the shear stress. Both increase with depth in the glacier (see Box 18.3).

18.2.7.1 Internal deformation of ice

Stress applied to ice causes it to deform, which is often known as creep (see Box 18.3). The rate of deformation depends on a wide variety of factors including temperature, crystal orientation and ice impurities. Creep is the dominant flow mechanism for cold glaciers, although the creep rate is much lower for cold ice than warm ice. Once the pressure melting point is reached at the glacier bed the processes of sliding and deformation of basal sediments can also occur.

18.2.7.2 Sliding at the glacier bed

If the bed of a glacier is cold, there is a strong bond between ice and bed that tends to prevent sliding. If, however, the bed is at the pressure melting point, this bond is not present and the presence of liquid water reduces friction. As a result the glacier can slide over its bed. Sliding is the process by which many of the geomorphic features we associate with glaciers are formed, and is the most efficient process for glacial erosion. The rate of sliding is controlled by the drag at the bed, which results from two factors. These are the bed roughness, which results in **form drag**,

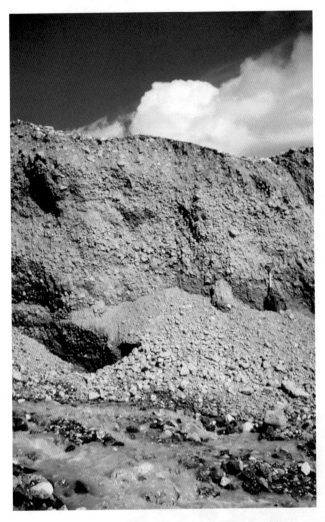

Figure 18.12 Extensive subglacially eroded sediments at the margin of Trapridge Glacier, Yukon Territory, Canada. These sediments used to lie beneath the ice. The presence of such sediments is thought potentially to affect both the glacier dynamics and the morphology of the water system.

in which case some form of additional drainage system must develop. Two drainage systems are postulated to form over soft sediments. The first is essentially the same as the R-channel system that forms over a hard bed. The second system is analogous to N-channels over bedrock; however, in this case the channels are eroded downwards into the basal sediments (Figure 18.9d). These channels are wide and shallow and are known as 'canals'. Sediment deformation will tend to close canals, which must be balanced by erosion of the bed if channels are to remain open. Unlike in R-channels, water within a large canal is at a higher pressure than in a small canal. There is therefore no tendency for flow to concentrate in a few large channels. As a result a canal drainage system is thought to consist

FUNDAMENTAL PRINCIPLES

STRESS, STRAIN AND ICE DEFORMATION

Stress is defined as a force acting per unit area and has units of pascals (Pa). For a glacier, the normal stress, σ (which is the stress exerted at right angles, or 'normal', to the slope), can be calculated from:

$$\sigma = \rho g h \qquad (18.1)$$

where ρ is the density of ice (Table 18.2), g is the acceleration due to gravity (9.8 m s^{-2}) and h is the ice thickness. For ice 100 m thick, the normal stress is about 880 kPa. The theoretical basal shear stress (the stress exerted at an angle parallel to the slope), τ, beneath an infinite parallel slab glacier is calculated from the bed slope (equal to the surface slope), α, and its thickness using:

$$\tau = \rho g h \, \sin \alpha \qquad (18.2)$$

For a typical glacier 100 m thick and with surface slope 4°, the shear stress will be about 57 kPa. Equation (18.2) allows us to predict certain geometric characteristics of a theoretical glacier. If we assume that the ice is perfectly plastic (does not deform until its **yield strength** but then has infinite **strain rate**) then we can set the basal shear stress in equation (18.2) to the yield strength of ice. This then tells us that the thickness is controlled by the

angle α, such that if α is large (the glacier is steep), then it will be thin. Conversely, if α is small and the glacier has a shallow surface and bed slope then the glacier will be thicker.

The stress applied to glacier ice causes it to deform, and the rate of deformation can be calculated from the flow law of ice (known as **Glen's law**). This flow law was determined in the laboratory and relates the strain rate, $\dot{\varepsilon}$, to the basal shear stress, τ:

$$\dot{\varepsilon} = A\tau^{n} \qquad (18.3)$$

The creep rate of ice is sensitively dependent on the temperature of the ice, and the value of A is about 1000 times greater at 0°C than it is at −50°C. The value of n determines how non-linear the response of ice is to an applied stress and is typically taken to be 3.

Glacier ice differs from the laboratory ice on which this result is based (Figure 18.13). Glacier ice is **polycrystalline** and the orientation of the crystals will vary. Glacier ice is inhomogeneous and can also contain water, air bubbles trapped as the snow becomes ice, and both soluble and insoluble impurities. These can all change the properties of the ice as can the **strain history** of the ice. In general, polycrystalline ice deforms less readily than a single crystal as reorientation must occur.

Figure 18.13 Thin section through glacier ice viewed through cross-polarizing filters. Each grain appears a different colour because of their different alignment. Also clearly seen are small air bubbles in the ice (seen as small round inclusions). These air samples provide a record of past atmospheric conditions. This thin section is from 333 m depth in the GISP2 core drilled at the summit of the Greenland ice sheet. (Source: photo courtesy of A. Gow)

BOX 18.3

and the rock–rock friction that results from the interaction between sediment that is lodged in the basal ice (Figure 18.14) and the bed, which is known as **frictional drag**. The morphology of the basal water system and the basal water pressure also strongly influence the rate of glacier sliding by decoupling the glacier from its bed. High basal water pressures enhance the rate of glacier sliding. Glacier beds are not smooth and basal sliding requires that ice is transferred around obstacles. Basal sliding occurs by two mechanisms under such rough conditions. These

are **regelation** and enhanced creep and are shown in Figure 18.15. If ice flows around obstacles at the bed, most of the shear stress is supported on the upstream side of the obstacles. This results in excess pressure upstream of the obstacle and lowered pressure on the downstream side. Increased pressure lowers the ice melting point upstream of the obstacle. The melted water then flows around the obstacle to the low-pressure downstream side where it refreezes because the melting point is higher. This mechanism therefore allows the ice to slide past the obstacle. The

Figure 18.15 Sliding of glacier ice over a rough bed by (a) regelation and (b) enhanced creep. Regelation occurs where increased pressure causes melting on the upstream side of the obstacle. Water flows around the obstacle and refreezes on the downstream side. Heat released by refreezing is conducted upstream through the obstacle. Enhanced creep results because of enhanced compressive and tensile stresses around the obstacle.

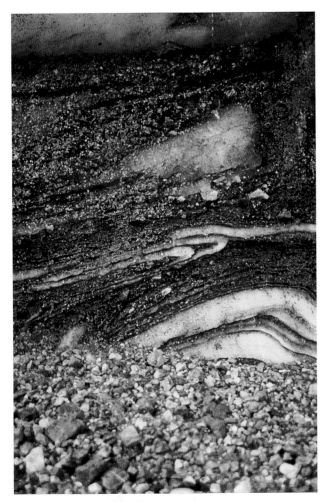

Figure 18.14 Laminated basal ice exposed at the margin of Trapridge Glacier, Yukon Territory. The sequence shown is ~60 cm in height. Ice flow was from right to left.

downstream refreezing releases latent heat (see Chapter 4). The pressure drop across the obstacle results in a downstream temperature rise, so that heat is conducted upstream through the obstacle where it assists in melting. This process of melting, transfer of water and refreezing is known as regelation (Figure 18.15a). Regelation is limited by the rate of heat conduction upstream, which is most efficient for small obstacles. The regelation process is important in the formation of basal ice and the entrainment of debris at the ice base (Figure 18.14).

The second mechanism of basal sliding is known as enhanced creep. The presence of the obstacle causes an increased compressive stress on the upstream side of the obstacle. Glen's flow law (see Box 18.3, equation (18.3)) shows that the creep rate is proportional to the third power of the stress. Hence increased stress greatly increases the local creep rate (Figure 18.15b). The magnitude of stress enhancement is thought to be related to the size of the obstacle so that the mechanism works most effectively for large obstacles.

Because regelation is most efficient for small obstacles and enhanced creep for large obstacles there is a critical obstacle size which provides the majority of the resistance to glacial flow. This critical obstacle size is thought to lie between 0.05 and 0.5 m in diameter. The total sliding velocity is equal to the sum of the velocity due to regelation and that due to enhanced creep.

18.2.7.3 Deformation of basal sediments

Sliding theories usually assume that the glacier overlies undeformable and impermeable bedrock. However, where an ice mass overlies soft sediments, deformation of these sediments can contribute to surface flow. At Whillans ice stream, West Antarctica, high-resolution seismic surveys undertaken in the mid-1980s showed a 5–6 m thick sediment layer that was laterally continuous for at least 8 km. This layer was thought to be deforming throughout its thickness and this helps explain why the ice stream flows so fast despite its low-slope angle (Alley *et al.*, 1986; Blankenship *et al.*, 1986). Furthermore, at Breiðamerkerjökull,

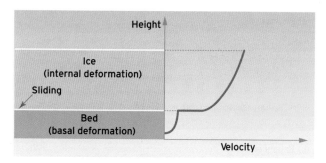

Figure 18.16 Mechanisms of glacial flow. The surface velocity of a glacier is made up from components due to internal deformation of the ice, sliding at the ice bed interface and deformation of subglacial sediments.

south-east Iceland, deformation of the bed was shown to cause 88% of the total surface motion (Boulton and Hindmarsh, 1987).

18.2.7.4 Relative importance of glacier flow mechanisms

Glacier motion is made up from contributions of ice deformation, sliding and sediment deformation (Figure 18.16). The relative importance of each at a particular site depends on the thermal regime, the availability and distribution of meltwater and the composition and morphology of the bed. At Trapridge Glacier, Yukon Territory, Canada, measurements have been performed using instruments installed through boreholes into the glacier bed designed to measure sliding and sediment deformation together with measurement of surface speed. The surface of Trapridge Glacier moves approximately 10 cm day^{-1}, of which sliding makes up about 4 cm day^{-1} and sediment deformation about 6 cm day^{-1}. The contribution at this site from internal ice deformation is less than 1 cm day^{-1} (Blake, 1992).

18.2.7.5 Glacier velocity, glacier surging and fast glacier flow

Glacier flow velocities vary considerably depending on the basal thermal regime, glacier mass balance, bed conditions and bed geometry (Clarke, 1987). Temperate valley glaciers typically flow at velocities of a few tens of metres per year, while cold-based continental glaciers may have maximum velocities of the order of 1–2 m yr^{-1}. The normal range of glacier velocities is considered to be 10–100 m yr^{-1}, typically associated with basal shear stresses of the order of

40–120 kPa. For example, White Glacier, in the Canadian Arctic, flows at about 30 m yr^{-1}. Often glaciers can be quite inaccessible and so it is difficult to determine glacier velocity in many locations. However, using satellite technology it may be possible to measure glacier flow and glacier topography. Box 18.4 provides information on one such technique. Even more inaccessible glaciers are found on other planets. However, it is still possible to examine glacier flow dynamics on other planets from space. An examination of polar ice on Mars is provided in Box 18.5.

A smaller class of fast-flowing glaciers exists, with velocities of the order of 100 to over 1000 m yr^{-1}, which includes ice streams, tidewater terminating glaciers and surging glaciers. Jakobshavn Isbræ, West Greenland, thought to be Earth's fastest continuously flowing ice mass, flows at nearly 7000 m yr^{-1}. Whillans ice stream, West Antarctica, flows at 827 m yr^{-1} and the Rutford Ice Stream, Antarctica, flows at 400 m yr^{-1}. Most fast-flowing glaciers are thought to do so by either fast sliding or fast sediment deformation. Several fast-flowing glaciers, such as Columbia Glacier, Alaska, and Whillans ice stream, West Antarctica, are discharging more ice than they accumulate. In contrast, Kamb ice stream, West Antarctica, which switched from a fast to slow flow mode approximately 250 years ago, flows at about 5 m yr^{-1}, and is currently gaining mass.

Surge-type glaciers switch between long periods of slow (or quiescent) flow and flow 10–1000 times faster (the surge). It is unclear whether the trigger that enables ice streams to switch between fast and slow flow is externally or internally controlled, but it appears that surging glaciers show internally driven switching, independent of external forcing. Most theories explain the change in velocity from slow to fast and vice versa as caused by a change from an efficient to an inefficient basal water system. Such a change can occur because the basal drainage system alters from a channelized to a linked cavity water system. If the glacier overlies a soft bed then the paths through sediments by which water is evacuated are thought to be destroyed by shear deformation, which traps pressurized water at the bed causing high water pressures and fast flow.

Glacier surging involves the release of ice stored in the upper portion of a glacier (reservoir zone), allowing it to flow down the glacier rapidly into the receiving zone. For example, Variegated Glacier, Alaska, flowed up to 65 m day^{-1} during its 1982–1983 surge (Kamb *et al.*, 1985). Glaciers surge repeatedly, with a periodic and relatively constant surge

TECHNIQUES

MEASURING GLACIER DYNAMICS FROM SPACE

Glaciers are often situated in remote and inaccessible regions, making remote sensing an attractive method for studying them. Satellite interferometry (see Chapter 23) uses two images taken at separate times and positions. Subtle differences in the phase of the back-scattered signal between the two images allow both the surface topography and velocity fields of ice masses to be measured from space (Goldstein *et al.*, 1993).

The limitations of the technique mean that coherence is lost if flow velocities exceed about 5 m day^{-1} between two images separated by one day or if surface characteristics change owing to precipitation or melting. However, the technique offers exciting prospects for monitoring inaccessible ice masses such as the Norwegian glaciers shown in Figure 18.17.

The satellites taking the two images from which this interferogram was formed were very close in space but separated by one day in time. This means that the fringes seen are mostly the effect of ice displacement rather than topography. Both of the glaciers have roughly the same surface slope so that the closely spaced fringes on Kronebreen show that it is flowing much faster than Kongsvegen. Steep slopes can also be picked out by more closely spaced fringes. The glaciers are discharging large amounts of sediment-laden basal water into the **fjord**.

(a) (b)

Figure 18.17 (a) Interferogram formed from two satellite images one day apart in December 1995. The Norwegian glaciers shown are Kronebreen (K), which is the fastest-flowing glacier in Svalbard with a flow speed of ~550–1500 m yr^{-1}, and Kongsvegen (V), which is a quiescent phase surge-type glacier (annual velocity ~3 m yr^{-1}). (Source: image courtesy of A. Luckman) The box shows area of enlargement shown in (b) which is an aerial photograph of these glaciers. The intense crevassing on Kronebreen is due to its fast flow. (Source: air photo (taken 1995), S95 1026 © Norwegian Polar Institute)

BOX 18.4

NEW DIRECTIONS

GLACIAL DYNAMICS ON MARS

The Martian polar regions are covered by extensive mantles of ice and dust that cover areas of about 1 million km^2 with ice masses up to 4 km thick such as that shown in Figure 18.18. These polar ice caps contain both carbon dioxide and water and might provide water required for a long-term human exploration of the planet (Byrne and Ingersoll, 2003; Titus *et al.*, 2002).

The most notable features of the ice caps are dark troughs that spiral outwards through generally white ice from the pole (Figure 18.18). It is thought that the white areas of the ice cap are probably accumulation areas and the dark areas, ablation zones. The ablation is thought to result from **sublimation** of the ice. Ice flow in these ice caps probably occurs from the higher white areas into the troughs by ice creep, albeit very slowly because of the extremely low temperatures (140-155 K or −133 to −118°C). It is suggested that the pattern of troughs and scarps migrates inwards from the edge of the ice cap towards the pole owing to the preferential sublimation from sunward-facing slopes. The different models of ice flow that explain the intriguing patterns will undoubtedly be tested by future space missions.

(a)

(b)

Figure 18.18 The Martian north polar ice cap: (a) the north polar ice cap (Source: NASA) and (b) topographic profile across the north polar ice cap. This ice cap is the larger of the two ice caps on Mars and is known to be at least partially comprised of water ice, although it is mainly frozen carbon dioxide. (Source: after Clifford *et al.*, 2000)

BOX 18.5

cycle between 20 and 500 years. The fast flow or active phase is short, typically 1–15 years. The surge often results in rapid advance of the front margin of the glacier as shown in Figure 18.19, sometimes by several kilometres in a few months or years. For example, the Hispar Glacier in the

Karakoram Himalayas is reported to have advanced '2 miles in 8 days' and the Hassanbad Glacier in the same region advanced '9.7 km in 2.5 months' (Hewitt, 1998). Strand lines are often left marking the former pre-surge ice surface in the reservoir zone (Figure 18.19c) and the ice surface typically

Figure 18.19 The effects of surging. (a) Aerial photograph showing the glacier Abrahamsenbreen, Svalbard, in 1969. The glacier surface is uncrevassed and tributary glaciers can be seen compressing the main trunk. (b) The same glacier in 1990 after a major surge. The glacier surface is highly crevassed and the moraine loops have been extended down the flow. The margin has advanced by ~3 km. (c) Glacier surges cause downdraw in the upper region of a glacier (the reservoir zone). At this glacier, Sortebrae in East Greenland, downdraw was measured to be up to 200 m, and (d) large crevasses opened in the lower part of the glacier. (Source: (a) air photo 569 1493 and (b) air photo 569 3134, © Norwegian Polar Institute; (c) and (d) Danish Lithospheric Centre)

becomes intensively and chaotically crevassed (Figure 18.19d). Tributary glaciers can be sheared off, resulting in looped patterns of medial **moraines** (Figure 18.19b and c). The slow-flow phase is often characterized by frontal retreat and wasting ice. During slow-flow periods the ice builds up in the reservoir zone ready for the next surge, which often results in the formation of a bulge of ice.

Although only about 1% of Earth's glaciers are surge type they tend to be more common in certain regions including Iceland, Svalbard, Karakoram, Pamirs, Tien Shan, Yukon–Alaska, Greenland and the Andes. These environments range from continental to maritime and the glacier thermal regimes from temperate to subpolar. Only cold-based glaciers have not been observed to surge.

Figure 18.20 Elevation change of the Antarctic and Greenland ice sheets measured using satellite altimetry. In general the central portions of the ice sheets are thickening because of increased precipitation whereas the peripheries are thinning. Overall, it is likely that both Antarctica and Greenland are contributing to sea-level rise. The altimeters used to make these maps cannot measure over steep slopes (more than a few degrees), such as on ice sheet margins, so these regions are missing from the figures. Other data sets, such as airborne altimetry, show the outlet glaciers of Greenland to be thinning rapidly. (Sources: (figure left) Davis *et al.*, 2005 from C.H. Davis, Y.H. Li, J.R. McConnell *et al.*, Snowfall-driven growth in East Antarctic ice sheet mitigates recent sea-level rise, *Science* 308 (5730): 1898-1901, 2005. Reprinted with permission from AAAS.; (figure right) Johanessen *et al.*, 2005, from O.M. Johannessen, K. Khvorostovsky, M.W. Miles *et al.*, Recent ice-sheet growth in the interior of Greenland, *Science* 310 (5750): 1013-1016, 2005. Reprinted with permission from AAAS.)

18.2.7.6 Glaciers in a warming climate

Global climate is warming, although the spatial pattern is not uniform (see Chapter 21) and this warming is causing widespread melting and retreat of glaciers. This loss of ice is resulting in sea-level rise (see Section 18.1). Mass loss is not happening everywhere, however. In some regions atmospheric warming has increased precipitation, resulting in mass gain. The major ice sheets are in general thickening in their central region, because of the increased snowfall, and thinning around their periphery (Figure 18.20). Some basins of the West Antarctic ice sheet which are grounded below sea level are thinning rapidly, as are many outlet glaciers in Greenland. Furthermore, a number of ice shelves on the Antarctic Peninsula have collapsed (see Box 18.6) and the glaciers feeding into them have accelerated. Many outlet glaciers in southern Greenland have also speeded up. In most

mountainous regions small glaciers are retreating and contributing to sea level despite the increase in precipitation in some areas.

Reflective questions

➤ A mining company suggests disposing of waste down crevasses of a temperate glacier, arguing that the material would be held for many thousands of years before being released into the environment. Do you agree with this assessment?

➤ Can you explain the major differences between a warm- and cold-bedded glacier in terms of thermal structure, hydrology and dynamics?

➤ A ski company decides that in order to prolong the ski season it will move snow from the accumulation zone of a glacier to the ablation zone. Can you comment on this proposal with a view to the long-term future of the ski resort?

➤ In the Delta River Valley, Alaska, the Alaska oil pipeline passes within hundreds of metres of glaciers, including several surge-type glaciers. Why might this be a problem?

18.3 Glacial geological processes and glacial sediments

The aim of this section is to develop an understanding of the main geomorphic processes operating beneath glaciers, concentrating on the themes of erosion and deposition, and emphasizing the characteristics of the sediments deposited.

18.3.1 Processes of glacial erosion

Glacial erosion has resulted in some of the most spectacular scenery on Earth. Deep, previously glaciated valleys and

CASE STUDIES

ICE SHELF COLLAPSE ON THE ANTARCTIC PENINSULA

The Antarctic Peninsula (Figure 18.21) is a chain of mountains with about 400 glaciers. The peninsula represents around 7% of the Antarctic land mass and contains sufficient ice to raise sea level by some 0.3 m. Warming on the Antarctic Peninsula has been approximately 3°C in the past 50 years, which is five times the global mean temperature rise and among the fastest measured (Vaughan *et al.*, 2003). As a result of this warming there have been profound changes due to greatly increased melting of snow and ice. About 75% of the glaciers on the peninsula are retreating, and a number of the floating ice shelves fed by these glaciers have melted and then sudden collapsed. The 1600 km² Larsen A ice shelf broke up in 1995, and the 1100 km² Wilkins ice shelf in 1998. The more southerly 3200 km² Larsen B ice shelf broke up over a few weeks in summer 2002 (Figure 18.21). While both the Larsen A and Wilkins ice shelves had collapsed previously in the last 2000–5000 years, geological

evidence shows Larsen B had existed for at least the past 10 000 years. After the ice shelves broke up, the glaciers that previously flowed into them speeded up by approximately

two to six times. These glaciers have also thinned. As a result of the break-up of ice shelves more ice is being delivered from the peninsula into the oceans.

Figure 18.21 This figure shows four MODIS images of the Larsen B ice shelf in summer 2002. On the first scene from 31 January there is clear evidence of surface melting and ponding of water on the ice shelf. The main collapse occurred between 23 February and 5 March 2002, when approximately 2600 km² of the ice shelf collapsed. (Source: National Snow and Ice Data Center, University of Colorado, Boulder)

BOX 18.6

fjords and lofty alpine ridges and horns are the typical scenery brought to mind. Enormous volumes of rock have been removed from these regions, transported by ice and water and deposited in lower-lying areas. This section outlines the processes by which these landscapes of erosion are formed. These processes are poorly understood, partly because of the inaccessibility of the subglacial environment and partly because of the complex interactions between the ice dynamics, thermal regime and hydrological system, which strongly affect the processes of erosion.

18.3.1.1 Glacial crushing

Glacial crushing is the direct fracturing of bedrock because of the weight of ice above it. The thickest ice that has existed on Earth was around 5 km thick, which results in a normal stress of about 44 MPa (Box 18.3). However, most rocks are stronger than this. For example, granites have unconfined compressive strengths of 140–230 MPa, sandstones between 70 and 210 MPa and even shales and tuffs have strengths greater than 10 MPa. It would appear that only the thickest glaciers overlying the weakest rocks will cause fracturing. Yet there is direct evidence that crushing does occur. This can result from the effects of stress concentration because of stones in the basal ice, the exploitation of pre-existing weaknesses and joints in the bedrock, and from repeated cycles of loading and unloading. Freeze–thaw weathering (see Chapter 11) in front of the glacier can fracture rock that the glacier subsequently advances over. Finally the removal of large amounts of rock by glacial erosion can cause fracturing due to pressure release. Bedrock crushing is enhanced by: (i) thick ice; (ii) the presence of particles entrained within the basal ice; (iii) cold patches at the bed; and (iv) large fluctuations in the basal water pressure. The products of bedrock crushing are typically large, angular rocks.

18.3.1.2 Plucking and quarrying

Once the bedrock is crushed it can be entrained (picked up) into the glacier ice. The process by which a glacier removes large chunks of rock from its bed is known as plucking or quarrying. Entrainment can result from freeze-on at the bed, by ice regelation around the rock or by incorporation into the ice along faults. Freeze-on can occur at the downstream side of obstacles at the bed where water refreezes as the glacier slides by regelation.

18.3.1.3 Glacial abrasion

Glacial abrasion occurs when glaciers slide relative to the material beneath them. Rock particles held within basal ice (Figure 18.14) are dragged over the glacier bed. This slowly scratches and wears the surface, rather like a piece of sandpaper wears wood. The rate of erosion due to abrasion is controlled by three factors. These are: the contact pressure between ice and its bed; the rate of sliding; and the concentration and nature of particles within the basal ice.

Abrasion rates increase as particle concentration within basal ice increases. Furthermore, fresh rough particles are more efficient tools of erosion than smoothed particles. Hence abrasion rates are higher in locations where basal melting brings a continual supply of fresh particles descending towards the bed. The relative hardness between entrained debris and the underlying bedrock will influence abrasion rates. For example, a glacier that flows from a hard to soft rock will have entrained hard rock pieces which will then efficiently abrade the softer rock downslope. Finally, erosion is more effective in locations where the fine sedimentary product is removed by meltwater flushing. Otherwise the fine sediment can clog or coat particles that are acting as abrasion tools. The sediment resulting from abrasion is very fine and when suspended in water is known as glacial flour. Glacial abrasion causes **striations** (see below) and smooths particles.

18.3.1.4 Mechanical and chemical erosion by basal meltwater

Meltwater at the bed of a glacier can cause erosion by mechanical or chemical processes. The rate of both is highly dependent on the nature of the glacial water system, the flux through it and the sediment or solute loading. Subglacial mechanical erosion occurs by the same processes as erosion in a surface channel flowing over bedrock. This operates by corrasion (mechanical wearing and grinding), cavitation (pressure changes due to bubble collapse in turbulent flow) and corrosion (chemical weathering) (see Chapter 14). Glacial meltwaters are efficient at erosion, because they typically flow at high velocities and often have high viscosity because of their large suspended sediment load and cold temperature. Mechanical erosion forms smoothed bedrock, potholes and N-channels. Chemical erosion by meltwater at the bed results in the decomposition of minerals into their ionic constituents. The processes operating include: solution, hydrolysis, carbonation, hydration, oxidation and reduction, and cation exchange (see Chapter 15). All of these processes act at the surface of particles. The rate of chemical erosion at the bed is enhanced by the influx of fresh surface water, by the presence of freshly abraded surfaces and by the presence of dissolved CO_2, which has an enhanced solubility at low temperatures.

18.3.1.5 Erosion rates

The rate at which glaciers and ice sheets denude bedrock varies greatly owing to changes in basal temperature, glacier velocity and properties of the bedrock. In one experiment, marble and metal plates were installed beneath valley glaciers to measure the rate at which they were abraded (Boulton, 1979). The plates were abraded at rates between 0.9 and 36 mm yr^{-1}. Other estimates are derived from measurements of the sediment output from glacial systems in streams and in basal ice. These measurements suggest total erosion rates of between 0.073 and 0.165 mm yr^{-1}. The best global estimate is approximately 1 mm yr^{-1} which is equivalent to 1 km per million years. However, these estimates are for valley glaciers and do not include ice sheets. It should be emphasized that in some situations ice may have a protective role and bedrock may be eroded at much slower rates than would have otherwise been the case. Indeed, recent evidence from cosmogenic isotope data suggests that denudation by large ice sheets was minimal for large areas at least during the advance of the last glacial (Fabel *et al.*, 2002).

18.3.2 Entrainment and transport

Freeze–thaw and other processes result in material falling from rock slopes onto the surface of glaciers where it is transported downstream by the ice (Figure 18.22). If the material falls onto the glacier in the accumulation zone it will become

buried, and either will be transported within the body of the glacier (englacially) or may descend to the bed. If material falls onto the surface in the ablation zone it will remain at the surface and be transported passively as supraglacial sediment. Material that is only buried partially will also be transported passively as englacial sediment and may emerge in the ablation zone as a result of ice melt. Sediment is also transported within the basal ice layer (Figure 18.14). Basal material may also become entrained within faults close to the margin or basal crevasses (Figure 18.23a). Medial and lateral moraines are made up of supraglacial material and the particles are typically coarse and angular (Figure 18.23b). Particles that are eroded from the bed, or descend to the bed, become rounded and worn through abrasion (Figure 18.23a).

The glacial system can be thought of as a conveyor belt transporting ice and sediment downslope, modifying the sediment particles as they are transported (Figure 18.22). As particles travel down the glacier they are abraded for progressively longer time periods and break down to finer and finer particles. However, experiments in grinding mills show that the breakdown of sediment particles ceases once particles reach their **terminal mode**, the size of which depends on the mineralogy (Haldorsen, 1981). As the material is further abraded more of the particles reach this terminal mode, but the material does not break down to finer particles. As a result of abrasion the particle size distribution of basal sediments often becomes bimodal (two peaks in the distribution; in this case there are lots of pebble-sized material and lots of fine-clay-sized material, but with less of all the other size fractions) (Dreimanis and Vagners, 1971) (Figure 18.24).

18.3.3 Deposition

Deposition of particles from the 'glacier conveyor belt' is an active process and alteration of the sediment properties continues as the particles are deposited. The sediment consequently develops characteristics that result from the particles' source, transport route through the system and their mode of deposition. The resulting sediments are often referred to by the process of their deposition, as this has the strongest influence on their properties. This is known as a **genetic classification**. The generic term for sediment deposited directly from a glacier is a **till**. It is important to understand the processes and nature of glacial deposits, because they cover about 70–80% of mid-latitude regions, which represents some 8% of the Earth's total land surface.

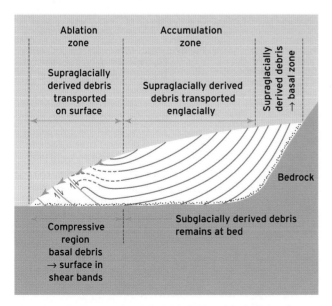

Figure 18.22 Transport paths of debris through glacial system. (Source: after Boulton, 1978)

(a)

(b)

Figure 18.23 (a) Basally derived material entrained in a fault in glacier ice. Basal material is actively transported and becomes rounded and finer. Ice flow was from right to left. (b) Supraglacial sediment is typically coarse and angular. Both supra- and englacial material are transported passively and so little modification occurs. Medial and lateral moraines are made up of supraglacial sediment.

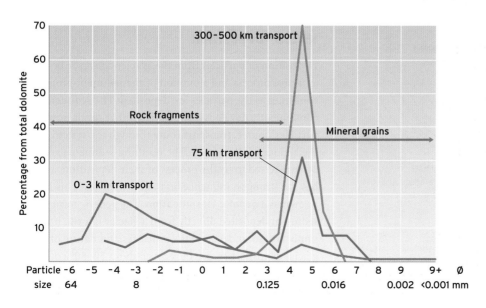

Figure 18.24 A typical particle size distribution for glacial till is bimodal or multimodal. The Φ index is shown (see Chapter 12) on the *x* axis along with values in mm. The two peaks in the distribution of the sediment sizes collected can be seen. There is more fine-grained (0.1 mm diameter) and pebble-sized (50 mm diameter) material than other particle sizes. This develops from the processes of glacial erosion that occur progressively with transport distance. With progressive distance from the ice mass centre successively higher percentages of rock fragments are reduced to a smaller particle size. (Source: after Dreimanis and Vagners, 1971)

18.3.3.1 Lodgement till

Lodgement of particles occurs when the frictional drag between a particle and the glacier bed exceeds the shear stress resulting from the moving ice (Box 18.3). The frictional drag on a particle depends on the contact pressure between it and the bed. This means that lodgement is most likely to occur beneath thick ice, where basal water pressures are low, on the upstream sides of obstacles, and in regions where basal melt rates are high. Boulder clusters will form because of preferential lodgement on the upstream side of obstacles at the bed. Lodgement tills form at the ice base from material that has been in transport in the basal zone. As a consequence the particles are typically rounded, are often striated and the sediment size distribution is often bimodal (see above).

The processes of lodgement may be size selective. A glacier slides over its bed by regulation and by enhanced creep. Since regulation is most efficient when obstacles are small, whereas enhanced creep is most efficient when obstacles are large, these processes will selectively keep particles of the critical size entrained within the glacier ice. The forces keeping small particles in motion within ice are relatively low because the glacier regelates around small particles easily, and the forces keeping large particles in motion are low because the ice creeps around the particles easily.

18.3.3.2 Deformation till

Many glaciers are thought to overlie deforming soft sediments. The processes operating beneath such glaciers are recorded in the tills that they deposit. Deformation rearranges particles and reorientates them. Examination of the tills can reveal structures that allow the strain history to be reconstructed (e.g. Figure 18.25). At low strain minor

Figure 18.25 Schematic diagram showing the effect of deformation on tills. (a) At low strain minor folding occurs. Photograph: small-scale faulting in clay layers that shows that this sediment was deformed only to very low strain. (b) At intermediate strain dramatic features can be formed, depending on whether the strain is compressive or extensive. (c) At high strain the till can appear to be completely homogenized and massive. Photograph: shear zones in sediment that has no features apparent on the macroscale. Two directions of shearing are evident. Matrix-supported till from Criccieth, North Wales. (Source: after Hart and Boulton, 1991; photos courtesy of (a) Sarah J. Fuller and (c) Andy J. Evans)

folding and faulting can occur. As the strain increases the features formed can be quite spectacular. Compressive features include folds and some faults, whereas extension results in the formation of **boudins** (Figure 18.25b). As the strain increases still further the features become progressively attenuated, and tectonic laminations can be formed. Finally at high strains, likely to be typical where deformation is a significant component of ice flow, the till can appear homogenized. However, the sediment may still have many structures visible at the microscopic scale (Figure 18.25). Deformation tills can have high porosity (% of void space) because **dilation** (expansion) occurs as the sediment deforms. Particle **fabrics** (the alignment of coarse particles) are typically stronger for intermediate strains than for low or high strain. Often the fabric has both a flow parallel and a transverse component. The sediment typically comprises basal material, containing rounded and striated pebbles and rocks, and if the sediment has experienced high strain there is likely to be a wide range of lithologies because material will have travelled long distances. However, it can be difficult to distinguish a deformation till from a lodgement till, and there is controversy about the origin of several deposits.

18.3.3.3 Meltout till

Meltout till forms when ice surrounding sediment melts. In general, the term is used to describe deposits from subglacial meltout. Supraglacial meltout usually results in high water contents and therefore reworking of the deposit often occurs. Meltout till usually forms beneath stagnant ice masses. Meltout till can inherit properties from the basal ice from which it forms, often being weakly stratified and maintaining the fabric (Figure 18.26). This is most likely to occur where the basal ice has a high-sediment and low-ice content, and where water produced when the ice melts can drain away freely. The extent of meltout till therefore depends on the distribution and concentration of sediment in basal ice, which is in turn controlled by the thermal regime and strain regime. Basal ice layers are thickest beneath polythermal glaciers and where a compressive flow regime occurs. Surging glaciers often develop thick basal ice layers.

18.3.3.4 Flow till

The proglacial environment is very active. In summer, there is often a continuous supply of water from melting ice. Because many glacial sediments have low permeability, they drain poorly. As a result sediment flows are common.

Figure 18.26 Meltout till at the margin of Trapridge Glacier, Yukon Territory.

Such flows alter sediment properties, realigning particles, and cause particle sorting (Figure 18.27). The resulting deposit is known as flow till. It is estimated that up to 95% of the marginal area in front of a glacier may comprise flow till.

18.3.3.5 Deposition from water

Sediments entrained into subglacial water flows may be transported considerable distances from the ice margin before deposition. As a result of the differences in time taken for particles of different sizes to settle through water (Table 18.3), there is potential for strong sorting of the deposits, both spatially and temporally. Spatial sorting occurs because the coarse particles settle close to the ice margin whereas the finer particles can remain in suspension and can travel long distances. Temporal sorting occurs because the outputs of both water and sediment vary on a

Figure 18.27 Mudflow (right to left) at the margin of Trapridge Glacier, Yukon Territory. Note that the flow has sorted the particles, with coarse particles migrating to the regions of least strain at the sides, surface and front of the flow. In the background a debris-rich basal ice sequence can be seen.

Table 18.3 Time taken for single grain to settle through 10 m of still water at 0°C. The very large differences in time between fine and coarse particles lead to strong sorting both spatially and temporally from a source

	Particle diameter (μm)	Time
Very fine sand	125	21 minutes
Coarse silt	31.25	5.7 hours
Fine silt	7.81	3.8 days
Coarse clay	0.98	7.7 months
Fine clay	0.06	175 years

variety of timescales (e.g. diurnally and seasonally). If we consider the seasonal variation, flows of sediment and water are highest in the early summer and lowest in the winter. The coarse particles are deposited first, in summer, and the finer particles progressively through the year. This forms repetitive sequences of deposited sediment that fine upwards; where such sequences represent annual inputs they are known as **varves**.

> ### Reflective questions
>
> ➤ What are the main processes of glacial erosion?
>
> ➤ Can you explain the differences between erosion beneath a warm- and a cold-bedded glacier in terms of the processes operating and their relative rates?
>
> ➤ What are the key differences that you expect to find in glacial sediments that allow the differentiation of: (i) supraglacial from basal material; and (ii) sediments deposited in water from those deposited on land?

18.4 The record of glacial change

The aim of this section is to develop an understanding of the effects of ice on the landscape at all scales from global to microscale, concentrating on the geomorphic features that result from the processes and environments discussed in Sections 18.2 and 18.3.

18.4.1 Ice sheet reconstruction

18.4.1.1 Geomorphology of regions of erosion

Regions dominated by the processes of glacial erosion form some of the most dramatic mountain scenery on Earth. Examples are the Southern Alps of New Zealand, the Rocky Mountains in North America and the Alps in Europe. In general erosion occurs most efficiently beneath a warm-based ice mass that is sliding over its bed. Furthermore, erosion requires that the glacier base is in contact with bedrock rather than sediments. This is more likely to be the case close to ice divides, at the centre of ice sheets or ice caps, and beneath local glaciers in alpine areas (Figure 18.28). The features formed by erosion are usually grouped by scale. At the macroscale, glaciers erode great **U-shaped valleys**, leaving ridges (or **arêtes**) and **horns** (Figure 18.29). Often side valleys and spurs are truncated, leaving **hanging valleys** (or hanging toughs) and **truncated spurs**. When the glaciers retreat, lakes often form in the overdeepened basins, which slowly fill with sediment over time. If the base of the glacial trough lies below sea level then after retreat it becomes inundated and forms a fjord. Less extensive glaciers often form **corries** (or cirques) which when they melt leave a depression that can be filled with water and are known as **tarns** (Figure 18.29). Figure 18.29 also shows some depositional features such as moraines that are discussed below.

Snowdonia, in North Wales, UK, is presently unglaciated but contains evidence of widespread erosion by glaciers. At around 23 000 years ago an ice sheet approximately

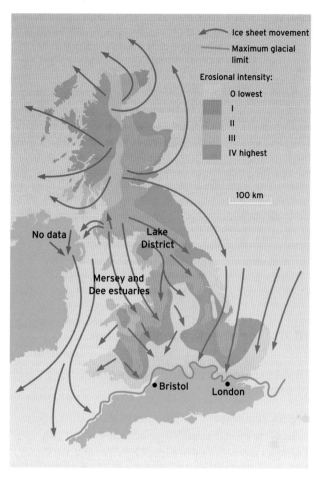

Figure 18.28 Pattern of glacial erosion in the United Kingdom. The erosion is greatest in the highest regions that formed the ice centres for major ice sheets and were glaciated by local glaciers during shorter or warmer stadials.

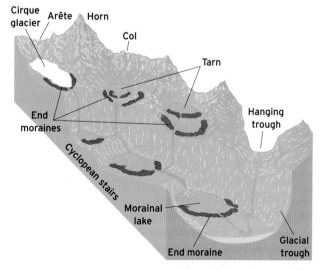

Figure 18.29 A landscape of glacial erosion. (Source: Easterbrook, D.J., *Surface processes and landforms*, 1st edition, © 1993. Reprinted with permission of Pearson Education, Inc., Upper Saddle River, NJ)

1000 m thick formed which was centred on North Wales. The accumulation area of this ice sheet was approximately 2000 km² at its maximum extent, about 18 000 years ago. This ice sheet lingered until about 12 000 years ago. Then a second short cold spell re-filled the area with ice between 11 000 and 10 000 years ago. Figure 18.30 shows the morphology of Cwm Idwal. This was one of the first sites where past glaciation was recognized in the 1840s. The corrie shows both erosional and depositional features. The overall corrie shape is governed by erosion, it is overdeepened and is currently occupied by a lake. Depositional moraines can be found within the valley (see below).

Mesoscale features are formed by the erosive power of both ice and water. Features formed by the sliding of debris-rich basal ice over the surface include **stoss-and-lee forms, roche moutonnées, whalebacks** and **crag and tail** features. Stoss-and-lee forms are streamlined features that have a gently sloped, glacially smoothed upstream side and a steeper, plucked, downstream lee side (Figure 18.31a). The stoss side is often striated. The features vary in size from centimetres to many metres in length. Small stoss-and-lee forms are often called roche moutonnées. Stoss-and-lee forms provide evidence of warm-based ice conditions and the formation of cavities downstream of obstacles at the bed. Whalebacks are a similar shape to stoss-and-lee forms but their steep side faces upstream and their tapered end downstream. Whalebacks probably form beneath warm-based ice where the sliding rate is slow enough that extensive cavitation does not occur behind obstacles at the bed. Crag and tail features form where resistant rock is left standing proud of the surface. Small cavities form behind the more resistant rock either protecting the tail of softer rock from erosion or allowing infilling to form a rock-cored **drumlin** (see Box 18.7 below). Steep-sided and smoothed channels are also found on bedrock, providing evidence of warm-based ice and a drainage system at least partly drained by N-channels.

Microscale features (Figure 18.31b) include those formed directly by erosion due to the ice, such as striations. These are formed by stones and rocks in the basal ice sliding over bedrock and leaving scratches on the surface. The presence of striations implies a warm-based ice mass, sliding over its bed, and provides a reliable indication of the direction of ice flow. Features related to striations include repetitive **chatter marks**, which show that the ice moved over its bed with a stick–slip motion. Larger **crescentric gouges**, scars and fractures also form, also typically concave down the glacier, but are less repetitive than chatter marks. All of these features form by

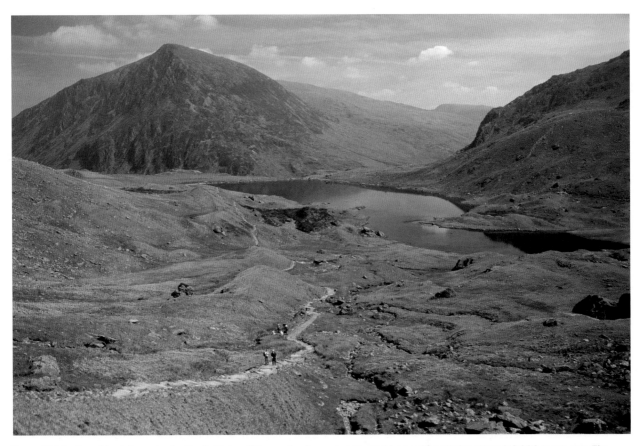

Figure 18.30 Cwm Idwal, Snowdonia, North Wales. This small valley was occupied by a valley glacier around 13 000 years ago. The glacier advanced to the terminal moraine (which forms at the front edge of a glacier). A complex suite of other moraines formed, which are discussed in the text. (Source: Mike Potts/Nature Picture Library)

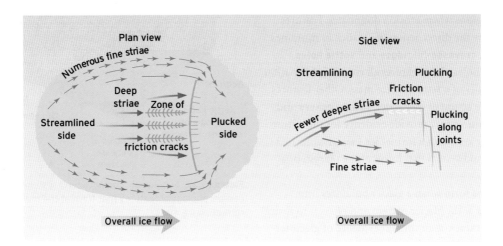

Figure 18.31(a) Medium- and small-scale ice erosional features, stoss-and-lee form; this example shows the typical features expected on a roche moutonnée, which is the name given to small versions of stoss-and-lee forms.

non-uniform slip over bedrock. Benn and Evans (1998) provide further details on these features. Glacial flow direction can also be determined using mini crag and tail features that form on inhomogeneous bedrock. They consist of tails of uneroded bedrock preserved behind small, more resistant, grains on the surface of a rock. Smooth channels (glacial grooves) and depressions are also found on bedrock surfaces.

Figure 18.31(b) Microscale features of glacial erosion. (Source: after Prest, 1983)

18.4.1.2 Geomorphology of areas of deposition

The geomorphology of areas of deposition is typically more subdued than regions where erosion dominates. Features may be formed by the direct action of ice such as moraines, drumlins, **flutes**, **crevasse-fill ridges** and **kettle holes**. Except for kettle holes, these features all encode information about the dynamics of a retreating ice mass. Other features are formed by the action of meltwater, such as **eskers** and **kames**. These features may provide evidence for the nature of the basal water system of a former ice mass. Figure 18.32 shows a typical land system that forms at the margin of a retreating glacier.

Push moraines form when a glacier bulldozes into existing sediment at the glacier margin and raises it to form a ridge. Push moraines often contain evidence of glaciotectonic deformation within them and consist of glacially derived material. If the material is not glacially derived, the term **ice-pushed ridge** is used. Major push moraines often mark the maximum extent of glacial advance and allow us to map ice extent. See Figure 20.7 in Chapter 20 for an example of a map of the extent of the Laurentide ice sheet that covered large areas of North America, which is based on push moraine evidence. Smaller annual moraines can

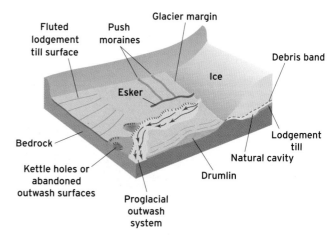

Figure 18.32 Subglacial-proglacial land system showing typical geomorphological features that form at the margin of a retreating ice mass.

be formed in front of a retreating glacier marking seasonal re-advances of the glacier front.

Dump moraines are ridges formed approximately transverse to flow from material delivered to the margin of a glacier by ice flow. They mark the stationary position of an ice margin. The size of dump moraine depends on the rate of ice flow and the debris content of the ice. **Lateral moraines**

(a)

(b)

(c)

Figure 18.33 (a) De Greer moraines, east of Hudson Bay, Quebec, Canada. Amplitude of ridges is 10-15 m. (b) Drumlin field south of Lake Athabasca, north-west Saskatchewan, Canada. Note the eskers that occur in close proximity to the drumlins. Ice flowed south-westward. (c) Drumlin field near Beverley Lake, District of Keewatin, NWT, Canada. Ice flowed north-westward. (Source: NAPL, Canada)

form parallel to the sides of glaciers from dumped material and frost-shattered material from the valley walls (Figure 18.32). **Hummocky moraines** form from the meltout of supraglacial or englacial material. These landforms tend to consist of irregular mounds of material.

Figure 18.33(a) shows arcuate ridges of sediment. The amplitude of these features, known as **de Geer moraine**,

is about 2–15 m in height. These features are formed approximately transverse to flow where a retreating ice mass borders on a glacial lake. The features are thought to form by deposition at the margin of active ice and mark the position of the winter margin.

Drumlins are streamlined features aligned along the direction of ice flow (Box 18.7). Figure 18.33(b) and (c)

CASE STUDIES

FORMATION OF DRUMLINS

Drumlins are streamlined, elongated hills that form approximately parallel to ice flow (Figure 18.33b and c). They are large-scale features and typical dimensions of a drumlin are 80–100 m high and 1–2 km long. Drumlins tend to have a blunt upstream end and a longer tapered downstream end. They are typically found in groups known as drumlin fields. Drumlins can be formed from either bedrock or sediment, and they may have a core of different composition at their upstream end. There are a number of theories as to the formation of drumlins. Not all theories can explain all types of drumlins. For example, theories that suggest that drumlins form by deposition of sediment cannot explain rock drumlins. Although both sediment drumlins and rock drumlins have similar morphologies it is possible that they form by different processes. Two contrasting theories of drumlin formation are the formation of drumlins by deformation of a soft bed and the formation of drumlins as a result of megafloods.

The idea that drumlins might form because of changes in the thickness and properties of a deforming bed was proposed by Boulton (1987) and Menzies (2000). They suggested that subglacial deformation can explain a continuum of forms from transverse features such as **rogen moraines**, through less elongate flow parallel forms such as drumlins, to very elongate forms such as flutes. The elongation ratio would increase with

distance downstream from an ice mass centre. The **Boulton-Menzies theory** suggests that a drumlin is formed by deposition in the lee of a slowly moving obstacle in the deforming layer and potentially by streamlining of the obstacle. This obstacle forms the core of the drumlin, and may be composed of bedrock, thermally frozen material or material that is better drained or less dilated than the surrounding material. The drumlin will slowly migrate down the glacier. As well as sedimentary evidence (Figure 18.34), proponents of this theory cite drumlins currently forming at sites where deformation is known to occur, such as at the margin of Breiðamerkerjökull, south-east Iceland. Furthermore, time-lapse

geophysical surveys have shown a drumlin forming from deforming sediments (Smith et al., 2007).

It has been suggested (J. Shaw, 1989, 1994) that all drumlins, including rock drumlins, were formed by a common agent, namely meltwater sheets during megafloods. Shaw used equivalence of form to argue that because drumlins have the same morphology as smaller-scale erosional scour features, they form by similar processes. He argued that horseshoe vortices form downstream of obstacles when megafloods occur (Figure 18.35a). These secondary flows cause erosion upwards into the ice and downwards into the bedrock. Rock drumlins are formed as the direct result of this

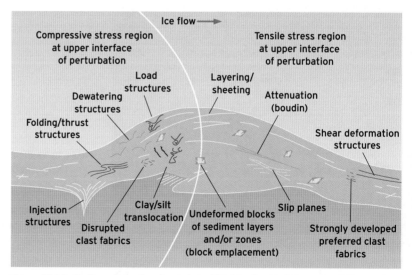

Figure 18.34 Sedimentary structures found in drumlins that would support their formation from an active deformable layer of sediment beneath a glacier via the Bouton-Menzies theory. Evidence thought to support this theory includes the presence of shear zones and other deformation structures in some drumlins. (Source: after Menzies, 1989)

BOX 18.7 ➤

(a)

Eroded furrow

Secondary flow

Obstacle

Direction of water flow

(b)

Ice bed with inverted erosional marks

Subglacial meltwater

Rogen moraine

Cavity fill

Drumlins

Lodgement till

Bedrock erosional marks

Figure 18.35 Shaw's theory of drumlin development as a result of megafloods. (a) During megafloods horseshoe vortices form downstream of obstacles resulting in erosion of the ice above and the rock or sediment below. (b) Drumlins form by this erosion or during the settling of the now drumlin-shaped ice sheet back onto soft sediments. Other bedforms such as rogen moraines may also form by the same process. (Source: (a) after Shaw and Sharpe, 1987; (b) after Shaw, 1994a)

pattern of erosion. Sediment drumlins then form either in a similar manner or at the end of flows when the flood waters become hyper-concentrated in sediment and the ice cap acts like a mould and settles back onto its bed (Figure 18.35b). Most of Shaw's work has considered drumlins that formed beneath the Laurentide ice sheet that covered Canada and parts of the United States. The megafloods that Shaw postulates involved a large amount of water. For example, for one drumlin field Shaw cited, 8×10^4 km^3 of water would be released over 16 to 162 days. This would have resulted in global sea-level rise of 0.23 m. Shaw suggests that the water required for this was probably stored in a depression beneath the ice sheet held back by frozen margins. There are currently about 70 lakes beneath the Antarctic ice sheet that store 4000–12 000 km^3 of water. Evidence to back up Shaw's theory includes oxygen isotope records (see Chapter 20), sedimentary cores from the ice sheet margin and flood myths in a number of human cultures.

BOX 18.7

shows two drumlin fields. The direction of ice flow can be determined from the shape of the drumlins. These features typically form with a blunt end pointing upstream. Note the large range of elongation ratios (ratio of length to width). Eskers can also be seen on Figure 18.33(b). These are typically thought to form in R-channels at the glacier bed, although they can also form in subaerial channels.

Glacial flutes form parallel to flow. Flutes have a much higher elongation ratio than the drumlins discussed above. Flutes are thought to form by the infilling of cavities on the downstream side of obstacles at the bed (Figure 18.36). The size of the flute is probably controlled by the ice velocity and the nature of the till material. Crevasse-fill ridges form when a glacier with deep crevasses sinks into soft, water-saturated sediments. Glaciers typically become heavily crevassed when they surge, and these features are common at the margins of surge-type glaciers.

Kettle holes are closed topographical depressions formed from melting stagnant ice, and are common in sediments downvalley from a glacier. Eskers are formed in ice-walled channels, either basal R-channels or in subaerial channels

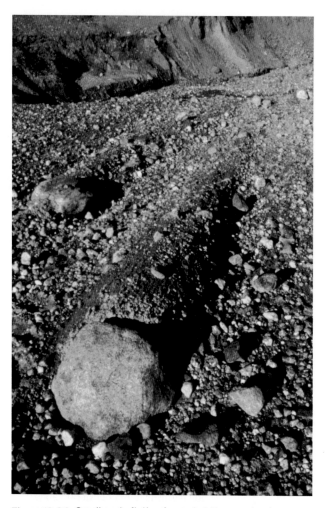

cut through ice blocks. An esker consists of a long, narrow ridge of sands and gravels which overlies till. Eskers are 20–30 m high, and range from a few to 500 km in length. Eskers formed in R-channels, which unlike subaerial channels may flow uphill as well as down, represent the relict form of a channelized water system. They are hence important in understanding the basal water system beneath past ice masses. Eskers are often quarried as a source of gravel and sand. A kame is a mound of sediment formed when a hole in a melting, stagnant ice mass becomes filled with sediment.

Figure 18.36 Small-scale fluting formed at the margin of a modern glacier. The features formed by the infilling of cavities formed at the downstream side of each boulder. The largest boulder is about 0.5 m in diameter. Ice flow is away from viewer. (Source: John Shaw)

Reflective questions

➤ What are the main landforms of glacial erosion and glacial deposition?

➤ Which glacial features allow the direction of former ice flow to be determined?

➤ It is suggested that the Martian ice caps have previously been much larger than they are at present. What features would you look for in order to reconstruct past glacial extent on Mars?

18.5 Summary

This chapter outlines the wide range of ice masses and their characteristics, and describes how they modify the landscape and how they are responding to climatic change. There are a range of feedbacks between the glacier water system, glacier dynamics and glacial thermal regimes. In summer, ice and snowmelt produce large volumes of water in many regions. This water must be removed on, within, or beneath the ice, often impacting on glacier dynamics by increasing rates of basal motion.

Glaciers move by three processes: ice deformation, basal sliding and the deformation of soft sediments beneath them. Of these, the last two are the most important in modifying the landscape: as the glacier sole scrapes over bedrock this bedrock is eroded by crushing, plucking and abrasion. The glacier then acts as a conveyor

➤

belt, moving this sediment downslope, modifying and finally depositing it. There are two transport routes through the glacial system, the high-level route via supraglacial and englacial transport, which is passive and results in coarse, angular deposits, and the low-level route via basal transport, which is active.

Sediments that have been transported in the base of the glacier are typically bimodal and the coarser material becomes rounded and sometimes striated. Upland areas are typically source areas for ice where erosion dominates and this forms dramatic alpine landscapes. Lowland areas become covered with thick deposits of the resulting sediments. Understanding of the effects of glaciation on the landscape can be used when reconstructing past glacial extent and conditions.

Further reading

Benn, D.I. and Evans, D.J.A. (1996) The interpretation and classification of subglacially-deformed materials. *Quaternary Science Reviews*, 15, 23–52.
This is a useful scientific review paper on glacial sediments which summarizes a lot of the research on the subject.

Benn, D.I. and Evans, D.J.A. (1998) *Glaciers and glaciation.* Edward Arnold, London.
This book provides excellent and encyclopaedic coverage of all aspects of glaciers, glacial sediments and glaciated landscapes. It is suitable for courses throughout an undergraduate degree.

Bennett, M.R. and Glasser, N.F. (1996) *Glacial geology, ice sheets and landforms.* John Wiley & Sons, Chichester.
This is an easy-to-access textbook covering most aspects of glacial systems.

Clarke, G.K.C. (1987) Fast glacier flow: ice streams, surging and tidewater glaciers. *Journal of Geophysical Research*, 92(B9), 8835–8841.
This is a review paper covering concepts of balance velocity and fast glacier flow.

Drewry, D. (1986) *Glacial geologic processes.* Edward Arnold, London.
This book is out of print but it is a unique approach to glacial geology. It contains many useful summary tables.

Fountain, A.G. and Walder, J.S. (1998) Water flow through temperate glaciers. *Reviews of Geophysics*, 36, 299–328.
This paper is the best and most up-to-date review of glacier hydrology in temperate glaciers.

Hambrey, M.J. (2003) *Glacial environments.* Routledge, London.
Hambrey provides a concise introductory text with excellent photographs.

Hooke, R.L. (2005) *Principles of glacier mechanics.* Cambridge University Press, Cambridge.
A more advanced text for those interested in the details of glacier dynamics.

Hubbard, B. and Glasser, N.J. (2005) *Field techniques in glaciology and glacial geomorphology.* John Wiley & Sons, Chichester.
This book is worth delving into if you are interested in how measurements are made in the field in glacial environments and will certainly be helpful if you are planning such fieldwork for a project.

König, M., Winther, J.-G. and Isaksson, E. (2001) Measuring snow and glacier ice properties from satellite. *Reviews of Geophysics*, 39, 1–27.
This is an excellent review of remote sensing in glaciology and of relevance to Box 18.4.

Murray, T. (1997) Assessing the paradigm shift: deformable glacier beds. *Quaternary Science Reviews*, 16, 995–1016.
This is the author's own review of deforming glacier beds.

Post, A. and LaChapelle, E.R. (2000) *Glacier ice.* University of Washington Press, Washington, DC.
There are beautiful photographs of a wide range of glacial features and landscapes in this book. It contains mainly Alaskan examples.

Willis, I.C. (1995) Interannual variations in glacier motion – a review. *Progress in Physical Geography*, 19, 61–106.
This is a review of variability in glacial flow and the interactions between the glacial water system and glacier dynamics.

Web resources

British Antarctic Survey

http://www.antarctica.ac.uk/

This site provides information on Antarctic research.

Geomorphology from Space: Glaciers and Glacial Landforms

http://disc.sci.gsfc.nasa.gov/geomorphology/GEO_9/index.shtml

There is an extensive chapter on glaciers and glacial landforms within the online NASA booklet on 'geomorphology from space'. A list of satellite remote sensing images of glacial features is provided with explanations behind the various observations.

Glacial Erosion

http://gsc.nrcan.gc.ca/surf/kivalliq/photo2_e.php

This site provide images of landforms of glacial erosion.

National Snow and Ice Data Center

http://nsidc.org/

This site provides a lot of information, including an introduction to glaciers. An online glossary is also available at:

http://nsidc.colorado.edu/glaciers/glossary

Natural Resources Canada

http://gsc.nrcan.gc.ca/landscapes/index_e.php

An image catalogue of Canadian landscapes that includes a wide variety of glaciated landscapes.

Outburst Floods

http://www.hi.is/~mmh/gos/

This site also contains links to other information about megafloods resulting from the mixture of volcanic eruptions and ice.

Scientific Committee on Antarctic Research

http://www.scar.org/

Information on Antarctic research currently underway is given here including on sub-Antarctic lakes.

US Geological Survey: Satellite Image Atlas of Glaciers of the World

http://pubs.usgs.gov/fs/2005/3056/

The USGS provides a downloadable pdf introduction to the image atlas, sample images and updated glacier areas.

US National Science Foundation: West Antarctic Ice Sheet

http://www.nsf.gov/pubs/1996/nstc96rp/sb4.htm

A page on the West Antarctic ice sheet and its stability.

World Glacier Monitoring Service

http://www.geo.unizh.ch/wgms/

The website of the World Glacier Monitoring Service provides news, data, general information and links particularly related to glacier mass balance.

Visit our website at **www.pearsoned.co.uk/holden** for further questions, annotated weblinks, interactive models and video-clips showing physical processes in action.

CHAPTER 19

Permafrost and periglaciation

Tavi Murray
Department of Geography, Swansea University

Learning objectives:

After reading this chapter you should be able to:

➤ explain the differences between permafrost and periglacial environments and describe their main characteristics

➤ describe how ground temperatures in permafrost zones vary with time and define the active layer, and continuous and discontinuous permafrost

➤ describe and explain the main geomorphological features that form in permafrost and periglacial regions and their relict forms, as well as those geomorphological features that allow differentiation between these two environments

19.1 Introduction

Periglacial environments are defined as those that are cold but non-glacial, regardless of their spatial proximity to glaciers. These are environments in which freeze–thaw processes drive geomorphic change. Therefore they develop distinctive geomorphological features. **Permafrost** refers to soil or bedrock that is perennially frozen over long timescales (greater than a minimum of two years). The presence of permafrost means that a periglacial environment

exists and periglacial processes are occurring. However, it is obviously possible to have a periglacial environment without permafrost. Permafrost usually contains ice within pore spaces (Figure 19.1). These ice features in the ground cause slow movement of the ground itself through expansion and contraction processes. Because the air temperature usually exceeds 0°C during summer in permafrost zones, a thin unfrozen layer (the **active layer**) can be present at the ground surface even when there are frozen layers below this. Currently permafrost affects approximately 26% of the Earth's surface (Washburn, 1979), while 35% is affected by freeze–thaw processes (Williams and Smith, 1989). The climate change that resulted in the advance and retreat of the major ice sheets during the past 2.4 million years (see Chapters 18 and 20) also resulted in past periglacial conditions far beyond the maximum extent of the ice sheets themselves. It is likely that an additional 20–25% of the Earth's surface has experienced periglacial conditions at some time in the past (French, 1996). These past periglacial conditions have left their mark on the landscape by producing periglacial landforms.

Freezing and thawing are important geomorphological processes as they drive fundamental changes in the ground's mechanical and hydrological properties. Both processes result in a volume change of the ground and there is typically a 9% increase in volume on freezing (Williams and

Figure 19.1 Ice layers from the top of the permafrost in Longyearbyen, Svalbard. (Source: Ole Humlum, University Centre, Svalbard)

water migrates to the freezing zone as the ground freezes. Frost heave is the vertical lifting of the soil surface and it disrupts structures and road surfaces in many parts of the world susceptible to permafrost conditions (Figure 19.2). Its understanding is therefore of great practical importance.

Most of the Earth's permafrost exists at temperatures just a few degrees below 0°C (IPCC, 2007a) and is thus highly susceptible to changes in climate or surface conditions. A slight warming or small changes to the ground surface, such as changes in vegetation cover, might cause dramatic change to permafrost zones. Particularly important for engineering in permafrost regions is the thawing that can occur owing to the increased insulation of the ground around a built structure, which can lead to subsidence (frozen ground often has 10–20% more water within it than unfrozen drained soil; see Figure 19.1). It is thus necessary for modern construction methods to take account of permafrost processes. This chapter is therefore of relevance to a large part of the Earth's land surface in terms of understanding landform development and human interactions with the landscape. It will begin by discussing permafrost processes before moving on to discuss permafrost and periglacial landforms and their formation.

Smith, 1989). This change in volume can result in frost shattering of rock. It is also partly the cause of **frost heave** although other processes are also responsible. In fact, most of the ground movement caused by frost heave results because

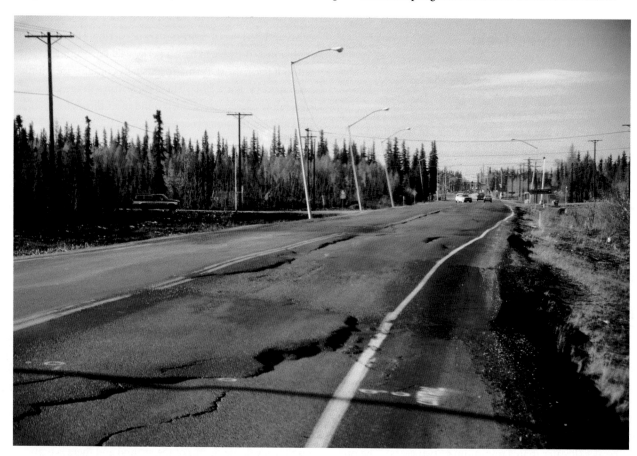

Figure 19.2 Road subsidence passing between leaning light poles in Alaska as a result of frost heave. (Source: Larry D. Hinzman, University of Alaska Fairbanks)

19.2 Permafrost processes

19.2.1 The distribution of permafrost

Today permafrost affects 82% of Alaska, 50% of Canada (Prowse and Ommaney, 1990; Trenhaile, 1990) and large parts of northern Siberia (Figure 19.3). This is known as **polar permafrost** and occurs because of low temperatures at high latitudes. **Alpine permafrost** occurs because of low temperatures at high altitudes. The major region of alpine permafrost occurs on the Tibetan Plateau, where some 2 million km^2 are affected (Figure 19.3). Alpine permafrost also occurs in the European Alps and the Rocky Mountains, North America. Permafrost is furthermore found beneath the sea, where it is known as **subsea permafrost**. The extent of subsea permafrost is not well known. Subsea permafrost usually occurs as a remnant of past colder temperatures and rising sea levels which drown frozen ground.

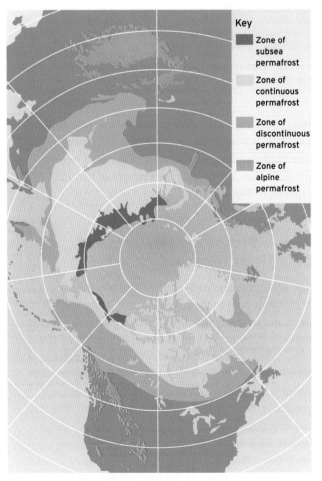

Key

- Zone of subsea permafrost
- Zone of continuous permafrost
- Zone of discontinuous permafrost
- Zone of alpine permafrost

Figure 19.3 The distribution of permafrost in the northern hemisphere. Permafrost covers approximately 26% of the Earth's surface, the large majority of which is in the northern hemisphere. (Source: after Péwé, 1991)

A transect south across Canada from the Arctic Ocean towards the United States, as shown in Figure 19.4, demonstrates differences in the thickness and extent of permafrost. These differences correlate approximately with latitude. In the far north, the permafrost is laterally continuous and 1000 m in thickness. This is the **continuous permafrost zone**, where the frozen ground is broken only beneath lakes, rivers, glaciers and other thermal disturbances. Further south in the **discontinuous permafrost zone** the permafrost is not laterally continuous, and it varies from 1 to 10 m or more in thickness (see also Figure 19.3). The transition between the two zones occurs approximately at the mean annual air temperature of the −6 to −8°C isotherm (line of equal temperature) and often occurs at about the same location as the **treeline**. The southernmost extent of discontinuous permafrost is often taken to coincide with the −1°C isotherm, although isolated relict patches of permafrost occur further south, often in peatlands; this is sometimes known as sporadic permafrost. In both the continuous and discontinuous permafrost zones the ground is subject to annual and sometimes diurnal freeze–thaw close to the surface. This forms the active layer (Figure 19.4). Thawing of the active layer in summer causes considerable difficulties for travel in permafrost regions when thawed and waterlogged terrain can become almost impassable. The active layer varies from a few tens of centimetres in the continuous permafrost zone to 15 m or more in the discontinuous permafrost zone. Unfrozen regions within the permafrost are known as **taliks**. Taliks are known as open if they are in contact with the active layer and closed if they are completely surrounded by permafrost.

Some measured permafrost depths are shown in Figure 19.5. At Resolute, Nunavut, Canada (74°N), the mean annual air temperature is −16.4°C. The permafrost thickness is greater than 400 m and the active layer is thin, only around 0.45 m thick. At Norman Wells, North West Territories, Canada (65°N), the permafrost is about 45 m thick with a moderate 1–2 m thick active layer. At Hay River, North West Territories, Canada (61°N), the mean annual air temperature is −4.4°C, the permafrost is discontinuous, approximately 12 m thick, and there is a 2–3 m thick active layer (Williams and Smith, 1989).

Fundamental changes in ground properties occur when it freezes. These changes include the ground mechanical properties (e.g. density, heat capacity, **thermal conductivity** and hydraulic conductivity; see Chapter 13) and electrical properties such as **electrical conductivity**. These changes mean that geophysical survey techniques can be successfully

Figure 19.4 Permafrost zones in Canada. The continuous permafrost zone corresponds approximately to the −6 to −8°C isotherm of mean annual air temperature. (Source: adapted from Williams and Smith, 1989)

used to map the extent of permafrost and massive ice within the ground. These techniques include seismic reflection surveying, ground-penetrating radar (see Chapter 23) and electromagnetic induction. These techniques are useful when trying to establish the best locations and methods for engineering structures. The importance of permafrost for large engineering projects is discussed using the example of the trans-Alaska pipeline in Box 19.1.

Figure 19.5 Permafrost in a north-to-south transect across Canada. The boundary between the continuous and discontinuous permafrost zones corresponds approximately with the treeline. Taliks are the unfrozen part of the ground within the permafrost. (Source: adapted from Williams and Smith, 1989)

19.2.2 Ground temperatures and permafrost thickness

Temperatures measured in permafrost boreholes are typically lowest close to the ground surface, excluding the active layer, and increase with depth, reaching the melting point at the base of the permafrost as shown in Figure 19.7 (Isaksen *et al.*, 2000). This trend may vary in the upper permafrost layer (see Box 19.2 for a discussion of how temperature varies in the ground in this upper layer). However, in general Figure 19.7 shows warming with depth. The main controls on permafrost thickness are the mean annual surface temperature, ground conductivity and the **geothermal heat flux**. Box 19.3 describes how to estimate permafrost thickness and the time taken to develop or degrade it. Surface albedo (see Chapter 4) can also play a role in permafrost processes. Dark vegetation cover may absorb more short-wave solar radiation and thus increase

CASE STUDIES

THE TRANS-ALASKA PIPELINE

A large portion of the oil and gas reserves in North America lies along the northern coast of Alaska. The Alyeska oil pipeline (trans-Alaska pipeline) was built to transport oil 1300 km from Prudhoe Bay to Valdez at a cost of over $7 billion (Figure 19.6a). The pipeline consists of a 120 cm steel pipe that carries oil at a temperature of about 65°C; oil at lower temperatures is too viscous to transport through a pipeline. About 75% of the route overlies permafrost, which is about 600 m thick near Prudhoe Bay. The hot temperature of the pipeline would result in thawing of the surrounding ground if the pipe were buried or at the ground surface and this would result in subsidence and damage to the pipeline itself. The design chosen was to elevate the pipeline above the ground, thus minimizing thawing of the permafrost (Figure 19.6b). The pipe itself expands as it warms and contracts as it cools. To allow the pipe to expand and contract as its temperature changes, the pipeline was built with bends. The pipe can move laterally by up to 4 m and vertically on its supports (Figure 19.6b). Furthermore, the vertical supports are equipped with thermal devices that help cool the permafrost during winter and prevent summer thawing. Provision was made for animals to cross the pipeline in certain places, which meant burying the pipeline for short distances with refrigeration units to prevent thawing of the ground. Despite its design the pipeline suffered two ruptures in 1977 and 1986 caused by unexpected settlement of the ground beneath (Williams, 1986).

(a)

(b)

Figure 19.6 The trans-Alaska pipeline (heavy line): (a) the route of the trans-Alaska pipeline crosses both the continuous and discontinuous permafrost zones; (b) the design of the trans-Alaska pipeline allows it to move laterally and vertically with expansion and contraction. The supports are designed to prevent melting of the permafrost by the pipeline (source: T & P Leeson/Photo Researchers Inc./Science Photo Library).

BOX 19.1

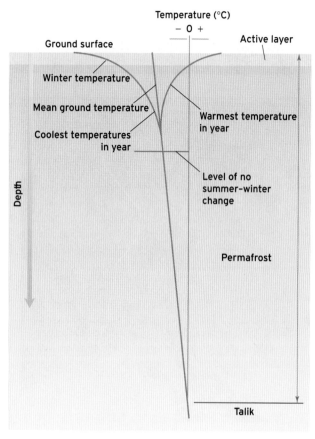

Figure 19.7 Theoretical temperature profile through material with constant thermal conductivity in a region in equilibrium with present conditions. (Source: after Isaksen *et al.*, 2000)

local ground temperature. Furthermore, vegetation cover or buildings can insulate the ground and cause thinning of permafrost.

The temperature at the Earth's surface varies between warmer summers and colder winters. However, deeper in the soil profile these variations are not so discernible. The **ground diffusivity** and the temperature fluctuation frequency control the depth to which surface temperature fluctuations are propagated. Box 19.2 describes the variation of ground temperature with depth in permafrost zones.

19.2.3 Reconstructing climate change from permafrost temperatures

Permafrost is ultimately a climatic phenomenon and there will be major changes in its distribution if climate warms as predicted over the twenty-first century. Because large quantities of carbon are stored in the permafrost, especially in peatlands and tundra regions, changes in distribution of frozen ground and the thickness of the active layer may result in the release of large amounts of greenhouse gas (e.g. methane) into the atmosphere (see below). Permafrost responds to climate change only slowly by aggrading or degrading over thousands of years (Molochuskin, 1973). It therefore damps out short-period oscillations and records major climatic change of several degrees over long

FUNDAMENTAL PRINCIPLES

SEASONAL VARIATION OF GROUND TEMPERATURE

Figure 19.8 shows the air and ground temperatures measured at Barrow, Alaska (71°N). Compare the variation in air temperature with the temperature at different depths in the ground. The air temperature is the forcing mechanism and the ground temperature is the response to this forcing. In general, the response is damped (the variation in temperature is less) at greater depths in the ground. For example, the air temperature shows a

double peak in July/August, which can also be seen at 0.6 m depth in the ground but delayed by approximately one month. Below this depth this detail cannot be seen. It becomes more difficult to see the maxima and minima at greater depth in the ground and at 18.2 m there is almost no variation in temperature from summer to winter. The graphs also show the progressive delay in timing of the maxima and minima as they propagate into the ground. For example, at 9.1 m depth the maximum temperature occurs in March delayed

by some 8 months from the air temperature change that caused it.

The wavelength of a temperature variation, λ (such as diurnal or annual temperature variation), is given by:

$$\lambda = \sqrt{4\pi D/f} \qquad (19.1)$$

where D is the diffusivity of the ground (Table 19.1) and f is the frequency of the perturbation. The thermal diffusivity varies with soil type as shown in Table 19.1. At a depth of one wavelength, the amplitude of the temperature change is reduced by a factor of about 0.002. Diurnal

BOX 19.2 ➤

➤ fluctuations are thus attenuated rapidly and affect only the upper 1 m or so, whereas annual fluctuations propagate to depths of 15 m or more. The timing of the maximum and minimum temperature becomes progressively lagged with depth and the lag increases with decreasing frequency; the velocity of propagation of maxima and minima into the subsurface is $4D\pi f$.

It is possible to estimate the depth of the active layer at the site shown in Figure 19.8. The base of the active layer occurs where the ground temperature remains below freezing throughout the year. The graphs show that at 0.6 m depth in the ground the temperature is above freezing during July and August. At 2.4 m depth the temperature remains below freezing throughout the year; the active layer is therefore

between 0.6 and 2.4 m in thickness. A final observation is that at 0.6 m depth the ground temperature remains constant at 0°C during October and November despite the air temperature becoming progressively colder during this period. This is because of the release of latent heat of freezing, which prevents the temperature of the ground dropping below 0°C until all the water at this depth in the soil is frozen.

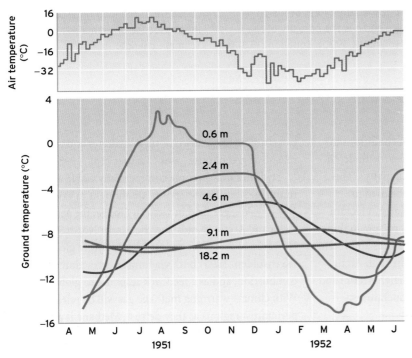

Figure 19.8 Typical air and ground temperatures at Barrow, Alaska. Surface temperature fluctuations take longer to have an impact for deeper layers and so the lag time between surface temperature increase and ground temperature increase is greater with depth. The magnitude of the change is also much greater near the surface and declines with depth. (Source: after Price, 1972)

Table 19.1 Typical thermal properties of soils and constituent components

Material	Water content (m^3 m^{-3})	Density (kg m^{-3})	Mass heat capacity (J kg^{-1} K^{-1})	Thermal conductivity (W m^{-1} K^{-1})	Thermal diffusivity ($\times 10^{-6}$ m^2 s^{-1})
Sandy soil	Dry	1600	800	0.3	0.2
	0.2	1800	1180	1.8	0.9
Clay soil	0.2	1800	1250	1.2	0.5
Water (0°C)		1000	4180	0.6	0.1
Ice (0°C)		917	2100	2.2	1.2
Air		1.2	1010	0.03	20.6

(Source: after Williams and Smith, 1989)

BOX 19.2

ESTIMATING PERMAFROST THICKNESSES

As discussed above, the controls on permafrost thickness are the mean annual surface temperature (T), ground conductivity (k) and the geothermal heat flux (G) and these parameters can be used to estimate the equilibrium permafrost thickness, H:

$$H = \frac{Tk}{G} \qquad (19.2)$$

An estimate of the time t taken to form a thickness of permafrost H from unfrozen sediment at 0°C can be obtained from:

$$H = \sqrt{\frac{2kTt}{\rho Ln}} \qquad (19.3)$$

where ρ is the density of the ground, L is the latent heat of fusion of ice and n is the porosity of the soil (Lock, 1990). Equation (19.3) is applicable to the early stages of growth of ice-rich permafrost from zero initial thickness. Large discrepancies in actual thickness compared with this simple model (equation (19.2)) can occur, mainly because of the presence of surface thermal disturbances such as water bodies and because of the long timescales for the response of permafrost to climate change.

BOX 19.3

timescales (Figure 19.9). Under equilibrium conditions the temperature profile within the ground is linear with depth (assuming a constant ground thermal conductivity). The ground's temperature profile can therefore be used to extrapolate past surface conditions, particularly in tectonically stable areas. Permafrost is generally increasing in thickness in those areas with low surface temperatures and recent tectonic uplift (e.g. Melville Island, Canadian High Arctic) or glacial retreat (e.g. Svalbard, Norwegian High Arctic) and degrading in areas with warmer surface temperatures.

19.2.4 Gas hydrates

A **gas hydrate** is a crystalline solid in which molecules of gas are combined with molecules of water. The hydrate of methane is stable at high pressures and low temperatures and so occurs commonly in ocean sediments with smaller amounts in permafrost regions. Global estimates of the methane stored in hydrate are around 10^{16} kg, which represents one of the largest sources of hydrocarbons on Earth: there is about twice as much carbon stored in gas hydrates than in all other fossil fuels put together. Increases in temperature or decreases in pressure may result in the hydrate becoming a mixture of gas and ice or water. The instability of hydrates may be a hazard during drilling operations, resulting in blow-outs due to gas build-up following warming via the drilling mechanism (Yakushev and Chuvilin, 2000). Hydrates of carbon dioxide are stable in the Martian ice caps (see Chapter 18). The flooding that appears to have scarred the landscape of Mars in the past may have resulted from

catastrophic breakdown of these hydrates and associated release of greenhouse gases (Kastner *et al.*, 1998).

Hydrates may become commercially viable as a natural resource. Significant deposits are thought to exist in the Messaryakha gas field in western Siberia, the Mackenzie Delta and Arctic Islands and the Alaska North Shore. However, hydrates may contribute to future global warming (Kvenvolden, 1995) both via human extraction and combustion, and as a feedback response to global warming. With climate warming methane gas will be released from permafrost regions as they defrost. Methane is a strong greenhouse gas, and so this may result in positive feedback, leading to warmer global temperatures and release of further gas to the atmosphere. Release of methane at the end of the last glacial period may have played an important role in the rapid melting of the major ice sheets (Buffet, 2000).

19.2.5 Hydrology in permafrost regions

19.2.5.1 Groundwater flow

Groundwater movement within permafrost regions is often restricted by the presence of frozen ground that acts as a barrier to flow. At temperatures significantly below 0°C, the hydraulic conductivity of the ground is greatly reduced, exerting a retarding influence on groundwater flow. In such regions the freezing of the active layer introduces a seasonal aquitard (see Chapter 13) and the underlying permafrost represents a perennial aquitard to vertical flow, restricting aquifer recharge and water flow. In these regions permafrost

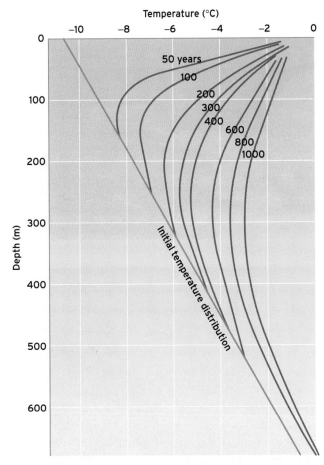

Figure 19.9 The effect of climate change on ground temperatures. The theoretical ground temperature curves show the effect of climate change (in this case a step climate warming) on the measured temperature in the ground. The figure ignores the active layer. (Source: reprinted with permission, after *Proceedings of the Second International Conference on Permafrost, Yakutsk, USSR* (1973) by the National Academy of Sciences, courtesy of the National Academies Press, Washington, DC)

can act as a barrier to contaminant transport. However, at temperatures close to zero there may be little difference between the conductivity of frozen and unfrozen regions of the ground (Anderson and Morgenstern, 1973), allowing water flow through frozen ground and aquifer recharge. In the discontinuous permafrost zone, water can flow through open taliks. Furthermore, in some regions seasonal freezing of the active layer does not reach the depth of the permafrost, allowing groundwater flow through a residual thaw layer above the permafrost confined below the upper frozen part of the active layer. Water can also flow beneath the permafrost in both discontinuous and continuous permafrost zones. The occurrence and flow of groundwater in permafrost regions are governed by the same physical processes as in more temperate regions (see Chapter 13).

19.2.5.2 River flow

River regimes in periglacial regions are typically very seasonal with large discharges resulting from the melt of winter snow cover during spring. This snowmelt usually occurs over a short period of time, typically 2–3 weeks, resulting in a short-lived flood event. In many areas between 25 and 75% of the total runoff is concentrated in a few days. Furthermore, because permafrost retards downward percolation, runoff is often rapid following a snowmelt or rainfall event. In glaciated catchments the large and variable flow continues into summer owing to ice melt and changes in the glacier hydrological system as shown in Figure 19.10.

Small rivers in periglacial regions flow only during the summer. However, despite cold winters major rivers continue to flow all year round under an ice cover. Thus, a distinction should be made between these large rivers and smaller rivers. The large rivers in the Arctic are often fed from a mixture of snowmelt plus input from non-periglacial regions and from deep springs within the discontinuous permafrost zone. Their discharge characteristics are therefore less peaked than smaller rivers which are fed solely from snow- and ice melt. The larger rivers, such as the Mackenzie River in Canada, and the Lena, Ob, Yenesei, Kolyna and Indigirka Rivers in Siberia, form an important transport network in the Arctic. Despite the fact that they flow for most of the year they are typically navigable for only a short annual period between May and October when the surface ice layer breaks up.

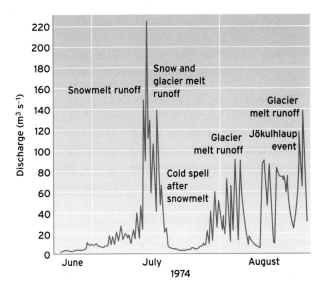

Figure 19.10 Hydrograph from Sverdrup River, Ellesmere Island, Nunuvut, Canada, showing a major peak from snowmelt and subsequent peaks from glacier melt and a glacier outburst flood known as a jökulhlaup. The basin area is 1630 km². (Source: after Woo, 1986)

The more peaked runoff characteristics of smaller rivers result in a high sediment transport rate compared with rivers with similar total discharges but which have discharge more evenly distributed during the year. For example, the River Mecham in Arctic Canada is fed largely by snowmelt. As a result 80–90% of its annual flow is concentrated into a 10 day period during which peak velocities may reach 4 m s^{-1} (Summerfield, 1991). Total sediment yield from the catchment is estimated to be 22.1 t km^{-2} yr^{-1} (French, 1996).

19.2.5.3 Seasonal features

Seasonal icings are mounds of ice that form in topographic lows during winter in locations where groundwater reaches the surface (Figure 19.11a). These are essentially zones where return flow (a component of saturation-excess overland flow, see Chapter 13) occurs and freezes. Icings also occur in river channels that freeze to their beds. Icings are common features downslope of warm-based glaciers (see Chapter 18) in Svalbard, Norway. Icings may form either below or above the ground surface. The ice forming these features is often stratified and may comprise **candle ice**, which consists of vertically orientated crystals of over 1 m in length (Figure 19.11b). In some regions, for example Yakutia, Siberia, large icings may form that do not fully melt in the summer and so survive from year to year.

(a)

(b)

Figure 19.11 (a) Icing formed downslope of a warm-based glacier in Yukon Territory, Canada. (b) The icing is made up of stratified ice and candle ice such as these elongated crystals. The black lens cap provides some idea of the scale of the crystals which can be over 1 m in length.

Reflective questions

➤ What are the controls on the distribution of permafrost?

➤ How are surface temperature changes reflected by ground temperatures in permafrost zones?

➤ Aklavik, a town in Canada situated in a permafrost region, was planned to be replaced in 1954 because of flooding, subsidence and difficulties in construction due to increasing depth of the active layer. Inuvik was built to replace Aklavik, although in fact many residents stayed in Aklavik. What considerations would you use to choose the most suitable site for a new town like Inuvik?

➤ What are the characteristics of river regimes in periglacial environments?

➤ Why are permafrost regions particularly sensitive to climate change and why might they contribute to further global warming?

19.3 Geomorphology of permafrost and periglacial environments

In this section we discuss the geomorphological features that form in periglacial and permafrost environments (see Section 19.1 for the distinction). In particular we will emphasize those features that allow differentiation between the two. Cold climate environments develop distinctive geomorphology because of three basic processes:

1. The 9% expansion of water on freezing which causes frost shattering which adds to scree development.
2. The contraction and cracking of rapidly freezing soils which forms **ice wedges** and polygonally shaped surface features.
3. The migration of water to the freezing front by suction which causes the formation of segregated ice.

Furthermore, permafrost and periglacial features form distinctive relict forms, collectively known as **thermokarst** because of their similarity to **karst** (cavernous limestone) features. Thermokarst features are extremely important in reconstructing past climate and the extent of former glaciers. They form in cold but non-glaciated terrain and are thus found at the margins of glaciers and can be used to demarcate former glacial boundaries. The presence of thermokarst features within the United Kingdom and Europe implies widespread permafrost conditions during the past (Figure 19.12). A range of active and relict permafrost and periglacial landforms are described in the following sections.

19.3.1 Ground ice features

Very large lenses of ice may slowly build up in soil that is frozen as a result of the migration of water to the freezing front in permafrost regions. These bodies of ice typically form only in the upper 5–6 m of ground and are known as **segregated ice**. Segregated ice may form thin bands. Where the bands are thick, sometimes up to several metres, they are known as **massive ice** (Figure 19.13). In both cases the concentration of ice can exceed 50% by volume. Clearly, the melting of such ground will produce a large volume of excess water. Such melting can cause seasonally impassable roads or the collapse of structures. Coarse deposits such as gravels and sands are highly permeable but have low potential to retain water, whereas finer deposits such as clay have low permeability but high water retention potential. This means that intermediate grain sizes, such as silt, have the greatest potential to form segregated or massive ice within the ground and are most susceptible to ice heave. When ice-rich ground thaws, **involutions** often form. Such features are shown in Figure 19.14. They are disruptions to the sedimentary structure of the ground and these features are often used as a diagnostic for past permafrost conditions. The melting of massive ground ice also affects the ground surface topography and produces thermokarst consisting of small irregularly shaped thaw lakes and depressions known as **alas**, which form when these thaw lakes drain. These features cover large areas in North America and Siberia. Similar features have been reported on the planet Mars (see Box 19.4).

Figure 19.12 The distribution of periglacial patterned ground in the United Kingdom. Widespread periglacial conditions occurred beyond the margin of the Quaternary ice sheets. (Source: after Sparks and West, 1972)

- Single ice wedges
- Ice wedge polygons
- Stone polygons and stripes
- Chalkland patterns
- Involutions
- Limit of ice during the last glacial maximum (see Chapter 20)

100 km

Figure 19.13 An example of banded massive ice, Peninsula Point, near Tuktoyaktuk, NWT, Canada. (Source: Photo was taken by Julian Murton at Peninsula Point near Tuktoyaktuk, NWT, Canada)

Figure 19.14 Involutions (frost-disturbed structures) in fluvial silt and sand, northern Belgium. The structures are related to loading and density differences in water-saturated sediments, probably during the thawing of ice-rich permafrost. (Source: from French, 1996)

NEW DIRECTIONS

PERMAFROST AND PERIGLACIATION ON MARS

On planet Mars, the mean annual temperature is −60°C and the planet has a dry periglacial-type climate. Permafrost currently extends over the planet's entire surface. Early in Martian history there is evidence for warmer climates and liquid water on the planet surface. At this time freeze-thaw processes were probably common. Certainly the surface of the planet provides evidence for polygonal-patterned ground similar, but at a larger scale, to that formed on Earth. Furthermore, the presence of apparent thermokarst features such as rampart craters, which resemble those resulting from the melting of ground ice on Earth, suggest high ice contents within the permafrost.

The dynamics of the Martian polar ice caps are discussed in Box 18.5 in Chapter 18. On Earth, microbial life has developed strategies to cope with extreme conditions and can survive within permafrost and basal ice beneath glaciers. It is thus possible that similar life forms have developed on Mars. Future Mars missions are likely to probe regions that may harbour such life forms.

BOX 19.4

(a)

(b)

Figure 19.15 Ice wedge landforms: (a) ice wedge in Svalbard (Source: Ole Humlum, University Centre, Svalbard); (b) ice wedge polygons in the Canadian Arctic seen from the air (Source: Peter Dunwiddle/Visuals Unlimited).

Figure 19.16 Formation of ice wedges. Soil will crack if it is cooled quickly (A). In summer this crack will fill with water (B), which subsequently freezes and expands the crack. Repeated thaw and freeze events (C) lead to the formation of an ice wedge (D). (Source: after Lachenbruch, 1962)

Ice wedges are V-shaped bodies of ground ice up to 1.5 m in width that can extend some 3–4 m into the permafrost (Figure 19.15a). Ice wedges develop because at low temperatures (less than approximately −15°C) frozen ground contracts as it is further cooled. If this occurs rapidly then cracking can occur as shown in Figure 19.16. The cracking of the feature is thought to occur in early winter, and the crack fills with water in spring and summer, which then freezes. The ice wedge, once developed, creates a weakness that tends to reopen annually as the ground contracts and hence the ice wedge grows. Ice wedges often exist in a network of **ice wedge polygons** (Figure 19.15b) which currently cover millions of square kilometres of the Earth's surface (Williams and Smith, 1989). Such polygons can also be identified on Mars (Box 19.4). Ice wedges actively grow in the continuous permafrost zone, forming only in perennially frozen ground, although some wedges may persist in the discontinuous permafrost zone. When ice wedges melt they often leave behind a landform known as an ice wedge cast, as shown in Figure 19.17. The ice is replaced by sediment, occasionally forming polygonal or linear troughs. Such features are easily recognized in sediment sections that were marginal to the ice sheets in the United Kingdom and Europe and they are reliable thermokarst features for identifying past permafrost conditions.

Pingos are ice-cored mounds up to 55 m high and 500 m in length which form in permafrost zones. Examples are shown in Figure 19.18. The mounds can be either conical or elongated and they contain some segregated ice and a core of massive ice described as a lens. The top of the mound often becomes cracked as ice core within the pingo grows. Two types of pingos are recognized: hydrostatic pingos (Figure 19.18a; formerly known as closed-system or Mackenzie Delta pingos) and hydraulic pingos (Figure 19.18b; formerly known as open-system or east Greenland pingos).

Hydrostatic pingos are caused by the doming of frozen ground as a result of the freezing of water expelled during talik elimination and the growth of permafrost beneath a former lake or other water body (Figure 19.19a). The features are usually isolated landforms found in continuous

Figure 19.17 Ice wedge cast, Belchatow, Poland. Note infilling from above. (Source: from French, 1996)

(a)

(b)

Figure 19.18 (a) Aerial view of a hydrostatic pingo in NWT, Canada. The Ibyuk pingo is 49 m high and 300 m long (Source: Paolo Koch/Science Photo Library Ltd.). (b) Hydraulic pingo in Svalbard. (Source: Ole Humlum, University Centre, Svalbard)

permafrost regions, predominantly in areas of low relief. Pingos formed over drained lakes are usually circular in shape, whereas pingos over old river channels may be linear in form. After the drainage of a water body, a closed talik may form within the ground, which traps water within the freezing sediments. Migrating water moves to the freezing front, forming pore and segregated ice (Figure 19.19b). Very high pore water pressures develop that force the ground upwards creating the pingo (Figure 19.19c). The growth rate of pingos in the Mackenzie Delta has been measured to be initially quite fast at approximately 1.5 m yr^{-1} (Mackay, 1973, 1979) and to decrease with age so that some of the largest pingos are probably around 100 years old and growing at a rate of 2.3 cm yr^{-1} (Mackay, 1986). Similar results have been determined in Siberia. As pingos grow, their sides become steeper, eventually opening radial cracks at their summits (Figure 19.18), which expose the ice core and the pingo begins to subside. Hydrostatic pingos are common on the Mackenzie Delta, where they are found across the delta following switches in the course of the river. Hydrostatic pingos have also been reported on the floor of the Beaufort Sea.

Hydraulic pingos form at the foot of slopes and result from the inflow and freezing of groundwater seeping from upslope (Figure 19.20). These features are most likely to occur in thin or discontinuous permafrost regions. In Svalbard, Norway, the water supply for pingos often flows from warm-based glaciers upslope as in the example given in Figure 19.20. Hydraulic pingos are usually round or elliptical in form and typically smaller than hydrostatic pingos, rarely exceeding 35 m in height. They often consist of multiple mounds (unlike hydrostatic pingos) because as one begins to subside the continuing water supply will tend to form another. These features are common in east Greenland, central Yukon, Alaska and Svalbard.

The melting of the ice core of a pingo initially leaves a small lake, but the final relict form is a central depression with sediment ramparts, sometimes termed a **pingo scar**. Such features have been identified at various locations in Europe including the United Kingdom and the Netherlands.

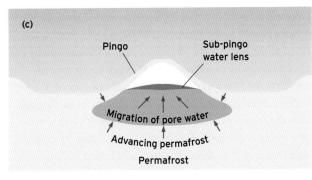

Figure 19.19 Formation of a hydrostatic pingo: (a) talik develops beneath a thaw lake; (b) lake drainage leads to refreezing of the talik; (c) progressive refreezing of the talik leads to pingo growth. (Source: after Mackay, 1983)

Figure 19.20 Formation of an hydraulic pingo. This figure shows a pingo forming from drainage from beneath a warm-based glacier in Svalbard. (Source: after Liestøl, 1977)

Palsas are low mounds, 1–10 m high, that form in peat in permafrost zones (Gurney, 2001). They form where snow is thin or discontinuous and have a core of segregated ice generated through suction of unfrozen water migrating to the freezing front. The lack of snow allows the ground to freeze to a greater depth than the surroundings and this ice then survives the subsequent summer, insulated by the peat above. Repeated winters result in frost heave, and a mound forms. This topography causes a positive feedback as snow is more likely to blow off, resulting in even deeper freezing depths. Changing vegetation cover can also play a role in palsa formation. When palsas melt, mounds and small lake-filled hollows are formed. **Frost blisters** are small ice-cored mounds that develop over just a single winter as a result of groundwater that freezes and uplifts the ground surface.

The ground surface in periglacial regions is often characterized by metre-scale organization of topography, vegetation or particle size in regular geometric patterns. There are two types of patterned ground. These are sorted patterned ground, such as circles, polygons or stripes, and unsorted patterned ground defined by topography, such as unsorted circles and stripes, or by alternation of vegetated and non-vegetated ground.

Sorted stone circles are typically arranged so that the fine material occurs in the centre of an area of lowered relief and coarse material forms an uplifted outer rim as shown in Figure 19.21(a) and (b) (Hallet *et al.*, 1988). Typically, polygons or circles occur on flat surfaces, and sorted stripes that are elongated downslope form on low-angle slopes (Figure 19.21c). On slopes greater than about 30°, mass movement prevents the formation of patterned ground. The features probably form by a variety of processes. One hypothesis involves convection within the ground. In summer, saturated soil close to the ground surface warms during the day whereas water at depth remains colder. Since water is densest at 4°C the colder water at depth is less dense than the water close to the surface, and this drives convection. Descending warm water can then melt the frozen surface below resulting in an undulating interface between frozen and unfrozen ground that is reflected in the surface topography. Sorting of the soil particles can occur if the soil particles convect with the soil water (Figure 19.22).

19.3.2 Slope processes

Many of the features that form in lowland areas also occur in alpine environments above the treeline. However, these alpine environments are often characterized by exposed, hard bedrock that has been eroded by the action of glacier ice, and steep slopes. Mountain tops are often covered by

(a)

(b)

(c)

Figure 19.21 Organized topographic features: (a) stone tilting near Rea Point, eastern Melville Island; (b) stone polygons, Svalbard; (c) stone stripes, Central Banks Island. (Source: from French, 1996)

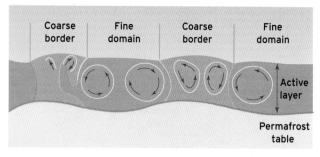

Figure 19.22 Inferred soil circulation patterns resulting in sorted circles. (Source: after Hallet *et al.*, 1988)

blockfields of frost-shattered material. Alpine permafrost may occur in these areas.

Periglacial activity, in particular repeated freeze–thaw, causes the formation and modification of slope deposits. Four types of processes cause **mass wasting** in periglacial environments: slopes evolve owing to fracture, debris (see Chapter 11) and **solifluction** flows, creep processes and **nivation**. Solifluction is especially likely to occur in regions underlain by permafrost. During the summer the active layer melts forming a mobile water-saturated layer. The process results in the formation of lobes and terraces. Slow creep of the active layer, at rates of only a few centimetres a year, occurs by processes of **frost creep** and a type of solifluction known as gelifluction. Frost creep occurs because freezing expands the soil normal to its surface but thawing results in settlement vertically resulting in a net downward movement. Gelifluction is this slow creep of water-saturated material. In seasonally frozen soil, movement typically occurs during spring as the ground thaws from the surface downwards. Because the underlying permafrost is largely impermeable, mass movement may occur on very low slopes (as low as 1°). Slow creep typically results in the formation of stepped ridges. Nivation is the localized erosion of a slope caused by a combination of frost action, gelifluction, frost creep and meltwater flow at the edges and underneath snow patches. Nivation commonly occurs in periglacial regions and is accentuated in permafrost-free areas. The combination of processes causes the development of nivation hollows as the snow patches sink into the hillside.

While most of these processes are not unique to periglacial environments and Chapter 11 provides details of their operation, mass wasting is probably most efficient in periglacial conditions (Table 19.2). The formation of screes from freeze–thaw fracturing of near-vertical rock faces occurs mainly in resistant rocks that are permeable only along fractures. Snow avalanches may move significant volumes of rock debris. Flows of unfrozen material are

Table 19.2 Rates of mass wasting by different processes at Karkevagge, Lapland, 1952-1960

Process	Rate (t km^{-1} yr^{-1})	Denudation (mm yr^{-1})
Transport of solutes by running water	26	0.010
Debris slides and flows	49.4	0.019
Slush avalanches, rock debris transport	14	0.005
Rock falls	8.7	0.003
Solifluction	5.4	0.002
Talus creep	1.5	0.001

(Source: adapted from Rapp, 1960)

Figure 19.23 Ploughing boulder. A prow has built up in front of the boulder and a trough has formed upslope of it.

promoted in permafrost regions because the seasonal melting of the active layer forms an upper layer of high water content.

Periglacial slope processes result in a range of landforms particular to periglacial slopes. **Protalus ramparts** are linear mounds of coarse sediment that form a small distance from the base of a slope. Snow persists at the foot of these slopes, particularly in the shade, which means that when a rock fall occurs boulders tend to slide across the snow and come to rest just beyond the snow bank. **Ploughing boulders** can also be seen on slopes in periglacial regions as shown in Figure 19.23.

The boulders move slowly downslope, leaving a trough upslope and forming a prow of sediment downslope. The movement of the boulders, typically a few millimetres per year, is thought to occur because of the different thermal conditions beneath the boulder compared with its surroundings.

Rock glaciers, as shown in Figure 19.24, typically consist of angular debris and have the form of a small glacier. They often show evidence of flow although there is no evidence of glacier ice at their surface. Within a rock glacier there is ice within the pore spaces (**interstitial ice**), and the slow flow of these features downslope is due to this. Some rock glaciers are probably formed from relict glaciers that have become debris covered, whereas other features are thought to

Figure 19.24 Rock glacier in Svalbard forming from glacier ice covered by debris. Clean ice can still be seen in the upper part of this rock glacier.

originate as snow and rock avalanches or ice-cemented rock debris in permafrost regions. Relict rock glaciers have been identified in the United Kingdom, including one in North Wales (Harrison and Anderson, 2001).

19.3.3 Loess and aeolian activity

Many permafrost regions are characterized by extreme aridity and are defined as polar deserts. In these regions there is abundant evidence for the transport of material by wind and for erosion by abrasion of the wind-blown material. Dust clouds are characteristic of many permafrost regions in autumn (Sparks and West, 1972). Material is deposited when the wind velocity drops or when precipitation falls, forming a silt deposit known as loess. During the past 2.4 million years, deposits of fine sediment formed on the outwash plains of the ice sheets, providing a plentiful source of material for transport. Intense frost action broke down larger material and provided additional material. Stronger winds occurred during this period. As a result erosion by wind-borne sediment and wind entrainment

has played a major role in the middle latitudes. Erosion formed wind-modified pebbles and blocks as well as **stone pavements**. Deposition resulted in loess which covers tens of thousands of square kilometres and is more than 100 m thick in some places. Loess deposits occur in large areas of North America and Europe that were situated on the southern margin of the ice sheets (see Chapter 20).

Reflective questions

➤ How can pingos and ice wedge casts be used to make climatic reconstructions?

➤ What are the main landforms associated with ground ice and how do they form?

➤ How do slopes evolve in periglacial conditions?

➤ What are the unique geomorphological features found on periglacial slopes?

19.4 Summary

This chapter describes both permafrost and periglacial environments, their distribution and the processes that occur within them. Periglacial processes currently affect about 35% of the Earth's surface, and during the past their impact was even more widespread. This past activity has left a range of landforms that are characteristic of such environments and that can therefore be used to reconstruct past climates. Understanding the processes operating and features formed in these is important for engineering in permafrost and periglacial regions. These regions are extremely sensitive to climate change as well as being the possible source of greenhouse gases that may drive further climate change through gas hydrate release.

Permafrost can be found in high latitudes and at high altitudes. In the coldest locations it can be over 1000 m deep but in more marginal climates it can be discontinuous and thinner. The upper ground layer often melts during the summer and this is known as the active layer.

Surface temperature fluctuations propagate down through the ground so that there is a delay in response. The lag time increases with depth and the magnitude of change decreases with depth. Deeper layers respond only very slowly to surface temperature change so that they can be used to identify former temperature regimes at that site.

Both permafrost and periglacial regions have a unique geomorphology with a range of landforms that develop owing to ice formation and melt at and below the ground surface. These melt and thaw processes cause expansion and contraction of the surface layers which results in surface collapse and cracking. As cracks fill with water and freeze they can expand and eventually large ice wedges can form. These are manifest at the surface by polygonal features. Freezing of water beneath thermal disturbances such as lakes can cause trapped water at high pressure which can result in features such as pingos. Slope features include rock glaciers and ploughing boulders. Periglacial slopes are dominated by mass wasting and gelifluction and frost creep processes.

Further reading

Ballantyne, C.K. and Harris, C. (1995) *The periglaciation of Great Britain.* Cambridge University Press, Cambridge.
This is an excellent book giving detailed explanations of relict periglacial features, their distribution and the processes that formed them. Although it is related to Great Britain the discussion can equally be applied to other sites.

Buffet, B.A. (2000) Clathrate hydrates. *Annual Reviews of Earth and Planetary Science*, 28, 477–507.
This is a paper providing further (but technical) details on hydrates.

French, H.M. (2007) *The periglacial environment*, 3rd edition. John Wiley & Sons, Chichester.
A new edition of this clearly written textbook with an excellent range of diagrams and good use of examples.

Summerfield, M.A. (1991) *Global geomorphology.* Longman Scientific, Harlow.

Refer in particular to chapter 19 which provides many results on the geomorphology of Mars and provides further information on Box 19.4.

Trenhaile, A.S. (1990) *The geomorphology of Canada: An introduction.* Oxford University Press, Oxford.
This is a short book describing the physical geography of Canada. Many of the descriptions are of features developed in permafrost and periglacial regions.

Williams, P.J. (1979) *Pipelines and permafrost: Physical geography and development in the circumpolar north.* Longman, Harlow.
This is an excellent short book describing the making of the Alaska pipeline and other related topics.

Williams, P.J. and Smith, M.W. (1989) *The frozen Earth: Fundamentals of geocryology.* Cambridge University Press, Cambridge.
This is the permafrost 'bible', although much is at quite a high level. If you want more information on processes in permafrost regions this is the book to look at.

Web resources

Arctic National Wildlife Refuge
http://www.r7.fws.gov/nwr/arctic/
The search engine of the US Fish and Wildlife Service websites provides articles and information on periglacial environments and animated views of periglacial features.

Intergovernmental Panel on Climate Change
http://www.ipcc.ch/
These are the International Panel on Climate Change (IPCC) pages. The latest reports contain material on permafrost and the effects of climate change in permafrost regions.

International Permafrost Association
http://www.geo.uio.no/IPA/
The International Permafrost Association provide news on permafrost research and a large list of links to organizations undertaking research on permafrost.

Natural Resources Canada
http://atlas.nrcan.gc.ca/site/english/maps/environment/land/permafrost
http://tsdmaps.gsc.nrcan.gc.ca/website/_permafrost/permafrost_e.htm
These sites provide permafrost maps for Canada.

Permafrost at the Geological Survey of Canada
http://gsc.nrcan.gc.ca/permafrost/index_e.php
A whole suite of pages covering many of the topics discussed in this chapter is given here.

US Geological Survey's Borehole Temperature Logs
http://esp.cr.usgs.gov/data/bht/alaska/
US Geological Survey's borehole temperature logs from Arctic Alaska, pre-1989, are given here. This website contains a rich resource of data. You may want to choose an east–west or north–south transect of boreholes and plot the subsurface temperatures. Then you can try to explain the variation in the plots that you have made. From this you will be able to produce a map of permafrost depths and active layer thicknesses in Alaska.

Visit our website at **www.pearsoned.co.uk/holden** for further questions, annotated weblinks, interactive models and video-clips showing physical processes in action.

Environmental change

Figure PVI.1 Global warming is leading to enhanced rates of ice melt which reduces albedo and raises sea levels. (Source: Seth Resnick/Getty Images)

Part contents

Scope

This final part deals with environmental change, its monitoring, modelling and management. It starts with Chapter 20 which describes the nature of global environmental change over the past 2.4 million years. A range of evidence suggests that the Earth's climate has warmed and cooled several times over that period, resulting in massive expansion and contraction of the world's ice sheets and an associated rise and fall of global sea level. These natural environmental changes are driven by a range of processes, some of which are related to external factors such as Earth's changing distance from the Sun and others driven by internal factors associated with surface-ocean-atmosphere-biosphere interactions. Such natural changes have been observed over long timescales (hundreds of thousands of years) and over centennial and decadal timescales. The evidence from history suggests that the Earth's climate system can sometimes change very rapidly. Studying the past can provide us with greater foresight and lessons for the future. Important and complex feedback mechanisms have been observed involving interactions between oceans, ice sheets, the biosphere and the atmosphere. The discovery of these interrelated adjustments helps explain more about present-day landforms, plant and animal distributions, and the climate system and how it may behave in the future. Chapter 20 therefore provides the context within which to view environmental changes over the timescale of human existence so that we can judge their importance and potential implications.

Chapter 21 examines contemporary climate change and demonstrates how this is occurring at an unprecedented rate with higher atmospheric CO_2 concentrations than at any other time in the past 2.4 million years. It describes how human populations have changed the landscape and altered the concentration of natural and human-made chemicals in the atmosphere, land and waters of the Earth. Much of the change has been a result of human requirements for food and for fuel. Over 6 billion humans now consume or dominate 40–50% of the land's biological production. They have progressively attained the capacity to alter the Earth's physical systems (such as the carbon and nitrogen cycles) in drastic and potentially permanent ways. The chapter discusses predictions of global climate models and how we might mitigate against climate change. The chapter also discusses the carbon cycle and the role of humans in altering the amounts of carbon stored in different parts of the cycle.

Chapter 22 deals with how climate change is driving changes in vegetation distributions and processes, but also establishes further feedbacks between vegetation and the environment. The chapter presents a series of case studies from different environments to provide evidence that vegetation is responding to human-induced climate change. Biodiversity loss and its importance are discussed in this chapter, which also shows how human modification of the world's biomes is having large impacts on climate and how integrated climate-vegetation modelling techniques are allowing us to predict what might happen in the future.

It is necessary to measure contemporary environmental change so that we can be aware of the global and local nature of such change. Many global and local-scale changes are difficult to measure from the ground and many areas remain inaccessible. However, types of remote sensing technology (such as satellite imagery) help provide quick, cost-effective and new methods of monitoring global change at all scales. Remote sensing now plays a pinnacle role in monitoring and detecting environmental change, such as alterations in vegetation cover and type, ice caps, global and oceanic temperatures, cloud cover and the ozone layer. The significance of this vital commodity will no doubt continue to grow, and for this reason it is important to understand the basic techniques involved. Chapter 23 gives a valuable overview of the methods employed.

A basic premise of environmental science is the desire to apply the knowledge we attain to help manage the environment. Elements of environmental change are described in every chapter in this book. In addition, environmental management issues are also discussed in every chapter in this book. Therefore it is necessary for the final chapter of this book to describe some of the general tools and pitfalls associated with managing environmental change that can be applied to all of the areas that physical geography touches. Arguably, the management of environmental change, species threatening as it is, is one of the most important aspects of science and management today. With improved physical geography it is often possible to foresee and perhaps to forestall the consequences of our actions. As demonstrated in the accompanying chapters of Part VI, change is not a single or simple concept: complex relationships exist between the factors that promote change and the nature of the change in terms of its direction, rate and reversibility. Accordingly, management of environmental change is not infallible and requires both scientific understanding and an ability to judge and make difficult decisions. The situation is further compounded by the economic and political imperatives that often take precedence. Nevertheless, Chapter 24 seeks to show how physical geography can be tailored to a form that is useful in supporting environmental management.

CHAPTER 20

Quaternary environmental change

Joseph Holden
School of Geography, University of Leeds

Learning objectives

After reading this chapter you should be able to:

➤ describe the major climate changes of the Quaternary

➤ understand the orbital forcing hypothesis and a range of feedback mechanisms

➤ describe the main types of evidence for Quaternary environmental change

➤ list the main dating techniques and explain issues surrounding correlation of different types of evidence from different locations

➤ evaluate the role of modelling in palaeoclimatic reconstruction

20.1 Introduction

The Quaternary is the period of geological time within which we live. It covers approximately the past 2.4 million years. The onset of the Quaternary is defined by a change in the Earth's climate from a period of general stability (a very slow cooling over 50 million years with some minor fluctuations) to a period of great instability. The Quaternary period is characterized by frequent fluctuations of warming and cooling of the Earth's atmosphere and the growth and retreat of major ice sheets. Defining the exact time when the Quaternary period started is problematic because the change from the warmer earlier Tertiary period to the colder Quaternary was gradual. Figure 20.1 provides the subdivision of Earth history within which we can place the Quaternary with approximate dates before present (BP). It should be noted that although 2.4 million years might seem a long time, in the context of Earth's history the Quaternary period is actually very short.

The Quaternary has often been split into two parts, the **Pleistocene** and the **Holocene**. The Holocene is the past 10 000 years when the Earth's climate has been relatively warm and the major ice sheets that covered large parts of the land surface retreated. We are currently in one of many interglacial periods during the Quaternary. Since we live in the Quaternary, we are actually living in the ice ages; this is evident by the fact that there are still glaciers and ice sheets on the planet today. Before the Quaternary the world was much warmer, although as Figure 20.1 shows, there were probably four other cold periods in Earth's history. The most recent of these was around 280 million years ago.

One of the major challenges that Quaternary scientists face is in putting together the evidence for past environmental change. Many natural systems are affected by

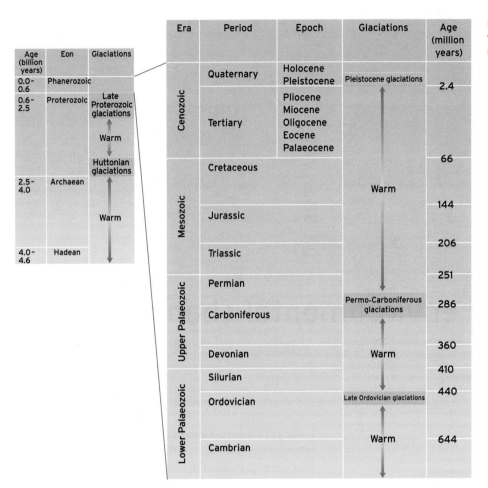

Figure 20.1 Earth's geological timescale. The Quaternary is the most recent period.

climate and so evidence of these past systems might give us information on past climate. There are many types of evidence for Quaternary environmental change including the constituents of gas bubbles trapped in ice cores, ocean floor sediments and their organic and inorganic constituents, glacial deposits and erosional features, periglacial and glacio-eustatic features, aeolian deposits (see Chapters 12, 16, 18 and 19), mineral deposits, pollen found in peat and lake deposits that have built up over time, and other plant and animal remains. There are two main challenges with these types of data: the first is in establishing how these types of evidence actually relate to a particular state of the environment; the second is in tying together one type of evidence from one place with another type of evidence from a different place. This is particularly problematic because there are often no appropriate absolute dating methods that can be employed.

It is useful to understand the environmental changes that have taken place during the Quaternary period so that we can explain more about present-day landforms and plant and animal distributions. It also helps us to explain how the climate system is behaving today and how it is likely to behave in the future. Understanding Quaternary environmental change allows us to place present-day climate changes into a much longer-term context so that we can judge their importance and potential implications. For example, we know that carbon dioxide (CO_2) concentrations have fluctuated between high and low in the past. By studying the past we can see how the Earth–ocean–atmosphere system behaved under naturally high CO_2 conditions and learn how the system might change if humans continue to increase atmospheric CO_2 concentrations. As this chapter will show, there are important and complex feedback mechanisms that must be taken into account. While some climatic changes can take thousands of years to take place, others can be sudden and occur over just decades. It is therefore crucial that we understand Quaternary environmental change as there are major implications for our present-day world.

20.2 Long-term cycles, astronomical forcing and feedback mechanisms

Several theories have been put forward in an attempt to explain why there have been phases of cooling (with ice advance) and warming (with ice retreat) during the Quaternary. These include plate tectonics which can result in changing ocean and atmospheric circulation patterns through mountain building, plate movements and releases of greenhouse gases such as methane and CO_2 into the atmosphere. Volcanic dust emissions can also cool the Earth by blocking out the Sun's energy. Plate tectonics (see Chapter 2) created the suitable conditions for the beginning of the ice ages by positioning Antarctica in a thermally isolated position over the South Pole. The northern hemisphere continents were also huddled around the Arctic Ocean. This then allowed other factors to become important (Imbrie and Imbrie, 1979). Several driving forces behind the Quaternary environmental changes have been proposed and many of the interactions between these forces will be discussed in this chapter. However, there is an important external driving factor that helps partly explain the glacial cycles that have occurred during the Quaternary. This is known as orbital forcing or the **Milankovitch theory**.

20.2.1 Orbital forcing theory

From 1912, Milutin Milankovitch, a mathematician, followed up the nineteenth-century work of astronomers Joseph Adhemar and James Croll by computing the radiation received at the top of the Earth's atmosphere over time as the Earth's orbit varied. The basis of the Milankovitch theory is that the amount of energy reaching different parts of the Earth from the Sun varies as the shape of Earth's orbit around the Sun, the angle of tilt of the Earth's axis, and the direction that the axis of rotation points change over time. These three parameters illustrated in Figure 20.2 vary in a way that determines the amount and distribution of solar radiation at the Earth's surface and in a way that is regular and predictable.

The first variable orbital parameter is called the **eccentricity of orbit** which is where gravitational forces cause the shape of the Earth's orbit around the Sun to change from more circular to more elliptical and back again over a period of 100 000 years. As the eccentricity increases, the difference in the Earth's distance from the Sun at the orbit's closest and furthest points also increases. This causes the seasons in one hemisphere to become more intense while the seasons in the other are moderated. The second orbital factor is the **tilt of the Earth**. The tilt of the Earth's

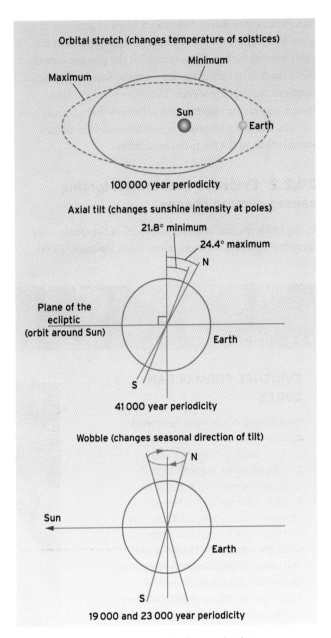

Figure 20.2 Milankovitch orbital forcing mechanisms.

axis varies from approximately 21° to 24° and back over a period of 41 000 years. The greater the tilt, the more intense the seasons in both hemispheres become; summers get hotter and winters colder. The third variable is axial 'wobble' which causes a phenomenon called **precession of the equinoxes**. The gravitational pull exerted by the Sun and the Moon cause the Earth to wobble on its axis like a spinning top, where the end of the axis describes a circle in space and determines where in the orbit the seasons occur, and most importantly the season when the Earth is closest to the Sun. This happens over two cycles of 19 000 and 23 000 years. This factor governs the interplay between the first two factors since precession determines whether summer in a given hemisphere occurs at a near or distant point in the orbit around the Sun. For example, at the present time the Earth reaches its furthest point from the Sun during the southern hemisphere winter. Therefore southern hemisphere winters are slightly colder than northern hemisphere winters. Southern hemisphere summers are also slightly warmer than in the northern hemisphere.

20.2.2 Evidence that orbital forcing causes climate change

In the 1950s the first continuous record of the Quaternary came from marine sediment cores, when Emiliani (1955)

analysed the isotopic composition of fossil calcium carbonate skeletons of small, single-celled marine organisms called **foraminifera**. This record consisted of a long core of ocean sediment that had been deposited over time. The sediment on the ocean floor builds up slowly as the remains of marine creatures sink to the bottom when they die. The nature of the oxygen isotope ratios from the skeletons of foraminifera taken from different layers of the core can tell us how much of the world's water was locked up in glaciers. This is because the chemistry of the ocean water changes depending on how much water is left in the oceans compared with that stored as ice. The mechanism is explained in Box 20.1.

The results from ocean core work showed that there were many fluctuations in the Earth's climate during the

FUNDAMENTAL PRINCIPLES

EVIDENCE FROM OCEAN CORES

On the deep ocean floor, sediments have been accumulating relatively undisturbed for millions of years. These sediments are a mixture of land-derived material and marine biogenic sediments. The terrestrial sediments are mainly sand, silt, clay and dust, which reach the ocean via wind, **ice rafting** and fluvial inputs. The biogenic sediment is composed of calcareous and siliceous skeletal remains of microorganisms that lived in the ocean waters. Figure 20.3 shows a typical section of sediment in a core being analysed while Figure 20.4 illustrates the remains of planktonic foraminifera which are surface-dwelling organisms and are indicators of the prevalent sea surface temperature at the time. Their shells are coiled left in temperatures below 7°C and right in temperatures greater than 7°C. The different species that make up foraminiferal assemblages also indicate the temperature, salinity and nutrient availability in the oceans.

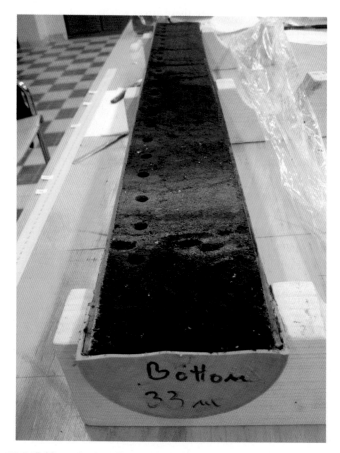

Figure 20.3 Evidence from sediments deposited over time being investigated in this part of a core. (Source: photo courtesy of Alice Milner)

BOX 20.1 ➤

Figure 20.4 Fossil shells of foraminifera. Species assemblages can tell us about sea temperatures and productivity and the chemistry of their shells can tell us about the chemistry of the ocean water when the microorganisms were alive. (Source: Thomas M. Gibson/USGS)

^{18}O isotope in the oceans decreases. During interglacials when the water is returned to the oceans the proportion of ^{16}O increases. Therefore, the ratio of ^{18}O to ^{16}O in the ocean sediment provides a long-term proxy for global ice volume during the Quaternary. This record has been divided up into stages, each of which corresponds to either a glacial or interglacial period (Figure 20.5; Figure 20.6e). Odd numbers represent warm periods and even numbers represent cold periods. More minor climatic events (stadials and interstadials) are sometimes given a letter coding or decimal place (e.g. 5e = 5.5) after a number (Figure 20.5).

Shackleton and Opdyke (1973) formalized the use of oxygen isotope stratigraphy to indicate changes in global sea level. Oxygen isotope analysis can be performed on foraminifera, and these are typically done on bottom-dwelling (**benthic**) species. Their skeletons contain some of the chemical constituents of the seawater when they were formed. During glacial periods more of the world's water is locked up in ice sheets and less is in the ocean. The lighter isotope of oxygen which is contained in water (H_2O) is evaporated more easily and more readily reaches polar regions to be stored as ice (the heavier isotope is also precipitated out more easily before it can get there). Therefore, during glacial periods the proportion of ^{16}O to the heavier

Figure 20.5 (a) The oxygen isotope curve and (b) sea-level change for the last glacial and interglacial period. Note that the ice seems to build up slowly and then very quickly melt away.

BOX 20.1

Quaternary that caused the growth and retreat of ice sheets. These results seemed partly to match the predictions provided by the Milankovitch model as shown in Figure 20.6 (Broecker and Denton, 1990). Figure 20.6(a), (b) and (c) shows the effect of eccentricity, tilt and precession on energy received at 65°N during the past 1.6 million years. Figure 20.6(d)

shows the cumulative effect of the three forcing mechanisms on radiation received at 65°N. Figure 20.6(e) shows the volume of the Earth's ice sheets determined from foraminiferal oxygen isotope ratios in ocean sediments for the same period. Significantly there is an approximate match between Figure 20.6(d) and (e). For example, there

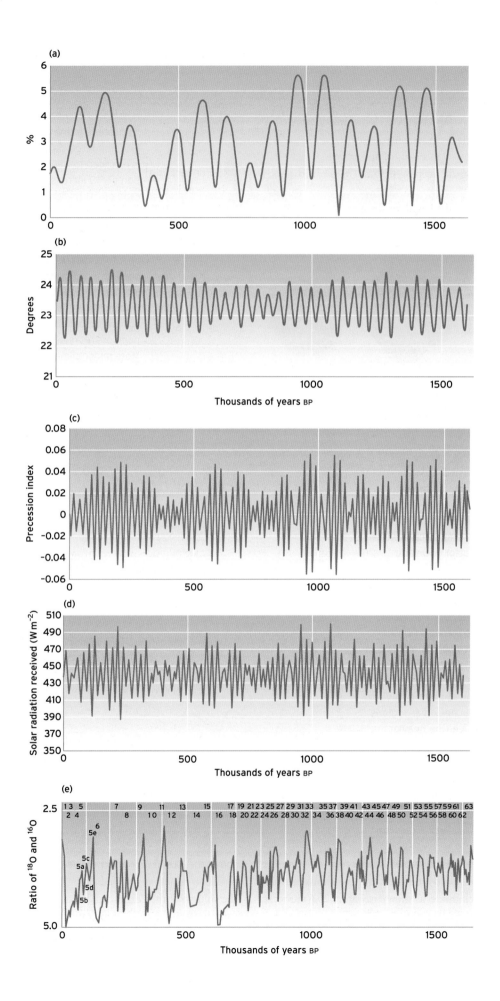

are eight large glacial build-ups over the past 800 000 years on an approximately 100 000 year cycle, each coinciding with minimum eccentricity. Smaller decreases or surges in ice volume have come at intervals of approximately 23 000 years and 41 000 years in keeping with the precession and tilt frequencies.

Other evidence for the length of the glacial cycles comes from the record of sea-level change provided, for example, by coral reefs in Barbados or Papua New Guinea (Chappell and Shackleton, 1986). Since water is stored on the continents as ice during glacials, and is released during interglacials when ice sheets melt, sea levels rise and fall by over 100 metres during climatic cycles. During the last glacial period (the peak of which was reached only 18 000 years ago) it was possible to walk from present-day England to mainland Europe as the North Sea and English Channel were dry land. Coral growth is linked to sea level and different growth periods for the coral can be dated up to about 160 000 years ago. This dating can be done by determining the ratio of uranium to thorium in the coral. Newly formed coral is high in uranium whereas old coral has less uranium and more thorium (see below). Again the evidence seems to match that from the ocean sediments and the predictions of the Milankovitch cycles as to the timing of glacial cycles (e.g. Chappell and Shackleton, 1986). High sea level coincides with the interglacial intervals and also coincides with the predicted timing of these from Milankovitch calculations.

Today it is recognized from the ocean sediments that there have been dozens of alternating cold and warm phases during the Quaternary. The cold phases are called **glacial intervals** and the warm phases are known as **interglacial intervals**. During glacial periods ice sheets and glaciers extended to lower latitudes from the poles. Less intense and shorter variations in global temperature that occur during these main cold and warm periods are called **stadial** (cold) intervals and **interstadial** (warm) intervals.

20.2.3 Problems with orbital forcing theory

There are a range of problems with orbital forcing theory and the use of marine oxygen isotope records. Firstly, it has recently emerged that benthic foraminifera oxygen isotope

records (Box 20.1) do not strictly represent a pure sea-level record and hence a record of ice advance and retreat (Skinner and Shackleton, 2005, 2006; Shackleton, 2006). There appears to be a signal from deep ocean temperature and this signal can vary between oceans as the local hydrology of the oceans reacts differently to changes in atmospheric temperature and ice growth or retreat. For example, Skinner and Shackleton (2005) showed that the isotope record for the end of the last glacial appears to be about 2200 years earlier than the likely change in ice volume, which is further 1700 years earlier than the deep Pacific isotope record. This offset in timing creates a problem because it is now no longer possible to simply link all sediments showing a change from cold to warm period from different oceans as being of the same age.

Secondly, the expected changes in temperature based on the changes in insolation via orbital forcing are not enough to explain temperature changes required for the vast ice expansions and retreats recorded. In fact there appears to be a 4–6°C shortfall. So something other than orbital changes must have been acting to cause changes in Quaternary environments.

Thirdly, the 100 000 year cycle, according to orbital calculations, should have a much weaker effect on incoming solar radiation than the shorter cycles. However, as Figure 20.6(e) shows, the 100 000 year cycle is dominant at least from about 800 000 years ago. Finally, the Milankovitch cycles should show a smooth rise and fall as in Figure 20.6(e) whereas the ice curve in Figure 20.6(e) is sawtooth in pattern. Over tens of thousands of years ice sheets built up several kilometres thick and scoured and scarred the landscape as far south as central Europe and Midwestern United States (e.g. Figure 20.7). But each cycle ended abruptly. Within a few thousand years the ice sheets melted back to present-day patterns.

So while orbital forcing mechanisms seem to be a good 'pacemaker' of Quaternary environmental change (Imbrie and Imbrie, 1979) it is necessary to look at other processes too. The next section details how internal feedback mechanisms may help explain some of the patterns that orbital forcing theory alone cannot account for.

Figure 20.6 Orbital forcing and global ice volume for the past 1.6 million years with calculated effect of orbital forcing parameters on solar radiation received at 65°N: (a) eccentricity; (b) tilt; (c) precesion; (d) cumulative effect of all orbital forcing parameters at 65°N during June; (e) the actual oxygen isotope record of global ice volume. (Source: (e) after Bridgland, 1994. Compiled from data published in Ruddiman *et al.*, 1989)

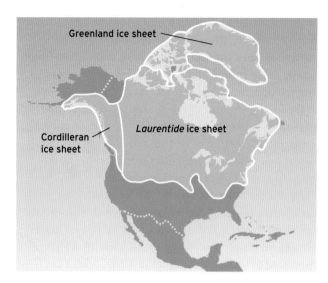

Figure 20.7 The Laurentide ice sheet grew very slowly over tens of thousands of years and then melted within a few thousand years.

20.2.4 Internal feedback mechanisms

20.2.4.1 Albedo and sea-level change

Once ice sheets start to form, the albedo of the Earth's surface increases because ice is highly reflective of the Sun's energy. This increased albedo results in a further drop in temperatures, allowing ice sheets to expand further. A further positive feedback mechanism comes from that of sea-level changes induced by ice sheet growth. The expanding ice sheets result in a fall of global volumetric sea level (this type of sea-level change is often called **eustasy** or eustatic sea level). This would make it easier for ice to flow out from the land further onto the continental shelves. Therefore the ice sheets can expand even further, and albedo would be greater, allowing the Earth to cool further. This might help explain some of the extra global cooling we have seen during the Quaternary that could not be explained by the changes in insolation predicted by the Milankovitch theory alone. However, general circulation models (GCMs) suggest that although albedo and sea-level change may be important they still would not account for the full magnitude of cooling seen during the Quaternary glacial periods (Broecker and Denton, 1990).

There are two potentially more important internal feedback processes that have been the focus of research since the 1990s: these relate to the combined effect of (i) changes in the ocean circulation and atmospheric CO_2 concentrations and (ii) the nature of ice sheet dynamics.

20.2.4.2 The missing CO_2 link: oceans and ice sheet dynamics

The centres of some very large ice sheets do not melt even during interglacials (that is why we still have ice over the poles today). Long ice cores from the centre of very old ice sheets in Greenland and Antarctica have been drilled and analysed. The ice cores contain bubbles of gas locked within them. These bubbles are representative of the air contents when the ice originally fell as snow (Alley *et al.*, 1993). Because ice layers build up during each year's snow season it is possible to calculate how old the air within the ice bubbles is by counting layers. This allows us to gain information about global air composition from thousands of years ago. Further information on ice cores and data available from these are provided in Box 20.2.

Results from ice cores have shown that CO_2 concentrations were lower during glacials and greater during interglacials. CO_2 is an important greenhouse gas and therefore would be expected partly to control the temperature of the Earth. However, this poses an interesting problem: why should CO_2 levels rise and fall during the Quaternary?

The oceans hold around 60 times as much CO_2 as the atmosphere. The gas readily diffuses between the ocean surface and the atmosphere. Therefore, its concentration in surface waters regulates the atmospheric concentration. If there are changes to CO_2 concentrations in the ocean surface waters this may affect atmospheric concentrations which may then cause the Earth's climate to warm or cool. Interestingly, however, the CO_2 concentration changes appear to lag behind the temperature changes.

While there are many ocean currents, often driven by surface trade winds (see Chapter 3), there is one very important deep ocean current. This is known as the thermohaline circulation system. The current is driven by temperature gradients (thermo) and salt concentration gradients (haline). This ocean circulation is described in detail in Chapter 3 which should be referred to in order to comprehend fully the remainder of this section. This strong deep current acts as a pump that can transfer CO_2 and nutrients from the surface of the oceans to the deeper waters and return them to the surface again. Today there are sensitive zones where such downwelling and upwelling occurs (see Chapter 3).

It is, however, living things in the oceans that partly control the concentration of CO_2 in the surface waters (Broecker and Denton, 1990). Tiny green plants (plankton) in the upper sunlit layers of the ocean take up CO_2 from the water as part of photosynthesis and this helps to form the plant tissue itself. When the plants die, their debris falls to the bottom of the ocean where bacteria oxidize it back to

EVIDENCE FROM ICE CORES

Ice builds up incrementally on the surface of ice sheets. Typically the winter layers of ice are light in colour while summer layers are darker because the partial melting and lower summer accumulation rate produces higher concentrations of impurities. This layering and X-ray analysis allow each seasonal and annual layer of ice to be identified. Figure 20.8 shows an ice core from Greenland being examined. It will contain annual increments that will allow a relative chronology of the ice to be developed. Each of the layers can then be analysed for their chemical properties. Major ice drilling programmes have been carried out in Greenland, the Canadian Arctic and the Antarctic as well in sites in Peru and Tibet. The longest cores are over 3 km deep.

Ice cores contain an abundance of highly detailed climate information. There are three main approaches to climatic reconstruction based on ice cores. These are analysis of: (i) oxygen isotopes; (ii) gases from air bubbles trapped in the ice; and (iii) dissolved and particulate matter in the ice.

- *Isotopic analysis:* this is partly based on the same approach as discussed in Box 20.1 for ocean sediments. The isotopic composition of the ice is partly controlled by the isotopic composition of seawater. In addition, atmospheric temperature controls the isotopic ratio in the precipitation and thus the relative amount of the heavier ^{18}O isotope reveals the temperature at the site when the snow fell. Figure 20.9 shows the stable isotope ratios during the past

160 000 years for the Greenland Ice Core Project (GRIP) summit core and the Vostok core from Antarctica. The present interglacial and last interglacial period some 120 000 years ago (the Eemian) are evident in the ice core records (Figure 20.9). These are separated by the last glacial period. The end of the glacial period before the Eemian can also be identified at the bottom of these plots.

- *Gas analysis:* the mixing time for atmospheric gases over Earth is around one to two years, meaning that changes are rapidly diffused throughout the lower atmosphere. Bubbles trapped in the ice contain records of past atmospheric composition. It is possible, by analysing the bubbles, to determine how the past concentrations of CO_2, methane and other gases have varied over time. The ice cores have revealed dramatic increases in CO_2, methane and nitrous oxide during interglacial episodes and decreases during glacial episodes. There are also many shorter-term fluctuations in gas concentrations over millennial and even centennial scales. However, a fundamental problem is that air in ice bubbles is always younger than the age of the surrounding ice. As snow becomes buried and transforms to firn and then ice, the air between the snow crystals remains in contact with the atmosphere until the pores of air become fully sealed upon ice formation. The sealed bubbles

Figure 20.8 Part of an ice core from Greenland. (Source: P. WIlliamson, University of Berne)

BOX 20.2 ➤

573

therefore contain air that is representative of the conditions long after snow deposition. This needs to be taken into account when producing high-resolution records.

- *Chemical content:* the presence of dust and trace chemicals in the ice can be determined. Evidence for volcanic episodes or periods of increased aridity can be traced. The types of dust can also be analysed to determine their sources and hence prevailing atmospheric circulation patterns.

Figure 20.9 Stable isotope variations during the past 160 000 years recorded in the Greenland summit core (GRIP) and Vostok core from Antarctica. (Source: after Peel, 1994)

BOX 20.2

CO_2. If the thermohaline circulation is acting as it is today then this CO_2 from the bottom of the oceans can be stirred up and taken back to the surface again. However, if the circulation slows, deep carbon stores will not be circulated back to the surface as quickly. Therefore, when the plants in the ocean waters die and fall to the bottom of the ocean, taking their absorbed CO_2 with them, this CO_2 will not be returned to the surface in as great a quantity as it is being sent down into the deep. If the thermohaline circulation system were to slow down and to lose its efficiency somehow, the surface waters of the oceans would have less CO_2 (since less will be pumped back to the surface by currents) and therefore less will be returned from the water

into the atmosphere. Hence, atmospheric concentrations of CO_2 would fall as the ocean plants continue photosynthesis. This would result in global cooling. Evidence for changes in ocean circulation has come from faunal and chemical studies of deep-sea sediments. These have indicated that the production of deep water in the Atlantic, the driving force of present-day circulation systems (see Chapter 3), was reduced greatly during past glacials.

Changes in the strength of the ocean circulation system would also change the way energy transfer took place between the equator and poles. In the North Atlantic, for example, a reduction in the strength of the thermohaline system today would cause western Europe to cool by several

Figure 20.10 Change in annual temperature 30 years after a collapse of the thermohaline circulation. (Source: Hadley Centre)

degrees changing local climates quite dramatically (Paillard, 2001). Figure 20.10 shows predicted changes in mean annual temperature 30 years after the collapse of the thermohaline circulation.

An explanation is needed for why the ocean circulation strength should change in the first place. One of the answers might lie in the formation of large ice sheets. Experiments with GCMs suggested that the topographical effects of ice sheets could explain a lot of the extra cooling not accounted for by the Milankovitch mechanism alone. A small amount of cooling caused by orbital forcing can cause the growth of some ice sheets. These thick ice sheets (several kilometres) could change local air currents that were deflected around the ice domes which might contribute further to cooling and ice sheet growth. Changing air currents may result in changes to ocean currents. For example, MacAyeal (1993) argued that the topography of the huge ice sheets altered the North Atlantic trade winds. This, in combination with a cooler climate, reduced the evaporation in sensitive areas of deep-water formation in the North Atlantic, thereby reducing the strength of the thermohaline circulation system. This occurred because evaporation controls the saltiness of the water left behind; increased evaporation leads to increased salinity. Saltier water is denser and will sink, allowing the circulation 'pump' to remain strong. A reduction in evaporation would result in a slowing of the thermo-

haline circulation as there would no longer be intense downwelling of dense salty water. Return of CO_2 to the atmosphere would slow, since the upwelling amounts are reduced. Hence the northern hemisphere ice sheets could cause changes to climate worldwide, allowing southern hemisphere ice sheets to advance and global cooling to be greater than if simply caused by orbital forcing alone.

One problem with this theory is that extra ice formation at the poles and particularly in the Antarctic Ocean would cause more downwelling. This is because as ice forms it excludes the salt, allowing the remaining ocean water to become denser and to sink as a strong downwelling pump. This would increase the strength of the circulation system and lead to a negative feedback as a strong thermohaline circulation brings heat from the equator to high latitudes. Furthermore, a colder ocean would mean lower biological productivity and thus more CO_2 in the atmosphere. A colder climate will also reduce the amount of precipitation available to supply glaciers. Again these are negative feedbacks which suggest that there may be some self-regulation to the Earth's climate system. However, it seems that the North Atlantic is the crucial and most sensitive part of the entire system.

The ocean–atmosphere circulation system is extremely important but very complex and as yet we do not fully understand all the processes involved. In addition to long-term Quaternary changes, the thermohaline circulation system may play a role in short-term climate changes including those that may occur in the very near future. These processes are discussed further below.

20.3 Short-term cycles

20.3.1 Glacial instability

Ocean cores provide long continuous records of climate change but because sedimentation rates are so slow they are of lower temporal resolution than ice cores. Ice cores provide high-resolution data through annual layering of snow accumulation (Box 20.2). However, the ice records do not extend back very far into the Quaternary. The Greenland and Antarctic ice cores, which are around 3 km deep, provide evidence only to around 130 000 and 400 000 years ago respectively. The most important evidence to emerge from the ice records is that Quaternary climate changes were often very rapid with significant warm interstadial episodes and cool stadial episodes occurring just a few centuries apart. Figure 20.9 shows that glacials were not very stable and there are several short warming and cooling episodes.

Some 20 interstadial events have been identified in the Greenland ice cores during the period 80–20 000 years BP during which temperatures fluctuated by 5–8°C. These events are known as **Dansgaard-Oeschger (D–O) events** and last for no more than 500–2000 years. The D–O events are characterized by abrupt changes in temperature (gradual cooling followed by abrupt warming), dust content, ice accumulation rate, methane concentrations and CO_2 concentrations (Broecker, 1994). The abruptness is of the order of a few decades. Bond *et al.* (1993) grouped these D–O events together into larger cycles which contain a long cooling trend followed by an abrupt warming (Figure 20.11). Ocean sediments from the North Atlantic revealed that at the

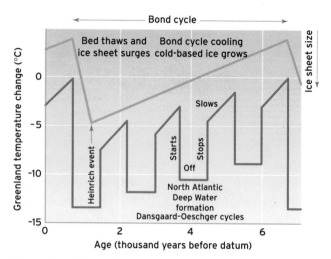

Figure 20.11 Schematic representation of the relationship between Dansgaard–Oeschger events, Bond cycles and Heinrich events.

coldest part of these longer **Bond cycles** vast discharges of icebergs floated southwards across the North Atlantic from the North American and European ice sheets. These releases of ice are known as **Heinrich events**. Heinrich events were immediately followed by abrupt warming. Evidence for Heinrich events comes from the fact that some distinctive layers of sediments taken from North Atlantic sediment cores are poorly sorted and angular and contain rocks that have come from Canada and Europe. They are named after Helmut Heinrich who described and numbered them and explained how the layers came to be there (Heinrich, 1988).

Several scientists (e.g. MacAyeal, 1993; Hunt and Malin, 1998) have proposed that Heinrich events were caused by inherent instabilities of large ice sheets. The great Laurentide ice sheet grew over North America so that it became very thick during glacials. This meant that under the sheer mass of the ice the sediments beneath could weaken and meltwater would be produced in greater quantities at the ice base owing to the extra pressure. This ice loading and failure of the substrate may then have caused parts of the ice sheet to flow much more quickly. This would result in the release of large quantities of icebergs from the Canadian coast. The icebergs would carry with them sediment from the Canadian land mass that had been plucked and scoured by the ice and then as the iceberg melted the sediment would be deposited on the ocean floor. Once the excess ice has been released, the ice sheet may stabilize again and start to grow once more; the ice rafting would cease. This type of mechanism is known as the **binge-purge model** of ice sheet development.

Given the sudden change in global climate following Heinrich events this would mean that ice rafting events in the North Atlantic somehow sent a signal to the entire world. MacAyeal

(1993) proposed that the decrease in ice sheet volume changed the wind action around the ice sheet. This change returned the wind to its pre-glacial formation with enhanced sea surface evaporation. This would result in enhanced downwelling and a return of the strong thermohaline circulation system. This reinvigorated thermohaline system would then result in rapid global warming through return of CO_2 to surface waters from the deep and enhanced heat transport.

20.3.2 Interglacial instability

We have seen that glacial periods were highly unstable and this was probably related to interactions between ice sheet dynamics, ocean circulation and biological productivity. Results from the GRIP shown in Figure 20.9 suggested that the Eemian (last) interglacial was also unstable. At the time when these results came out in the early 1990s this caused great excitement as it had previously been considered that interglacials such as the Holocene were relatively stable. The Vostok core, for example, also shown in Figure 20.9, had previously revealed a stable Eemian period. However, it is now widely accepted that the lower layers of the GRIP ice core were disturbed by ice deformation processes. Evidence from nearby cores and other environmental proxies for the period could not corroborate the GRIP findings for the Eemian. Figure 20.12 illustrates how this deformation could take place. Ice does not simply build up in one place over time; it also flows and deforms. When ice deforms it does not do so uniformly. An ice core taken from the centre of the ice dome is likely to suffer less deformation than cores taken away from the dome. However, despite the fact that the GRIP core was taken from the summit of the ice dome in order to minimize ice flow disturbance through the core, the lower parts of the core were disturbed. This was probably because the summit of the ice dome had moved over the 160 000 year period and local geology may have interrupted the ice flow. The general consensus for instability during interglacial periods is that millennial-scale variability continues but it is more subdued than during glacial stages.

20.3.3 The Younger Dryas

The last short, major, cold event is known as the 'Younger Dryas'. It occurred during the transition from the last glacial into the present Holocene and lasted from around 11 000 to 10 000 years ago. The Earth was warming from about 18 000 years ago. This warming trend was interrupted by several cold reversals, the most pronounced of which was the Younger Dryas. A suggested explanation is that the meltwater of retreating continental ice masses was released into the sensitive parts of the North Atlantic (from the melting North American ice sheet) where it substantially reduced the density of the ocean surface water (Broecker and Denton, 1990). As the ice sheets melted, a switching of drainage of the Laurentide ice sheet from the Caribbean towards the North Atlantic, via the Gulf of St Lawrence, led to an input of fresh meltwater to surface ocean layers. Being fresh (less dense than seawater) this meltwater input slowed the downwelling of water in the North Atlantic and thereby slowed down the deep-water formation. The result of this was to slow down the thermohaline ocean circulation system which had carried warm tropical waters to the north. Without this source of heat Europe and North America began to cool again (Figure 20.10) and the ice sheets started to re-advance. Many of the glacial landscapes that we can see in northern Europe today are those created by that sudden cooling and re-advance of glaciers. This series of events demonstrates that climatic feedback effects can be strongly non-linear; global warming led to a sudden cooling.

> **Reflective questions**
>
> ➤ Why might glacial periods be more unstable than interglacial periods?
>
> ➤ Why do results from Younger Dryas research have implications for predictions of the future impacts of current global warming caused by humans?

20.4 Further evidence for environmental change

Evidence from deep-sea and ice core sediments has already been discussed in this chapter but a range of other types of evidence can be used to reconstruct Quaternary environments. Table 20.1 summarizes the main evidence types used.

Figure 20.12 Ice flow and deformation can cause the lower layers of ice cores to be unreliable in palaeoclimatic reconstruction.

Table 20.1 Main sources of data used to reconstruct Quaternary environments

Proxy data source	Variable measured	Proxy data source	Variable measured
Geology and geomorphology – continental:		**Biology and biogeography – continental:**	
Relict soils	Soil types	Tree rings	Ring-width anomaly, density
Closed-basin lakes	Lake level		
Lake sediments	Varve thickness		Isotopic composition
Aeolian deposits – loess, desert dust, dunes, sand plains	Composition, depth, layering	Fossil pollen and spores	Type, relative abundance and/or absolute concentration
Lacustrine deposits and erosional features	Age	Plant macrofossils	Age, distribution
Evaporites, tufas	Stable isotope composition	Plant microfossils	Age, distribution
		Vertebrate fossils	Age, distribution
Speleothems	Composition, layering, dated layers	Invertebrate fossils: mollusca, ostracods	Age, distribution
Geology and geomorphology – marine:		Diatoms	Type, assemblage abundance
Ocean sediments	Ash and sand accumulation rates	Insects	Refuges
	Fossil plankton composition	Modern population distributions	Relict populations of plants and animals
	Isotopic composition of planktonic and benthic fossils	**Biology and biogeography – marine:**	
Continental dust	Mineralogical composition and surface texture	Diatoms	Age, distribution, geochemistry
Biogenic dust: pollen, diatoms, phytoliths	Geochemistry	Foraminifera	Faunal and floral abundance
Marine shorelines	Coastal features	Coral reefs	Morphological variations
Fluviatile inputs	Reef growth	**Archaeology:**	
Glaciology:		Written records	
Mountain glaciers, ice sheets	Terminal positions	Plant remains	Age, distribution
Glacial deposits and features of glacial erosion	Composition, orientation, shape	Animal remains, including hominids	Age, distribution
Periglacial features	Composition	Rock art	Age, distribution, content
Glacio-eustatic features	Shorelines	Hearths, dwellings, workshops	Age, distribution, geochemistry
Layered ice cores	Oxygen isotope concentration	Artefacts: bone, stone, wood, shell, leather	Composition, age, distribution
	Physical properties (e.g. ice fabric)		
	Trace element and microparticle concentrations		

(Source: from Williams *et al.*, 1998)

The value of using multiple lines of evidence lies in the richness of the information that they bring. For example, if data from one site are mapped with data from other locations then a regional synthesis can be produced, providing greater insight into palaeoclimate and former circulation patterns. Local environmental gradients and information on precipitation and aridity, for example, can be reconstructed. This information helps us understand more about the processes involved in environmental change and shows us which factors need to be manipulated in models predicting future global environmental change. Multiproxy studies provide a diversity of information, spatial coverage, temporal coverage, a diversity of environments recorded and a completeness of records. They enable us to obtain information about a range of different environments and within each environmental setting they enable us to obtain a record of different aspects of that environment. A selection of evidence types and their uses is discussed in the following sections and in Chapters 12, 16, 18, 19, 21 and 22.

20.4.1 Landforms

Over 20 major variations in climate are recognized in the undisturbed records of the deep oceans, but the terrestrial (land-based) record of climatic change preserves far fewer, because processes of erosion make the record much more fragmentary. For example, the last glacial period allowed ice sheets to expand across much of northern Europe, eroding earlier sedimentary records of Quaternary landscapes, plant and animal life. Nevertheless there are still many geomorphological features providing ample evidence for climatic change. These include landforms of former glacial erosion and deposition, former periglacial landforms, river terraces, cave deposits and wind-blown sediments. Data can be obtained from the sediments themselves by relating observations of present depositional environments to features present in the stratigraphic record. Furthermore, since many deposits contain fossils, inferences can be made about the type of environment and climate experienced at that point in the sedimentary sequence.

20.4.1.1 Glacial and periglacial landforms

Sedimentary evidence for cold climates includes glacial tills, fluvioglacial sands and gravels, and glaciolacustrine (from water bodies associated with glaciers) silts and clays. Such proxy evidence can be augmented by other landform

evidence, including, for example, moraines, eskers, kames, drumlin fields and meltwater channels and periglacial landforms (see Chapters 18 and 19). The unvegetated landscape of periglacial areas is often subject to intense wind erosion and the abundant supply of fine-grained sediment often leads to the deposition of significant thicknesses of fine-grained silt (termed loess) and sand deposits (termed **coversands**).

Mapping the spatial distribution of these sediments and landforms (by both remote sensing techniques (see Chapter 23) and ground mapping) allows the reconstruction of glacial limits (e.g. Clarke *et al.*, 2000). Figure 20.13 shows an example of geomorphological mapping in northern Europe, allowing ice sheets to be reconstructed through analysis of landforms. The direction of ice movements can also be determined through analysis of morphological features such as striations and moraines. This type of mapping work, however, requires understanding about the genesis of different landforms and the complexity of such landscapes and is often subjective. Therefore different people may produce different reconstruction maps depending on the way they view landscape features.

20.4.1.2 River terraces

River valleys throughout the world contain evidence of past climates. River terraces often exist along valley sides (Figure 20.14) or on the floodplain and sometimes exist in vertical steps such as on the River Meuse (Figure 20.15). Terrace sequences reflect both incision and lateral migration of a river channel. Often the highest terrace is the oldest, and lower terraces are younger. Sometimes the sequence is more complex than this where, for example, older terraces have become buried by younger alluvial deposits. Terraces can develop through a range of processes including: changes in sea level where a fall in sea level causes a river to incise further; changes in precipitation regimes and vegetation cover resulting in changes to water and sediment supply; tectonic processes; and human activity (e.g. forest clearance can result in increased runoff and erosion and increased sedimentation in downstream areas).

River terraces are of great interest because they contain sediments that are often rich in plant and animal remains. Often the nature of the environment under which the terraces evolved can be determined from the deposits. The terraces of the River Thames in England are some of the most studied in the world and fossil assemblages have

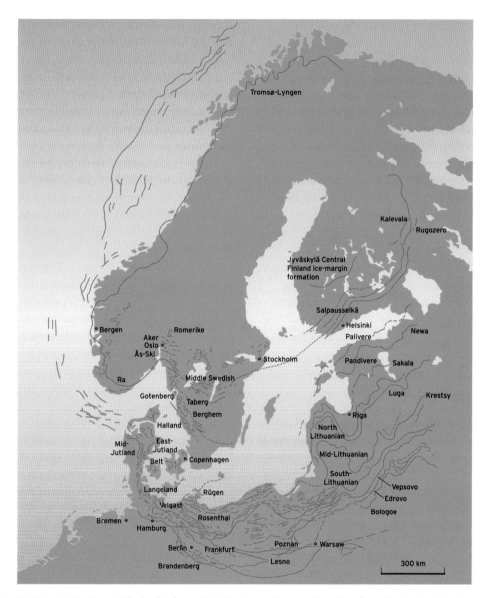

Figure 20.13 Ice sheet reconstruction on the basis of morphological mapping over Scandinavia during the last glacial maximum. (Source: after Lowe and Walker, 1997)

allowed glacial, interstadial and interglacial sequences to be determined and correlated with terrace sequences throughout lowland England (Gibbard, 1994; Bridgland, 1994). This allows a comprehensive picture of local environmental change to be built up involving changes in vegetation and animal communities through time.

20.4.1.3 Cave deposits

Cave systems and **rock shelters** (shallow niches in the hillside) can also yield useful information about palaeoclimates. Caves form excellent traps for sediment as they are protected from surface weathering and erosion. Cave detritus can often contain skeletal parts of animals and many caves and rock shelters were occupied by humans resulting in rich archaeological deposits. For example, the rock shelters of northern Greece have yielded evidence of human tools and debris dating back to 50 000 years ago. Interspersed with these human artefacts are sediments of faunal, floral, fluvial and mineralogical processes providing rich detail on local palaeoenvironments (Woodward and Goldberg, 2001).

These cave systems can also contain **speleothems** which are mineral deposits formed in limestone regions by water

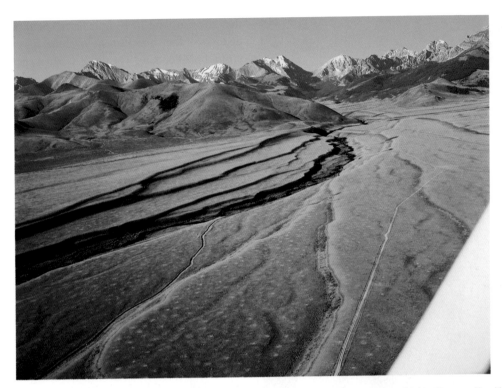

Figure 20.14 River terrace series formed by river incision on the Pahoemeroi River, Idaho. (Source: Ken Hamblin)

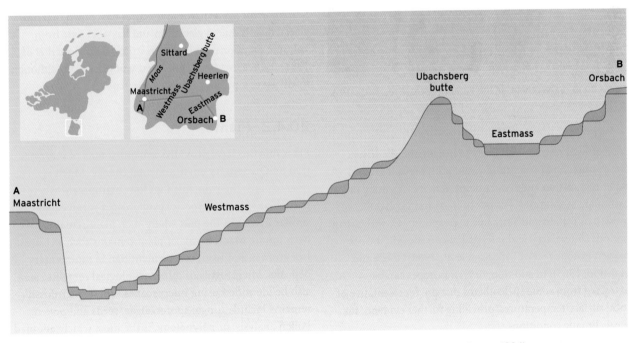

Figure 20.15 Terrace sequence on the River Meuse, near Maastrict, the Netherlands. (Source: after Ruegg, 1994)

dripping from the ceiling or walls of a cave (e.g. Figure 20.16). Often these are formed as stalactites and stalagmites. They are primarily composed of calcium carbonate which is precipitated from groundwater that has percolated through the adjacent carbonate rock. Certain trace elements such as uranium can be used to determine the ages of layers of speleothems. Cessation of speleothem growth can be detected where adjacent layers yield very different ages.

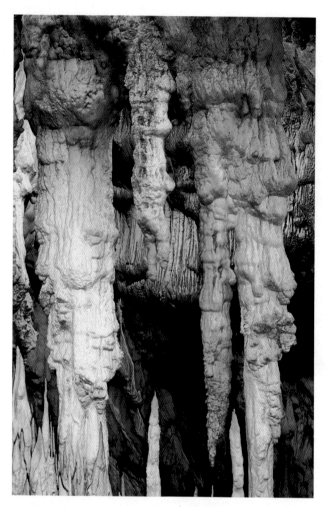

they cover around 10% of the Earth's land surface. The deposits of the Loess Plateau in China, for example, cover around 276 000 km^2. The deposits often contain evidence of climate change such as layers of soil formed during warm periods, found between layers of wind-blown sediments deposited during cold phases. The sediments also provide useful evidence of prevailing wind direction, atmospheric circulation patterns and gradients in the past (e.g. Figure 20.17).

Increased atmospheric dust is often cited to indicate a glacial epoch. However, we must be careful to distinguish between cause and effect. Extra atmospheric dust emitted by volcanic action could cause cooling by reducing the amount of solar radiation reaching the Earth's surface. The dust could also be a consequence of glaciation. With more water locked up as ice the world would be more arid. The thermal gradient between the equator and poles would also be greater during glacial periods because the ice sheets extended towards the equator. This would result in a more intense atmospheric circulation. Enhanced aridity combined with stronger winds led to more dust in the atmosphere. Useful evidence to solve this problem comes from analysis of the type of dust in ice cores deposited during glacials. Volcanic dust contains glass shards. Searching for glass shards in ice cores has shown that most glacial dust is due to glaciation and has not come from volcanic eruptions but is of terrestrial origin (Jouzel *et al.*, 1993). Hence the dust is not a cause of glaciation (although it will act as a positive feedback mechanism).

20.4.2 Plants

Plant remains have provided a major source of information on Quaternary environmental change. Much research uses the principle that the environment where particular species or species assemblages survive today is similar to that in the past where such plant remains have been found. Macrofossils of plant remains such as leaves, tree stumps and seeds can be identified in sedimentary deposits. Microfossils of minute biological remains can also be identified using microscopes. These microfossil remains include fungal spores, algae, seeds and pollen. Pollen analysis, or **palynology**, is the most widely adopted method of reconstructing Quaternary environmental change.

Pollen grains are formed by seed-producing plants and are the male fertilization units. They aim to meet with the stigma of the female part of the flower to allow fertilization. Pollen grains are released from plants in large

Figure 20.16 Speleothems in the Perama Caves, Greece. (Source: Dorling Kindersley)

Such growth restriction is likely to be climatically driven. For example, cold conditions could both prevent groundwater percolation and reduce biotic activity (resulting in less carbonate in solution). Periods of maximum speleothem growth in Britain and Tasmania correspond well with warm periods (interglacial and interstadial episodes) in the ocean oxygen isotope record (Atkinson *et al.*, 1986; Goede and Harmon, 1983). In addition, oxygen isotopes can be extracted from some speleothems that are representative of local surface temperatures, providing further evidence for local climatic conditions.

20.4.1.4 Wind-blown sediments

Deposits in many arid parts of the world are dominated by wind-blown sediment (see Chapters 12 and 16). **Loess** and coversands can accumulate to great thicknesses and

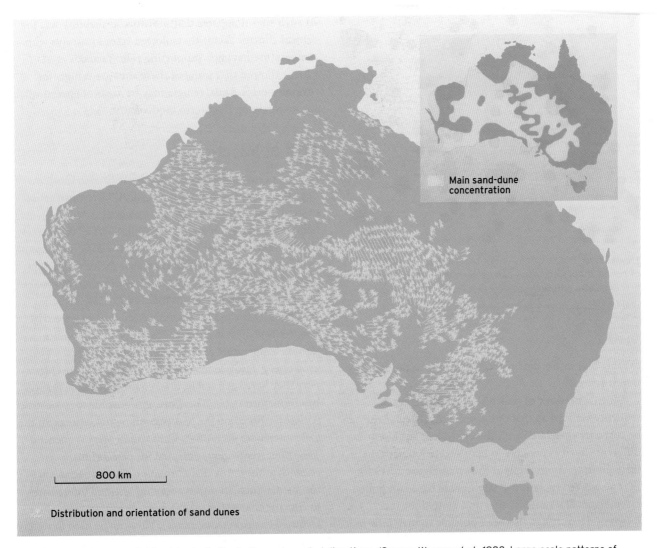

Figure 20.17 Fossil dune fields of Australia illustrating palaeowind directions. (Source: Wasson *et al.*, 1988, Large-scale patterns of dune type, spacing and orientation in the Australian continental dunefield, *Australian Geographer* (**19**) 89-104, Taylor & Francis Ltd and http://www.informaworld.com)

numbers to increase the chances of successful pollination and many accumulate on the ground surface or in water bodies. Pollen grains are extremely resistant to decomposition and so some are preserved in the sediments and become fossilized. Particular plant types produce highly characteristic pollen grains making it possible, using a microscope, to identify the plants that produce them (Figure 20.18). The extraction, identification and counting of these preserved pollen grains have yielded much useful data on Quaternary environmental change. Spores, which are small reproductive structures released by lower plants such as ferns and mosses, are often analysed in the same way alongside pollen grains.

If samples are taken from a sedimentary sequence then an analysis of the pollen found within each horizon can show changes in pollen content over time. This indicates changes in the vegetation cover in the area adjacent to the study site. The data are usually plotted as pollen diagrams and an example is shown in Figure 22.4 in Chapter 22. Here either the percentage of pollen counted for each plant type is plotted for each horizon in the stratigraphic sequence or the number of pollen grains for each species per volume of sediment for each layer, the concentration, might be calculated. A range of issues must be dealt with in order to interpret pollen diagrams. These include the fact that some types of pollen are preserved more readily than others and that pollen may travel long distances to the deposition site and hence not be representative of the local vegetation. Moreover, not all plants produce the same quantities of pollen,

Figure 20.18 Pollen grains under the microscope. A range of pollen shapes and sizes can be seen, each indicating the presence of a different species. (Source: photo courtesy of John Corr)

causing bias in the pollen spectra relative to the vegetation it represents.

When a network of pollen sites exists, and each sequence is dated so that one layer at a particular site can be correctly correlated with the equivalent layer at a different site, it is possible to map vegetation patterns through time. An example is shown in Figure 20.19 for Europe where vegetation groupings are based on pollen assemblages at a range of sites and plotted for four different dates. It has also been possible to map how individual species have migrated over time and Figure 20.20 presents information for deciduous oak trees in Europe. During the last Glacial Maximum at 18 000 years BP, the oaks survived only in isolated pockets in the south and were not present elsewhere in Europe. During the post-glacial period, as the ice melted, the oaks spread northwards. However, the sudden onset of the Younger Dryas lasting about 1000 years caused them to retreat southwards again before further warming allowed full expansion. It seems that the isolated pockets in southern Europe where the oak trees persisted during glacial periods allowed the species to survive and provided a reservoir of oak trees which expanded again to cover northern Europe once the climate warmed. These sites, known as **refugia**, are crucial to the preservation of particular species through cold episodes. Indeed, Tzedakis (1993) showed for a 430 000 year record at Ioannina, in north-west Greece, that this site has acted as a long-term refugium for the oak trees throughout the Quaternary.

At such sites it appears that moisture availability was the critical climatic factor that allowed trees to survive with temperature having a supporting role. Tzedakis *et al.* (2002) showed that some of these sites are refugia for several species. Thus, refugia may be areas of special value for maintaining long-term biodiversity.

20.4.3 Insects

A range of insects have been found fossilized in Quaternary deposits and as with vegetation these allow reconstruction of local temperature and moisture regimes based on comparison with contemporary habitats. Of all the insects it is the remains of beetles that have proved the most useful in Quaternary research. This is because their exoskeletons are well preserved and allow good identification of species. Beetles (**Coleoptera**) account for around 25% of all the species on Earth and they occupy almost every terrestrial and freshwater zone. Many species survive only within narrow temperature ranges or very specific habitats. They evolve very slowly so that over the timescale of the Quaternary they have been stable; their environmental tolerances have remained stable. Therefore, it is possible to reconstruct with some precision the past environmental settings for the Coleopteran fossils found within sedimentary sequences. They offer tremendous potential for palaeoclimatic research as they provide information on short-term environmental fluctuations because they respond immediately to climatic change (by migrating to remain in their favoured environmental range). They can be used to reconstruct seasonal and mean annual temperature maps for different periods. Indeed, they can also be used to tell us about likely vegetation assemblages that they were associated with.

20.4.4 Other animal remains

A range of other fossil remains are used in reconstructing Quaternary environmental change, such as terrestrial and marine molluscs, diatoms, non-biting midges (chironomids), bivalved crustaceans (ostracods), foraminifera, animal bones and teeth, coral polyps, fungal remains and testate amoebae. The last, for example, are found in many peats and are controlled by hydrological variations that affect peat formation. They can be used to reconstruct past moisture regimes. Care should be taken when interpreting animal and plant remains and inferring climate because some species may or may not be present owing to ecological processes rather than climatological ones. For example,

Figure 20.19 Vegetation communities in northern Europe for four periods during the past 9000 years. The figure shows how main communities have changed over time at four time slices. (Source: after Huntley and Prentice, 1993, Holocene vegetation and climates of Europe. In: Wright, H.E. Jr *et al.* (eds), *Global climates since the Last Glacial Maximum*, University of Minnesota Press, Copyright 1993 by the regents of the University of Minnesota)

Figure 20.20 Distribution of deciduous oaks during the past 18 000 years. The figure shows that oaks were only found in isolated pockets in southern Europe during the Glacial Maximum and then migrated northwards during the warming period. The changes during the Younger Dryas can also be seen by comparing (b) with (c). (Source: after Taberlet, P. and Cheddadi, R. 2002, Quaternary refugia and persistence of biodiversity, *Science*, **297**, pp. 2009-10. Copyright 2002 AAAS. Illustration: Katharine Sutcliff. Reprinted with permission from AAAS)

species competition or migrational isolation might mean that some species are not present when at other times with a similar climate they were present.

Reflective questions

➤ Why is it beneficial to use several types of evidence when reconstructing Quaternary environmental change at any given location?

➤ What are the advantages and disadvantages of using pollen grains to infer environmental change?

20.5 Dating methods

Reliable dating is highly desired by Quaternary scientists. There are three types of technique: age estimation, age equivalence and relative dating methods (Lowe and Walker, 1997). Some examples of these techniques are listed in Figure 20.21. Often the techniques used are valid for only limited parts of the Quaternary. There are no perfect techniques and each has a range of problems which lead to difficulties when interpreting data.

20.5.1 Age estimation techniques

There are two types of age estimation techniques. The first comprise the **incremental methods** which involve

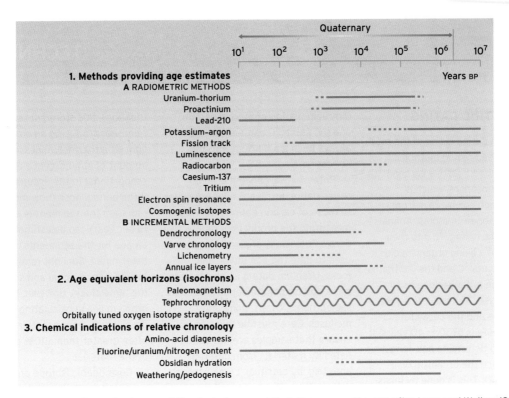

Figure 20.21 Some Quaternary dating techniques and their time range. (Source: after Lowe and Walker, 1997)

measurements of regular accumulations of sediment or biological matter through time. These include the use of tree rings for dating (**dendrochronology**), annual layers of ice on a glacier and analysis of varves (layered seasonal accumulations of sediment often found in glaciofluvial settings). The layers can be counted back through time. The second types of estimation technique comprise the **radiometric methods** that are based on the natural radioactive properties of particular isotopes. Some isotopes decay over time to produce a more stable atomic form. Because radioactive decay is time dependent and the rate of decay over time is constant, the age of a sample can be determined. Some elements decay in a few seconds whereas others transform slowly over millions of years. Typical methods include potassium–argon dating, uranium series dating, carbon dating and *in situ*-produced cosmogenic isotope analysis (Box 20.3).

20.5.2 Age equivalent labels

Distinctive horizons can often be found in Quaternary deposits which are found in more than one place. If the horizon can be dated at just one of the locations where it is found (e.g. by **radiometric** or incremental methods) then this allows dates to be extended to other places where this

distinctive horizon occurs. Distinctive horizons are produced by layers of volcanic ash spread across a large proportion of the planet after a major volcanic eruption. They are also produced by reversals and strength changes of the Earth's magnetic field (see Chapter 2).

One commonly used age equivalence technique uses the marine oxygen isotope curve as the basis of a globally applicable stratigraphic scheme. The oxygen isotope stages (shown in Figure 20.6e), established in deep-ocean sediment records, can be correlated with sediment sequences (both marine and terrestrial) throughout the world. The ages of each stage are assigned by matching the stage with the predicted age for that stage as determined by Milankovitch orbital forcing mechanisms. This 'tuning' of course only provides a relative age and assumes the deep-sea record of climate change should be perfectly tuned to the orbital cycles. This is unlikely to be the case because of a large number of feedback mechanisms operating in the climate system (Elkibbi and Rial, 2001).

20.5.3 Relative chronology

Sediments and rocks are affected by chemical reactions that take time to occur. The amount of weathering or organic decomposition may provide a basis for relative dating.

TECHNIQUES

RADIOMETRIC DATING

Radiocarbon dating is one of the most frequently used dating methods. Plants absorb radioactive carbon (^{14}C) from the atmosphere, which is produced when cosmic rays from space hit nitrogen-14 atoms. Animals eat the plants, taking ^{14}C into their bodies. When a living organism dies it stops absorbing ^{14}C and the ^{14}C that is already in the organism begins to decay. Because ^{14}C decays at a slow, constant rate and the decay rate is known (half-life is 5730 ± 40 years) it is possible to determine how long it has been since the organism died (Figure 20.22). This is done by measuring the ^{14}C left in the sample. It should be noted that radiocarbon dating is only really valid for the past 50 000 years because by that age the amount of ^{14}C left is so small that it is impossible to measure accurately making dating impossible. Furthermore, there are problems with ^{14}C dating because of (i) temporal variations in ^{14}C production, (ii) contamination with ancient or modern carbon,

(iii) recirculation of marine carbon and (iv) plant ^{14}C content may not always be in equilibrium with the atmosphere.

Potassium-argon and uranium series dating involve similar principles to those of carbon dating involving measuring the products of radioactive decay. Potassium-argon dating is used to date the age of volcanic rocks. Uranium dating is used to date rocks and carbonate deposits such as speleothems, corals, bone and molluscs. Care must be taken to ensure that samples are not contaminated by materials from different ages (e.g. by detritus). Luminescence dating allows the amount of time sediment has been buried to be calculated. Materials exposed to radiation accumulate a **thermoluminescent** property over time. Sediments everywhere contain low concentrations of uranium, thorium and potassium which produce, over geological time periods, a constant flux of ionizing radiation. The ionizing radiation is absorbed and stored by surrounding sediments and, with stimulation

(heating), this stored dose can be released, producing an emission of light (the thermoluminescence). The amount of luminescence emitted is proportional to the accumulated dose. If the annual dose (how much ionizing radiation the sediment is exposed to per year) can be estimated then an age for the sediments can be determined. Sunlight removes the luminescence signal and so resets the 'time clock'. However, the major problem lies in calculating the annual dose, which means that errors are often greater than ±10% of the age of the material.

Cosmogenic isotope analysis produced *in situ* is now becoming the standard technique for dating landforms and some types of terrestrial deposits (Cockburn and Summerfield, 2004). The Earth is under constant bombardment by cosmic radiation. Some of this radiation comes from the Sun, some from other locations within our galaxy, while the highest-energy particles originate from outside our galaxy. These high-energy particles which consist mainly of neutrons interact with elements in a shallow surface layer when they reach the Earth's surface. This interaction produces extremely small quantities of cosmogenic nuclides. Measurements of the amounts of these cosmogenic nuclides accumulated over time can provide valuable information on the age and rate of change of the land surface. They are particularly useful because, unlike ^{14}C, the cosmogenic isotopes have very long half-lives, ranging from thousands to millions of years. For more information on these techniques see the further reading at the end of this chapter.

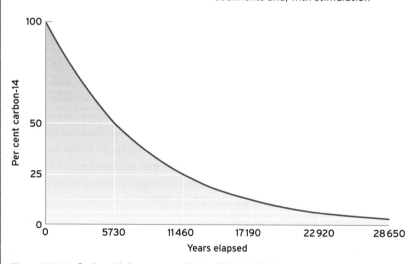

Figure 20.22 Carbon-14 decay curve. The half-life is 5730 years.

BOX 20.3

Bones and shells contain proteins even hundreds of thousands of years after death. The protein undergoes transformations over time, which changes, for example, the ratios of certain amino acids. Thus a relative chronology can be developed by examination of the types and amount of transformation (**diagenesis**) that has taken place. Commonly this involves analysis of amino acids. The rate of the chemical reactions concerned is directly related to temperature. Consequently, amino acid diagenesis will proceed more slowly at cooler sites than at warmer sites. Samples from some mid-latitude regions, for example, provide a resolution of 20 000–30 000 years with a useful range of only approximately 2 million years since diagenesis proceeds rapidly; Arctic samples provide a resolution of 100 000 years with a useful range of 5 to 6 million years, because diagenesis is slower here (Bradley, 1999). Fossil bones can also be assessed for their content of elements absorbed into them from the surrounding sediments over time. For example, bones often absorb fluorine and uranium from groundwater over time.

Reflective question

➤ Why do we need to develop and use several different dating techniques?

20.6 Quaternary stratigraphy and correlation

In order to assess environmental change through time it is necessary to analyse sequences of sediments preserved in a range of contexts on land, in ice and beneath the ocean floor. Two aspects of this work are ordering the record at any one location into a time sequence (stratigraphy) and determining how the evidence at one site relates to the evidence at another (correlation). Since only a short period of time is recorded in the terrestrial deposits at any one site it is rare to find even one complete cycle of glacial and interglacial sediments on the land surface of northern Europe or North America. Therefore sequences from different sites must be pieced together to form a complete picture. Only in deep oceans, deep lake deposits and thick loess are long stratigraphic sequences preserved. At most places, where the record is fragmentary, careful analysis is required before sequences can be ordered at one place and then related to those at another.

(a)

(b)

Figure 20.23 Stratigraphic deposits: (a) a short terrestrial deposit which dates back approximately 18 000 years; (b) deposits made visible by cliff erosion on the Norfolk coast, England, dating back 25 000 years.

Often stratigraphic methods are descriptive as it is the visible features that allow formal subdivision of the sequence (Figure 20.23). Classification of subdivisions can be done in a number of ways. For example, sequences can be classified on the basis of fossil evidence found within them, with each **biostratigraphic** unit having a distinctive fossil assemblage. Traditionally, pollen assemblages have been used in the subdivision of interglacial stages of North America and northern Europe. Other methods involve the relative dating of landforms present or inferring changes in climate from the sedimentary structure. Where a particular stratigraphic unit is very clear and well recorded, and where its lower boundary can be well defined, this site may be designated a **stratotype** or typesite. This site becomes the reference point and the place where a particular

Table 20.2 Quaternary stratigraphic correlations of the northern hemisphere

Time-scale (Myr BP)	Marine oxygen isotope stages	Northern Europe	The Netherlands	British Isles	European Russia	Northern Alps	North America	C/T†
0.01	1	Holocene	Holocene	Flandrian	Holocene	Holocene	Holocene	T
0.08	2–4d	Weichselian	Weichselian	Devensian	Devensian	Würm	Wisconsinan	C
0.13	5e	Eemian	Eemian	Ipswichian	Mikulino	Riss-Würm	Sangamon	T
0.19	6	Warthe			Moscow	Penultimate Glacial Late Riss?	Late	C
0.25	7	Saale/Drenthe		'Wolstonian'	Odintsovo / Dneipr Glaciation		Illinoian	T
0.30	8	Drenthe			Dneipr	Antepenultimate Glac. Early Riss/Mindel?	Early	C
0.34	9	Domnitz [Wacken]			Romny			T
0.35	10	Fuhne [Mehleck] / Holsteinian Interglacial		Hoxnian	Pronya	Pre-Riss?	Pre-Illinoian A	C
0.43	11	Holsteinian [Muldsberg]			Lichvin			T
0.48	12	Elster 1		Anglian		Late Mindel? /Donau	B	C
0.51	13	Elster 1/2	Elster		Oka			T
0.56	14	Elster 1		Cromerian		Early Mindel? /Donau	C	C
0.63	15	Cromerian IV	Cromerian IV [Noordbergum]					T
0.69	16	Glacial C	Glacial C				D	C
0.72	17	Interglacial III	Interglacial III [Rosmalen]					T
0.78	18	Glacial B	Glacial B				E	C
0.79	19	Interglacial II	Interglacial II [Westerhoven]					T
	20	Helme [Glacial A]	Glacial A			Early Günz?	F	C
	21	Astern Interglacial I	Interglacial I [Waardenburg]					T
0.90	22		Bavelian / Dorst				G	C
			Bavelian / Leerdam					T
0.97			Bavelian / Linge					C
			Bavelian / Bavel					T
			Menapian					T/C
			Waalian					T
								C
			Eburonian				H	T/C
								T
				Beestonian			I	C
1.65			Tiglian / C5-6	Pastonian				T
			Tiglian / C-4c	Pre-Pastonian/Baventian				C
			Tiglian / Cl-4b	Bramertonian/Antian				T
			Tiglian / B	Thurnian			J	C
	103		Tiglian / A	Ludhamian				T
	104		Praetiglian	Pre-Ludhamian				C
2.60			Pliocene	Pliocene				

†C, cold; T, temperate.

(Source: Lowe and Walker, 1997)

stratigraphic subdivision is officially defined. Then other sites where the record of that equivalent unit is only partly present or poorly preserved can be compared with the stratotype that is the standard reference site. Often the location of these stratotype sites provides the names for apparent events in the Quaternary record. However, because of the spatial and temporal variation of Quaternary environments often the stratotypes are only locally important.

Nevertheless, it is important to try to correlate evidence at one site with that of others so that the extent and spatial variability of environmental change can be determined and so that we can piece together a long continuous record. However, the repeated cooling and warming during the Quaternary has meant that similar depositional features may be preserved that are actually of very different ages. This makes correlation difficult. Table 20.2 presents how stratigraphic schemes established using evidence from different regions across the northern hemisphere have been related to each other on a large scale. There is often a great deal of debate about whether a particular sequence really represents similar changes recorded elsewhere. In some places there is no evidence in the stratigraphic record for events that have been

recorded elsewhere. For example, Table 20.2 shows that six climatic stages were preserved in northern European records during the two climatic stages of the Wolstonian and Hoxnian periods for the British Isles. The evolution of the water vole *Mimomys savini* to *Arvicola catniana* took place during only one apparent interglacial stage in the British sequence but over four interglacial stages in Europe (Gibbard, 1994). This is further evidence to suggest that there is evidence missing in Britain. It is now common practice to try to relate terrestrial stratigraphic units to the marine isotope stratigraphy (Box 20.4), but many of the marine stages have yet to be identified in the terrestrial record. This may be because the event did not have an impact on local processes at a given site or region, but it is more likely to be because sedimentary evidence of the event has not been preserved.

Even at a small scale problems can arise in correlating sedimentary sequences. This may happen because during one glacial period the ice may extend over a particular site. However, even if the next glacial period was just as cold and ice volumes just as great, the same site may not be subject to the same glacial action. It may, for example, be part of a fluvioglacial outwash plain during the next glacial period.

NEW DIRECTIONS

DIRECT MARINE AND TERRESTRIAL STRATIGRAPHIC CORRELATION

In order to get around the problems of making sure that a land sequence really was deposited at the same time as an apparently equivalent ocean record some workers have used deep-ocean sediment cores obtained near land margins (e.g. Heusser *et al.*, 2006). This is because the sediment deposited at these sites often contains pollen and dust from local terrestrial sources. Marine pollen records of the vegetation that grew on land can be correlated directly with proxy evidence of the marine environment and the oxygen isotope stratigraphy preserved in the same

sediment (e.g. Heusser and Oppo, 2003). This also makes sedimentary sequences on land with similar pollen distributions easy to correlate with the marine sequences.

Kershaw *et al.* (2003), for example, analysed pollen and charcoal records from marine cores off the northern coast of Australia. The radiocarbon dates for the charcoal and the oxygen isotopes from the ocean sediment allowed the pollen records to be more rigorously correlated. Vegetation change not only reflected Milankovitch orbital forcing patterns but a 30 000 year fluctuation reflected changes in the intensity of El Niño induced fire frequency (which produced the charcoal washed into the ocean sediments).

Analysis of deep-sea cores from the western Portugese margin

provide continuous, high-resolution records of millennial-scale climatic oscillations and work has been done on cores dating between 9000 and 65 000 years BP (Roucoux *et al.*, 2001) and between 180 000 and 345 000 years BP (Roucoux *et al.*, 2006). Pollen analysis of the same cores allows direct assessment of the lags between the North Atlantic climate system and the vegetation changes on the adjacent landmass. The pollen was transported into the ocean by the Douro and Modego Rivers which flow into the Atlantic. Work on the cores has shown that variability in tree population size closely tracked both millennial-scale climate variability and Milankovitch-scale variability.

BOX 20.4

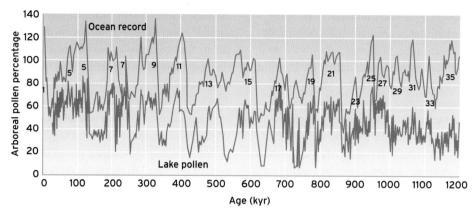

Figure 20.24 Correlation of the Funza pollen record with the marine oxygen isotope record over the past 1.2 million years. (Source: after Hooghiemstra and Sarmiento, 1991)

There are also problems caused by the erosion or reworking of old sediments which are then redeposited on top of a younger layer of sediment. Thus, fossils and other sediments may be incorrectly associated with a younger period of time than they actually belong to because of their erroneous stratigraphic position.

Deep-ocean records appear to be relatively undisturbed, unlike many terrestrial sequences, and therefore represent a good long-term global timeframe of events. In fact the ocean record is now used as the standard of reference for most other stratigraphic sequences. Producing a good correlation between terrestrial and marine sequences is one of the key areas of Quaternary research (Box 20.4). However, some terrestrial sequences preserve temporal detail that is as good as the deep-ocean sedimentary records and reveal similar patterns of change. These include undisturbed lake sediments (e.g. Funza, Colombian Andes; Lake Biwa, Japan; Lake George, Australia) or tectonic basins (e.g. Carpathian Basin, Hungary) where sediments have been consistently accumulating for millions of years. For example, Hooghiemstra and Sarmiento (1991) showed that tree pollen at Funza in the Colombian Andes correlated well with the marine isotope record (Figure 20.24).

Shorter-term changes such as D–O events and even shorter decadal to century-scale events recorded in the ice core record are more difficult to identify in the terrestrial sequences. This may be because vegetation and animal responses to such rapid climate changes are too slow and/or because the resolution of terrestrial sequences is rarely good enough to be able to pick out such short events (e.g. sedimentation rates are not fast enough). Furthermore, other effects such as changes in moisture regimes, atmospheric circulation and environmental gradients may not be identified in the record until the vegetation has had time to respond and migrate or *in situ* decline can take place. However, the vegetation surrounding tree population refugia (e.g. in southern Europe) seems to respond rapidly and therefore registers the D–O cycles in its pollen records (Tzedakis *et al.*, 2002).

Reflective questions

➤ What are the main problems and benefits of correlating evidence for Quaternary environmental change?

➤ Why is using deep-sea cores close to land margins an exciting prospect for future Quaternary research?

20.7 Palaeoclimate modelling

The use of data from a number of sources and use of high-speed numerical simulation models has benefited Quaternary research enormously. At a simple level, wave modelling (or spectral analysis) has allowed the fluctuations of the ocean oxygen isotope record to be analysed in more detail to determine more exact matches and deviations from the Milankovitch predictions (Elkibbi and Rial, 2001). However, it is concern over the consequences of human activity on atmospheric greenhouse gases that has driven the development of models of the climate system. GCMs have been run with pre-industrial levels of CO_2 (as indicated by ice core evidence) and then with double CO_2 levels, or with sequential changes in CO_2, to examine projected climatic changes.

However, such models, which are based on the present state of the Earth system, should be questioned for reliability in light of what we have learned about the Quaternary. For example, it is questionable how well such a model can work for a future climate state in which boundary conditions (sea level, solar radiation, extent of ice, etc.) are very different from today (just as different as some of the climatic states that occurred during the Quaternary). In order to get around this problem many of these models have been used to predict past climates using boundary conditions that are known to have occurred in the past. If models can reproduce past climatic conditions that we know about from their proxy environmental record we may be more confident in their ability to predict future climates.

This modelling approach, which is concerned with improving future predictions, has also benefited Quaternary research. The models provide insight into potential forcing mechanisms and the interactions between different parts of the climate system (oceans, ice sheets, atmosphere, biosphere, land surface). They can also indicate potentially unreliable data or areas where further research is needed. For example, many models are unable to produce glaciations with relatively warm tropical sea surface temperatures and this goes against some data from ocean sediments.

Several types of models are used for palaeoclimatology, including energy balance models, radiative convective models and GCMs. Energy balance models simply consider energy exchange between zones of the Earth and can incorporate latitudinal, longitudinal and altitudinal transfers. These simple models allow the effects of feedbacks such as albedo, CO_2 concentrations and solar input to be investigated relatively quickly. Radiative convective models examine atmospheric radiation processes and have been used to examine the effects of aerosols and clouds on temperature. Often these models have been used to look at changes in one parameter at a time (e.g. methane concentration), manipulating them to see which factors are most important and how individual factors might act as feedbacks in the system (Bradley, 1999).

For GCMs the Earth's surface is divided into a grid of boxes which also extend vertically into the atmosphere. The verticals are sliced up into boxes for each ground-based cell (Figure 20.25). Equations involving conservation of energy, mass and momentum are then solved at each grid point and for each vertical level for every time interval required. This requires an enormous amount of computer time to run, and the higher the spatial and temporal resolution the more calculations that have to be performed and the longer

Figure 20.25 Putting grid cells over the Earth. The atmosphere and oceans are split into columns and each calculation is performed for each box. (Source: after McGuffie and Henderson-Sellers, 1997)

it takes. Nevertheless, research is demonstrating that even more complex models that combine a series of smaller models are required to simulate the climate system. Ocean circulation models are often coupled to atmospheric circulation models. Ice dynamic models are also added in to demonstrate the impacts on ocean circulation and atmospheric circulation of ice sheets growing and melting. Coupling these models to biosphere (Chapter 22) and land surface models will add complexity but may aid understanding of the whole system behaviour. GCMs can also be used to trace the pathways of materials within the climate system. For example, desert dust pathways have been modelled and predictions tested against observed changes in the dust content of ice cores in order to determine the direction of prevailing circulation patterns (Mahowald *et al.*, 1999). Sources of moisture supplying precipitation to ice sheets have also been modelled in order to help understand how source regions differed in the past, which helps with interpretation of ice core geochemistry.

Given the strong evidence in the climate record for sudden and dramatic changes in the climate system indicating thresholds in the system, it is important that such thresholds are incorporated into palaeoclimate models. However, it is often difficult to work out what these thresholds should be. For example, we do not really know how great a change in evaporation or meltwater input is required in the North Atlantic to shut down the thermohaline circulation system, or whether the amount required changes through time as other processes operate (e.g. changes in ocean circulation due to Mid-Atlantic Ridge formation changes). Some models are able to simulate key features of the climate record but uncertainties about the processes, the role of feedbacks and boundary conditions still need to be resolved. Even the most complex models today are still too simple to represent climate processes and many of the feedbacks in detail. Nevertheless, even simple models can produce interesting results that suggest new avenues for research. Such an example is given at the start of this book in Chapter 1 (Box 1.3).

Reflective questions

➤ What are the possibilities and limitations of palaeoclimate models?

➤ Why are palaeoclimate models useful for making predictions about future climate change?

20.8 Humans and Quaternary environmental change

Humanity's impact on the Earth's environment was minor until the Holocene. Prior to the Holocene, human groups were involved in hunting and gathering food and they were not widespread or great in number across the planet. Humans responded to glacial advance and retreat through migration of communities. During the Holocene, however, humans managed to domesticate certain plants and animals for food. These cultivation practices and the development of tools and techniques for improving such practices have left their mark on the landscape and the biosphere. Evidence of human activity can be found in Quaternary deposits and must be taken into account when interpreting such deposits. For example, the large-scale forest clearances in northern Europe by humans during the late Holocene should not be mistaken, in the pollen records, for a change in climate. Rather it was a human-induced (anthropogenic) change in the environment. These sorts of environmental change are

of particular concern for the long-term preservation of biodiversity on Earth because, as we have seen, certain areas act as refugia for a range of species during glacial times. If these refugia are disturbed then there may be no place for survival for a range of species during the next glacial period. These issues are discussed in more detail in Chapter 22, which describes the interactions between vegetation and climate.

Urbanization, industrialization and the degradation of biodiversity are major contributors to recent Quaternary environmental change and these are discussed in more detail in Chapters 21 and 22. However, human activity has also aided Quaternary research by producing datable layers (such as radionuclide fallout on the Earth's surface following atomic bomb testing in the 1950s) and by the recording of temperatures and other climatic parameters using instruments over the past two centuries. There are major concerns surrounding the action of humans for future Quaternary environmental change. As the world warms owing to increased atmospheric CO_2 caused by fossil fuel burning and deforestation, more of the ice sheets that exist today might melt. If all the Earth's land ice were to melt then global sea levels around the planet would increase by a massive 75 m. This would cause huge problems for the Earth's population, most of whom live in low-lying coastal areas (see Chapter 17).

However, through studies of past climate we have learnt that a slight warming can cause negative feedback mechanisms to operate. For example, warming at the end of the last interglacial episode resulted in fresh meltwater entering the North Atlantic, which caused the thermohaline circulation to slow down, resulting in global cooling. Warming can also result in greater precipitation in Antarctica allowing more snow and ice to build up. So a change in one direction can cause a very sudden flip of the climate system in the opposite direction because of negative feedback effects. Alternatively, positive feedback effects can exaggerate climatic change. For example, the increasing albedo of a growing ice sheet can cause further cooling. Therefore, it is difficult to establish just how human-induced increases in atmospheric CO_2 levels may affect global climate. The world may begin to warm, with extra greenhouse gases, but this may trigger a negative feedback effect and ultimately cause rapid climatic cooling.

Whether or not the thermohaline circulation will be affected by human-induced global warming is strongly dependent on the future temperature distribution and freshwater supply over the North Atlantic region. Models predict an increase in precipitation in high latitudes and warming over the North Atlantic, using a scenario of doubling CO_2

within the next 70 years. Models also predict a decrease in the strength of the thermohaline circulation. However, the reduction estimates vary from 30% to only 10%. Few studies have so far explored the long-term effects (more than 100 years) of these changes but those studies that have been done illustrate that global warming can affect the climate system in a non-linear fashion. Importantly, evidence for recent changes to the North Atlantic is now being provided. For example, Dickson *et al.* (2002) showed that the North Atlantic has in fact reduced in salinity owing to the extra meltwater inputs and river influxes caused by global warming over the past 40 years.

The twentieth century was exceptionally warm in the context of the last thousand and perhaps several thousand years. However, the coldest decades of the nineteenth century were among the coldest times in the late Holocene. Thus, within 200 years we have seen the coldest and warmest extremes of the late Holocene (Bradley, 1999). It may only be a matter of a decade or two before another rapid change in climate occurs as we force the system to flip over a threshold.

Reflective question

➤ Thinking about future climate change over the next century, what has been achieved by doing research into Quaternary environmental change that might be of use?

20.9 Summary

The Quaternary is a dynamic period of Earth's history which has seen oscillation upon oscillation in climatic conditions. Huge ice sheets have expanded and retreated over tens of thousands of years. Short-term, but dramatic, changes in global temperature superimposed on top of this have occurred on a timescale of just a few decades. Evidence for Quaternary environmental change comes from a range of sources on land, in ice and on the ocean floor. The terrestrial evidence is detailed but fragmentary, the ocean evidence is continuous but low resolution (and is restricted in that it tells us only about the ocean and ice volume) and the ice core evidence is high resolution but relatively short in comparison with the ocean sedimentary record.

There are several problems in correlating different types of evidence from different locations. One problem is that it is often difficult to date a piece of evidence for past environmental change precisely. Other problems include the fragmentary nature of some records and the fact that in some places a climatic change may not have a local impact whereas in other places it does.

The evidence for climatic change is always indirect evidence (a proxy, e.g. vegetation or animal assemblages). Therefore, even if proxy data are well dated they will be of little use for climatic reconstruction unless the climatic dependency of that proxy can be clearly established. There is room for considerable improvement in this field because it is often not appropriate to use modern analogues. For example, species can adapt to new environments (Chapters 8, 9 and 10) and environmental gradients can vary over time. Nevertheless a rich amount of detail can be obtained on ecological responses to climate, the physical extent of former ice sheets and former atmospheric and ocean circulation patterns, using proxy environmental and climatic data.

This chapter has shown that there are a range of factors exerting strong controls on Quaternary climate. These factors include external orbital forcing processes and a range of internal positive and negative feedback effects. The complex interactions between oceans, ice sheets, the biosphere and the atmosphere mean that there are no simple cause–effect relationships. The Earth's environment appears to have certain thresholds which, when crossed, can result in sudden changes to the climate system. Quaternary research is of use to those involved in predicting future climate change. It has suggested that slow changes of climate in one direction can result in sudden changes in a different direction. Modelling of climate change allows us to make predictions of future climates while modelling past climates serves to test the models' sensitivity. It also allows us to determine where we need more detailed information in order to improve the quality of these predictions. It seems evident that a great deal more research is required before our predictions include the full complexity of real environmental processes and the feedbacks between them.

Further reading

Bradley, R.S. (1999) *Paleoclimatology: Reconstructing climates of the Quaternary.* Harcourt Academic Press, London.
This is a very detailed (and sometimes technical) textbook on techniques and the advantages and disadvantages of different types of evidence.

Ehlers, J. (1996) *Quaternary and glacial geology.* John Wiley & Sons, Chichester.
Translated by Phil Gibbard, the latter part of this book contains useful examples of stratigraphic analysis and provides regional synopses of local environmental change inferred from the evidence.

Imbrie, J. and Imbrie, K.P. (1979) *Ice ages: Solving the mystery.* Macmillan, London.
This book is rather dated now but is very relaxed bed-time reading about the development of Quaternary research.

Lockwood, J.G. (2001) Abrupt and sudden climatic transitions and fluctuations: a review. *International Journal of Climatology*, 21, 1153–1179.
John Lockwood presents a straightforward review summarizing the main debates and issues surrounding rapid climate oscillations.

Lowe, J.J. and Walker, M.J.C. (1997) *Reconstructing Quaternary environments.* Pearson Education, Harlow.
This is a very comprehensive overview of types of evidence and methods of their analysis. This book contains lots of useful examples and explanations and is an excellent textbook.

Paillard, D. (2001) Glacial cycles: toward a new paradigm. *Reviews of Geophysics*, 39, 325–346.
Paillard's paper is a discussion about the problems with orbital forcing and the need to think about feedback mechanisms and thresholds.

Walker, M. (2005) *Quaternary dating methods.* John Wiley & Sons, Chichester.
This book clearly explains the basics of dating, the techniques available and the strengths and weaknesses of each technique. While the title might sound daunting, the book is very accessible and the techniques are described in a very understandable way.

Walker, M. and Bell, W. (2005) *Late Quaternary environmental change: Physical and human perspectives.* Pearson Education, Harlow.
Another clearly written book, but this time concentrating on the more recent period of the Quaternary over the past 20 000 years during which time the ice sheets have retreated.

Web resources

Canadian Institute for Climate Studies: Paleoclimate Models
http://www.cics.uvic.ca/climate/crn/reports95-96/paleo95.htm
A good summary of palaeoclimate modelling is offered; the various inputs from oceans, the atmosphere, ice and solid earth are outlined.

Colby College: Beetles (Coleoptera) in Quaternary Studies
http://www.colby.edu/geology/Beetles.html
The use of fossil beetles in environmental reconstruction is described. This is a useful site for introducing the value of Coleoptera.

Ice Core Projects
http://www.gisp2.sr.unh.edu/GISP2/
http://www.ncdc.noaa.gov/paleo/icecore/greenland/summit/document/gripinfo.htm
Websites relating to ice coring projects (GISP2, GRIP and Vostok).

Minnesota State University: Dating Techniques
http://www.mnsu.edu/emuseum/archaeology/dating/index.shtml
An online museum exhibition by Minnesota State University is provided here; the pages on dating techniques discuss relative and absolute dating techniques in detail. The site provides good accounts of the preparation and application of each technique used for the reconstruction of Quaternary environments.

NOAA Paleoclimatology Program: Astronomical Theory of Climate Change

http://www.ncdc.noaa.gov/paleo/milankovitch.html
This site provides a good summary of Milankovitch theory.

Quaternary Websites

http://www.colby.edu/geology/Quatresources.html
Extensive links to other Quaternary sites on the WWW.

Remembrance of Things Past: Greenhouse Lessons from the Geologic Record

http://www.gcrio.org/CONSEQUENCES/winter96/geoclimate.html
These are online articles that analyse the palaeorecord of historic climate change and its relevance to current concerns

over enhanced greenhouse warming, discussing the potential for abrupt transitions and the comparison of model results to palaeodata.

University of Wales, Aberystwyth: Quaternary Environmental Change

http://www.aber.ac.uk/~qecwww/index.htm
This site contains detailed information on luminescence, palaeoecology and volcanic impact data.

 Visit our website at **www.pearsoned.co.uk/holden** for further questions, annotated weblinks, interactive models and video-clips showing physical processes in action.

CHAPTER 21

Climate change: an unprecedented environmental challenge

John Grace

School of GeoSciences, University of Edinburgh

Learning objectives

After reading this chapter you should be able to:

➤ discuss the evidence for anthropogenic global warming

➤ understand how the use of fossil fuels has impacted upon the climate

➤ describe how the carbon cycle has been perturbed

➤ appreciate how humankind has created environmental problems and perceive how they may be solved

21.1 Introduction

Environmental change on a global scale first became a matter of public concern in the 1960s. Before then, the perceived environmental problem was urban pollution, which affected human health and the quality of life of many people. Although urban pollution became acute during the **Industrial Revolution**, it was not new. The **smelting** of toxic metals such as copper and lead was a health hazard in ancient Rome, as revealed by analysis of hair samples from the preserved corpses of Roman soldiers found in bogs, and

from traces of metal in Greenland ice cores. Coal was used in London in the thirteenth century. Coal contains not only carbon but also 1–4% sulphur and traces of heavy metals, and therefore its combustion releases a multitude of pollutants as well as carbon dioxide (CO_2). With the onset of the Industrial Revolution in western Europe, around 1780, the use of coal increased dramatically and cities such as London became heavily polluted with smog, a mixture of fog and smoke. Domestic coal burning was a major contributor to smog, and the industrial regions around Birmingham in England became known as the Black Country; even non-industrial Edinburgh was known as Auld Reekie, referring to the smell of coal burning.

Diseases such as bronchitis and tuberculosis were widespread following the Industrial Revolution, and nearly a quarter of deaths in Victorian Britain (1837–1901) were from lung diseases. In one week of December in 1952, 4000 Londoners were killed by a particularly severe episode of smog. The ensuing public outcry resulted in the Clean Air Act of 1957, which restricted coal burning and resulted in the use of cleaner energy sources such as oil, gas and electricity. Other coal-burning cities of the world such as Pittsburgh in the United States have a similar history. Problems were greatly exacerbated by the growth in use of the motor car, especially in regions receiving high solar radiation, such as Los Angeles, Athens and Mexico City, where the ultraviolet radiation reacts with uncombusted hydrocarbons from exhausts of cars to yield photochemical smog, irritating the eyes, nose and throat. An important milestone in the awakening of environmental concern was prompted by the widespread use of the persistent pesticides that were introduced after the Second World War and the publication of Rachel Carson's book *Silent spring* in 1962. *Silent spring* warned against the dangers of pesticides, especially to songbird populations, indicating how persistent chemicals might spread in food chains as well as in the atmosphere, and ultimately damage non-target species. At the same time, other scientists were demonstrating that the pesticide DDT could be found in snow in the Antarctic, and that pesticides were responsible for eggshell-thinning in wild birds, threatening especially those species at the end of food chains such as raptors. Thus, the idea of environmental change *on a global scale* soon became a permanent part of the western culture, and part of the international research agenda.

The global scale of human influence on the planet is today felt even more strongly, but not because of fears of widespread pollution of the land and sea by pesticides. The global environmental challenge that we face now is climate change, and that is the main focus of this chapter.

21.2 Climate change

21.2.1 Long-term change

The climate has always fluctuated, but usually over very long timescales. There are many sources of information that help in the reconstruction of past climates. These include historical records, evidence from the annual growth rings of trees, deposits of pollen in lakes and bogs, isotopes and fossils. The picture that emerges is quite complex, showing cyclic fluctuations on several scales (Figure 21.1). The long-term cyclic trends in the Earth's temperature, seen in Figure 21.1(a), associated with periods of glaciation known as 'the Ice Ages' were attributed by Serbian astronomer Milutin Milankovitch to the irregularities in the orbit and tilt of the Earth, which influence the energy received from the Sun (see Chapter 20). In contrast to these gradual changes, there have also been catastrophic events causing mass extinctions on a global scale. For example, in the Late Permian (245 million years ago) about half the families of marine animals were lost. At the boundary of the Cretaceous and Tertiary (known as the KT boundary, some 65 million years ago) 15% of marine families were

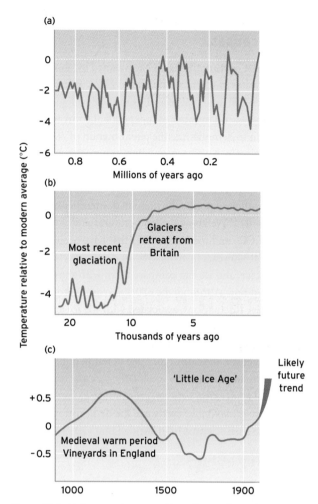

Figure 21.1 Air temperature over three timescales, relative to modern records: (a) the past million years; (b) the past 20 000 years; (c) the past millennium.

lost, perhaps 75% of plant species, and (most famously) dinosaurs became extinct. Such events are now usually attributed to the impact of comets, asteroids or large meteorites, which would have thrown up debris into the atmosphere and greatly reduced the penetration of solar radiation, causing widespread cooling, a reduction in **photosynthesis** and collapse of food chains. The KT boundary is considered to have been caused by an asteroid 10 km in diameter, which impacted at Chicxulub, northeast Yucatan, Mexico. Geologists recognize five such mass extinction events in the fossil record, all of them global in extent, taxonomically broad and most evident in marine invertebrates (for which the fossil record is relatively complete). It is against this background that we examine the changes in the climate system which are currently occurring, and their link to anthropogenic activity.

21.2.2 Recent climate change and its causes

Over the past century the Earth has warmed by about 0.7°C, with rapid rates in recent years (Figure 21.2). Apart from this modern instrumental record from meteorological stations, there are a number of independent sources of information to demonstrate the phenomenon of climate warming: glaciers have been receding, snow cover has declined, polar ice has been melting, sea levels have been rising and spring has been earlier.

Many authors describe the present-day temperatures as 'unprecedented'. We know the temperatures and the concentrations of CO_2 and methane (CH_4) that have occurred over the past 650 000 years from analysis of deep cores taken from polar ice, and we can compare them with those being experienced now. Although the 650 000 year record does contain large fluctuations, associated with the ice ages and the warm interglacial periods, we can observe that today's temperatures and concentrations of CO_2 and CH_4 are much higher. Moreover, the rate of increase in temperature is faster now than previously.

A causal association between greenhouse gases and temperature is inevitable, ever since the demonstration in 1859 by the Irish scientist John Tyndall that CO_2 absorbs infrared radiation. In the Earth's atmosphere, CO_2 and a range of other gases, including water vapour, absorb some of the infrared radiation that would otherwise stream directly out to space, thus causing a heating effect known as the greenhouse effect (the name arises because glass also absorbs infrared radiation and so the glass panes in a greenhouse have exactly the same effect on a local scale). The amounts of three of these gases, CO_2, CH_4 and N_2O, have risen sharply in recent times, and the extent of warming to be expected from these rises can be calculated (see the right hand axis of Figure 21.3). The rise in heat supply to the Earth's surface, known as radiation forcing (see Chapter 4), amounts to 2–3 W m^{-2}. This adds only about 1% to the heat supply from the incoming solar radiation (averaging about 230 W m^{-2} at the Earth's surface) and this is enough to increase the global temperatures.

We know for sure that humans have emitted vast quantities of CO_2 by burning fossil fuels and biomass, thus interfering with the global carbon cycle. The rise in CH_4 concentrations can be attributed to increases in various types of human activity. Only about 45% of all CH_4 emissions are produced naturally: from wetlands, termites, the ocean and from the decomposition of **gas hydrates**. The remainder is anthropogenic: from energy production, rice fields, landfills, ruminant livestock, waste treatment and biomass burning. The rate of increase in CH_4 has in fact been falling in the past few years. As for N_2O, the causes of its increase are somewhat less clear. It is produced naturally by microbial activity in the nitrogen cycle (see Chapter 22), and at a much faster rate when land is 'improved' by the use of nitrogen fertilizer. It is estimated that about one-third of the global emissions of N_2O are anthropogenic.

Other processes influence global temperatures (see Figure 4.23 in Chapter 4). Some are less well understood, and are the subject of current research. One such case is the influence and general behaviour of aerosols. These are particles in the atmosphere, including fumes and smoke from industrial processes and transport, and naturally produced particles such as pollens and spores. To some extent, they shield the planet from solar radiation, absorbing, scattering and reflecting solar radiation. The aerosol 'haze' which we see in the clear sky (especially in the northern hemisphere) effectively reflects part of the incident solar energy back into space, contributing to a cooling effect and therefore offsetting the warming effect of greenhouse gases (see Section 4.9 of Chapter 4). Periodic changes in the aerosol content of the atmosphere, for example by major volcanoes and by periods of heavy industrialization or biomass burning, have the capacity to change the temperature of the planet. Marine **phytoplankton** and the vegetation itself contribute to the

Figure 21.2 Observed trends on (a) global average temperature, (b) global average sea level and (c) northern hemisphere snow cover for the past century. (Source: IPCC 2007a)

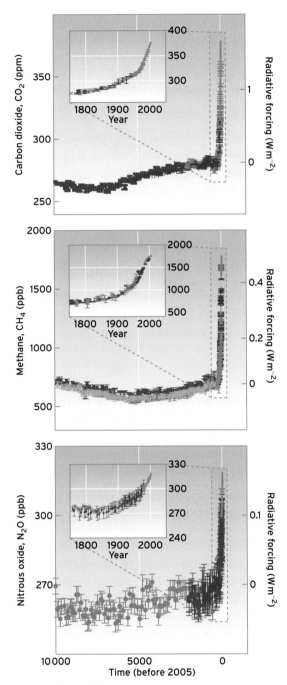

Figure 21.3 Trends in three global atmospheric greenhouse gases over the past 10 000 years. Different colours denote different studies. The inset box shows the period since 1750 in more detail. The left hand axis shows the concentration of each gas and the right hand side shows the radiative forcing that the concentration implies. (Source: IPCC, 2007a)

aerosol content of the atmosphere, by emitting certain volatile organic compounds which form aerosols (Meir *et al.*, 2006).

The uncertainty inherent in estimating the radiative forcing of aerosols comes from the recognition that not all

of them behave in the same way. Aerosols from biomass burning (black carbon) and the 'brown clouds' that come from urban sources may have the opposite effect. Ramanathan *et al.* (2007) flew small unmanned aircraft in brown clouds over the Indian Ocean and showed that these low-elevation clouds had a warming effect.

Variation in the Sun's energy output is sometimes proposed as a possible cause of global warming. Although we talk of the average energy incident on the Earth as measured outside the atmosphere as the 'solar constant' (and assign it the value of 1366 W m^{-2}), it is not quite constant. The most conspicuous variations are associated with sunspots, which appear as dark marks on the solar surface and arise because of variations in the magnetic properties of the Sun. They occur in an 11 year cycle, but there is a possibility of less conspicuous longer-term trends. However, according to estimates in IPCC (2007a) the changes in solar irradiance since 1750 have caused a radiative forcing of only +0.12 W m^{-2}, and recently Lockwood and Fröhlich (2007) have shown that the changes in solar radiation over the past 20 years have been in the opposite direction to that required to explain the observed rise in global mean temperatures.

Volcanic eruptions eject aerosols into the atmosphere, causing more radiation to be reflected back into space. They are sometimes large enough to have a short-term impact on the climate. The June 1991 eruption of Mount Pinatubo injected large amounts of aerosols into the stratosphere. Over the following months, the aerosols formed a reflective layer of sulphuric acid haze and global temperatures dropped by about 0.5°C. Likewise, the April 1815 eruption of Mount Tambora in (modern-day) Indonesia is believed to have been the cause of the exceptionally cold conditions everywhere in the world in the following year: 1816 is known as 'the year without a summer'. Significant volcanic eruptions in recent times were Mount Agung in Bali in 1963 and El Chichonal in Mexico in 1981. Volcanoes that are large enough to eject massive quantities of aerosols into the atmosphere have not become more frequent over the last 100 years and so cannot be the cause of the trend in global warming.

The scientific consensus is, overwhelmingly, that the production of greenhouse gases by humans is the primary cause of recent global warming, as outlined in reports from the IPCC (**http://www.ipcc.ch**). One of the most compelling lines of evidence is that global climate models (GCMs), in which production of these gases is simulated, show the same pattern of global warming as that observed, and in all parts of the world (Figure 21.4). When run without adding anthropogenic production of greenhouse gases, GCMs show no appreciable global warming.

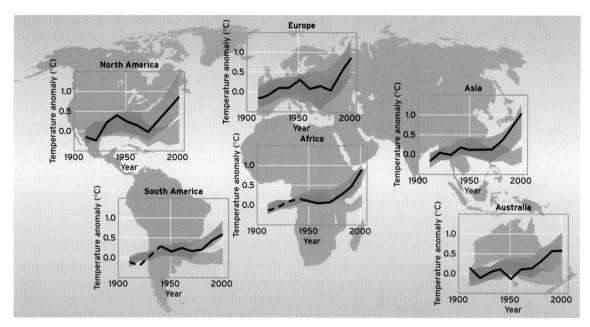

Figure 21.4 Observed warming rates in different regions of the world. Black lines are meteorological observations, blue bands represent the range of model results when no anthropogenic effects are included, red bands represent the range of model results when anthropogenic effects are included. (Source: IPCC, 2007a)

Understanding the causes of climatic variation is still an area of intense research, drawing upon expertise from many scientific disciplines (see Chapters 4 and 20). One important issue is the behaviour of a myriad of negative feedbacks which tend to dampen any instability, and the extent of the influence of positive feedbacks which might cause run-away warming. For example, as the climate warms and the polar ice melts, the overall albedo of the planet will decline (land and sea absorb more energy than ice and snow). A decline in albedo will make the planet's surface absorb more solar radiation, and thus further warming will occur. In this way, warming may give rise to further warming, an example of a positive feedback loop. Some examples of positive and negative feedbacks are given in Box 21.1.

21.2.3 Predictions from global climate models (GCMs)

Global climate models have developed from global circulation models (both abbreviated GCMs), which in turn sprang from the application of numerical methods to weather forecasting. Predictions of the climate for the next century are made by running GCMs with specified prior assumptions about the pattern of greenhouse gas emissions. In reality, these patterns will depend on social, political and economic development in the world, and they are patterns we cannot foretell. So researchers define them as 'scenarios' or

'storylines', each storyline having a particular pattern of greenhouse gas emissions, and use GCMs to investigate the consequence of each scenario. The scenarios are defined exactly in the Special Report on Emission Scenarios, SRES (IPCC, 2001) and are summarized here.

A1: In this scenario there is rapid economic growth, an increasing human population until mid-century and thereafter a decline, and the rapid introduction of more efficient technologies. Three A1 groups are distinguished: A1FI is fossil-fuel intensive, A1T uses non-fossil energy and A1B uses a mixture of the two. A1B corresponds to what most traditionalists imagine will happen. A1FI leads to the CO_2 concentration rising from its present 380 ppm to around 960 ppm while A1B results in 710 ppm by 2100.

A2: Here, the world develops in a more heterogeneous way with emphasis on self-reliance. Fertility patterns have regional characteristics and converge slowly; economic growth and technological uptake are more fragmented. In this scenario, the CO_2 concentration rises from its present 380 ppm to 860 ppm by 2100.

B1: Like A1, scenario B1 is a convergent world with a population that peaks in the mid-century but with a strong evolution of a service and information technology, with reductions in material intensity and clean technologies. In B1 global solutions are found to economic, social and environmental sustainability. In

ENVIRONMENTAL CHANGE

CLIMATE FEEDBACK

Anthropogenic activity is believed to be enhancing climate change and encouraging the planet to warm. However, there are a range of feedbacks that result in different responses to human activity. Positive feedbacks on the climate system will accelerate global warming, while negative feedbacks will suppress warming. There are a whole range of interlinked processes that suggest we need to look at environmental change taking a whole-system viewpoint. This box lists some of the hypothesized climate feedbacks that global modellers are investigating.

Positive feedbacks

- Warming will cause release of CO_2 from increased biomass decomposition, primarily in the forest regions of the world but also in the tundra, thus accelerating warming.

- Warming will melt snow and ice, decreasing albedo and thus increasing warming, melting even more snow and ice.

- Tropical deforestation will cause warming and drying, itself causing a decline in the rainforests of the world.

- Increased cover of woody vegetation in the high latitudes, caused by warming, will decrease the reflectance of the land surface, and thus accelerate warming.

- Warming will increase the decomposition rate of gas hydrates (see Chapter 19), leading to a release of the potent greenhouse gas methane; this will increase warming.

Negative feedbacks

- Deforestation will lead to an increase of soil erosion,

atmospheric aerosols will increase and solar radiation at the surface will decline, causing cooling.

- Replacement of coniferous forest by warmth-loving broadleaved forests and by agriculture will decrease planetary reflectance, causing cooling.

- Increased transpiration in a warm world will lead to more clouds, cooling the planet.

- Increased precipitation and ice melt will result in increased runoff into sensitive parts of the oceans altering the balance between freshwater and saline water, thereby resulting in a slowing of ocean circulation and allowing northern high latitudes to cool (see Chapter 20).

BOX 21.1

---

this scenario, the CO_2 concentration rises from its present 380 ppm to 540 ppm by 2100.

B2: In B2 local solutions to economic, social and environmental sustainability are found; the population growth rate is slower than A2, with intermediate levels of economic development, and there is less rapid technological change than in A1 and B1. In this scenario, the CO_2 concentration rises from its present 380 ppm to 615 ppm by 2100.

When the GCM are run, we see a warming by 2100 ranging from 1.8°C in the B1 scenario to nearly 4°C in the A1FI scenario (Figure 21.5). There are associated changes in rainfall. For example, in the A1B scenario (Figure 21.6) the rainfall patterns are substantially different from those today, with more rain falling in the polar regions while the mid-latitudes will become drier. The Mediterranean regions of Europe and Central America will become especially dry according to this prediction.

These changes are profound, especially so as the models suggest a more variable climate with an increasing frequency of extreme events. News reports of storms, droughts and hurricanes are increasingly shown in the media but these alone should not be taken as evidence of a link between global warming and extreme events. Analyses of reliable long-term records and model predictions are the proper evidence that must be considered. Emanuel (2005) investigated data on hurricanes and found the total power dissipated (longer storms and more intense storms) has increased markedly since the mid-1970s. Moreover, model predictions do indeed show an increased variability with an increased frequency of extreme events (IPCC, 2007b). Heat-waves, for example, are expected to increase (Figure 21.7). Results such as these prompt economic analysis and receive the attention of the public and of politicians. Insurance companies can no longer base their premiums on the analysis of past data when the climate system is so clearly

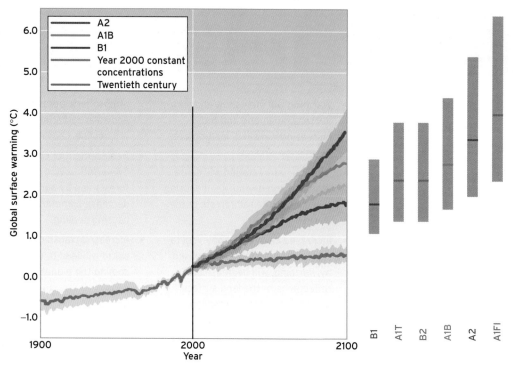

Figure 21.5 Predictions of global warming for different scenarios (see text for an explanation). The coloured lines show the assumed socio-economic scenarios and the bands around the lines show the range of model behaviours. Note: the orange line shows the effect of keeping the concentration constant from the year 2000. (Source: IPCC, 2007a)

changing its behaviour. Such 'extremes' in temperature, rain and wind will all cause appreciable damage, and the cost of repairing the damage will ultimately consume much of the wealth of the world, as emphasized by the report of Sir Nicholas Stern made to the UK Government (Stern, 2006). Box 21.2 provides some examples of potential impacts of global warming.

21.2.4 Critical evaluation of the state of the art in GCMs

Global climate models have significant weaknesses, which are frequently highlighted by sceptics. However, according to the IPCC, most climatologists agree that better models would not materially influence the conclusions of the model

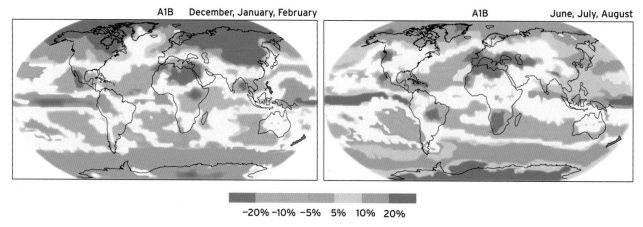

Figure 21.6 Relative changes in precipitation for the period 2090-2099, relative to 1980-1999, assuming the A1B scenario for December to February (left) and June to August (right). White areas are where less than 66% of the models agree on the sign of change. (Source: IPCC, 2007a)

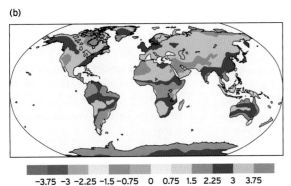

Figure 21.7 Increase in heatwaves over the rest of the century, expressed as the standard deviation of temperature: (a) trends from runs of the GCM for three scenarios; (b) global pattern for A1B. (Source: IPCC, 2007a)

runs. Some of these perceived weaknesses are mentioned for consideration here.

21.2.4.1 Resolution

Spatial resolution may be too coarse. For example, in the HADCM3 model, the GCM used at the Hadley Centre in the United Kingdom, the grid cells for the global runs made for the IPCC are 2.5 × 3.75 degrees in latitude × longitude, and the time steps are half-hour. There is a practical limitation on spatial and temporal resolution imposed by the speed of the supercomputer as it takes a very long time to run a GCM for a 100 year or more simulation. As computing power increases over the next few years there will be improvements in resolution. Some features of the climate system of course have a characteristic size which is small in relation to the grid cells (hurricanes and even clouds have to somehow be represented).

21.2.4.2 Biological and chemical coupling

Attempts to represent the impact of the biology and chemistry in GCMs are in their infancy. To some extent this

bottleneck relates to the lack of process understanding. For example, how should we model the effect of warming on the respiratory production of CO_2 by the soil microbes? In general, our understanding of the carbon cycle is incomplete, and arguably we are not yet ready to represent it in GCMs, yet its behaviour clearly has the potential to generate 'surprises' in the form of new and substantial sinks and sources of carbon. This is touched upon in Section 21.3. Similar remarks could be made in the realm of atmospheric chemistry and aerosol science.

21.2.4.3 The behaviour of ice

It is very difficult to represent properly the melting of ice, and the existing GCMs fail to deal specifically with the consequences of the possible melting of the Greenland ice sheet. This would cause a massive influx of meltwater into the North Atlantic, changing the ocean circulation patterns and therefore profoundly altering the distribution of heat over the Earth's surface. Such events may have occurred before, as the Heinrich events, which are evident during the last glacial period (Rahmstorf, 2006; see Chapter 20).

21.2.4.4 The human dimension

No one has attempted to incorporate models of social and economic life into this type of climate model. The most significant change in land-use is currently tropical deforestation. When forest is replaced by pasture, as for example in Amazonia, the land surface becomes more reflective, the pattern of evapotranspiration becomes more seasonal, the cloud cover is reduced, and the surface becomes aerodynamically smoother (Figure 21.8). These changes, on the scale of the Amazon Basin (over 4 million km^2), have the potential to change the climate not only in Brazil but elsewhere.

> **Reflective questions**
>
> ➤ Why do most scientists believe that most of the recent global warming is caused by human actions?
>
> ➤ Can you list some positive and negative climate feedbacks and some of the impacts of climate change that could be associated with increased greenhouse gas concentrations?
>
> ➤ What are the current limitations of GCMs?

IMPACTS OF GLOBAL WARMING

Most current models suggest a 2–4°C increase in global temperature over the next century. This rate is 10 times faster than the warming experienced over the past 10 000 years, and substantial impacts on human societies are anticipated. The following is a summary of some of the main effects that have been widely discussed in recent publications.

- Extinction rates, already very high, will be increased even more. Species would have to be capable of very fast migration to keep up with the isotherms which would move northwards at about 10 km per year. A recent study suggests that 13–37% of species will be 'committed to extinction' by 2050 (Thomas *et al.*, 2004).

- Although cold countries such as Russia may enjoy an increase in agricultural productivity as a result of warming, at least in the short term, many of the world's main food-producing regions may become too hot and dry for crops to grow. This would include major 'breadbasket' regions such as central and southern Europe and North America.

- Low-lying ground, including many major cities of the world and some entire small island states, may be inundated as a result of thermal expansion of the ocean and the melting of ice. Rates of rise in sea level are likely to be in the range 0.20–0.86 m from 1990 to 2100 (IPCC, 2001).

- Some geographical regions will suffer more than others. Temperature rises are currently especially high in the high latitudes (leading to melting of ice). Models show an increase in the extent of El Niño, with high rates of warming and drying in some of the tropical regions, causing replacement of the rainforest of Brazil by savanna (see Chapter 22).

- Diseases are likely to spread from the tropics to the temperate and northern regions as the climate warms. Outbreaks of pests may become more extreme, as the natural biological control processes may not always be present. Of particular concern is the northward spread of insect pests which damage crops or transmit disease. One such example is Lyme disease, a life-threatening disease carried by ticks and found to be more prevalent in the United Kingdom during warm years.

- The cost of repairing damage caused by extreme events will escalate and occupy a major proportion of the world's economic production.

Although there is considerable uncertainty in model predictions, partly because the models do not incorporate many of the likely feedbacks that derive from the vegetation itself, there is now agreement that the countries of the world must cooperate to reduce the emissions of greenhouse gases.

BOX 21.2

Figure 21.8 The effect of land-use change (tropical forest to pasture) on the energy and water balance of the landscape. Forests absorb more solar radiation and have a higher evapotranspiration rate than pastures.

21.3 The carbon cycle: interaction with the climate system

The carbon cycle is a natural biogeochemical cycle whereby carbon as CO_2 is transferred from the atmosphere to the land and ocean, where it resides in another form in water, soil or living material, before returning to the atmosphere as CO_2. The principal processes involved in transfer from the atmosphere are the dissolution of CO_2 in the oceans and the uptake of CO_2 by the photosynthesis of green plants. The processes involved in the return to the atmosphere are the release of CO_2 from the ocean in regions where the ocean upwells (see Chapter 3), and the breakdown of organic matter by respiration or fire. We can thus envisage the

carbon cycle as a set of fluxes between major pools as shown in Figure 21.9. The pools differ in magnitude and in the average time a carbon atom resides in them, so the dynamic behaviour is likely to be complex. For example, an 'average' carbon atom can be expected to reside in the atmosphere for about 5 years, while in the ocean the corresponding residence time will be about 400 years.

An understanding of the carbon cycle is fundamental to our understanding of life itself, as all biomass is carbon-based. The principal biochemical constituents of cells (carbohydrates, proteins, lipids) have a high carbon content, and the overall carbon content of dry biomass is in the range 45–55%. The carbon cycle has become especially topical in recent years, since the realization that warming is

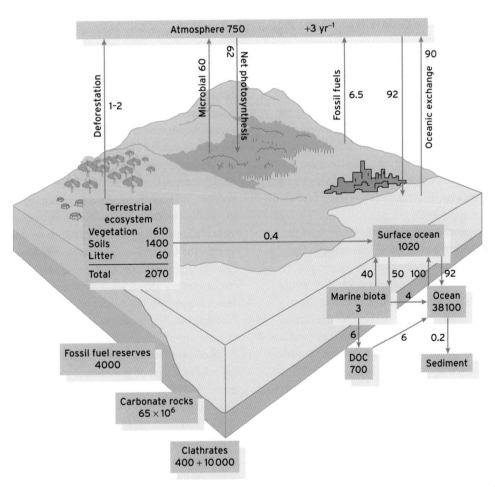

Figure 21.9 The carbon cycle, c. 2000. Stocks of carbon are shown in blue (units are gigatonnes, i.e. billions of tonnes, 10^9 tonnes or 10^9 petagrams). The +3 in the atmosphere box refers to the mean annual increase of C as CO_2 in the atmosphere. 'Net photosynthesis' is the net between photosynthesis and plant respiration. 'Microbial' refers to the heterotrophic respiration, dominated by soil microbes but also including animal respiration. 'DOC' is an abbreviation for dissolved organic carbon. Some of the stocks and fluxes are uncertain: soil stocks and ocean stocks and fluxes are not well determined. Stocks in fossil fuels, carbonate and clathrates are uncertain. 'Clathrates' refer to methyl clathrates, hydrated forms of methane, and the two numbers 400 and 10 000 refer to known quantities and possible upper limit, respectively.

large enough to force the cycle out of equilibrium, whereby the concentration of the gas in the atmosphere is rising. As we have seen above, CO_2 is by far the most important of the several gases that absorb infrared radiation emitted from the planetary surface, and its continuing rise is capable of causing additional global warming.

The quantity of carbon emitted by burning fossil fuels is about 6.5 Gt C yr^{-1} and clearing forests in the tropics releases 1–2 Gt C yr^{-1}, making a total anthropogenic burden on the atmosphere of 7.5–8.5 Gt C yr^{-1}. However, the concentration of CO_2 in the atmosphere is increasing by only about 3 Gt yr^{-1}. The 'missing' carbon is being taken up by terrestrial photosynthesis or dissolving in the ocean. The search for the so-called 'missing sink' has occupied scientists' attention for the past 20 years. It seems that the terrestrial vegetation may be absorbing about half of the missing carbon, and the rest is dissolving in the ocean (Gurney *et al.*, 2002). This is reflected in the fluxes shown in Figure 21.9. These fractions do vary from year to year, as some years are more favourable for plant growth. However, the ocean and terrestrial sinks might not continue. According to some models, the terrestrial sink will diminish and then become a source as a result of the impact of high temperatures and droughts in the tropics (Cox *et al.*, 2000). On the other hand, as warming occurs there is a reasonable expectation that the sinks in the northern regions will strengthen, perhaps enough to compensate for the loss of sink strength in the southern regions. This theme is discussed further in Chapter 22.

There are prospects of enhancing the strength of the sinks in order to slow the rate of climate warming. On the terrestrial side, this might be done by planting more forests, or by protecting existing tropical forests. It could also be done by modifying agricultural practices (less ploughing, for example) in order to conserve the carbon stocks in the soil or to protect peatlands and other wetlands. The potential of enhancing the terrestrial sink by land-use changes of this kind is considerable (IPCC, 2001). The ocean sink might also be managed by fertilizing the ocean. The scientific basis for this proposition is the observation that phytoplankton in the deep ocean are short of the micronutrient iron. When iron is added as ferric ions the productivity of phytoplankton is increased. This provides more food for zooplankton, and more food for the fish that eat the zooplankton, so there should be an enhanced stream of dead biota and carbonate shells that sink to deeper layers of the ocean. This, in turn, should enable more CO_2 to dissolve in the surface waters. No one really knows whether this will work on a large scale. Environmentalists have generally opposed all suggestions of increasing the strength of ocean sinks, arguing that the mechanisms and processes are imperfectly understood, and the sinks cannot be depended upon in the long term. They argue that there is no alternative but to reduce emissions by reducing consumption and finding alternative 'clean' sources of energy.

> **Reflective question**
> ➤ Can you explain the carbon cycle and what is meant by the 'missing sink'?

21.4 Mitigation

The prospect of a three or even four degree rise in global temperatures by 2100 constitutes 'dangerous climate change'. The question of 'what can be done?' to avoid dangerous climate warming has to be addressed at a global scale. If one country alone were to apply stringent measures at great expense to reduce fossil fuel emissions while other countries go ahead and increase theirs, then that country would be at an economic disadvantage and the world as a whole would scarcely benefit. Clearly, countries must engage in debate and decide on the actions required before it is too late. Following the 1992 'Rio Summit' to discuss environmental change, many countries joined an international treaty – the United Nations Framework Convention on Climate Change (UNFCCC) – to consider what might be done to reduce global warming. Later, in 1997, a number of nations approved an addition to the treaty: the Kyoto Protocol, which imposes powerful (and legally binding) measures to reduce emissions.

The Kyoto Protocol is an international agreement to limit the emission of greenhouse gases. Six gases are mentioned in the protocol, of which CO_2 is the most important contributor to warming (Table 21.1). They differ greatly in their residence time in the atmosphere, and in the extent to which they are effective in absorbing infrared radiation. These two factors together are incorporated into an index called the global warming potential (GWP), which measures the relative effectiveness of the gas, on a per kilogram basis, in causing global warming over a century.

Only Parties to the Convention that have also become Parties to the Protocol (by ratifying, accepting and approving) are bound by the Protocol's commitments; 174 countries have ratified the Protocol to date. Of these,

Table 21.1 Greenhouse gases in the Kyoto Protocol, lifetime in the atmosphere and global warming potential (GWP) on a scale where $CO_2 = 1$ (see text for definition) (Source: Woodward *et al.* 2004)

Gas	Sources	Lifetime (years)	GWP	Comment
Carbon dioxide (CO_2)	Fossil fuel burning, cement manufacture	100	1	On the increase still, but Annex 1 countries[*] likely to reduce emissions
Methane (CH_4)	Wetlands, rice fields, burning, oil wells, ruminants, termites	12	21	
Nitrous oxide (N_2O)	Land disturbance, use of nitrogen fertilizers	120	310	
Hydrofluorocarbons	Industry, refrigerants	1-300	140-11 700	Substituting for CFCs. They contain only hydrogen, fluorine and carbon, and do not damage the ozone layer (CF_3CFH_2, CF_3CF_2H, CHF_3, CF_3CH_3 and CF_2HCH_3)
Perfluorocarbons (CF_4, C_2F_6)	Industry, electronics, firefighting, solvents	2600-50 000	6 500	Also substituting for CFCs
Sulphur hexafluoride	Electronic and electrical industries, insulation	3200	23 900	Increasing; by-product of aluminium smelting

[*]See Box 21.3 for a list of Annex 1 countries.

36 countries and the countries of the European Union are required to reduce greenhouse gas emissions below levels specified for each of them in the treaty. The individual targets for these Annex I countries are listed in the Kyoto Protocol (see Box 21.3 and the Protocol itself at **http://unfccc.int/2860.php**). These add up to a total cut in greenhouse gas emissions of 5.2% by 2012 from 1990 levels, but for these countries only.

Unfortunately, only a few countries are on course to meet their Kyoto targets, and even if they were on course, the 5.2% reduction in GHG emissions would not be sufficient to make a substantial difference. Moreover, the United States, the world's largest GHG emitter, opted out of the procedure in 2001, while countries such as India and China, with fast-growing economies and rapidly increasing emissions, were never included in the Protocol.

At an individual level, some people opt to take personal responsibility for their carbon emissions. In a developed western society it is possible to make substantial savings in this way, as Reay (2006) has pointed out. But only a few people have so far taken direct control over their 'carbon footprints' by changing their lifestyles. For example, most people in the developed world, and an increasing number in the developing countries, have a substantial component of emissions from travel; they often ignore options to reduce these by selecting low-carbon-emitting modes of transport (train not aeroplane, bicycle not car). Currently, the cheapest and most convenient mode of transport is usually not the one with the lowest carbon emissions. In a recent survey of the travelling habits of the people of Edinburgh a linear relationship was found between income and transport emissions (Figure 21.10), suggesting a fatal inevitability between income and carbon emissions. The challenge for governments is now to enable modes of travel that are efficient and affordable yet that do not incur such high emissions of greenhouse gases.

At national level, some countries have set themselves carbon emission reduction targets which go well beyond those prescribed by the Kyoto Protocol. In the United Kingdom, for example, there is a target for a CO_2 reduction of 20% by 2010 and 60% by 2050. These targets will be achieved only by quite radical and unpopular changes which

CASE STUDIES

SOME OF THE MAIN COMPONENTS OF THE KYOTO PROTOCOL

The aim of the Kyoto Protocol is to obtain international agreement to limit the emission of greenhouse gases, and thus reduce the extent of global warming. In the words of The Framework Convention on Climate Change the ultimate aim is: 'to achieve stabilisation of greenhouse gas concentrations . . . at a level that would prevent dangerous anthropogenic interference with the climate system'.

The Protocol mainly concerns actions of 'Annex 1 Parties', which are the following countries: Australia, Austria, Belgium, Bulgaria, Canada, Croatia, Czech Republic, Denmark, Estonia, Finland, France, Germany, Greece, Hungary, Iceland, Ireland, Italy, Japan, Latvia, Liechtenstein, Lithuania, Luxembourg, Monaco, Netherlands, New Zealand, Norway, Poland, Portugal, Romania, Russian Federation, Slovakia, Slovenia, Spain, Sweden, Switzerland, Turkey, Ukraine, United Kingdom, United States of America.

- Each Party in Annex 1 pledges to: enhance energy efficiency, protect and enhance sinks for carbon, promote sustainable agriculture, promote and develop renewable energy and CO_2 absorption technologies; phase out any subsidies on greenhouse gas (GG) emitting

sectors, to limit emissions in the transport sector, and to limit methane emissions.

- Parties should meet, cooperate to share experiences and discuss all measures and policies. This is achieved through a Conference of the Parties which is held periodically. The 13th meeting, called COP13, was held in Bali in December 2007.

- Annex 1 Parties should reduce GG emissions by 5% below 1990 levels in the commitment period 2008-2012. Sinks achieved by land-use changes including afforestation, reforestation can be set against emissions. Carbon stocks (including soil, biomass, fuel stores) in the baseline year 1990 should be stated.

- Parties can reach commitments jointly by making alliances. Thus, the European Union has declared itself as a 'bubble' in which some of its countries (Greece, Portugal and Ireland) do not have to restrict their emissions because other countries within its territory (United Kingdom, Germany) have agreed to make savings that exceed the overall target of 5%.

- In estimating the stocks, the methodologies defined by the Intergovernmental Panel on Climate Change (IPCC) will be used.

- Any Annex 1 Party may transfer emission reductions to another Annex 1 Party. Trading can therefore

occur. If one country finds it too difficult to meet its target for emission reduction, it can pay another country to do this on its behalf.

- There are ways to assist poor countries to achieve sustainable development, while assisting Annex 1 countries to achieve their emission targets. This is called the Clean Development Mechanism (CDM). For example, a tropical country might decide to accept payment from an Annex 1 country to reduce C emissions on its behalf. This is one of the most controversial aspects of the Protocol as there are ethical as well as operational difficulties. The tropical country could also absorb the CO_2 through growing new forest. It is unclear how, and by whom, the uptake of carbon would be monitored.

The Kyoto Protocol came into effect on 16 February 2006, when 141 countries, accounting for 55% of the greenhouse gas emissions, ratified the treaty and thus pledged to reduce their greenhouse emissions according to agreed targets. If this were to be effective, these emissions would be reduced by 5.2% (from 1990 levels) by the year 2012. Unfortunately, the agreement does not include the world's top polluter, the United States, or countries with rapidly expanding economies such as China and India. Moreover, very few countries seem able to meet their emission reduction targets.

BOX 21.3

will have to include: carbon taxes, new technologies (especially, the burial of CO_2 in geological strata) and a move back to nuclear power. The use of renewable energy sources such as wind power and biomass energy can make only a

small contribution. Countries such as China that are developing economically at a very fast rate, and have a large reserve of easily accessible coal, may take several decades to control their emissions effectively.

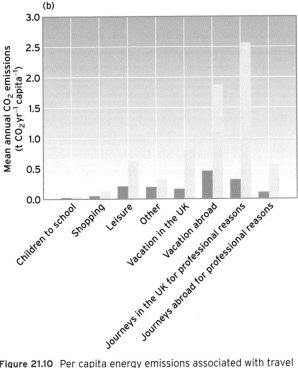

Figure 21.10 Per capita energy emissions associated with travel for people of Edinburgh, Scotland, in 2005: (a) relationship with income; (b) type of travel (blue bars are deprived social group, grey bars are affluent social group). (Source: from Korbetis *et al.*, 2006)

Reflective questions

➤ Using Internet tools work out your own personal carbon footprint and reflect how you might make it smaller. Would this impact on your quality of life?

➤ How can we mitigate against climate change?

21.5 Destruction of the ozone layer by chlorofluorocarbons (CFCs)

Electromagnetic radiation from the Sun reaches the Earth's atmosphere at a rate of about 1366 W m^{-2}. Much of it is scattered back into space, and only a small fraction reaches the surface; it drives photosynthesis and evaporation, and warms the planetary surface. The radiation contains ultraviolet (waveband 100–400 nm) as well as visible radiation, which happens also to be the photosynthetic waveband (waveband 400–700 nm), and near-infrared (waveband 700 nm to a few micrometres). Ultraviolet radiation is absorbed by the DNA of all organisms, causing damage to the genetic code and consequently interfering with protein synthesis and the control of cell division. In humans the most common effects include reduction in the immune system (all races), skin cancer in Caucasian-type humans and damage to the eyes.

For the past billion years, the Earth has been shielded from damaging ultraviolet radiation by ozone (O_3) in the stratosphere. This protection has enabled life to develop on the land. Now, the ozone layer is diminishing as a result of a chain of chemical reactions that begins with totally human-made chemicals, the chlorofluorocarbons (CFCs). These CFCs are synthetic non-reactive gases and liquids first made in 1930 and used as refrigerants (later as propellants in spray cans). Being inert under normal conditions, they persist in the atmosphere, and slowly make their way to the stratosphere. Laboratory studies in 1974 established that CFCs could catalytically break down ozone in the presence of ultraviolet radiation to form highly reactive radicals such as ClO and OClO. It is these radicals that catalyse the breakdown of O_3 to O_2. A ground-based survey of stratospheric ozone was started in Antarctica in 1956, and surveys continued using satellites in the early 1970s. In 1985 a British team based in Antarctica reported a 10% drop in the ozone level during the spring, which they attributed to CFCs and oxides of nitrogen. A similar decline was also seen in data from NASA's Nimbus 7 satellite carrying TOMS (Total Ozone Mapping Spectrometer; see Chapter 23), and it is now evident that a steady decline is occurring over Antarctica and a decline has been detected over the Arctic (Figure 21.11). In the 1980s Australians sunbathed much less than before, and sales of sunhats and skin creams to protect against ultraviolet radiation increased. Plants and animals are less able to take protective measures.

The Montreal Protocol of 1987, in which nations agreed to phase out the use of CFCs, has undergone several modifications. Trade sanctions on CFCs have been imposed and a total phasing out is due in 2030. In March 1989

1980

1991

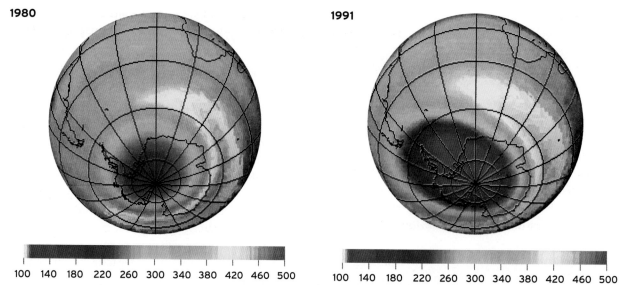

| 100 | 140 | 180 | 220 | 260 | 300 | 340 | 380 | 420 | 460 | 500 |

| 100 | 140 | 180 | 220 | 260 | 300 | 340 | 380 | 420 | 460 | 500 |

Figure 21.11 Images of the thickness of the stratospheric ozone layer over the southern hemisphere. The units are Dobson units named after G.M.B. Dobson, an early investigator. Normal thickness is 300 and the scale is linear. (Source: Centre for Atmospheric Science, University of Cambridge, UK)

environmental ministers of the European Union announced a total phase-out of CFCs in Europe by the year 2000. More recently, related chemicals which do not significantly destroy ozone have been introduced: these are the hydro-fluorocarbons (HFCs) and the perfluorocarbons (PFCs). Unfortunately, these gases are powerful greenhouse gases, although their concentration in the atmosphere is very low and so they do not presently contribute much to global warming.

Reflective questions

➤ Why is the ozone layer crucial to life on Earth?

➤ Has the Montreal Protocol of 1987 been more successful than the Kyoto Protocol of 1997?

21.6 The future

Fossil fuel continues to be the main source of energy. More-over, the developing world, which consists of about five-sixths of humankind, will continue to increase its population and its fossil fuel burning for many years after the rich countries have stabilized and decreased their dependency on fossil fuels. Some poor countries have neither fossil fuels nor any other supply of energy, and so cannot develop. Even fuel-wood is in short supply.

Nuclear power was developed enthusiastically by many countries in the 1950s, and 29 countries were running 437 nuclear power plants by 1998. Early optimism about devel-opment of an energy economy from nuclear fission faded following nuclear accidents and leakages such as Chernobyl in the USSR (now in the Russian Federation) in 1986. Many environmentalists believe that the risks that are inherent in nuclear fission are quite unacceptable. Power from nuclear fission is very expensive, once the costs of handling radio-active waste and decommissioning old power stations are taken into account. Despite all this, many governments are in favour of continuing and even expanding their nuclear power programmes, and for many it is the only practical way to reduce carbon emissions.

There are, however, some reasons for optimism. In the period since 1960 considerable progress has been made towards developing alternative sources of energy to replace carbon-based fuels (coal, oil and gas). Governments in the richer countries are setting ambitious targets to decarbonize their energy economies and are pushing for investment in renewable forms of energy such as wind and biofuels. Solar cells are likely to become increasingly important (see Figure 24.15 in Chapter 24). This technology was first developed during the exploration of space in the 1960s. Silicon solar cells convert solar energy to electrical energy with an efficiency of 20%, and the energy may be stored and transported in fuel cells. The construction cost of a solar cell is rather low, as the main elemental ingredient, silicon, is abundant. Unlike wind turbines and wave-power generators there are no moving parts and consequently maintenance costs are extremely low.

Even in winter in northern countries, solar cells can provide useful quantities of solar energy. In the future, roof-tiles incorporating solar cells may be used in all new housing construction, and solar energy 'farms' may cover large areas of deserts.

The major problems facing the world are related to each other: climate change, energy supply, poverty, disease and the tensions and hostility arising from a disparity of living standards between different countries. It is difficult to foresee what kinds of environmental change are just around the corner, and therefore it is hard to plan for the future. A hundred years ago, the problems of today were invisible and quite unpredicted. But there is some evidence from recent history (Box 21.4) that we are at last beginning to grasp the nature of the human–environment interaction. For 200 years humans have been inadvertently damaging the life-support system of the planet. In the past 30–40 years we have realized what is happening, and governments are beginning to take remedial action.

Reflective question

➤ What cause for optimism is there in thinking about future climate change and human response to it?

ENVIRONMENTAL CHANGE

IMPORTANT RECORDED EVENTS

The most important dates in the relationship between humans and the global environment are listed below. These events and discoveries changed our perceptions of the world we inhabit, and contributed to global environmental change. From 1500 to 2007 the human population increased from 0.5 billion to 6.6 billion, and the world gross domestic product (GDP) rose from $240 billion to $30 000 billion.

1400–1450   Chinese fleets arrive in Middle East and East Africa; Arab fleets cross Indian Ocean. Intercontinental trade begins.

1450   Johannes Gutenberg, Germany, establishes printing press, enabling humans to communicate efficiently and to learn from each other more effectively.

1490–1500   Decade of European maritime exploration of Asia, Africa and America, paving way for settlement, slavery, trading, further spread of economically important plant and animal species.

1610   Galileo Galilei, Italy, uses the telescope to observe behaviour of the Moon and planets. Other scientific instruments for Earth observation were developed: microscope (1618), thermometer (1641) and barometer (1644).

1780–1820   Industrial Revolution. Dramatic increase in the use of coal. Western Europe sees rapid technological, social and economic transformation, driven largely by the steam engine fuelled by coal. Widespread urban pollution, exploitation of workforce, occupational diseases. Humans begin to alter the composition of the global atmosphere.

1796   First blast furnace opens in Gleiwitz, Poland. Manufacture of iron and steel followed in urban centres: Belgium, Germany, Great Britain. Respiratory diseases increase.

1798   Thomas Malthus, English clergyman, writes *Essay on the Principle of Population*, pointing out the natural tendency of human populations to grow exponentially, outrunning the food supply.

1821   William Hart obtains natural gas (methane) from a 9 m well in New York, and provides street lighting.

1827   Jean Baptiste Joseph Fourier (France) first proposed that 'light finds less resistance in penetrating the air, than in repassing into the air when converted into non-luminous heat'. Possibly the first articulation of the greenhouse effect.

1838   Birth of John Muir at Dunbar, Scotland. First person to call for conservation of wilderness areas, arguably the father of the modern conservation movement.

BOX 21.4 ➤

➤

1839 Antoine-Cesar Becquerel, France, discovers the photoelectric effect, demonstrating that sunlight can be converted into electricity. But practical solar cells were not developed until 1954.

1842 John Lawes, England, invented superphosphate fertilizer.

1851 James Young, Scotland, discovers how to extract hydrocarbons from oil shale, and develops process of refining oil. He establishes a paraffin industry in Scotland (paraffin is called kerosene in the USA) and is nicknamed 'Paraffin Young'.

1859 Edwin Drake strikes oil at 20 m in Pennsylvania, USA. Oil was soon discovered in North and South America, Mexico, Russia, Iran, Iraq, Romania, Japan, Burma and elsewhere. Oil soon plays its part in the industrialization of the world.

1859 Charles Darwin publishes *The origin of species,* proposing the theory of evolution by natural selection.

1859 Irish scientist John Tyndall discovers that H_2O and CO_2 absorb selective wavebands of infrared radiation, and suggested a role for these gases in the regulation of the Earth's temperature.

1864 James Croll, Scotsman, proposed theory of long-term climate change to account for the Ice Ages (see also the reference to Milutin Milankovitch 1895)

1866 German engineers Langen and Otto patent the internal combustion engine. The manager in Otto's factory, Daimler, makes the first petrol engine in 1884.

1868 In Japan, the Meiji Restoration. Japan opens to the west and large-scale industrialization spreads to Asia.

1885 German chemist Robert Bunsen discovers how to make a very hot flame from gas, by mixing it with air before combustion. Gas burners were thereafter much more efficient.

1895 Serbian astrophysicist Milutin Milankovitch describes theory of long-term climate change. Essentially it is the same theory that Croll had proposed in 1864, but Croll's work was ignored by his peers.

1896 Arrhenius, Swedish chemist, advances theory that CO_2 emissions will lead to global warming, and postulates the ocean as a CO_2 sink.

1901 Italian Guglielmo Marconi invents radio, achieves first transatlantic radio message.

1903 Henry Ford, USA (1863–1947), establishes the Ford Motor Company, makes Model-T Ford cars in 1908. Others would follow Ford's idea of mass production, and car ownership would increase rapidly.

1903 Orville and Wilbur Wright, USA, demonstrate a flying machine based on the internal combustion engine. Rapid intercontinental travel by air would follow in 50 years.

1907 Henry Ford completes first tractor, a machine that was to revolutionize agriculture.

1909 Fritz Haber, Germany, shows how to synthesize ammonia from N_2 in the atmosphere, and Karl Bosch uses this process for mass production of nitrogen fertilizer. This was to greatly increase the capacity of humans to grow crops.

1915 German geophysicist Alfred Wegener publishes his controversial hypothesis of continental drift, in a book entitled *The origin of continents and oceans.*

1917 Chainsaw manufactured for the first time. Enables rapid deforestation.

1926 John Logie Baird, Scotland, invents television, but TV broadcasts did not start until 1936 (UK) and 1941 (USA), and TV sets were not widespread until the 1950s. Television has allowed ordinary people to see how others live, and thus to understand better their own place in the world.

1927 Alexander Fleming, Britain, discovers the antibiotic effect of the fungus *Penicillium*. Much later (1940) Florey and Chain, working in the USA, discover how to make the antibiotic penicillin in bulk. This launches golden age of medicine. Infant mortality declines and people live longer.

1928 Mohandas Karamchand Gandhi, India leader, questions the sustainability of the industrial age: 'God forbid that India should ever take to industrialism after the manner of the West. If it took to similar economic exploitation, it would strip the world bare like locusts.'

1930 Thomas Midgely, USA, invents the gas freon. It was the first of the chlorofluorocarbons (CFCs) which much later (1970s) became widely used and caused thinning of the stratospheric ozone layer.

BOX 21.4 ➤

1938 Guy Stewart Callendar (UK) predicted global warming at a rate of 0.03°C per decade.

1943 Primitive electronic computers constructed: Harvard Mk I (USA) and Colossus (UK).

1944 Pilotless planes and rockets developed in Germany for use in warfare. Later, the technology formed the basis for space exploration and Earth observation.

1945 Atomic bomb dropped on Hiroshima. *Guardian* newspaper (UK) comments 'man is well on the way to mastery of the means of destroying himself utterly'.

1948 Agricultural efficiency increases dramatically in the developed world, as a result of mechanization, fertilizers, pesticides, plant breeding and managerial skill. Crop yields increase. Later (1960s and 1970s) the new agricultural technology is taken to the developing countries, and becomes known as the Green Revolution.

1948 First operational stored-program computer, known as Manchester Mk I (Williams, Kilburn and Wilkes, UK).

1951 Age of nuclear power starts with first commercial nuclear reactor at Idaho, USA. Later there are significant accidents: fire at Windscale, UK, in 1957; meltdown at Three Mile Island, Michigan, USA, in 1979; meltdown and large release of fission products at Chernobyl, USSR, in 1986.

1952 British jet airliner, De Havilland Comet, begins regular intercontinental travel.

1954 Chapin, Fuller and Pearson develop silicon solar cell capable of converting solar energy into electrical energy with a conversion efficiency of 15%.

1957 First spacecraft, Sputnik, USSR; to be followed by first man in space, Yuri Gagarin, USSR, in 1961 and first man on the Moon, Neil Armstrong, USA, 1969.

1958 Charles Keeling, of the Scripps Institute in the USA, begins the first reliable measurements of atmospheric CO_2 at Mauna Loa in Hawaii.

1960 Soviet engineers begin large-scale irrigation using rivers flowing to the Aral Sea, the world's fourth largest lake. Within 40 years the lake would almost disappear, possibly the greatest hydrological change yet engineered by humankind.

1962 *Silent spring* by Rachel Carson, USA, warns of dangers of pesticides to wildlife. This best-seller inspired a whole generation of environmentalists.

1968 Satellite remote sensing starts. Pictures of Earth from deep space, Apollo 8 mission, USA; followed in 1972 by Earth Resources Satellite ERTS-1 carrying multispectral sensors later called Landsat.

1969 In the USA, the Advanced Research Projects Agency (ARPA) begins the ARPANET. Soon, global communication by e-mail and Internet would become possible.

1970 Establishment of Environmental Protection Agency (EPA), USA.

1971 Formation of Greenpeace. A group of activists sail their small boat into a US bomb-test zone near Alaska to draw attention to the environmental dangers of nuclear war.

1971 Swedish scientists demonstrate long-range transport of sulphur as the cause of acidification of Swedish lakes, and predict that acid rain will damage fresh water ecosystems and forests.

1972 In the UK, publication in *The Ecologist* of 'A blueprint for survival', warning of the extreme gravity of the global situation and criticizing governments for failing to take corrective action.

1972 Publication of *The limits to growth* by the Club of Rome, dealing with computer simulation of global environmental change. Fails to identify the threat of global warming; points to resource depletion and pollution as the major threats.

1972 First international conference on the environment, Stockholm, leading to the establishment of the United Nations Environment Programme (UNEP). Acid rain is widely publicized, especially in relation to forest decline, but since then the developed world has been moving to low-sulphur fuels.

1972 The anchovy fishery of Peru collapses because of overfishing and bad weather. Other fish stocks decline sharply, and management of marine resources becomes an important issue.

1973 Organization of Petroleum Exporting Countries (OPEC) restricts the supply of oil, forcing its price to rise five-fold and threatening the global economy.

BOX 21.4 ➤

➤

1979	James Lovelock proposes the Gaia hypothesis (see Chapter 23).
1985	Farman, Gardiner and Shanklin, a British team working in the Antarctic, report thinning of stratospheric ozone, attributable to CFCs.
1986	Nuclear accident at Chernobyl, USSR, creates radioactive fallout everywhere in the northern hemisphere, reminding people that environmental problems cross political boundaries. The expansion of nuclear power in the West falters.
1987	First appearance of the word *biodiversity* in the scientific literature (by E.O. Wilson, USA).
1987	Ice core from Antarctica, taken by French and Russian scientists, reveals close correlation between CO_2 and temperature over the past 100 000 years.
1987	Montreal Protocol signed, an agreement to phase out CFCs.
1987	United Nations World Commission on Environment and Development produces the Brundtland Report, dealing with definitions of sustainability.
1988	Intergovernmental Panel on Climate Change (IPPC) is established.
1990	IPPC's first Scientific Assessment Report, linking greenhouse gas emissions to warming.
1992	Implementation of the International Geosphere Biosphere Programme (IGBP) to predict the effects of changes in climate, atmospheric composition and land-use on terrestrial ecosystems; and to determine how these effects lead to feedbacks to the atmosphere.
1992	Earth summit, Rio de Janeiro. Leaders of the world's nations meet in Rio and set out an ambitious agenda to address the environmental, economic and social challenges facing the international community. Heads of state sign the UN Framework Convention on Climate Change. The UNFCCC was one of three conventions adopted. The others – the Convention on Biological Diversity and the Convention to Combat Desertification – involve matters strongly affected by climate change.
1997	Kyoto Protocol is drafted, the first international agreement to limit greenhouse gas emissions.
1997-1998	Particularly severe El Niño causes drought and widespread forest fires in Indonesia, Malaysia, Brazil and Mexico. In SE Asia the fires affect 10 000 km^2 of forest.
1998	The warmest year of the century, and probably of the millennium.
2000	International Coral Reef Initiative reports that 27% of the world's coral reefs are lost, mainly a consequence of climate warming.
2001	US President George Bush announces that the USA will not ratify the Kyoto Protocol.
2002	As warming of Antarctica proceeds, some 3200 km^2 of the Larsen B ice shelf collapses.
2002	Schools in Seoul, South Korea, are closed when a dust storm originating from China sweeps over the country.
2003	Gates of China's Three Gorges Dam are shut, and the world's largest hydropower reservoir is created, destroying archaeological sites and forcing the relocation of nearly 2 million people.
2003	European heat wave causes premature death of 35 000 people.
2004	Indian Ocean earthquake causes large tsunami, killing a quarter of a million people.
2005	Kyoto Protocol comes into force, 16 February.
2005	Hurricanes sweep the US Gulf Coast, causing widespread damage and loss of life. New Orleans evacuated.
2006	The film *An Inconvenient Truth* (Director Davis Guggenheim, Presenter Al Gore) presents the science of global warming in a manner accessible to non-scientists. It is a box-office success and wins awards.
2006	Nicholas Stern, an economist, in a report for the UK government, suggests that global warming will cause the greatest and widest ranging market failure ever seen, and proposes environmental taxes as the best remedy.
2007	IPCC Fourth Assessment Report predicts that temperatures will rise by 1.8–4.0°C by the end of the century.
2007	UN meeting on climate change in Bali agrees on 'road map' towards post-2012 agreement on climate change.

BOX 21.4

21.7 Summary

World population growth has been associated with increased utilization of the land for agriculture. Increased domestication of plants and animals and use of wood for fuel have resulted in vast amounts of deforestation, particularly over the past 200 years when the human population has increased from 1 billion to 6.6 billion. Humans now appropriate 40–50% of the land's biological production. Large-scale change in the land surface from forest to farm influences regional and global climates in many ways that are still not properly understood. Conversion of forest to pasture involves release of CO_2 to the atmosphere, changes to albedo, air movements, and the water, carbon and nitrogen cycles.

In addition, the burning of fossil fuels and the creation of human-made chemicals such as CFCs and techniques of creating nitrogen fertilizer all have an impact on the environment. The enhanced greenhouse effect appears to be causing accelerated climate change which will have major impacts on humans and global ecosystems. However, there are a number of positive and negative feedback effects that global circulation models are only just being able to predict and model. Atmospheric pollution from industry, combustion engines and agricultural practice is impacting on human health and biodiversity. Cities often experience harmful smogs, and the ozone layer which protects the Earth from vast amounts of harmful ultraviolet radiation is suffering severe damage due to CFC use.

However, humans are an inventive species and technological improvements are continually being made that may help us cope with and mitigate environmental change. The development of safe nuclear fusion and increased use of solar cells may allow us to harness the world's resources in a more sustainable manner. In addition, the international recognition that global environmental change is taking place has been achieved and there is a willingness around the world to try to combat environmental problems.

Further reading

Houghton, J.T. (2004) *Global warming: The complete briefing*, 3rd edition. Cambridge University Press, Cambridge.
Highly readable account of global warming, written by the former Chairman of the Intergovernment Panel on Climate Change.

Leggett, J.K. (2000) *The carbon war: Global warming and the end of the oil era*. Penguin, London.
This is excellent bed-time reading, a first-hand story about international negotiations surrounding reductions in carbon emissions.

Lovelock, J.L. (1979) *Gaia: A new look at life on Earth*. Oxford University Press, Oxford.
This is a very readable book, and seminal work, proposing that the biosphere behaves as a homeostatic system. In his recent book, *The revenge of Gaia*, published in 2006, the author argues that it is now too late to avoid substantial global heating which will make large regions of the Earth inhospitable.

McNeill, J. (2000) *Something new under the Sun*. Allen Lane & Penguin, London.
This is a brilliant and readable account of environmental history.

Moore, P.D., Challoner, W. and Stott, P. (1996) *Global environmental change*. Blackwell, Oxford.
This provides a good overview of the subject and is well illustrated.

Reay, D., Hewitt, N., Smith, K. and Grace, J. (2007) *Greenhouse gas sinks*. CABI, Wallingford, UK.
Further reading, to find out more about the subject of sinks.

Wilson, E.O. (2002) *The future of life*. Abacus, London.
This is a thoughtful and highly informed set of speculations on what will happen next.

Web resources

Climate Change Prediction

http://www.iop.org/activity/policy/Publications/file_4147.pdf

Institute of Physics website holding Alan Thorpe's easy-to-read article about climate prediction.

IPCC 2001 and 2007 reports

http://www.ipcc.ch/

This is the IPCC website from which the various reports on climate change may be downloaded (they are very long), and also graphics for teaching purposes. For research-level downloads of data see

http://www.ipcc-data.org/

News about Greenhouse Gases

http://www.ghgonline.org

This is Dave Reay's greenhouse gas website with news and teaching resources.

Online Trends

http://cdiac.ornl.gov/trends/trends.htm

This site provides trends in CO_2 emissions, country by country.

The Stern Report

http://www.hm-treasury.gov.uk/media/3/6/
Chapter_1_The_Science_of_Climate_Change.pdf

At this site you can download the report prepared for the UK government on the economics of climate change; it will soon be available as a book from Cambridge University Press.

UK Hadley Centre

http://www.metoffice.gov.uk/research/hadleycentre/

This is a good place to view climatic predictions, and to obtain a view of the Hadley Centre's research.

The United Nations Framework Convention on Climate Change

http://unfccc.int/2860.php

It has the Kyoto Protocol and data on progress of countries in meeting their Kyoto commitment.

University of Cambridge ozone hole tour

http://www.atm.ch.cam.ac.uk/tour/

Images of the ozone hole are provided here by the University of Cambridge.

 Visit our website at **www.pearsoned.co.uk/holden** for further questions, annotated weblinks, interactive models and video-clips showing physical processes in action.

CHAPTER 22

Vegetation and environmental change

John Grace

School of GeoSciences, University of Edinburgh

Learning objectives

After reading this chapter you should be able to:

➤ understand the way vegetation responds to climatological variables, and appreciate some of the underlying mechanisms of this response

➤ outline the evidence that shows how temperature and water supply have a leading role in determining the global patterns in vegetation

➤ appreciate how researchers are using models to predict future vegetation patterns

➤ discuss how human activities interact with climatic impacts, in both 'good' and 'bad' ways

➤ understand the relationship between climate models, vegetation models and observations, and realize that much remains to be understood about the impact of climate on vegetation and vice versa

22.1 Introduction

Vegetation responds to a large number of factors in the physical environment, such as temperature, the available solar energy, and the supply of nutrients and water via the soil. Humans have been discovering the nature and extent of this response for a very long time, at least since biblical times when people first began to grow crops. The impact of drought, for example, is woven into the history of tribes and whole civilizations. In modern times, the study of the relationship between plant life and the climate falls within several scientific disciplines. **Plant physiology** is concerned with understanding the functioning of plants, and this includes the response of plants to their environment and the acquisition of resources by plants. From plant physiology we learn how species have different environmental requirements as a result of differing structural and biochemical make-up. **Agronomy, horticulture** and **silviculture** are applications of plant physiology in the service of humankind to provide field crops, garden plants and wood products. Ecology, on the other hand, looks more broadly at vegetation, and has a focus on plant distribution; here we learn that plant distribution is not only constrained by climate and soil, but also influenced by factors such as **competition**, herbivory, fire and disturbance. **Palaeoecology** is about ancient distributions (see Chapter 20), often inferred from deposits of pollen in a few places where material is well-preserved, such as peat bogs and lake sediments. Every year, the surface of the peat or the bed of a lake is added to by a 'rain' of pollen grains. Cores can be extracted from the material, and slices can be acid-digested to remove

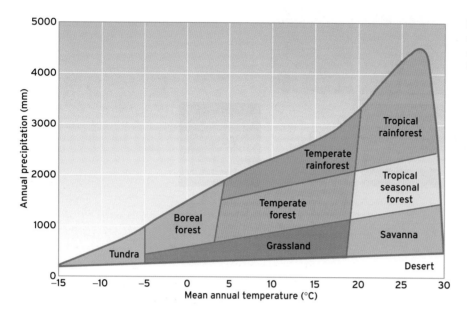

Figure 22.1 Approximate distribution of biomes in relation to precipitation and mean annual temperature. (Source: from Woodward *et al.*, 2004)

organic debris leaving pollen grains. The temporal patterns (with some assumptions) can be correlated with changes in temperature. Likewise, patterns of growth over long periods can also be measured in some organisms, such as corals and trees. **Dendrochronology** is the study of climate patterns from the annual growth rings of trees, and the related **dendroecology** is the study of tree rings to investigate ecological processes. In old trees these growth records extend over hundreds of years, and by joining together the records of past generations of trees, for example using trees found preserved in bogs or the timbers of ancient buildings, it is possible to obtain a record over thousands of years. Finally, **phenology** is the study of the annual variations in the timing of key events in the life cycle of plants, such as the date when specific tree species open their leaves, the date when flowers appear, or the date of autumn colouring when leaves lose their chlorophyll and photosynthesis ceases. There are long-term records of these phenological events, and they can be related to trends in the climate.

In this chapter we will draw upon work from all these disciplines, noting at the same time that to synthesize knowledge over a range of disciplines often requires some kind of mathematical model that incorporates knowledge and understanding, and that can be run from historical climate data or from data generated by climate models. In fact, we use predictions of what the climate may be in the future to estimate how the vegetation may change in the future.

We may also note in passing that there is an inevitable relation between climate and the native vegetation, as discussed in classic work by Wladimir Köppen, a German climatologist, in a 1931 paper, and Holdridge, an ecologist, in 1947. We emphasize 'native' vegetation, because much of the world's vegetation is affected by humans and transformed or removed so that the land can be used for agriculture and forestry. It turns out that for the native vegetation, the most important variables determining distribution on a global basis are average annual and monthly temperatures, and precipitation. A modern expression of these relationships can be seen in a paper by Woodward *et al.* (2004), reproduced here as Figure 22.1. Here, the land cover is represented as only nine 'biomes', each biome occupying a particular region of climate space. From such relationships we see how a warmer world might shift particular locations from one biome to another.

22.2 Fundamentals of how plants respond to climatic variations

22.2.1 Light

About 50% of the dry weight of plants is carbon, which has been accumulated through the process of **photosynthesis** (from the Greek *phos* = light and *synthesis* = combination) summarized as follows:

$$6CO_{2(gas)} + 12H_2O_{(liquid)} + photons \rightarrow C_6H_{12}O_{6(aqueous)} + 6O_{2(gas)} + 6H_2O_{(liquid)} \quad (22.1)$$

carbon dioxide + water + light energy → glucose + oxygen + water

Green leaves achieve this by capturing photons (energy as light from the Sun) with a set of pigments, of which the most important are the chlorophylls (green pigments); then they use the captured energy to drive a series of chemical reactions that result in CO_2 being absorbed from the atmosphere and converted into simple sugars, such as glucose. Subsequently, glucose is made into storage compounds, the most common being starch, and structural compounds such as cellulose (of which cell walls of leaves and roots are made) and lignin (a component of the cell walls of woody stems). From this discussion, it is clear that green plants need light, and that the more light they have, the more growth can be expected. It is worth pointing out that the process of photosynthesis has been going on for some 4 billion years, and that the by-product of photosynthesis, oxygen, enables all aerobic life including our own.

From a biochemical perspective, different types of photosynthesis are recognized. The majority of plant species are found to fix CO_2 into 3-carbon compounds, triose phosphates. However, some other species have a different enzyme system and they make a 4-carbon compound instead, oxaloacetic acid. The former condition is known as C_3 photosynthesis and the latter as C_4 photosynthesis. C_4 plants are relatively recent. We see them in the fossil record only 20–25 million years ago and they spread remarkably only 8 million years ago. Most of them are grasses, some are sedges, a few are herbs and shrubs, and only one of them is a tree. When they evolved from C_3 ancestors, fire was on the increase because of a drying climate, and browsing increased with the evolution and spread of large mammals in the Oligocene and Miocene.

Only a few per cent of species are currently C_4 but they can utilize high levels of solar radiation more effectively and may account for as much as 30% of global photosynthesis, mostly in the tropics and mainly in savannas. There are many differences between C_3 and C_4 but here we need note only a few of them. C_4 plants utilize water more efficiently and so they tolerate periods without rain. C_4 photosynthesis works best in warm climates, and so we can reasonably expect C_4 plants to be favoured in the future, warmer climate (Figures 22.2 and 22.3). They are among the faster growing crops of the world, including maize, sorghum, millet and sugar-cane. They dominate some of the most productive natural **ecosystems** in the world, from the floodplains in Brazil where we find a C_4 aquatic grass *Echinochloa polystachya* to the lakes in Africa where monospecific **stands** of *Cyperus papyrus* occur, the sedge used by ancient civilizations to make 'papyrus'. They are widespread as the C_4 grasses in tropical savannas, where they co-exist with C_3 trees. A third type of photosynthesis, first discovered in the plant family Crassulaceae, is associated with leaves that are fleshy and accumulate malic acid during the night. This condition is termed crassulacean acid metabolism (CAM). In contrast to all other plants, CAM plants

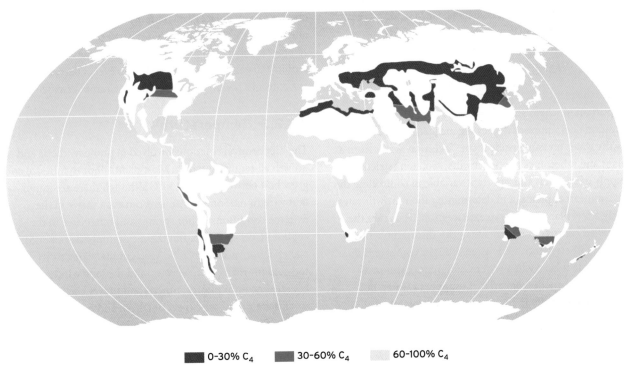

0-30% C_4 30-60% C_4 60-100% C_4

Figure 22.2 Grasslands of the world, showing the percentage of the grasses that are C_4. (Source: from Ehleringer *et al.*, 2005)

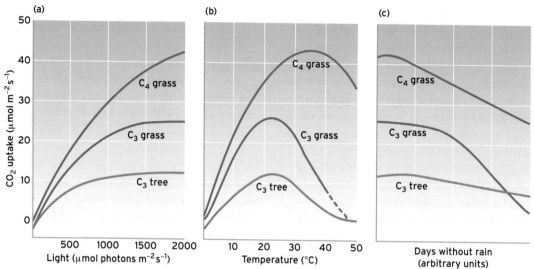

Figure 22.3 Environmental responses of photosynthetic CO_2 uptake by leaves of three hypothetical species: a C_4 grass, a C_3 grass and a C_3 tree. (a) Under different amounts of light; the C_4 grass has higher rates of CO_2 uptake and is able to continue to exploit the incident solar energy even in bright sunlight; C_3 grasses and herbs often have a higher rate of photosynthesis than C_3 trees. (b) The performance of the three species at high temperature; we see how the C_4 grass has a higher optimum temperature for photosynthesis. (c) How the hypothetical species might respond to a period with no rain: C_4 species have a higher water use efficiency than C_3 species (they lose only 250–300 g of water for every gram of CO_2 fixed, while C_3 lose 400–500 g water per g CO_2); grasses have near-surface roots and they die down in drought but trees often have deep roots and they can exploit stored water more effectively (grasses will regrow leaves when the rain comes).

keep their stomata closed in the day and hence avoid water loss; but they open them at night using a different enzyme system to capture CO_2 and form malic acid. Then, during the day the malic acid is broken down to yield CO_2 which is fixed into glucose in the same way as we find in C_3 plants. As well as members of the Crassulaceae, CAM is found in cacti, the pineapple family (Bromeliaceae) and orchids.

22.2.2 Water

Inside the plant cell, water is the medium in which all the important biochemistry takes place. For land plants water is additionally important as they are mechanically supported by the water pressure inside them. Take away water from a plant and it becomes flaccid (it wilts). So, rain (strictly, the supply of water) is an important variable in controlling the distribution of plants, and determining their growth rates. Only a few plants can survive desiccation. These are the so-called resurrection plants than can disassemble the photosynthetic machinery when drought comes and then reassemble it later when water is abundant. Almost all land plants benefit from an extra supply of water, but species vary hugely in how well they can tolerate dry periods. Some tolerate drought by shedding leaves or by having special characteristics such as a thick and water-proof coating (cuticle) on their leaf surfaces, or by having small thick leaves and special organs to store water. Plants

with adaptations to dry conditions are called **xerophytes**, and are found especially in deserts. Changes in water supply are expected as a component of climate change: as we saw in Chapter 21 we expect some regions to become wetter and others to be drier; we may therefore expect vegetation to change in some places and patterns of food production to alter. For example, it has been predicted that the Amazon rainforest will be replaced by a more xerophytic vegetation within 100 years from now as El Niño events become more extreme and possibly more frequent (Cramer *et al.*, 2001).

22.2.3 Temperature

The biochemical reactions involved in photosynthesis and growth all require warmth. Most plants photosynthesize and grow best between 10 and 30°C although there are important variations. Some organisms can even thrive in the extreme conditions of hot springs and others live in the coldest places on Earth. But these extremophiles are mostly bacteria, not vascular plants. Much of the planet's surface is too cold for many plant species, and the long winters of the boreal zone limit photosynthesis and prevent cell division. As climate warms we expect northern and mountain regions to become greener, to photosynthesize more rapidly and to grow faster. Indeed, there is evidence from historical photography and satellite imagery that this is already happening (Nagy *et al.*, 2003). We expect that in the north especially, there will be widespread changes in the distribution of plants, as the length of the 'growing season' increases.

22.2.4 Other climatic variables

Other climatic variables are important too, especially on a local basis. Photosynthesis responds strongly to changes in humidity, as leaves tend to close their stomata in dry air. Wind is an important factor also. It plays a direct role in some regions of the world, where it may limit the extent to which forest or woodland can develop. In all environments wind is important because it determines the relation between the climate and the microclimate at the surface of leaves. In this regard, the influence of wind on heat transfer between leaves and the atmosphere determines surface temperatures, and plants that are sheltered are generally a few degrees warmer by day than those that are exposed in the same location. Wind may also convey adverse materials, including salt spray and pollutants.

Reflective questions

➤ Should we take into account *extremes* of climatological variables or do we work merely with mean values?

➤ Is the temperature of a leaf the same as the temperature of the air?

➤ How will global food production change with global warming?

22.3 Observational studies: how we know for sure that vegetation responds to a changing climate

In this section we review examples of field observations that show how natural vegetation frequently changes as a result of a climate change.

22.3.1 The forest/savanna boundary in southern Amazonia

The Amazon rainforest occupies about 4 million km² and holds around one-third of the global **biodiversity**. However, the world's most famous rainforest may not have been as extensive just a few thousand years ago. Records obtained

from lake sediments in Bolivia in the southern part of Amazonia suggest that the boundary with the savanna may have undergone substantial changes over the past few thousands of years (Mayle *et al.*, 2000). The results are presented as a pollen diagram (Figure 22.4), in which we see the fluctuations of some major groups of plants found as pollen from a 3 m core taken from the sediment at the bottom of a deep lake. In such studies, the age of the samples is determined by the ^{14}C dating (see Box 20.3 in Chapter 20) of the strata from which the samples have been taken. At present, the lake is surrounded by rainforest and the pollen 'rain' into the lake is dominated by pollen of the families Moraceae and Urticaceae which contain predominantly trees. There is also a signal from *Cecropia*, a tree genus. Grass pollen from members of the Poaceae is relatively uncommon, and so are *Mauritia* and *Mauritiella*, both palms of wet places.

However, just 4000 years ago the situation was very different. The site was evidently much drier. We can tell the water level in the lake must have been much lower because we see pollen of *Isoetes* (quillwort), a plant that lives in the shallows at lake margins, Figure 22.4). Most significantly, there is a strong component of grass pollen and relatively little tree pollen, showing this area to have been savanna at the time. There is also charcoal, an indication of dry conditions and human presence. The conclusion, that rainforest is quite recent, and preceded by grassland vegetation, can be supported by other evidence. Mayle *et al.* (2000) point to a regional increase in rainfall at the time tree pollen began to be common: it is known that the water level in Lake Titicaca (located in the Andes on the border of Peru and Bolivia) rose after 3200 BP, reaching modern levels by 2100 BP. This change in climate was not anthropogenic, although the savannas were probably occupied and fire was used as a tool, but rather the consequence of a southward shift in the intertropical convergence zone (see Chapter 4), itself triggered by Milankovitch forcing. A few similar studies have been conducted at other tropical and subtropical regions, for example the Sahara Desert which was formerly much wetter and vegetated, with a considerable human population.

The work is relevant to ideas of how tropical vegetation will change in the future. Some climate models predict substantial drying in Amazonia and elsewhere, and it is thought likely that part of the region will revert to savanna. Today, the changes might well be accelerated by humans who are inclined to use fire for clearing. When the forest is dry there can be large-scale destruction of forests by fire as we have seen in many parts of the world in recent years. This removal of forest by fire is one of the mechanisms by which savanna replaces forest.

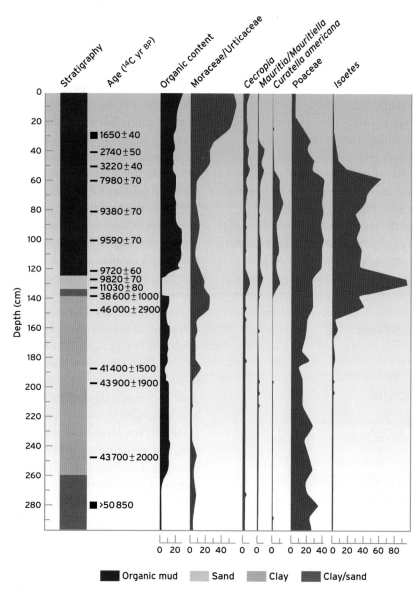

Figure 22.4 Vegetational history of southern Amazonia seen in the pollen record. (Source: Mayle *et al.*, 2000, from F.E. Mayle, R. Burbridge & T.J. Killeen, Millennial-scale dynamics of Southern Amazonian rain forests, *Science* 290: 2291-2294, 2000. Reprinted with permission from AAAS.)

22.3.2 The northern tree line

Trees are excluded from cold places. In most older texts it is claimed that trees are absent from sites wherever the temperature of the warmest month is less than 10°C, following a line of thinking by Köppen. This is not entirely true: for some Andean sites the tree line coincides with a maximum summer temperature of only 6°C. However, there is a consensus that summer temperatures are important and that they must be sufficient for adequate rates of photosynthesis and cell division. The consequence of this is the existence of a phytogeographical boundary where trees give way to dwarf shrubs, found at northern latitudes and on mountains all over the world. Around the boundary, known as the

tree line, the trees grow slowly and are often stunted with contorted stems. The German word to describe the woodland composed of these trees *is krummholz* (crooked wood).

Climate warming has so far been more rapid in the extreme northern regions than elsewhere, and so it is in the north that we might expect to see a sign of warming in the vegetation, for example increases in the rates of growth, and northerly advances in the distribution of trees. In fact, observations from several regions of the world suppport this view. Gamache and Payette (2004) sudied subarctic spruce–tundra in northern Québec, Canada. Here, the black spruce (*Picea mariana*) is normally rather shrubby. The authors measured growth rates of the plants along a 300 km

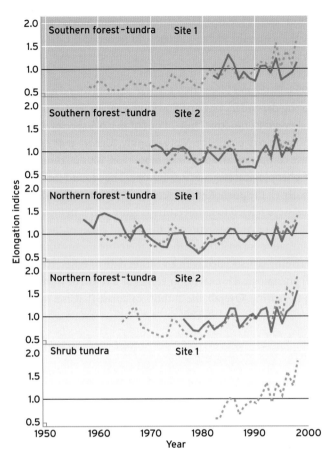

Figure 22.5 Recent spurts of growth of black spruce in northern forest and tundra. Growth is measured as elongation of the shoots. Data are for trees or saplings found at the tree line (solid lines) and above the tree line (dashed lines). (Source: from Gamache and Payette, 2004)

latitudinal transect. Height growth decreased with increasing latitude, and they found that height growth had increased in the northern plants since the 1970s, so that the northern trees were growing almost as well as the southern trees (Figure 22.5). Similar changes in tree line species have been recorded in other parts of the world too. In Sweden, Kullman (2005) has been studying *Pinus sylvestris* since 1973. He finds the size of the population has increased by 50%. In the first period (1973–1987) summer temperatures did not increase at all and there was a decline in the species; however, in the period 1988–2005 there was a strong increase in growth and reproduction, and also a decrease in winter damage.

22.3.3 Upward march of vegetation in mountains

In some places it is possible to obtain photographic evidence of shifts in vegetation, either by repeatedly revisiting the same place, comparing aerial photographs (often,

they go back to the 1940s), or simply as photographs taken by ecologists and naturalists on the ground. One such case is the Montseny Mountains of Catalonia in north-east Spain (Peñuelas and Boada, 2003), where ground-based images were available from 1945. In these mountains the beech forest has moved up the slope by 70 m at the highest altitudes. At medium altitudes the existing beech forest is becoming degraded, partially defoliated and is not regenerating itself as it formerly did. Nearby, heathland is being replaced by the more drought and heat-tolerant holm oak (Figure 22.6). Changes like this have been reported from the European Alps as well. There, the existence of many small alpine species is threatened with local extinction as their particular habitat declines in area as a result of global warming (Pauli *et al.*, 2007). This general phenomenon, first recognized two decades ago, has prompted the establishment of long-term sample plots, which are revisited every few years for enumeration of the plants. The project has

Figure 22.6 Rapid vegetational change attributed to climate warming. Panels on the left hand side are from photographs in 1969, those on the right hand side are from 2001. (Source: from Peñuelas and Boada, 2003)

established a global network of sites in mountain regions, known as GLORIA (Global Observation Research Initiative in Alpine Environments, **http://www.gloria.ac.at**).

22.3.4 Changes in the timing of flowering

The recording of the annual cycle of growth and development of plants has interested amateur naturalists and professional meteorologists for over 100 years (Jeffree, 1960; Sparks *et al.*, 2000). In some countries, 'phenological gardens' have been established to monitor such things as the date of bud-break every spring, the date of first flowering and the date on which trees shed their leaves. These data sets are a rich archive of information, linking biology and meteorology. In some cases these records are very long ones, broken only by interruptions by war or failure of funding. In other cases, they are the result of one person's painstaking

efforts over a lifetime. One of the best examples is provided by the naturalist Richard Fitter, who recorded the data of first flowering for a set of 557 wild plant species over a period of 47 years in south-central England (Fitter *et al.*, 1995; Fitter and Fitter, 2002). These authors found a very strong dependency of first flowering date on temperature in the previous months. Only 24 out of the 243 species selected showed no dependency. More recently, Menzel *et al.* (2006) have completed an analysis of more than 100 000 time series of phenological stages across Europe. They found that nearly all the phenological events, and also farmers' activities, were related to temperature (Figure 22.7). In Figure 22.7, the recordings represent the date of onset of the event. A negative correlation coefficient therefore means that warming is associated with an earlier date of onset. Nearly all phenological events occur earlier as a result of warming in the spring. Overall, the authors calculate that as a result of

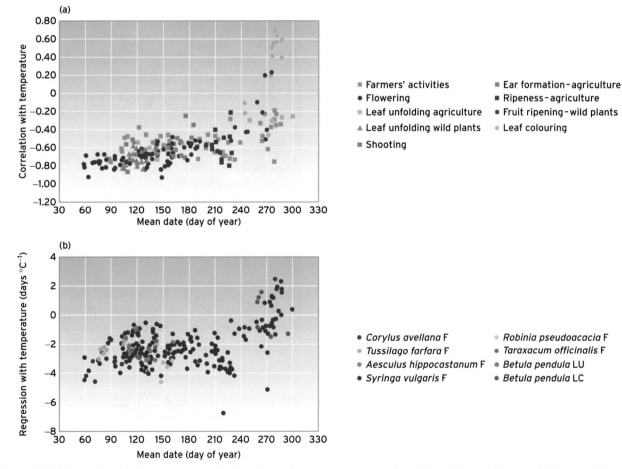

Figure 22.7 The relationship between phenological events and temperature measured on different days of the year: (a) the correlation coefficient (negative coefficient means high temperatures on that day of the year produce earlier events); (b) the sensitivity of the bud opening to temperature on that day of the year for a range of six trees and two perennial herbs. For example, a regression of −4 on day 120 means that one degree of warming stimulates early bud-opening by 4 days. F = flowering, LU = leaf unfolding, LC = leaf colouring. (Source: from Menzel *et al.*, 2006)

warming, spring in Europe is arriving 2.5 days earlier per decade. A few events have positive correlations with temperature, meaning that warming is associated with a later occurrence of the event. For example, warm summers are associated with later leaf colouring in the autumn.

We may safely conclude that, at least for temperate perennial plants, the life cycle is to a large extent set by the temperature, within certain limits. The cycle of growth and development may be advanced or retarded, and the growth period shortened or prolonged, according to the temperature pattern of a particular year.

From such data, it is possible to model what might happen with a few degrees of warming. To go further and to predict the ecological consequences is somewhat more difficult. If insect-pollinated plants flower early, for example, they will not be fertilized unless the relevant insect pollinators are available at the same time. Also, if flowering is too early there might be some increased risk of damage by frost. Clearly, this work has enormous economic relevance and it assists growers to plan for the future and also to evaluate risks.

Today, satellite sensors can be brought to bear on the problem of relating climate change to the cycles of plant life, especially in relation to the greening up of the land in spring. However, satellites look at huge swathes of landscape and it is usually hard to disaggregate the signal of natural vegetation from the frequent changes in agricultural practice. So, there remains a role for amateur phenologists, and interest in phenological gardens is currently as high as it has ever been.

Reflective question

➤ What evidence is there that vegetation is responding to climate change?

22.4 Models for prediction

In the preceding sections we saw something of how plants respond to climatic variables. There are substantial differences between species: how long they live, their growth rates and physiology, and this in general is why different species occupy different regions of the land surface. In principle, it would be possible to determine a set of parameters for each species, and then to use a model to determine where the species could possibly occur within a global climatic envelope like the example shown in Figure 22.1. The task would be large, as even closely related species respond differently to

environmental variables, and in any case there are important other influences, to which we have already alluded, such as competition between species and the tendency for dominant species to modify the conditions for other species. It would not be practical to screen the characteristics of all extant species of vascular plants, as there are about 400 000 of them. The size of this task is one of the bottlenecks in the development of predictive models. The research community is still discussing the best way to link knowledge of species to knowledge of biomes; however, there is a consensus that some coarse classification of species is required so as to capture the essence of what makes one vegetation type different from another. Ideally, this description should have a 'meaning' across several disciplines, so that remotely sensed data on reflectance of the land cover can be related to what ecologists can discern as a more or less homogeneous entity. To this end, and after a number of international conferences on the subject, the idea of plant functional types (PFTs) emerged.

A PFT is a group of plants with similar traits and which are similar in their association with environmental variables. Each PFT is defined by a variety of optical, morphological and physiological parameters. At present there is no real consensus beyond this, but in Table 22.1 we show one set of PFT definitions from Woodward *et al.* (2004). From this starting point we can make mathematical models of how vegetation responds to a changing climate. Most models keep track of the flow of carbon as well as the PFTs and biomes. The important processes which need to be modelled in that case are as follows.

1. *The rate of photosynthesis, and its dependency on environmental factors and the supply of nutrients.* The rate of photosynthesis on an ecosystem scale is termed gross primary productivity (GPP); it is the CO_2 uptake in the daytime period, adjusted by an estimate of how much carbon is simultaneously lost by plant respiration.

2. *Allocation patterns.* The allocation of the carbon acquired in photosynthesis to different plant parts (leaf, stem and root) is required and so is the respiration rate required to build complex molecules such as proteins and lignin from the simpler ones like glucose. If we know how this process of allocation depends on environment, and we also know the rate of photosynthesis, then we can compute the growth rate. On an ecosystem scale this is known as net primary productivity (NPP), related to GPP as follows:

$$GPP = NPP + R_a \qquad (22.2)$$

where R_a is the respiration that the plant expends, known as autotrophic respiration. There are some established theoretical

Table 22.1 A plant functional type classification adopted by Woodward *et al.* (2004)

Classification	Comment
Evergreen needleleaf forests	Lands dominated by trees with a canopy cover of more than 60% and height exceeding 2 m. Almost all trees remain green all year. Canopy is never without green foliage.
Evergreen broadleaf forests	Lands dominated by trees with a canopy cover of more than 60% and height exceeding 2 m. Almost all trees remain green all year. Canopy is never without green foliage.
Deciduous needleleaf forests	Lands dominated by trees with a canopy cover of more than 60% and height exceeding 2 m. Consists of seasonal needleleaf tree communities with an annual cycle of leaf-on and leaf-off periods.
Deciduous broadleaf forests	Lands dominated by trees with a canopy cover of more than 60% and height exceeding 2 m. Consists of seasonal broadleaf tree communities with an annual cycle of leaf-on and leaf-off periods.
Mixed forests	Lands dominated by trees with a canopy cover of more than 60% and height exceeding 2 m. Consists of tree communities with interspersed mixtures or mosaics of the other four forest cover types. None of the forest types exceeds 60% of the landscape.
Closed shrublands	Lands with woody vegetation less than 2 m tall and with shrub canopy cover more than 60%. The shrub foliage can be either evergreen or deciduous.
Open shrublands	Lands with woody vegetation less than 2 m tall and with shrub canopy cover between 10 and 60%. The shrub foliage can be either evergreen or deciduous.
Woody savannas	Lands with herbaceous and other understorey systems, and with forest canopy cover of between 30 and 60%. The forest cover height exceeds 2 m.
Savannas	Lands with herbaceous and other understorey systems, and with forest canopy cover between 10 and 30%. The forest cover height exceeds 2 m.
Grassland	Lands with herbaceous types of cover. Tree and shrub cover is less than 10%.

relationships that help us to find R_a. For example, the synthesis of cellulose requires the energy obtained from the breakdown of glucose as well as a supply of glucose molecules, so the formation of 1 g cellulose is associated with the release of a specific quantity of respiratory CO_2. These are 'classical' issues in plant physiology, discussed by Amthor (2000).

3. *Birth, death, phenology.* Plant parts are formed and they are shed, often on an annual cycle, and plants as a whole are 'born' every time a seed germinates and they eventually die. Some are annuals (programmed to die after a few weeks or months), others are biennials (the life cycle takes two years, sometimes a little longer) and perennials (most grasses and all trees are good examples). Generally, in global models we have to provide some 'rules' for each PFT. As in all models, we have to ignore the exceptions and abide by simple rules. For conifer trees, for example, leaves are retained for several years whereas for deciduous trees they are shed on an annual cycle. The resulting 'litter' from shedding and death is incorporated into

the soil, and decomposed by microbes. The microbial respiration gives rise to further efflux of CO_2 from the ecosystem, R_h, known as heterotrophic respiration. The resulting net flux of CO_2 between the ecosystem and the environment is the net ecosystem production (NEP), related to the previous terms as:

$$GPP = NEP + R_a + R_h \qquad (22.3)$$

In a constant environment, all the GPP would be respired by plants or microbes and so NEP would be zero.

In the real world there are additional losses of carbon caused by natural and human influences. To take these into account we need to insert a new carbon loss, the disturbance flux R_d.

4. *Population-level processes.* Models need to represent the process of plant **succession**, whereby species colonize the land and develop towards an equilibrium or near-equilibrium state. To do this, the attributes of each PFT are needed, and some further rules are required to determine in what circumstances one species succeeds another.

Models are run using a sequence of climatological data, in time steps that vary from months to minutes, according to the scheme of Figure 22.8. How good are such models? They can be tested in several ways: (i) by examining whether the model calculations can reproduce today's vegetation by running the models from historical climate data; (ii) by investigating whether they produce the carbon fluxes that are measured in field studies; and (iii) by comparing them with other models, developed in other laboratories more of less independently.

An important comparison of six models was reported by Cramer *et al.* (2001). They computed the 'expected' vegetation on the basis of a world divided into regions of 3.75° longitude and 2.5° latitude, given the climatic scenario. In this scenario of climate change, temperature increases by 4.5°C over the next 100 years, and the CO_2 concentration rises from 380 to 800 ppm by the end of the century. Although the six models do not all agree in detail, and although they do not faithfully reproduce the current vegetation, there is a general consensus on how vegetation change will occur (Figure 22.9). The principal changes are:

- Transitions from forest to savanna. Such transitions are predicted to occur in the Amazon, the central part of the American continent and in South-East Asia.
- Evergreen forest will replace 'grassland' in parts of North America and North Europe.
- Parts of the Mediterranean grasslands will become savannas.

As for the carbon balance, the models produce an interesting trend, and there is some agreement between models. The NPP is stimulated by warmer conditions (felt especially in the cold northern regions) and also by the elevated CO_2 (Figure 22.10). However, the heterotrophic respiration is increased by warming to an even larger extent, and the consequence is a large rise in global respiration over several decades. It is interesting to put these numbers into perspective as follows. The heterotrophic respiration from all the microbes (heterotrophic respiration of all terrestrial ecosystems) is currently around 50–60 Gt of carbon per year, which completely dwarfs the fossil fuel emissions of about 6.5 Gt of carbon per year.

Large though R_h is, it is more than offset by photosynthetic production, so the NEP is positive. In other words the terrestrial ecosystems are collectively a 'sink' for carbon. In fact, the current terrestrial sink strength of 1–2 Gt of carbon per year is predicted to rise for a few decades before taking a downward turn as the effect of temperature on respiration is increasingly felt. There is a substantial predicted downturn in NEP so that by 2100 the sink has diminished and, in one of the models, has turned into a source.

The most significant weakness of these models is that they are not coupled to a model of the climate system, so the CO_2 effect of a vegetational source or sink on the temperature is not apparent. In the 'real world' the atmospheric CO_2 and therefore the global warming rate would be influenced by the vegetation. Such feedbacks may well be important.

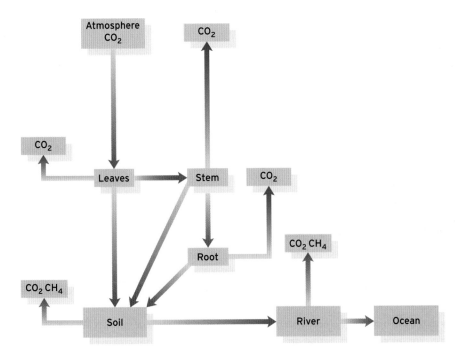

Figure 22.8 Structure of a very basic model of the carbon flows associated with photosynthesis, respiration, growth and storage of carbon. 'Losses' of CO_2 from leaves, stems and roots (autotrophic respiration) are associated with maintenance and growth of the plant tissues; losses from the soil are either from microbial respiration (heterotrophic respiration) or in the drainage water. Such a model might be configured for an individual plant or for a set of PFTs in an ecosystem. More advanced models include flows of water and nutrients, and their interaction with carbon flows.

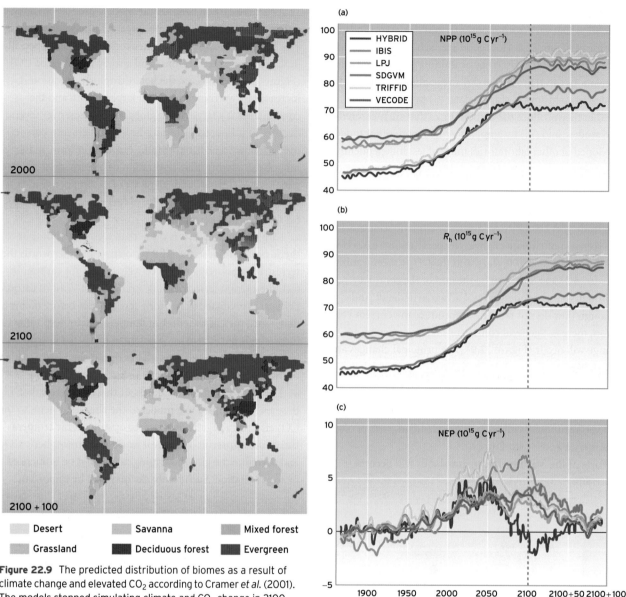

Figure 22.9 The predicted distribution of biomes as a result of climate change and elevated CO_2 according to Cramer et al. (2001). The models stopped simulating climate and CO_2 change in 2100. Therefore changes from the middle to bottom panel are related to vegetation dynamics as the system equilibrates. See text for further interpretation. (Source: from Cramer et al., 2001)

Figure 22.10 Predicting the carbon balance of terrestrial vegetation for the next century by means of six models: (a) the net primary productivity (NPP); (b) the heterotrophic respiration (R_h microbes and animals); (c) the overall carbon flux made by subtracting R_h from NPP. Each line is a different model, as shown in (a). (Source: from Cramer et al., 2001)

In recent years, simple vegetation models have, however, been coupled to global circulation models (GCMs) in an attempt to capture the essential feedbacks. Here, we refer to a synthesis study (Friedlingstein et al., 2006) in which 11 models were compared. The results are quite variable and the models do not all show the same general trend (Figure 22.11). They clearly differ in their sensitivity to climate change. The most sensitive result of all is shown by the Hadley Centre Model where the vegetation becomes a progressively weakening carbon sink in the next few decades and then moves into carbon deficit rather in the same way as the uncoupled models discussed in the previous paragraph. For the other

models in the Friedlingstein study, there is a weakening of the sink except in one case where the sink continues to intensify. Further work is clearly needed before we can make further estimates of the future based on models.

One of the more difficult aspects of predicting the future from models is that humans interact with climate change and this interaction is hard to predict. In truly managed ecosystems the economically relevant outputs are the ones that are recognized, monitored and increased largely as a

Figure 22.11 Carbon uptake from the atmosphere by terrestrial ecosystems (units: Gt = gigatonnes = 10^9 tonnes = 1 billion tonnes). The calculations are made after coupling several GCMs to simple models of the terrestrial biosphere. Each line is the result from a different model. The current terrestrial carbon 'sink' is somewhere between 0 and 4 Gt of carbon per year. Some models predict that the land sink will become a source. (Source: from Friedlingstein *et al.*, 2006)

result of applying science. In modern UK agriculture, for example, the agriculture has become more intensive since World War II, and has enjoyed substantial government subsidies, yet a recent study showed widespread recent carbon losses from the soil in England and Wales, thus contributing inadvertently to global warming.

> ## Reflective questions
>
> ➤ Models are only as good as the understanding built into them. What are the gaps in our understanding of vegetation–climate models?

> ➤ How well have the vegetation models been tested?
>
> ➤ What is a plant functional type?
>
> ➤ What important processes need to be modelled in PFT, biome and carbon flow vegetation models?

22.5 The complex interaction between human activities and climate change

22.5.1 Does atmospheric pollution sometimes benefit plants?

As a result of human activities, notably agriculture and the driving of motor vehicles, much more nitrogen in a chemically active form (ammonium and nitrate especially) is deposited to the land surface than hitherto (Box 22.1). If there is too much N-deposition, ecosystems may show signs of 'nitrogen saturation', a condition whereby the land surface may 'leak' nitrogen to the drainage water and give off nitrous oxide, another greenhouse gas. If, on the other hand, the deposition rate is below a certain threshold level, there may be a fertilizer effect for many ecosystems, especially those that are otherwise N-deficient. This would be seen as a stimulation of photosynthesis and possibly an increase in growth rates and a strengthening of the carbon sink. Indeed, many model calculations of the impact of environmental change on global vegetation contain a term to allow for this. The N-effect and climate change interact in ways that are not understood very well, as highlighted by Magnani *et al.* (2007). Data on the C-fluxes over forests in Europe and North America were collected, and attempts were made to relate the GPP and R_h to temperature. In fact, both showed a remarkable linear relation with temperature; but the NEP (i.e. GPP − R_{eco} where R_{eco} is the total ecosystem respiration) is rather a weak function of temperature (Figure 22.12). However, NEP is strongly related to the deposition of anthropogenic nitrogen from the atmosphere. Hence, we may suggest that pollution of the atmosphere with nitrogen (principally from vehicles, agricultural systems) is at a level that stimulates production. This is of course a somewhat controversial claim, as most people link N-deposition to nitrogen saturation of ecosystems or, worse, with the production of acid rain that has deleterious impacts on forests.

Figure 22.12 Measured carbon fluxes over 18 forests in Europe and North America: (a) ecosystem respiration, R_{eco}; (b) gross primary productivity, GPP; (c) and (d) the overall carbon flux as directly measured as NEP, showing a carbon sink of between 0 and 5 t C ha^{-1} yr^{-1}. Note that the overall carbon flux NEP is a fairly weak function of temperature but a strong function of the deposition of nitrogen from the atmosphere. (Source: Magnani *et al.*, 2007, reprinted by permission of MacMillan Publishers Ltd.: *NATURE*, F. Magnani, M. Mencuccini, M. Borghetti, P. Berbigier, F. Berninger, S. Delzon, A. Grelle, P. Hari, P.G. Jarvis, P. Kolari, A.S. Kowalski, H. Lankreijer, B.E. Law, A. Lindroth, D. Loustau, G. Manca, J.B. Moncrieff, M. Rayment, V. Tedeschi, R. Valentini, J. Grace, The human footprint in the carbon cycle of temperate and boreal forests, vol. 447: 848-850. Copyright 2007.)

22.5.2 How does fire interact with climate change?

Many predictions regarding climate change suggest that some areas of the tropics will become dryer as a result of an increased frequency and harshness of El Niño events. As we saw above, this increase in drought is one of the causes of the expected conversion from forest to savanna. However, as humans encroach upon the rain forest, some researchers believe the effect will be amplified by the creation of forest edges in a fragmented forest, and the use of fire (Laurance and Williamson, 2001). Dense forests have a microclimate characterized, for example, by high humidities and daytime temperatures that are usually lower than those measured at the top of the canopy. At forest edges the situation is different, with free horizontal ventilation and mixing of canopy air with air from outside. When drought occurs, relatively dry air penetrates the canopy, and reductions in humidity and increases in plant mortality have been measured at up to 100 m from the canopy's edge. Humans light fires, and these fires are likely to ignite more easily and spread more rapidly in the dry conditions of the forest edge. Hence, the forest is damaged and possibly destroyed at a faster rate to what would occur in the absence of humans. The processes involved are quite complex and collectively amount to a positive feedback (Figure 22.15). In the

Laurance–Williamson model, deforestation causes less evaporation, which in turn leads to less rainfall and hence droughts are exacerbated. Logging can also be important as it thins the canopy and increases vulnerability to combustion.

Reflective questions

➤ How might atmospheric pollution benefit plants?

➤ How much N-deposition is required before N-saturation occurs? (Research the literature and see what you can find.)

➤ How can the MODIS fire product (**http://modis-fire.umd.edu/**) be used to understand how climate and fires are related? You need to visit the website to address this question.

22.6 Loss of biodiversity

Biodiversity is defined as the number and variety of species in ecological systems, at local, regional and global scales (see Chapters 9 and 10). There is concern that both human

ENVIRONMENTAL CHANGE

THE NITROGEN CYCLE AND ANTHROPOGENIC PERTURBATIONS

Nitrogen is an important constituent of proteins and nucleic acids, and so is essential for life. Nitrogen exists primarily as an unreactive gas in the atmosphere as dinitrogen, N_2. It constitutes 79% of the air we breathe. Very important reactive forms of nitrogen also exist as gases and ions. The gases are ammonia (NH_3) and the oxides NO, NO_2 and N_2O, all present in trace concentrations. The main ions, found in soils and water, are ammonium NH_4^+, nitrate NO_3^- and nitrite NO_2^- ions (see Chapter 7). Two natural processes convert nitrogen to reactive forms that can be taken up by the roots of plants: lightning and biological nitrogen fixation (BNF). Very small amounts of N_2 are reacted with O_2 during lightning, to form the gas nitric oxide, NO, which eventually reaches the ground as nitrate. BNF is quantitatively more important than lightning as an agent of nitrogen fixation: bacteria living in the soil fix N_2 to make reactive forms ammonium NH_4^+ and nitrate NO_3^- which can be used by plants. Some of these nitrogen-fixing bacteria are free-living, but others form **symbiotic** relationships with plants, especially those of the pea family Leguminosae. Many members of this family are used in agricultural systems as a 'free' source of nitrogen fertilizer (examples are clover, lucerne, groundnuts, soybeans, alfalfa and lupins). Such plants have root nodules containing populations of the nitrogen-fixing bacteria. As a consequence of the free-living nitrogen-fixers and the symbiotic nitrogen-fixers, 130–330 Gt N yr^{-1} are made available in the soil solution as ammonium NH_4^+ and nitrate NO_3^- and can thus be taken up by plants to make protein and other biochemical constituents. Herbivorous animals obtain their protein by consuming plant material in prodigious quantities. Dead plants and animals decompose in the soil, and some of the nitrogen is acted upon by denitrifying bacteria. Nitrogen is thereby returned to the atmosphere as N_2 or N_2O. The process is thus cyclic (Figure 22.13).

Humans perturb the nitrogen cycle in fundamental ways. The industrial fixation of N_2 was developed in 1909 by Haber and Bosch, and provides a supply of nitrogen fertilizer, estimated to be 78 Gt N yr^{-1}. This is applied to

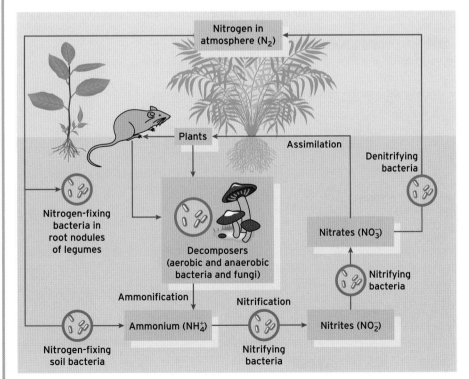

Figure 22.13 The nitrogen cycle.

BOX 22.1 ➤

➤

the soil, but much of it is released to the atmosphere as nitrous oxide N_2O, a greenhouse gas. Some of it enters drainage water, and causes excessive growth of algae in streams and rivers (see Chapter 15). There are other anthropogenic sources of nitrogen. The internal combustion engine and some other fuel-burning devices are responsible for emissions of oxides of nitrogen to the atmosphere, through the combination of atmospheric oxygen and nitrogen inside the combustion chamber. Moreover, cultivation and disturbance of the land results in emissions of nitrous oxide. Finally, animal rearing is associated with the emission of NH_3, produced by the decomposition of urine and faeces.

The consequence of increased formation of NH_3 and oxides of nitrogen is an enhanced rate of ammonium and nitrate deposition to land and waters. Nitrogen deposition rates are now much higher in populous regions of the world than they were in pre-industrial times (Figure 22.14). This deposition occurs as dry deposition of the gases themselves, and also

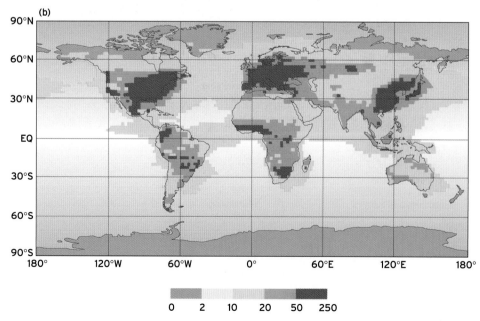

Figure 22.14
Deposition of reactive nitrogen to the Earth's surface (in mmol $N\,m^{-2}\,yr^{-1}$): (a) pre-industrial levels; (b) current atmospheric deposition. Note the enhanced deposition in densely populated areas. (Source: after Galloway *et al.*, 1995)

BOX 22.1 ➤

> as nitric acid, HNO₃, a contributor to acid rain. The growth of plants is generally limited by the availability of nitrogen in the soil, and so the enhanced deposition may be increasing plant growth, and contributing to a widespread increase in the rate at which trees grow. On the other hand, in some areas, the imbalance in nutrients in acid rain may cause damage to forests (see Chapter 7).

BOX 22.1

activities and climate change are causing a decline in the number of species. Given global warming, the importance of climate change is likely to become progressively more important, as pointed out in the recent report by the IPCC (IPCC, 2007b).

Although it is impossible to know the total number of species in the world (only 1.5 million are known but many are yet undiscovered), extinctions themselves are generally well documented. Since 1600, a minimum of 490 plant and 580 animal species have become extinct. Some groups, such as mammals and birds, have suffered more than others. In geological time there have of course been catastrophic mass extinctions. The natural or background extinction rate can be estimated from the fossil record. For example, in mammals the background rate is about one in 400 years, but this is much lower than the observed rate. The number of species that are threatened far exceeds our capacity to protect those species, and so conservationists have recently identified 'hot spots' where conservation effort and resources should be greatest (Myers *et al.*, 2000). It is particularly important to protect areas with a high degree of endemism (an endemic species is one restricted to a particular region). When this exercise was carried out there were some surprises, as shown in Box 22.2. For example, the natural vegetation of the tropical Andes heads the list. Its vegetation has been reduced to 25% of its original extent, yet it contains 6.7% of all plant species in the world and 5.7% of all vertebrate animals.

Reflective question

> Why is biodiversity important?

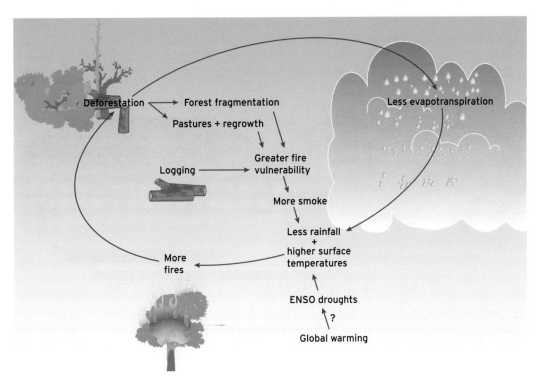

Figure 22.15 The Laurance-Williamson model of how droughts and human impacts combine to degrade and destroy rainforests. (Source: Laurance and Williamson, 2001)

ENVIRONMENTAL CHANGE

HOT SPOTS AND CLIMATE CHANGE

Myers *et al.* (2000) reported that 44% of all species of vascular plants are confined to 25 hot spots constituting only 1.4% of the land surface. The authors urged these to be singled out for the attention of conservationists, to attempt to protect them. The Myers *et al.* (2000) map of hot spots is reproduced here, along with projections of temperatures from the IPCC (2007b) 4th Assessment Report (Figure 22.16). Figure 22.16(b) shows the warming predicted from the A2 scenario (see Chapter 21). Note: all hot spots will be warmer.

Over the next century there will be further extinctions, and climate change will be an important driver, along with land-use change. Everyone agrees that species should be protected, and that natural environments have an inherent value. Indeed, the idea that humans are the guardians of nature is deeply

(a)

(b)

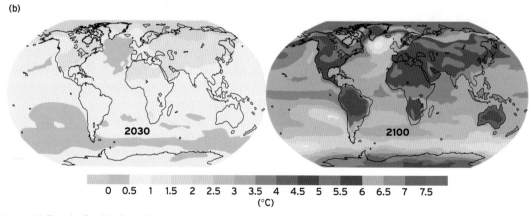

Figure 22.16 (a) Twenty-five biodiversity hot spots identified where 44% of all species of vascular plants and 35% of all species in four invertebrate groups are confined to the hot spots comprising only 1.4% of the land surface of the Earth. (b) Projected temperature increases for 2030 and 2100 from model scenario A2. (Source: (a) from Myers *et al.*, 2000, reprinted by permission of MacMillan Publishers Ltd.: *NATURE*, Norman Myers, Russell A. Mittermeier, Christina G. Mittermeier, Gustaro, A.B. da Fonseca & Jennifer Kent, Biodiversity hotspots for conservation priorities, vol. 403: 853–858: Copyright 2000; (b) from IPCC, 2007a)

BOX 22.2 ➤

embedded in Christian and other religions. There is also much folklore that evokes the conservation ethic. The author's grandmother used to say:

if you wish to live and thrive let a spider run alive.

This appears to be a reference to a 'keystone species', as spiders are voracious predators, required to control populations of small flying insects which often carry disease. When keystones are removed the ecosystem is in trouble.

At the moment there is a plethora of international agreements to protect species and habitats and many individuals subscribe to organizations such as Friends of the Earth or Greenpeace. The challenge for nature conservation is to protect natural ecosystems but traditional conservation and protection against human encroachment is clearly not sufficient: it does not protect species and ecosystems from the impact of human-made climate change.

BOX 22.2

22.7 Summary

Plant growth is influenced by climatological variables, especially light, temperature and moisture. For natural vegetation it is predominantly temperature and moisture that determine the type of land cover on a global scale. Both temperature and moisture are changing and they are expected to change rapidly in the next 100 years. Observations in the field, conducted over the last few decades, support the general view that the vegetation is changing in response to global climate change. Over the next century, it is likely that substantial changes in vegetation will result, and that these are likely to be so important as to impact upon the carbon cycle and the climate system.

Specifically, regions of the world such as the humid tropics, which are now believed to be a carbon sink, may become a source; conversely, cold northern regions will become warmer and therefore more favourable for the growth of plants and especially for trees. They may become a sink for carbon. Predictions are, however, based on state-of-the-art model calculations, and there are still many uncertainties. One of the largest unknowns in the system is the behaviour of humans. Their behaviour determines the rate of global warming and the nature of the land cover, and also modifies the response of vegetation on a global scale through the use of agents such as cultivation, fertilizers and fire.

Further reading

Beerling, D.J. and Woodward, F.I. (2001) *Vegetation and the terrestrial carbon cycle. Modelling the first 400 million years.* Cambridge University Press, Cambridge.
Original and insightful if you are interested in long (geological) timeframes.

Lovejoy, T.E. and Hannah, L. (2005) *Climate change and biodiversity.* Yale University Press, New Haven, CT.
Broader and deeper than this chapter can possibly be.

Malhi, Y. and Phillips, O. (2005) *Tropical forests and global atmospheric change.* Oxford University Press, Oxford.
A research-level enquiry into one of the most important issues raised in this chapter.

Schulze, E.D., Beck, E. and Müller-Hohenstein, K. (2001) *Plant ecology.* Springer-Verlag, Berlin.
An excellent compendium of plant physiology as it relates to ecology; a good place to learn about photosynthesis.

Web resources

IPCC report

http://www.ipcc.ch/SPM13apr07.pdf
This downloadable document from the IPCC website has a lot of important information about impacts of climate change on agriculture, ecology and forestry.

Mountain Vegetation Change

http://www.gloria.ac.at/
An initiative to record vegetation changes in mountains.

Nature's Calendar

http://www.naturescalendar.org.uk/
A UK website charting current phenological events. Excellent in the spring.

Phenology on Wikipedia

http://en.wikipedia.org/wiki/Phenology
An overview of the field of phenology.

 Visit our website at **www.pearsoned.co.uk/holden** for further questions, annotated weblinks, interactive models and video-clips showing physical processes in action.

CHAPTER 23

Remote sensing of environmental change

Timothy D. James
Department of Geography, Swansea University

Learning objectives

After reading this chapter you should be able to:

➤ appreciate the role of remote sensing as an indispensable tool in physical geography for monitoring and understanding the processes of environmental change

➤ understand the nature of electromagnetic radiation and how it interacts with the physical environment

➤ evaluate different methods of remote sensing for collecting environmental change data

➤ describe methods of digital image processing that enhance the quality and interpretation of remotely sensed data

23.1 Introduction

In physical geography, a great many subjects are studied that cover a wide range of scales. Geographers study processes that occur over the macroscale (e.g. global climates, Chapter 5), the mesoscale (e.g. glaciers and ice sheets, Chapter 18) and the microscale (e.g. soils, Chapter 7). At each of these levels, scientists need to collect suitable data as quickly and efficiently as possible using cost-effective methods and without damaging or interfering with the environments they seek to

understand. Whatever the scale, many of these data collection requirements are met through the methods of remote sensing. Over the last few decades, data collected using remote sensing have significantly increased our ability to measure the physical, chemical, biological and cultural characteristics of the Earth's surface.

The American Society of Photogrammetry and Remote Sensing defines remote sensing as any technique whereby information about objects and the environment is obtained from a distance. Remote sensing data can be acquired from one of three platforms: terrestrial, airborne and spaceborne. Terrestrial sensors may include simple handheld instruments such as a regular 35 mm camera handheld or mounted on a gantry, or sensors that are placed close to the ground for detecting features beneath the Earth's surface. Airborne sensors are mounted in helicopters or small aircraft for low-altitude missions, and more specialized aircraft for higher-altitude missions. Spaceborne sensors are mounted on orbiting vehicles such as satellites or the Space Shuttle.

Most often, remotely sensed information comes in the form of images that can be interpreted and analysed. There are several characteristics of remote sensing that make it ideal for use in physical geography and detecting environmental change. Firstly, it minimizes the need for field visits. This is important when studying environments that are either dangerous (e.g. natural hazards, political strife), isolated and difficult to reach (e.g. remote islands or glaciers), or fragile (e.g. periglacial and dune ecosystems). Secondly, within some limitations remote sensing instruments can be positioned as far from, or as near to, a surface as required. Thus, they can collect information over a very large area as in satellite imagery or over a very small area such as a camera mounted on a microscope. Finally, remote sensing offers the ability to repeat data collection over relatively short periods thus adding a fourth dimension to geographical studies and allowing environmental change to be detected. As a result, the methods of remote sensing have become essential for research in physical geography.

The subject of remote sensing is exceptionally extensive as it covers a wide range of instruments and applications. To address the subject in one chapter can scarcely do it justice. Where necessary, you are therefore encouraged to consult the further reading and references for more detailed coverage of the topics herein. With this in mind, the aim of this chapter is two-fold. The first is to provide you with a general introduction to the various components of remote sensing. The second is to act as a starting point from which you can explore more comprehensive references on any of the subjects discussed in the following sections. In this chapter you will be introduced to important image characteristics, the foundations of remote sensing, the various data sources and digital image processing.

23.2 Image characteristics

23.2.1 Types of image

Although remote sensing data are not exclusively in the form of imagery, this is by far the most common format and the most useful to physical geography. Therefore, this section discusses the various characteristics of imagery produced by remote sensing techniques. There are basically two types of images in remote sensing: analogue and digital. **Analogue images** include photographic negatives, **diapositives** and prints, each of which can be described as continuous tone images. Even in this digital age, analogue images still represent a common image format. However, in the face of improving technology and decreasing costs, digital equivalents are slowly replacing their analogue predecessors.

Wolf and Dewitt (2000) defined a **digital image** as a computer-compatible, pictorial rendition divided into a fine, two-dimensional grid of **pixels**. The term pixels comes from a contraction of 'picture elements' and each represents a finite area in the image. A **digital number (DN)** is assigned to each of these pixels to summarize the average reflection

Table 23.1 Bit scale and grey levels

Bit scale (depth)	Range of DNs	Number of grey levels
1-bit (2^1)	0-1	2
6-bit (2^6)	0-63	64
7-bit (2^7)	0-127	128
8-bit (2^8)	0-255	256
24-bit (2^{24})	0-16 777 215	16 777 216
32-bit (2^{32})	0-4 294 967 296	4 294 967 297

that was recorded for that unit area of surface. In the displayed imaged, the DNs determine the colour of each pixel according to the **bit scale** of the recorded image and the **colour palette** used for its display. A colour palette is simply a record of predefined colours each linked to one or more DNs. Colour palettes can range from only two colours (i.e. black and white) to billions of colours. The bit scale (or depth) of the image determines the number (or range) of DNs that are used in an image. A simple example is a 1-bit image, called a binary image, composed only of the numbers 1 and 0. Thus, the number of colours used in an image is determined by the bit scale as shown in Table 23.1. An example of the DNs in an image is given in Figure 23.1. Here, a simple 8-bit image using 256 grey levels is shown with small sections of increasing magnification to show how the DNs combine to form an image.

23.2.2 Image orientation, scale and resolution

For both analogue and digital images, there are a variety of image characteristics that are essential for their use in geographical research. An appreciation of these characteristics is central to understanding remote sensing and therefore they are addressed here. The first three characteristics discussed are variables that are determined prior to data collection. They are **image orientation**, **scale** and **resolution**.

A general assumption is that remote sensing instruments are always pointed downwards (vertical imagery) from some position above the surface of interest. However, this is not necessarily the case. Although vertical imagery is more common, the orientation of a remote sensing instrument can theoretically be in any direction: vertically (up or

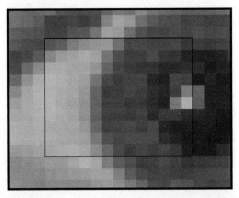

112	122	128	153	163	169	172	142	108	110	106	106	108	97
125	159	176	181	188	180	145	114	102	101	111	96	94	84
162	187	188	182	182	168	132	107	99	93	94	88	87	91
191	196	191	184	177	146	110	102	106	107	88	83	109	102
195	192	186	182	171	132	100	103	108	103	92	86	103	151
193	195	185	178	166	131	105	104	98	98	89	82	147	195
193	193	186	181	167	131	106	105	100	106	88	98	97	87
191	186	185	184	173	137	107	109	111	104	111	85	79	80
191	190	185	183	179	150	118	113	116	113	112	92	84	76
182	184	190	193	185	164	137	114	102	109	103	109	99	82

Figure 23.1 Construction of a digital image. As the magnification increases, what appears to be an image of continuous tone breaks down into units called pixels, each of which is represented by a digital number. In this case the image is 8-bit greyscale which has 256 shades of grey.

down), horizontally or obliquely (at an angle other than vertical or horizontal). The orientation of the imagery depends on the characteristics of the available sensor and the data requirements of each application. Although more difficult to analyse quantitatively, oblique imagery can be more cost effective for qualitative applications as a greater area can be covered than in one vertical image. This is illustrated in Figure 23.2, which shows a vertical and oblique image of the flooded Ouse River in York, England, in November 2000. Mapping water extent and flood risk management (Box 23.4 below) is an important application of remote sensing.

The scale of an image describes the relationship between a linear distance on the image and the corresponding linear horizontal distance on the ground. Thus, scale is usually expressed as a ratio such as 1 : 50 000 which describes an image–ground relationship such that 1 cm on the image equals 50 000 cm (or 500 m) on the ground. It is important to note that where images contain relief or are captured obliquely, image scale will change across an image. In any case, images or maps can be described as being small (least detail), medium or large scale (most

detail) and although the limits are not standardized, a rough guide is 1 : 50 000 and smaller for small scale and 1 : 12 000 and larger for large scale (Lillesand and Kiefer, 2003). The importance of the scale of an image lies in that it largely dictates the usefulness of an image for a particular application. For example, a weather satellite image covering all of western Europe would be of little use for studying hillslope processes, whereas an aerial photograph of the English Channel would not be useful for studying global ocean circulation.

The resolution of an image also plays an important role in determining its suitability for a given application. In general it refers to the ability of a system to separate a scene into constituent 'parts'. In remote sensing, resolution is divided into three components: spatial, temporal and spectral. Spatial and temporal resolution are discussed here, whereas spectral resolution is more appropriately addressed later in this chapter.

In most cases, the resolution of an image describes its spatial resolution. This refers to the degree to which a system can isolate a unit of area on a surface as being separate from its surroundings. Resolution is expressed

(a)

(b)

Figure 23.2 Vertical (a) and oblique (b) aerial photographs of the November 2000 flood of the River Ouse in the city of York, England. These images were used to map the extent of the flood waters for Britain's Environment Agency. Notice how much more of the flood-plain is visible in the oblique image compared with the vertical image. (Source: photos © the UK Environment Agency)

as the size of the smallest individual component of an image in surface measurement units. Although there are many factors that influence resolution during image capture, ultimately the resolution is determined by the size of individual film grains in photographs or by the size of individual pixels in digital images. These in turn are largely dependent on scale. Typically, the smaller the scale of the imagery, the lower the resolution. For example, the size of the smallest unit in a satellite image taken from 800 km above the Earth's surface will be much larger (smaller scale, lower resolution) than the smallest unit in an aerial photograph taken from 1000 m (larger scale, higher resolution).

With temporal resolution, the ability of the system to separate a scene into constituent parts refers not to parts over space but over time. Thus, a system that can capture many images over a unit period of time will have a higher temporal resolution than one that captures only a few. Clearly temporal resolution is highly dependent on the sensor used to collect the data. For terrestrial systems, the resolution can be very high. For example, with a 35 mm camera on a tripod or gantry, photographs can be taken separated only by a few tenths of a second. Such instruments are useful for recording events of environmental change that occur very quickly, such as avalanches and volcanic activity. However, with satellites that have to orbit the Earth several times before revisiting the same scene, the temporal separation of images will be of the order of days. Coarse temporal resolution can be very useful for recording environmental changes that occur over longer periods of time, such as deforestation, sea ice dynamics and coastal processes.

23.2.3 Characteristics of image content

There are a number of image characteristics that form the basis of image interpretation. They do not describe characteristics of the imagery as a whole but rather the characteristics of image content and thus are used to extract information from the images through human image interpretation and computer-aided analysis. They are dependent on the illumination of the scene and the characteristics of the surface being imaged. They include shape, size, pattern, association, tone and texture.

The first four elements, shape, size, pattern and association, are fairly self-explanatory. They describe the spatial characteristics of features in an image and their relationship to each other. The tone or colour describes the relative brightness of the surface as detected and interpreted by the sensing instrument. Given the context of an image, many features can be identified based largely on their tone. For example, green tones in an image are associated with parks and trees, brown tones with crops and soil, and grey tones with buildings and roads. However, tone is of little use in isolation.

Texture can be described as the tonal variation in an image as a function of scale. Thus, where tone describes the spectral information in an image, texture describes the spatial variation of the spectral information in an image. In essence, texture is the effect created by an agglomeration of features in an image that are too small to be detected individually. The size of these texture features determines the coarseness or smoothness of the texture and it is one of its most important defining characteristics of features in an image. It might be argued that tone is more important, but why then is it so easy for humans to identify objects in a greyscale image where the role of tone is greatly diminished? This shows how central the role of texture is in human vision and in our ability to identify objects around us and, therefore, in image interpretation.

Reflective questions

➤ How do digital images differ from analogue images?

➤ What do you think are the most important image characteristics that govern the suitability of imagery for detecting environmental change? Why?

23.3 Foundations of remote sensing

All objects on the Earth's surface are capable of reflecting, absorbing and emitting energy called **electromagnetic radiation**. The foundations of remote sensing lie in the nature of this energy; how it interacts with our atmosphere and the surface of the Earth; and in our ability to detect and record it using remote sensing instruments. These are discussed in the following sections.

23.3.1 Electromagnetic radiation

The objective of remote sensing is to detect, measure and analyse electromagnetic radiation. When a sensing instrument is pointed at a surface from a remote location, it is the electromagnetic radiation intercepting the sensor from the surface of interest that is being measured and recorded. Sometimes called **electromagnetic energy** or just simply radiation, electromagnetic radiation refers to a form of energy in transit where both the electric and magnetic fields vary simultaneously. Visible light is just one form of electromagnetic radiation. In a vacuum, this energy travels in a straight line at the speed of light (c = approximately 300 000 km s^{-1}) and can only be detected when it interacts with some form of matter.

Electromagnetic radiation can be described either as a wave or as a stream of particles travelling through space. For the applications of physical geography, the characteristics of electromagnetic radiation are best described using the wave model (Avery and Berlin, 1992). However, the particle model is important for understanding and describing how electromagnetic radiation interacts with objects. Therefore, both models should be explored and these are addressed in Box 23.1.

23.3.2 Electromagnetic spectrum

Equations (23.1) to (23.3) in Box 23.1 describe functions where wavelength and frequency are continuous rather than discrete variables. This implies that electromagnetic radiation exists over a continuum. This continuum is called the **electromagnetic spectrum**. As shown in Box 23.1, the way in which radiation interacts with the atmosphere and with objects on the Earth's surface is dependent on wavelength and thus on its position in the spectrum. Therefore, the electromagnetic spectrum has been divided up into discrete categories of wavelengths sharing similar properties called **spectral bands**. The spectral bands and their location on the electromagnetic spectrum, shown in

FUNDAMENTAL PRINCIPLES

WAVE AND PARTICLE MODELS OF ELECTRO-MAGNETIC RADIATION

When electromagnetic radiation is described in terms of the wave model, reference is made to its wavelength (λ) and **wave frequency** (*f*). Wavelength describes the linear distance between successive peaks or troughs of the energy wave and frequency refers to the number of peaks (or troughs), called cycles, that pass a fixed point in space in a given period of time (Figure 23.3). The relationship between the wavelength, frequency and velocity of the radiation is described as:

$$c = f\lambda \quad \text{or} \quad f = \frac{c}{\lambda} \quad (23.1)$$

where *c* is velocity, equal to the speed of light, $\sim3.00 \times 10^8$ m s^{-1}, λ is wavelength, in metres (m), and *f* is frequency, in cycles per second or hertz (Hz). This inverse relationship between wavelength and frequency is illustrated in Figure 23.4.

The particle model takes a different approach by describing radiation in terms of a discrete unit called a **photon** which is a quantum of radiation. The energy of a quantum of radiation varies directly with its frequency (*f*) and inversely with its wavelength. This is given by:

$$E = hf \quad (23.2)$$

where *E* is the energy of one quantum, in joules (J), and *h* is Planck's constant, 6.626×10^{-34} J s^{-1}.

Using equation (23.1), we can link together the wave and particle models with:

$$E = \frac{hc}{\lambda} \quad (23.3)$$

This shows that the energy content of radiation is dependent on its wavelength and therefore has important consequences on how the radiation will interact with and affect any object it encounters. This will become increasingly obvious as this chapter progresses. Further details about radiation are also provided in Chapter 4.

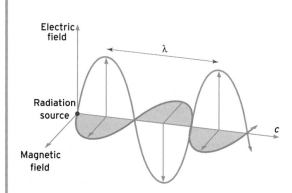

Figure 23.3 Illustrating the wavelength (λ) of the magnetic and electric fields of electromagnetic radiation. *c* denotes the speed of light equal to $\sim3.0 \times 10^8$ m s^{-1}.

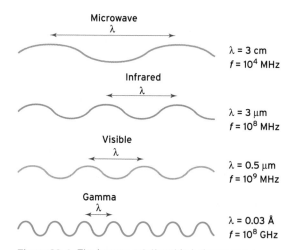

Figure 23.4 The inverse relationship between wavelength (λ) and frequency (*f*).

BOX 23.1

Figure 23.5, provide a convenient system of reference for describing electromagnetic radiation of similar properties (Lillesand and Kiefer, 2000).

From Figure 23.5, it can be seen that radiation ranges from lethal gamma rays (short wavelength, high frequency, high energy) to harmless television and radio waves (long wavelength, low frequency, low energy). The most familiar range of the spectrum for humans is the visible band that quite literally includes all the colours of the rainbow. This can be found between the ultraviolet wavelengths at 0.4 μm and the near-infrared wavelengths at 0.7 μm. The visible band is the range of radiation to which human vision has adapted. However, important information can also be obtained about objects by measuring and analysing radiation from other parts of the spectrum. This is discussed in later sections of this chapter.

Figure 23.5 The electromagnetic spectrum showing a magnified section of the visible bands to which human vision is adapted. (Source: after Lillesand and Kiefer, 2000)

23.3.3 Atmospheric and terrestrial interactions

Because the characteristics of electromagnetic radiation vary from one end of the spectrum to the other, it follows that the way it interacts with matter will also vary. The implications of this are that not all bands of the spectrum are suitable for remote sensing applications. One property of energy that is constant throughout the spectrum is that the energy that intersects an object, whether it be an atmospheric molecule, a leaf or a road surface, must be in balance with the energy that leaves the object. This conservation of energy is called the first law of thermodynamics, which states that the energy that enters a system must equal the energy that leaves the system. This can be represented by:

$$E_i = E_o \tag{23.4}$$

where E_i is incident energy and E_o is output energy.

When energy intercepts matter in the atmosphere such as gases, water molecules or particulate matter, it will be scattered, absorbed or transmitted (Figure 23.6; see also Chapter 4). Therefore, we can write equation (23.4) as:

$$E_i = E_s + E_a + E_t \tag{23.5}$$

where E_s is scattered energy, E_a is absorbed energy and E_t is transmitted energy.

Scattering is divided into three types: Rayleigh, Mie and non-selective. **Rayleigh scattering** is caused by atmospheric molecules and particles whose diameters are much smaller than the wavelength of the incident radiation. As there is an inverse relationship between wavelength and the degree of scatter, shorter wavelengths are most affected by Rayleigh scattering. This accounts for the blue colour of the sky. **Mie scattering** is caused by atmospheric molecules that are about equal in diameter to the wavelength of the

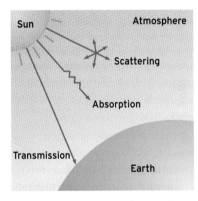

Figure 23.6 Schematic diagram illustrating the division of incoming solar radiation into the three components: scattered, absorbed and transmitted energy.

incident radiation. This includes water molecules and dust particles and most strongly affects longer wavelengths. Conversely, **non-selective scattering** is not dependent on wavelength and scatters all wavelengths between the visible and mid-infrared. Non-selective scattering is caused by atmospheric particles such as water droplets and ice crystals that are much larger in diameter than the incoming wavelength.

Matter in the atmosphere that scatters and absorbs energy prevents the energy from the Earth's surface from reaching the scanner and is therefore a hindrance to remote sensing. As this **interference** is often wavelength-dependent, some wavelengths of radiation will pass through the Earth's atmosphere virtually unobstructed while others will be almost completely scattered or absorbed. This yields bands of the spectrum called **transmission bands** (or atmospheric windows) and **absorption bands**. The various transmission and absorption bands of the atmosphere and the various contributing atmospheric molecules are shown in Figure 23.7.

Figure 23.7 Atmospheric transmission and absorption bands of electromagnetic radiation. Orange shaded area indicates transmitted energy while green indicates absorbed energy. (Source: after Scherz and Stevens, 1970)

Equations (23.4) and (23.5) must also hold true for terrestrial interactions with incoming electromagnetic radiation. The way in which natural and artificial objects on the Earth's surface distribute incident energy varies greatly. When incident radiation strikes a surface, it is reflected, absorbed or transmitted. In terrestrial interactions the broader term 'reflection' is used rather than scattering. In this context, reflection is subdivided into two types: specular and diffuse (scattering) reflection. **Specular reflection** occurs when radiation hits a surface that is smooth relative to the radiation's wavelength. In this case, the radiation is redirected in a predictable direction such as when the Sun glints off a body of water. In contrast, **diffuse reflection** occurs when radiation encounters a surface that is rough relative to its wavelength. Here, the radiation is reflected randomly in many directions like the soft light on a cloudy day. Thus equation (23.5) can be expressed as:

$$E_i = E_r + E_a + E_t \qquad (23.6)$$

or

$$E_i = (E_{st} + E_{dr}) + E_a + E_t \qquad (23.7)$$

where E_r is reflected energy, E_{st} is specular reflection and E_{dr} is diffuse reflection. Albedo, discussed in Chapter 4, is the term used to describe the ability of an object to reflect incoming radiation. Some typical albedo values for some common surfaces are given in Table 4.1 in Chapter 4.

Equation (23.7) suggests that specular and diffuse reflection, called directly reflected radiation, is the only source of radiation that reaches the sensor. However, there is also indirectly reflected radiation to consider. Indirectly reflected radiation occurs after incoming radiation is absorbed (E_a) by an object, converted to internal heat energy and then

subsequently emitted at longer wavelengths. Since the emitted radiation has a longer wavelength than its solar source, this energy is called long-wave radiation, whereas the incoming solar radiation and directly reflected energy is called short-wave radiation. Therefore, equation (23.7) holds true either only for short-wave radiation or if the term E_r includes both types of reflected radiation. In any event, both directly and indirectly reflected radiation are important in remote sensing for characterizing the reflectance characteristics of a surface.

The reflectance characteristics of an object or surface across the electromagnetic spectrum are called its **spectral signature**. A surface's spectral signature describes to what extent electromagnetic radiation is transmitted, absorbed and reflected at different wavelengths and, therefore, to some extent is an indication of its chemical composition and physical state. Thus, in many ways a spectral signature is analogous to a fingerprint. Every object reflects natural and artificial radiation in different ways. Just as fingerprints can be used to identify people, we can use spectral signatures to identify surface features. Figure 23.8(a) gives the average spectral signatures of various types of common surface covers. Notice the high reflection of vegetation and the absorption of water in the near-infrared. Figure 23.8(b) shows how different types of vegetation reflect differently. Notice how much overlap there is between the curves except in the near-infrared. This is a good example of the value of performing remote sensing in bands outside the visible spectrum.

Having introduced the concept of spectral signatures, it is now possible to revisit the issue of spectral resolution. Consider that at every wavelength in the spectrum a surface will interact in a predictable way. The spectral resolution of

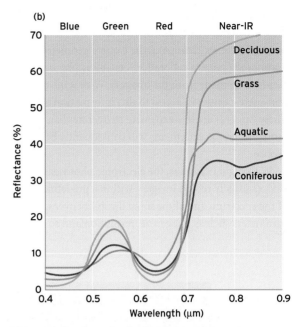

Figure 23.8 Spectral signatures of: (a) several common surface covers; and (b) vegetation types for the blue to near-infrared (0.4 and 0.9 μm) portion of the spectrum. (Source: after Avery and Berlin, 1992)

a sensor refers to its ability to define sections of the spectrum by wavelength and to provide a measurement of the radiation at each section. For example, a camera using black and white film takes one measurement for the entire visible portion of the spectrum (0.4–0.7 μm). From this no information about how the surface reflects in the rest of the spectrum will be available. Such a system has a low spectral resolution. However, a camera using colour film essentially takes a measurement for three separate bands (one for each of blue, green and red) thereby providing more information about the reflectance characteristics of the surface. Therefore, this system would have higher spectral resolution. If you were to split up the entire spectrum into small slivers and take a measurement for each sliver you would have a very high spectral resolution indeed.

In the following sections, two classifications of sensing instruments will be discussed: passive and active. Passive sensors measure the naturally occurring radiation that is reflected or emitted from a surface. A common 35 mm camera, when used without a flash, is a good example of a passive system. Alternatively, active sensors provide their own source of illumination. They emit a radiation pulse and measure any radiation that is reflected back to the source. A 35 mm camera used with a flash is a good example of an active sensing system. The main divisions of remote sensing instruments as they are addressed in the following sections are: (i) camera sensors; (ii) electro-optical sensors; and (iii) microwave and ranging sensors.

Reflective questions

➤ How does scattering affect our ability to detect environmental change through remote sensing?

➤ How is a spectral signature like a fingerprint and what does it reveal about a surface?

23.4 Camera sensors

One of the most versatile remote sensing instruments is the analogue photographic camera and its digital equivalent. The difference between analogue and digital cameras lies in how each system captures and stores images. Whereas traditional analogue cameras use photographic film to capture images, and diapositives, glass plates and prints to store them, digital cameras use electronic photosensitive devices to capture images and electronic media for storage (e.g. computer hard disk). These technological advances have necessitated the expansion of the traditional definition of 'camera' to include digital imaging devices. However, although the performance and cost of digital cameras are improving, analogue cameras are still heavily used, especially for aerial applications.

Cameras used in remote sensing range from standard single lens reflecting (SLR) cameras to highly specialized, large-format aerial cameras. Whether analogue or digital, these are divided into non-metric, semi-metric and metric cameras. In metric cameras, the camera geometry is closely monitored (camera calibration), enabling precise measurements to be made from the imagery. With semi-metric and non-metric cameras, access to this information is limited. These cameras are less expensive to operate than metric cameras, but measurements will consequently be less precise.

With digital cameras, the electronic photosensitive devices usually record imagery within a predefined portion of the spectrum and allow some flexibility over collection parameters such as image size, resolution and so on.

Conversely, analogue cameras depend on different types of film. Thus, as modern image processing methods require digital imagery as input, analogue images are digitized using scanners which range in quality from standard desktop scanners to highly specialized metric scanners, which aim to preserve the geometric and radiometric qualities of the analogue photographs.

Common film formats for analogue cameras include standard 35 mm, medium-format and large-format film. The last is the most common format used in airborne remote sensing and is designed for both high-resolution and high image quality. An example of a standard, large-format aerial photograph is given in Figure 23.9, which

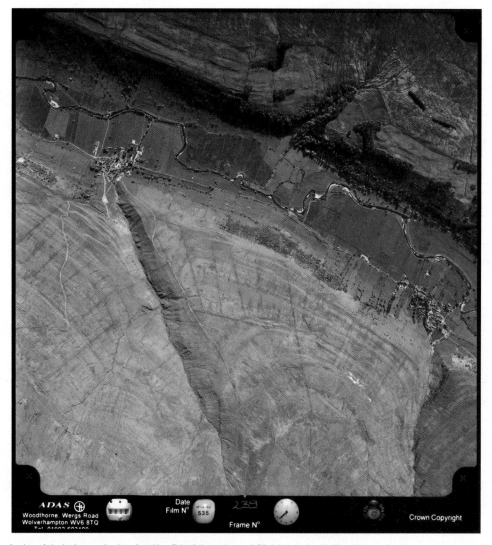

Figure 23.9 Typical aerial photograph showing the fiducial marks and flight data including the contractor contact details, date and time of flight, aircraft elevation, image number, film and frame number and level indicator. This image was taken over Upper Wharfedale in the Yorkshire Dales National Park in northern England on 17 May 1992. North is roughly to the right and the image measures 3.2 km across. (Source: photo © the UK Environment Agency)

shows the **fiducial marks** in the four corners (used to define a coordinate system in image space) and flight data recorded around the border of the image. This image was taken using black and white **panchromatic** film, which is sensitive to all the colours of the visible spectrum and uses shades of grey between black and white to record them. In addition to panchromatic film, black and white film that is also sensitive to infrared light is in common use. Figure 23.10 gives a comparison of a panchromatic and black and white infrared photograph. Take care to note the differences between these two images and how the vegetation and water reflect differently.

The obvious alternative to black and white film is colour film to which human vision is better adapted. However, the processing of colour film tends to be more expensive. With colour film, the printed colours need not necessarily correspond to the real colour of the scene as this can be controlled by film type and processing method. As a result, infrared-sensitive colour film can be displayed in a false-colour image where colours are assigned so that green light is recorded as blue, red light as green and near-infrared as red. This is much easier to interpret over black and white infrared images. Like Figure 23.10, Figure 23.11 provides a comparison between a normal colour and colour infrared image. Again, take note of the differences between the images. As predicted by Figure 23.8, the bright vegetation in both infrared images and the dark water in Figure 23.10(b) indicate how strongly vegetation reflects and how much water absorbs in the near-infrared.

(a)

(b)

Figure 23.10 Examples of (a) a panchromatic and (b) a black and white infrared photograph. Notice the difference in reflection of the water and the vegetation between the two images. These 1 : 9000 images show flooding of Bear Creek in north-west Alabama, USA. (Source: from Lillesand and Kiefer, 2000)

(a)

(b)

Figure 23.11 Examples of (a) a normal colour and (b) a colour infrared aerial photograph. These images show the campus and stadium of the University of Wisconsin. Notice that the vegetation appears red in the colour infrared image and how the artificial turf in the stadium is green in both. (Source: from Lillesand and Kiefer, 2000)

23.4.1 Photogrammetry

A common application of camera-based aerial imagery and increasingly of satellite imagery (discussed in the next section) is in a process called photogrammetry. The American Society for Photogrammetry and Remote Sensing defines photogrammetry as the recording, measurement and interpretation of both photographic images and recorded radiant electromagnetic radiation and other phenomena (Wolf and Dewitt, 2000). However, the main application of photogrammetry is the generation of **digital elevation models (DEMs)**. A DEM is a digital file that stores the three-dimensional coordinates of an array of points that correspond to a real surface. They are often oriented in a regular grid so they can be easily manipulated in standard image processing packages where they can be useful for modelling Earth surface processes that are dependent on surface gradient topography such as erosion (Chapter 11) and for monitoring and measuring environmental change such as glacier mass balance (Chapter 18).

To derive three-dimensional (3D) topographic coordinates from remote sensing data requires stereo imagery. **Stereo images** are images that have been captured in overlapping pairs, strips or blocks (Kasser and Egels, 2002). Overlapping images allow a surface to be observed from two different positions. This produces a phenomenon called **parallax**, which refers to the apparent change in position of a stationary object when viewed from two different positions. To demonstrate, hold your thumb up in front of you at arm's length and look at it with one eye shut. Now switch to the other eye and observe the change of position of your finger relative to the background. The magnitude of this change in position is dependent on several factors including the distance between your eyes (i.e. between the points of observation), the distance from your finger to your eyes and the distance from your finger to the background. Because of the parallax, the position of any point that has been imaged in at least two aerial photographs can be determined if information such as the flying height and attitude of the aircraft and the distance between the photographs is known (Wolf and Dewitt, 2000). Therefore, the goal of photogrammetry is to reconstruct the exact geometry of the film, camera (sensor), **platform** (usually an aircraft) and ground at the time each photo was captured. The geometrical relationship between these components is shown in Figure 23.12. Traditionally, this was accomplished using large, highly specialized instruments called stereoplotters. However, nowadays digital photogrammetry is the norm where the re-creation of these geometrical relationships is carried out entirely in a digital environment. In either case,

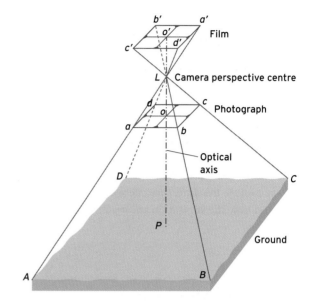

Figure 23.12 Geometrical relationship between the four imaging components of photogrammetry: film, camera perspective centre, photograph and ground. (Source: after Wolf, P.R. and Dewitt, B.A., *Elements of Photogrammetry with Applications in GIS*, 3rd edition, 2000, McGraw-Hill, Fig. 6.1, reproduced with permission of the McGraw-Hill Companies)

the variety of high-precision measurements made in the camera, on the photographs and on the ground are used to produce a model of the system geometry (Figure 23.12) at the time each photo was taken using a process called a **least-squares adjustment**. This is a mathematical method for fitting a model to data so as to minimize error between the observed values and the model predicted values. With a good fit of data to the model, 3D positional measurements can be made from the photographs.

As a result, photogrammetry represents an important source of topographic data for a variety of applications. For example, Lane *et al.* (2000) applied digital photogrammetry on 1 : 3000 scale aerial photography to an area of complex topography in the coniferous-forest-covered Glen Affric catchment in Scotland. The results showed that the precision of defining the surface topography was largely governed by photogrammetric data quality (camera calibration, base : distance ratio, ground control), combined with either scanning density or digital image resolution. The effect of the vegetation itself could be digitally removed using the stereo images, allowing the ground surface beneath to be mapped. Another example of the use of photogrammetry in the monitoring of environmental change is the work of Andreassen *et al.* (2002) who used historical aerial photography to track volume changes of a series of Norwegian glaciers.

23.5 Electro-optical scanners

Scanners use electro-optical detectors to measure incoming radiation. An important difference between camera-type instruments and scanners is the higher spectral resolution that can be achieved. Scanning sensors are capable of measuring reflectance over the entire electromagnetic spectrum but most commonly operate between about 0.3 and 14 μm (blue, green, red, near-, mid- and thermal infrared) and can measure radiation in numerous very narrow bands of the spectrum, thus providing high spectral resolution. Conversely, cameras tend to operate in one broad band between about 0.3 and 0.9 μm (ultraviolet, blue, green, red and near-infrared) although they tend to provide a higher spatial and temporal resolution than scanners.

Scanning instruments form a family of sensors that produce two-dimensional (2D) digital images by collecting data continuously under a **swath** beneath a moving platform. This differs from cameras, which use a lens and a shutter to capture an entire image simultaneously. However, the data can similarly be described in terms of resolution, bit scale and so on. What is very different between these images is how they are acquired. There are two main types of scanning instruments: **across-track** and **along-track sensors**. Although both types of instruments measure the incoming radiation from a swath below a moving platform, in across-track sensors the scanner's line of sight (also called the instantaneous field of view, IFOV) is directed in a sweeping motion at right angles to the direction of travel by a rotating or oscillating mirror. The forward motion of the platform causes the field of view to move forward, thus covering the entire 2D swath beneath the platform. The operation of these sensors is depicted in Figure 23.13(a). Because of this motion, across-track sensors are often referred to as whiskbroom scanners.

Along-track sensors produce 2D images using a linear array of charge-coupled devices (CCDs) oriented perpendicularly to the direction of travel covering one side of the swath to the other. As the platform moves forward, the field of view of the sensor array moves forward, thus producing a continuous 2D image of the swath. The operation of along-track sensors, often referred to as pushbroom scanners, is illustrated in Figure 23.13(b).

Scanners are most often mounted on airborne or spaceborne platforms and there are a variety of different types and configurations, which are discussed in the following sections.

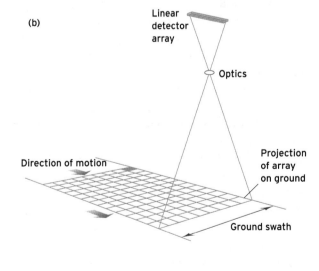

Figure 23.13 The operation of (a) an across-track (whiskbroom) sensor and (b) an along-track (pushbroom) sensor. (Source: after Lillesand and Kiefer, 2000)

23.5.1 Multispectral, thermal and hyperspectral instruments

Using either whiskbroom or pushbroom mechanisms, **multispectral scanners** produce digital images with more bands than is possible with camera-type sensing instruments. With technological advances, these instruments have given way to **hyperspectral scanners**, also called **imaging spectrometers**, which are similar in principle to multispectral scanners except that they can image with hundreds of bands, each of which is very narrow (approximately 2 nm). This produces images over virtually a continuous spectrum between the visible and thermal infrared regions of the spectrum. This provides very high spectral resolution and therefore offers a very detailed spectral signature for surface materials in the image. In hyperspectral imagery, surface objects can be more accurately identified and characterized remotely. These types of data have innumerable applications such as detecting changes in land-use (Section 23.2), water quality monitoring, and assessing vegetation health (Box 23.3 below). To show that they are made up of hundreds of spectral bands, hyperspectral images are often displayed graphically as an image stack or image cube as shown in Figure 23.14.

Thermal scanners measure radiation in the same way as multispectral and hyperspectral scanners but target the thermal infrared band of the spectrum. All objects with a temperature above 0 K (−273.15°C) emit thermal or radiant energy as a function of their internal temperatures (see Chapter 2). This energy can be measured, interpreted and analysed in the same way as in multispectral and hyperspectral scanners. Some applications of thermal remote sensing are qualitative, thus requiring only interpretation of relative differences between surface objects. Other applications may require absolute temperature difference to be measured. To do so requires that the **emissivity**, which is measured by the sensor, is converted to absolute temperatures. Thermal remote sensing is useful for a variety of applications in physical geography including studying ocean circulation patterns, geological structure, soil mapping and for assessing volcanic activity. Figure 23.15 shows a good comparison of the Chiliques Volcano in northern Chile imaged in the visible and thermal wavelengths.

23.5.2 Spaceborne instruments

Remote sensing of the Earth from space has undergone many developments since the days of the Mercury, Gemini and Apollo missions in the early 1960s. Today, there are numerous satellite programmes providing endless volumes

Figure 23.14 This image is an example of a hyperspectral image stack. The instrument used to capture the data was the Airborne Visible/Infrared Imaging Spectrometer (AVIRIS) which collects data in 224 bands of about 10 nm width between 0.40 and 2.45 mm. This particular image was captured over Konza Prairie, Kansas, in late August. (Source: from Lillesand and Kiefer, 2000)

of imagery over many regions of the spectrum. In this section an overview is given of the two established Earth observation programmes, Landsat and SPOT, as well as an introduction to some more recent spaceborne sensors.

23.5.2.1 Landsat

The Landsat programme has been very prominent in Earth observation research since the early days of remote sensing. The programme was launched in 1967 and has so far included seven satellites, of which two are still in use. The characteristics of each Landsat mission are summarized in Table 23.2. Overall, five sensor instruments have been used in these missions. They are the Return Beam Vidicom (RBV), the Multispectral Scanner (MSS), the Thematic Mapper (TM), the Enhanced Thematic Mapper (ETM) and the Enhanced Thematic Mapper Plus (ETM+). Table 23.3 summarizes some of the specifications of these five sensors.

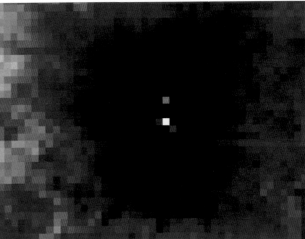

Figure 23.15 The Chiliques Volcano in northern Chile has been dormant for the past 10 000 years and, as shown in this pair of images, is now coming back to life. These images were captured by the Advanced Spaceborne Thermal Emission and Reflection Radiometer (ASTER) and show the volcano in the visible and near-infrared (left) and in the thermal infrared (right). Note the hot spots around the volcano's crater caused by magma appearing just below the surface. (Source: images courtesy of NASA/GSFC/MITS/ERSDAC/JAROS, and US/Japan ASTER Science Team)

Enhancements included improvements to spatial resolution (80 m to 30 m), temporal resolution (18 versus 16 days to complete coverage) and spectral resolution (four to eight bands).

The MSS, TM and ETM+ sensors aboard Landsat 5 and 7 are all based on the whiskbroom scanner design and, unlike their predecessors, they are still very much in use and represent an important source of Earth observation data. Landsat 5 was launched in a circular, Sun-synchronous, near-polar orbit at an altitude of 705 km. The use of **Sun-synchronous orbits** means that satellites pass over any given latitude at the same local time each day. This ensures that the same solar illumination conditions prevail for each pass. The repeat period, the time it takes for a satellite to

Table 23.2 Landsat mission characteristics (Source: Lillesand and Kiefer, 2000)

Satellite	Launched	Decommissioned	RBV bands	MSS bands	TM bands	Repeat period (days)	Altitude (km)
1	23 July 1972	6 January 1978	1-3	4-7	None	18	900
2	22 January 1975	25 February 1982	1-3	4-7	None	18	900
3	5 March 1978	31 March 1983	A-D	4-8*	None	18	900
4	16 July 1982	15 June 2001	None	1-4	1-7[†]	16	705
5	1 March 1984	TM operational, MSS powered off August 1995	None	1-4	1-7	16	705
6	5 October 1993	Failed on launch	None	None	1-7, ETM	16	705
7	15 April 1999	Still operational	None	None	1-7, ETM	16	705

[*]Band 8 failed after launch.
[†]TM data transmission failed in August 1993.

Table 23.3 Landsat programme sensor

Sensor	Mission	Sensitivity (μm)	Resolution (m)	Swath width (km)	Bit scale
RBV	1, 2	0.475–0.575	80	185	n/a
		0.580–0.680	80		
		0.690–0.830	80		
	3	0.505–0.750	30	183	
MSS	1–5	0.5–0.6	79/82*	185	6-bit
		0.6–0.7	79/82*		
		0.7–0.8	79/82*		
		0.8–1.1	79/82*		
	3	10.4–12.6	240		
TM	4, 5	0.45–0.52	30	185	8-bit
		0.52–0.60	30		
		0.63–0.69	30		
		0.76–0.90	30		
		1.55–1.75	30		
		10.4–12.5	120		
		2.08–2.35	30		
ETM	6	TM bands	30 (120 in thermal)	185	8-bit
		0.50–0.90	15 (panchromatic)		
ETM+	7	TM bands	30 (120 in thermal)	185	8-bit
		0.50–0.90	15 (panchromatic)		

*79 m for Landsat 1 to 3 and 82 m for Landsat 4 and 5.

(Source: from Lillesand and Kiefer, 2000)

revisit a point on the ground, is completed every 16 days but the orbits were offset to reduce this period to 8 days when Landsat 4 was still operational. As Landsat 7 was designed in order to maintain data continuity with Landsat 4 and 5, the same specifications (orbit, bands, resolution and swath width) were used with the exception of an added high-resolution 15 m panchromatic band. A sample of the five non-thermal bands that are the most frequently used TM bands (Bands 1 to 5) is given in Figure 23.16. The information that can be extracted from images has been used in innumerable applications of physical geography and in detecting environmental change, including studies of vegetation change (Chapters 8 , 9, 10 and 22), soil and coastal erosion (Chapters 7 and 17), sediment and pollution movements in watercourses (Chapters 12, 13, 14 and 15) and glacial studies (Chapter 18).

23.5.2.2 SPOT

The SPOT programme, an acronym of the Système Pour l'Observation de la Terre, was undertaken by the French Government in 1978 with early collaboration with Belgium and Sweden and has now become a successful international endeavour. The SPOT satellites, of which there are five, use the pushbroom sensor design. Their mission characteristics are given in Table 23.4. As the goal of the SPOT programme was long-term data continuity, the characteristics of various satellites are similar. All the SPOT satellites are in identical circular, Sun-synchronous, near-polar orbits at an altitude of 832 km. Although the SPOT satellites take much longer to cover the globe than Landsat (i.e. 26 days), the optical devices in the sensors are movable to permit off-vertical imaging. This decreases the 26-day repeat period to less

Figure 23.16 This figure provides a sample of the five non-thermal Landsat TM bands, each in 8-bit greyscale: (a) Band 1 (0.45–0.52 mm blue); (b) Band 2 (0.52–0.60 mm green); (c) Band 3 (0.63–0.69 mm red); (d) Band 4 (0.76–0.90 mm near-infrared); and (e) Band 5 (1.55–1.75 mm mid-infrared). This Landsat-5 image was captured over the Bering Glacier, Alaska, on 25 September 1986 at 11:03 local time. Notice the large difference in reflectance between the vegetation along the bottom in the image and the glacier in Bands 4 and 5. North is up and the images are roughly 53 km across. (Source: NASA Landsat Program, 11/08/1999, Landsat TM scene, p064r18_5t860925. Geocover Ortho, MDA Federal, Rockville, 09/25/1986. Data courtesy of Global Land Cover Facility, www.landcover.org, images courtesy of the Global Land Cover Facility, and Adrian Luckman)

Table 23.4 SPOT mission characteristics

SPOT satellite	Launched	Decommissioned	Repeat period (days)	Altitude (km)	High-resolution stereoscopic
1	22 February 1986	31 December 1990	26	832	No
2	22 January 1990	Still operational	26	832	No
3	25 September 1993	14 November 1997	26	832	No
4	23 March 1998	Still operational	26	832	No
5	3 May 2002	Still operational	26	832	Yes

than 5 days. It also enables stereoscopic viewing, enabling the imagery to be used easily in photogrammetric applications. Sensor systems also include vegetation monitoring instruments that provide 1 km resolution multispectral imagery and daily global coverage. On the newer SPOT 5, imagery is provided at 5 m resolution. For more detailed information about the SPOT instruments including band sensitivity visit the SPOT technical information link at **www.spot.com**.

23.5.2.3 New systems

Although the resolutions of the Landsat and SPOT sensors were cutting edge in their time, they now are some way from matching that of the new high-resolution satellites that have recently been launched. Plenty of information on these types of sensors is available on the Internet, but noteworthy commercial undertakings include:

- IKONOS, launched by Space Imaging on 27 April 1999, can capture 4 m resolution multispectral and 1 m resolution panchromatic imagery;
- QuickBird, launched by DigitalGlobe on 18 October 2001, can collect 0.61 m resolution panchromatic and 2.5 m resolution multispectral imagery;
- MODIS, the Moderate Resolution Imaging Spectroradiometer from NASA, boasts a high radiometric sensitivity (12 bits) in 36 spectral bands which range in wavelength from 0.4 μm to 14.4 μm and with a spatial resolution of 250 and 1000 m. The first instrument (Terra) was launched on 18 December 1999 while the second (Aqua) was launched on 4 May 2002.

High-resolution (spatial, temporal and spectral) instruments such as these significantly increase the potential of remote sensing techniques for detecting environmental change. Their ability to provide low-cost, flexible and high-resolution data will make them increasingly indispensable in the future.

Interpreting multispectral data depends on many factors including time of day, specific bandwidths, imaging conditions (e.g. weather) and experience of the interpreter. However, some general guidelines exist for interpreting the different bands of the spectrum. These are highlighted in Table 23.5.

Reflective questions

➤ What types of electro-optical scanners are there?

➤ What is the advantage of high spectral resolution instruments over lower spectral resolution instruments such as the Landsat instruments?

➤ How do sensor mechanics affect the output image?

23.6 Microwave and ranging sensors

In this section we investigate the sensors of the microwave wavelengths as well as the ranging sensors including radar, sonar and laser altimetry (lidar). Although the ranging sensors use vastly different sources of illumination, they have many operational similarities and therefore it is appropriate to discuss them here.

23.6.1 Microwave sensors

The microwave portion of the electromagnetic spectrum can be found between the wavelengths of 0.001 and 1 m.

Table 23.5 Principal applications of bands commonly used in remote sensing

Spectral band	Principal applications
Blue	Good water penetration so suitable for coastal mapping and bathymetry. Also useful for soil/vegetation/forest-type discrimination, cultural feature identification. Sensitive to atmospheric haze
Green	Has some ability to penetrate water but sensitive to turbidity. High reflectance from vegetation so useful for vegetation discrimination and vigour assessment. Also cultural feature identification. Sensitive to atmospheric haze
Red	High chlorophyll absorption in vegetation and high reflection in soils so good for differentiating between soil and vegetation. Also good for delineating snow cover and cultural features
Near-IR	High reflection from vegetation and absorption in water. Best band for discriminating between vegetation types and vigour. Good for delineating water bodies and soil moisture
Mid-IR	In shorter wavelengths, good for vegetation and soil moisture content and for discriminating between snow and clouds. In longer wavelengths, good for discriminating between mineral and rock types and also moisture content
Thermal	Vegetation stress analysis, soil moisture and thermal mapping

(Sources: after Avery and Berlin, 1992; Lillesand and Kiefer, 2000)

Radiation in this portion of the spectrum is more commonly characterized in terms of its frequency (3.0×10^5 to 3×10^9 Hz). Microwave remote sensing has two important advantages. Firstly, it is well suited for penetrating through the atmosphere in conditions that typically interfere with other wavelengths, such as cloud, dust, rain, snow and smoke. Secondly, the interaction of microwaves with surfaces tends to be quite different from those of other common wavelengths such as the visible and infrared bands. This provides researchers with unique insights into the surfaces under investigation.

There are two main types of remote sensing instruments that operate in the microwave portion of the electromagnetic spectrum. The first is a passive instrument and is called a microwave radiometer. The second, a more widely used instrument, is an active microwave sensor called radar. Remember that, unlike an active sensor, a passive sensor does not provide its own source of radiation. Thus, microwave radiometers detect only the very low levels of naturally occurring microwave radiation that are emitted from all objects on the Earth's surface. Because all objects emit these microwave signals, the interpretation of passive microwave data can be very difficult since the signal for any given point is the sum of four component sources of radiation: (i) emitted radiation from the object of interest; (ii) emitted radiation from the atmosphere; (iii) reflected radiation from another source such as the Sun; and (iv) transmitted energy from the subsurface. Thus, the applications of passive systems are somewhat limited.

Radar, however, is very widely used and has many applications. It is an acronym of **ra**dio **d**etection **an**d **r**anging, although microwaves are now used in place of radio waves. As an active sensor, it generates a microwave pulse that is aimed at the surface of interest and any returned energy, called the echo or **backscatter**, is measured by an antenna. Two common non-imaging radar systems are Doppler radar and the plan position indicator (PPI). Doppler radar is used to measure the velocity of remote objects such as cars. PPI is used to plot the planimetric position of large objects such as aircraft around an airport or ships at sea. It involves the use of a rotating antenna that continuously updates measured positions of objects on a circular screen.

Imaging radar systems were developed after World War II and were based on PPI technology primarily for peering over enemy lines. Radar systems were mounted in aircraft to look sideways deep into enemy territory while flying safely in friendly skies. This technology was called side-looking radar (SLR) or side-looking airborne radar (SLAR) for airborne systems. Like the Doppler and PPI radar, the instrument sends out a short, high-energy pulse as shown in Figure 23.17. The energy from this pulse reaches the ground and reflects off the surfaces it encounters (buildings, trees, soil and so on). The first object the pulse encounters will produce a return signal that will be the first to reach the antenna. The travel time of this return is recorded. All subsequent objects that the pulse encounters will produce returns that will be detected by the antenna. Since the pulse travels at the speed of light, c, and the travel time is known (t),

Figure 23.17 Operation of a side-looking radar system. Inset shows a profile of the current image line.

the distance (d) of the return producing objects from the aircraft is given by:

$$d = \frac{ct}{2} \qquad (23.8)$$

As the aircraft moves forward a series of pulse returns can be combined to form an image. The tone of a point in a radar-generated image measures the intensity of the backscatter from that point.

The resolution of an SLR image is determined by several factors. One is the length of the pulse and the other is the **beamwidth** (Figure 23.17). Two ways to improve the resolution of the system are to use shorter pulses and narrower beamwidths. Shorter pulses can be achieved by shortening the time the instrument emits the outgoing signal. Instruments can control beamwidths either by increasing the physical length of the antenna or by simulating a longer antenna length. The former are called brute force or real aperture systems. Those that simulate antenna length are called synthetic aperture radar (SAR). Thus, SAR systems enable the use of very narrow beamwidths without the physical requirements of a long antenna. These instruments represent one of the most important sources of remote sensing data in the microwave portion of the spectrum.

Overall, the reflection characteristics of the radar pulse can be attributed to a surface's geometric properties such as surface roughness and orientation, and electrical properties, which are largely influenced by the presence of water. Although these interactions are complex, they have enabled the successful application of radar to a wide variety of applications such as geological mapping, discriminating among vegetation types and sea ice studies. Many applications take

advantage of the number of radar satellites that operate from space. Two noteworthy examples include the European Space Agency's (ESA's) ERS 1 and 2 and Canada's RADARSAT. An example of a radar image is given in Figure 23.18 used for examining the Monaco Glacier in Svalbard, Norway.

Another use of radar technology, which has many environmental uses, is ground-penetrating radar (GPR). It is a technology that has been used since the 1920s when it was first applied to a glacier in Austria (Stern, 1929, 1930). Unfortunately, the method was virtually forgotten until the late 1950s. Since then, GPR has become an invaluable tool for a whole host of applications including locating water tables, buried objects and delineating stratigraphy and internal structures of glaciers, bedrock, concrete and sediment.

Essentially GPR is used to explore, characterize and monitor subsurface structures. It can be operated on the ground by hand or in some type of vehicle, airborne by an aircraft or helicopter and even from a satellite. Operating in the same way as the radar systems described above, a microwave pulse is generated and directed into the subsurface. Structures in the subsurface cause some of the waves of energy to be reflected back to the instrument's receiving antenna as echo or backscatter where they are detected and recorded. Figure 23.19 gives an example of the resulting image from a peat bog in northern England. Peat depths and soil pipes (see Chapters 13 and 15) across the survey catchment could easily be measured, producing a continuous profile (Holden *et al.*, 2002). This is advantageous over the traditional point data that would have been produced

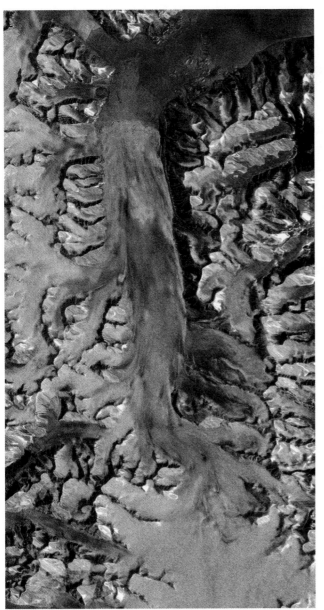

23.6.2 Sonar

Although sonar employs a very different type of energy to 'illuminate' a surface, it is operationally very similar to radar. Sonar is an acronym of **so**und **n**avigation **a**nd **r**anging and is a form of active remote sensing that uses sound waves or acoustic energy propagated through water to detect objects and surfaces. There are two groups of sonar systems. The first are non-imaging systems used to measure water depths and include echo-sounding profilers or bathymetric sonar (for measuring underwater topography). The second produce images or sonographs and are called sidescan imaging sonar.

Figure 20.20 illustrates how a sonar system is deployed. A torpedo-like vessel called a towfish is towed at a known depth behind a survey ship joined by a cable. Within this cable is the link between the recording instruments on board the ship and the sending and receiving instruments (called transducers) in the towfish. The transducers send out fan-shaped acoustical pulses perpendicular to the direction of travel. Upon receiving and recording the echo from surface features, a sonograph is produced. The topographic features on the sea floor, for example, must be thought of as reflectors in the same way features are with microwave energy (Avery and Berlin, 1992). Features that face the transducers will be better reflectors than those that face away from the transducers. Furthermore, smooth surfaces (e.g. sand and mud) will tend to act as specular reflectors of acoustic energy and thus will return no signal. Alternatively, rough surfaces such as gravel and boulders will reflect much of the energy back to the sensor. The main application of sonar in physical geography is the mapping of the ocean sediments, river and lake beds, and the ocean floor topography.

23.6.3 Lidar

Lidar technology has been used since the 1970s and is also called laser scanning or laser altimetry. It is an acronym of **li**ght **d**etection **a**nd **r**anging. Instead of using microwaves or acoustical energy to 'illuminate' a subject, lidar employs a highly focused beam of light called a laser. The primary function of lidar is to produce DEMs. A laser scanner measures the time it takes for the laser pulse to travel from the scanner to the surface and back to the scanner. If the direction, velocity and travel time of the laser pulse are known, the distance and position of the surface point relative to the platform (e.g. aircraft or helicopter) can be determined. However, in order to turn this relative position into absolute coordinates, the exact position and attitude of the platform

Figure 23.18 The radar image shown here shows Monacobreen (the Monaco Glacier) and surrounding mountains in north-west Svalbard, Norway. This image was captured in June 1995 by the ESA's Earth Resources Satellite (ERS). Notice the distortion in the mountain peaks due to the look angle of the radar instrument. North is up and the image is roughly 23 km across. (Source: image courtesy of Tavi Murray and Adrian Luckman)

with rods or by coring. This type of work has been carried out in many parts of Finland and Norway and allows us to establish the amount of peat on Earth. This is invaluable data because peatlands are a very important store of carbon, which if released may enhance global warming (Holden, 2005b).

Distance (m)

Peat

Figure 23.19 A GPR radar-gram from a peatland survey in the North Pennines, England. The image clearly shows the peat depth and some of the layers within the peat, and how it changes across the transect. The 'substrate reflector' is the bedrock below the peat.

Stratigraphic layering   Substrate reflector   N

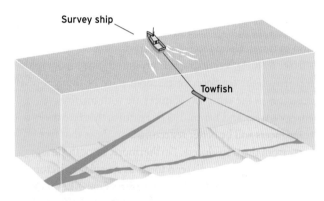

Survey ship

Towfish

Figure 23.20 Operation of a sonar instrument. (Source: after Avery and Berlin, 1992)

must be precisely known at a frequency that matches that of point measurement (approximately 15 000 pulses s⁻¹). Until very recently this locating technology was not available but today laser scanning systems rival and often exceed the performance and efficiency of digital photogrammetric systems in producing DEMs.

A modern system requires three components as illustrated in Figure 23.21. The first is the laser scanner to determine the distance and direction of each measured point from the platform. The second is a **global positioning system (GPS)** used to determine the position of the aircraft using a constellation of satellites that orbit the Earth. This requires a static reference GPS receiver (called a **base station**) on a known point on the ground, and a mobile GPS receiver (called a **rover**) in the aircraft. This system is called **differential GPS**. The third component is an inertial navigation system (INS), which measures any unwanted

rotation of the aircraft. With these three systems operating in synchronization with each other, a very accurate 3D position of a point on the ground can be achieved. Modern instruments now also record the strength or intensity of each laser return signal. The strength of the return is heavily dependent on surface characteristics in the same way the strength of reflected light is heavily dependent on the surface from which it was reflected. Therefore, this intensity can be used in the form of an image, like a photograph, to assist in the interpretation for the lidar data. This is demonstrated in Box 23.2.

One of the great benefits of lidar systems is their ability to record multiple returns for the same point. As the laser beam travels towards the surface, it experiences some divergence such that it produces a **footprint** on the ground whose diameter (usually around 25 cm) is dependent on the height of the aircraft. The first object within this footprint that the beam encounters will produce a return. Likewise, all objects within this footprint that the beam encounters will produce a return. Ideally, the first return will represent the top of a vegetation canopy and the last return will represent the ground. This enables the modelling of canopy heights while at the same time offering an accurate representation of the surface. However, sufficiently dense vegetation and solid objects such as buildings do present a problem for accurate representation of the ground surface. Thus, sometimes DEMs that include non-surface objects are called digital surface models (DSMs) and those representing only the ground surface are called digital terrain models (DTMs). Lidar segmentation, the process by which lidar points (often called hits) that represent non-surface objects are separated from hits which represent the surface, is one of the principal

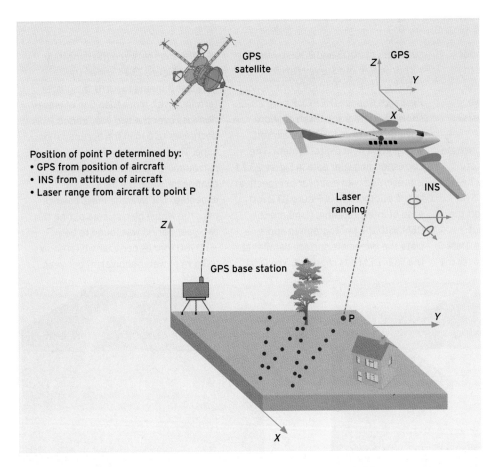

Position of point P determined by:
• GPS from position of aircraft
• INS from attitude of aircraft
• Laser range from aircraft to point P

Figure 23.21 Operation of a lidar system. The combination of an accurate differential GPS position of the platform/ instrument, an INS measurement of the attitude of the platform/instrument (rotation on X, Y and Z axes) and the laser range gives an accurate position in X, Y and Z of point P.

CASE STUDIES

AIRBORNE LIDAR FOR MEASURING CHANGES IN GLACIER GEOMETRY, EXTENT AND MASS BALANCE IN SVALBARD, NORWAY

Mountain glaciers constitute only about 3% of the glacierized area on Earth. However, under current climate change predictions, they are expected to contribute significantly to sea-level rise because of their heightened sensitivity to climate change (IPCC, 2007a). Unfortunately, estimates of the contribution of these glaciers to future sea-level rise are uncertain due largely to the global shortage of long-term glacier mass balance observations (see Chapter 18). Glacier mass balance is an important measure of the response of a glacier to changes in climate variables, especially summer temperatures and winter precipitation. Of more than 160 000 glaciers worldwide only about 40 have mass balance records of more than 20 years. In the high latitudes mass balance records are even worse because glaciers in these areas tend to be very remote. Unfortunately, it is at these high latitudes where climate change is expected to have the greatest impact and where long-term mass balance measurements are needed most.

There are several challenges to measuring glacier mass balance:

• Glaciers are usually found in remote locations.

• They are often associated with extreme topography.

• They can be very large, which makes it difficult to ensure that a small number of measurements are representative of the whole glacier.

BOX 23.2 ➤

➤

Airborne remote sensing methods such as lidar (Section 23.6.3) can provide an ideal solution to these challenges and, as a result, have become important tools for measuring glacier mass balance.

The figures in this box demonstrate one recent application of lidar for mass balance monitoring in the heavily glaciated Arctic archipelago of Svalbard, which lies about 700 km off the north coast of Norway and about 1100 km south of the North Pole. Data presented here were collected over the summers of 2003 and 2005 by the Airborne Remote Survey Facility (ARSF) of the UK Natural Environment Research Council (NERC). Figure 23.22 shows a photograph taken of Midre Lovénbreen from the NERC aircraft with a hand held digital camera. The corresponding lidar data in Figure 23.23 is presented with a similar perspective to the photograph in Figure 23.2 for ease of interpretation. Compare the DEM with the photograph to appreciate the incredible surface detail that the lidar DEM has captured. The DEM of difference between the two study years in Figure 23.24 shows the greatest thinning towards the glacier terminus with more than 5 m of elevation loss over the two year period. This decreases to 0 m in the accumulation zone. The blue colour of the surrounding bedrock indicates no change between the data sets as expected. The average mass balance over the whole glacier area during this two year period was found to be −121 180 tonnes per year which equals −0.6 m of water equivalent per year.

Figure 23.22 Hand held digital photograph of the Svalbard glacier Midre Lovénbreen looking south. The glacier is about 4 km long and 1 km wide.

Elevation loss

0 m

5 m

Figure 23.24 DEM of difference between the 2005 lidar data shown in Figure 23.23 and a similar lidar data set that was collected (also by NERC) during summer 2003. North is up and the DEM of difference is approximately 4 km across.

(a)

(b)

Figure 23.23 The 2005 lidar DEM of Midre Lovénbreen: (a) with intensity overlay; and (b) in shaded-relief. The latter clearly highlights the surface detail that can be captured by this high-resolution instrument. Both DEMs were interpolated from the raw lidar point cloud (inset) to a 5 m grid. Thanks to 3D imaging software, QT Modeler, courtesy of Applied Imagery for use of the software to produce this image.

BOX 23.2

challenges of laser altimetry. For example, lidar data used in mass balance modelling (Box 23.2) and coastal management (Chapter 17) must only model points that represent the ground surface whereas land-use and vegetation mapping (Chapter 22) may require that all surface features are modelled. Compare this to a more complex scenario such as floodplain mapping for flood management which requires the modelling of the ground surface but also of any surface feature that would act as a barrier to slow flood waters (i.e. walls and hedges but not buildings or trees). With lidar becoming increasingly important for monitoring environmental change, the lidar segmentation issue has become an important area of research (Sithole and Vosselman, 2006).

Reflective questions

➤ What are the advantages of microwave remote sensing over optical sensing?

➤ When collecting lidar data, if any of the three instruments (scanner, GPS and INS) are in error or out of synchronization, are the resulting lidar measurements reliable?

23.7 Digital image processing

Regardless of the source of remotely sensed data, photographs and images are rarely in a state suitable for immediate analysis. Most often, owing to radiometric or geometric errors, images require some form of manipulation

or interpretation before they are ready for analysis. The manipulation and interpretation of digital images are called digital image processing. There are many procedures in digital image processing that are frequently used for detecting environmental change. Important procedures include image rectification, image enhancement and image classification.

23.7.1 Digital images

Digital images are made up of a 2D array of DNs that usually represent the brightness of the corresponding area on the surface. A pixel can also be defined in terms of its position in a coordinate system. In the simplest case, pixel space coordinates are used where the top left corner is defined as the origin (0, 0). To link an image to an absolute coordinate system (called **georeferencing**), a pixel can be assigned an x and y coordinate that corresponds to some ground reference system (e.g. latitude/longitude). Images that are georeferenced can be joined together to create an **image mosaic**.

As digital images are made up of a series of numbers, their characteristics can be expressed in terms of a **histogram** and summary statistics. Figure 23.25 provides a histogram of the greyscale range of DN values for Figure 23.1. Images can be subjected to a variety of processes based on the manipulation of their DNs and histograms.

23.7.2 Image rectification

The purpose of image rectification is to attempt to compensate for any image distortions or degradations that may have resulted from the image acquisition process. The potential

Figure 23.25 Histogram and descriptive statistics of the image in Figure 23.1.

sources of such artefacts are many, as are the procedures for their correction. Image rectification processes are divided into two types: geometric and radiometric.

When an image requires geometric correction, inconsistencies in the relationship between the sensor and the surface have occurred during data collection. These inconsistencies depend greatly on the sensor used but in general they can be caused by changes in the sensor's altitude, attitude or velocity, changes in the surface such as the rotation and curvature of the Earth, and effects of the atmosphere. Some of these errors are predictable (e.g. Earth's rotation and curvature) and can be modelled mathematically to remove their effects. These are called **systematic errors**. Other errors that are unpredictable cannot be modelled mathematically. These are called **random errors**. To account for the effects of random errors, **ground control points** (GCPs) are used to tie the imagery to known points on the ground. The location of many GCPs in the distorted images is measured. Then, the coordinates in both ground and image systems are used to determine the transformation equations to correct the position of each pixel. Orthorectification is a specific type of rectification for aerial photography where the effects of elevation are removed from the imagery using not only GCPs but also information about the internal geometry of the camera to compensate for these errors.

Unlike geometric correction, radiometric correction does not change the position of the pixels, only the value of their DNs. However, like geometric correction, sources of radiometric errors are also largely dependent on the sensor used. There is a long list of factors that can interfere with the radiance measured by an instrument. Some examples include atmospheric conditions, illuminance (time of year/day), instrument sensitivity and viewing geometry. To correct an image radiometrically, the DNs are adjusted based on some model to correct the pixel brightness to represent the true radiance characteristics of that scene. The three most common adjustments are: noise removal, Sun-angle and haze correction.

Noise removal is necessary when electronic noise is recorded in an image that is unrelated to the radiance of the scene. It appears either as random or periodic errors found throughout the image. In many cases, these errors are caused by the malfunction of the sensing instrument. The errors can be removed by interpolating new DN values from unaffected pixels around the errors. The effect of a noise removal routine can be seen in Figure 23.26.

Remote sensing imagery is collected at various times of the year and different parts of the day. The position of the Sun at any given point in time will have an influence on how

(a)

(b)

Figure 23.26 Image showing the effects of noise removal. (a) An image containing noise which can be seen as a random white speckle distributed throughout the image. In (b) a noise removal algorithm has been applied. Although the noise has been removed, notice how the image appears less sharp than previously.

much sunlight reaches the surface and therefore on the radiance that is recorded by the instrument. This only becomes a problem when working with groups of images with different **illuminations**. Thus, compensation for solar position, called Sun-angle correction, is often required. This is easily achieved by applying a correction factor to each DN based on the illumination angle. Another problem is that atmospheric scattering (called haze) can be recorded by the sensing instrument. Haze has the effect of making an image appear foggy and it tends to mask the radiance characteristics of the surface. Haze correction can be applied when a feature in the scene is known to have a reflection at or near zero, such as water or shadows. Haze makes dark features appear more grey rather than black. Thus, an estimate of the effects of haze can be made and subtracted from the rest of the image.

23.7.3 Image enhancement

Image enhancement refers to any form of image manipulation that attempts to redisplay the information in the image in a way that better represents image characteristics or features of interest. Image enhancement applications are infinite. However, the most common applications include contrast stretching, spatial filtering and band ratioing.

Contrast stretching refers to a group of processes that are used to redistribute the range of DNs of an image to make better use of the image's bit scale. The image in

Figure 23.27(a) and its histogram in Figure 23.27(b) show a rural scene dominated by pastures all of fairly uniform tone and very little texture. In this image, only the grey levels between 55 and 230 are being used to display the image, leaving many unused grey levels on either end of the histogram. To improve the interpretability of the information available in this image, it is sometimes beneficial to make use of all the grey levels available on the bit scale (in this case 0–255) and thus we use contrast stretching. A linear stretch is simplest and uniformly expands the range of DNs that are used in the image to include the entire range of values. A histogram equalization stretch also makes use of the entire range of DN values. However, unlike a linear stretch, a histogram equalization takes into account the frequency of DN occurrences. This means that in the output image, more DNs are reserved for brightness values in the image that occur frequently. For Figure 23.27, for example, this histogram equalization stretch will increase the contrast of the pasture pixels more than the urban pixels since the former occur more frequently in the image. The histogram equalization stretch is well suited for improving the interpretability of this image and the results of its application are given in Figure 23.28.

Another type of commonly used image enhancement is spatial filtering. Spatial filtering is a broad term used to refer to a variety of applications for the enhancement of the spatial variation in image tone (texture). In spatial filtering, texture is often described in terms of frequency. A smooth texture (small variations in tone) has a low frequency, whereas a coarse texture (large variations in tone) has a high frequency. This type of operation is called filtering because,

(a)

Figure 23.27 (a) Small section of an aerial photograph showing a relatively poor image contrast in the pastures. (b) The histogram and descriptive statistics of the image.

like a filter, it selectively passes or preserves certain spatial frequencies, thereby enhancing some spatial frequencies and suppressing others. Low-pass filters produce a smoother image by enhancing low-frequency features and suppressing high-frequency features, whereas high-pass filters do the reverse and produce a sharper image. To illustrate this Figure 23.29 shows the results of a low- and high-pass filter when applied to the image in Figure 23.27.

Spatial filtering is a local operation, which means that the pixels of the filtered image only reflect the conditions in the immediate neighbourhood of each pixel. This is accomplished using a process called convolution. Convolution

describes the process where a small moving window, usually 3 × 3, 5 × 5 or 7 × 7 pixels in size, passes over each pixel in an image. For each pixel some operation is computed using only the pixels within the window to arrive at a new DN value for the central pixel. For the low-pass filter described previously, the convolution filter (also called a kernel) determines the mean of the values in the moving window and assigns this new value to the central pixel of the window before moving to the next pixel to repeat the calculation. This process is illustrated in Figure 23.30. Thus, it can be seen that by changing the values in the kernel, a great number of spatial filtering operations can be carried out.

Another set of spatial filters are edge enhancement filters. These are also conducted using convolution and are used to exaggerate abrupt changes of DNs in an image that are usually associated with an edge or boundary of some feature of interest. Because edges are an example of a high-frequency feature, these filters tend to produce sharper images similar to the high-pass filter. Edge enhancement filters can be used simply to highlight all edge pixels, to remove non-edge pixels or to identify edges that lie only in a certain direction.

Finally, band ratioing is a process where the DN values of one band are divided by those in another to reveal the subtle variations between bands that would otherwise be masked by differences in illumination across the scene. One very common example of band ratioing is the calculation of a normalized difference vegetation index (NDVI) image which provides an important tool for monitoring vegetation. It is calculated using the red and near-infrared bands of a sensor

(a)

Figure 23.28 (a) This figure shows the image in Figure 23.27 after a histogram equalization stretch has been applied. Notice how the improved contrast highlights the small variance in reflection in the pastures. (b) The histogram and descriptive statistics of this image are given. The image now uses all the 256 available grey levels and the mean and standard deviation have increased accordingly.

(a)

(b)

Figure 23.29 The image in Figure 23.27 is shown after applying (a) a low-pass filter which suppresses high-frequency variation in the image and (b) a high-pass filter which highlights the high-frequency variation in the image.

which represent wavelengths of high absorption and high reflection, respectively, and is the subject of Box 23.3.

23.7.4 Image classification

When humans look at an image, the brain automatically orders the scene into categories (e.g. trees, buildings and roads) based on the colours, textures and other spatial relationships of pixels in an image. This process is called pattern recognition. However, many applications such as mapping and land-use change detection require a more numerical approach where the results can be stored and manipulated digitally. Therefore we use computers to identify relationships with surrounding pixels based on image tones and patterns (Lillesand and Kiefer, 2000) and to sort patterns into meaningful categories called classes. This process is called image classification. Since a classification usually involves the use of several bands of data, it is often referred to as multispectral classification and it is really the sum of a three-component process: training, defining signatures and the decision rule.

Before the computer can classify an image, it has to be trained to recognize data patterns. This training can be carried out either automatically (unsupervised training) or under the guidance of the user (supervised training). In unsupervised training the computer identifies intrinsic statistical clusters in the data that often do not correspond to typical land-use classes. In supervised training the user defines the clusters of data by choosing groups of pixels in the image that are representative of the desired classes. These pixel groups are called training samples while their

physical locations in the image are the training sites. Supervised training tends to yield more accurate results than unsupervised training since there are many tests that can be applied to the training samples to determine how representative they are of the classes they characterize and how spectrally differentiable they are from the other classes.

The goal of the training process is to provide spectral information about classes in the image from which a representative spectral signature can be developed for each class. If each class in the image has a unique signature, then the class to which each pixel belongs can be easily determined. Unfortunately, this is rarely the case since many land cover types have overlapping signatures. Take the visible spectrum, for example. Trees and pasture both have a high reflection in the green portion of the spectrum and a low reflection in the red. Fortunately, the input into a classification is not limited to the visible portion of the spectrum. In fact, multispectral classification is not even limited to bands of surface reflectance but can include elevation, and statistical and map data layers. Additional layers added to a classification are called ancillary layers and the number of layers used in a classification is referred to as its dimensionality. The results of a classification can be considerably improved by increasing its dimensionality (Tso and Mather, 2001) but there is a trade-off because the inclusion of redundant layers can strain storage and processing resources (Ohanian and Dubes, 1992).

The third stage of the classification process is the decision rule, which is the mathematical algorithm that uses the signatures to assign each pixel to a class. In an unsupervised

(a)

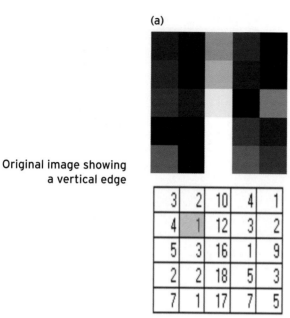

Original image showing
a vertical edge

(b)

-1	2	-1
-1	2	-1
-1	2	-1

Vertical edge
detection kernel

(c)

(-1*3)+(2*2)+(-1*10)+(-1*4)+(2*1)+(-1*12)+(-1*5)+(2*3)+(-1*16) = -38

Since the value is negative the new pixel is assigned a value of 0.

(d)

Image after convolution showing
a highlighted vertical edge

Figure 23.30 This figure provides an example of the application of convolution. For illustration purposes, a vertical edge detection kernel is applied to a simple 5 × 5 image. The original image is shown in (a) and the edge detection kernel is given in (b). In essence, the kernel is applied by centring the kernel window over each pixel in the image and multiplying the overlapping pixels. The value of the new pixel is achieved by summing the results of the nine cells. In this case, a value of 0 is given to all negative results. (c) An example of the calculation performed when the window is centred over the highlighted pixel in (a). (d) The results of the vertical edge detection kernel applied to the whole image. Notice how the vertical edge in the original image has been enhanced in the filtered image.

ENVIRONMENTAL CHANGE

LANDSAT TM FOR MEASURING AND MONITORING THE EARTH'S GREEN VEGETATION

The use of remote sensing data has proved very successful for assessing the type, extent and condition of vegetation over the Earth's surface. Such data have allowed researchers to estimate the area of different crop types, to assess the impact on vegetation from natural or human-made stresses (e.g. pests, fire, disease and pollution) and to delimit boundaries between various types of vegetation cover. It has also been an important tool in assessing and publicizing the extent of desertification (Chapter 16) and deforestation.

Satellite imagery can be enhanced to improve its interpretability through a variety of image processing techniques. One such technique that is applicable to vegetation monitoring is a band ratioing application called the normalized difference vegetation index (NDVI). Like most other vegetative indices, it is calculated as a ratio of measured reflectivity in the red and near-infrared portions of the electromagnetic spectrum. These two spectral bands are chosen because they are most affected by the absorption of chlorophyll in green leafy vegetation and by the density of green vegetation on the surface. Also, in the red and near-infrared bands, the contrast between vegetation and soil is at a maximum.

The NDVI is a simple ratio of the difference between these two bands to their sum and it has been applied to the study of seasonal vegetation variations, leaf area index measurements and tropical deforestation. It can be calculated by:

$$NDVI = \frac{NIR - RED}{NIR + RED} \qquad (23.9)$$

where NIR is the reflectivity in the near-infrared and RED is the reflectivity in the red portion of the spectrum.

For Landsat data the red band is Band 3 and the near-infrared band is Band 4. Figure 23.31 shows the red and near-infrared bands of a Landsat scene of the Wirral Peninsula in north-west England, which is separated from the City of Liverpool by the River Mersey. Applying the NDVI to these bands gives the image in Figure 23.32. Notice how the vegetated parts of the Landsat scene stand out in the NDVI image compared with either the red or infrared bands independently.

(a)

Figure 23.31 A magnified view of (a) Band 3 (0.63-0.69 mm red) and (b) Band 4 (0.76-0.90 mm near-infrared). The image was captured over Liverpool, England, on 19 December 2000 at 12:15 in the afternoon. The city of Liverpool can be clearly seen in the top-centre portion of the image on the north side of the Mersey. (Source: images (a) and (b) courtesy of MIMAS)

BOX 23.3 ➤

➤
(b)

Figure 23.31 *Continued.*

Figure 23.32 Landsat NDVI image which is generated using Landsat Bands 3 and 4 shown in Figure 23.31. Notice how clearly the city of Liverpool stands out against the parks and surrounding farmland as well as the brightness of the marshlands on the north shore of the Dee in the south-west corner of the image. (Source: image courtesy of MIMAS)

BOX 23.3

classification, the three stages (training, signatures and the execution of the decision rule) are usually carried out automatically in one continuous process. Thus, it is the most automated form of image classification and requires minimal input from the user. The most common decision rule used in unsupervised classification is called the Iterative Self-Organizing Data Analysis Technique (ISODATA) and

is based on the spectral distance of each pixel from the cluster means as determined automatically by the computer from the statistics of the image (Tou and Gonzalez, 1974). The number of classes used is defined by the user. Initially, arbitrary means are assigned to each class, and a classification is performed. The means are adjusted to the resulting classes and the process is repeated until the means closely fit

(a)

(b)

(c)

	Trees
	Buildings
	Roads
	Pasture
	Water

200 m

Figure 23.33 Image classification: (a) a subsection of an aerial photograph that was taken over Upper Wharfedale, northern England, in May 1992; (b) the same image classified by an unsupervised classification using five arbitrary classes; (c) a supervised classification applied using five predefined classes: trees, buildings, roads, pasture and water. Notice the confusion in the supervised classification image between walls, roads and buildings owing to their similar reflective characteristics. Also, notice how the unsupervised classification identified the heterogeneity of the pastures and trees while the supervised classification did not. Each image is about 440 m across.

the classes in the data. As a result, the final classes rarely coincide with land-use classes on the ground. Nonetheless, such information is useful when little is known about the surface in question or as a guide for subsequent supervised classification.

Unlike unsupervised classification, supervised classification requires much more input from the user and the decision rule stage is no exception. There are a variety of methods available but these are the three most common methods:

- Minimum distance – assigns the pixel to the class whose spectral mean is closest to that of the pixel.
- Mahalanobis distance – the same as minimum distance except that it takes into account the variability of each class. A highly variable class will have pixels assigned to it with greater spectral distances than one with low variation.
- Maximum likelihood – assigns pixels to classes based on the probability of their membership (Tso and Mather, 2001).

To illustrate the types of classification results that can be expected, Figure 23.33 gives an image that has been classified using an unsupervised classification (Figure 23.33b) and a maximum likelihood supervised classification (Figure 23.33c), both of which used five classes. Notice the difference between the results for the two classifications. For Figure 23.33(b) the intrinsic statistical divisions in the image can be seen whereas Figure 23.33(c) shows land cover divisions.

Reflective questions

➤ Which image processing routine would you apply if you wanted to highlight the location of roads, hedges or fences?

➤ Which image processing routine would you apply to remove the effects of atmospheric scattering?

23.8 Summary

This chapter has introduced remote sensing in the context of its application to physical geography and environmental change. There is not one major component of physical geography covered by the chapters in this textbook that has not benefited from remote sensing. Digital images are made up of pixels, each of which has a digital number. This number is a code for the colour of that pixel. Important image characteristics include orientation, scale and resolution as well as content characteristics such as tone, texture, shape, size and pattern. These characteristics determine the quality and type of information that can be gained from the image. Photogrammetry can be used to produce digital elevation models and digital terrain models from photographs and if repeated images are taken over time then environmental change such as landscape erosion can be measured remotely.

The properties of electromagnetic radiation are fundamental to remote sensing because it is not only visible light that can be detected and recorded by sensors but the full range of the spectrum. Microwave sensors and ranging sensors such as radar, sonar and lidar provide additional information on environmental change. Reflection, transmission, absorption and scattering processes are important in determining the type and amount of energy received by sensors. Every object reflects radiation in different ways and thus has a characteristic spectral signature. These signatures can be detected by passive and active sensors. Active sensors use their own source of radiation and measure the quantity of emitted radiation reflected back. Scanners can be used to detect radiation across narrow wavelength bands or over the full spectrum. A range of spaceborne, airborne and ground scanners are used to create a wealth of spatial and temporal information on environmental change.

➤

> Often it is not sufficient to use raw remote sensing images and it is necessary to process these images in some way. This processing may involve rectification which allows us to compensate for any image distortions or degradations that may have resulted during image acquisition. Image enhancement attempts to redisplay image information in a way that better represents image characteristics or features of interest. Methods include contrast stretching, spatial filtering and band ratioing. Computer classification of image features is also an important tool and aids automation of remote sensing data analysis.

Further reading

Kasser, M. and Egels, Y. (2002) *Digital photogrammetry.* Taylor & Francis, London.
This is a good photogrammetry reference that specifically addresses issues of digital photogrammetry.

Lillesand, T.M., Kiefer, R.W. and Chipman, J. (2003) *Remote sensing and image interpretation,* 5th edition. John Wiley & Sons, Toronto.
This well-respected reference provides an in-depth look at the principles of remote sensing. It also explores the links between remote sensing and geographical information systems. The Appendix on p. 744 provides an extensive list of remote sensing data sources with web addresses.

Russ, J.C. (2001) *The image processing handbook.* CRC Press, Boca Raton, FL.
This is a well-respected user manual of digital image processing. It presents a high-quality discussion of the techniques of image processing and when/why they are applicable. It is up to date and draws on examples from a variety of real-world applications such as medicine, microscopy and remote sensing.

Wolf, P.R. and Dewitt, B.A. (2000) *Elements of photogrammetry with applications in GIS.* McGraw-Hill, New York.
This new edition of the principal photogrammetry reference provides a thorough and modern perspective on photogrammetry in the twenty-first century. It includes sections on principles of photography, coordinate systems, GIS, topographic mapping, digital image processing and project planning.

Web resources

American Society of Photogrammetry and Remote Sensing
http://www.asprs.org/
This is the American organization dedicated to advancing knowledge and improving understanding of mapping sciences and to promote the responsible applications of photogrammetry, remote sensing, geographic information systems (GIS) and supporting technologies. The website provides a variety of remote sensing resources.

Canada's National Air Photo Library
http://airphotos.nrcan.gc.ca/index_e.php
Canada's national archive of over 6 million aerial photographs covering all of Canada dating back 70 years can be accessed here.

Canadian Centre for Remote Sensing
http://www.ccrs.nrcan.gc.ca/
This is the Canadian remote sensing agency website that provides learning resources, image samples and research and development information for sustainable development and environmental protection.

International Society for Photogrammetry and Remote Sensing
http://www.isprs.org/
This is the site of the international organization devoted to the development of international cooperation for the advancement of knowledge, research, development, education and training in the photogrammetry, remote sensing and spatial information sciences. The site provides a variety of resources, access to publications as well as career postings.

LANDSAT Program
http://landsat.usgs.gov/index.php
This is the website of the Landsat program at the United States Geological Survey (USGS).

NASA Earth Observatory
http://earthobservatory.nasa.gov/
An impressive database of a variety of Earth observation images from a variety of sensors is given here. The site provides interesting write-ups on all the archived images.

NASA LANDSAT 7

http://landsat.gsfc.nasa.gov/

This site is specifically devoted to the Landsat 7 satellite and its applications.

National Oceanic and Atmospheric Administration (NOAA): Remote Sensing

http://www.csc.noaa.gov/crs/rs_apps/

These are the US National Oceanic and Atmospheric Administration web pages on remote sensing within the context of predicting changes in the Earth's environment and to conserve and manage coastal and marine resources. Links to other useful sites can also be found here.

Natural Environmental Research Council Airborne Remote Sensing Facility

http://arsf.nerc.ac.uk/

The British remote sensing facility web page offers information about a variety of instruments and applications from a practical perspective.

Ordnance Survey

http://www.ordsvy.gov.uk/

This is Britain's national mapping agency website that provides information on British map and aerial photography products.

Remote Sensing and Photogrammetry Society

http://www.rspsoc.org/

This is a British organization devoted to coordinating and promoting activities in remote sensing and photogrammetry. The website provides access to publications and information on remote sensing opportunities.

SPOT Programme

http://www.spotimage.fr/home/

This is the website of the French remote sensing satellite programme.

University of Cambridge Unit for Landscape Modelling

http://www.uflm.cam.ac.uk/

This is the site of the Cambridge-based group that uses and develops remote sensing techniques for landscape ecology applications. It is also a provider of aerial photography and lidar images.

 Visit our website at **www.pearsoned.co.uk/holden** for further questions, annotated weblinks, interactive models and video-clips showing physical processes in action.

CHAPTER 24

Managing environmental change

Adrian T. McDonald

School of Geography, University of Leeds

Learning objectives

After reading this chapter you should be able to:

➤ understand environmental change and define and categorize such change

➤ appreciate the dependence of environmental management on good physical geography and describe how physical geography can be applied

➤ understand basic environmental management tools and management processes

➤ evaluate problems and processes involved in implementation of environmental interventions

24.1 Introduction

Even five years ago the management of the environment was viewed as important by only a small number of resource-dependent companies. The B & Q do-it-yourself stores stood out from the pack as a supplier of certified timber (from well-managed forests to internationally agreed standards), for example. Today, however, most of our biggest companies are addressing environmental issues. BP, one of the world's major oil companies, is looking 'Beyond Petroleum'. Tesco (a major grocery store in Europe) has three key aims of:

- helping our customers by making green choices easier and more affordable;
- setting an example by measuring and making big cuts in Tesco's greenhouse gas emissions around the world;
- working with others to develop new low-carbon technology throughout the supply chain.

It has made a number of specific commitments, for example:

- to use a 50 : 50 bio-diesel mix in its fleet;
- to add 5% bio-ethanol to its petrol mix in 300 stations;
- to halve the price of energy-efficient bulbs;
- to label air-freighted food and to reduce this to 1% of products.

Now compare this situation with that of 2004 when the first edition of this book was written and when it expressed concern that environmental management was all talk and no action. Perhaps this was driven by pessimism derived from the outcomes of the major Johannesburg conference on climate change in 2003 which seemed to have very few actions resulting from it. However, the growing strength of a market-led approach to environmental performance seems to have created optimism so that progress may be made. However, activists such as George Monbiot continue to argue that the rate and scale of a changed lifestyle are too slow to make a difference and so we still need to imagine a world without life. This might be a world in which the biosphere had been silenced by excessive and careless use of bioaccumulating pesticides (see Chapters 3, 7 and 8). This was the future offered by the courageous ecologist and author Rachel Carson when she published, in 1962, her landmark book *Silent spring* in an effort to communicate the dangers of DDT (see also Chapter 22). This was a pesticide sprayed on crops and used around the world from the 1930s. The problem with this pesticide is that it contained toxins that stayed in the food chain and accumulated in soils, plants and animals. These toxins became more and more dangerous as they built up over time. Therefore DDT was banned over most of the world in the late 1970s.

Silent spring was probably the warning that finally awoke the world to the significance of the environment and the degradation that was in progress. People began to realize that humans were degrading the landscape and damaging the environment. Surprisingly, the subsequent increase in environmental awareness, investigation and popular reporting has been slow to impact on our thinking. When environmental degradation first began, with such actions as forest clearance for timber and agricultural activities, it was at a small scale and the overall effects of the actions were not collated or understood. That cannot be the case today. We have a global picture of change (see Chapter 23) and some appreciation of the ramifications of such changes (see Chapter 21). We also have the tools and management skills with which to start to cope or respond and it is the purpose of this chapter to describe these management tools.

A fundamental question must be answered before we continue in this chapter. Why, in a book about physical geography, is there a chapter on environmental management? We study physical geography for two basic reasons. The first relates to intellectual curiosity, which drives us to understand better our world in all its diversity and the processes through which the landscapes in which we live have become as they are. The second relates to the effect we have on our environment by living in it and drawing natural resources from it. To manage the way in which we impact on the environment, to foresee and perhaps to forestall the consequences of our actions require that we understand the processes which govern our environment. After all, Rachel Carson was first and foremost a practising scientist who understood the processes operating in the environment. Society needs good physical geographers so that process-based science can provide the information necessary for environmental management. Table 24.1 provides, for each of the other chapters in this book, just two examples of environmental

management issues that were discussed within those chapters. It is clear that all areas of physical geography have relevant application to environmental management.

Arguably, the management of environmental change, species threatening as it is, is one of the most important aspects of science and management today. Managing and promoting change in the environment are full of uncertainties. Yet action needs to be taken, often at an early stage before the picture is clear. This is known as the **precautionary principle**. This is action taken as a precaution before we understand the outcome, but being aware that some aspects of the possible outcomes cannot be tolerated. However, taking such action also involves risk, both a physical risk of the wrong outcome developing and an economic risk of taking expensive action that is not required. The scientific research discussed in the earlier chapters of this book provides information to help reduce uncertainty about how the environment will respond to a management action. It is, however, the management tools themselves that will be discussed in this chapter. Environmental management is the practical application of physical geography. It has become a discipline in its own right. This chapter seeks to show how physical geography informs environmental management and to outline some of the key techniques used to tailor understanding of physical geography to a form that is useful for management.

24.2 Understanding environmental change

The previous four chapters of this book outline issues surrounding environmental change over different timescales. Chapter 20 discusses environmental changes over the past 2.4 million years, whereas Chapter 21 provides information on more recent human-induced environmental change and related problems. Chapter 22 describes interactions between vegetation and climate change while Chapter 23 discusses contemporary techniques for monitoring present global and local environmental change. The following section, however, provides more general ideas about the characteristics of change of relevance to environmental management.

24.2.1 Characteristics of change

In order to manage environmental change we need to have some understanding of change itself, both the 'upstream' drivers and the 'downstream' consequences. This more holistic approach is gathered in the DPSIR process. This is

Table 24.1 Two example environmental management issues from each of the other chapters in this textbook

Chapter	Subject	Two example management issues
1	Approaching physical geography	Channel avulsion, rock weathering on building materials
2	Earth geology and tectonics	Earthquakes, volcanic eruptions
3	Oceans	Overfishing, pollution
4	Atmospheric processes	Drought, aerosols
5	Global weather and climate	Hurricanes, monsoons
6	Regional and local climates	Haze, shelter belts
7	Soils	Soil erosion, soil pollution
8	The biosphere	Fire, deforestation
9	Biogeography	Alien introductions, recreation needs
10	Ecological processes	Conservation, bioaccumulation
11	Erosion, weathering, hillslopes	Mass movements, erosion/landscape degradation
12	Sediments	Dams, mining
13	Hydrology	Flooding, salinization
14	Rivers	Bank erosion, biodiversity
15	Solutes	Water pollution/contamination, lake acidification
16	Drylands	Desertification, water resources
17	Coasts	Coastal erosion, sea-level rise
18	Glaciers and ice sheets	Water resources and energy, skiing/tourism
19	Periglaciation	Infrastructure stability, gas hydrates
20	Quaternary change	Greenhouse gases, global climate model predictions
21	Climate change	Mitigation, pollution
22	Vegetation change	Biodiversity, food supply
23	Remote sensing	Land-use change, flooding

a framework that assumes a cause and effect relationship between the various interrelated parts (spatial, temporal and component) of social, economic and environmental systems:

Driving forces of environmental change (e.g. greater agricultural production)

Pressures on the environment (e.g. release of chemicals to the atmosphere)

State of the environment (e.g. urban air quality)

Impacts on population, economy, ecosystems (e.g. health impacts, respiratory problems)

Response of the society (e.g. clean air acts, vehicle emission control)

DPSIR is extremely flexible and although created with developed country economies and systems in mind, it has already proven effective in rural environmental management in northern Ghana. Box 24.1 summarizes this application.

Perhaps the most important aspect of change concerns the relationship between factors that promote change and the nature of the change. The nature of change is summarized in Table 24.2. If the change is irreversible or quick and unstable, for example, it is a much more problematic type of change than one that is reversible, slow and stable. Change can be considered to be driven by a factor or group of factors. For example, increased acid rain may be driven by the sulphur released in the burning of coal for energy

CASE STUDIES

DPSIR APPLICATION TO DESERTIFICATION IN NORTHERN GHANA

The DPSIR framework is a five-partite assessment indicator formulated by the European Environmental Agency with the legal basis of the European Union Environmental Policy Acts 95, 174, 175 and 176 of the consolidated version of the treaty of the European Union. The framework was carefully designed to recognize the complexity of environment-related problems and provides a means of analysing them. Each indicator conveys its own distinctive meaning and application. *Driving forces* are social processes that cause either the increase or mitigation of pressures on the environment. *Pressures* are represented by direct human activities on the environment beyond its carrying capacity. *State* relates to the current state and trends of the environment and determines the extent and magnitude of degradation. *Impacts* are direct human consequences resulting from the change on the environment and the *Responses* are what the affected individuals perceive needs to be done to improve the quality of the environment.

The results of the adopted DPSIR framework to desertification in northern Ghana (Figure 24.1) indicated that most areas originally covered with savanna woodland have recently been replaced by a barren environment with a potential threat of desertification. Economic policies transformation, demographic factors, changing tenure system, economic hardships (poverty) and behavioural and attitudinal changes are some of the social processes that drive people in northern Ghana to engage in various activities such as small-scale mining that bring about severe deterioration of the environment. The observed impacts included threats of desertification through severe savanna vegetation loss, declining crop yield, cross-cultural tensions, health risk and reduction in living standards. Individual responses for a better environment included a reduction in the driving forces, effective environmental management practices and environmental awareness campaigns to be considered in the Ghanaian environmental policy formulation. Lessons learnt through the adoption of the DPSIR framework in northern Ghana's threat of desertification was that it allowed the interconnection of the various indicators to be well understood and applied to policy formulation. The DPSIR framework is therefore seen as an effective means of organizing complex environmental information for policy formulation.

The information in this box was kindly provided by Isaac Agyemang, University of Leeds.

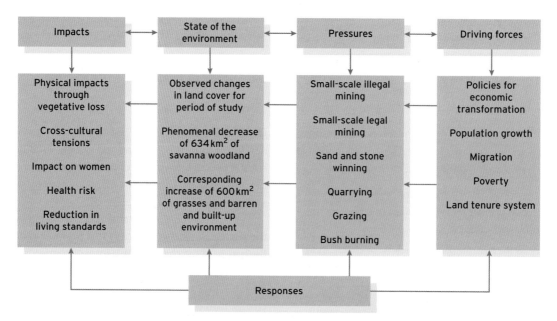

Figure 24.1 DPSIR framework analysis results from northern Ghana. (Source: courtesy of Isaac Agyemang)

BOX 24.1

Table 24.2 The nature of change. Change is not simple. If a circumstance or situation involves all the attributes listed as 'good' then it is certainly simpler than a situation involving some or all of the 'problematic' attributes. It is a good management exercise to consider where, in the spectrum of change, a particular case lies

Good attributes	Problematic attributes
Stable	Unstable
Reversible	Irreversible
Slow transformation	Rapid transformation
Limited	Unlimited
Within tolerances	Beyond tolerances
Linear	Non-linear
Small scale	Global

(see Chapters 7 and 21). However, the nature of the relationship between coal burning and acid rain and the resulting management implications will be constrained by the mix of good and problematic attributes of the relationship. This is generalized in Figure 24.1 which shows the range of relationships that exist. These can be reversible or irreversible, simple (linear) or complex (non-linear) or even contain hysteresis (see Chapter 15). If we assumed that a relationship was linear and reversible then we might put in place a management strategy to reverse this change. However, it may be the case that the change was not linear or reversible and so this management strategy might make the problem worse.

A good example of this comes from many peatlands which have been drained (Figure 24.2). The digging of ditches into a peatland often causes the soil to dry out and to crack (the change) as shown in Figure 24.2. The ditches themselves often lead to faster water movement when it rains, causing more flooding, and also to more land erosion because of the channelling of flow along the unvegetated drains (associated change). The management solution is to block up the drains by placing dams into them. This fills up the drains with water when it rains, rather than letting water run along their floors. However, in many cases this sends more water through the newly created cracks on the floors and sides of the drains, which results in more erosion of the soil below the surface (Holden *et al.*, 2006). The cracking is an irreversible change because when peat dries out it changes its properties so that it can no longer hold as much water (Holden and Burt, 2002b). Therefore the management

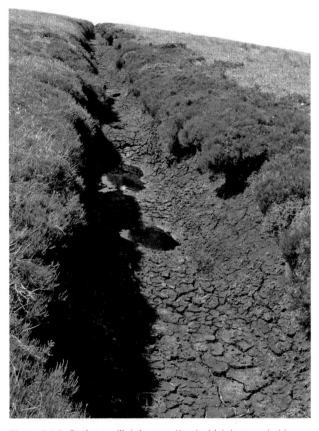

Figure 24.2 Drainage ditch in a peatland which has eroded to become a wide channel. The exposure of the bare peat to the Sun has caused it to dry out and crack. (Source: photo courtesy of Alona Armstrong)

solution may result in even faster change in the landscape by encouraging more turbulent flow through soil cracks causing erosion. So the simple definition of change, a transition from one state to another, can hide the risks, uncertainty and complexity of change and it is therefore important to be aware of the attributes of change as described in Table 24.2 and Figure 24.3.

24.2.2 Rate of change

The rate of onset of change is a vital consideration because, if slow, (i) it may offer species the opportunity to adapt or shift their range, (ii) it may provide the environmental manager with the time required to devise and implement a remediation strategy, and (iii) it may offer the politicians sufficient time to perceive the potential damage and provide

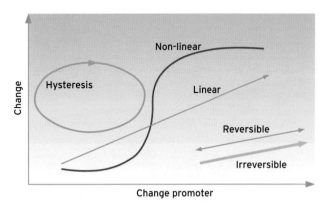

Figure 24.3 Some characteristics of change represented graphically.

the resources needed by the manager. Of course in this last case, slow onset may also be a disadvantage in that it might not develop to be a serious problem within the limited time horizon that some politicians possess.

Non-linearity of, and rapid alteration in, rate of onset of change is a further, and in many ways frightening, issue. If we consider Figure 24.4 we can see that it consists of a surface which slowly curves from one stable location (look at either the upper or lower part of the surface). Forecasts based on a statistical assessment of the trend may indicate a slow onset change (relatively flat surface) allowing plenty of time for a considered response or perhaps an equilibrium

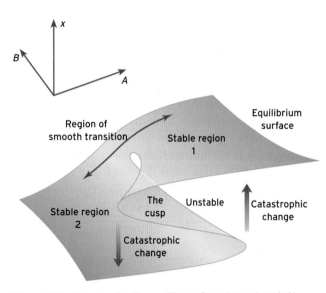

Figure 24.4 Catastrophe theory. The surface shown here is in three dimensions. A and B are the changes in environmental conditions and x is the change in the response. In the real environment there are many more than two variables but for simplicity just two have been drawn. Catastrophe occurs at the two arrows, the region of instability at the cusp. Rapid change occurs which is not reversed by a reversal in system conditions.

situation in which a large change in the driver causes little response in the factor of concern. Unfortunately, the next section of the surface, the 'cusp', shows a very rapid response to a new stable state. In other words, a very slow change suddenly becomes a very rapid change as if the system has reached a threshold and it suddenly jumps out of one stable state. This figure summarizes **catastrophe theory** (Thom, 1968). A sudden catastrophic change occurs and then the system reaches a new stable (or equilibrium; see Chapter 1) position. So the theory describes cases when we have a sudden jump from one stable mode of operation to another, but where any changes at first may be very slow or hardly noticeable.

A good example of such a problem is provided by the thermohaline circulation system discussed in Chapters 3 and 20. This is the deep ocean circulation system that is today a very strong system for transporting heat away from the equator towards the poles. It is a strong system because of the saltiness of the water in certain sensitive locations. This salty water is dense and so sinks, thereby forcing water at the bottom of the oceans up in a large circulation system. However, with global warming, more of the world's glaciers are melting and so rivers flowing into oceans such as the North Atlantic are producing more freshwater (Dickson *et al.*, 2002). As river water is not very salty, these inputs may prevent the deep sinking that previously took place. This may shut down the ocean circulation system (Paillard, 2001). This change may be very sudden, and there is evidence from the past that the climate system has suddenly jumped from one stable warm mode to a stable very cold mode because of changes in ocean circulation (Broecker and Denton, 1990). The immediate result would be a dramatic cooling of Europe by several degrees and a consequent growth of glaciers. These glaciers would increase albedo and thereby cool the planet even further and we would enter a glacial period. Thus, while we are in a stable warm period we may have a very slow change (slow warming) that eventually results in a rapid and catastrophic shutting down of the ocean circulation system which then sends the system into a stable cold phase. This of course would be irreversible on human timescales. Thus the concept of catastrophe theory, of a non-linear switch between stable states without the prospect of a reverse, is real and possible.

24.2.3 Environmental tolerance

Many definitions of the environment are somewhat limited in that they focus primarily on an ecological definition of the environment relating to organisms. However, environmental change relates to Earth surface–ocean–atmosphere–biosphere environments at any point on the global to local scale. The environmental manager needs to encompass the ecological perspective but must also include the physical characteristics and processes of the Earth's surface, the oceans and the atmosphere. At a global scale this perspective merges with the concept of **Gaia** (Box 24.2) at least to the extent of recognizing highly complex interactions of all elements of the Earth surface and that the response to a stimulus will be complex and unpredictable, almost like a giant organism (Lovelock, 1979).

An environmental tolerance is a valuable concept within the ecological community. Each species has a range of environmental conditions within which it can operate. This might, for example, be a temperature range such as $-2°C$ to $+30°C$. In reality all species will have a series of 'ranges' within which they can survive, controlled for example by soil moisture content, oxygen saturation, temperature and pH. When conditions fall outside any *one* of these values for a critical period of time (seconds or hours depending on the species and determinant) the organism cannot survive. That just *one* of a series of environmental factors is crucial at a particular time is known as the **law of limiting factors**. The same law or concept appears in other areas of management under titles such as 'bottleneck theory' and 'threshold theory'. The generalized concept of environmental tolerance is measured and made specific for an individual determinant through measures such as the $LD_{50/24}$ measure, which is the dose of a determinant which would be lethal to 50%

ENVIRONMENTAL CHANGE

GAIA HYPOTHESIS

James Lovelock (1979) presented his Gaia theory which suggested that the Earth is a superorganism. It was argued that the maintenance of conditions suitable for life and particularly the maintenance of the Earth's atmospheric chemistry was derived from the self-regulation and feedback mechanisms that exist on the planet. The biological systems and the landscape-ocean-atmosphere systems are highly coupled and operate as a single living entity. A justification for this hypothesis is that the composition of the Earth's atmosphere would be radically different if there were not life on the surface. Without flora and fauna, the atmosphere would be mostly carbon dioxide, with very little nitrogen or oxygen. However, with the addition of life, in combination with the Earth's other subsystems, the Earth's various aspects constitute a feedback system that seeks an optimal environment to sustain life. In its most basic form, the Earth acts to regulate flows of energy and cycling of materials. The unlimited input of energy from the Sun is captured by the Earth as heat or photosynthetic processes, and returned to space as long-wave radiation. At the same time, the material possessions of the Earth are limited. Thus, while energy flows through the Earth (Sun to Earth to space), matter is recycled within the Earth.

The idea of the Earth acting as a single system as put forth in the Gaia hypothesis has stimulated a new awareness of the connectedness of all things on our planet and the impact that humans have on global processes. It means we cannot think of separate components or parts of the Earth; there are too many interconnections for this to be the case. Therefore what humans do on one part of the planet can affect other parts of the planet. The most difficult part of this idea is how to determine whether these effects are positive or negative. If the Earth is indeed self-regulating, then it will adjust to the impacts of humans. However, these adjustments may act to exclude humans, much as the introduction of oxygen into the atmosphere by **photosynthetic bacteria** millions of years ago acted to exclude **anaerobic bacteria**. This is the crux of the Gaia hypothesis.

Lovelock argued that the Earth is not 'alive' in the normal sense but that seeing the Earth as a living organism is just a convenient way of organizing facts about the Earth. He argued that a tree processes sunlight and water to grow and change, but this happens so imperceptibly that in practical terms it often appears unchanged, and wondered if the Earth might have the same characteristics. Many people will reject Gaia as one step too far. However, if we accept that Earth is an integration between all elements of a system at all scales and if we seek to understand how all components of the system relate to each other and avoid excluding any factors, then we start to approach Gaia.

BOX 24.2

of the organisms under consideration within 24 h
(Lawrence *et al.*, 1998).

An ecosystem contains a set of species which have individual tolerances but there is also a tolerance of the
ecosystem as a whole. The tolerance of the ecosystem will be
greater than that of the individual species that form it. Thus
the scale of environmental change will have to be greater to
destroy ecosystems than to destroy an individual species. Of
course, ecosystems may be more degraded if the individual
species that is destroyed is a keystone species (see Chapters 10
and 22). Figure 24.5 shows that the tolerance range of the
ecosystem (the range beyond which none of the ecosystem
species survive) is considerably greater than that of individual component species (McDonald, 2000). Under
'normal' conditions nine of the species survive, although
species number 8 is at the extreme of its range. About half
the species will survive changes to unusual conditions but
note that the species that survive an upward unusual change
are different from those that survive a downward unusual
change. Only two species are shown to survive swings to
either extreme. This is one reason why the promotion of
diversity is important. Diverse ecosystems tend to be more
resilient (Bradbury, 1998). Box 24.3 provides a case study of

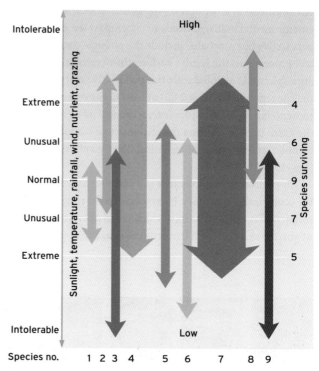

Figure 24.5 Species and ecosystem tolerance. (Source: after McDonald, 2000)

CASE STUDIES

MANAGING THE CONSEQUENCES OF CLIMATE CHANGE - THE PINE BEETLE IN PRINCE GEORGE, BRITISH COLUMBIA

Prince George lies in the geographical centre of British Columbia. It is a small city on the banks of the Upper Fraser River and is the administrative centre of northern British Columbia. Its origins are as a logging town and it is surrounded by coniferous forest for hundreds of kilometres in every direction. What is strange about Prince George is that it appears from the trees as if it is autumn all year. The trees are brown and light filters

through to the forest floor. However, these are coniferous trees that should stay green all year round. This is wholesale destruction by the pine beetle (Figure 24.6) in epidemic proportions. Seven million trees are infected (Figure 24.7) with areas where the tree kill is 100%.

So there are three (generic) questions that any manager will pose in such a situation:

1. What is the cause?
2. What are the consequences?
3. What can be done?

Causes

The beetle is a natural grazer and occupier of conifers. Normally it is in

balance with the growth of the vegetation. However, beetle numbers are increasing substantially and damaging the tree cover. There appear to be two basic causes of the spread of this beetle. The first relates to climate change. The occurrence of long-duration, severe cold periods in winter has declined. Temperatures reach only −20°C for a few days at a time whereas in the past such temperatures were exceeded for weeks at a time. Beetles are therefore surviving the winter in much larger numbers than before. An added factor may be the drier years that have occurred since 1995 which may have weakened the mature pines, making

BOX 24.3 ➤

Figure 24.6 Pine trees damaged by the pine beetle. (Source: Mickey Gibson/Animals Animals/Photolibrary.com)

them more susceptible to pine beetle attack. As is so often the case it is the larvae and the fungal infections carried by the adult that destroys the trees – not the beetle itself.

The second cause is more directly human induced. Fire is an integral part of the ecological processes that operate in the forests of British Columbia. The fires occur naturally from lightning strikes and are usually relatively low-temperature burns that reduce the numbers of beetles in an area. However, as forests become more important commercially, as fires start to arise from deliberate or care-less human actions and as more homes and communities develop in the forests, there has been a growing pressure to provide a forest fire-fighting service. This service is now

extremely effective. Forest fire-fighting is coordinated between the United States and Canada, there are rapid response times and effective action. The result is that there are fewer fires and fewer beetle controls.

Consequences

There are many consequences, some direct and tangible, some indirect and intangible:

- The immediate result is that there is a growing area of dead forest.
- Timber not extracted in a five year period will have limited value.
- A rapid increase in timber volume reaching market has an influence on price.

- Replanting with pine is a high-risk option.
- Loss of leaf cover alters snow capture, trunk climate and snowmelt.
- Loss of cover means direct rainfall to the forest floor, increasing erosion.
- Loss of habitat and loss of biodiversity.
- Raised fire risk from dead trees (and accumulated unburnt brash).
- Loss of forest as a cultural resource to first nations.

Intervention

Many companies offer advice on detecting the pine beetle and treating the infection at a suburban or

BOX 24.3 ➤

Figure 24.7 A map showing the total proportion of pine trees killed by pine beetle between 1999 to 2005 in British Columbia. (Source: Courtesy of BC Ministry of Environment)

Pine killed
100%
0.5%

300 km

small-plot scale. However, the economic treatment of very large swathes of landscape by chemicals is unrealistic. The options are:

- species change occurring naturally;

- the exploitation of the trees at a younger age (as it is a disease of mature trees);

- the relaxation of the burning control regimes to promote large areas of younger trees;

- harvesting of the damaged trees (luckily wood prices are high at the moment) followed by, legally required, replanting.

Alternatively we may opt to accept this as a natural occurrence and so accept:

- changed flow regimes in the rivers;

- more flooding downstream;

- more erosion;

- loss of habitat;

- loss of employment;

- increased risk of wildfires.

An excellent video of the pine beetle problem can be found on the British Columbia government website **http://www.for.gov.bc.ca/hfp/ mountain_pine_beetle/video.htm** and the substantial costs of replanting and community fire protection are given in **http://www2.news.gov.bc.ca/ news_releases_2005-2009/ 20050TP0108-000832.htm**

BOX 24.3

managing beetle infestations around the Upper Fraser River, British Columbia, which have come about due to climate change and change in local forest management.

24.2.4 The 'duty' and need to manage change

Some ecologists take comfort in ecosystem resilience to change. However, the environmental manager can take much less comfort because economic and political imperatives always appear to have the highest priorities. In other words, money and power often come before environmental management issues. The human species is nearly always placed at the top in any environmental management decision. However, there are both ancient and modern values that do not place the human species higher than others in the biosphere. These ideas form part of 'environmental ethics' and suggest that we might have a duty to care for other species and the Earth itself.

Environmental ethics is the idea that different people might have different viewpoints (e.g. religious, cultural) on environmental management. One viewpoint is that we should treat other species and environmental processes as equal to ourselves. It is a philosophy that places humans on an equal level with other species. Therefore, advocates of environmental ethics would suggest that we should think carefully about managing the environment for the environment's sake and not just for our own selfish uses of the environment.

Environmental ethics is not a new concept but it has been dismissed in industrialized western societies over the past few centuries. These societies have tended to have a technocentric perspective on the environment. This is a view that suggests that with technological developments we can 'conquer' the environment. In western society today an appreciation of the importance of environmental ethics has awakened. Many people around the world believe that we have a duty to look after and care for the environment. It is therefore vital that the environmental manager appreciates the very different values and perspectives that can be held by different communities (Hargrove, 1989).

When managing the environment we must carefully think about the actions of others and not just about our own actions. A good example of this is what has been termed the **tragedy of the commons** (Harding, 1968). This involves a field that anyone can use (common land). It is to be expected that each farmer will try to keep as many cattle as possible on the field. However, the logic of the commons will bring tragedy. This is because, as a rational being, each farmer seeks to maximize their gain. The farmer will ask

'what is the benefit or disadvantage of adding one more animal to my herd?' However, this action has one negative and one positive component. The positive component is a function of the increment of one animal. Since the farmer receives all the proceeds from the sale of the additional animal, the positive utility is nearly +1. However, the negative component is a function of the additional overgrazing created by one more animal. Since, however, the effects of overgrazing are shared by all the farmers using the field, the negative utility for any particular decision-making farmer is only a small fraction of −1.

Adding together the component parts, the rational farmer will conclude that the only sensible course to pursue is to add another animal to the farmer's herd on the field. This will be followed by the addition of another and maybe many more. However, this is also the conclusion reached by each and every rational farmer sharing a field. Therein is the tragedy. Each farmer exists within a system that encourages them to increase their herd without limit but in a world with limited resources. The field will be massively overgrazed and all of the vegetation will be removed. Therefore none of the animals will survive and the farmers will be ruined. So a society that believes in freedom of the commons will be a society that brings its own downfall.

This analogy, of course, is the same as allowing all people the freedom to pump pollutants into the atmosphere or oceans when and where they want and in whatever quantity they want as if the atmosphere and oceans were unlimited resources. It is also the same as individuals thinking 'what will it matter if I throw just one more plastic carton in the bin; surely one more will not do any harm?' The analogy fits environmental problems throughout the world. We have finite resources and we must therefore manage those resources. Allowing people freedom to do what they want may bring only degradation. Management is therefore required in order to identify the range of feasible alternative outcomes in response to changing environmental conditions, the identification of the costs, risks and uncertainties associated with pursuing each alternative, and the organization and mediation of the resources needed to implement a chosen alternative (Lawrence et al., 1998).

24.2.5 Types of change to be managed

There are broadly three types of environmental change that can be 'managed':

1. Responding to natural environmental change.
2. Controlling anthropogenic environmental change.
3. Implementing local change.

Table 24.3 Some east coast flooding responses in England

Flood response	Advantages	Disadvantages
Sea walls and similar hard engineering responses	Rapid implementation	Ugly
	Defined action	Promotes flood potential
	Retain existing use	Costly
Beach maintenance (adding sand)	Rapid implementation	Regular action needed
	Retains scenic value	Questionable sustainability
Land-use	Long-term viability	Change in use
	Recognizes risk	Loss of existing business
Insurance	No capital cost	Availability?
	Not a government expenditure	Cost
	Individual responsibility	
Managed retreat (allowing the sea to advance inland)	A sustainable defence	Change in use
	Provides biodiversity	Outcome not easy to predict
	Reduced expenditure?	
No active response	No capital cost	Lack of coordination
	Individual responsibility	Lack of expertise

We do not have the ability to alter or reverse natural environmental change and certainly not in a sustainable manner (see below) and not at any scale beyond the local. For example, the isostatic rebound of the land masses in the north of Britain following the thousands of years under the weight of the ice and the resultant settling of the southern half of Britain as part of that response has increased coastal flood risk in southern Britain and will continue to do so. We cannot raise the south-east of England and so all our responses are forms of '**coping strategies**' (see Chapter 17). Table 24.3 lists some of the coping strategies to flooding on the east coast of England and lists the advantages and disadvantages of each option. The lists of options, advantages and disadvantages are by no means complete but the table provides examples to show the sorts of responses we may be able to perform.

When considering the management of environmental change we must consider how long we want to manage the system for. An action taken today to cope only with the current situation may not be adequate in the long term as, for example, sea levels continue to rise. This necessity for long-term management is often subsumed within the term

'**sustainable development**' incorporating a requirement to consider not just fair outcomes for the present population but outcomes that are satisfactory for future generations: 'Sustainable development is development that meets the needs of the present without compromising the ability of future generations to meet their own needs' (World Commission on Environment and Development, 1987). Sustainable development also incorporates the broader vision of the social and economic environment.

Anthropogenic change, or change resulting from human activity, is usually more rapid than similar processes operating through natural phenomena. Examples of anthropogenic change are widespread (acid rain, overfishing, deforestation, river diversion, and so on) and further cases involve a blend of anthropogenic and natural change such as desertification. In most cases the change is one directional rather than cyclical, with broadly irreversible consequences. For the manager there are two main sets of options: (i) using the coping strategies available to manage natural change; and (ii) limiting the drivers that are creating the environmental change. Of course to limit the drivers assumes that the process is reversible or at least reversible at the stage to which the situation has reached. It is also assumed, more importantly, that the political, legal and economic will is present to commit to curtailing the activity that is giving rise to the environmental change. This 'will' is only forthcoming if the environmental scientist can prove clearly and simply that the relationship between the initiating activity and the consequent change exists and that the change is very serious and must be addressed. Thus the disciplines of the physical geographer and the environmental manager must never become competitive. Each must be aware of the contribution of the other.

At a local scale we now have the power to make changes to the environment. For example, for a hundred years or more (thousands in the case of the Rivers Amu Darya, Syr Darya, Tigris, Euphrates and Nile at least) we have 'trained' rivers to follow the course most convenient to our developing use of land. However, such power has often not been accompanied by an equal measure of knowledge and so, as discussed in Chapter 14, river channelization techniques have often made flood problems worse by sending the water downstream in larger volumes over shorter periods of time (e.g. by shortening the length of a river by removing meander bends). They have also resulted in a series of other problems related to geomorphological processes. For example, straightening a section of river causes it to erode its bed and banks in the straightened section because the average slope of the river along that section has increased (shorter river channel distance over the same fall in altitude) and so stream power is greater. This undermines the river channel engineering structures themselves and also adds more sediment to the downstream part of the river. This can cause the downstream river channel to infill with sediment and so its capacity to carry water is reduced and it will overflow more easily, causing increased flooding. It is therefore important to improve and incorporate process understanding into environmental management plans.

Reflective questions

➤ What is the DPSIR approach?

➤ Why are diverse ecosystems more resilient than individual species?

➤ What is catastrophe theory?

➤ Can you think of an example of human action that is analogous to the tragedy of the commons?

24.3 Tools for management

The needs of the environmental manager are simple. The manager needs to be able to identify the nature, scale and timing of the impacts of any actions. This can be the action of others or humankind in general. The manager must be able to demonstrate that remedial action is required either to pre-empt or reverse other actions or to create change for which there is demonstrably no adverse reaction. Hence the first need of managers is to be able to forecast. Statistical and numerical modelling can help make predictions about environmental change and the impacts of management strategies. Such modelling tools are discussed in Chapter 1 and elsewhere in this book.

24.3.1 Hazard assessment

Studies of slope processes (Chapter 11), solutes (Chapter 15), tectonics (Chapter 2) and many other areas of physical geography covered in this book provide us with information that allows us to gain some idea of the risk of an environmental hazard. This risk is measured in terms of the likely recurrence interval (e.g. one in 1000 years event) or when the event might occur and how much damage the hazard might cause. Thus, for environmental management it is often necessary to be aware of environmental hazards and be able to predict the risks that those hazards pose. There are a range of tools for assessing and predicting hazards but most of these are site or hazard specific such as models for diagnosing urban areas against the risk of earthquakes, or slope stability equations for assessing when slope failure might occur (Figure 24.8).

These techniques are often spatial in their approach and it may be of great benefit to map areas of risk associated with a given environmental problem. For example, Figure 24.9 shows a map of soil erosion risk in Europe. Once these hazard assessments have been made it is then possible to show the results to local or international governors in order to gain the money (and legislation if necessary) to spend on management strategies for those areas considered to be most risky. For example, those areas with a high risk should be protected by reducing grazing levels or planting trees, or using some other management technique (Kirkby, 2001). These hazard assessments can then justify environmental management policies such as spending money on trying to prevent soil erosion in certain sensitive areas, or by moving people out of an area that is very risky (e.g. in terms of slope failure or contaminants).

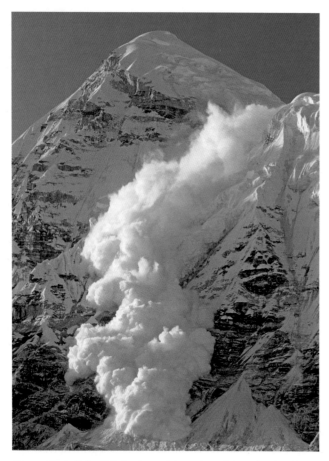

Figure 24.8 Scientists try to explain and predict avalanches in order to protect people from the hazard. This avalanche is occurring on Pumori, Nepal. (Source: Christopher Boisvieux/ Corbis)

24.3.2 Impact assessment

Making an environmental management decision is rarely simple. Environmental systems are complex. Small changes in one component may have unforeseen consequences elsewhere. To reflect the complexity of possible impacts, managers have developed a series of assessment techniques that seek to report fully the impacts of an action. Initially these were called **environmental impact assessments (EIAs)**. They are vital tools in the arsenal of the manager and are legally required in many countries depending on the characteristics of a proposed action (Petts, 1999). However, the focus on the physical environment to the exclusion of other elements in early forms of EIAs and related legislation has led to other forms of impact assessment such as **environmental technology assessment** (a new technology is assessed for its potential environmental, economic and social impacts), **social impact assessment** and **health impact assessment (HIA)**. Since the manager must seek to identify, quantify and aggregate the impacts regardless of the 'arena' in which they occur, it is clear that a scoping study is required to identify the type of impact assessment required.

All forms of impact assessment, environmental, social, strategic and health, have a generic similarity. They all have screening and scoping elements to determine whether the assessment is required in the first place and to define the boundaries of the investigation. They should all be prospective (occur before the proposed management activity) in nature and are therefore a form of forecasting tool. They are relatively simple and are driven by checklists and matrices (e.g. Table 24.4 in Box 24.4) to try to ensure that all elements are considered and that the significance of the impact in terms of scale, intensity, cost, irreversibility, longevity and so on are addressed. The key common stages are shown in Figure 24.10. Screening assesses whether an EIA is required and, if so, at which scale (short/outline or full). Scoping involves defining the boundaries of the investigation in spatial, regulatory and ramification terms. The third stage requires the definition of the type of impact assessment (strategic, social, environmental, health, etc.) while the fourth step is the implementation of the EIA. Implementation is usually promoted through some form of checklist which prompts the analyst to consider each possible impact element. For example, a prompt could concern the nature of any pollutant: liquid, gas, particulates and so on. The fifth step is the reporting and monitoring stage. In total all these form a series of steps that provide a consistent framework for making environmental management decisions. It allows results between sites to be compared and for decisions to be made in a clear way so that other people can see how a given decision has been arrived at. A case study of an EIA using these steps is provided in Box 24.4.

HIAs are a newer form of EIA and feature more prominently on the public agenda than normal EIAs. They have gained wide acceptance by governments and are a combination of procedures, methods and tools by which a policy, programme or project may be judged as to its potential effects on the health of a population, and the distribution of those effects within the population (World Health Organization, 1999). These methods have global endorsement. Policies to address environmental change at anything beyond the local scale are almost, by default, major policies and so are likely to require an HIA. However, 'health' in this context is not limited to the absence of disease but includes the presence of well-being: mental, physical and social. HIA has applied impact analysis to a broadly interpreted health agenda in a

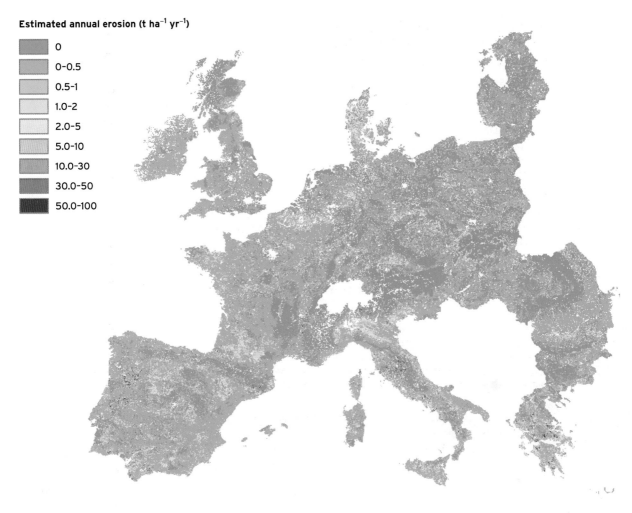

Estimated annual erosion (t ha^{-1} yr^{-1})

- 0
- 0–0.5
- 0.5–1
- 1.0–2
- 2.0–5
- 5.0–10
- 10.0–30
- 30.0–50
- 50.0–100

Figure 24.9 Soil erosion map for Europe showing estimated rates of erosion. (Source: courtesy of Brian Irvine and the EU PESERA Project)

CASE STUDIES

FLOOD RISK MANAGEMENT STRATEGY FOR THE RIVER SEVERN AT GLOUCESTER, UK

In Autumn 2000, there was extensive flooding in the River Severn catchment at Gloucester. It was therefore decided that a study was required in order to help determine a 50 year flood risk management strategy for the site. At the screening stage it was decided an EIA was needed. An EIA was performed during 2003 for flood risk management at the site. A scoping study identified 10 possible options:

1. Do nothing.
2. Do minimum (maintain existing defences and keep flood warning at current level of service).
3. Improve flood warning.
4. Retreat or lower existing local railway embankments.
5. Remove/modify floodplain obstructions.
6. Construct new defences.
7. Maintain the channel (e.g. dredging).
8. Remove existing weirs.
9. Reduce fluvial flows (e.g. by changing rural land management).
10. Flood flow routes (e.g. build flood diversion channels).

BOX 24.4 ➤

689

For each of these options an EIA was performed. This is normally done using a matrix which consists of around 10 factors which include human beings, flora and fauna, cultural heritage, and so on (Table 24.4). The matrix is completed describing the impact on each factor and potential significance of that impact. For example, the 'do nothing' option will result in increased flooding. Table 24.4 illustrates completed matrices for three of the options as examples. The impact columns identify the likely impact resulting from the effect. Impacts can be positive, negative or both, depending on the factor. The evaluation is qualitative (e.g. for the 'do nothing' option the impact of increased flooding on human beings is considered to be disruption, damage to and loss of property and quality of life). The 'potential significance' column provides an indicator, based on qualitative evaluation, of the potential significance of the impact brought about by the option selected. The evaluation broadly takes into account the geographical extent, likelihood of occurrence, timescale (temporary/permanent, gradual/sudden), magnitude and cumulative or secondary impacts.

In order to determine the impact and significance of the impact it is often necessary to do a great deal of research into the site and the processes operating at the site. Modelling can assist prediction of what will happen under different option scenarios and so many EIAs are associated with model results and descriptions. For example, at Gloucester, there is a disused railway embankment near the river. Modelling showed that removing the embankment would lower water levels at Gloucester and would markedly improve the conveyance of flood water across the floodplain.

Table 24.4 Environmental impact assessment matrices for three management options for the River Severn at Gloucester: (a) do nothing; (b) lower disused railway embankments; (c) construct new defences

Receptor	Description of possible effect	Impact	Potential significance	Notes
(a) Do nothing				
Human beings	Increased flooding	Disruption Damage loss	Major	Existing defences would eventually breach (owing to lack of maintenance). Increased flooding would result in disruption to quality of life, the significance of which could be major (far-reaching, temporary and permanent effects).
Flora and fauna	Increased flooding	Change in habitats	Moderate	Habitats could change with changing flood regimes – could be positive or negative – depending on what habitats were lost and what established instead. There might be opportunities to promote wetlands in areas that flood more frequently.
Air and climate	Increased flooding	No impact	None	
Landscape and visual amenity	Increased flooding	Change	Moderate	The existing landscape could change (positively or negatively).

BOX 24.4 ➤

Receptor	Description of possible effect	Impact	Potential significance	Notes
Water	Increased flooding	Possible risk of decrease in quality	Major	Contaminated land, e.g. Hempsted landfill site, may flood more regularly, affecting water quality significantly.
Land-use	Increased flooding	Change	Major	Increased flooding could alter the existing land-use and economic value (positively or negatively). Development may be restricted. Using agri-environment schemes to promote flood management may help to offset some adverse change.
Cultural heritage, archaeology and material assets	Increased flooding	Damage loss	Major	Surface assets may be flooded more frequently and could be damaged as a result. Buried assets may be adversely or positively affected by changes in the duration and frequency of flooding.
Traffic and transport	Increased flooding	Disruption	Major	Existing transport routes may experience more frequent flooding – with associated disruption to traffic.
Soil, geology and hydrogeology	Increased flooding	Reduction in groundwater quality	Major	Contaminated land, e.g. Hempsted landfill site, may flood more regularly, affecting groundwater quality significantly.
Use of natural resources	Increased flooding	No impact	None	

(b) Retreat or lower existing embankments

Receptor	Description of possible effect	Impact	Potential significance	Notes
Human beings	Managed increases in flooding	Flood alleviation for certain areas	Moderate to major	Benefits are proportional to the amount of additional flood storage created. Areas would be selected to minimize effects on buildings.
Flora and fauna	Some areas flooded more frequently	Change	Moderate to major	Opportunities for major wetland habitat creation. Significance depends on area. Existing terrestrial habitats may be adversely affected.
Air and climate	Managed increases in flooding	None		

BOX 24.4 ➤

691

Receptor	Description of possible effect	Impact	Potential significance	Notes
Landscape and visual amenity	Some areas flooded more frequently	Change	Moderate	Wetland creation or similar could bring about gradual change to the existing landscape.
Water	Managed increases in flooding	No effect on water quality	Minor	Areas selected to avoid release of harmful substances into flood water.
Land-use	Managed increase in flooding	Change	Moderate	Land management regimes could experience change. Agri-environment schemes may support the transition. Scope for development in retreated areas could reduce. Areas would be selected to minimize impact.
Cultural heritage, archaeology and material assets	Some areas flooded more frequently	Change	Moderate	Changes in wet/dry cycles may lead to deterioration of some artefacts. Areas selected to minimize impact on known terrestrial and buried remains. Watching brief employed during construction to mitigate potential loss where considered appropriate.
Traffic and transport	Managed increases in flooding	No effect on traffic and transport	Minor	Areas could be selected to avoid impact.
Soil, geology and hydrogeology	Managed increases in flooding	Possible increase in groundwater recharge	Minor	Areas could be selected to avoid release of harmful substances into groundwater.
Use of natural resources	Build set-back defences	Use of resources	Minor	Existing embankment material could be reused.

(c) Construct new defences

Receptor	Description of possible effect	Impact	Potential significance	Notes
Human beings	Reduction in flooding locally	Reduction in disruption, stress, etc., caused by flooding	Major	Selected if other options not effective. Impacts on flooding elsewhere would be investigated.
Flora and fauna	Reduction in flooding / Construction of defence	Change in habitats / Damage loss	Major	Reduction in flood frequency may affect existing habitats (positively or negatively). Existing habitats likely to be damaged during construction of a new defence – could be mitigated to an extent by re-creation elsewhere.
Air and climate	Reduction in flooding	None		
Landscape and visual amenity	Reduction in flooding / Construction of defence	Change	Major	Existing long views, local landscape and river views likely to be altered by new defences. Gradual change in landscape likely if land flooded less frequently.

BOX 24.4 ➤

Receptor	Description of possible effect	Impact	Potential significance	Notes
Water	Reduction in flooding	Reduced risk to water quality	Moderate	Sensitive/contaminated sites could experience less flooding so the risk of surface water quality contamination could be reduced. Drainage through the defences to be adequate so that surface water flooding of defended areas does not increase.
Land-use	Reduction in flooding	Reduced risk to existing land-use	Major	Existing land-use could experience less flooding, so long-term impacts, e.g. insurance blight, may be mitigated. There may be opportunities for diversification of land-use if land flooded less frequently; however, the Environment Agency policy regarding strict control on development in the floodplain will be maintained even if the area was defended.
Cultural heritage, archaeology and material assets	Reduction in flooding Construction of defence	Change Damage	Major	Existing assets (buried) likely to experience change in wetting/crying cycle, which may affect their preservation. Terrestrial assets could experience less frequent flooding. Defence construction could damage assets (buried or terrestrial).
Traffic and transport	Reduction in flooding	Improvement	Major	Disruption to traffic and transport caused by flooding could be reduced.
Soil, geology and hydrogeology	Reduction in flooding	Change	Minor	Flood waters can deposit fertile silts on agricultural land. If flooding were reduced, deposition of these silts could also reduce – this may be considered an adverse effect. Conversely, soil structure can be damaged when it is overworked during wet periods. Reducing flooding could avoid such damage.
Use of natural resources	Construction of defence	Use of resources	Minor	Recycled and secondary materials should be used where possible.

BOX 24.4

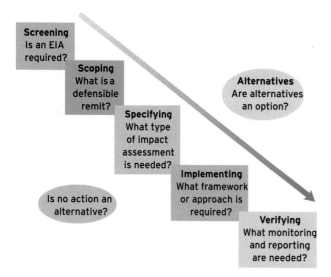

Figure 24.10 The main stages in an environmental impact assessment.

process that incorporates the principles of sustainable development. Because HIA has only recently been developed it incorporates a more explicit reflection of sustainable development principles. Many economic, social and environmental (as well as inherent biological) factors influence the well-being of individuals and communities. For example, personal circumstances, lifestyles and the social and physical environment influence health (e.g. education, income, employment, behaviour, culture, social support networks, community participation, air quality, housing, crime, civic design and transport). In addition HIAs tend to have a moral and ethical dimension with an aim of trying to make sure that everyone can enjoy the same health quality of life.

24.3.3 Life costing

Putting impact assessment and forecasting together allows the development of **life-cycle analysis (LCA)** and raises an interesting question about the perceptions of environmental managers. Do managers have the vision to see all the ramifications of an action or do they tend to see only the immediate consequences and limited future effects irrespective of the perception they may hold? In the light of many studies which identify the shortcomings of a manager's vision, LCA has become an important tool in allowing a full 'cradle to grave' assessment of management actions or intervention. LCA is the examination of everything that happens in the manufacture, use and disposal of a product, from the time the

raw materials are taken from the Earth to the time the product is thrown away and added to the ecosystem (Figure 24.11). The basic idea of LCA is to identify and evaluate all the environmental impacts of a given product (Ciambrone, 1997).

This is similar to producing an ecological footprint, which is where a calculation is performed as to what area of biosphere is required to sustain an individual, a company/ organization or a country. The challenge of LCA, however, lies in the word 'evaluate'. We are aware that many natural systems have a capacity to absorb impacts. What is needed is a measure of this capacity. Thus we need detailed physical geography science to be carried out to provide us with such information. Sometimes we can use, as in the case of the ecological footprint, a measure of land area to indicate the capacity of a system. We can also use other indices such as the amount of carbon used. Most of the time, however, we are using multiple resources (fossil fuels, nutrients, etc.) and can create substances that may last for thousands of years without being recycled into the ecosystem (e.g. nuclear waste). Environmental managers, like managers in all other sectors, are seldom free to do what they like. The actions that they take need resources, and these need to be justified. Furthermore, the boundaries within which managers can exercise choice must be agreed. Since all actions take place from within a limited budget and since money spent on the environment will inevitably divert resources away from other deserving causes, it is often helpful to express impact and effects in monetary values to promote communication with decision-makers. This has expanded to a major discipline in its own right, known as **environmental economics**. When we merge LCA with the financial expression of consequences the resulting approach is called **whole-life costing**.

> **Reflective questions**
>
> ➤ What is an environmental impact assessment?
>
> ➤ What is life-cycle analysis?

24.4 Implementation

Without effective implementation, environmental science and management approaches discussed above are futile. There are two key elements to implementation: (i) building

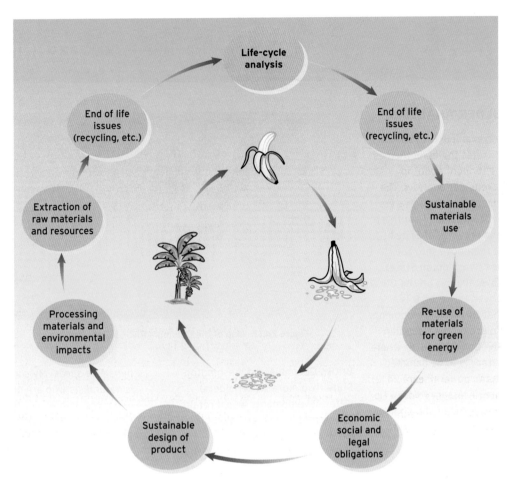

Figure 24.11 Life-cycle analysis.

a consensus and (ii) managing the project. The following section deals with these two elements.

24.4.1 Stakeholder involvement

Building a consensus for action is vital but is usually time consuming. It requires that everyone with an interest is consulted or feels consulted. The identification of precisely 'everyone with an interest' is not simple. Each discipline has its own perspective and, certainly in the past, engineers would tend to consult engineers and politicians would negotiate with other politicians. Today we seek to identify '**stakeholders**' through a process called stakeholder analysis. Stakeholder analysis is simply a sequence of brainstorming events at which a group of project managers seek to identify everyone with a reasonable interest in a project. It is important to appreciate that

this goes far beyond simply the list of statutory bodies that a politician might first identify and beyond the owners, businesses, consumers and liability holders that a lawyer might identify. Box 24.5 provides an example of stakeholder analysis.

Making an adequate definition of stakeholders is the foundation of consensus building. Agreement cannot arise from people who are not part of the agreement process. This may seem self-evident but the Chinese Government, in November 2003, paid compensation to contractors who were unable to proceed with the development of the Hong Kong harbour. This was after the courts decided that protestors who disagreed with the proposed development had not had an opportunity to register their disagreement and had not been able to engage in a proper debate about the proposals with those in favour of the scheme.

TECHNIQUES

STAKEHOLDER ANALYSIS

The following example is hypothetical and illustrates the main themes that must be thought about in terms of stakeholder analysis. The Columbia River has its headwaters in the Rocky Mountains of southern Canada in the province of British Columbia. It then flows into the United States (Washington State). There are several dams on the US part of the river. However, it is proposed that further developments are to take place on the Canadian section of the Columbia River where a large tributary is to be dammed for hydroelectric power (Figure 24.12). It is therefore necessary to draw up a comprehensive list of stakeholders for this situation. We would start by listing the different groups that might have an interest. It is usually better to attempt this systematically.

Often it is good to start at the largest scale and work down and thus we might identify governments, regional administrations, businesses, pressure groups, native first peoples and local communities. Note that each of these can be subdivided. For example, governments subdivide into US Government and Canadian Government and the International Joint Commission. Pressure groups might subdivide into global interest groups such as Greenpeace, Friends of the Earth, the Sierra Club and local pressure groups. (You might like to complete this process through additional web research.) The list then has to be tested to determine whether, for example, pressure groups that had an interest in some of the earlier

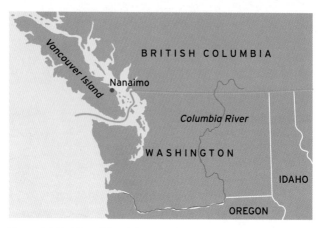

Figure 24.12 Map of the Columbia River, North America.

British Columbia and Washington State (two more stakeholders) water developments would retain an interest in further Columbia River developments.

When you have completed the list, look at Table 24.5 which lists some of the stakeholders involved with the nearby Vancouver Island power generation project at Nanaimo (a smaller project; see Figure 24.12). You may now feel that you need to add more stakeholders to your list. It is very important to be inclusive of stakeholders and to make sure they are adequately represented as many people become very sensitive about their involvement in decision-making. For example, the following extract is from a letter sent in May 1998 to the British Columbia Premier about membership of a local watershed council from members of the Federation of British Columbian Naturalists (FBCN):

Your letter was addressed to Denis and June Wood, with no mention of the groups on whose behalf we speak. My constituency is naturalists who are organised under the umbrella of the Federation of B.C. Naturalists. The FBCN is comprised of over 50 naturalist clubs province-wide, representing over 5300 members. The Federation has provided a unified voice for naturalist clubs since 1969 and our input is valued and respected throughout British Columbia as reasoned and informed.

The FBCN resigned from the watershed council because it felt it was not being respected by the council as a full stakeholder, and this quote illustrates how people can become upset if they feel their organizations are not given sufficient recognition as a stakeholder in an environmental management decision. Without people being fully involved with project decisions it is then very difficult to reach a consensus and therefore to gain agreement about an environmental management decision.

BOX 24.5 ➤

➤
Table 24.5 Some of the stakeholders contacted regarding the Vancouver Island generation project at Nanaimo in British Columbia, Canada

Government	Businesses and community groups
Municipal/regional government	Vancouver Island Real Estate Board
Islands Trust	Fire departments
Fire Department	Harmac Mill
City councillors	General hospitals
Recreation directors	E&N Railway Company
Tourism directors	Chambers of Commerce
Environmental service director	Community Futures Development Corporation of Central Island
Trails coordinator	Nanaimo Rotary
Liquid waste manager	Vancouver Island building trades
Planner manager	Nanaimo Port Authority
Public works officer	Tourism Association of Vancouver Island
Roads and traffic office	Nanaimo fisheries
Airport manager	
Mayor	
Economic development officer	
Provincial government	
Environmental assessment office	
Aboriginal relations advisor	
Ministry of Health services	
Ministry of Health Planning	
Ministry of Sustainable Resource Management	
Ministry of Transportation	
Ministry of Water, Land and Air Protection	
Federal government	
Mortgage and Housing Corporation	
Department of Fisheries and Oceans	
Environment Canada	
Royal Canadian Mounted Police	

BOX 24.5

However, even if you do manage to identify all of the stakeholders it is very unlikely that all the stakeholders will agree with a single inflexible proposal. Indeed such an approach is almost inevitably going to cause opposition. If, as an environmental manager, this 'single solution' approach is adopted it will need to be resolved through the legal system or through some national or international arbitration process. Of course any such approach is costly and may have unfortunate consequences:

- For example, a project such as a runway development, abandoned for environmental considerations, may reduce employment potential. The same project permitted to go ahead will result in claims of loss of property value.
- Future opposition to similar projects is likely to be stronger and more organized. This will occur because opponents will have learned key lessons and will be strengthened by their past success or hardened by their previous failures.
- The legal process is costly.
- Arbitration and legal processes are lengthy. Such processes can result in problems where local development opportunities are lost because people do not know which way a region, area or project will develop.
- International disagreement can result in stalemate. Within a country there is likely to be a mechanism for dispute resolution. In the international arena, however, there are few such mechanisms and even if apparent agreement is reached, one side can simply decide to ignore the agreement, which is effectively how the United States views the Kyoto Carbon Treaty.

Therefore the process of offering a single option destroys rather than builds on the prospects of future consensus. In place of such an approach there should be an intelligent negotiation of alternatives. In reality, however, these two approaches (offering a single solution and offering a choice of options) are simply two points on a spectrum of types of participation in decision-making identified by van Ast and Boot (2003). The full spectrum is given in Table 24.6 which shows how the style of management and role of stakeholder may interact. The two approaches discussed above correspond to 'open authoritative' and 'participating/delegating' categories within the table.

Alternative dispute resolution is a process that avoids any winners and losers. It does this by presenting a range of choices and uses the stakeholder groups to participate in the analysis of which options are viable and which are not. The process of encouraging all stakeholders to take an active part in the decision process is called **participatory analysis** while the identification of key criteria through which to evaluate options is called **options analysis**. Box 24.6 presents a recent options analysis for a river management project in Yorkshire. The options were analysed by the participants and one option was agreed upon and has been implemented (McDonald *et al.*, 2004).

24.4.2 Project management

The second part of implementation involves project management with an extended period of monitoring. Project management is the delivery of an agreed objective to an agreed timescale. It follows then that there are two requirements that are fundamental to effective project

Table 24.6 Levels of governance and styles of participation

Style of management	Role of stakeholder	Form of interaction
Facilitative	Initiator	Interactive
Cooperative	Cooperating partner	
Delegating	Co-decision-maker	
Participating	Adviser	
Consultative	Consultant	Non-interactive
Open authoritative	Information receiver	
Closed authoritative	None	

(Source: from van Ast and Boot, 2003)

CASE STUDIES

RIVER RESTORATION CHOICES

Flood protection on the River Wharfe, northern England, was being compromised by sediment deposition that was reducing channel capacity. As the river channel was filling up with sediment it was flooding more frequently because there was less room in the channel for water. In an attempt to control this problem a gravel trap was installed upstream of the problem sites (Figure 24.13). The gravel trap consisted of a large excavation in the base of the river sufficient to hold many tonnes of sediment, and was expected to be emptied every five years. However, the gravel trap actually required to be emptied:

- much more frequently;
- at a higher cost than anticipated; and
- with more disposal problems in a national park setting.

In addition the large cobbles trapped and retained preferentially in the gravel trap captured further large cobbles to create a shoal in the river which in turn promoted bank erosion, loss of valuable land and threatened infrastructure. Initially only two (extreme) management options were considered, namely (i) highly engineered physical protection and (ii) complete

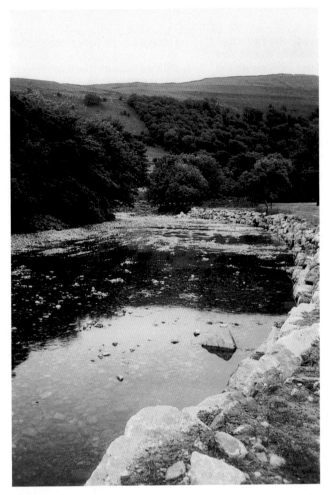

Figure 24.13 The gravel trap on the River Wharfe with bank protection.

removal of management to allow the river to find its own course at an unknown future time. This caused a polarization of views. To resolve the dispute the seven options shown in Table 24.7 were offered

and the merits of each considered. A consensus towards an integrated compromise solution was found that gave time for the dynamics of the longer-term sediment budget to be better understood.

BOX 24.6 ➤

management: (i) a clearly specified, agreed, feasible objective and (ii) the resources to reach that objective. Project management requires the subdivision of the project into a series of discrete tasks. Each task will have a clear specification,

resources allocated, a time duration to complete the task and a person or agency responsible for the delivery. There must also be an identification of the relationships between tasks because in some cases one task may not be completed

➤

Table 24.7 Seven options for management of a gravel trap on the River Wharfe, England

Option	Summary
1. An engineered structure to consolidate eroding bank, protect the footpath and protect against flooding at that site	Enhance current bank protection works on the true right bank by providing an engineered structure capable of consolidating the bank thus enabling the long-distance footpath (the Dales Way) to remain in its current position. Acknowledges that the Wharfe channel, local to a tributary, Cray Beck, has narrowed because of the northern (left) bank walling, gravel accumulations from Cray Beck, plus a large gravel shoal opposite the confluence that has accumulated in the last 130 years. The accumulated right bank gravel shoal opposite Cray Beck would be removed, allowing free passage of water and further reducing the erosion stress on the new river bank walling upstream. This section of channel seems to be developing into a tight meander; maps produced over the past 130 years illustrate this process. The option therefore seeks to reverse this trend in order to protect the right bank and current footpath. The planform would be designed so as to enhance sediment transport through the reach.
2. Reduce left bank protection and encourage movement on the north side of the river	Combines elements of bank protection on the right bank with elements of removing bank protection on the left bank. Acknowledges that the meander within this reach has become tighter, and allows this process to continue on the left bank (currently heavily protected), local to Cray Beck, although it may reduce erosion pressure upstream on the right bank. By allowing some river migration, there is the possibility of floodplain construction, and this begins to address the issue of downstream impacts. Option 2 seeks to use more vegetation to protect the banks. This would extend upstream and downstream of the meander. The islands associated with the Cray Beck confluence would be restored.
3. Bypass channel	This option tries to overcome one of the key concerns with option 2, namely that the area for sediment accumulation is marginally smaller. This aspect is of concern since ideally we are trying to protect downstream river sections through the management of the gravel trap reach. The proposal is therefore to establish a 'bypass' channel on the northern floodplain (left bank) that allows some water and sediment to bypass the right bank defences thereby reducing erosion pressure. The area of woodland increases to protect the new channel through soft bank protection methods.
4. Realignment and planned retreat	This proposal is to realign the River Wharfe to the lowest point in the floodplain. Centuries of river training, gravel use and levee construction have resulted in levee levels well above the floodplain elevation and river bed levels that are approaching the general floodplain level. If continued and in the light of the cessation of gravel exploitation, the river bed level may well be significantly above the floodplain as has occurred in some Lake District rivers. Following realignment, management would allow normal flooding and sediment to spread laterally over the floodplain. This would involve the restoration of natural processes and enable the river to define its own channel. It would also restore the crucial process of floodplain construction associated with sediment storage.
5. Gravel extraction from gravel trap and manage the gravel deposition	This is the 'do nothing' option now required to be evaluated in much US environmental legislation. However, the 'do nothing' option can, in this case, take two forms. The first, 'do nothing', would be to make no change to the planned current management regime. The gravel trap scheme would be maintained as intended with the gravel trap emptied at regular intervals. This is treated as option 5 here. The second 'do nothing' option is the 'hands-off' option. In effect then this maintains the *de facto* status quo of the last decade, namely maintain neither the gravel trap, nor any of the gravel shoals. But, and this is a major change, it also requires non-intervention with current bank erosion and allows the river to adjust its form and position from its current position in the valley. This is treated as option 6 below.

BOX 24.6 ➤

700

Option	Summary
6. No intervention. Unplanned retreat	No intervention or maintenance. Allow the river to erode and eventually change course and form with a return to natural processes of sediment erosion, deposition and storage.
7. Working with the river	This proposal takes elements of option 6, namely do nothing and allow the river to develop its own form, with elements of options 1 and 2. The main elements are to remove the gravel trap bank and river bed protection measures. Allow Cray Beck Island to be reformed and the natural sinuosity of this river reach to develop, on both the left and right bank. The flood risk to fields south of the meander would be protected by strengthening the current levee. In the medium to long term, and with no gravel clearance within this section, the meander should be expected to tighten on the basis of current changes. The levee will be reformed behind the eroding bank at intervals to reduce the risk of the river breaching this part of the bank and flowing southwards over the lower-lying floodplain (blister repair). Farm access to the field and the public footpath would follow the newly formed levee. The proposal therefore seeks to allow the river to define its bed form, allow gravels to accumulate and banks to erode until a quasi-stable river form is reached where erosion rates become lower. Bank protection works will 'retreat' with the eroding bank or 'advance' with the revegetation and soil development of older gravel areas. This option seeks to avoid an uncontrolled breach of the levee while promoting some natural processes, including sediment storage, with beneficial downstream impacts.

BOX 24.6

without another task being completed. For example, a structure such as a weir could not be added to a river without permission from the planning agency and that permission might not be forthcoming without a formal EIA. Not all tasks must be performed in sequence and many might be performed at the same time. However, in any large project a network of tasks is necessary and the track through the network which has the highest cumulative duration is the critical path. This is the path that will take the longest to complete as it relies on one thing being completed before the next can be taken forward. A delay in any of the tasks on this track will delay the whole project and conversely a saving on this track will speed up the whole project. Larger projects also need definitions of milestones and deliverables. In effect these are simply key stages at which major steps are realized and at which identifiable items can be 'delivered' to the end user or next developer.

For the management of environmental change, however, the project management as outlined above is too simple or

at least too clearly defined. Any project that involves managing environmental change will have a large element of uncertainty and so there will be a need to monitor outcomes to ensure that the anticipated outcomes are being realized. However, there are usually too many outcomes to report effectively to a monitoring group, particularly as many members of that group will not be familiar with the science or the terminology.

For example, we may want to restore urban air quality that has been degraded by pollution (Figure 24.14). Urban air quality will be measured in several ways such as particulate concentrations, low-level ozone, sulphur dioxide concentrations, visibility and so on, and each of these will be related to one or more sources of pollution. For each pollution source there may be several options for solution or reduction, and the implementation of these options would constitute a project. Some sources, causes and solution options are given in Table 24.8. The environmental manager must reflect on the solution options, such as declaring a smoke-free zone, or incentivize use

Figure 24.14 Urban pollution in Shenyang, China. (Source: Fritz Hoffmann/The Image Works/TopFoto)

Table 24.8 Options to limit urban air quality degradation. SO_x and NO_x are forms of sulphur oxide and nitrous oxide respectively and PM_{10} is the concentration of particles that are less than 10 μm in diameter

Measure of degradation	Cause of degradation	Solution option
SO_x	Domestic emissions	Smoke-free zone Punitive taxation on domestic fuel Home insulation support Combined heat and power facilities
NO_x	Transport emissions	Congestion charging Odd- and even-day access Parking restrictions Subsidized public transport Pedestrianization Zero-emission vehicles
PM_{10}	Industrial emissions	More aggressive regulation Changed process Resiting Incentivize use of renewable energy

of renewable energy such as wind power or solar panels (Figure 24.15) and consider the best way forward. Nevertheless, the question still remains as to how a non-technical project management or oversight group can measure progress. At the simplest level an 'improving' trend is the clearest signal of progress. However, this would ignore the rate of improvement unless an annual target had been set or a satisfactory concentration had been defined. In addition, a large number of 'trends' with no recognition of differences in the significance or meaning of the trends will make the derivation of information from data difficult.

(a)

(b)

Figure 24.15 Renewable energy generation: (a) wind power and (b) a building clad in solar panels.

Therefore, oversight groups need indicators. These may be selected key indicators or may be aggregate indicators. The indicators may be 'normalized' so that all measures are within similar numeric values or they may be transformed to word format (e.g. improving, satisfactory, deteriorating, unsatisfactory, unchanged) or symbol (e.g. tick/cross, traffic light symbol) to improve ease of communication to a diverse group.

Reflective questions

➤ What are the main issues associated with gaining agreement on environmental management?

➤ What are the main general problems associated with project management?

24.5 Summary

Environmental management is about the management of change. The objective may be to stop a detrimental change or to encourage a responsible change. However, change is not simple and predictable, yet it is vital to be able to forecast change. Environmental managers therefore rely on process understanding provided by physical geographers to offer scenarios of the possible changes that might take place and the uncertainty that is associated with each scenario. Change may be linear or non-linear, reversible or irreversible, and hysteretic. Slow change may cause a system eventually to cross a threshold and then dramatically alter (e.g. river channels suddenly switching their course) and such changes may be illustrated by catastrophe theory.

Environmental impact assessments are important environmental management tools and allow us to determine the relative merits of a variety of management options. It is often possible to estimate the financial advantages and disadvantages of management options but most of the time such schemes do not incorporate full life-cycle analysis or costing of the wider effects of a given management activity or technology.

The money to pay for environmental management will be provided only if politicians are willing to provide such finance. The political stance has to reflect the consensus for support for a given action and such consensus can only be found through all stakeholders finding common ground. Implementation of the actions that are agreed requires monitoring. The monitoring results eventually need to be reported to the non-technical (e.g. non-scientist) groups that are paying for the management or that are stakeholders within the project. Therefore the results must often be expressed as a series of non-technical indicators.

Further reading

Asante-Duah, D.K. (1998) *Risk assessment in environmental management.* John Wiley & Sons, Chichester.
This book describes the nature of contaminant problems, and demonstrates how you can calculate risk in different ways. There is occasionally some complex theory and a little maths.

Barrow, C.J. (1997) *Environmental and social impact assessment: An introduction.* Arnold, London.
This book describes the development, role, processes and methods of EIAs. There are good boxes with case studies.

Barrow, C.J. (2006) *Environmental management for sustainable development.* Routledge, London.
This is a clear text which also covers law, standards and economics in addition to the topics covered in this chapter.

Erickson, S.L. and King, B.J. (1999) *Fundamentals of environmental management.* John Wiley & Sons, Chichester.
This book relates a lot of its discussion to regulations in North America. There is a very useful appendix which contains checklists for environmental management audit exercises and a good glossary.

Nath, B., Hens, L., Compton, P. and Devuyt, D. (eds) (1998) *Environmental management in practice.* Routledge, London.
This book has three volumes and is very detailed. Volume 2 is good for information on the management of physical components such as soils, water quality, transport, sustainable agriculture, tourism, fisheries, and so on, whereas volume 3 deals with ecosystems of the world.

Owen, L. and Unwin, T. (eds) (1997) *Environmental management: Readings and case studies.* Blackwell, Cambridge, MA.
This book is aimed at undergraduate level and is a compilation of important papers and chapters by different authors. It is an excellent resource for case studies which could be used in essays and exams.

Owens, S. and Owens, P.L. (1991) *Environment, resources and conservation.* Cambridge University Press, Cambridge.
While this is now a bit dated, many of the general problems associated with environmental management are described in this text. There is also a good theoretical discussion about management of resources.

Web resources

Catastrophe Teacher: An Introduction for Experimentalists
http://perso.orange.fr/l.d.v.dujardin/ct/engr_index.html
For an outstanding teaching package on catastrophe theory including explanations, models, source Java script and examples visit this site.

Center for Environmental Philosophy
http://www.cep.unt.edu/
Here there is access to Internet resources that relate to environmental ethics and environmental philosophy, in particular a brief history for the novice. The only concept omitted is the idea of environmental stewardship that links the environmental manager to the responsibilities that the discipline should carry.

Environment Canada
www.ec.gc.ca/envhome.html
This site provides one of the largest sets of online resources about the environment and environmental decision-making (including impact assessments); information is given on the various action plans to deal with environmental issues such as climate change.

European Environment Agency
http://www.eea.eu.int/
The home page of the European Environment Agency provides a mass of information on a series of environmental themes (from waste disposal, biodiversity issues and desertification risk), and environmental news from throughout Europe. It aims to achieve significant improvement in Europe's environment through provision of relevant and reliable information to policy-makers and the public. Details on environmental management strategies can be found on this site.

Google Directory: Impact Assessment
http://directory.google.com/Top/Science/Environment/
Impact_Assessment/
Google maintains an extensive directory with links to many sites (government, academic and private organizations) that deal with environmental impact assessment (EIA) and also strategic impact assessment (SIA).

SD Gateway
http://sdgateway.net
This site integrates the online information developed by members of the Sustainable Development Communications Network. It is one of the best organized and authoritative sources on sustainable development and the Bruntland Report.

UK Environment Agency

http://www.environment-agency.gov.uk/

This site provides excellent but rather disaggregated information on environmental management (you need to search through the site). There is a large amount of information, including facts, figures and brief discussion on a range of conservation and environmental issues such as air quality, endangered species, contaminated lands and so on.

US Environmental Protection Agency

http://www.epa.gov/ems/index.htm

Information and resources related to environmental management systems (processes and practices that enable an organization to reduce its environmental impacts) for a variety of users are presented here.

Whole Systems

http://www.worldtrans.org/whole.html

This web page discusses many aspects of whole-system thinking. Its information ranges from hard mathematics and fuzzy logic as well as softer philosophical holistic considerations. There is a good introduction to whole-systems cybernetic thinkers such as Buckmaster Fuller, Stafford Beer and Pierre Teilhard de Chardin (the predecessor of James Lovelock).

Visit our website at **www.pearsoned.co.uk/holden** for further questions, annotated weblinks, interactive models and video-clips showing physical processes in action.

A

Ablation The action of removing and carrying away a superficial material; it primarily relates to the *sublimation*, melting and evaporation which remove snow and ice from the surface of a glacier or snowfield. It can also be applied to the wearing away of rock by water and the removal of salt or sand from a surface by the action of wind.

Ablation zone The region of a glacier that experiences net loss of mass throughout the year (i.e. ablation exceeds accumulation).

Absolute zero The temperature at which molecules have no internal energy and are at a complete rest. This is equivalent to zero kelvin or −275.15 degrees Celsius.

Absorption bands Sections of the electromagnetic spectrum that interfere with the passage of radiation to the Earth's surface.

Abyssal plain The flat deep-sea floor extending seawards from the base of the continental slope and continental rise, reaching depths of 4–6 km between mid-ocean ridges and trenches.

Accumulation zone The region of a glacier that experiences net gain of mass throughout the year (i.e. accumulation of snow, *firn* and ice).

Across-track sensors Scanning instruments that collect data by directing the line of sight in a sweeping motion at right angles to the direction of travel by a rotating or oscillating mirror. These are also called whiskbroom scanners.

Active layer The thin top layer of the ground surface that is seasonally frozen and unfrozen above *permafrost*.

Adiabatic The expansion of a body of air without loss or gain of heat.

Adsorbed When a specific gas, liquid or substance in solution adheres to the exposed surface of a material with which it is in contact, usually a solid, e.g. the adsorption of anions to the surface of clay minerals by electrostatic attractions.

Adsorption Attachment of a substance in solution to a solid.

Advection The horizontal circulation of an ocean body or air mass.

Advective solution The process in which solutes are removed and transported through the soil in a flow of water (e.g. runoff and percolating water); it is very effective at removing solutes from near the surface; similar to leaching.

Aeolian Pertaining to the processes, Earth materials and landforms that involve the role of wind.

Aeration The ventilation of soil.

Aerosols Minute particles suspended in the atmosphere that interact with the Earth's radiation budget and climate.

Aggradation The raising of a surface caused by the accumulation of material deposited by various geomorphological agents (e.g. wind, water or wave).

Aggregate A grouping of soil particles adhered together and separated from surrounding aggregates by voids in the soil. Also known as *ped*.

Agronomy Subject of utilizing plant processes for crop growth.

Air mass An extensive body of air possessing relatively uniform conditions of temperature and moisture that is in contact with the ground.

Alas Small irregularly shaped lakes and depressions caused by the melting of massive ground ice.

Albedo The proportion of radiation reflected from a surface. Surfaces such as snow have a high albedo.

Alkalinity The capacity of water to neutralize acid, determined by the quantity of base *cations* (Na^+, K^+, Mg^{2+}, Ca^{2+}) in the substance. Measured by titration of the sample with a strong mineral acid.

Allogenic Pertaining to a change in system dynamics caused by the influence of an external environmental factor, i.e. in relation to river channel adjustment, an allogenic change involves a move away from equilibrium conditions in response to an alteration in the sediment and water regime of the river.

Alluvial fan A fan-shaped landform composed of alluvium, deposited where a tributary stream loses momentum on entering a more gently sloping valley.

Along-track sensors Scanning devices that collect data using a linear array of instruments oriented perpendicularly to the direction of travel covering one side of the *swath* to the other. These are also called pushbroom scanners.

Alpine permafrost Permafrost that occurs locally owing to low temperatures at high altitudes, e.g. the Rocky Mountains.

Alternative dispute resolution The process of settling an environmental dispute through mediation and arbitration (avoiding litigation); it avoids strict winners and losers by presenting a range of choices and encouraging *participatory analysis*.

Amphidrome Points in the oceans where there is zero tidal range due to cancelling out of tides. Tides radiate out from amphidromes.

Anaerobic Functioning in the absence of oxygen.

Anaerobic bacteria Microorganisms that survive in environments containing no free or dissolved oxygen; they obtain oxygen through the decomposition of chemical compounds, such as nitrates.

Analogue image An image composed of continuous tone.

Anastomosing Pertaining to the tendency for certain rivers to divide and reunite, producing a complex pattern of channels with large, stable islands between the channels.

Angle of repose The maximum slope gradient at which unconsolidated material will remain stable without collapse.

Anions Negatively charged *ions*, i.e. an atom which has gained one or more negatively charged electrons, e.g. the chloride ion (Cl^-).

Anoxic Depleted of oxygen; in water usually a result of bacterial oxygen consumption and other respiration in areas of restricted circulation.

Antarctic Bottom Water (ABW) A body of water formed along the edge of the Antarctic continent. Very dense water created by the very cold, saline conditions is forced to sink and flow north underneath the *North Atlantic Deep Water (NADW)*; together they power the *thermohaline circulation* of the world's oceans.

Anticyclogenesis A condition in which a zone of descending air results in high pressure at ground level and air to circulate slowly outwards from the descending zone. This results in anticyclonic conditions.

Anti-dune A type of small-scale cross-bedding feature formed from a sand deposit on a river bed. It develops from a 'normal' dune when the flow velocity increases in a highly loaded river; erosion from the downstream slope throws material into saltation and suspension more easily than it can be replenished from upstream, causing upstream migration of the bedform feature.

Aquifer A layer of rock with sufficient porosity to absorb and store water and permeable enough to allow water to pass freely through as groundwater.

Aquitard An *aquifer* which has been confined between impermeable rock layers and only open for recharge and discharge at certain locations.

Archimedes' principle Any object wholly or partially immersed in a fluid will experience a buoyant force (or upthrust) equal to the weight of the fluid displaced.

Arête A steep knife-edge ridge that divides the steep walls of two adjacent cirques in a mountainous region.

Argillic A term to describe a soil horizon characterized by clay accumulation.

Aridity A state of lacking in moisture, when evapotranspiration exceeds precipitation. It can be defined by the annual overall net negative moisture balance of a particular environment.

Armoured layer The coarser stoned surface layer of a mixed gravel-bed river, protecting the finer material beneath.

Artificial neural network (ANN) A type of parallel computing in which memory is distributed across a number of smaller processing units that process information in a parallel manner.

Aspiration The act of drawing air.

Asthenosphere The ductile layer of the Earth's mantle located 100–400 km below the surface, on which the rigid lithospheric plates glide.

Atmometer An instrument for taking direct measurements of evaporation; by connecting a water supply to a porous surface the amount of evaporation over a given time is measured by the change in water stored.

Atolls Coral reefs that surround a central lagoon; most are found in the Indian and Pacific Oceans.

Autogenic Pertaining to a change in system dynamics caused by the influence of an internal, self-produced factor, i.e. in relation to river channel adjustments, an autogenic change involves a fluctuation about an equilibrium condition.

Autotrophs Those life forms that acquire their energy from the Sun via the process of photosynthesis. They form the first *trophic level* by creating a source of energy for other animals, birds and insects.

Available water The water available in soil for plant growth after excess water has drained owing to gravitational forces, i.e. the water retained between the states of *field capacity* and *wilting point*.

Avulsion The process whereby a channel shifts, abandoning its old course for a new course, and leaving an intervening area of floodplain intact.

B

Back-scar The upslope section of the wall from which a landslide has occurred, creating a scar.

Backscatter The return signal of a radiation pulse from an active remote sensor. This is also called the echo.

Backwash The seaward return pulse of water from a breaking wave along the shoreline, moving under the force of gravity.

Badlands A deeply eroded barren landscape characterized by very irregular topography with ridges, peaks and mesas resulting from wind and water erosion of sedimentary rock. Badlands originally referred to the heavily eroded arid region of south-west South Dakota and north-west Nebraska in the United States but is now a more generic term.

Balance velocity The flow velocity required by a glacier to maintain the ice in equilibrium; the mass transferred down the glacier should equal that lost by melting in the ablation zone.

Bank-full Condition when the river channel is full of water.

Bank-full discharge The level of discharge at which any more water would cause the river to spill out of the channel onto adjacent low-lying land.

Barchan dunes Isolated crescentic sand dunes with a shallow windward and a steep lee side whose horns point in the direction of dune movement (usually forming under conditions of limited sand supply).

Barrier beaches Elongated offshore banks of coarse granular debris (sand, gravel) lying parallel to the coastline that are not submerged by the tide (see *barriers*).

Barrier islands Elongated offshore islands of coarse granular material, lying parallel to the coastline, similar to barrier beaches but larger in scale and forming behind barrier beaches (see *barriers*).

Barriers Elements of a beach planform located just offshore that involve the accumulation of landward-migrating sand shoals running parallel to the coastline that achieve surface elevation as they roll inland (see *barrier beaches* and *barrier islands*).

Bars Ridges of coarse sediment deposited on a stream bed where the stream velocity drops, especially mid-stream and on the inside of meanders.

Base Pertaining to substances with a pH above 7 (notably calcium, magnesium, potassium and sodium) or substances that release hydroxide ions (OH$^-$).

Base saturation The percentage of base cations that make up the total exchangeable cations in soil.

Base station A stationary *global positioning system* receiver positioned over a known point that continuously collects data from satellites used to correct the recorded positions of the roving global positioning system receivers.

Baseflow The stable portion of a river's discharge, contributed by *groundwater* transfers.

Bathymetry The study and mapping of ocean floor topography.

Bay-head delta A delta at the head of an estuary or a bay into which a river discharges. They typically occur if the river carries large amounts of sediment, or where the coastline is being submerged.

Beach cusps Crescent-shaped accumulations of sand or shingle surrounding a depression on a beach; they are always found in combination and formed when outgoing *rip currents* and incoming waves combine to create circular water movements.

Beamwidth The width of a radar pulse in the direction of travel. A narrow beamwidth means higher resolution.

Bed load Sediment grains transported in water by rolling along the bed surface or through *saltation*.

Bedform A morphological feature developed by fluid flow across the surface of soft sediment, involving the entrainment or deposition of sediment.

Benthic Pertaining to organisms dwelling on the sea floor.

Bergeron process The formation of precipitation described by the Bergeron–Findeison theory. Ice crystals fall from the upper part of a cloud, leading to aggregation of crystals and accretion of supercooled water. Ice crystals grow preferentially by *sublimation* at the expense of surrounding water droplets because the relative humidity above an ice surface is greater than a liquid surface and hence the saturation vapour pressure over water is greater than ice, causing a pressure gradient towards the ice.

Berms Prominent ridges at the back of a beach with a steep seaward face and flat top, marking the limit of the swash zone.

Best Management Practices (BMPs) Methods of minimizing diffuse pollution. BMPs consist of two types: structural, e.g. wetlands, and procedural, e.g. handling methods for polluting chemicals.

Bimodal distribution A statistical term signifying that the frequency curve of a distribution of data has two maxima (two modal classes).

Binge-purge model A model of ice sheet development related to inherent instabilities of large ice sheets; ice loading due to the growth of an ice sheet increases basal pressure causing substrate failure and greater meltwater production, thereby increasing ice flow velocity and ice rafting. The release of this excess ice would then stabilize the ice sheet again. Proposed as a possible cause of *Heinrich events*.

Bioaccumulation The accumulation of toxins in specific parts of the ecosystem (usually at the higher levels of food chains) due to the greater ability of some chemicals to accumulate in zones where they become bioavailable, and are taken up and stored by producers and consumers. The materials must be stored in those parts of the individual that will be consumed. It may result in severe adverse effects on an ecosystem once a threshold level of toxin storage is reached.

Biochemical oxygen demand (BOD) A measure of the amount of biochemically degradable organic matter in water that is widely used in water pollution assessments.

Bioconcentration The level of concentration of accumulated toxins found in plant or animal tissues (via the process of *bioaccumulation*) compared with background natural levels.

Biodiversity The number and variety of taxonomic groups (usually species) of plants and animals at a site or within a region.

Biogenous Pertaining to material derived from organisms.

Biogeochemical Pertaining to the chemical relationships between the geology of an area and its plant and animal life.

Bioherm An ancient mass of rock formed by sedentary organisms, such as corals.

Biological sediments Sediments derived from organic materials, either remains of dead organisms (e.g. shells, plants) or framework organisms (e.g. coral reefs).

Biomass The total dry weight of living organic matter, usually measured per unit area over a particular time interval. Tends to include dead parts of organisms when referring to soils.

Biome A coarse unit of ecosystem classification based on what the land cover and ecosystems look like.

Bioregional theory A concept first put forward by Brunckhorst (1995) as a framework for regional planning based on a firm understanding of the sustainable and interconnected elements of both the geophysical and the cultural indicators that unify an *ecosystem*, therefore valuing both the natural *landscape ecology* and the conservation values developed from local cultural beliefs.

Biosphere All the organisms on the planet, viewed as a system of interacting components making a thin film on the planet's surface, and including parts of the atmosphere, hydrosphere and lithosphere.

Biostratigraphic Pertaining to the division of sedimentary deposits based upon their fossil evidence, each biostratigraphic unit having a distinctive fossil assemblage, e.g. the use of fossilized pollen assemblages for studying European *interglacial intervals*.

Bit scale Refers to the number of colours used to quantify brightness values in a digital image.

Black body An ideal body or surface that absorbs and emits all radiant energy dependent on its absolute temperature.

Black smokers Hydrothermal springs lying along the rift valley of *mid-ocean ridges*. Seawater that seeps into fissures in the basaltic lava becomes superheated and chemically interacts with the basaltic rock to create a black precipitate of metal sulphides.

Blockfield A continuous spread of angular rock fragments across a high mountain or plateau in a periglacial environment; formed *in situ* by frost shattering (occasionally transported and deposited by saturated material in *gelifluction*).

Bond cycles A grouping of *Dansgaard-Oeschger (D-O) events* together into a larger cycle with a long cooling trend followed by an abrupt warming.

Bottomset beds Horizontally layered sediment beds deposited in front of a delta as it prograades seawards. They become covered and end up at the bottom of a stack of deltaic sediments (below the *foreset beds* and *topset beds*).

Boudins Bands of connected debris-rich ice lenses in a glacier, once connected but broken under pressure (forming a sausage shape; *boudin* is French for sausage).

Boulder-controlled slope A scree slope at the base of a cliff in which the scree material is removed as quickly as new material is added to it by rockfall creating a thin covering maintained at the angle of repose. The landform (cliff and boulder-controlled slope) retreats at an almost constant ratio.

Boulton-Menzies theory A theory of *drumlin* formation suggesting drumlins are formed by deposition in the lee of a slowly moving obstacle in the deforming layer of a glacier, therefore streamlining the deposition.

Boundary conditions The physical conditions at the boundaries of a system. They are particularly used in modelling work and for example would refer to the impermeable nature of the floor of an aquifer in the model.

Boundary roughness The roughness of the river channel bed and the submerged bank.

Braided channel A river channel consisting of separate, but interlinked, migrating channels flowing either side of active unvegetated bars that change position owing to bed load transport.

Breakwater A coastal management feature in which a submerged artificial barrier offshore acts to break incoming waves or create new diffraction patterns, protecting the shoreline from wave action in the process.

Bulk density The weight per unit volume of a solid particulate as it is normally packed (including solids and pore spaces). Usually expressed as lb/ft^3 or $g\ cm^{-3}$.

C

Calcareous ooze Fine-grained deep-ocean *biogenous* sediment containing at least 30% skeletal remains of marine organisms based on calcium carbonate ($CaCO_3$).

Calcrete A *duricrust* composed mainly of calcium carbonate.

Caldera A large, steep-sided, land surface depression containing volcanic vents. Formed by large-scale subsidence as the parent magma chamber cools and contracts following a major volcanic eruption.

Calibre The size of sediment particles.

Candle ice Ice consisting of vertically orientated crystals often over 1 m in length.

Capillary water Water that remains in small pores in the soil against the forces of gravity; the major source of water available for plant uptake.

Carbon sequestration The uptake of carbon by a system. Carbon dioxide can be absorbed by plants from the atmosphere and then this is converted into solid plant material. The carbon is then part of the terrestrial system and has been 'taken up' from the atmosphere.

Carnivores Organisms (usually an animal) that consume only meat and therefore occupy a high *trophic level* in the *ecosystem*.

Catastrophe theory A theory in which non-linear interactions within a system cause a threshold to be crossed which then leads to a sudden and dramatic change to a new stable model of operation. Before the threshold is crossed changes may be slow and barely noticeable.

Catchment An open system defined as the area of land drained by a particular stream or river; it represents a fundamental unit in hydrology and is usually topographically well defined. A catchment may be composed of a series of subcatchments.

Catena The sequence of soils occupying a slope transect from the topographical divide to the bottom of the adjacent valley that have developed from similar parent material but vary in profile characteristics owing to the differing topographical and drainage conditions under which they formed.

Cation exchange The process of interchange between a *cation* in soil solution and another on the surface of a soil colloid.

Cation exchange capacity (CEC) The overall net negative charge of clay minerals per unit mass of soil, usually expressed as milliequivalents (meq) per kg of oven-dried soil.

Cations Positively charged *ions*, e.g. the sodium ion (Na^+).

Causal inference The process in which a cause is linked to observations under the assumption that every event must have a cause. It is a key element in the scientific method.

Cavitation A process of fluvial erosion, characteristic of waterfalls and rapids. Constriction of channel flow raises the flow velocity, thereby reducing water pressure and leading to the formation of air bubbles. As the stream widens again and the velocity decreases the air bubbles collapse and the shock waves place considerable stress on the channel walls.

Channel planform The form of channels when viewed from above.

Channel sinuosity A measure of the degree of curvature in channels with meandering *channel planforms*.

Channelization The artificial modification of natural river channels for the purposes of flood alleviation, land drainage or relocation. It may involve channel widening, deepening, straightening, stabilizing (using concrete or piling) or embanking.

Chatter marks Microscale erosional features that appear as crescentic scars.

Chelates A stable compound formed between organic molecules and metallic cations in which more than one bond links the two components (also see *complex*). Chelates are especially important for the behaviour of aluminium and iron in the soil.

Chemical mixing model A chemical mass balance model in which it is assumed that the concentration of an *ion* in solution consists of the mixture of concentrations and flows from different sources.

Chemical sediments Sediments produced by chemical processes, formed predominantly as a result of precipitation of minerals directly from a water body.

Chinook A warm, dry local wind that blows east down the lee slopes of the Rocky Mountains. The wind is subject to warming by adiabatic compression on descent and is warmer in absolute terms at any given altitude than on its windward ascent.

Chroma A measurable variable of soil colour describing the purity or strength of the colour (a chroma of 0 is natural grey).

Clastic sediments Sediments composed of grains of rock which have been weathered and eroded from a pre-existing bedrock material; they are dominated by those grains most resistant to weathering.

Clay A soil mineral particle within the *fine earth* fraction, having an upper limit of 2 μm (two-millionths of a metre) in diameter; very important in determining soil properties.

Clay skin A thin film of clay which has lined an area of soil.

Climax communities A community of plants and animals in steady-state equilibrium with prevailing conditions in the physical environment, seen as the self-perpetuating terminal stages of *ecological succession*.

Coefficient of friction The ratio of the frictional force between two surfaces sliding across one another to the normal force acting perpendicular to the surfaces; it depends primarily on the nature of both surfaces in contact.

Cohesion The force by which a homogeneous substance is held together owing to attraction between like molecules.

Cold glacier A glacier in which ice remains at very low temperatures, tens of degrees below freezing, with no appreciable surface melting. The absence of meltwater causes the glacier to remain largely frozen to the underlying substrate.

Coleoptera A large and important order of insects, distinguished by anterior wings converted into hard sheaths covering the other pair when not in use, i.e. beetles.

Colloid A substance in which very small particles (1–10 μm (millionths of a metre) in diameter) are held in a state midway between a solution and a suspension.

Colour palette Colour reference table for displaying the *digital number* of an image.

Community (in ecological terms) The total living biotic component of an ecosystem (plants, animals and microbes).

Competition (in ecological terms) Negative interaction between organisms caused by the need for a common resource such as light, water or nutrients.

Complex A compound formed between organic molecules and metallic cations by a single bond (also see *chelates*).

Concentration The mass of substance of interest per unit volume, for solutes normally expressed as mg l^{-1}.

Conditionally unstable Instability in the atmosphere that is conditional upon an air parcel becoming saturated, which leads to a shift from cooling via the dry adiabatic lapse rate to the saturated adiabatic lapse rate. This causes the air to become warmer than the surrounding air and ascend more rapidly, leading to strong upward convection.

Conduction Transfer of energy between two bodies in contact.

Continental shelf The zone bordering a continent extending from the line of permanent immersion to the depth at which there is a marked increase in the downward slope which descends to the deep ocean floor.

Continental shield The ancient, stable, low-relief interior of continents; composed primarily of Precambrian crystalline rocks, some as old as 2 to 3 billion years.

Continuous permafrost zone A region in which permafrost occurs everywhere in the ground surface except beneath large bodies of water or ice.

Control A single test performed within a larger set of experiments, whereby no variables are altered from the norm; it acts to monitor the quality of the experimentation and ensure that no unaccounted variables are influencing the results.

Convection Transfer of energy through a fluid (liquid or gas) by molecular motions.

Convergent plate boundary The boundary between two lithospheric plates in which one plate descends below the other, resulting in the consumption of *lithosphere* via the process of *subduction*.

Coping strategies Strategies used to reduce stress by mitigating the influences of an event/situation as opposed to attempting to change the situation.

Coriolis effect As a result of the Earth's rotation any moving object or fluid is deflected towards the right in the northern hemisphere and to the left in the southern hemisphere.

Corrasion The mechanical breakdown of rock due to wearing and grinding caused by material carried in transport across the rock surface.

Corries Basins excavated into a mountainside by the erosive power of a cirque glacier possessing defining features: steep retaining rock walls, a gently inclined rock basin, abundant signs of glacial scour and a terminal moraine. Also known as cirques.

Corrosion A weathering process involving the breakdown of solid rock by means of chemical reactions.

Cosmogenous Pertaining to material that originated extraterrestrially, e.g. meteor fragments and cosmic dust.

Coversands An extensive sand sheet (generally thin and lacking bedforms), formed by wind action in an unvegetated periglacial environment adjacent to an ice sheet. Similar to *loess* but coarser grained.

Crag and tail A streamlined ridge consisting of a resistant rock mass (the 'crag') with an elongated body of less resistant glacial till (the 'tail') on the lee side; a result of preferential erosion around a ridge below a glacier.

Crescentric gouges Crescentric fractures formed by irregular rolling of boulders carried at the base of a glacier; similar to *chatter marks* but larger in scale and less repetitive.

Crevasse-fill ridge A short, linear ridge of glaciofluvial material deposited as meltwater debris in a crevasse of a previous glacier (similar to an *esker* but shorter and less sinuous).

Critical rationalism A form of *positivism* involving the use of *deductive* reasoning. A theory is first adopted, leading to the formation of a hypothesis; this hypothesis is then tested in an attempt to falsify it.

Crusting The process in which a crust (a hard coating) is formed on the ground surface, caused either by concentration of minerals in surface layers due to high evaporative rates, or through high-intensity raindrop impact. It is most common in dryland environments.

Cryptozoic Pertaining to organisms which seek shelter as the preferred ecological niche.

Curie point The temperature above which a metal is no longer attracted to a magnet.

Current ripples Small unidirectional ridges (up to 5 cm in height and 30 cm in wavelength) formed on a beach or on a sandy river bed by the motion of water.

Cyclogenesis A condition in which high-level air divergence is greater than low-level convergence such that air is able to rise. This results in low pressure and the development of cyclonic conditions.

D

Dansgaard–Oeschger (D–O) events *Interstadial* episodes during the Quaternary lasting no more than 500–2000 years involving an abrupt change in temperature of the order of 5–8°C as quickly as a few decades (gradual cooling followed by rapid warming). Large differences are also recorded in atmospheric dust content, ice accumulation rate, methane concentration and CO_2 concentrations. They are interpreted to be the response of internal feedback mechanisms.

Darcy's law A physical law to determine the flow of water through the matrix of a porous medium (e.g. groundwater through an aquifer). The relationship holds only for laminar (non-turbulent) flow of fluids in homogeneous porous media and does not hold well for well-jointed limestone with numerous fissures. According to Darcy's law the discharge of water through permeable material is equal to the cross-sectional area times the slope of the water surface times the coefficient of permeability.

Davisian cycles of erosion A theoretical sequence of 'weardown' processes and forms that occur between the initial uplift of a land mass and its erosion to a *peneplain* (first codified by William Davis in 1899). Its utility within real situations has often been challenged, particularly in relation to the effect of frequent global climatic changes that render any simple sequence of forms improbable.

de Geer moraine An arc-shaped ridge of glaciofluvial material (2–15 m in height) formed transverse to flow at the margin of a retreating glacier where the ice mass borders a glacial lake.

Debris avalanche A type of sudden and very rapid mass movement common on steep mountain slopes mobilized by gravity and commonly originating from a rockslide. It is composed of an unsorted mass of rock and soil that disintegrates during movement into a range of fragment sizes. Movement is characterized by flowage in a dry or wet state; momentum is maintained by a bouncing layer of grain-to-grain collisions at the ground surface.

Debris flow A highly destructive mass wasting process involving a slurry-like flow composed of rock grains, sediment and water with a wide range of sediment size grades and little internal stratification on deposition.

Deductive Pertaining to the process of drawing a conclusion via observational and measurement testing of a principle/hypothesis that is already assumed, i.e. inference by reasoning from general laws to particular instances (opposed to *inductive*).

Degrees of freedom Pertaining to the capability of variation within a system. The number of degrees of freedom in a particular system refers to the number of independent variables that can be freely changed to bring the system to a new equilibrium without altering the phases of the system.

Delta A sedimentary landform where the mouth of a river reaches another water body such as an ocean or lake.

Delta front The limit of the accumulation zone of a delta; it forms from the settling of finer-grained sediment carried furthest into the basin by the decelerating currents.

Delta plain The flat surface of the delta over which the river channel migrates and channel sands accumulate.

Delta switching The process whereby the position of a river delta changes from one site to another. This can occur when the delta becomes so infilled with sediment that the river changes its course dramatically to allow the water to drain more freely into the ocean.

Dendrochronology The study of annual growth rings in certain tree species for dating of the recent past.

Dendroecology The study of tree rings to understand ecological processes.

Denudation chronology The process of attempting to determine the history of a landscape by establishing which stage of the *Davisian cycles of erosion* it represents.

Desert varnish A tough dark layer covering *scree slopes* found in arid areas, produced by weathering of the interior of the scree slope boulders.

Desertification The spread of desert-like conditions and land degradation in arid and semi-arid environments as a result of mainly human influence or climatic change.

Detritus Waste from living organisms, including dead organisms and cast-off fragments.

Deuterium A stable isotope of hydrogen containing two neutrons in the nucleus.

Diagenesis Minor, non-destructive changes in the mechanical or chemical properties of rock shortly after deposition, particularly cementation and compaction (associated with the final stages of lithification (see *lithified*)).

Diapositives A positive photograph developed on plastic or glass with high dimensional stability, rather than on paper, to minimize distortions with shrinkage or expansion of the photographic media.

Diatoms Microscopic single-celled marine or freshwater plants with silica skeletons that contribute to the formation of sedimentary deposits when they die.

Differential GPS A global positioning system that uses one or more roving receivers along with one stationary global positioning system receiver positioned over a known point that continuously collects data from the satellites. This information can be used to correct errors in the global positioning system signals received by the rovers to produce high-quality positional measurements.

Differentials Those species present as a subtype of an ecosystem community that is dominated by another species.

Diffuse pollution The release of contaminants over a large area, e.g. leaching of nitrates from cultivated fields, wash-off of oil from highway surfaces.

Diffuse reflection The redirection of radiation off a rough surface such that the radiation is redirected in many random directions.

Digital elevation model (DEM) A digital representation of a three-dimensional surface, where pixel *digital numbers* represent elevation rather than brightness.

Digital image An image composed of an array of discrete pixels with a numerical assignment to define its tone.

Digital number (DN) The numerical value assigned to a pixel in a digital image, the range of which is defined by the image depth or *bit scale*.

Dilation In general terms, the expansion of material.

Dinoflagellate Unicellular organisms which exhibit a great diversity of form; the most dramatic effect on surrounding life is in marine ecosystems during 'bloom' periods.

Dipterocarp A family of large tropical trees that dominate South-East Asian rainforest ecosystems.

Discontinuous permafrost zone A region in which frozen ground occurs but is not laterally continuous.

Disjunct When two related groups of organisms are separated geographically by a large distance.

Dispersive grain stress A force acting to lift particles in a debris avalanche, caused by grain-to-grain collisions which bounce the particles along the base of the flow.

Displacement flow The method in which soil water at the bottom of a slope is rapidly pushed out of the soil by new infiltrating water entering at the top of a slope, contributing directly to storm hydrographs.

Distributaries Separate river channels that are created when a river splits and does not rejoin the main channel.

Divergent plate boundaries The boundary between two lithospheric plates which are moving apart, resulting in the formation of new *lithosphere*.

Drainage basin That part of the landscape which is drained by a unitary river system.

Drainage density The measure of total stream channel length per unit area of drainage basin (stream length divided by drainage area).

Drainage divide The perimeter boundary of a *drainage basin*.

Drumlin A depositional bedform characterized by elongated accumulations of till streamlined in the direction of ice flow (they can reach 1 km in length, 500 m in width and 50 m in height). Debate exists as to the process of their formation; most accept they are deposited when the competence of a glacier overloaded with sediment reduces.

Dry adiabatic lapse rate The rate at which rising air is cooled as it expands when no condensation is occurring: 9.8°C km^{-1}.

Dry-snow zone A zone within the accumulation zone of a glacier in which there is no surface melt, even in summer.

Dump moraines Ridges formed at the margin of a glacier from material delivered by the ice flow; they lie transverse to the flow direction.

Dunes Migrating ridges of sediment, either as terrestrial deposits of sand formed by aeolian processes, or as small stream bed deposits of sand and clay common in streams of high velocity. The steeper front of the dune is termed the 'stoss' side and the gentler the 'lee' side.

Duricrust A hard, crystalline crust found on arid land surfaces. Evaporation and limited flushing by rains can lead to the accumulation of minerals on the surface or subsurface as capillary action transports minerals from underlying soils and rocks towards the surface.

E

Eccentricity of orbit The shape of the Earth's orbit around the Sun changes from more circular to more elliptical and back again over a 100 000 year period owing to gravitational forces. An increase in eccentricity causes the seasons in one hemisphere to become more intense while the seasons in the other are moderated.

Ecocline A biogeographical term introduced by Whittaker (1953) to describe the combination of environmental factors changing together through space along a gradient, e.g. the simultaneous change in temperature, exposure and soil type resulting from a change in altitude.

Ecological succession The mixture of processes that produce a gradual directional change in ecosystem structure and community at a given site over time, involving progressive habitat modification. Clements (1916) first described ecological succession as a sequence of plant communities characterized by increasing complexity of life form.

Ecosystem An organized open system consisting of biotic (plants and animals) and abiotic (environmental) components interconnected through flows of energy and materials.

Ecotone A zone of transition between two plant communities that is generally characterized by plant competition and can have special significance for more mobile animals due to edge effects.

Edaphic Pertaining to the characteristics of soil, i.e. those environmental conditions influencing a terrestrial ecosystem that are determined by the physical, chemical and biological properties of the soil.

Eh–pH stability field A plot showing the Eh–pH conditions in which the aqueous species of an element occur in a system at equilibrium (see also *redox potential*).

El Niño Southern Oscillation (ENSO) A reduction in the trade wind strength over the equatorial Pacific Ocean causes the westward-driven equatorial ocean current to falter. This leads to the cessation of the typical upwelling of cold deep water off the South American Pacific. The result is the appearance of unusually warm weather and the disturbance of pressure and precipitation systems throughout the southern hemisphere. Literally, 'The Christ Child' for its periodic occurrence every few years commencing during the Christmas season.

Elastic creep Deformation caused by strain forces; the method of movement of solid mantle rocks.

Electrical conductivity The degree to which a substance conducts an electric current.

Electromagnetic energy/radiation A type of energy in transit (or radiation) in which electric and magnetic fields vary simultaneously.

Electromagnetic spectrum The continuum comprising the entire range of wavelengths of electromagnetic energy.

Eluviation The removal of solid or dissolved material from one soil horizon.

Emissivity The rate of emission of energy from a surface per degree of temperature difference between the surface and surrounding substances.

Encapsulated countryside A type of urban ecosystem that is likely to suffer ecosystem degradation in the absence of human management, either ancient habitats or previously managed land.

Endemic Referring to a plant or animal species that is indigenous to one particular location or region, i.e. a consequence of geological isolation and allopatric (populations become more isolated from one another) *speciation*.

Endoreic Pertaining to an inland drainage system which does not terminate at the coastline; predominantly found in dryland environments ending in salt pans or *playas*.

Englacial Inside a glacier between the surface and the bed.

Entrainment The process in which small sediment particles are mobilized from a bed surface and transported in fluid suspension.

Entropy The degree of disorder or uncertainty in a system.

Environmental economics A branch of economics that involves the analysis and expression of the impacts of an activity on the environment in monetary values in order to promote communication with decision-makers and providing a more straightforward view of the impacts.

Environmental gradient The change in an environmental variable that acts as a control on plant and animal communities along a transect from one location to another, e.g. altitude, moisture, temperature, soil acidity. Environmental gradients vary in steepness, direction and the severity of their influence upon the community; they can lead to evolutionary branching and hence *speciation*.

Environmental impact assessment (EIA) An evaluation designed to identify and predict the impact of proposed action or project on the environment in order to ensure all possible impacts are considered before implementation. They are legally required in many countries.

Environmental lapse rate The actual rate at which temperature falls with increasing altitude in the local atmosphere.

Environmental technology assessment An evaluation designed to identify and predict the potential environmental, economic and social impacts of a new technology.

Ephemeral Short-lived.

Epilimnion The surface layer of water in a water body which is warmer and less dense than the water layer below which remains trapped as it is cooler and more dense than the water above.

Epiphyte A plant growing above the ground surface that is not rooted in the soil but uses other plants for support; commonly associated with tropical rainforests.

Equilibrium line The boundary between the *accumulation zone* and the *ablation zone* of a glacier where the net mass balance is zero.

Ergodic method Studying the development of a process or object over time (i.e. the sequence of landforms) by evaluating areas that represent different stages of advance in the process, therefore substituting space for time. For example, the sequence of successional stages of a salt marsh over time can be observed by studying a horizontal transect through the marsh from the youngest to the oldest section.

Ergs Another term for *sand seas*.

Esker A narrow winding ridge of glaciofluvial sand and gravel deposited by a meltwater stream flowing at the bed of a glacier.

Essential elements Elements found in the soil without which, or in the wrong proportions, green plants cannot grow normally. There are 16 essential elements consisting of *micronutrients* and *macronutrients*.

Estuary The mouth of a river where it broadens into the sea and within which the tide ebbs and flows, leading to an intermixing of freshwater and seawater. Estuaries are usually sites of deposition, especially if the river charges more sediment than can be removed by the tidal current or wave action.

Eustasy Global change in ocean water level due to change in the volume of water in the oceans.

Eutrophication Enrichment of freshwater and marine water bodies with plant nutrients to the extent that plants in the water bloom at the expense of other aquatic organisms.

Evaporite A mineral or sedimentary rock composed of soluble salts resulting from the evaporation of a body of water.

Evapotranspiration The transfer of liquid water from the Earth's surface to water vapour in the atmosphere by means of evaporation and plant *transpiration*.

Exchangeable cations Cations attracted to the surface of clay minerals and adsorbed by electrostatic attractions which can be displaced by cations in the soil solution through *cation exchange*.

F

Fabric The orientation of particles within a rock or sediment.

Falling limb The section of a storm hydrograph depicting the decrease in river discharge after rainfall has ceased following a storm event.

Fecundity The faculty of reproduction. An organism has a high fecundity if it reproduces quickly and in large numbers and can therefore recover its population quickly after a problem.

Feeder–seeder mechanism A type of *orographic* enhancement of precipitation. Adiabatic cooling of air forced to rise over mountains causes saturation of water vapour and cloud formation. The water vapour of this 'feeder' cloud is swept into the precipitation of a frontal 'seeder' cloud aloft, increasing the overall precipitation on the mountain.

Fiducial marks Marks exposed on imagery that act as a frame of reference for the *x,y* coordinate system of the image.

Field capacity A term to describe the state of the soil when all *gravitational water* has drained away, leaving only the *capillary water*.

Field drains Subsurface drainage system installed under agricultural land to reduce the soil moisture content.

Fine earth The fine fraction of soil mineral particles consisting of sand, silt and clay, less than 2 mm in diameter.

Firn Compacted granular snow with interconnecting air spaces in at least its second accumulation season, in the process of being transformed into glacier ice (with a density usually greater than 0.4 but less than 0.8 kg m^{-3}).

Fjord A long, deep basin, previously excavated by a glacier, and has since become inundated by the sea as a result of sea-level rise during deglaciation.

Flashy regime A term used to describe a stream with a fast hydrological response to precipitation events involving a rapid rise in channel water levels.

Flow duration curve A graphically plotted curve used in the analysis of river flow frequency. The frequency distribution of the mean flow of a river at a particular site is calculated and the percentage of time any particular discharge rate is reached and exceeded is then plotted. The slope of the curve indicates the magnitude of the flow.

Flow traction A type of wash process that involves the transport of sediment along the ground surface due to the stress applied by rainwater flowing downslope and the friction maintaining the rolling and bumping of particles within the moving water; a type of *rillwash*.

Flows A type of rapid mass movement in which different parts of the mass move over each other with differential levels of movement; see *debris flow*.

Fluidization The process in which water detaches an entire fine-grained sediment deposit and converts it into a mixture of water and sediment that takes on most of the properties of a fluid.

Flute An elongated ridge formed by the infilling of cavities on the downstream side of obstacles in the bed (more elongated than drumlins); the depositional equivalent of crag and tail features.

Föhn wind The European equivalent of the *Chinook* wind.

Footprint The area on the ground 'seen' by a sensor at an instantaneous moment in time.

Foraminifera An order of Rhizopoda: small, single-celled marine organisms with a shell of calcium carbonate.

Foredune ridge A ridge of sand on the seaward side of a dune system where several smaller dunes which are forming have coalesced.

Foreset bed A seaward-sloping sediment bed deposited at the advancing edge of a growing delta; the sediment accumulates underwater and constitutes the bulk of the delta (overlain by the *topset bed*).

Form drag A drag force that slows and curves a flow path over surface topography, e.g. the slowing of glacier flow due to obstacles on the glacier bed, or the slowing of air flow over a mountain range.

Frequency distribution A method of plotting numerical data to show the number of times different values of a variable occur within a sample.

Friction cracks A variety of rock fractures caused by the action of glacier ice passing over bedrock (including *crescentic gouges*).

Frictional drag The retarding drag force associated with the interaction between a particle moving across the surface of another particle.

Frost blisters Small ice-cored mounds that develop over a single winter as a result of *frost heave*.

Frost creep The slow downslope creep of the *active layer* on a slope due to freezing of the soil causing it to expand normal to the surface; subsequent thawing then permits the vertical settlement of the soil, resulting in a net downward movement.

Frost heave The vertical lifting of the soil surface into doming frost hillocks as a result of pressures caused by the freezing of groundwater under periglacial conditions.

Fundamental niche A *niche* containing the ideal conditions for the species requirements, only realized in simple situations where no other competitors for any of the resources exist.

G

Gabbro A type of intrusive igneous rock that crystallizes slowly at depth; its minerals (plagioclases and pyroxenes) are large and visible to the naked eye.

Gabions Wire rock-filled cages used to prevent river bank erosion and stabilize slumping hillsides.

Gaia A conceptual theory proposed by James Lovelock (1979) that proposes that the Earth maintains conditions suitable for life by self-regulation and feedback mechanisms whereby all elements of the Earth are interlinked at all scales; the Earth acts almost as a conscious biological organism.

Gas hydrate An ice-like crystalline solid formed from a mixture of water and a gas, often methane.

Gelifluction A type of *solifluction* only occurring in areas of permafrost. When seasonally frozen soil in the *active layer* thaws during spring, water cannot percolate down owing to lower layers of permafrost, thereby creating a lubricating effect for the slow downslope creep of the water-saturated material.

Genetic classification A term used in association with the classification of sediment types by the mode of their deposition.

Geographical information system A system of hardware, software and procedures designed to support the capture, management, manipulation, analysis, modelling and display of spatially referenced geographical data for solving complex planning and management problems.

Georeferencing Correcting a remote sensing image to remove geometric distortions caused by the motion of the sensor or by the motion of the Earth beneath the sensor so that objects in the image correspond to their positions on the ground.

Geostrophic wind High-level wind blowing parallel to isobars, in which the pressure gradient and Coriolis force are in balance.

Geothermal heat Energy derived from the heat in the interior of the Earth. This energy is found everywhere beneath the Earth surface, although the highest temperatures are concentrated in regions of active or geologically young volcanoes.

Geothermal heat flux The amount of heat given off by the Earth's crust; generated by radioactive decay, chemical processes and friction at plate boundaries.

Glacial intervals An extended cold phase during the Quaternary in which ice sheets and glaciers extended widely from the poles to lower latitudes.

Glen's law A physical law referring to the rate of deformation of glacier ice due to shear stress; fundamental in the understanding of glacier flow.

Gleying The formation of a gley soil; a blue-grey soil or soil layer caused by the reduction of iron and manganese compounds in stagnant saturated conditions.

Global positioning system (GPS) A system consisting of a constellation of orbiting satellites and one or more global positioning system receivers on the ground (or in an air- or watercraft) that are used for precise positioning.

Gondwanaland A large ancient continent of the southern hemisphere made up of present-day South America, Africa, Australia and Antarctica.

Graben A structural rock mass that is downdropped by parallel faults on both sides, often forming a structural valley.

Grain roundness A textural property of sediments associated with the degree of angularity/roundness that provides environmental information, e.g. very rounded grains are formed via considerable abrasion and are likely to have undergone extended transport.

Gravitational water The water that drains from *macropores* in the soil following a period of soil saturation due to gravitational forces and is replaced by air.

Greenhouse effect - enhanced Human disturbance within the last few centuries has caused the concentration of the major greenhouse gases to increase. Thus, the presence of higher concentrations of such gasses may enhance the greenhouse effect, making the planet warmer. The Earth's average global temperature has been estimated to have risen by 0.3–0.6°C in the past 100 years, as a result of human influence.

Greenhouse effect - natural The atmosphere traps heat energy at the Earth's surface and within the atmosphere by absorbing and re-emitting long-wave radiation; 90% of long-wave radiation emitted back to space is intercepted and absorbed by greenhouse gases such as carbon dioxide and water vapour. Without the greenhouse effect the Earth's average global temperature would be −18°C.

Ground control points Points on the ground of known coordinates that can be identified in the imagery and can be used to remove distortions in a process called geocorrection.

Ground diffusivity Pertaining to the ability of the soil to propagate fluctuations in surface temperature to greater depths in the ground.

Ground-penetrating radar A geophysical technique involving the propagation of high-frequency electromagnetic waves into the ground which are then reflected back to the surface from boundaries at which there are electrical property contrasts. It allows high-resolution mapping of subsurface features such as soil pipes, bedrock and other geophysical anomalies.

Groundwater The portion of subsurface water stored in both soils and aquifer rocks below the water table (in the *saturated zone*).

Groynes Artificial structures positioned across a beach at right angles to the shore to trap sediment being transported by longshore currents, therefore inhibiting loss of sediment by longshore drift.

Gypcrete A *duricrust* composed of calcium sulphate.

Gypsum An *evaporite* mineral composed of calcium sulphate with water.

Gyre A large circular movement of water. It usually refers to the large oceanic gyres in the subtropical high-pressure zones where geostrophic currents rotate clockwise in the northern hemisphere and counterclockwise in the southern hemisphere.

H

Halocline A layer of water in which the salinity (saltiness) changes rapidly with depth.

Hanging valley A tributary valley which, at convergence with the trunk valley, has a higher ground level, resulting in a sharp drop in elevation (the result of glacial activity in the main valley).

Hard engineering Pertaining to coastal management practices that involve the construction of large-scale structures to protect the coastline (e.g. sea walls, breakwaters and groynes); most hard engineering practices change the local sediment dynamics.

Health impact assessment (HIA) An evaluation designed to identify and predict the health impacts on society of a proposed activity.

Heinrich events Referring to events occurring during the coldest points of *Bond cycles* in which vast amounts of icebergs were discharged into the North Atlantic, immediately followed by abrupt warming.

Helicoidal flow A process by which water flows in an outward direction when approaching a meander bend, causing water levels on the outside of meander bends to become super-elevated. Water then flows inwards along the channel bed as a return flow. This results in individual water molecules cork-screwing around the meander bend.

Herbivore An organism that feeds off plants (primary producers) and therefore occupies the second *trophic level*.

Heterotrophs Organisms that cannot photosynthesize and therefore feed directly on *autotrophs* and other heterotrophs for their energy supply. They are described as first-order consumers and form the second *trophic level*.

Histogram A column diagram where values are divided into equal parts and the frequency of occurrences within each subdivision are summed and plotted.

Hjulström curve An empirical curve defining the threshold flow velocities (i) to initiate motion of sediment grains of different sizes on a stream bed, (ii) necessary to keep the sediment grains in transport and (iii) the depositional velocity.

Holocene The second epoch of the Quaternary, in which we live now, preceded by the *Pleistocene*. It began approximately 10 000 years ago. The Holocene is an interglacial period.

Horizontal interception Formation of water droplets by condensation of atmospheric moisture on vegetation surfaces, contributing significantly to catchment precipitation inputs under vegetation canopies in conditions of high atmospheric humidity.

Horn A high, spire-shaped mountain summit with steep sides formed by the convergence of intersecting walls of several cirques.

Horticulture Subject of using plant processes for the purposes of garden development.

Hot spots A centre of volcanic activity and igneous rock production located away from plate margins, thought to be positioned over a rising mantle plume and related to convection processes which originate at the core-mantle boundary deep in the Earth.

Hue A measurable variable of soil colour that describes the dominant colour of the pure spectrum (usually redness or yellowness).

Hummocky moraine A type of *moraine* characterized by considerably undulating terrain; caused by deposition from meltout of *supraglacial* and *englacial* material in *kettle holes* and crevasses.

Humus A type of soil organic matter which is very resistant to decomposition.

Hydraulic geometry Referring to the river flow characteristics of a channel (such as discharge, depth, width and velocity) and their relationship to one another.

Hydraulic radius The ratio of the cross-sectional area of water flowing through a channel to the length of the *wetted perimeter*; it represents a measure of the efficiency of the channel at conveying water and the proportion of water subject to bed surface friction.

Hydraulic sorting The process by which particles of river bed material are sorted into sections of near uniform particle size due to the change in river competence throughout its journey from source to mouth.

Hydrogenous Pertaining to sediments derived from *ions* in seawater through geochemical processes, e.g. metal ions of iron and manganese are release from hydrothermal vents and oxidize or combine with silica to form metal-rich sediments.

Hydrograph A graph showing river discharge plotted against time for a point on the river channel network, displaying a characteristic shape during rainfall events.

Hydrophyte A plant that has adapted to grow in wet or waterlogged conditions.

Hygroscopic water Soil water held as a tight film around individual soil particles and unavailable to plants because of the very strong attraction between the water and soil particle.

Hyper-arid Pertaining to extremely dry areas ('true deserts') that may go as long as 12 months without rainfall (e.g. central Sahara).

Hyper-concentrated flow Similar to a debris flow but with a higher water content, acting more like a liquid and with less viscosity; these flows behave like a sediment-rich stream maintained by forces of turbulence.

Hyperspectral scanners Scanning remote sensing instruments that record digital images using multiple narrow bands. Similar to *multispectral scanners* but with higher spectral resolution. These are also called imaging spectrometers.

Hypolimnion A cooler, lower layer of water in a water body which does not readily mix with the upper warmer layer as it is more dense and hence remains below the warmer, less dense layer.

Hysteresis A process whose progress is determined by the direction in which the reaction is occurring. It is normally described by a bivariate plot in which the value of one variable is dependent on whether the other variable is increasing or decreasing.

I

Ice creep A slow, continuous movement of ice involving non-recoverable deformation of the ice owing to intergranular motion caused by internal pressure and the force of gravity.

Ice rafting The process by which glacially eroded debris is transported by floating ice (ice floes or icebergs); it may be transported great distances and deposited either on the sea floor when the ice melts or on beaches.

Ice shelves Thick floating sheets of ice extending over the sea from a landward ice sheet, fed by the ice sheet and snow accumulation.

Ice streams Fast-flowing 'rivers' of ice within more slowly moving ice sheet walls.

Ice wedge polygons *Ice wedges* that have joined together owing to the annual reopening and expansion of the ice wedge.

Ice wedges V-shaped bodies of ground ice that extend into permafrost (up to 1.5 m in width and 3–4 m in depth). Under very low temperatures frozen ground contracts as it is further cooled, causing the ground to crack; water enters during spring and summer and then freezes into an ice wedge.

Iceberg calving The process in which a large mass of floating ice breaks away from an ice shelf; a major method in which mass is lost from ice sheets.

Ice-pushed ridge A ridge of material accumulated by the ploughing action of a glacier but composed of material that is not glacially derived (i.e. similar to a *push moraine* but not consisting of glacially derived debris).

Igneous rock Rock that has originated from a molten state such as lava from a volcano.

Illumination The degree to which a scene or object is lit, in this case by the Sun.

Illuviation The deposition of solid or dissolved material into a soil horizon.

Image mosaic A composite of remote sensing images to produce an image of greater coverage.

Image orientation The direction in which a sensor is pointed to capture an image. Images can be orientated vertically, horizontally or obliquely.

Imaging spectrometers Another term for *hyperspectral scanners*.

Imbrication The wedging of particles among others. Often small particles become trapped by larger ones so that even though the flow is great enough to entrain them, they cannot move until the larger particles are entrained.

Incremental methods Techniques for estimating the age of deposits based on the measurements of regular accumulations of sediment or biological matter through time, e.g. *dendrochronology*, analysis of *varves* and ice cores.

Inductive Pertaining to the process of inferring a general law or principle from the observation of particular instances; by classifying and ordering unordered knowledge, regularities may be identified and general laws discovered (opposed to *deductive*).

Indurated Pertaining to soils and sedimentary rocks which have become hardened and compacted by post-depositional chemical and physical alterations.

Industrial Revolution A major shift of technological and cultural practices in the late eighteenth century and early nineteenth century in some western countries. It began in Britain and spread throughout the world and consisted of an engagement with energy generation through fossil fuel burning, construction, invention and mass transport systems.

Infiltration capacity The maximum rate at which water can enter soil under specified conditions.

Infiltration rates The volume of water passing into the soil per unit area per unit time (i.e. the rate at which water added to the surface enters the soil).

Infiltration-excess overland flow A form of *overland flow* occurring when rainfall intensity exceeds the infiltration capacity and excess water is stored and transported on the surface (also known as Hortonian overland flow).

Interception The process by which precipitation is prevented from reaching the ground by the vegetation layer.

Interception storage The storage of water on leaves and tree trunks when precipitation has been intercepted by vegetation en route to the ground surface.

Interference The fading, disturbance or degradation of a signal (in this case surface reflectance) caused by signals from unwanted sources (i.e. the atmosphere).

Interglacial intervals A long, distinct warm phase between glacial stages during the Quaternary; the Earth's glaciers become severely diminished owing to climatic amelioration (restricted to very limited locations with sufficient conditions).

Interstadial A short period of climatic amelioration and ice retreat within a glacial stage, less pronounced than an *interglacial interval*.

Interstitial ice Ice crystals (individual or fused together) occupying the pore spaces of a soil or rock.

Involutions Features caused by the deformation of unconsolidated surface materials (i.e. disruption to the sedimentary structure and soil profile) due to thawing of ice-rich ground; often used as a diagnostic for past permafrost conditions.

Ionic diffusion The upward movement of ions through the soil without the aid of water, due to the difference in concentration of ions from the base to the surface of the soil. The close proximity to parent material at the bottom of the soil profile results in a greater quantity of ions; the random movements of the ions will then form a general upward movement to an area of fewer ions.

Ions Positively or negatively charged atoms.

Island biogeography The study of the distribution and evolution of organisms on islands or even 'virtual islands' (resulting from some barrier other than the sea). More narrowly, island biogeography is the examination of MacArthur and Wilson's (1967) equilibrium theory of *speciation* in geographically isolated areas, whereby a relationship is identified between species richness of an island and its size and isolation, among other characteristics.

Isobars Lines on a map joining points of equal atmospheric pressure.

Isohyets Contour lines connecting points of equal rainfall.

Isomorphous substitution During the formation of a clay mineral, the process in which one atom in the crystal lattice is replaced by another of similar size without disrupting the crystal structure. The replacing ion is generally of a lower positive charge, causing the clay mineral to become electrically negative.

Isostasy The principle by which the Earth's crust 'floats' upon the denser *mantle*, following Archimedes' law of hydrostatics. The thicker, more buoyant crust (continental regions) stands topographically higher than the thinner, denser crust (under the oceans) to create an equilibrium situation.

Isovels Contour lines connecting points of equal velocity.

J

Jet streams High-speed long, narrow winds in the upper atmosphere. These currents meander and reach speeds of 400 km h^{-1}.

Jetties *Hard engineering* coastal management structures built along the banks of a tidal inlet at a river mouth in order to stabilize unpredictable shifting channels for navigation purposes.

K

Kames Steep-sided isolated conical hills of bedded glaciofluvial materials deposited by meltwater along the sides or margins of a glacier.

Karst Referring to the ground surface depressions and extensive underground drainage network created by limestone solution.

Katabatic drainage Radiative cooling at night causes the air close to the ground to cool; this cooler air is slightly denser and slowly moves downslope to lower ground and into depressions. It is greatest in cloud-free and dry conditions with light winds (limited mechanical mixing of the air).

Kettle hole A closed depression found in glacial till deposits, formed by the melting of a large mass of ice that became incorporated and preserved in glacial till.

Keystone species Species that are highly connected to the entire food web; their loss may result in *ecosystem* collapse and huge loss of biodiversity.

Kinematic viscosity The ratio between the density and viscosity of a fluid.

L

Lagoon A coastal bay totally or partially enclosed and cut off from the open sea by a *barrier beach*, *spit*, shingle ridge or an offshore reef.

Lahars Flows of loose soil, rock, ash and water following a volcanic eruption.

Laminar flow One of the two ways in which water can flow; it involves all water molecules flowing in the same direction parallel to one another resulting in no mixing of water.

Landscape corridors Narrow strips of land that differ from the *landscape matrix* existing on either side; the key characteristic relates to their function in connecting different environments and the often sharp microclimatic and soil gradients from one side of a corridor to another.

Landscape ecology A concept used in exploring regional and small-scale biogeographical distributions. It refers to the analysis of the cause–effect relationships between the living community and the immediate environmental conditions, which have created the specific landscape pattern observed. The theory suggests the landscape consists of a matrix of patches and corridors providing oases and pathways for species dispersal and movement.

Landscape matrix The element of the landscape that contains within it other landscape components (patches and corridors) into a complex system that controls the local biogeography. The stability of the matrix is dependent upon the extent and development of *landscapes patches* and *landscape corridors*.

Landscape patches Distinctive elements within the wider land-scape, such as ponds, woods or towns. Their analysis involves the influence of patch characteristics (shape, frequency, origin and stability) upon the local ecosystem. The community of a landscape patch may vary substantially to the surrounding landscape and be very vulnerable to its influences, thus having particular relevance to conservation ecology.

Landslides A mass movement process whereby a large coherent mass of material moves down a slope under the influence of gravity, remaining undeformed.

Lapse rate Rate at which temperature decreases with altitudinal increase.

Latent heat The amount of heat required to change the state of a substance, e.g. from a liquid to a gas, or vice versa.

Lateral moraine A ridge of glacial debris lying parallel to the sides of a glacier or lying along the sides of a valley formerly occupied by a glacier, consisting of dumped material and frost-shattered material from the valley walls.

Laterization The process in which high temperatures and heavy rainfall cause intense weathering and leaching of the soil, producing horizons depleted in base cations and enriched in silica and oxides of aluminium and iron.

Laurasia A large ancient continent of the northern hemisphere made up of present-day North America, Europe and Asia.

Law of limiting factors Pertaining to a species, the necessity for all the environmental factors that control its survival to be maintained within a range that the organism can tolerate; if just one of these controlling variable falls outside of the toler-ance range the organism will not survive.

Laws of thermodynamics Laws pertaining to the conservation of energy. The first law of thermodynamics states that energy cannot be created or destroyed, only transformed from one form into another; thus energy is conserved. The second law of thermodynamics states that isolated systems become more disorganized over time.

Leaching Downward transport of soluble soil material in solution through the soil profile by percolating surplus water, depositing some in lower layers but removing the most soluble entirely.

Least-squares adjustment A mathematical method for fitting a model to data so as to minimize error between the observed values and the estimated values.

Liana A woody vine supported on the trunk or branch of trees, usually tropical.

Life-cycle analysis (LCA) The evaluation of all the environ-mental impacts of a product from the time the raw materials are taken from the Earth to the time the product is thrown away and added to the ecosystem (including its manufacture, use and disposal).

Linear wave theory Main theory of ocean surface waves used in ocean and coastal engineering from which important equa-tions are derived.

Lithified Pertaining to the transformation of unconsolidated sediments into a cohesive sedimentary rock mass through cementation, compaction and crystallization (lithification).

Lithogenous Pertaining to material derived from the physical and chemical breakdown of rocks and minerals.

Lithosphere The rigid outermost layer of the Earth, consisting of the crust and upper section of the mantle above the asthenosphere; characterized by brittle behaviour.

Litter A type of soil organic matter consisting of decomposing residues of plant and animal debris.

Littoral drift The transport of beach material along the coast, sometimes referred to as longshore drift. Waves surging along the beach at an oblique angle transport sediment up and along the beach in the *swash* followed by transport more perpendicular to the coast in the *backwash* (creating a zig-zag movement of sediment along the beach).

Lobate Characterized by having a tongue-like shape, e.g. the ice lobe of an alpine glacier.

Loess A fine-grained (less than 50 μm (fifty-millionths of a metre)), commonly non-stratified and unconsolidated sedi-ment. It is composed of quartz, feldspar, carbonate and clay minerals that have been transported by wind from arid land surfaces and deposited elsewhere, sometimes thousands of kilometres away.

Logical positivism A form of *positivism* in which *inductive* reasoning is used to form theory and acquire knowledge from experimentation.

Long profile A graphical curve displaying the longitudinal alti-tude profile of a river from source to mouth (height of the river plotted against distance from stream source). It illus-trates the change in river gradient downstream.

Longshore currents A net movement of water parallel to a coastline. This occurs because waves surging along beaches at oblique angles are followed by more perpendicular trans-port out to sea resulting in a net water movement along the coastline.

Longshore drift Another term for *littoral drift*.

Long-wave radiation Radiation that has been emitted by a surface at a longer wavelength than its solar source. This is also called terrestrial radiation.

Lumped model A catchment model in which catchment charac-teristics are assumed to be uniform across space.

Luvisols A group of soils produced by clay *eluviation* (also known as acid brown earths).

Lysimeter An instrument for taking direct measurements of *evapotranspiration*; by isolating a block of soil (with its vegeta-tion cover), the weight of the block can be used to represent the quantity of water and its change over time can be calculated.

M

Macronutrients The group of *essential elements* found in high concentration in plants (carbon, oxygen, hydrogen, nitrogen, phosphorus, sulphur, calcium, magnesium, potassium and chlorides).

Macropore Infrequent large opening or void in the soil (greater than 0.1 mm in diameter) that can promote rapid, preferential transport of water and chemicals, formed by structural cracks and fissures or by biological activity, e.g. earthworms, burrowing creatures and plant roots.

Macropore flow The movement of water through the soil within larger pores (*macropores*).

Magnetometer An instrument for measuring the strength of the Earth's magnetic field.

Main stream length The distance of the main river channel in a catchment from source to mouth (equating to the length of the *long profile*). Given in kilometres.

Mangroves A term applying to the variety of trees and shrubs which grow on saline mudflats in tropical coastal areas to form a dense swamp forest. Their roots trap silt which accumulates to form a swamp.

Mantle The zone within the Earth's interior lying between the partially molten core and the thin surface crust, containing 70% of the earth's total mass and composed principally of magnesium-iron silicates.

Mass balance The difference between the total accumulation and *ablation* of a glacier with time, i.e. a positive mass balance exists when accumulation exceeds ablation for a given period.

Mass movement The downslope movement of sediment, soil and rock material as a single unit (the individual fragments are in close contact); a number of mass movement processes can be identified including *debris flows*, *debris avalanches*, *slumping* and *landslides*.

Mass wasting The spontaneous downhill movement of surface materials (soil, *regolith* and bedrock) under the influence of gravity, without the active aid of fluid agents.

Massive ice Very thick bands of *segregated ice*, up to several metres thick.

Matrix flow The movement of water through the soil within very fine pores.

Meandering rivers Sinuous river channels that migrate downstream owing to river bank erosion on the outside of meander bends and deposition of bed material on the inner bank. Excessive meandering leads to oxbow lake formation.

Mesophyte A plant that requires a moderate climate in terms of temperature and precipitation in order to survive.

Metamorphic rock Rock which has altered its form through structural and mineralogical change due to heat and pressure from the surrounding conditions.

Metamorphosis (biol.) A change in the form, function or habits of a living organism by a natural process of growth or development, e.g. the change of a caterpillar into a butterfly.

Micronutrients The group of *essential elements* found in small concentration in plants (iron, manganese, zinc, copper, boron and molybdenum).

Micropores Very small pores in the soil that can hold soil water (less than 0.1 mm in diameter).

Mid-ocean ridges The zones in which oceanic lithosphere is created by the spreading of *divergent plate boundaries*.

The relative buoyancy of the newly formed oceanic crust causes the topography to be raised, creating a high-relief ridge.

Mie scattering The wavelength-dependent redirection or scattering of electromagnetic radiation at wavelengths of about the same magnitude as the size of the particles.

Milankovitch theory A theory describing the external driving force behind the glacial cycles of the Quaternary. The amount of solar radiation reaching different parts of the Earth from the Sun varies as the eccentricity of the Earth's orbit, the obliquity of the axis of rotation and the precession of the equinoxes change over time in a regular and predictable way.

Mineralization The process of forming a mineral by combination of a metal with another element.

Mohorovičić discontinuity (Moho) The contact surface between the crust and the mantle; the zone in which seismic waves are significantly modified.

Mole The quantity of a substance that contains the same number of chemical units as there are atoms in exactly 12.000 grams of carbon-12.

Molten rock Rock in a state of a liquid; the rock has melted and flows as any liquid.

Monoclimax A theory of vegetation requiring that all sequences of ecological succession within a given climatic region converge on a single uniform stable climax community depending solely on regional climate.

Monsoon A system of winds that switch direction from ocean-continent to continent-ocean between summer and winter in response to the northerly and southerly movements of the intertropical convergence zone (ITCZ). The characteristics of a monsoon climate are most apparent in India and South-East Asia; the *jet stream* reverses from westerly to easterly, causing the north-east and south-west monsoon seasons that are responsible for the majority of inter-annual climatic change in the region.

Montmorillonite A soft mineral that forms as very small plate-shaped crystals. Two silicon tetrahedral sheets enclose an aluminium octahedral sheet in the structure. Considerable expansion can occur when water moves between the silica sheets.

Moraine An accumulation of glacial till that has been transported and deposited by a glacier or ice sheet; classifications of moraines are usually based on the mode of their formation: see *de Geer moraine*, *dump moraines*, *hummocky moraine*, *lateral moraine*, *push moraine* and *rogen moraine*.

Moulin A rounded vertical or steeply inclined hole within a glacier down which meltwater travels.

Multispectral scanners Scanning remote sensing instruments that record digital images using several, moderately narrow bands, typically between the ultraviolet and infrared portions of the spectrum. Similar to *hyperspectral scanners* but with lower spectral resolution.

Munsell colour chart A standard system for the description of soil colour based on three measurable variables (*hue*, *value* and *chroma*).

Mycorrhizal Pertaining to the nature of mycorrhiza; a fungus growing in or on a plant root involving a symbiotic relationship between the two.

N

Neap tide A tide that occurs at the first and third quarters of the Moon when the gravitational force of the Moon is opposed to that of the Sun, thereby producing a relatively small tidal range, and causing lower than average high tides and higher than average low tides. The velocity of tidal currents is slowed at this time.

Nearshore A process-based term for the area comprising the swash, surf and breaker zone; the area in which waves are forced to break owing to the shallowing of water closer to the shoreline.

Negative feedback An event or process resulting from another event that counteracts its effects.

Net primary productivity (NPP) The amount of energy (carbohydrate) fixed by plants during photosynthesis subtracting that used in respiration; it represents the growth of the plant/ecosystem, measured in unit area per unit time.

Net radiation The difference between the total incoming radiation and the total outgoing radiation.

Niche The position or role of an animal or plant species within its community in relation to its specific requirement of habitat resources and microclimatic conditions (i.e. climate, shelter, food, water). No two species with identical resource requirements can occupy the same niche (the principle of competitive exclusion applies).

Nitrifying bacteria Bacteria that oxidize ammonium to nitrite and thence to nitrate.

Nivation Localized erosion of a slope caused by the combination of frost action, gelifluction, frost creep and meltwater flow at the edges and beneath a snowpack; accentuated in permafrost-free zones during periodic freezing and thawing of constantly moistened ground.

Nivometric coefficient The percentage of precipitation falling as snow within a given area.

Non-selective scattering The wavelength-independent redirection or scattering of electromagnetic radiation caused by atmospheric particles that are much larger than the wavelengths of the light they scatter.

North Atlantic Deep Water (NADW) A body of water formed in the North Atlantic Ocean. Relatively saline water from the Gulf Stream cools when it moves rapidly north into the Norwegian Sea; it becomes denser and sinks, flowing back south to form a major component of the *thermohaline circulation* of the oceans.

Nuée ardente A cloud of superheated gas-charged ash that develops into a pyroclastic gravity flow following a very explosive volcanic eruption.

O

Occluded front The process in which a cold front of a depression overtakes a warm front. The occluded front is classified as warm or cold depending on whether the air ahead of the warm front is colder or warmer than the air following the cold front.

Offshore A morphological term for the area below the wave base, just beyond the shoreline and foreshore.

Omnivore Organism that feeds on both plants and animals.

Ophiolites A layer of oceanic crust created at mid-oceanic ridges and uplifted at convergent plate boundaries, now lying exposed above the water at continental margins.

Options analysis The identification of key criteria through which to evaluate management options when dealing with an environmental issue.

Organic Pertaining to any compound containing carbon, except simple compounds such as oxides and carbonates (which are considered inorganic).

Orographic Pertaining to mountains; for example, orographic precipitation is caused by the forced ascent of air over high ground/mountain barrier.

Oscillatory flow Currents that oscillate backward and forward such as wave currents.

Overland flow The motion of a surface layer of water as sheet flow (unchannelled).

Oxidation A chemical weathering process involving the combination of free oxygen with minerals to form oxides with a positive electrical charge.

Oxidation state The electronic state of an atom in a particular compound; equal to the difference between the number of electrons it has compared with a free atom, e.g. in calcium chloride ($CaCl_2$), calcium has the oxidation state of +2 (Ca^{2+}) and chlorine has the state -1 (Cl^-).

Oxisols A soil order found in the tropics consisting of old, extremely weathered soils which have been highly leached and consequently become infertile with a low base status.

P

Palaeoecology The study of ancient plant and animal distributions and processes.

Palaeomagnetism The study of the magnetism of igneous rock; the strongly magnetic particles of magnetite in igneous rock become permanently orientated in the direction of the Earth's magnetic field at the time of the lava cooling.

Palsas Low permafrost-cored mounds, 1-10 m high, formed in peat of both continuous and discontinuous permafrost zones; caused by differential frost heaving linked to the thermal conductivity of peat.

Palynology The branch of science concerned with the study of living or fossil pollen and spores; often used in the reconstruction of palaeoenvironments via analysis of pollen types preserved in peat, organic soils and lake muds.

Panchromatic Sensitive to all colours of the visible spectrum.

Pangaea The Earth's most recent supercontinent formed during the Permian by the coalescence of most continental plates (Gondwanaland and Laurasia, among other smaller continents) and rifted apart in the Jurassic.

Parallax The apparent change in position of a stationary object when viewed from two different positions.

Parent material The material upon which soil is developed and constitutes the main input of soil material through the process of weathering. It may be the weathered surface of exposed unconsolidated *in situ* rock surfaces, or unconsolidated superficial material transported and deposited by gravity, water, ice and wind.

Partial contributing area concept The idea that *infiltration-excess overland flow* will often occur only in spatially localized parts of the hillslope as opposed to the entire catchment (as originally postulated by Horton).

Partially mixed estuaries Estuaries that are highly influenced by tidal currents, causing greater mixing (*advection* and *diffusion*) of fresh- and saltwater and a more gradual salinity gradient in the water column.

Participatory analysis The process of encouraging all stakeholders to take an active part in the decision-making process when deciding on an environmental management strategy where conflicting interests are involved.

Particle movement The physical transportation of material down a hillslope where grains move one, or a few, at a time and do not significantly interact with one another, as opposed to a *mass movement*.

Peat A type of predominantly dark organic soil derived from partially decomposed compacted plant materials that accumulate under waterlogged conditions.

Pedogenesis The process of soil formation.

Peds Clumps or structural units of soil separated by small natural voids.

Pelagic sedimentation Sediments formed in an open-ocean environment by the slow background sedimentation of fine-grained material (usually marine organisms and red clays) falling through the water column to the seabed.

Peneplain A low-relief plain that is the theoretical end product of erosion in the absence of tectonic activity (following *Davisian cycles of erosion*).

Perennating system A group of plants that persist for several years, usually with new growth from a perennating part of the plant, e.g. bulbs, *rhizomes* or tubers.

Permafrost A condition existing below the ground surface, in which the soil or bedrock material remains perennially frozen, below 0°C for a minimum of two years. Currently permafrost affects approximately 26% of the Earth's surface.

Phanerophytes Woody perennials (trees and shrubs) with visible buds on upright perennial stems high above the ground, e.g. the palm family.

Phenological Pertaining to the timing of recurring natural phenomena, such as the timing of events such as leaf fall and buds appearing on plants.

Phenology The study of the timing of natural phenomena in relation to climate. For example, the appearance of the first flower of spring.

Photic zone The layer of the surface ocean which receives enough sunlight to enable photosynthesis to occur.

Photon A quantum of electromagnetic radiation. Units of light or other electromagnetic radiation, the energy of which is proportional to the frequency of the radiation.

Photoreduction Chemical reduction of a substance caused by ultraviolet radiation, e.g. in sunlight.

Photosynthesis The process of converting light energy to chemical energy and storing it as sugar. This process occurs in plants and some bacteria and algae. Plants need only light energy, CO_2 and H_2O to make sugar.

Photosynthetic bacteria Bacteria that are able to carry out photosynthesis (light is absorbed by bacteriochlorophyll), e.g. blue-green algae.

Phytoplankton Microscopic plants that live in the ocean and are the foundation for the marine food chain.

Pillow lava The name for lava that erupts from vents underwater and cools rapidly forming rounded structures surrounding the vent.

Pingo An ice-cored mound (up to 55 m high and 500 m long) found in permafrost areas; derived from an Inuit word meaning 'hill'.

Pingo scar A relict periglacial feature formed by the melting of the ice core of a *pingo*, leaving a central surface depression with sediment ramparts.

Pipeflow The movement of water through the soil within *soil pipes*.

Pixel A contraction of 'picture element' that refers to the smallest unit of a digital image.

Plant physiology The study of the functioning of plants, and this includes the response of plants to their environment and the acquisition of resources by plants.

Platform The stationary (i.e. gantry) or moving (i.e. aircraft) position on which remote sensors are mounted.

Playa A depression in the centre of an inland desert basin; the site of occasional temporary lakes; high levels of evaporation often create alluvial flats of saline mud.

Pleistocene The first epoch of the Quaternary, preceded by the Pliocene and succeeded by the *Holocene*. Lasting from approximetly 1.8 million to 10 000 years before the present (when the Earth was most extensively glaciated).

Ploughing boulders Boulders found on periglacial slopes that slowly move downslope owing to different thermal conditions beneath the boulder compared with the surroundings. They leave a trough upslope and form a sediment prow downslope.

Point source pollution Release of contaminants from a clearly identified point, e.g. a pipe from a factory.

Polar front The surface of contact between a cold polar air mass and a warm tropical air mass.

Polar permafrost Extensive permafrost that occurs owing to low temperatures in high-latitude areas, e.g. Alaska.

Polder Land reclaimed from the sea via the development of embankments.

Polyclimaxes A theory of vegetation allowing the co-existence of several final *climax communities* for a given type of area, all of which rank equally rather than being subordinate to a single climatic climax community (as required by the *monoclimax* theory). Instead of total convergence into a single community type, succession therefore produces partial convergence to a mosaic of different stable communities in different habitats.

Polycrystalline Referring to a crystalline structure in which there is a random variation in the orientation of different parts.

Polythermal glacier A glacier composed of both warm and cold ice.

Pore spaces The voids between solid soil particles.

Pore water pressure The pressure exerted by water in the pores of soil and aquifer rocks which may force particles apart during saturated conditions.

Porosity The pore space of a substrate (i.e. the factor controlling soil and rock permeability).

Positive feedback An event or process resulting from another event or process which exacerbates or magnifies the original effect.

Positivism A traditional philosophy of science, originally attempting to distinguish science from religion by ensuring the application of a unitary scientific method of observation, involving direct and repeatable experimentation on which to base theory. The underlying premise is that a firm empirical basis will lead to the identification of scientific laws which become progressively unified into a system of knowledge and 'absolute truth' about the natural world.

Potential evapotranspiration The *evapotranspiration* that would occur from a vegetated surface with an unlimited water supply.

Potholes Circular depressions found on bedrock surfaces. In reference to rivers they are scoured out by the effect of a pebble rotating in an eddy.

Precautionary principle An approach to decision-making which states that where there are threats of serious or irreversible damage, lack of scientific certainty should not be used as an excuse to preclude preventative action. Action should be taken at an early stage before victims or negative impacts occur; 'better safe than sorry'.

Precession of the equinoxes The gravitational pull exerted by the Sun and the Moon cause the Earth to wobble on its axis like a spinning top, determining where in the orbit the seasons occur, and the season when the Earth is closest to the Sun.

Precipitation The deposition of water in a solid or liquid form on the Earth's surface from atmospheric sources (including dew, drizzle, hail, rain, sleet and snow).

Precipitation deficits The lack of precipitation in a water balance when considering the losses in the form of evaporation or losses through gravity-driven movement of water to riverflow or groundwater.

Primary endemism When a species occurrence is unique to one specific area alone and unknown to any other region, e.g. Australian marsupials.

Primary minerals Minerals that have not changed from their original state since they were formed in magma (e.g. quartz, feldspars and micas).

Primary productivity The amount of biological material (biomass) produced by photosynthesis per unit area and unit time by plants.

Primary succession *Ecological succession* beginning on a newly constructed substrate previously devoid of vegetation (e.g. a new volcanic island); the recently exposed land is colonized by animals and plants.

Pro-delta The shelf area offshore of a river mouth which marks the intersection between the delta sediments and the adjacent basin.

Protalus ramparts Linear ridges of coarse sediment found a small distance away from a slope base, formed from the accumulation of frost-shattered debris that, once fallen from a backwall, slides down a snowpack to its lower margin.

Push moraine A ridge of material accumulated at the glacier margin by the bulldozing action of a glacier front and consisting of glacially derived material.

R

'r' and 'K' selection A theory of two life strategies to cope with competition and stress. Natural selection may favour either individuals with high reproductive rates and rapid development ('r' selection) or individuals with low reproductive rates and better competitive ability ('K' selection).

Radiation Emitted electromagnetic energy.

Radiometric Of or pertaining to the measurement or representation of radiation.

Radiometric methods Techniques for estimating the age of deposits based on the time-dependent radioactive decay of particular radioactive isotopes found in sediments.

Raindrop impact The force exerted by a falling raindrop on a soil surface. The impact of the raindrop causes a shock wave which detaches grains of soil or small aggregates up to 10 mm in diameter and projects them into the air in all directions; the rate of detachment is roughly proportional to the square of rainfall intensity.

Rainflow In shallow overland flow, the transport of water resulting from a combination of detachment by raindrop impact and transportation by rainwater flowing downhill.

Rainsplash A type of soil erosion caused by *raindrop impact* in which sediment is transported through the air.

Rainwash The erosion of soil by overland flow processes; normally occurs in concert with *rainsplash*.

Random errors Non-systematic errors that are unpredictable and cannot be removed from data and can only be estimated.

Rating equation An equation used to infer river discharge values from measured water levels at particular points along a river. The known discharges of the river at various different water levels are plotted and the equation for the line of best fit is calculated; the discharge at any water height can then be inferred (although there are inevitable errors in this process).

Rayleigh scattering The wavelength-dependent redirection or scattering of *electromagnetic radiation* caused by atmospheric particles that are much smaller than the wavelengths of the light they scatter.

Realized niche A term to describe the *niche* more commonly utilized by most species whereby competitive interaction between several species attracted to the same resource has inhibited attainment of the *fundamental niche*.

Recharge Replenishment of groundwater stores.

Redox potential The reducing or oxidizing intensity of a system, measured with an inert platinum half-cell and a reference half-cell calibrated against the hydrogen electrode. A measurement conducted in this manner is known as the Eh.

Reduction A chemical weathering process in which oxygen is dissociated from minerals creating a negative electrical charge; it usually occurs in anaerobic conditions.

Reductionist The assumption that the system under study is 'closed', i.e. all other variables within the system are held constant, allowing the direct relationship between two variables to be ascertained, and thereby eliminating reference to the potential influence of extraneous variables (e.g. *positivism* is reductionist).

Reflection The process by which a wave approaching a vertical or near-vertical object (e.g. sea cliff or sea wall) is rebounded from the object. If the angle of wave approach is parallel to the object, the wave will be reflected in the opposite direction to the line of approach. If the wave strikes at an angle of incidence other than parallel the wave is reflected in the tangent to the angle of approach.

Reflectivity The ability of a body to return energy.

Refraction The process by which a wave front bends and changes direction owing to a reduction in velocity as the wave enters shallow water.

Refugia Isolated habitats with distinctive ecological, geological, geomorphological or microclimatic characteristics that allow formerly widespread species to survive following a period of climatic change.

Regelation A two-fold process involving the melting of ice under pressure (the melting point of ice under pressure is lower than 0°C) and its subsequent transport and refreezing where the pressure is reduced; a major factor in the mechanism of downslope movement of a glacier.

Regime The seasonal variation in river flow which tends to be repeated each year is the river regime.

Regolith The basal layer of soil overlying the bedrock composed of loose, unconsolidated weathered rock and gravel debris; it is the raw material from which soils are developed.

Regressive barriers Large mounds of sediment that have developed under the influence of a falling sea level and/or excess sediment supply. Landward sediments are deposited on top of more seaward ones.

Relative sea level Level of the sea relative to the land determined by *eustasy* and *isostasy*.

Relaxation time The amount of time that an environment/landscape takes to recover from a major event (e.g. a flood or landslide).

Relay floristics A concept relating to species invasion and disappearance from a local community; it suggests that as one group of species establishes itself it is replaced by another which is then replaced until a stable state is achieved.

Renaissance A period of change in culture in Europe when classical art and learning was re-examined and embraced. It began in the late fifteenth century in Italy and then spread to other European countries.

Reptation A method of sand transport in which grains are set into a low motion due to the high-velocity impact of a descending saltating grain.

Residual tidal current Net movement of water due to tides occurring over long time periods. Tidal movements do not necessarily balance out over time thereby creating an overall water movement in a particular direction.

Resilience The ability of a system to recover from an event, change or shock.

Resolution Describes the ability of a system to separate a scene into constituent parts whether these parts be spatial, temporal or spectral.

Return flow Subsurface flow in the soil, either throughflow or macropore flow, that encounters a zone of soil saturation or lower hydraulic conductivity and is forced up through the soil profile to flow over the ground surface.

Rhizomes The underground lateral stems of certain plants that send up the new shoots (the rootstock).

Riffle A *bar* deposit found on the bed of river channels, usually spaced between 5 and 10 times the channel width apart. The height above the average bed surface causes fast-flowing, shallow and broken water under low- and medium-flow conditions.

Rill A small channel, formed by the merging of sheet wash into channelized flow, that acts as a conduit for water and sediment and is liable to collapse and change location between each runoff event.

Rillwash A hillslope erosion process that occurs when rain flow is deeper than 6 mm (generally in small channels carved out of the hillslope), rendering raindrop detachment ineffective; sediment detachment occurs when the downslope component of gravity and fluid flow traction overcome the frictional resistance of the soil.

Rip currents A strong seaward-directed current associated with water returning to the sea after being brought onshore by wave-breaking activity; an accumulation of water develops which pushes down the beach via a line of least resistance.

Rising limb The increase in river discharge in response to a rainfall event, as depicted in a storm hydrograph.

River reach A length of river channel.

Roche moutonnées Small *stoss-and-lee forms*.

Rock flour Fine-grained rock particles pulverized by glacial erosion.

Rock glacier A tongue-like body of angular debris resembling a small glacier but with no ice evident at the surface and only *interstitial ice* in the pore spaces between the debris. Their movement downvalley is very slow and many appear stagnant.

Rock shelters Shallow, sheltered niches in a hillside, smaller and less pronounced than a cave. Ancient human occupation often results in rich archaeological findings, in addition to other deposits indicative of past environmental conditions.

Rockfall A mass-wasting process whereby consolidated material falls and breaks up into a jumble of material at the base of a cliff or steep slope.

Rogen moraine A moraine characterized by a series of ribs of sediments lying transverse to the direction of ice advance, approximately 10–30 m in height.

Rossby waves Upper-air waves that undulate horizontally in the flow path of the jet streams and the westerlies.

Roughness length An indicator of the roughness of the ground surface and its impact upon surface winds, i.e. an urban surface has a much greater roughness length (up to 10 m for tall buildings) than agricultural crops (approximately 5–20 cm).

Rover A non-stationary global positioning system receiver that is used to collect three-dimensional position data over an area.

Ruderal Pertaining to species with a good colonizing ability, capable of growing on new or disturbed sites, e.g. weeds (also described as 'r' strategists).

S

Sabkha A salt-encrusted tidal flat environment; evaporation of groundwater draws in seawater which upon evaporation precipitates *gypsum* (e.g. the coasts of the Persian Gulf).

Safety factor The ratio of the sum of forces resisting movement to the sum of forces promoting movement of material down a slope; a value below 1 means movement will begin.

Salcrete A *duricrust* predominantly composed of sodium chloride (rock salt), a halite.

Salinization A process involving the accumulation of soluble salts of sodium, magnesium and calcium in the soil to the extent that the soil fertility is severely reduced.

Salt marshes Coastal marshes that develop on low-lying sheltered sections of coastlines (primarily in a lagoon, behind a spit or in an estuary). Specialized salt-tolerant vegetation (halophytes) trap silt particles and consolidate the environment through processes of vegetation succession.

Saltation A mechanism of sediment transport involving sediment grains being bounced along a bed surface.

Sand Sediment particles between 0.06 and 2 mm in diameter.

Sand seas Large areas of sand accumulations characterized by sand sheets and dunes; sediment grains are well rounded and typically quartz (e.g. the Sahara and Namibian Deserts). Also known as *ergs*.

Saprolite A soft, clay-rich, disintegrating rock found in its original place, formed by chemical weathering of igneous or metamorphic rock in humid, tropical or subtropical climates.

Saprovores An organism that survives on dead organic matter.

Saturated A term to describe the state of the soil when all soil pores are filled with water.

Saturated adiabatic lapse rate The rate at which temperature decreases in a rising parcel of saturated air.

Saturated hydraulic conductivity The rate of water movement through a porous medium when it is saturated (calculated using *Darcy's law*).

Saturated zone The zone under the surface, which lies beneath the *water table*, in which all pores of the aquifer rock or soil are filled with groundwater (i.e. saturated).

Saturation-excess overland flow A form of overland flow that occurs when all available soil pore spaces become full (i.e. the soil is saturated). Excess water is forced to flow over the surface.

Scale Describes the linear relationship between a linear distance on an image and the corresponding distance on the ground which determines how much detail is captured in the image.

Sclerophyllous Refers to plants with small, tough evergreen leaves which maintain a rigid structure at low water potentials thereby avoiding wilting. They are usually found in low-rainfall areas since the tough leaves help to reduce water loss.

Scree Loose, angular, rocky material that has been loosened from a slope through weathering and deposited further down the slope.

Scree slope The area at the base of a hillside where loose angular sediment (*scree*) accumulates.

Sea walls Massive concrete, steel or timber structures built along the coastline, with a vertical or sometimes curved face. A *hard engineering* coastal management technique employed to protect local infrastructure from flooding or erosion.

Seamounts Individual volcanoes on the ocean floor whose origin is distinct from the plate boundary volcanic system of mid-ocean ridges or subduction zones, i.e. usually formed as a plate moves over a *hot spot*.

Sea-salt events Enrichment of precipitation with sea salts incorporated from sea spray in windy conditions.

Seasonal icings Mounds of ice formed in winter in topographic lows where groundwater reaches the surface, i.e. in areas where *return flow* occurs and freezes.

Secondary endemic A species becomes endemic through the extinction of those species occurring in other places where they once survived (e.g. mammals of the West Indies).

Secondary minerals Minerals formed by the breakdown and chemical weathering of less resistant primary minerals (e.g. clays and oxides of iron and aluminium).

Secondary succession *Ecological succession* beginning on a previously vegetated site that has been recently disturbed by natural agents (e.g. fire, flood and hurricanes) or by human activities (e.g. deforestation). Remnant seed banks and root systems may influence the character of the resulting community.

Sediment budget An account of the inputs, outputs and stores of sediment for a given system.

Sediment yield The amount of sediment, both in suspension and transported as bed load, that is lost from a catchment. Usually measured as tonnes per year or tonnes per year per unit catchment area.

Sedimentary rock Rock which has formed by the gradual accumulation of sediment through time which has then solidified.

Sedimentation The process in which sediment is deposited leading to its accumulation (e.g. at deltas).

Segregated ice Very large lenses of ice that have slowly built up in frozen soil as a result of the migration of water to the freezing front (typically only in the upper 5–6 m of ground).

Seif dunes Linear dunes formed where two dominant wind directions are present at approximately right angles to each other.

Sensible heat Heat that can be measured by a thermometer and felt by humans.

Seral stage A stage within the process of *ecological succession* which is characterized by a particular biotic community. A series of seral stages (and their associated biotic communities) successively follow one another in the path to the climax community; each community creates conditions more favourable for a succeeding community.

Sesquioxides An oxide containing three atoms of oxygen to two atoms (or radicals) of some other substance.

Sessile A term to describe benthic organisms attached to a substrate and hence immobile (fixed to the ocean bottom).

Shadow dunes Small wind-blown dunes that develop in coastal or dryland areas around obstacles such as driftwood, a rock or a dead animal.

Shear A condition or force causing two contacting layers to slide past each other in opposite directions parallel to their plane of contact.

Shear stress A stress that acts upon a particle in the same plane as the surface the particle is resting upon (i.e. opposed to normal stress acting in the direction of gravity), resulting in either movement or strain of the particle. In the context of river systems, shear stress is the velocity of flowing water; when a critical flow velocity (and hence a critical shear stress) is reached, frictional forces may be overcome and a particle lifted from the bed.

Shield volcanoes Large, dome-like rarely explosive volcanoes with gentle slopes of 6–12° formed by alternate layers of runny basalt, e.g. the Hawaiian shield volcanoes.

Shifting cultivation A form of plant cultivation in which seeds are planted in the fertile soil prepared by cutting and burning the natural growth. Relatively short periods of cultivation are followed by longer periods of fallow to allow soil rejuvenation, returning to the site years later.

Shoaling A gradual shallowing of the seabed.

Shore The land bordering the sea between the water's edge at low tide and the upper limit of effective wave action.

Shoreline The water's edge where the shore and the water meet; it varies over time.

Shore platform An erosional surface of horizontal or gently sloping rock in the intertidal zone that has developed following erosion of a rocky coast.

Short-wave radiation Incoming radiation whose wavelength is unchanged from its solar source.

Significant wave height The mean height of the top tenth of all wave heights recorded at a given location (used as an approximate measure of wave energy for that location).

Silcrete A *duricrust* predominantly composed of silicates.

Silt Sediment particles between 0.004 and 0.06 mm in diameter.

Silviculture Subject of utilizing plant processes to grow trees for harvesting.

Sinusoidal The mathematical shape of a curve of sines, i.e. a wave consists of a simple sinusoidal form.

Slaking A process that involves raindrops striking a soil surface and water being forced into a soil aggregate therefore compressing the air inside and causing the aggregate to explode into its constituent grains.

Slantwise convection Convection (vertical rise in an air parcel) is inhibited when the prevailing lapse rate is less than the appropriate adiabatic lapse rate. However, a poleward horizontal movement of an air mass may bring the air parcel into an environment denser than itself, thereby allowing the air parcel to rise through slantwise convection.

Slide *Mass movements* which involve a large mass of earth or rock essentially moving as a block as opposed to *flows*.

Slumping A mass movement process whereby saturated slope material moves downslope under the force of gravity and deforms upon movement.

Smelting The process of extracting a metal from its ores by heating.

Snowline The altitude marking the lower limit of permanent snow in upland or high-latitude areas, i.e. the line where the winter snowfall exceeds the amount removed by summer melting and evaporation.

Social ecology The study of the dynamics and diversity of social behaviour and social systems of animals; social ecological variables include measures of group composition, intermale competition, and habitat preference.

Social impact assessment An evaluation of the impact of a proposed activity on all the social aspects of the environment including: people's *coping strategies* (economic, social and cultural); use of the natural environment; the way communities are organized through social and cultural institutions; and the identity and cultural character of a community. It involves characterizing the existing state of these aspects of the social environment in addition to predicting how they might change.

Soft engineering Pertaining to more 'sensitive' management practices that involve methods more closely associated with geomorphological processes and local sediment dynamics; large '*hard engineering*' types of structures are avoided.

Soil A complex medium consisting of inorganic materials, organic matter (living and dead), and water and air variously organized and subject to dynamic processes and interactions. It forms the natural terrestrial surface layer that is the supporting medium for the growth of plants.

Soil colloid Very small mineral particles (less than 0.002 mm in diameter) that stay suspended in water, the most important being clay minerals capable of remaining suspended in water indefinitely.

Soil colour A visible characteristic of the soil that allows the determination of soil properties such as organic matter content, iron content, soil drainage and soil aeration.

Soil creep The very slow, imperceptible, movement of material downslope under the force of gravity.

Soil horizons Distinctive horizontal layers within a *soil profile*, created primarily by the translocation of materials with water moving through the soil.

Soil pipes Horizontal tube-like subsurface cavities within the soil; special forms of macropores greater than 1 mm in diameter. They are continuous in length such that they can transmit water, sediment and solutes through the soil and bypass the soil matrix.

Soil profile A vertical section through the soil from the ground surface down to the parent material; the profile characteristics determine the soil type.

Soil solution The water held in the soil pores that contains dissolved organic and inorganic substances and hence is not pure.

Soil structure The shape, size and distinctiveness of soil *aggregates*, divided into four principal types (blocky, spherical, platy and prismatic).

Soil texture The relative proportions of sand, silt and clay-sized fractions of a soil.

Solifluction Form of slow mass movement in environments that experience freeze–thaw action or highly variable warming and cooling of the surface. This results in a slow movement of soil material downslope.

Solum The portion of the soil where soil-forming processes are active and plant and animal life are mostly confined; the A, E and B horizons.

Solute load The total mass of material transported in solution by a flow.

Sorting A measure of the spread, or standard deviation, of grain sizes within a sediment. In general the further a sediment deposit has been transported from its source, the greater the sorting of grains.

Speciation The evolution of new species involving the relatively gradual change in the characteristics of successive generations of an organism, ultimately giving rise to species different from the common ancestor. Most biologists accept Darwin's basic hypothesis of speciation from a common ancestor as a result of natural selection of those attributes best suited to survival in a given habitat with limited resources. Speciation can take a number of forms, whether sympatric (populations overlapping) or allopatric (populations become isolated from one another).

Specific conductance The ability of water to conduct an electric current, dependent on the concentration of ions in solution.

Specific heat The energy required to change the temperature of 1 gram of a substance by 1 degree Celsius. Water has a higher specific heat than air, requiring more energy to be absorbed for any given temperature change.

Spectral band A division of the electromagnetic spectrum that groups energy according to similarities.

Spectral signature Describes the reflectance characteristics of a surface across the electromagnetic spectrum.

Specular reflection The redirection of radiation off a smooth surface such that the radiation is otherwise unchanged.

Speleothems Structures formed in a cave by the deposition of minerals from water, e.g. a stalactite, stalagmite. They are primarily composed of calcium carbonate precipitated from groundwater percolating through carbonate rock, e.g. limestone.

Spits Narrow and elongated accumulations of sand and shingle projecting into the sea, usually with a curved seaward end caused by wave action. They grow out from the coastline when the shore orientation changes but longshore currents do not deviate and continue to transport and deposit along a projected coastline, e.g. at the mouth of an estuary.

Spring tides A tide that occurs at or near the new moon and full moon when the gravitational pull of the Sun reinforces that of the Moon producing a large tidal range, causing higher than average high tides and lower than average low tides.

Stable isotope Isotopes of an element possess the same number of protons in their nuclei but have different numbers of neutrons. A stable isotope does not break down by radioactive decay. For example, ^{12}C is a stable isotope and the most widespread form of carbon in the environment, but the radioactive isotope ^{14}C and the stable isotope ^{13}C also occur.

Stadial A short period of climatic deterioration within an interglacial period; glaciers advanced and periglacial conditions extended but in a less pronounced way than during a *glacial interval*.

Stakeholders A person or group who can affect or is affected by an action and therefore has a vested interest in the outcomes.

Responsible decision-making requires consideration of the effects on all stakeholders.

Stand An area of more or less homogeneous vegetation.

Standard deviation A measure of how spread out the data are around the mean.

Stellate dunes Star-shaped dunes formed under conditions of variable wind direction with no one prevailing wind direction. These dunes do not migrate.

Stemflow The flow of water down the trunk of a tree or stems of other vegetation allowing water to reach the hillslope.

Stereo images Aerial photographs that have been obtained such that each photo overlaps another by a prescribed amount. Overlapping coverage provides two points of observation to provide *parallax* required for *digital elevation model* generation.

Stochastic A model that contains some random element in the operation or input data so that more than one, and usually a very large number of, outcomes are possible.

Stone pavements Accumulations of flat-lying boulders in a mosaic pattern at the ground surface in periglacial environments. Some argue they are formed by aeolian removal of fine surface particles, but it is more commonly argued they are displaced upwards as small particles fall into ground cracks created during freeze–thaw cycles while the larger boulders cannot.

Stormflow The peak flow that occurs during or immediately following a rainfall event occurring as a result of overland flow and rapid subsurface flow (e.g. pipeflow contributions may also be high).

Stoss-and-lee forms Streamlined elongated rock exposure formed by the sliding of debris-rich basal ice over the bedrock surface under a glacier; characterized by a gently sloping glacially smoothed upstream side and a steeper plucked downstream side (centimetres to metres in length).

Strain history The amount of deformation of a substance that has occurred owing to previous stress impact; it can affect present and future stress–strain relationships.

Strain rate The amount of deformation occurring over time for a given material (i.e. the rate of deformation). For glaciers, the strain rate for a given shear stress is determined by *Glen's law*.

Stratification Division of water in deep lakes, reservoirs and stable water bodies into layers of differing density.

Stratified estuaries Estuaries with limited saltwater and freshwater mixing (via advection and diffusion) causing a lower layer of denser and saltier water with an upper layer of less dense freshwater; a salt wedge develops.

Stratigraphy The layering of sediments.

Stratosphere A layer of the atmosphere lying above the *troposphere* about 50 km above the Earth's surface.

Stratotype A particular stratigraphic unit with clear and well-recorded characteristics and boundaries. This site can become the point of reference for comparison with a more poorly preserved record. Also known as typesite.

Stream competence The maximum particle size a stream can transport.

Stream order Numbering of the drainage network according to the number of tributaries and stream network linkages.

Stream power The rate of energy supply in a river that is available for work to be done at the stream bed, measured in $W\,m^{-2}$.

Stress The force per unit area acting on a plane within a body due to application of an external load; six values are required to characterize the stress at a point completely (three normal components and three shear components).

Striations Microscale erosional features on rock surfaces, resembling a scratch.

Subduction The process in which one lithospheric plate descends beneath another into the asthenosphere when the two plates converge.

Subglacial Pertaining to the environment at the base of a glacier.

Sublimate A change in the physical state of a substance directly from solid to gaseous form.

Sublimation The chemical process in which a solid changes directly into a gas.

Subsea permafrost *Permafrost* found beneath the sea; sometimes due to low temperatures at the bed, more usually a remnant of past colder temperatures and rising sea levels (drowning frozen ground).

Subsurface flow Pertaining to *throughflow* that occurs through *micropores*, *macropores* and *soil pipes*.

Succession Changes over time in the structure or composition of an ecological community. These changes often follow a predictable pattern.

Sun-synchronous orbit The orbit of a satellite travelling around the Earth which is timed such that it passes over any given latitude at the same time at each pass so as to ensure that illumination conditions remain constant between subsequent images.

Super-adiabatic A term used for localized steep *lapse rates* that are greater than even the *dry adiabatic lapse rate* causing rapid local convection, e.g. strong radiational heating of the ground surface.

Superimposed ice The ice formed when water from melting snow comes into contact with the cold surface ice of a glacier at the base of the snowpack and refreezes.

Supply limited A transport process that is limited by the lack of sediment supply, not the capacity to transport sediment since more force is available than is being utilized. For example, rockfalls are limited by the amount of material that is loose enough to fall.

Supraglacial Pertaining to the environment at the surface of a glacier.

Surf zone A process-based term for the area within the *nearshore* zone where breaking waves approach the shore usually over a wide, low gradient.

Surface boundary layer A layer extending upwards from the Earth to a height that ranges anywhere between 100 and

3000 m. Here, almost all interactions between the atmosphere and humans take place.

Suspended load The sediment transported in water when lifted from the bed surface and kept in suspension by turbulent fluid flow.

Sustainable development Development (any form of development from an action, project, strategy or legislation) that meets the needs of people today without compromising the ability of future generations to meet their own needs.

Swash The thin sheet of water that travels up the beach following the breaking of a wave.

Swash zone Process-based term for the area within the *nearshore* zone where broken waves travel up the beach as *swash* and return as *backwash*.

Swath The area on the ground covered by the motion of a remote sensing instrument.

Symbiotic Living together in a mutually beneficial relationship.

Systematic errors Predictable errors that can be modelled and removed from the data.

T

Talik A Siberian word for an unfrozen pocket within permafrost; for example, beneath a lake or warm-bedded glacier.

Talus An accumulation of angular rock debris from rockfalls found at the base of a slope.

Tarn A depression located at the site of a melted *corrie* glacier; a lake usually forms in the centre.

Taxonomy The study, description and systematic classification of living organisms (plant and animal) into groups based on similarities of structure or origin. Synonymous with systematics.

Telemetry The process of obtaining measurements in one place and relaying them for recording or display at a different site.

Temperate glacier A glacier formed in temperate climates where the temperature of the entire glacier is at the pressure melting point except for the surface 10–20 m (which fluctuates with the season); considerable quantities of meltwater are generated causing high rates of glacier movement and erosion.

Temperature inversion A reversal of the normal environmental temperature *lapse rate*; air temperature increases with altitude.

Terminal mode The final form of a particle of glacial sediment in which the particle will not break down into a finer form even with prolonged transport in the glacial system.

Terminal velocity The velocity at which the frictional drag forces acting on a falling object are equal to the driving forces of gravity, resulting in a constant fall rate (neither accelerating nor slowing down).

Thalweg The line of maximum water velocity down the path of a river.

Thermal conductivity The degree to which a substance transmits heat.

Thermal scanner A remote sensing instrument similar to a *multispectral scanner* but that can only sense radiation in the thermal infrared portion of the spectrum.

Thermocline The depth at which the temperature gradient of the water column rapidly changes in the vertical dimension, marking the contact zone between water masses of markedly different temperatures.

Thermohaline circulation Large-scale circulation of the world's oceans, involving the vertical movement of large bodies of water, driven by water density differences (a result of variations in temperature and salinity).

Thermokarst A term referring to the ground surface depressions which are created by the thawing of ground ice (and subsequent water erosion) in periglacial areas, e.g. *pingos*.

Thermoluminescent Pertaining to luminescence (an emission of light) resulting from exposure to high temperature; used as a means of dating ancient material.

Thiessen polygon The spatial influence of a particular data point calculated using arithmetic spatial averaging techniques on a network of data points.

Thixotropic Pertaining to the property of becoming fluid when agitated but recovering its original condition upon standing; viscosity decreases as the rate of shear (relative movement) increases.

Throughfall Water reaching the ground surface after dripping or bouncing off overlying vegetation.

Throughflow The downslope movement of water draining through the soil.

Through-wash A wash process involving the movement of regolith particles through the pores between grains in the regolith; the particles must be at least 10 times smaller than the grains they are passing between, and the process is therefore only significant in washing silt and clay out of clean sands.

Tidal currents Currents produced by the rise and fall of the tide; either the movement in and out of an estuary or bay, or the movement of water between two points affected by different tidal regimes (especially common in straits).

Tidal prism The volume of water that moves in or out of an area such as an estuary during a tidal cycle.

Tidal range The difference in water level between high and low water during a tide.

Till The generic term for sediment deposited directly by glacier ice.

Tillage erosion An anthropogenic soil erosion process (similar to creep) which is the result of ploughing either up- and downslope or along the contour. The turning over of soil produces a direct downhill movement. Whatever the ploughing direction, the process is faster than natural soil creep.

Tilt of the Earth The Earth's axis lies at an angle that varies from approximately 21° to 24° and back again in a 41 000 year cycle. The greater the tilt, the more intense the seasons in both hemispheres become.

Topset bed A horizontal bed of coarse sediment deposited by braided streams crossing a *delta plain*; it represents the subaerial part of the delta.

Topsoil The upper section of the soil that is most important for plant growth (usually the A horizon or plough horizon).

Total stream length The combined length (km) of all components of the channel network.

Trade winds The prevailing winds in the tropics blowing from high pressure at the tropics to low pressure at the equator. The winds do not blow directly north–south because the rotation of the Earth deflects them to the left in the southern hemisphere and to the right in the northern hemisphere.

Tragedy of the commons A term coined by Garret Harding in 1968 that refers to the excessive exploitation of a communal resource to a point of degradation due to the selfish nature of rational people who will use more than their fair share of the resource; no one person will take responsibility for something owned by all. It is often used to demonstrate the mistake in allowing a growing population to increase steadily its exploitation of the ecosystem which supports it.

Transform faults Major strike–slip faults occurring where two plates slide past each other in the horizontal plane. They are capable of causing major destructive earthquakes, e.g. the San Andreas Fault.

Transgressive barriers Accumulations of sediment just offshore running parallel to the coastline which have formed under the influence of rising sea level and/or a negative sediment budget. They tend to consist mainly of tidal delta and/or washover deposits, and are underlain by estuarine or lagoonal deposits. In this instance sediments deposited in seaward environments end up on top of sediment that originated in more landward environments.

Translocation The transport of dissolved ions and small particles through the soil within the *soil solution*, to surface water and groundwater.

Transmission bands Sections of the electromagnetic spectrum that allow radiation to pass unobstructed. These are also called atmospheric windows.

Transpiration The loss of water to the atmosphere through the process of evaporation from leaf pores and plants.

Transport limited A transport processes that can only move material a limited distance from the source despite the plentiful supply of material, e.g. rainsplash.

Transporting capacity The maximum amount of material which the transport process can carry.

Transverse dunes Linear dunes with a shallow windward side and steep lee slope (similar in structure to dunes and ripples formed below water).

Treatments A single test within a larger experiment where a single variable has been altered from the control situation by a known quantity and applied to the principal substrate in order to ascertain its effect.

Treeline The altitudinal upper limit of tree growth; affected by latitude and local factors such as slope, soil, aspect and exposure.

Trophic level A functional or process category describing the position of an organism or group of organisms in a food chain. Primary producers are at the first trophic level, those that feed on primary producers are at the second trophic level.

Tropopause The boundary between the *troposphere* and the *stratosphere*.

Troposphere The lowermost layer of the atmosphere extending to approximately 11 km above the Earth's surface.

Truncated spur A valley side spur that has been abruptly cut off and steeped at its lower end by the erosive action of a glacier.

Tsunami A large sea wave generated by submarine seismic activity (earthquakes, slides, volcanic activity) or meteor impact in the ocean. These waves can be extremely destructive, especially in the Pacific Rim, contributing to the coastline development of these areas.

Turbidite current A density current involving mixtures of sediment and water which, owing to their increased density relative to seawater, flow down and along the bottom of the oceans transporting sand and clay-sized sediment from shelf slopes to deeper oceanic environments.

Turbulent flow One of the two ways in which water can flow. It involves water molecules moving in many directions, with an overall net flow in one direction; as a result the flow is well mixed.

U

Unidirectional flow Currents flowing in one dominant direction (e.g. rivers and wind).

Uniformitarianism A practical principle of modern science concerning the method in which scientists explain phenomena; it advocates the use of the simplest explanation which is consistent with both evidence and known scientific laws. It is primarily related to James Hutton's demonstration in 1788 that the simplest explanation of the development of the Earth's landscape is through the observed processes of erosion and uplift acting gradually over time, rather than catastrophic landform development through divine intervention.

Urban boundary layer The section of the atmosphere directly overlying the *urban canopy layer*, subject to *urban heat island* effects through the entrainment of air from the urban canopy layer and anthropogenic heat from roofs and chimneys.

Urban canopy layer The section of the atmosphere immediately overlying an urban area, subject to heavy localized *urban heat island* effects, including greater daytime heat storage, anthropogenic heat release from buildings and decreased evaporation.

Urban heat island Urban landscapes adjust the local microclimatic processes, resulting in an 'island' of warmer air surrounded by cooler rural air; two distinct regions of atmospheric modification are observed (the *urban canopy layer* and the *urban boundary layer*). The effect is most apparent at night, owing to slower cooling of the urban landscape, and during light winds.

U-shaped valley A wide valley with steep sides formed by glacial erosion of a V-shaped valley, involving the formation of truncated spurs during the valley straightening.

V

Value A measurable variable of soil colour describing the degree of darkness or lightness of the colour (a value of 0 represents black).

Variable source area concept The idea that the area of a catchment that produces *saturation-excess overland flow* will vary through time, i.e. during a rainfall event a greater proportion of the catchment will begin to contribute saturation-excess overland flow as time progresses, and the catchment becomes more saturated.

Varve A thin laminar bed of glacial sediment deposited by a proglacial stream in a repetitive annual sequence; coarse particles are deposited in summer and the finer particles progressively throughout the year.

Ventifacts The smoothed surfaces of individual stones eroded by sand and dust particles entrained in the wind.

Venturi effect Wind forced to funnel between two buildings causes the local wind speed to increase.

Vital attributes The critical physical characteristics of plants that determine their ability to survive disturbance, including their methods of persistence, conditions for establishment and timing of life stages.

W

Wadati–Benioff zone The band of rock (20 km thick) which dips from the trench region under an overlying plate in a subduction zone. It is the location of earthquake foci that are associated with descending lithospheric plates.

Wandering gravel-bed river A gravel-bed river channel characterized by an irregularly sinuous thalweg that is frequently split around vegetated islands and low-order braiding within complex bar deposits where the river is laterally unstable.

Water balance Pertaining to the cyclical movements of volumes of water within a drainage basin per unit time. It relates to the various inputs, storage and outputs of water within the drainage basin system and controls the nature of river discharge.

Water table The upper boundary of the zone of groundwater saturation (the *saturated zone*). Its level varies with the amount of precipitation, evapotranspiration and percolation.

Wave asymmetry The nature of waves that is not symmetrical on either side of the wave crest. As waves enter shallower water they develop peaked crests and flat troughs. In addition to the asymmetrical shape, the water flow velocities also become asymmetric with the onshore side of the wave being stronger but of shorter duration than the offshore side of the wave.

Wave base The point of a wave below which there is no orbital movement of water.

Wave breaking The destruction of a wave when it becomes too steep and disintegrates.

Wave convergence The focusing of wave rays so that they come together increasing in energy and height.

Wave crest The peak of the curve of a wave.

Wave divergence The separation of wave rays so that waves move apart. Typically waves will become shorter and less energetic.

Wave energy flux The rate of transfer of energy by waves.

Wave frequency The number of *wave crests* which pass a fixed point over a set timescale.

Wave height The vertical distance between the *wave trough* and the *wave crest*.

Wavelength The distance between a *wave crest* to the next wave crest, or between trough to trough.

Wave period A measure of wave speed; the time taken for two successive *wave crests* to pass a fixed point.

Wave set-up Wave breaking results in water piling up against the shore. This results in a slope of water with higher water pushed nearer the shore and this 'set-up' is sufficient to oppose the shoreward wave stresses.

Wave steepness Wave height divided by wavelength.

Wave trough The base of the curve of a wavelength.

Weathering The breakdown of rocks and minerals by the physical and chemical processes of erosion.

Well-mixed estuary An estuary in which mixing is so effective that the salinity gradient in the vertical direction vanishes entirely. If the estuary is wide enough then the Coriolis force pushes the flow of the outflowing river to the margin of the estuary and may result in a horizontal separation of river-water and seawater.

Wet-snow zone The region of a glacier in which the entire snowpack becomes saturated at the end of the summer.

Wetted perimeter The contact area between the channel bed and water when viewed in cross-section. Bank-full wetted perimeter is calculated as the estimated contact zone when the channel is completely full with water.

Whaleback A streamlined elongated rock exposure formed by the basal sliding action of a glacier; similar in shape to *stoss-and-lee forms* but the steep side faces upstream and the tapered end downstream.

Whole life costing Expressing the results of *life-cycle analysis* in financial terms, i.e. placing a monetary value on all the environmental impacts of a product from its manufacture to disposal.

Wilting point The condition of a soil when plants cannot withdraw the necessary water for growth as the only remaining soil water is that held tightly to soil particles by hygroscopic forces and is unavailable for plant use.

Wind shear A change in wind speed or direction with altitude in the atmosphere.

X

Xeromorphic The possession, by a plant, of features adapted to conditions of limited moisture availability.

Xerophyte A plant that has adapted to grow in very arid conditions with restricted water availability by minimizing water loss and maximizing water efficiency.

Xerophytic Pertaining to having the character of a *xerophyte*, i.e. an organism adapted to growth in conditions of limited water availability.

Y

Yardangs Large-scale dryland features; the erosive power of dust carried in the wind leads to the smoothing of entire hills streamlined in the direction of sediment transport.

Yield strength The stress at which a material exhibits a deviation from the proportionality of stress to strain, to produce a specified amount of plastic deformation, i.e. below the yield strength the material acts as an elastic and above as a viscous material.

A

Aggard, T. and Masselink, G. (1999) The surf zone. In: Short, A.D. (ed.) *Handbook of Beach and Shore Morphodynamics*, pp. 72–118. Wiley Interscience.

Abrol, I.P., Yadav, J.S.P. and Messoud, F.I. (1988) Salt affected soils and their management. *FAO Soils Bulletin*, **39**.

Achard, F., Eva, H.D., Stibig, H.J., Mayaux, P., Gallego, J., Richards, T. and Malingeau, J.P. (2002) Determination of the deforestation rate of the world's humid tropical forests. *Science*, **297**, 999–1002.

Agnew, C. and Anderson, E. (1992) *Water resources in the arid realm*. Routledge, London.

Ahnert, F. (1976) Brief description of a comprehensive three-dimensional process-response model of landform development, *Zeitschrift für Geomorphologie, Supplement*, **25**, 29–49.

Ahrens, C.D. (2000) *Meteorology today*, 6th edition. West Publishing, Mineapolis, MN.

Ahrens, C.D. (2003) *Meteorology today – An introduction to weather, climate, and the environment*. Brooks/Cole, Pacific Grove, CA.

Allen, J.R.L. (1985) *Principles of physical sedimentology*. George Allen and Unwin, London.

Allen, P.A. (1997) *Earth surface processes*. Blackwell Science, Oxford.

Alley, R.B. (1991) West Antarctic collapse – How likely? *Episodes*, **13**, 231–238.

Alley, R.B., Blankenship, D.D., Bentley, C.R. and Rooney, S.T. (1986) Deformation of till beneath ice stream B, West Antarctica. *Nature*, **322**, 57–59.

Alley, R.B., Meese, D.A., Shuman, A.J. *et al.* (1993) Abrupt increase in Greenland Snow Accumulation at the end of the Younger Dryas Event. *Nature*, **362**, 527–529.

American Meteorological Society (1997) Meteorological drought – policy statement. *Bulletin of the American Meteorological Society*, **78**, 847–849.

Amthor, J.S. (2000) The McCree–de Wit–Penning de Vries–Thornley respiration paradigms: 30 years later. *Annals of Botany*, **86**, 1–20.

Anderson, D.M. and Morgenstern, N.R. (1973) Physics, chemistry and mechanics of frozen ground: a review. *North American contribution, 2nd Permafrost International Conference*, US National Academy of Sciences, Washington, DC, 257–288.

Anderson M.G. (ed.) (1988) *Modelling geomorphological systems*. John Wiley & Sons, Chichester.

Anderson, M.G. and Burt, T.P. (eds) (1985) *Hydrological forecasting*. John Wiley & Sons, Chichester.

Anderson, R.S. and Haff, P.K. (1988) Simulation of eolian saltation. *Science*, **241**, 820–823.

Andoh, R.Y.G. (1994) Urban runoff – nature, characteristics and control. *Journal of the Institution of Water and Environmental Management*, **8**, 371–378.

Andreassen, L.M., Elvehøy, H. and Kjøllmoen, B. (2002) Using aerial photography to study glacier changes in Norway. *Annals of Glaciology*, **34**, 343–348.

ANRA (2001) *Fast facts 21 – dryland salinity in Australia*. Australian National Resources Atlas, Canberra.

Archibold, O.W. (1995) *Ecology of world vegetation*. Chapman and Hall, London.

Arnell, N. (2002) *Hydrology and global environmental change*. Pearson Education, Harlow.

Asante-Duah, D.K. (1998) *Risk assessment in environmental management*. John Wiley & Sons, Chichester.

Atkins, P., Roberts, B. and Simmons, I. (1998) *People, land and time: An historical introduction to the relations between landscape, culture and environment*. Arnold, London.

Atkinson, T.C., Briffa, K.R., Coope, G.R. *et al.* (1986) Climatic calibration of coleopteran data. In: Berglund, B.E. (ed.), *Handbook of Holocene palaeoecology and palaeohydrology*. John Wiley & Sons, Chichester, 851–858.

Avery, B.W. (1990) *Soils of the British Isles*. CAB International, Wallingford.

Avery, T.E. and Berlin, G.L. (1992) *Fundamentals of remote sensing and airphoto interpretation*. Macmillan, Toronto.

B

Bagnold, R.A. (1941) *The physics of blown sand and desert dunes*. Methuen, London.

Baillie, M.G.L. and Munroe, M.A.R. (1988) Irish tree rings, Santorini and volcanic dust veils. *Nature*, **332**, 344–346.

Ballantyne, C.K. and Harris, C. (1994) *The periglaciation of Great Britain*. Cambridge University Press, Cambridge.

Barber, D.C., Dyke, A., Hillaire-Marcel, C. *et al.* (1999) Forcing of the cold event of 8,200 years ago by catastrophic drainage of Laurentide lakes. *Nature*, **400**, 344-348.

Barrow, C.J. (1997) *Environmental and social impact assessment: An introduction*. Arnold, London.

Barrow, C.J. (2003) *Environmental management: Principles and practice*. Routledge, London.

Barry, R.G. (1992) *Mountain weather and climate*. Routledge, London.

Barry, R.G. and Chorley, R.J. (2003) *Atmosphere, weather and climate*. London, Routledge.

Bass, I.G. (1982) Ophiolites. *Scientific American*, 108-117.

Batjes, N.H. (1996) Total carbon and nitrogen in the soils of the world. *European Journal of Soil Science*, **47**, 151-163.

Battarbee, R.W., Flower, R.J., Stevenson, A.C. and Rippey, B. (1985) Lake acidification in Galloway: a palaeoecological test of competing hypotheses. *Nature*, **134**, 350-352.

Battjes, J.A. (1974) Surf similarity. *Proceedings 14th International Conference on Coastal Engineering*, ASCE, Reston, VA, 466-480.

Beatley, T., Brower, D.J. and Schwab, A.K. (2002) *An introduction to coastal zone management*. Island Press, Washington, DC.

Beeby, A.H and Brennan A.M. (1997) *First ecology*. Chapman and Hall, London.

Bencala, K.E., Kennedy, V.C., Zellweger, G.W., Jackman, A.P. and Avanzino, R.J. (1984) Interactions of solutes and streambed sediment 1. An experimental analysis of cation and anion transport in a mountain stream. *Water Resources Research*, **20**, 1797-1803.

Benn, D.I. and Evans, D.J.A. (1996) The interpretation and classification of subglacially-deformed materials. *Quaternary Science Reviews*, **15**, 23-52.

Benn, D.I. and Evans, D.J.A. (1998) *Glaciers and glaciation*. Edward Arnold, London.

Bennett, M.R. and Glasser, N.F. (1996) *Glacial geology, ice sheets and landforms*. John Wiley & Sons, Chichester.

Berner, E.K and Berner, R.A. (1996) *Global environment: Water, air and geochemical cycles*. Prentice Hall, Englewood Cliffs, NJ.

Betson, R.P. (1964) What is watershed runoff? *Journal of Geophysical Research*, **69**, 1541-1552.

Beven, K.J. (1989) Changing ideas in hydrology – the case of physically-based models. *Journal of Hydrology*, **105**, 157-172.

Beven, K.J. and Germann, P. (1982) Macropores and water flow in soils. *Water Resources Research*, **18**, 1311-1325.

Bigg, G.R. (2001) Back to basics: the oceans and their interaction with the atmosphere. *Weather*, **56**, 296-304.

Bird, E. (2000) *Coastal geomorphology: An introduction*. John Wiley & Sons, Chichester.

Bird, E.C.F. (1985) *Coastline changes: A global review*. Wiley-Interscience, Chichester.

Bird, E.C.F. (1993) *Submerging coasts: The effects of a rising sea level on coastal environments*. John Wiley & Sons, Chichester.

Bird, E.C.F. (1996) *Beach management*. John Wiley & Sons, Chichester.

Birkeland, P.W. (1999) *Soils and geomorphology*, 3rd edition. Oxford University Press, Oxford.

Bjørnsson, H., Gjessing, Y., Hamran, S.-E. *et al.* (1996) The thermal regime of sub-polar glaciers mapped by multi-frequency radio-echo sounding. *Journal of Glaciology*, **42**, 23-32.

Blake, E.W. (1992) The deforming bed beneath a surge-type glacier: measurement of mechanical and electrical properties. PhD thesis, University of British Columbia, Vancouver.

Blankenship, D.D., Bentley, C.R., Rooney, S.T. and Alley, R.B. (1986) Seismic measurements reveal a saturated porous layer beneath an active Antarctic ice stream. *Nature*, **322**, 54-57.

Bodman, G.B. and Colman, E.A. (1943) Moisture and energy conditions during downward entry of water into soils. *Proceedings of the Soil Science Society of America*, **8**, 116-122.

Bond, G., Broecker, W., Johnsen, S. *et al.* (1993) Correlations between climate records from North Atlantic sediments and Greenland ice. *Nature*, **365**, 143-147.

Boulton, G.S. (1978) Boulder shapes and grain size distributions of debris as indicators of transport paths through a glacier and till genesis. *Sedimentology*, **25**, 773-799.

Boulton, G.S. (1979) Processes of glacier erosion on different substrata. *Journal of Glaciology*, **23**, 15-38.

Boulton, G.S. (1987) A theory of drumlin formation by subglacial sediment deformation. In: Menzies, J. and Rose, J. (eds), *Drumlin Symposium*, Balkema, Rotterdam, 25-80.

Boulton, G.S. and Hindmarsh, R.C.A. (1987) Sediment deformation beneath glaciers: rheology and geological consequences. *Journal of Geophysical Research*, **92** (B9), 9059-9082.

Boyd, P.W., Watson, A.J., Law, C.S. *et al.* (2000) A mesoscale phytoplankton bloom in the polar Southern Ocean stimulated by iron fertilization. *Nature*, **407**, 695-702.

Boyd, R., Dalrymple, R. and Zaitlin, B.A. (1992) Classification of clastic coastal depositional environments. *Sedimentary Geology*, **80**, 139-150.

Bowen, A.J., Inman, D.L. and Simmons, V.P. (1968) Wave 'setdown' and setup. *Journal of Geophysical Research*, **73**, 2569-2577.

Bradbury, I.K. (1998) *The biosphere*. John Wiley & Sons, Chichester.

Bradley, R.S. (1999) *Palaeoclimatology: Reconstructing climates of the Quaternary*. Harcourt Academic Press, London.

Brady, N.C. and Weil, R.R. (2002) *The nature and properties of soil*, 13th edition. Prentice Hall, Upper Saddle River, NJ.

Brasher, R.E. and Zheng, X. (1995) Tropical cyclones in the south-west Pacific. *Journal of Climate*, **8**, 1249-1260.

Breckle, S.-W. (2002) *Walter's vegetation of the Earth: The ecological systems of the geo-biosphere*, 4th edition. Springer-Verlag. Berlin.

Bridgland, D.R. (1994) *Quaternary of the Thames*. Chapman and Hall, London.

Briggs, D., Smithson, P., Addison, K. and Atkinson, K. (1997) *Fundamentals of the physical environment*, 2nd Edition. Routledge, London.

Broecker, W.S. (1994) Massive iceberg discharges as triggers for global climate change. *Nature*, **372**, 421-424.

Broecker, W.S. (1998) Paleocean circulation during the last glaciation. A bipolar seesaw? *Paleoceanography*, **13**, 119-121.

Broecker, W.S. and Denton, G.H. (1990) What drives glacial cycles? *Scientific American*, **262**, 43-50.

Brookes, A. (1985) River channelisation, traditional engineering methods, physical consequences and alternative practices, *Progress in Physical Geography*, **9**, 44-73.

Browning, K.A. and Hill, F.F. (1981) Orographic rain. *Weather*, **36**, 326-329.

Brunckhorst, D.J. (1995) Sustaining nature and scociety - a bioregional approach. *Inhabit*, **3**, 5-9.

Brutsaert, W. (1982) *Evaporation into the atmosphere: theory, history and application*. D. Reidel, Dordrecht.

Bryan, K. (1922) Erosion and sedimentation in the Papago County, Arizona. *USGS Bulletin*, **730**.

Bryant, E. (2001) *Tsunami: The underrated hazard*. Cambridge University Press, Cambridge.

Budyko, M.I. (1974) *Climate and life*. Academic Press, New York.

Buffet, B.A. (2000) Clathrate hydrates. *Annual Reviews of Earth and Planetary Science*, **28**, 477-507.

Burke, E.J., Brown, S.J. and Christidis, N. (2006) Modeling the recent evolution of global drought and projections for the twenty-first century with the Hadley Centre Climate Model. *Journal of Hydrometeorology*, **7**, 1113-1125.

Burt, J., Barber, K. and Rigby, D.L. (2006) *Elementary statistics for geographers*, 3rd edition. Guilford, New York.

Burt, T.P. (1986) Runoff processes and solutional denudation rates on humid temperate hillslopes. In: Trudgill, S.T. (ed.), *Solute processes*. John Wiley & Sons, Chichester, 193-249.

Burt, T.P. (1988) A practical exercise to demonstrate the variable source area model. *Journal of Geography in Higher Education*, **12**, 177-186.

Burt, T.P. (1996) The hydrology of headwater catchments. In Petts, G.E. and Calow, P. (eds), *River flows and channel forms*. Blackwell, Oxford, 6-31.

Burt, T.P., Heathwaite, A.L. and Trudgill, S.T. (1993) *Nitrate: Process, patterns and management*. John Wiley & Sons, Chichester.

Byrne, S. and Ingersoll, A.P. (2003) A sublimation model for Martian South Polar Ice features. *Science*, **299**, 1051-1053.

C

Caissie, D., Pollock, T.L. and Cunjak, R.A. (1996) Variation in stream water chemistry and hydrograph separation in a small drainage basin. *Journal of Hydrology*, **178**, 137-157.

Campos, E.F., Fabry, F. and Hocking, W. (2007) Precipitation measurements using VHF wind profiler radars: measuring rainfall and vertical air velocities using only observations with a VHF radar. *Radio Science*, **42**, Art. No. RS3003.

Carson, M.A. and Kirkby, M.J. (1972) *Hillslope form and process*. Cambridge University Press, Cambridge.

Carson, R. (1962) *Silent spring*. Houghton Mifflin, Boston, MA.

Carter, R.W.G. (1988) *Coastal environments*. Academic Press, London.

Carter, R.W.G. and Woodroffe C.D. (1994) Coastal evolution: an introduction. In: Carter, R.W.G. and Woodroffe, C.D. (eds), *Coastal evolution*. Cambridge University Press, Cambridge, 1-31.

Chandler, T.J. (1965) *The climate of London*. Hutchinson, London.

Chang, H.-G.H., Eidson, M., Noonan-Toly, C. *et al.* (2002) Public health impact of reemergence of rabies, New York. *Emerging Infectious Diseases*, **8**, 909-913.

Chapman, D.V. (ed.) (1996) *Water quality assessments: A guide to the use of biota, sediments and water in environmental monitoring*, 2nd edition. Chapman and Hall, London.

Chapman, P.J., Reynolds, B. and Wheater, H.S. (1993) Hydrochemical changes along stormflow pathways in a small moorland headwater catchment in Mid-Wales, UK. *Journal of Hydrology*, **151**, 241-265.

Chappell, J. (1980) Coral morphology, diversity and reef growth. *Nature*, **286**, 249-252.

Chappell, J.M.A. and Shackleton, N.J. (1986) Oxygen isotopes and sea-level. *Nature*, **324**, 137-138.

Chengrui, M. and Dregne, H.E. (2001) Silt and the future development of China's Yellow River. *The Geographical Journal*, **167**, 7-22.

Cherrett, J.M (1989) *Ecological concepts*. Blackwell Scientific, Oxford.

Childs, E.C. (1969) *Introduction to the physical basis of soil water phenomenon*. John Wiley & Sons, London.

Chow, W.T.L. and Roth, M. (2006) Temporal dynamics of the urban heat island of Singapore. *International Journal of Climatology*, **26**, 2243-2260.

Church, J.A. and White, N.J. (2006). A 20th century acceleration in global sea-level rise. *Geophysical Research Letters*, **33**, L01602, doi:10.1029/2005GL024826.

Ciambrone, D.F. (1997) *Environmental life cycle analysis*. Lewis, Boca Raton, FL.

CLAG (Critical Loads Advisory Group) (1994) *Critical levels of acidity in the United Kingdom*. Summary report prepared at the request of the UK Department of the Environment, Transport and Regions.

Clark, M.J., Cresser, M.S., Smart, R. *et al.* (2004) The influence of catchment characteristics on the seasonality of carbon and nitrogen species concentrations in upland rivers of Northern Scotland. *Biogeochemistry*, **68**, 1-19.

Clarke, C.D., Knight, J.K. and Gray, J.T. (2000) Geomorphological reconstruction of the Labrador sector of the Laurentide Ice Sheet. *Quaternary Science Reviews*, **19**, 1343-1366.

Clarke, G.K.C. (1987) Fast glacier flow: ice streams, surging and tidewater glaciers. *Journal of Geophysical Research*, **92**(B9), 8835-8841.

Clements, F.E. (1916) *Plant succession*. Carnegie Institute of Washington, Washington, DC.

Clements, F.E. (1928) *Plant succession and indicators*. H.W. Wilson, New York.

Clifford, N.J. (1993) Formation of riffle-pool sequences: field evidence for an autogenetic process. *Sedimentary Geology*, **85**, 39-51.

Clifford, N.J. and Valentine, G. (eds) (2003) *Key methods in geography*. Sage Publications, London.

Clifford, S.M., Crisp, D., Fisher, D.A. *et al.* (2000) The state and future of Mars polar science and exploration. *Icarus*, **144**, 210-242.

Cockburn, H.A.P. and Summerfield, M.A. (2004) Geomorphological applications of cosmogenic isotope analysis. *Progress in Physical Geography*, **28**, 1-42.

Collier, C.G. (1999) Precipitation measuring weather radar. In: Herschy, R.W. (ed.), *Hydrometry: Principles and practice*. John Wiley & Sons, Chichester, 181-198.

Collinson, A.S. (1997) *Introduction to world vegetation*. Allen and Unwin, London.

COM (2002) *Towards a thematic strategy for soil protection*. Commission of the European Communities, Brussels.

Connell, J.H. and Slatyer, R.D. (1977) Mechanisms of succession in natural communities and their role in community stability and organisation. *American Naturalist*, **111**, 1119-1144.

Cosby, B.J., Hornberger, G.M. and Galloway, J.N. (1985) Modelling the effects of acid deposition: assessment of a lumped parameter model of soil water and streamwater chemistry. *Water Resources Research*, **21**, 51-63.

Cotton, C.A. (1922) *Geomorphology of New Zealand*. Dominion Museum, Wellington.

Cowell, P.J. and Thom, B.G. (1994) Morphodynamics of coastal evolution. In: Carter, R.W.G. and Woodroffe, C.D. (eds), *Coastal evolution*. Cambridge University Press, Cambridge, 33-86.

Cowles, H.C (1899) The ecological relations of the vegetation on the sand dunes of Lake Michigan. *Botanical Gazette*, **27**, 95-117, 167-202, 281-308, 361-391.

Cox, B. and Moore, P. (1993) *Biogeography: an ecological and evolutionary approach*. Blackwell Scientific, Oxford.

Cox, P.M., Betts, R.A., Jones, C.D. *et al.* (2000) Acceleration of global warming due to carbon-cycle feedbacks in a coupled climate model. *Nature*, **408**, 184-187.

Cramer, W., Bondeau, A., Woodward, F.I. *et al.* (2001) Global response of terrestrial ecosystem structure and function to CO_2 and climate change: results from six dynamic global vegetation models. *Global Change Biology*, **7**, 357-373.

Cresser, M.S., Kilhma, K. and Edwards, A.C. (1993) *Soil chemistry and its applications*. Cambridge University Press, Cambridge.

Critchfield, H.J. (1983) *General climatology*, 4th edition. Prentice Hall, Englewood Cliffs, NJ.

Croke, J. (1997) Australia. In: Thomas, D.S.G. (ed.), *Arid zone geomorphology: Process, form and change in drylands*. John Wiley & Sons, Chichester, 563-573.

Culling, W.E.H. (1963) Soil creep and the development of hillsides slopes. *Journal of Geology*, **71**, 127-161.

Curran, J.C. and Robertson, M. (1991) Water quality implications of an observed trend of rainfall and runoff. *Journal of the Institution of Water and Environmental Management*, **5**, 419-424.

Curtis, L.F., Courtney, F.M. and Trudgill, S.T. (1976) *Soils in the British Isles*. Longman, London.

D

Dai, A., Trenberth, K.E. and Qian, T. (2004) A global dataset of Palmer Drought Severity Index for 1870-2002: relationship with soil moisture and effects of surface warming. *Journal of Hydrometeorology*, **5**, 1117-1130.

Dalrymple, R.W., Zaitlin, B.A. and Boyd, R. (1992) Estuarine facies models: conceptual basis and stratigraphic implications. *Journal of Sedimentary Petrology*, **62**, 1130-1146.

D'Arcy, B.J., Ellis, J.B., Ferrier, R.C., Jenkins, A. and Dils, R. (eds) (2000) *Diffuse pollution impacts: The environmental and economic impacts of diffuse pollution in the UK*. Terence Dalton Publishers, London.

Darwin, C. (1859) *The origin of species by means of natural selection*. John Murray, London.

Davies, J.L. (1980) *Geographical variation in coastal development*, 2nd edition. Longman, London.

Davies, T.A. and Gorsline, D.S. (1976) Oceanic sediments and sedimentary properties. In: Riley, J.P. and Chester, R. (eds), *Chemical oceanography*, 2nd edition. Academic Press, Orlando, FL.

Davies, T.D., Tranter, M., Wigington, P.J. Jr and Eshleman, K.N. (1992) 'Acidic episodes' in surface waters in Europe. *Journal of Hydrology*, **132**, 25-69.

Davis, C.H., Li, Y.H., McConnell, J.R. *et al.* (2005) Snowfall-driven growth in East Antarctic ice sheet mitigates recent sea-level rise. *Science*, **308**, 1898-1901.

Davis, W.M. (1889) The rivers and valleys of Pennsylvania. *National Geographic Magazine*, **1**, 183-253.

Davis, W.M. (1954) *Geographical essays*, ed. D.W. Johnson. Dover, New York.

Dawson, A.G. (1992) *Ice Age Earth*. Routledge, London.

Dekker, L.W. and Ritsema, C.J. (1996) Preferential flow paths in a water repellent clay soil with grass cover. *Water Resources Research*, **32**, 1239-1249.

Delworth, T.L. and Dixon, K.W. (2000) Implications of the recent trend in the Arctic/North Atlantic Oscillation for the North Atlantic thermohaline circulation. *Journal of Climate*, **13**, 3721-3727.

Deosthali, V. (2000) Impact of rapid urban growth on heat and moisture islands in Pune City, India, *Atmospheric Environment*, **34**, 2745-2754.

de Wit, M. and Stankiewicz, J. (2006) Changes in surface water supply across Africa with predicted climate change. *Science*, **311**, 1917-1921.

Diamond, J. (1997) *Guns, germs and steel*. Vintage, London.

Diamond, J.M. (1975) The island dilemma: lessons of modern biogeographic studies for the design of nature reserves. *Biological Conservation*, **7**, 129-146.

Diamond, J.M. and May, R.E. (1976) Island biogeography and the design of nature reserves. In: May, R.E. (ed.), *Theoretical ecology*. Sinauer Associates, Sunderland, MA.

Diamond, J.M. and Veitch, C.R. (1981) Extinctions and introductions in the New Zealand avifauna - cause and effect? *Science*, **211**, 499-501.

Dickson, B., Yashayaev, I., Meincke, J. *et al.* (2002) Rapid freshening of the deep North Atlantic ocean over the past four decades. *Nature*, **416**, 832-837.

Dickson, R.R., Gmitrowicz, E.M. and Watson, A.J. (1990) Deep-water renewal in the northern North Atlantic. *Nature*, **344**, 848-850.

Dietrich, W.E. and Dunne, T. (1978). Sediment budget for a small catchment in mountainous terrain. *Zeitschrift für Geomorphologie, Supplement*, **29**, 191-206.

Dietrich, W.E. and Dunne, T. (1993) The channel head. In: Beven, K.J. and Kirkby, M.J. (eds), *Channel network hydrology*. John Wiley & Sons, Chichester, 175-219.

Dils, R.M. and Heathwaite, A.L. (1996) Phosphorus fractionation in hillslope hydrological pathways contributing to agricultural runoff. In: Anderson, M.A. and Brooks, S.M. (eds), *Advances in hillslope processes*, Vol. 1. John Wiley & Sons, Chichester, 229-251.

Doran, J.W. and Parkin, T.B. (1994) Defining and assessing soil quality. In: Doral, J.W. *et al.* (eds), *Defining soil quality for a sustainable environment*, Special Publication 35. Soil Science Society of America, Madison, WI, 3-21.

Dougill, A.J., Reed, M.S., Fraser, E.D.G. *et al.* (2006) Learning from doing participatory rural research: lessons from the Peak District National Park. *Journal of Agricultural Economics*, **57**, 259-275.

Douglas, I. (1967) Man, vegetation and the sediment yield of rivers. *Nature*, **215**, 925-928.

Dreimanis, A. and Vagners, U.J. (1971) Bimodal distribution of rock and mineral fragments in basal tills. In: Goldthwait, R.P. (ed.), *Till, a symposium*. Ohio State University Press, Columbus, OH, 237-250.

Drewry, D. (1986) *Glacial geologic processes*. Edward Arnold, London.

Drury, W.H. and Nisbet, I.C.T. (1973) Succession. *Journal of the Arnold Arboretum*, **53**, 331-368.

Dunbar, R.B. (2000) Clues from corals. *Nature*, **407**, 956-957.

Dunne, T. (1978) Field studies of hillslope flow processes. In: Kirkby, M.J. (ed.), *Hillslope hydrology*. John Wiley & Sons, Chichester, 227-293.

Dunne, T. and Aubry, B.F. (1986) Evaluation of Horton's theory of sheetwash and rill erosion on the basis of field experiments. In: Abrahams, A.D. (ed.), *Hillslope processes*. Allen & Unwin, London, 31-53.

Dyer, K.R. (1998) *Estuaries: A physical introduction*. John Wiley & Sons, Chichester.

Dyurgerov, M.B. and Meier, M.F. (2000) Twentieth century climate change: evidence from small glaciers. *Proceedings of the National Academy of Sciences, USA*, **97**, 1406-1411.

E

Easterbrook, D.J. (1993) *Surface processes and landforms*. Macmillan, London.

Edgren, G. (1993) Expected economic and demographic developments in coastal zones world-wide. *World Coast '93*, National Institute for Coastal and Marine Management, Coastal Zone Management Centre, Noordwijk, The Netherlands, 367-370.

Egler, F.E. (1954) Vegetation science concepts I. Initial floristic composition, a factor in old-field vegetation development. *Vegetatio*, **4**, 412-417.

Ehleringer, J.R., Cerling, T.E. and Dearing, M.D. (eds) (2005) *A history of atmospheric CO_2 and its effect on plants, animals, and ecosystems*. Springer-Verlag, New York.

Ehlers, J. (1996) *Quaternary and glacial geology*. John Wiley & Sons, Chichester.

Eicher, D.L. (1976) *Geologic time*. Prentice-Hall, Englewood Cliffs, NJ.

Elkibbi, M. and Rial, J.A. (2001) An outsider's review of the astronomical theory of climate: is the eccentricity-driven insolation the main driver of the ice ages? *Earth-Science Reviews*, **56**, 161-177.

Emanuel, K. (2005) Increasing destructiveness of tropical cyclones over the past 30 years. *Nature*, **436**, 686-688.

Emiliani, C. (1955) Pleistocene temperatures. *Journal of Geology*, **63**, 538-575.

Emmett, W.W. and Wolman, M.G. (2001) Effective discharge and gravel-bed rivers. *Earth Surface Processes and Landforms*, **26**, 1369-1380.

English Nature (2002) *Urbanities urbio: Urban diversity and human nature*. English Nature, Peterborough, Issue 1, 14.

Erickson, S.L. and King, B.J. (1999) *Fundamentals of environmental management*. John Wiley & Sons, Chichester.

Evans, J.G. (1999) *Land and archaeology: Histories of human environment in the British Isles*. Tempus Publishing, Stroud.

Evans, M.G., Burt, T.P., Holden, J. and Adamson, J. (1999) Runoff generation and water table fluctuations in blanket peat: evidence from UK data spanning the dry summer of 1995. *Journal of Hydrology*, **221**, 141-160.

F

Fabel, D., Stroeven, A.P., Harbor, J. *et al.* (2002) Landscape preservation under Fennoscandian ice sheets determined from *in situ* produced ^{10}Be and ^{26}Al. *Earth and Planetary Science Letters*, **201**, 397-406.

FAO/UNESCO (1974). *Soil map of the world, scale 1 : 5 000 000*: Volume 1, *legend*. UNESCO, Paris.

Ferguson, R.I. (1981) Channel form and channel changes. In: Lewin, J. (ed.), *British rivers*. George Allen and Unwin, London, 90-125.

Fisher, O. (1866) On the disintegration of a chalk cliff. *Geological Magazine*, **3**, 354-356.

Fitter, A.H. and Fitter, R.S.R. (2002) Rapid changes in flowering time in British plants. *Science*, **296**, 1689-1691.

Fitter, A.H., Fitter, R.S.R., Harris, I.T.B. and Williamson, M.H. (1995) Relationships between first flowering date and temperature in the flora of a locality in central England. *Functional Ecology*, **9**, 55-60.

FitzPatrick, E.A. (1983) *Soils: Their formation, classification and distribution*, 1st edition. Longman Scientific & Technical, Harlow.

Forman, R.T.T. (1995) *Land mosiacs - The ecology of landscapes and regions*. Cambridge University Press, Cambridge.

Fountain, A.G. and Walder, J.S. (1998) Water flow through temperate glaciers. *Reviews of Geophysics*, **36**, 299-328.

Fourier, J. (1824) Remarques generales sur les temperatures du globe terrestre et des espaces planetaires. *Annales de chimie et de physique*, **27**, 136-167.

Freeman, C., Fenner, N., Ostle, N.J. *et al.* (2004) Export of dissolved organic carbon from peatlands under elevated carbon dioxide levels. *Nature*, **430**, 195-198.

French, H.M. (1996) *The Periglacial environment*, 2nd edition. Longman, Harlow.

French, H.M. (2007) *The Periglacial environment*, 3rd edition. John Wiley & Sons, Chichester.

French, P.W. (1997) *Coastal and estuarine management*. Routledge, London.

French, P.W. (2001) *Coastal defences - Processes, problems and solutions*. Routledge, London.

Friedlingstone, P., Cox, P., Betts, R. *et al.* (2006) Climate-carbon cycle feedback analysis: results from the C^{4}MIP model inter-comparison. *Journal of Climate*, **19**, 3337-3353.

Friedrichs, C.T. and Aubrey, D.G. (1988) Non-linear tidal distortion in shallow well-mixed estuaries: a synthesis. *Estuarine, Coastal and Shelf Science*, **27**, 521-545.

G

Gaillardet, J., Negrel, P., Dupre, B. and Allegre, C.J. (1997) Chemical and physical denudation in the Amazon River Basin. *Chemical Geology*, **142**, 141-173.

Galloway, J.N., Schlesinger, W.H., Levy, H. *et al.* (1995) Nitrogen fixation: anthropogenic enhancement and environmental response. *Global Biogeochemical Cycles*, **9**, 235-252.

Galloway, W.E. (1975) Process framework for describing the morphologic and stratigraphic evolution of deltaic depositional systems. In: Broussard, M.L. (ed.), *Deltas, models for exploration*. Houston Geological Society, Houston, 87-98.

Gamache, I. and Payette, S. (2004) Height growth response of tree line spruce to recent climate warming across the forest-tundra of eastern Canada. *Journal of Ecology*, **92**, 835-845.

Ganderton, P. and Coker, P. (2005) *Environmental Biogeography*. Prentice Hall, Harlow.

Gedzelman, S.D. and Austin, S. (2003) Mesoscale aspects of the urban heat island around New York City. *Theoretical and Applied Climatology*, **75**, 29-42.

Geiger, R. (1965) *The climate near the ground*, 2nd edition. Harvard University Press, Cambridge, MA.

Geiger, R. and Pohl, W. (1953) *Revision of Koppen-Geiger climate maps of the Earth*. Justes Perthes, Darmstadt.

Geiger, R., Aron, R.H. and Todhunter, P. (2003) *The climate near the ground*. Rowman and Littlefield, Oxford.

Gellert, P.K. (1998) A brief history and analysis of Indonesia's forest fire crisis. *Indonesia*, **65**, 63-85.

Gerrard, J. (2000) *Fundamentals of soils*. Routledge, London.

Gibb, J. (1954) Feeding ecology of tits, with notes on treecreeper and goldcrest. *Ibis*, **96**, 513-543.

Gibbard, P.L. (1994) *Pleistocene history of the lower Thames*. Cambridge University Press, Cambridge.

Gilbert, G.K. (1909) The convexity of hilltops. *Journal of Geology*, **17**, 344-351.

Gildor, H. and Tziperman, E. (2001) A sea-ice climate switch mechanism for the 100-kyr glacial cycles. *Journal of Geophysical Research*, **106**, 9117-9133.

Gilvear, D.J. (2003) Patterns of channel adjustment due to impoundment of the Upper River Spey, Scotland (1942-2000). *River Research and Applications*, **19**, 1-17.

Gilvear, D.J. and Black, A.R. (1999) Flood induced embankment failures on the River Tay: implications of climatically induced hydrological change in Scotland. *Hydrological Sciences Journal*, **44**, 345-362.

Gilvear, D.J. and Bradley, S. (1997) Geomorphic adjustment of a newly constructed ecologically sound river diversion on an upland gravel bed river, Evan Water, Scotland. *Regulated Rivers*, **13**, 1-13.

Gilvear, D.J., Winterbottom, S.J. and Sichingabula, H. (2000) Mechanisms of channel planform change on the actively meandering Luangwa River, Zambia. *Earth Surface Processes and Landforms*, **25**, 421-436.

GLASOD (1990) *World map of the status of human-induced soil degradation*. Winand Staring Centre, Wageningen.

Gleason, H.A. (1917) The structure and development of the plant association. *Bulletin of the Torrey Botanical Club*, **43**, 463-481.

Gleason, H.A. (1926) The individualistic concept of plant association. *Bulletin of the Torrey Botanic Club*, **53**, 7-26.

Goede, A. and Harmon, R.S. (1983) Radiometric dating of Tasmanian speleothems - evidence of cave sediments and climate change. *Journal of the Geological Society of Australia*, **30**, 89-100.

Goldich, S.S. (1938) A study of rock weathering. *Journal of Geology*, **46**, 17-58.

Goldstein, R.M., Engelhardt, H., Kamb, B. and Frolich, R.M. (1993) Satellite radar interferometry for monitoring ice stream motion: application to an Antarctic ice stream, *Science*, **262**, 1525-1530.

Goodbred, S.L. and Kuehl, S.A. (1999) Holocene and modern sediment budgets for the Ganges-Brahmaputra river system: evidence for highstand dispersal to flood-plain, shelf, and deep-sea depocenters. *Geology*, **26**, 559-562.

Goudie, A.S. (1973) *Duricrusts of tropical and subtropical landscapes*. Clarendon Press, Oxford.

Goudie, A.S. (1997) Weathering processes. In: Thomas, D.S.G. (ed.), *Arid zone geomorphology: Process, form and change in drylands*. John Wiley & Sons, Chichester, 25-39.

Gowing, G. (2003) Salinisation - one of our biggest environmental problems: **http://www.amonline.net.au/factsheets/salinisation.htm**

Grace, J. and Unsworth, M.H. (1988) Climate and microclimate of the uplands. In: Usher, M.B. and Thompson, D.B.A. (eds),

Ecological Change in the Uplands, Special Publication Number 7, British Ecological Society. Blackwell, Oxford, 137-150.

Graumann, A., Houston, T., Lawrimore, J. *et al.* (2005) Hurricane Katrina – a climtaological perspective. NOAA National Data Center Technical Report 2005-01, Ashville, NC.

Gregory, K.J. (1985) *The nature of physical geography.* Arnold, London.

Gregory, K.J. (2000) *The changing nature of physical geography.* Arnold, London.

Gregory, K.J., Davis, R.J. and Downs, P.W. (1992) Identification of river channel changes due to urbanisation. *Applied Geography,* **12**, 299-318.

Grime, P. (1997) Climate change and vegetation. In: Crawley, M.J. (ed.), *Plant ecology.* Blackwell Scientific, Oxford.

Grimes, J.P. (1979) *Plant strategies and vegetation processes.* John Wiley & Sons, Chichester.

Guinotte, J.M., Buddemeier, R.W. and Kleypas, J.A. (2003) Future coral reef habitat marginality: climate change in the Pacific basin. *Coral Reefs,* **22**, 551-558.

Gupta, A. (2007) *Large rivers: Geomorphology and management.* John Wiley & Sons, Chichester.

Gurney, K.R., Law, R.M., Denning, A.S. *et al.* (2002) Towards robust regional estimates of CO_2 sources and sinks using atmospheric transport models. *Nature,* **415**, 626-630.

Gurney, S.D. (2001) Aspects of the genesis and geomorphology and terminology of palsas: perennial cryogenic mounds. *Progress in Physical Geography,* **25**, 249-260.

H

Hack, J.T. (1960) Interpretation of erosional topography in humid temperate regions. *American Journal of Science,* **258**, 80-97.

Haines-Young, R.H. and Petch, J.R. (1986) *Physical geography; Its nature and methods.* Harper and Row, London.

Haldorsen, S. (1981) Grain-size distribution of subglacial till and its relation to glacier crushing and abrasion. *Boreas,* **10**, 91-105.

Hallam, A. (1973) *A revolution in earth sciences.* Clarendon Press, Oxford.

Hallet, B., Anderson, S.P., Stubbs, C.W. and Gregory, E.C. (1988) Surface soil displacements in sorted circles, Western Spitzbergen. In: *Proceedings 5th International Conference on Permafrost,* Vol. 1. Tapir Publishers, Trondheim, 770-775.

Hambrey, M.J. (1994) *Glacial environments.* UCL Press, London.

Hammer, C.U., Clausen, H.B., Freidrich, W.L. and Tauber, H. (1987) The Minoan eruption of Santorini dated to 1645 BC? *Nature,* **328**, 517-519.

Hansen, M., DeFries, R., Townshend, J.R.G. and Sohlberg R. (1998) *UMD Global Land Cover Classification, 1 Kilometer, 1.0, 1981-1994.* Department of Geography, University of Maryland, College Park, MD.

Harding, G. (1968) The tragedy of the commons. *Science,* **162**, 1243-1248.

Harding, R.J. (1978) The variation of the altitudinal gradient of temperature within the British Isles. *Geografiska Annalar* A, **60**, 43-49.

Harding, R.J. (1979) Radiation in the British uplands. *Journal of Applied Ecology,* **16**, 161-170.

Harding, R.J., Hall, R.L., Neal, C., Roberts, J.M., Rosier, P.T.W. and Kinniburgh, D.G. (1992) *Hydrological impacts of broadleaf woodlands: Implications for water use and water quality,* National Rivers Authority R & D Note 56, National Rivers Authority, Bristol.

Hargrove, E.C. (1989) *Foundations of environmental ethics.* Prentice Hall, Englewood Cliffs, NJ.

Harré, R. (2002) *Great scientific experiments: Twenty experiments that changed our view of the world.* Dover Publications, New York.

Harrison, S. and Anderson, E. (2001) A Late Devensian rock glacier in the Nantlle Valley, North Wales. *Glacial Geology and Geomorphology:* **http://boris.qub.ac.uk/ggg/papers/full/ 2001/rp012001/rp01/html**

Hart, J.K. and Boulton, G.S. (1991) The interrelation of glaciotectonic and glaciodepositional processes within the glacial environment. *Quaternary Science Reviews,* **10**, 335-350.

Harvey, A.M. (1997) The role of alluvial fans in arid zone fluvial systems. In: Thomas, D.S.G. (ed.), *Arid zone geomorphology: Processes, forms and change in drylands.* John Wiley & Sons, Chichester, 231-259.

Haslett, S.K. (2000) *Coastal systems.* Routledge, London.

Haycock, N.E. and Muscutt, A.D. (1995) Landscape management strategies for the control of diffuse pollution. *Landscape and Planning,* **31**, 313-321.

Heinrich, H. (1988) Origin and consequences of cyclic ice rafting in the Northeast Atlantic Ocean during the past 130,000 years. *Quaternary Research,* **29**, 142-152.

Henderson-Sellers, A. and Robinson, P.J. (1986) *Contemporary climatology.* Longman, London.

Herschy, R.W. (ed.) (1999) *Hydrometry, principles and practice.* John Wiley & Sons, Chichester.

Hess, H.H. (1962) History of ocean basins. In: Engel, A.E.J., James, H.L. and Leonard, B.F. (eds), *Petrologic studies: A volume in honor of A.F. Buddington.* Geological Society of America, Boulder, CO.

Heusser, L. and Oppo, D. (2003) Millennial- and orbital-scale variability in southeastern United States and in the subtropical Atlantic during Marine Isotope Stage 5: evidence from pollen and isotopes in ODP Site 1059. *Earth and Planetary Science Letters,* **214**, 483-490.

Heusser, L., Heusser, C., Mix, A. and McManus, J. (2006) Chilean and Southeast Pacofoc palaeoclimate variations during the last glacial cycle: directly correlated pollen and delta ^{18}O records from ODP Site 1234. *Quaternary Science Reviews,* **25**, 3404-3415.

Hewitt, K. (1998) Recent glacier surges in the Karakoram himalaya, south central Asia: **http://www.agu.org/eosrelec/ 97l06e.html**

Hewlett, J.D. (1961) Watershed management. In: *Report for 1961 Southeastern Forest Experiment Station*, US Forest Service, Ashville, NC, 62–66.

Hey, R. (1996) Environmentally sensitive river engineering. In: Petts, G.E. and Calow, P. (eds), *River restoration*. Blackwell Scientific, Oxford, 80–105.

Hoffmann, A.A., Hallas, R.J., Dean, J.A. and Schiffer, M. (2003) Low potential for climatic stress adaptation in a rainforest *Drosophila* species. *Science*, **301**, 100–102.

Hofstede, J. (2005) Danish-German-Dutch Wadden Environments. In: E.A. Koster (ed.), *The physical geography of Western Europe*. Oxford University Press, Oxford, 185–205.

Holden, J. (2005a) Darcy's law. In: Lehr, J.H. and Keeley, J. (eds), *Water encyclopedia*, Volume 5. John Wiley & Sons, New York, 63–64.

Holden, J. (2005b) Peatland hydrology and carbon cycling: why small-scale process matters. *Philosophical Transactions of the Royal Society A*, **363**, 2891–2913.

Holden, J. and Burt, T.P. (2002a) Piping and pipeflow in a deep peat catchment. *Catena*, **48**, 163–199.

Holden, J. and Burt, T.P. (2002b) Laboratory experiments on drought and runoff in blanket peat. *European Journal of Soil Science*, **53**, 675–689.

Holden, J. and Wright, A. (2004) A UK tornado climatology and the development of simple prediction tools. *Quarterly Journal of the Royal Meteorological Society*, **130**, 1009–1022.

Holden, J., Burt, T.P. and Cox, N.J. (2001) Macroporosity and infiltration in blanket peat: the implications of tension disc infiltrometer measurements. *Hydrological Processes*, **15**, 289–303.

Holden, J., Burt, T.P. and Vilas, M. (2002) Application of ground penetrating radar to the identification of subsurface piping in blanket peat. *Earth Surface Processes and Landforms*, **27**, 235–249.

Holden, J., Evans, M.G., Burt, T.P. and Horton, M. (2006) Impact of land drainage on peatland hydrology. *Journal of Environmental Quality*, **35**, 1764–1778.

Holden, J., Shotbolt, L., Bonn, A. *et al.* (2007) Environmental change in moorland landscapes. *Earth-Science Reviews*, **82**, 75–100.

Holling, C.S. (1973) Resilience and stability of ecological systems. *Annual Review of Ecology and Systematics*, **4**, 1–23.

Holt-Jensen, A. (1980) *Geography: History and concepts*. Sage Publications, London.

Hooghiemstra, H. and Sarmiento, G. (1991) New long continental pollen record from a tropical intermontane basin: Late Pliocene and Pleistocene history from a 540 m core. *Episodes*, **14**, 107–115.

Hooke, J.M. (1980) Magnitude and distribution of rates of river bank erosion. *Earth Surface Processes and Landforms*, **5**, 143–147.

Hooke, R.L. (2005) *Principles of glacier mechanics*. Cambridge University Press, Cambridge.

Hornberger, G.M., Cosby, B.J. and Wright, R.F. (1989) Historical reconstructions and future forecasts of regional surface water acidification in southernmost Norway. *Water Resources Research*, **25**, 2009–2018.

Horton, R.E. (1933) The role of infiltration in the hydrological cycle. *Transactions of the American Geophysical Union*, **14**, 446–460.

Horton, R.E. (1945) Erosional development of streams and their drainage basins; hydrophysical approach to quantitative morphology. *Bulletin of the Geological Society of America*, **56**, 275–370.

Houghton, J. (1994) *Global warming: The complete briefing*, 1st edition. Oxford: Lion Publishing.

Houghton, J.T. (1997) *Global warming: The complete briefing* (paperback). Cambridge University Press, Cambridge.

Howard, A.D. (1994) A detachment limited model of drainage basin evolution. *Water Resources Research*, **30**, 2261–2285.

Hubbard, B. and Glasser, N.J. (2005) *Field techniques in glaciology and glacial geomorphology*. John Wiley & Sons, Chichester.

Hudson-Edwards, K.A., Macklin, M.G., Miller, J.R. and Lechlar, P.J. (2001) Sources, distribution and storage of heavy metals in the Rio Pilcomayo, Bolivia. *Journal of Geochemical Exploration*, **72**, 229–250.

Hudson-Edwards, K.A., Heather, E., Jamieson, H.E., Charnock, J.M. and Macklin, M.G. (2005) Arsenic speciation in waters and sediment of ephemeral floodplain pools, Rios Agrio-Guadiamar, Aznalcóllar, Spain. *Chemical Geology*, **219**, 175–192.

Huggett, R.J. (1995) *Geoecology: An evolutionary approach*. Routledge, London.

Huggett, R.J. (2004) *Fundamentals of biogeography*. Routledge, London.

Hulme, M. and Kelly, M. (1993) Exploring the links between desertification and climate change. *Environment*, **35**, 5–45.

Hunt, A.G. and Malin, P.E. (1998) Possible triggering of Heinrich events by ice-load induced earthquakes. *Nature*, **393**, 155–158.

Huntley, B. and Prentice, C.I. (1993) Holocene vegetation and climates of Europe. In: Wright, H.E. Jr. *et al.* (eds), *Global climates since the Last Glacial Maximum*. University of Minnesota Press, Minneapolis, 136–168.

Hutton, J. (1795) *Theory of the Earth*. Edinburgh, 2 volumes.

I

ICONA (1991) *Plan national de lutte contre l'érosion. Ministère de l'Agriculture, de la Peche et de l'Alimenation*. Institut National pour la conservation de la Nature, Madrid.

Ilaiwi, M., Abdelgawad, G. and Jabour, E. (1992) Syria: human induced soil degradation. In: *UNEP world atlas of desertification*. Edward Arnold, London, 43–45.

Imbrie, J. and Imbrie, K.P. (1979) *Ice ages: Solving the mystery*. Macmillan, London.

Imeson, A.C. and Kwaad, F.J.P.M. (1990) The response of tilled soils to wetting by rainfall and the dynamic character of soil erodibility. In: Boardman, J., Foster, I.D.L. and Dearing, J.A. (eds), *Soil erosion on agricultural land*. John Wiley & Sons, Chichester, 3–14.

Inkpen, R. (2004) *Science, philosophy and physical geography*. Routledge, London.

IPCC (1995) *IPCC second assessment - Climate change 1995,* A Report of the Intergovernmental Panel on Climate Change. IPCC, Geneva.

IPCC (2001) *Climate change 2001: The scientific basis.* Contribution of Working Group 1 to the Third Assessment Report of the Intergovernmental Panel on Climate Change, ed. J.T. Houghton *et al.* Cambridge University Press, Cambridge.

IPCC (2007a) *Climate change 2007: The physical science basis.* Contribution of Working Group 1 to the Fourth Assessment Report of the Intergovernmental Panel on Climate Change. Solomon, S., Qin, D., Manning, M. *et al.* (eds). Cambridge University Press, Cambridge.

IPCC (2007b) *Climate change 2007: Climate change impacts, adaptation and vulnerability.* Contribution of Working Group 2 to the Fourth Assessment Report of the Intergovernmental Panel on Climate Change. Adger, N. *et al.* (eds). Cambridge University Press, Cambridge.

Isacks, B., Oliver, J. and Sykes, L.R. (1968) Seismology and the new global tectonics. *Journal of Geophysical Research,* **73,** 5855-99.

Isaksen, K., Mühll, D.V., Gubler, H., Kohl, T. and Sollid, J.L. (2000) Ground surface temperature reconstruction based on data from a deep borehole in permafrost at Janssonhaugen, Svalbard. *Annals of Glaciology,* **31,** 287-294.

J

Jarvis, M.G., Allen, R.H., Fordham, S.J., Hazelden, J., Moffat, A.J. and Sturdy, R.G. (1984) *Soils and their use in south east England.* Soil Survey of England and Wales. London.

Jauregi, E. (1990-91) Influence of a large urban park on temperature and convective precipitation in a tropical city. *Energy and Buildings,* **15-16,** 457-463.

Jeffree, E.P. (1960) Some long-term means from *The Phenological Reports* (1891-1948) of the Royal Meteorological Society. *Quarterly Journal of the Royal Meteorological Society,* **86,** 95-103.

Jeffries, M.J. (1997) *Biodiversity and conservation.* Routledge, London.

Jenny, H. (1941) *Factors of soil formation: A system of quantitative pedology.* McGraw-Hill, New York.

Johannessen, O.M., Khvorostovsky, K., Miles, M.W. *et al.* (2005) Recent ice-sheet growth in the interior of Greenland. *Science,* **310,** 1013-1016.

Jones, A., Duck, R., Reed, R. and Weyers, J. (2000) *Practical skills in environmental science.* Prentice Hall, Harlow.

Jones, J.A.A. (1981) *The nature of soil piping: A review of research,* British Geomorphological Research Group Monograph Series 3. Geo Books, Norwich.

Jones, J.A.A. (1997a) *Global hydrology: Processes, resources and environmental management.* Pearson Education, Harlow.

Jones, J.A.A. (1997b) Pipeflow contribution areas and runoff response. *Hydrological Processes,* **11,** 35-41.

Jouzel, J., Barkov, N.I., Barnola, J.M. *et al.* (1993) Extending the Vostok ice-core record of palaeoclimate to the penultimate glacial period. *Nature,* **364,** 407-412.

K

Kamb, B., Raymond, C., Harrison, W. *et al.* (1985) Glacier surge mechanism: 1982-1983 surge of Variegated Glacier, Alaska. *Science,* **227,** 469-479.

Kaplan, M. (2002) Let the river run. *New Scientist,* **175,** 32-35.

Karavayeva, N.A., Nefedova, T.G. and Targulian, V.O. (1991) Historical land use changes and soil degradation on the Russian Plain. In Brouwer, F.M., Thomas, A.J. and Chadwick, M.J. (eds), *Land use changes in Europe: Processes of change, environmental transformations and future patterns.* Kluwer Academic, Dordrecht, 351-377.

Kasser, M. and Egels, Y. (2002) *Digital photogrammetry.* Taylor & Francis, London.

Kastner, M., Kvenvolden, K.A. and Lorenson, T.D. (1998) Chemistry, isotopic composition and origin of a methane-hydrogen sulfide hydrate at the Cascadia subduction zone. *Earth Planetary Science Letters,* **156,** 173-183.

Kay, R. and Alder, J. (1999) *Coastal planning and management.* Routledge, London.

Kellerhals, R. and Church, M. (1989) The morphology of large rivers: characterization and management. *Canadian Special Publication of Fisheries and Aquatic Sciences,* **106,** 31-48.

Kent, M. and Coker, P. (1992) *Vegetation description and analysis - a practical approach.* Belhaven Press, Chichester.

Kershaw, A.P., van der Kaars, S. and Moss, P.T. (2003) Late Quaternary Milankovitch-scale climatic change and variability and its impact on monsoonal Australasia. *Marine Geology,* **201,** 81-95.

King, C. (1991) *The Depositional Environments,* Sedimentology Book 2. Longman, Harlow.

Kirkby, M.J. (1971) Hillslope process-response models based on the continuity equation. *Transactions,* Institute of British Geographers, Special Publication No. 3.

Kirkby, M.J. (2001) Modelling the interactions between soil surface properties and water erosion. *Catena,* **46,** 89-102.

Kirkby, M.J., Naden, P.S., Burt, T.P. and Butcher, D.P. (1993) *Computer simulation in physical geography,* 2nd edition. John Wiley & Sons, Chichester.

Kocurek, G., Townsley, M., Yeh, E. *et al.* (1990) Dune and dunefield development on Padre Island, Texas, with implications for interdune deposition and water table controlled accumulation. *Journal of Sedimentary Petrology,* **62,** 622-635.

Kolpin, D.W., Furlong, E.T., Meyer, M.T. *et al.* (2002) Pharmaceuticals, hormones, and other organic wastewater contaminants in U.S. streams, 1999-2000: a national reconnaissance. *Environmental Science and Technology,* **36,** 1202-1211.

Komar, P.D. (1998) *Beach processes and sedimentation,* 2nd Edition. Prentice Hall, Englewood Cliffs, NJ.

Kondolf, G.M., Piegey, H. and Landon, N. (2002) Channel response to increased and decreased bedload supply from land use change: contrasts between two catchments. *Geomorphology*, **45**, 35-51.

Kondolf, M. and Piegey, H. (2003) *Tools in fluvial geomorphology*. John Wiley & Sons, Chichester.

König, M., Winther, J.-G. and Isaksson, E. (2001) Measuring snow and glacier ice properties from satellite. *Reviews of Geophysics*, **39**, 1-27.

Korbetis, M., Reay, D.S. and Grace, J. (2006) New directions: rich in CO_2. *Atmospheric Environment*, **40**, 3219-3220.

Koren, I., Kaufman, Y.J., Washington, R. *et al.* (2006) The Bodélé depression: a single spot in the Sahara that provides most of the mineral dust in the Amazon forest. *Environmental Research Letters*, **1**, 014005.

Koster, M.J. and Hillen, R. (1995). Combat erosion by law: coastal defence policy for The Netherlands. *Journal of Coastal Research*, **11**, 1221-1228.

Kraus, U. and Wiegand, J. (2006) Long-term effects of the Aznalcóllar mine spill – heavy metal content and mobility in soils and sediments of the Guadiamar river valley (SW Spain). *Science of the Total Environment*, **367**, 855-871.

Krinsley, D.P. (1970) *A geomorphological and palaeoclimatological study of the playas of Iran*. US Geological Survey, Washington, DC.

Kullman, L. (2002) Rapid recent range-margin rise of tree and shrub species in the Swedish Scandes. *Journal of Ecology*, **90**, 68-77.

Kullman, L. (2005) On the presence of Lateglacial trees in the Scandes. *Journal of Biogeography*, **32**, 1499-1500.

Kumar, K.K., Rajagopalan, B., Hoerling, M. *et al.* (2006) Unraveling the mystery of Indian monsoon failure during El Niño. *Science*, **314**, 115-119.

Kvenvolden, K.A. (1995) A review of the geochemistry of methane in natural gas hydrate. *Organic Geochemistry*, **23**, 997-1008.

L

Lachenbruch, A. (1962) *Mechanics of thermal contraction cracks and ice-wedge polygons in permafrost*, Special Report 70. Geological Society of America, Boulder, CO.

Lack, D. (1971) *Ecological isolation in birds*. Blackwell Scientific, Oxford.

Lal, R. (2004) Soil carbon sequestration impacts on global climate change and food security. *Science*, **304**, 1623-1627.

LaMarche, V.C. and Hirschboeck, K.K. (1984) Frost rings in trees as records of major volcanic eruptions. *Nature*, **307**, 121-126.

Lane, S.N. (2001) Constructive comments on D. Massey 'Space-time, "science" and the relationship between physical geography and human geography'. *Transactions of the Institute of British Geographers*, **26**, 243-256.

Lane, S.N. (2003) Numerical modelling in physical geography: understanding, explanation and prediction. In: Clifford, N.J. and Valentine, G. (eds), *Key methods in geography*. Sage Publications, London, 263-290.

Lane, S.N., James, T.D. and Crowell, M.D. (2000) Application of digital photogrammetry to complex topography for geomorphological research. *Photogrammetric Record*, **16**, 793-821.

Lane, S.N., Brookes, C.J., Kirby, M.J. and Holden, J. (2004) A network-index-based version of TOPMODEL for use with high-resolution topographic data. *Hydrological Processes*, **18**, 191-201.

Langbein, J.B. and Schumm, S.A. (1958) Yield of sediment in relation to mean annual precipitation. *Transactions of the American Geophysical Union*, **39**, 1076-1084.

Langmuir, D. (1997) *Aqueous environmental geochemistry*. Prentice Hall, Englewood Cliffs, NJ.

Laurance, W.F. and Williamson, B. (2001) Positive feedbacks among forest fragmentation, drought and climate change in the Amazon. *Conservation Biology*, **15**, 1529-1535.

Laurance, W.F., Cochrane, M.A., Bergen, S. *et al.* (2001) Environment – the future of the Brazilian Amazon. *Science*, **291**, 438-439.

Lauscher, F. (1976) Weltweite Typen der Höhenabhängigkeit des Niederschlags, *Wetter und Leben*, **28**, 80-90.

Lawrence, E., Jackson, A. and Jackson, J. (1998) *Dictionary of environmental science*. Longman, Harlow.

Leatherman, S.P. (1983) Barrier dynamics and landward migration with Holocene sea-level rise. *Nature*, **301**, 415-418.

Leeder, M.R. (1999) *Sedimentology and sedimentary basins*. Blackwell Scientific, Oxford.

Leeks, G.J., Lewin, J. and Newson, M.D. (1988) Channel change, fluvial geomorphology and river engineering: the case of the Afon Trannon, Mid Wales. *Earth Surface Processes and Landforms*, **13**, 207-223.

Leggett, J.K. (2000) *The carbon war: Global warming and the end of the oil era*. Penguin, London.

Leith, H. (1964) *Versuch einer kartographischen Darstelung der Produktivitat der Pflanzendecke auf der Erde*. Geographische Taschebuch, Weisbaden.

Leonard, J., Perrier, E. and de Marsily, G. (2001) A model for simulating the influence of a spatial distribution of large circular macropores on surface runoff. *Water Resources Research*, **37**, 3217-3225.

Leopold, L.B. and Maddock, T. Jr (1953) The hydraulic geometry of stream channels and some physiographic implications. *US Geological Survey Professional Paper* **252**.

Leopold, L.B. and Wolman, M.G. (1957) River channel patterns – braided, meandering and straight. *US Geological Survey Professional Paper* **282B**.

Li, X., Zhou, J. and Dodson, J. (2003) The vegetation characteristics of the 'Yuan' area at Yaoxian on the Loess Plateau in China over the last 12 000 years. *Review of Paleobotany and Palynology*, **124**, 1-7.

Liestøl, O. (1977) Pingos, springs and permafrost in Spitsbergen. *Norsk Polarinstitutt Årbok*, **1975**, 7-29.

Likens, G.E. and Bormann, F.H. (1995) *Biogeochemistry of a forested ecosystem*, 2nd edition. Springer-Verlag, New York.

Lillesand, T.M. and Kiefer, R.W. (2000) *Remote sensing and image interpretation*. John Wiley & Sons, Toronto.

Lillesand, T.M., Kiefer, R.W. and Chipman, J. (2003) *Remote sensing and image interpretation*, 5th edition. John Wiley & Sons, Toronto.

Linacre, E. and Geerts, B. (1997) *Climates and weather explained*. Routledge, London.

Livingstone, I. and Warren, A. (1996) *Aeolian geomorphology*. Longman, London.

Lock, G.S.H. (1990) *The growth and decay of ice*. Cambridge University Press, Cambridge.

Lockwood, J.G. (1962) Occurrence of Föhn winds in the British Isles. *Meteorological Magazine*, **91**, 57-65.

Lockwood, J.G. (1965) The Indian Monsoon - a review. *Weather*, **20**, 2-8.

Lockwood, J.G. (1979) *Causes of climate*. Edward Arnold, London.

Lockwood, J.G. (1993) Impact of global warming on evapotranspiration. *Weather*, **48**, 291-299.

Lockwood, J.G. (1999) Is potential evapotranspiration and its relationship with actual evapotranspiration sensitive to elevated atmospheric CO_2 levels? *Climatic Change*, **41**, 193-212.

Lockwood, J.G. (2001) Abrupt and sudden climatic transitions and fluctuations: a review. *International Journal of Climatology*, **21**, 1153-1179.

Lockwood, M. and Fröhlich, C. (2007) Recent oppositely directed trends in solar climate forcings and the global mean surface air temperature. *Proceedings of the Royal Society A*. doi: 10.1098/rspa.2007.1880.

Lovelock, J.L. (1979) *Gaia: A new look at life on Earth*. Oxford University Press, Oxford.

Lowe, J.J. and Walker, M.J.C. (1997) *Reconstructing Quaternary environments*. Pearson Education, Harlow.

M

MacArthur, R.H. and Wilson, E.O. (1967) *The theory of island biogeography*. Princetown University Press, Princeton, NJ.

MacAyeal, D.R. (1993) Binge-purge oscillations of the Laurentide Ice Sheet as a cause of the north Atlantic's Heinrich events. *Palaeoceanography*, **8**, 775-784.

Mackay, J.R. (1973) The growth of pingos, western Arctic coast, Canada. *Canadian Journal of Earth Sciences*, **10**, 979-1004.

Mackay, J.R. (1979) Pingos of the Tuktoyaktuk Peninsula area, Northwest Territories. *Canadian Journal of Earth Sciences*, **15**, 461-462.

Mackay, J.R. (1983) Pingo growth and subpingo water lenses. In: *Permafrost: Fourth International Conference, Proceedings, Fairbanks, Alaska*. National Academy Press, Washington, DC, 762-766.

Mackay, J.R. (1986) Growth of Ibyuk Pingo, Western Arctic Coast, Canada and some implications for environmental reconstructions. *Quaternary Research*, **26**, 68-80.

Macklin, M.G., Brewer, P.A., Balteanu, D. *et al.* (2003) The long term fate and environmental significance of contaminant metals released by the January and March 2000 mining tailings dam failures in Maramures County, upper Tisa Basin, Romania. *Applied Geochemistry*, **18**, 241-257.

Magnani, F., Mencuccini, M., Borghetti, M. *et al.* (2007) The human footprint in the carbon cycle of temperate and boreal forests. *Nature*, **447**, 848-850.

Mahowald, N., Kohfeld, K.E., Hansson, M. *et al.* (1999) Dust sources and deposition in the last glacial maximum and current climate: a comparison of model results with paleodata from ice cores and marine sediments. *Journal of Geophysical Research - Atmospheres*, **104**, 15895-15916.

Maier, H.R. and Dandy, G.C. (1996) The use of artificial neural networks for the prediction of water quality parameters. *Water Resources Research*, **32**, 1013-1022.

Makepeace, D.K., Smith, D.W. and Stanley, S.J. (1995) Urban stormwater quality: summary of contaminant data. *Critical Reviews in Environmental Science and Technology*, **25**, 93-139.

Malkus, J.S. (1958) Tropical weather disturbances: why do so few become hurricanes? *Weather*, **13**, 75-89.

Mannion, A.M. (2002) *Dynamic world land-cover and land-use change*. Arnold, London.

Marshall, J., Kushnir, Y., Battistf, D. *et al.* (2001) North Atlantic climate variability: phenomena, impacts and mechanisms. *International Journal of Climatology*, **21**, 1863-1898.

Mason, R.G. and Raff. A.D. (1961) Magnetic surveys off the west coast of North America, 32°N to 42°N latitude. *Geological Society of America Bulletin*, **72**, 1259-1266.

Masselink, G. and Hughes, M.G. (2003) *Introduction to coastal processes and geomorphology*. Hodder Arnold, London.

Matthews, J.A. (1992) *The ecology of recently deglaciated terrain: A geoecological approach to glacier forelands and primary succession*. Cambridge University Press, Cambridge.

May, R.E. (1975) Island biogeography and the design of nature reserves. *Nature*, **254**, 277-278.

May, R.M., Lawton, J.H. and Stork, N.E. (1995) Assessing extinction rates. In: Lawton, J.H. and May, R.M. (eds), *Extinction rates*. Oxford University Press, Oxford, 1-24.

Mayle, F.E., Burbridge, R. and Killeen, T.J. (2000) Millennial-scale dynamics of Southern Amazonian rainforests. *Science*, **290**, 2291-2294.

McClatchey, J. (1996) Spatial and altitudinal gradients of precipitation in Scotland. In: Merot, P. and Jigorel, A. (eds), *Hydrologie dans les pays celtiques*. INRA, Paris, 45-51.

McClatchey, J., Runcres, A.M.E. and Collier, P. (1987) Satellite images of the distribution of extremely low temperatures in the Scottish Highlands. *Meteorological Magazine*, **116**, 376-386.

McDonald, A.T. (2000) Ecosystems and their management. In: Kent, A. (ed.), *Reflective practice in geography teaching*. Paul Chapman, London, 11-25.

McDonald, A.T. (2002) Environment at the crossroads. *Regional surveys of the world: The USA and Canada*. Europa Press, London, 22-28.

McDonald, A.T., Lane, S.N., Haycock, N.E. and Chalk, E.A. (2004) Rivers of dreams: on the gulf between theoretical and practical aspects of an upland river restoration. *Transactions of the Institute of British Geographers*, **29**, 257-281.

McGuffie, K. and Henderson-Sellers, A. (1997) *A climate modeling primer*. John Wiley & Sons, Chichester.

McIlveen, J.F.R. (1998) *Fundamentals of weather and climate*. Stanley Thornes, Cheltenham.

McKenzie, D.P. and Parker, D.L. (1967) The North Pacific: an example of tectonic on a sphere. *Nature*, **216**, 1276-1280.

McKnight, D.M. and Bencala, K.E. (1990) The chemistry of iron, aluminium, and dissolved organic material in three acidic, metal-enriched, mountain streams, as controlled by watershed and in-stream processes. *Water Resources Research*, **26**, 3087-3100.

McManus, J. and Duck, R.W. (eds) (1993) *Geomorphology and sedimentology of lakes and reservoirs*. John Wiley & Sons, Chichester.

McNeill, J. (2000) *Something new under the sun*. Allen Lane & Penguin, London.

McRae, S.G. (1988) *Practical pedology; Studying soils in the field*. Ellis Horwood, Chichester.

Meadows, D.H., Meadows, D.I., Randers, J. and Behrens, W.W. III (1972) *The limits to growth*. Earth Island, London.

Meigs, P. (1953) World distribution of arid and semi-arid homoclimates. *Arid zone hydrology*, UNESCO Arid Zone Research Series 1. UNESCO, Paris, 203-209.

Meir, P., Cox, P. and Grace, J. (2006) The influence of terrestrial ecosystems on climate. *Trends in Ecology and Evolution*, **21**, 254-260.

Melton, M.A. (1965) Debris-covered slopes of the southern Arizona desert. *Journal of Geology*, **73**, 715-729.

Menzel, A., Sparks, T.H., Estrella, N. *et al.* (2006) European phenological response to climate change matches the warming pattern. *Global Change Biology*, **12**, 1969-1976.

Menzies, J. (1989) Drumlins - products of controlled or uncontrolled glaciodynamic response? *Quaternary Science Reviews*, **8**, 151-158.

Menzies, J. (2000) Micromorphological analyses of microfabrics and microstructures indicative of deformation processes in glacial sediments. In: Maltman, A.J., Hubbard, B. and Hambrey, M.J. (eds), *Deformation of glacial materials*. The Geological Society, London, 245-257.

Meybeck, M. (1980) Pathways of major elements from land to ocean through rivers. In: *River Inputs to Ocean Systems*, UNEP/UNESCO Report, 18-30.

Micheli, E.R. and Kirchner, J.W. (2002) Effects of wet meadow riparian vegetation on streambank erosion: remote sensing measurements of streambank migration and erodibility. *Earth Surface Processes and Landforms*, **27**, 627-639.

Middleton, G.V. and Wilcock, P.R. (1994) *Mechanics in the Earth and environmental sciences*. Cambridge University Press, Cambridge.

Middleton, N.J. and Thomas, D.S.G. (1997) *World atlas of desertification*, 2nd edition. Edward Arnold, London.

Millennium Ecosystem Assessment (2005) *Ecosystems and human well-being: Biodiversity synthesis*. World Resources Institute, Washington, DC.

Mills, M.G.L. (1989) The comparative behavioural ecology of hyenas: the importance of diet and food dispersion. In: Gittleman, J.L. (ed.), *Carnivore behaviour, ecology and evolution*. Chapman and Hall, London.

Milne, G. (1935) Composite units for the mapping of complex soil associations. *Transactions of the Third International Congress of Soil Science*, **1**, 345-347.

MISR (1984) *Organisation and methods of the 1:250 000 Soil Survey of Scotland*. The Macaulay Institute for Soil Research, Aberdeen.

Mitchell, T.L. (1837) Account of the recent exploring expedition to the interior of Australia. *Journal of the Royal Geographical Society*, **7**, 271-285.

Moffitt, B.J. and Ratcliffe, R.A.S. (1972) Northern hemisphere monthly mean 500 millibar and 1000-500 millibar thickness charts and some derived statistics (1951-66). *Geophysical Memoirs*, 117, Meteorological Office, London.

Molochuskin, E.N. (1973) The effect of thermal abrasion on the temperature of the permafrost in the coastal zone of the Laptev Sea. In: *Proceedings of the Second International Conference on Permafrost, Yakutsk, USSR*, USSR Contribution. National Science Academy, Washington, DC, 90-93.

Monteith, J.L. and Unsworth, M.H. (1990) *Principles of environmental physics*, 2nd edition. Edward Arnold, London.

Montgomery, D.R. and Buffington, J.M. (1997) Channel-reach morphology in mountain drainage basins. *Geological Society of America Bulletin*, **109**, 596-611.

Moore, P.D., Challoner, W. and Stott, P. (1996) *Global environmental change*. Blackwell, Oxford.

Morgan, J.J. and Stumm, W. (1965) The role of multivalent metal oxides in limnological transformations, as exemplified by iron and manganese. In: *Advances in water pollution research, Proceedings of the Second International Conference on Water Pollution Research*, Pergamon Press, Oxford, **1**, 103-118.

Morgan, R.P.C. (ed.) (1986) *Soil erosion and conservation*. Longman, Harlow.

Murray, J. (1968) *The world of life: The biosphere*. University of Colorado/Jarrold and Sons, Norwich.

Murray, T. (1997) Assessing the paradigm shift: deformable glacier beds. *Quaternary Science Reviews*, **16**, 995-1016.

Myers, N., Mittermeier, R.A., Mittermeier, C.G. *et al.* (2000) Biodiversity hotspots for conservation priorities. *Nature*, **403**, 853-858.

N

Nägeli, W. (1946) Weitere Untersuchungen uber die Windverhaltnisse im Bereich von Windschutzanglen. *Mitteilungen Schweizerischen Anstalt für das forstliche Versuchswesen*, **24**, 659-737.

Nagler, P.L., Glenn, E.P. and Huete, A.R. (2001) Assessment of spectral vegetation indices for riparian vegetation in the Colorado River delta, Mexico. *Journal of Arid Environments*, **49**, 91-110.

Nagy, L., Grabherr, G., Körner, C. and Thompson, D.B.A. (2003) *Alpine biodiversity in Europe*. Springer-Verlag, Berlin.

Nath, B., Hens, L., Compton, P. and Devuyt, D. (eds) (1998) *Environmental management in practice*. Routledge, London.

NEGTAP (2001) *Transboundary air pollution: Acidification, eutrophication and ground-level ozone in the UK*. Report by the National Expert Group on Transboundary Air Pollution

(NEGTAP) on behalf of the UK Department for Environment, Food and Rural Affairs.

Netzband, A., Reincke, H. and Bergemann, M. (2002) The River Elbe: a case study for the ecological and economical chain of sediments. *Journal of Soils and Sediments*, **2**, 112-116.

Neuman, A.C. and MacIntyre, I. (1985) Reef response to a sea level rise: keep-up, catch-up or give-up. *Proceedings, 5th International Coral Reef Congress*, 105-110.

Nicholls, N. (1991) The El Niño–Southern Oscillation and Australian vegetation. *Vegetatio*, **91**, 23-36.

Nicholls, R.J. and Mimura, N. (1998) Regional issues raised by sea-level rise and their policy implications. *Climate Research*, **11**, 5-18.

Nichols, G. (1999) *Sedimentology and stratigraphy*. Blackwell Scientific, Oxford.

Nicholson, S.E., Ba, M.B. and Kim, J.Y. (1996) Rainfall in the Sahel during 1994. *Journal of Climate*, **9**, 1673-1676.

Nilsson, J. and Grennfelt, P. (1988) *Critical loads for sulphur and nitrogen* (Report 1988 : 15). Nordic Council of Ministers, Copenhagen.

Noble, I.R. and Slatyer, R.O. (1980) The use of vital attributes to predict successional changes in plant communities subject to recurrent disturbances. *Vegetatio*, **43**, 5-21.

Norris, R.M. (1966) Barchan dunes of Imperial Valley, California. *Journal of Geology*, **74**, 292-306.

Novotny, V. and Harvey, O. (1994) *Water quality: Prevention, identification and management of diffuse pollution*. Van Nostrand Reinhold, London.

O

Ohanian, P.P. and Dubes, R.C. (1992) Performance evaluation for four class of texture features. *Pattern Recognition*, **25**, 819-833.

O'Hara, S.L., Wiggs, G.F.S., Wegerdt, J. *et al.* (2001) Dust exposure and respiratory health amongst children in the environmental disaster zone of Karakalpakstan, Central Asia: preliminary findings of the ASARD project. In Brebbia, A. and Fayzieva, D. (eds), *Environmental health risk*. WIT Press, Southampton, 71-82.

O'Hare, G., Sweeney, J. and Wilby, R. (2005) *Weather, Climate and Climate Change*. Prentice Hall, Harlow.

Oke, T.R. (1976) The distinction between canopy and boundary-layer urban heat islands. *Atmosphere*, **14**, 268-277.

Oke, T.R. (1987) *Boundary layer climates*. Routledge, London.

Oke, T.R. and East, C. (1971) The urban boundary layer in Montreal. *Boundary Layer Meteorology*, **1**, 411-437.

Oldeman, L.R., Wakkeling, R.T.A. and Sombroek, W.G. (1991) *World map of the status of human induced soil degradation*. Volume 2: *Global assessment of soil degradation*. International Soil Reference and Information Centre, Wageningen, and UNEP.

Open University (1989) *Waves, tides and shallow water progress*. Pergamon Press, Oxford.

Owen, L. and Unwin, T. (eds) (1997) *Environmental management: readings and case studies*. Blackwell, Cambridge, MA.

Owens, P.N. (2005). Conceptual models and budgets for sediment management at the river basin scale. *Journal of Soils and Sediments*, **5**, 201-212.

Owens, P.N., Walling, D.E. and He, Q. (1996). The behaviour of bomb-derived caesium-137 fallout in catchment soils. *Journal of Environmental Radioactivity*, **32**, 169-191.

Owens, S. and Owens, P.L. (1991) *Environment, resources and conservation*. Cambridge University Press, Cambridge.

P

Paillard, D. (2001) Glacial cycles: toward a new paradigm. *Reviews of Geophysics*, **39**, 325-346.

Parker, D.E. and Alexander, L.V. (2002) Global and regional climate in 2001. *Weather*, **57**, 328-340.

Paterson, W.S.B. (1994) *The physics of glaciers*, 3rd edition. Pergamon Press, Oxford.

Pauli, H., Gottfried, M., Reiter, K., Klettner, C. and Grabherr, G. (2007) Signals of range expansions and contractions of vascular plants in the high Alps: observations (1994-2004) at the GLORIA master site Schrankogel, Tyrol, Austria. *Global Change Biology*, **13**, 147-156.

Pearce, A.J., Stewart, M.K. and Sklash, M.G. (1986) Storm runoff generation in humid headwater catchments 1. Where does the water come from? *Water Resources Research*, **22**, 1263-1272.

Peel, D.A. (1994) The Greenland Ice-Core Project (GRIP): reducing uncertainty in climate change? *NERC News*, 26-30 April.

Penck, W. (1924) *Morphological analysis of landforms*, trans. from the Czech, 1953. Macmillan, London.

Peñuelas, J. and Boada, M. (2003) A global change-induced biome shift in the Montseny mountains (NE Spain). *Global Change Biology*, **9**, 131-140.

Perry, C.T. and Taylor, K.G. (eds) (2007) *Environmental sedimentology*. Blackwell Publishing, Oxford.

Perry, M.C., Prior, M.J. and Parker, D.E. (2007) An assessment of the suitability of a plastic thermometer screen for climatic data collection. *International Journal of Climatology*, **27**, 267-276.

Pethick, J. (1984) *An introduction to coastal geomorphology*. Arnold, London.

Pettijohn, F.J., Potter, P.E. and Siever, R. (1987) *Sand and sandstone*. Springer-Verlag, New York.

Petts, G.E. (1979) Complex response of river channel morphology subsequent to reservoir construction. *Progress in Physical Geography*, **3**, 329-369.

Petts, G.E. (1984) Sedimentation within a regulated river. *Earth Surface Processes and Landforms*, **9**, 125-134.

Petts, G.E. and Calow, P. (eds) (1996a) *River flows and channel forms*, Blackwell Scientific, Oxford.

Petts, G.E. and Calow, P. (eds) (1996b) *River restoration*. Blackwell Science, Oxford.

Petts, J. (ed.) (1999) *Handbook of environmental impact assessment*. Blackwell Scientific, Oxford.

Péwé, T.L. (1991) Permafrost. In: Kiersch, G.A. (ed.), *The heritage of engineering geology: The first hundred years*, Geological

Society of America, Boulder, CO, Centennial Special Volume 3, 277-298.

Philander, G. (1998) Learning from El Niño. *Weather*, **53**, 270-274.

Pilkey, O.H. (2003) *A celebration of the world's barrier islands*. Columbia University Press, New York.

Pimentel, D., Harvey, C., Resosudarmo, P. *et al.* (1995) Environmental and economic costs of soil erosion and conservation benefits. *Science*, **267**, 1117-1123.

Pinder, G.F. and Jones, J.F. (1969) Determination of the groundwater component of peak discharge from the chemistry of total runoff. *Water Resources Research*, **5**, 438-445.

Pinet, P.R. (1996) *Invitation to oceanography*, 1st edition. West Publishing, St Paul, MN.

Pinet, P.R. (2000) *Invitation to oceanography*, 2nd edition. Jones and Bartlett, Sudbury, MA.

Pirrie, D., Power, M.R., Wheeler, P.D. *et al.* (2002) Geochemical signature of historical mining: Fowey Estuary, Cornwall, UK. *Journal of Geochemical Exploration*, **76**, 31-43.

Playfair, J. (1802) *Illustrations of the Huttonian Theory of the Earth*. Reprinted 1964, Dover, New York.

Post, A. and LaChapelle, E.R. (2000) *Glacier ice*. University of Washington Press, Seattle, WA.

Press, F. and Siever, R. (1986) *Earth*. W.H. Freeman, New York.

Press, F., Siever, R., Grotzinger, J. and Jordan, T.H. (2004) *Understanding Earth*. W.H. Freeman, New York.

Prest, V.K. (1983) *Canada's heritage of glacial features*. Miscellaneous Report 28, Geological Survey of Canada.

Price, J.S. (1992) Blanket bog in Newfoundland: Part 1. The occurrence and accumulation of fog-water deposits. Hydrological processes. *Journal of Hydrology*, **135**, 87-101.

Price, L.W. (1972) *The Preglacial environment, permafrost and man*. Commission on College Geography Resource Paper No. 14, Association of American Geographers, Washington, DC.

Price, M. (2002) *Introducing groundwater*. Nelson Thornes, Cheltenham.

Prowse, T.D. and Ommaney, C.S.L. (1990) *Northern hydrology, Canadian perspectives*. NHRI Science Report 1.

Pye, K. (1987) *Aeolian dust and dust deposits*. Academic Press, London.

Q

Quammen, D. (1996) *The song of the dodo: Island biogeography in an age of extinctions*. Hutchinson, London.

R

Rahmstorf, S. (2006) Thermohaline ocean circulation. In: Elias, S.A. (ed.), *Encyclopedia of Quaternary science*. Elsevier, Amsterdam.

Ramanathan, V., Ramana, M.V., Roberts, G. *et al.* (2007) Warming trends in Asia amplified by brown cloud solar absorption. *Nature*, **448**, 575-578.

Ranasinghe, R., Symonds, G. and Holman, R.A. (2004) Morphodynamics of intermediate beaches: a video imaging and numerical modelling study. *Coastal Engineering*, **51**, 629-655.

Rapp, A. (1960) Recent development of mountain slopes in Karkevagge and surroundings, northern Sweden. *Geografiska Annaler*, **42**, 71-200.

Raschke, K., Vonder Haar, T.H., Bandeen, W.R. and Pasternak, M. (1973) The annual radiation balance of the earth-atmosphere system during 1969-70 from Nimbus 3 measurements. *Journal of Atmospheric Sciences*, **30**, 341-364.

Raunkiaer, C. (1934) *The life forms of plants & statistical plant geography*. Oxford University Press, Oxford.

Readers Digest Universal Dictionary (1987) Readers Digest, London.

Reading, H. (1998) *Sedimentary environments: Processes, facies and stratigraphy*, 3rd edition. Blackwell Scientific, Oxford.

Reading, H.G. and Collinson, J.D. (1996) Clastic coasts. In: Reading, H.G. (ed.), *Sedimentary environments: Processes, facies and stratigraphy*. Blackwell Science, Oxford, 154-231.

Reay, D.S. (2006) *Climate change begins at home: Life on the two-way street of global warming*. Macmillan Science, Basingstoke.

Redfield, A.C., Ketchum, B.H and Richards, F.A. (1963) The influence of organisms on the composition of seawater. In: Hill, M.N. (ed.), *The Sea*, Vol. 2. Interscience, New York.

Reid, J.M., McLeod, D.A. and Cresser, M.S. (1981) Factors affecting the chemistry of precipitation and river water in an upland catchment. *Journal of Hydrology*, **50**, 129-145.

Reiser, D.W., Ramey, M.P., Beck, S., Lambert, T.R. and Geary, R.E. (1991) Flushing flow recommendations for maintenance of salmonid spawning gravels in a steep regulated stream. *Regulated Rivers*, **3**, 267-276.

Renfrew, I.A. and Anderson, P.S. (2002) The surface climatology of an ordinary katabatic wind regime in Coats Land, Antarctica. *Tellus, Series A: Dynamic Meteorology and Oceanography*, **54**, 463-484.

Rhoads, B.L. and Thorn, C.E. (eds) (1996) *The scientific nature of geomorphology: Proceedings of the 27th Binghampton Symposium in Geomorphology, 27-29 September*, John Wiley & Sons, Chichester.

Richards, K.S. (1996) Samples and cases: generalisation and explanation in geomorphology. In: Rhoads, B.L. and Thorn, C.E. (eds), *The scientific nature of geomorphology: Proceedings of the 27th Binghampton Symposium in Geomorphology, 27-29 September*, John Wiley & Sons, Chichester, 171-190.

Richards, K.S., Brookes, S., Clifford, N.J., Harris, T.R.J. and Lane, S.N. (1997) Theory, measurement and testing in 'real' geomorphology and physical geography. In: Stoddart, D.R. (ed.), *Process and form in geomorphology*. Routledge, London, 265-292.

Riehl, H. (1965) *Introduction to the atmosphere*. McGraw-Hill, New York.

Robertson, D.J., Taylor, K.G. and Hoon, S.R. (2003) Geochemical and mineral magnetic characterisation of urban sediment particulates, Manchester, UK. *Applied Geochemistry*, **18**, 269-282.

Robinson, P.J. and Henderson-Sellers, A. (1999) *Contemporary climatology*. Pearson Education, Harlow.

Root, T.L., Price, J.T., Hall, K.R, *et al.* (2003) Fingerprints of global warming on wild animals and plants. *Nature*, **421**, 56-60.

Rosenberg, N.J., Blad, B.L. and Verma, S.B. (1983) *Microclimate: The biological environment*. John Wiley & Sons, New York.

Rosenqvist, I.T. (1990) From rain to lake: water pathways and chemical changes. *Journal of Hydrology*, **116**, 3-10.

Rotty, R.M. and Mitchell, J.M.J. (1974) *Man's energy and world climate*. Institute for Energy Analysis Report, Oak Ridge Associated Universities.

Roucoux, K.H., Shackleton, N.J., Schonfeld, J. and Tzedakis, P.C. (2001) Combined marine proxy and pollen analyses reveal rapid Iberian vegetation response to North Atlantic millennial-scale climate oscillations. *Quaternary Research*, **56**, 128-132.

Roucoux, K.H., Tzedakis, P.C., de Abreu, L. and Shackleton, N.J. (2006) Climate and vegetation changes 180 000 to 345 000 years ago recorded in a deep-sea core off Portugal. *Earth and Planetary Science Letters,* **249**, 307-325.

Rowell, D.L. (1994) *Soil science: Methods and applications.* Longman, London.

Rowley, R.J., Kostelnick, J.C., Braaten, D. *et al.* (2007) Risk of rising sea level to population and land area. *Eos*, **88**, 105-107.

Roy, P.S., Cowell, P.J., Ferland, M.A. and Thom, B.G. (1994) Wave-dominated coasts. In: Carter, R.W.G. and Woodroffe, C.D. (eds), *Coastal evolution: Late Quaternary shoreline morphodynamics.* Cambridge University Press, Cambridge, 121-186.

Royal Commission on Environmental Pollution (1996) *Nineteenth Report – Sustainable Use of Soil*. HMSO, London.

Ruddiman, W.F., Raymo, M.E., Martinson, D.G., Clement, B and Backman, J. (1989) Pleistocene evolution: northern hemisphere ice sheets and North Atlantic ocean. *Paleoceanography*, **4**, 353-412.

Ruegg, G.H.J. (1994) Alluvial architecture of the Quaternary Rhin-Meuses river. *Geologie en Mijnbouw*, **72**, 321-330.

Russ, J.C. (2001) *The image processing handbook*. CRC Press, Boca Raton, FL.

S

Sánchez Goñi, M.F., Eynaud, F., Turon, J.L. and Shackleton, N.J. (1999) High resolution palynological record off the Iberian margin: direct land-sea correlation for the Last Interglacial complex. *Earth and Planetary Science Letters*, **171**, 123-137.

Sarnthein, M. (1978) Sand deserts during glacial maximum and climatic optimum. *Nature*, **272**, 43-46.

Sarnthein, M. and Walger, E. (1974) Der aolische Sandstrom aus der W-Sahara zur Atlantikkuste. *Geologische Rundschau*, **63**, 1065-1087.

Scheidegger, A.E. (1973) On the prediction of the reach and velocity of catastrophic landslides. *Rock Mechanics*, **5**, 231-236.

Scherz, J.P. and Stevens, A.R. (1970) *An Introduction to Remote Sensing for Environmental Monitoring,* Report No. 1, University of Wisconsin Madison.

Schlosser, P., Bayer, R., Bonisch, G. *et al.* (1999) Pathways and mean residence times of dissolved pollutants in the ocean derived from transient tracers and stable isotopes. *Science of the Total Environment*, **238**, 15-30.

Schmugge, T.J. and Jackson, T.J. (1996) Soil moisture variability. In: Stewart, J.B., Engman, E.T., Feddes, R.A. and Kerr, Y. (eds), *Scaling up in hydrology using remote sensing*. John Wiley & Sons, Chichester, 183-192.

Schultz. J. (1995) *Die Okozonen der Erde*, 2. Aufl. UTB 1514. Ulmer, Stuttgart.

Schumm, S.A. (1956) Evolution of drainage systems and slopes on badlands at Perth Amboy, New Jersey. *Geological Society of America Bulletin*, **67**, 597-646.

Schumm, S.A. (1964) Seasonal variations of erosion rates and processes on hillslopes in western Colorado. *Zeitschrift für Geomorphologie Supplement*, **5**, 215-238.

Schumm, S.A. (1977) Applied fluvial geomorphology. In: Hails, J.R. (ed.), *Applied geomorphology*. Elsevier, Amsterdam, 119-156.

Schumm, S.A. (1985) *The fluvial system*. John Wiley & Sons, New York.

Schumm, S.A. (1991) *To diagnose the earth: Ten ways to be wrong*. Cambridge University Press, Cambridge.

Schumm, S.A. and Lichty, R.W. (1965) Time, space and causality in geomorphology. *American Journal of Science*, **263**, 110-119.

Segar, D.A. (1998) *Introduction to ocean sciences*. Wadsworth, Belmont, CA.

Selby, M.J. (1993) *Hillslope materials and processes*, 2nd edition. Oxford University Press, Oxford.

Sellers, W.D. (1965) *Physical climatology*. University of Chicago Press, Chicago.

Shackleton, N.J. (2006) Formal Quaternary stratigraphy – what do we expect and need? *Quaternary Science Reviews*, **25**, 3458-3461.

Shackleton, N.J. and Opdyke, N.D. (1973) Oxygen isotope analyses and palaeomagnetic stratigraphy of equatorial Pacific core V28-238: oxygen isotope temperatures and ice volumes on a 10^5 and 10^6 year scale. *Quaternary Research*, **3**, 39-55.

Shaw, E.M. (1994) *Hydrology in practice*. Stanley Thornes, Cheltenham.

Shaw, J. (1989) Drumlins, subglacial meltwater floods and ocean responses. *Geology*, **17**, 853-856.

Shaw, J. (1994a) A qualitative view of sub-ice-sheet landscape evolution, *Progress in Physical Geography*, **18**, 159-184.

Shaw, J. (1994) Hairpin erosional marks, horseshoe vortices and subglacial erosion. *Sedimentary Geology*, **91**, 269-283.

Shaw, J. and Sharpe, D.R. (1987) Drumlin formation by subglacial meltwater erosion. *Canadian Journal of Earth Sciences*, **24**, 2316-2322.

Shaw, P.A. (1997) Africa and Europe. In: Thomas, D.S.G. (ed.), *Arid zone geomorphology: Process, form and change in drylands*. John Wiley & Sons, Chichester, 467-485.

Shaw, P.A. and Thomas, D.S.G. (1997) Pans, playas and salt lakes. In: Thomas, D.S.G. (ed.), *Arid zone geomorphology: Process, form and change in drylands*. John Wiley & Sons, Chichester, 294-317.

Shaw, W.W., Harris, L.K. and Livingston, M. (1998) Vegetative characteristics of urban land covers in metropolitan Tucson. *Urban Ecosystems*, **2**, 65-73.

Sherman, D.J. and Bauer, B.O. (1993) Dynamics of beach-dune systems. *Progress in Physical Geography*, **17**, 413-447.

Shimda, A., Evenari, M. and Noy-Meir, I. (1986) Hot desert ecosystems: an integrated view. In: Evenari, M., Noy-Meir, I. and Goodall, D.W. (eds), *Ecosystems of the world*, Vol. 12B, *Hot deserts and arid shrublands*. Elsevier, Amsterdam, 378-388.

Shinn, E.A., Smith, G.W., Prospero, J.M. *et al.* (2000) African dust and the demise of Caribbean coral reefs. *Geophysical Research Letters*, **19**, 3029-3032.

Short, A.D. (1999) Global variation in beach systems. In: Short, A.D. (ed.), *Beach morphodynamics*. John Wiley & Sons, Chichester, 72-118.

Sibbesen, E. and Runge-Metzger, A. (1995) Phosphorus balance in European agriculture – status and policy options. In: Tiessen, H. (ed.), *Phosphorus in the global environment*. John Wiley & Sons, Chichester, 43-60.

Simberloff, D.S. (1983) When is an island community in equilibrium? *Science*, **220**, 1275-1277.

Simpson, C.J. and Smith, D.G. (2001) The braided Milk River, northern Montana, fails the Leopold Wolman discharge gradient test. *Geomorphology*, **41**, 337-353.

Sithole, G. and Vosselman, G. (2006) Bridge detection in airborne laser scanner data. *ISPRS Journal of Photogrammetry and Remote Sensing*, **61**, 33-46.

Skempton, A.W. and DeLory, F.A. (1957) Stability of natural slopes in London Clay. *Proceedings of the Fourth International Conference on Soil Mechanics and Foundation Engineering*, Vol. 2, 378-381.

Skinner, L.C. and Shackleton, N.J. (2005) An Atlantic lead over Pacific deep-water change across Termination I: implications for the application of the marine isotope stratigraphy. *Quaternary Science Reviews*, **24**, 571-580.

Skinner, L.C. and Shackleton, N.J. (2006) Deconstructing Terminations I and II: revisiting the glacioeustatic paradigm based on deep-water temperature estimates. *Quaternary Science Reviews*, **25**, 3312-3321.

Sklash, M.G., Stewart, M.K. and Pearce, A.J. (1986) Storm runoff generation in humid headwater catchments 2. A case study of hillslope and low-order stream response. *Water Resources Research*, **22**, 1273-1282.

Slingo, J. (1998) The 1997/98 El Niño. *Weather*, **53**, 274-281.

Smith, A.M., Murray, T., Nicholls, K.W. *et al.* (2007) Rapid erosion and drumlin formation observed beneath a fast-flowing Antarctic ice stream. *Geology*, **35**, 127-130.

Smith, C.L., Richards, K.J. and Fasham, M.J.R. (1996) The impact of mesoscale eddies on plankton dynamics in the upper ocean. *Deep-Sea Research I*, **43**, 1807-1832.

Smith, F.D.M., May, R.M., Pellew, R. *et al.* (1993) Estimating extinction rates. *Nature*, **364**, 494-496.

Smith, P. (2004) Soil as carbon sinks: the global context. *Soil Use and Management*, **20**, 212-218.

Smith, P., Mile, R., Powlson, D.S. *et al.* (2000) Revised estimates of the carbon mitigation potential of UK agricultural land. *Soil Use and Management*, **16**, 293-295.

Smithson, P., Briggs, D., Addison, K. and Atkinson, K. (2002) *Fundamentals of the physical environment*, 3rd edition. Routledge, London.

Snead, R.E. (1972) *Atlas of world physical features*. John Wiley & Sons, New York.

Solé, R.V. and Montoya, J.M. (2001) Complexity and fragility in ecological networks. *Proceedings of the Royal Society of London B*, **268**, 1-7.

Soulsby, C. (1997) Hydrochemical processes. In: Wilby, R.L. (ed.), *Contemporary hydrology*. John Wiley & Sons, Chichester, 59-106.

Sparks, B.W. and West, R.G. (1972) *The Ice Age in Britain*, Methuen, London.

Sparks, T.H., Jeffree, E.P. and Jeffree, C.E. (2000) An examination of the relationship between flowering times and temperature at the national scale using long-term phenological records from the UK. *International Journal of Biometeorology*, **44**, 82-87.

Spellerberg, I.F. and Sawyer, J.W.D. (1999) *An introduction to applied biogeography*. Cambridge University Press, Cambridge.

Spellerberg, I.F., Sawyer, J.W.D and Whitten, T. (1999) *An introduction to applied biogeography*. Cambridge University Press, Cambridge.

Stamp, L.D. (1949) *The world; A general geography*. Longman, London.

Stanley, D.J. and Sheng, H. (1986) Volcanic shards from Santorini (Upper Minoan ash) in the Nile delta, Egypt. *Nature*, **320**, 733-735.

Stanley, D.J. and Warne, A.G. (1998) Nile delta in its destruction phase. *Journal of Coastal Research*, **14**, 794-825.

Stanners, D. and Bourdeau, P. (eds) (1991) *The DOBRIS Report: Europe's environment*. Earthscan, London.

Stanners, D. and Bourdeau, P. (1995) *Europe's environment: The Dobris assessment*. European Environment Agency, Copenhagen.

Stephenson, W.J. (2000) Shore platforms: remain a neglected coastal feature. *Progress in Physical Geography*, **24**, 311-327.

Stern, N. (2006) *The economics of climate change: The Stern Review*. Cambridge University Press, Cambridge.

Stern, W. (1929) Versuch einer elektrodynamischen Dickenmessung von Gletschereis. *Gerlands Beiträge zur Geophysik*, **23**, 292-333.

Stern, W. (1930) Uber Grundlagen, Methodik und bisherige Ergebnisse elektrodynamischer Dickenmessung von Gletschereis. *Zeitschrift für Gletscherkunde*, **15**, 24-42.

Stewart, M. (2003) Using woods, 1600-1850 (i) the community resource. In: Smout, T.C. (ed.), *People and woods in Scotland – A history*. Edinburgh University Press, Edinburgh, 82-104.

Stoddart, D.R. (1966) Darwin's impact on geography. *Transactions of the Institute of British Geographers*, **56**, 683-698.

Strahler, A.N. (1957) Quantitative analysis of watershed geomorphology. *Transactions of the American Geophysical Union*, **38**, 913-920.

Strakhov, N.M. (1967) *Principles of lithogenesis*, Vol. 1. T. M. Thomas, London.

Strangeways, I.C. (2006) *Precipitation: theory, measurement and distribution*. Cambridge University Press, Cambridge.

Sugden, D.E. and John, B.S. (1976) *Glaciers and landscape*. Edward Arnold, London.

Summerfield, M.A. (1991) *Global geomorphology*. Longman Scientific, Harlow.

Sunamura, T. (1992) *Geomorphology of rocky coasts*. John Wiley & Sons, Chichester.

Sutherst, R.W. and Maywald, G.F. (1985) A computerised system for matching climates in ecology. *Agriculture, Ecosystems and Environment*, **13**, 281-299.

Sverdrup, H.U., Johnson, M.W. and Fleming, R.H. (1942) *The oceans*. Prentice Hall, Englewood Cliffs, NJ.

Sverdrup, K.A., Duxbury, A.C. and Duxbury, A.B. (2004) *An Introduction to the World's Oceans*. McGraw-Hill, London.

T

Taberlet, P. and Cheddadi, R. (2002) Quarternary refugia and persistance of biodiversity. *Science*, **297**, 2009-2010.

Takhtajan, A. (1969) *Flowering plants, origins and dispersal*. Oliver and Boyd, London.

Tansley, A.G. (1920) The classification of vegetation and the concept of development. *Journal of Ecology*, **22**, 554-571.

Tansley, A.G. (1935) The use and abuse of vegetational concepts and terms. *Ecology*, **43**, 614-624.

Tarbuck, E. and Lutgens, F. (1991) *Sedimentary petrology: An introduction to the origin of sedimentary rocks*. Blackwell Scientific, Oxford.

Taylor, K.G., Boyd, N.A. and Boult, S. (2003) Sediments, porewaters and diagenesis in an urban water body, Salford, UK: impacts of remediation. *Hydrological Processes*, **17**, 2049-2061.

Terzaghi, K. (1962) Stability of steep slopes on hard unweathered rock. *Géotechnique*, **12**, 251-270.

Thom, R.F. (1968) Une théorie dynamique de la morphogénèse. In: C.E.H. Waddington, *Towards a theoretical biology*, Vol. I. Edinburgh University Press, Edinburgh, 152-166.

Thomas, A.D. and Dougill, A.J. (2007) Spatial and temporal distribution of cyanobacterial soil crusts in the Kalahari: implications for soil surface properties. *Geomorphology*, **85**, 17-29.

Thomas, C.D., Cameron, A., Green, R.E. *et al.* (2004) Extinction risk from climate change. *Nature*, **427**, 145-148.

Thomas, D.S.G. (1997a) Arid environments: their nature and context. In: Thomas, D.S.G. (ed.), *Arid zone geomorphology: Process, form and change in drylands*. John Wiley & Sons, Chichester, 3-12.

Thomas, D.S.G. (1997b) Sand seas and aeolian bedforms. In: Thomas, D.S.G. (ed.), *Arid zone geomorphology: Process, form and change in drylands*. John Wiley & Sons, Chichester, 373-412.

Thomas, D.S.G. (ed.) (1997c) *Arid zone geomorphology: Process, form and change in drylands*. John Wiley & Sons, Chichester.

Thomas, D.S.G. and Middleton, N.J. (1994) *Desertification: Exploding the myth*. John Wiley & Sons, Chichester.

Thomas, D.S.G. and Shaw, P.A. (1991) *The Kalahari environment*. Cambridge University Press, Cambridge.

Thomas, D.S.G., Knight, M. and Wiggs, G.F.S. (2005) Remobilization of southern African desert dune systems by twenty-first century global warming. *Nature*, **435**, 1218-1221.

Thompson, L.G., Mosley-Thompson, E., Davis, M.E. *et al.* (1995) Late Glacial Stage and Holocene tropical ice core records from Huascarán, Peru. *Science*, **269**, 46-50.

Thorne, C.R., Hey, R.D. and Newson, M.D. (1997) *Applied fluvial geomorphology for river engineering and management*. John Wiley & Sons, Chichester.

Thornes, J.B. (1994) Catchment and channel hydrology. In: Abrahams, A.D. and Parsons, A.J. (eds), *Geomorphology of desert environments*. Chapman and Hall, London, 257-287.

Thurman, H.V. (2004) *Introductory oceanography*. Prentice Hall, Upper Saddle River, NJ.

Thurow, C. (1983) *Improving street climate through urban design*, Planning Advisory Service, No. 376. American Planning Association, Chicago.

Tilman, D. (1985) The resource-ratio hypothesis of plant succession. *The American Naturalist*, **125**, 439-464.

Titus, T.N., Kieffer, H.H. and Christensen, P.R. (2002) Exposed water ice discovered near the South Pole of Mars. *Science*, **299**, 1048-1051.

Tou, J.T. and Gonzalez, R.C. (1974) *Pattern recognition principles*. Addison-Wesley, Reading, MA.

Toy, T.J., Foster, G.R. and Renard, K.G. (2002) *Soil erosion: Processes, prediction, measurement and control*. John Wiley & Sons, Chichester.

Trenhaile, A.S. (1990) *The geomorphology of Canada, an introduction*. Oxford University Press, Oxford.

Trenhaile, A.S. (1987) *The geomorphology of rock coasts*. Oxford University Press, Oxford.

Troll, C. (1939) Luftbildplan und ökologische Bodenforschung. *Zeitschrift der Gesellschaft für Erdkunde zu Berlin*, **7/8**, 241-298.

Trudgill, S.T. (ed.) (1986) *Solute processes*. John Wiley & Sons, Chichester.

Tso, B. and Mather, P.M. (2001) *Classification methods for remotely sensed data*. Taylor & Francis, London.

Tucker, C.J., Dregne, H.E. and Newcomb, W. (1991) Expansion and contraction of the Sahara Desert from 1980 to 1990. *Science*, **253**, 299-301.

Tucker, G.E., Lancaster, S.T., Gasparini, N.M. and Bras, R.L. (2001) The channel-hillslope integrated landscape development model (CHILD). In: Harmon, R.S. and Doe, W.W.I. (eds), *Landscape erosion and evolution models*. Kluwer Academic, New York, 349-388.

Tucker, M.E. (1981) *Sedimentary petrology: An introduction*. Blackwell Scientific, Oxford.

Tucker, M.E. and Wright, V.P. (1990) *Carbonate sedimentology*. Blackwell Scientific, Oxford.

Turnbull, D.A. and Bevan, J.R. (1995) The impact of airport de-icing on a river: the case of the Ouseburn, Newcastle upon Tyne. *Environmental Pollution*, **88**, 321-332.

Tzedakis, P.C. (1993) Long-term tree populations in northwest Greece through multiple Quaternary climatic cycles. *Nature*, **364**, 437-440.

Tzedakis, P.C., Lawson, I.T., Frogley, M.R. *et al.* (2002) Buffered tree population changes in a Quaternary refugia: evolutionary implications. *Nature*, **297**, 2044-2047.

U

UK Department of Energy (1979) *Heat loads in British Cities*, Energy Paper No. 34. HMSO, London.

Ulén, B., Bechmann, M., Fölster, J., Jarvie, H.P. and Tunney, H. (2007) Agriculture as a phosphorus source for eutrophication in the north-west European countries, Norway, Sweden, United Kingdom and Ireland: a review. *Soil Use and Management*, **23**, 5-15.

UNEP (1997) *World Atlas of Desertification,* 2nd Edition, Edward Arnold.

UNEP (2000) *Guidelines for erosion and desertification control management*. United Nations Environment Programme.

United Nations (1992) *United Nations Framework Convention on Climate Change*. UN, New York.

Urban, F.E., Cole, J.E. and Overpeck. T. (2000) Influence of mean climate change on climate variability from a 155-year tropical Pacific coral record. *Nature*, **407**, 989-993.

V

van Ast, J.A. and Boot, S.P. (2003) Participation in European water policy. *Physics and Chemistry of the Earth*, **28**, 555-562.

Van der Berg, J.H. (1995) Prediction of alluvial channel pattern of perennial rivers. *Geomorphology*, **12**, 259-279.

van der Maarel, E. (1997) *Biodiversity: From Babel to biosphere management*. Opulus Press, Grangarde, Sweden.

van der Maarel, E. (ed.) (2000) *Succession and zonation on mountains, particularly on volcanoes*. Opulus Press, Grangarde, Sweden.

Van Tyne, J. (1951) *Proceedings of the International Ornithological Congress*, **10**, pp. 537-44.

Van Vuuren, D.P., Sala, O.E. and Pereira, H.M. (2006) The future of vascular plant diversity under four global scenarios. *Ecology and Society,* **11**, article 25.

Vaughan, D.G., Marshall, G.J., Connolley, W.M. *et al.* (2003) Recent rapid regional climate warming on the Antarctic Peninsula. *Climatic Change*, **60**, 243-274.

Velicogna, I. and Wahr, J. (2006) Acceleration of Greenland ice mass loss in spring 2004. *Nature,* **443**, 329-331.

Viles, H. and Spencer, T. (1995) *Coastal problems*. Arnold, London.

Vincent, P. (1990) *The biogeography of the British Isles*. Routledge, London.

Vine, F.J. and Matthews, D.H. (1963) Magnetic anomalies over oceanic ridges. *Nature*, **199**, 947-949.

Vitousek, P.M., Ehrlich, P.R., Ehrlich, A.H. and Matson, P.A. (1986) Human appropriation of the products of photosynthesis. *Bioscience*, **36**, 368-373.

W

Walder, J.S. and Hallet, B. (1979) Geometry of former subglacial water channels and cavities. *Journal of Glaciology*, **23**, 335-346.

Walker, B. and Steffan, W. (1997) An overview of the implications of global change for natural and managed terrestrial ecosystems. *Conservation Ecology*, **1**, 2.

Walker, M. (2005) *Quaternary dating methods*. John Wiley & Sons, Chichester.

Walling, D.E. and Webb, B.W. (1986) Solutes in river systems. In: Trudgill, S.T. (ed.), *Solute Processes*. John Wiley & Sons, Chichester, 251-327.

Walter, H. (1985) *Vegetation of the Earth and ecological systems of the geo-biosphere*, 3rd edition. Springer-Verlag, New York.

Walter, H. (1990) *Vegetationszonen und Klima*, 6th edition. Ulmer, Stuttgart.

Ward, R.C. and Robinson, M. (2000) *Principles of hydrology*. McGraw-Hill, London.

Waring, R.H. (1989) Ecosystems: fluxes of matter and energy. In: Cherrett, J.M. (ed.), *Ecological concepts*. Blackwell Scientific, Oxford, 17-41.

Washburn, A.L. (1979) *Geocryology: A study of periglacial processes and environments*. Arnold, London.

Wasson, R.J., Fichett, K., Mackay, B. and Hyde, R. (1988) Large-scale patterns of dune types, spacing and orientation in the Australian continental dunefield. *Australian Geographer*, **19**, 80-104.

Watson, A. and Nash, D.J. (1997) Desert crusts and varnishes. In: Thomas, D.S.G. (ed.), *Arid zone geomorphology: Process, form and change in drylands*. John Wiley & Sons, Chichester, 69-107.

Waugh, D. (1995) *Geography: An integrated approach*. Thomas Nelson, London.

Webster, P.J. (1981) Monsoons. *Scientific American*, **245**, 109-118.

Wegener, A.L. (1915) *Die Entstehung der Kontinente und Ozeane*. Sammlung Vieweg, No. 23. F. Vieweg und Sohn, Braunschweig.

Wegener, A.L. (1966) *The origin of continents and oceans*, trans. from the 4th revised German edition of 1929 by J. Biram, with an introduction by B.C. King. Methuen, London.

Wemelsfelder, P.J. (1953) The disaster in the Netherlands caused by the storm flood of February 1, 1953. *Proceedings of the 4th Coastal Engineering Conference*, ASCE, 256-271.

Werner, B.T. (1999) Complexity in natural landform patterns. *Science*, **284**, 102-104.

Werner, B.T. and Fink, T.M. (1993) Beach cusps as self-organized patterns. *Science*, **260**, 986-970.

Westhoff, V. (1983) Man's attitude towards vegetation. In: Holzner, W., Werger, M.J.A. and Ikusima, I. (eds), *Man's impact on vegetation*. Dr. W. Junk, The Hague, 7-24.

Wheater, C.P. (1999) *Urban habitats*. Routledge, London.

White, I.D., Mottershead, D.N. and Harrison, S.J. (1992) *Environmental systems: an Introductory Text,* 2nd edition, Taylor and Francis Books Ltd, Routledge.

White, R.E. (1997) *Principles and practice of soil science: The soil as a natural resource*, 3rd edition. Blackwell Scientific, Oxford.

Whitehead, P.G., Beck, M.B. and O'Connell, P.E. (1981) A systems model of flow and water quality in the Bedford Ouse River System – II. Water quality modelling. *Water Resources Research*, **15**, 1157-1171.

Whitehead, P.G., McCartney, M.P., Williams, R.J. *et al.* (1995) A method to simulate the impact of acid mine drainage on river systems. *Journal of the Chartered Institute of Water and Environmental Management*, **9**, 119-131.

Whittaker, R.H. (1953) A consideration of climax theory: the climax as a population and pattern. *Ecological Monographs*, **23**, 41–78.

Whittaker, R.H. (1975) *Communities and ecosystems*. Macmillan, New York.

Whittaker, R.J. and Fernández-Palacios, J.M. (2007) *Island biogeography: Ecology, evolution, and conservation*. Oxford University Press, Oxford.

Wiggs, G.F.W. (1993) Desert dune dynamics and the evaluation of shear velocity: an integratede approach. In: Pye, K. (ed.), *The dynamics and environmental context of aeolian sedimentary systems*, Special Publication 72. Geological Society of London, 37–46.

Wild, A. (1993) *Soils and the environment: An introduction*. Cambridge University Press, Cambridge.

Willgoose, G.R., Bras, R.L. and Rodriguez-Iturbe, I. (1991) Results from a new model of river basin evolution. *Earth Surface Processes and Landforms*, **16**, 237–254.

Williams, G.P. and Wolman, M.G. (1983) Downstream effects of dams on alluvial rivers. *US Geological Survey Professional Paper* 1286.

Williams, M.A.J. and Balling, R.C. (1995) *Interactions of desertification and climate*. Edward Arnold, London.

Williams, M.A.J., Dunkerley, D., Deckker, P. de *et al.* (1998) *Quarternary environments*. Edward Arnold, London.

Williams, P.J. (1986) *Pipelines and permafrost: Physical geography and development in the circumpolar north*. Longman, Harlow.

Williams, P.J. and Smith, M.W. (1989) *The frozen Earth, fundamentals of geocryology*. Cambridge University Press, Cambridge.

Williams, R.S. and Ferrigno, J.G. (1999) Satellite image atlas of glaciers of the world, Chapter A: Introduction. *US Geological Survey Professional Paper* 1386-A.

Willis, I.C. (1995) Interannual variations in glacier motion – a review. *Progress in Physical Geography*, **19**, 61–106.

Wilson, E.O. (1992) *The diversity of life*. Harvard University Press, Cambridge, MA.

Wilson, E.O. (1994) *Naturalist*. The Penguin Press, London.

Wilson, E.O. (2002) *The future of life*. Abacus, London.

Wilson I. (1985) *The exodus enigma*. Weidenfeld and Nicolson, London.

Withers, P.J.A., Edwards, A.C. and Foy, R.H. (2001) Phosphorus cycling in UK agriculture and implications for phosphorus loss from soil. *Soil Use and Management*, **17**, 139–149.

Wodzicki, K.A. (1950) *Introduced mammals of New Zealand*. Department of Scientific and Industrial Research, Wellington, New Zealand.

Wolf, P.R. and Dewitt, B.A. (2000) *Elements of photogrammetry with applications in GIS*. McGraw-Hill, New York.

Wolman, M.G. (1967) A cycle of sedimentation and erosion in urban river channels. *Geografiska Annaler*, **49**, 385–395.

Wolman, M.G. and Miller, J.P. (1960) Magnitude and frequency of forces in geomorphic processes. *Journal of Geology*, **68**, 54–74.

Woo, M.-K. (1986) Permafrost hydrology in north America. *Atmosphere-Ocean*, **24**, 201–234.

Woodroffe, C.D. (2003) *Coasts, form, process and evolution*. Cambridge University Press, Cambridge.

Woodward, F.I., Lomas, M.R. and Kelly, C.K. (2004) Global climate and the distribution of plant biomes. *Philosophical Transactions of the Royal Society B*, **359**, 1465–1476.

Woodward, J.C. and Goldberg, P. (2001) The sedimentary records in Mediterranean rockshelters and caves: archives of environmental change. *Geoarchaeology: an International Journal*, **16**, 327–354.

Wooldridge, S.W. and Linton, D.L. (1939) *Structure, surface and drainage in South East England*. Philip, London.

Woolgar, S. (1988) *Science: The very idea*. Methuen, London.

World Commission on Environment and Development (1987) *Our common future*. Oxford University Press, Oxford.

World Health Organization (1993) *Guidelines for Drinking Water Quality, Vol. 1 Recommendations*, 2nd edition. WHO, Geneva.

World Health Organization (1999) *Gothenburg Consensus Paper*. WHO, Geneva.

Worrall, F. and Burt, T.P. (2004) Time series analysis of long-term river dissolved organic carbon records. *Hydrological Processes*, **18**, 893–911.

Worrall, F. and Burt, T.P. (2007) Flux of dissolved organic carbon from U.K. rivers. *Global Biogeochemical Cycles*, **21**, GB1013, doi: 10.1029/2006GB002709.

Worrall, F., Harriman, R., Evans, C.D. *et al.* (2004) Trends in dissolved organic carbon in UK rivers and lakes. *Biogeochemistry*, **70**, 369–402.

Wright, L.D. and Short, A.D. (1984) Morphodynamic variability of surf zones and beaches: a synthesis. *Marine Geology*, **56**, 93–118.

Wright, L.D. and Thom, B.G. (1977) Coastal depositional landforms: a morphodynamic approach. *Progress in Physical Geography*, **1**, 412–459.

Wright, L.D., Chappell, J., Thom, B. *et al.* (1979) Morpho-dynamics of reflective and dissipative beach and inshore systems, South Australia. *Marine Geology*, **32**, 105–140.

Y

Yakushev, V.S. and Chuvilin, E.M. (2000) Natural gas and gas hydrate accumulations within permafrost in Russia. *Cold Regions Science and Technology*, **31**, 189–197.

Yalden, D. (1999) *The history of the British mammals*. Academic Press, London.

Yatsu, E. (1966) *Rock control in geomorphology*. Sozosha, Tokyo.

Yatsu, E. (1988) *The nature of weathering: an introduction*. Sozosha, Tokyo.

Yatsu, E. (1992) To make geomorphology more scientific. *Transactions of the Japanese Geomorphological Union*, **13**, 87–124.

Young, I.R. and Holland, G.J. (1996) *Atlas of oceans, wind and wave climate*. Elsevier, Oxford.

Z

Zhu, T.X. (1997) Deep-seated, complex tunnel systems – a hydrological study in a semi-arid catchment, Loess Plateau, China. *Geomorphology*, **20**, 255–267.

INDEX

MORE KEY TEXTBOOKS FROM PEARSON EDUCATION

2000
ISBN 13: 9780582328730
ISBN 10: 058232873X

2005
ISBN 13: 9780130283191
ISBN 10: 0130283193

2006
ISBN 13: 9780131293175
ISBN 10: 0131293176

2002
ISBN 13: 9780582369368
ISBN 10: 0582369363

2005
ISBN 13: 9780582318298
ISBN 10: 0582318297

2003
ISBN 13: 9780130888907
ISBN 10: 0130888907

1991
ISBN 13: 9780582301566
ISBN 10: 0582301564

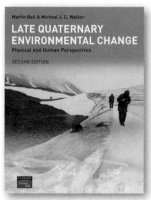

2005
ISBN 13: 9780130333445
ISBN 10: 0130333441

2002
ISBN 13: 9780582369849
ISBN 10: 0582369843

For further information or to order these books, please visit:
www.pearsoned.co.uk
*Prices are correct at time of going to press.